AF370785

# Land Use Transitions under Rapid Urbanization

# Land Use Transitions under Rapid Urbanization

Editors

**Hualou Long**
**Xiangbin Kong**
**Shougeng Hu**
**Yurui Li**

MDPI • Basel • Beijing • Wuhan • Barcelona • Belgrade • Manchester • Tokyo • Cluj • Tianjin

*Editors*
Hualou Long
School of Public Administration,
Guangxi University
China

Xiangbin Kong
College of Land Science
and Technology,
China Agricultural University
China

Shougeng Hu
Department of Land Resources
Management, China University
of Geosciences
China

Yurui Li
Institute of Geographic Sciences
and Natural Resources Research,
Chinese Academy of Sciences
China

*Editorial Office*
MDPI
St. Alban-Anlage 66
4052 Basel, Switzerland

This is a reprint of articles from the Special Issue published online in the open access journal *Land* (ISSN 2073-445X) (available at: https://www.mdpi.com/journal/land/special_issues/ LandUseTransitions_Urbanization).

For citation purposes, cite each article independently as indicated on the article page online and as indicated below:

LastName, A.A.; LastName, B.B.; LastName, C.C. Article Title. *Journal Name* **Year**, *Volume Number*, Page Range.

**ISBN 978-3-0365-2113-8 (Hbk)**
**ISBN 978-3-0365-2114-5 (PDF)**

Cover image courtesy of Hualou Long

National Natural Science Foundation of China, No. 41731286

# Contents

# About the Editors

**Hualou Long** Professor at School of Public Administration, Guangxi University, Nanning 530004, China; Adjunct Professor of human geography at the University of Chinese Academy of Sciences and the Institute of Geographic Sciences and Natural Resources Research, Chinese Academy of Sciences. Director of the Working Committee on Youth of China Land Science Society and the Commission on Agricultural Geography and Rural Development of the Geographical Society of China and a Steering Committee Member of the IGU Commission on the Sustainability of Rural Systems. He received his Ph.D. degree from Peking University in 1999. In 2005, he worked as Research Scholar at the department of Sustainable Rural Development, International Institute for Applied Systems Analysis (IIASA), Austria. His research interests include the following: land use transition; sustainable land use; rural restructuring; urban–rural development. He was awarded Highly Cited Researcher in the field of Social Sciences in both 2019 and 2020. Currently, he is serving as an Associate Editor of Habitat International and is a Member of the Editorial Board of Land, Land Use Policy, the Journal of Rural Studies, and the Journal of Urban and Regional Analysis.

**Xiangbin Kong** Distinguished Professor in the College of Land Science and Technology, China Agricultural University (CAU), Beijing 100193, China; Vice President of the National Academy College of Agricultural Science and Technology Strategy, CAU. Committee member of China Land Science Society (CLSS) and Sino-Europe Panel on Land and Soil (SEPLS). Special expert of science and technology strategy for rural modernization in 2040 of the Chinese Academy of Engineering, professional consultant of Surface Survey and Evaluation of Ministry of Natural Resource. He received his Ph.D. degree from CAU and was the visiting scholar at Ohio State University and the University of California. His research interests include the following: sustainable arable land use; arable land quality evaluation; food security; low-carbon food systems. His work has been published in journals such as Nature and Advances in Agronomy. He is serving as Associate Editor of Transactions of the Chinese Society of Agricultural Engineering, Guest Editor of Land, Land Use Policy, and the Scientific World Journal.

**Shougeng Hu** Professor at Department of Land Resources Management, China University of Geosciences, Wuhan 430074, China; Dean of School of Public Administration, China University of Geosciences (Wuhan). His research interests include the following: land use transition; urban–rural development; land space optimization; cultivated land protection; natural resource utilization policy; spatial analysis and modeling.

**Yurui Li** Associate Professor at the Institute of Geographic Sciences and Natural Resources Research, Chinese Academy of Sciences, Beijing 100101, China; Deputy director and Secretary general of the Commission on Agricultural Geography and Rural Development of the Geographical Society of China. He received his Ph.D. degree from the Graduate School of Chinese Academy of Sciences in 2011. His research interests include the following: agricultural geography; land use; land consolidation; rural development; poverty alleviation; regional development. He was awarded Highly Cited Researcher in the field of Social Sciences in 2020. Currently, he is serving as Member of Editorial Board of Habitat International, Land, Scientia Geographica Sinica, and the Journal of Agricultural Resources and Environment.

# Preface to "Land Use Transitions under Rapid Urbanization"

Recently, tremendous land use transitions have been triggered in the world and have brought about a series of challenges to sustainable urban and rural development, especially in the developing world where is experiencing a rapid urbanization. The research on land use transitions has become a hot topic, as evidenced by an average annual growth rate of 16% of the documents found in Scopus via searching 'land use transition' within the article title, abstract, and keywords in the last five years. Chinese scholars have made an important contribution and have authored 16% of the searched documents, which ranks second only to the United States, with a proportion of 29%. The research of land use transitions has sparked great concern in both academic communities and the governmental authorities since it was introduced into China at the beginning of the new century (Long et al., 2020). Owning to the Key Program of National Natural Science Foundation of China entitled "Land Use Transitions Driven by Urbanization and Their Effects on Local Environment in the Farming Areas of China: The Case of Huang-Huai-Hai Plain" (Grant No. 41731286), the concepts and connotations of land use transitions, as well as the research paradigm, have been further developed, and abundant research results have emerged (Long et al., 2020). This inspired us to edit this Special Issue of *Land* titled "Land Use Transitions under Rapid Urbanization".

In the book *Land Use Transitions and Rural Restructuring in China*, published by Springer (Long, 2020, pp. 5–6), the concept and connotations of land use morphology were developed and expanded into two kinds, i.e., dominant morphology and recessive morphology. The former refers to the land use structure of a certain region over a certain period of time, with features such as the quantity (area, proportion) and spatial pattern of land use types. Recessive morphology is a special morphology which relies on the dominant morphology but can only be observed by the means of analyzing, testing, monitoring, and surveying, and it includes the land use features in the aspects of quality (nutrient, pollution, degradation), property rights (state-owned, collectively-owned), management mode (individual, joint-stock system, transfer, and large-scale management), input (capital, technology, labor), output (yield, output value, input–output ratio), and function (production, living, ecology, culture). Most papers included in this book as well as the Special Issue "Land Use Transitions under Rapid Urbanization" follow the research paradigm of land use transitions focusing on the changes of dominant morphology and recessive morphology.

Although gratifying achievements have been made in the research on land use transitions in developing countries under rapid urbanization in the last two decades, ongoing changes in land use and urban–rural development will trigger more theoretical and practical research needs. We hope this book will be helpful for enriching the research of land use transitions, as well as attracting more scholars to carry out further studies on land use transitions.

## References


Long H. Land Use Transitions and Rural Restructuring in China. Springer Nature, Singapore, 2020.
Long Hualou, Qu Yi, Tu Shuangshuang, Zhang Yingnan, Jiang Yanfeng. Development of land use transitions research in China. Journal of Geographical Sciences, 2020, 30(7): 1195-1214.


**Hualou Long, Xiangbin Kong, Shougeng Hu, Yurui Li**

*Editors*

*land*

# Land Use Transitions under Rapid Urbanization: A Perspective from Developing China

Hualou Long [1,2,*], Xiangbin Kong [3], Shougeng Hu [4] and Yurui Li [2]

[1] School of Public Administration, Guangxi University, Nanning 530004, China
[2] Institute of Geographic Sciences and Natural Resources Research, Chinese Academy of Sciences, Beijing 100101, China; liyr@igsnrr.ac.cn
[3] College of Land Science and Technology, China Agricultural University, Beijing 100193, China; kxb@cau.edu.cn
[4] Department of Land Resources and Management, China University of Geosciences, Wuhan 430074, China; hushougeng@cug.edu.cn
[*] Correspondence: longhl@igsnrr.ac.cn

**Citation:** Long, H.; Kong, X.; Hu, S.; Li, Y. Land Use Transitions under Rapid Urbanization: A Perspective from Developing China. *Land* **2021**, *10*, 935. https://doi.org/10.3390/land10090935

Received: 27 August 2021
Accepted: 1 September 2021
Published: 6 September 2021

**Publisher's Note:** MDPI stays neutral with regard to jurisdictional claims in published maps and institutional affiliations.

## 1. Introduction

Land use transition is a manifestation of land use and land cover change (LUCC) and is also a major research focus of the Global Land Project (GLP), as well as land system science (LSS). Land provides essential resources to society and its changes have large consequences for the local and global environment and human well-being. The past, current, and projected state and dynamics of land use represent the major focus of land use science, which is influenced by long-term anthropogenic changes. The concept of land use transition highlights the fact that land use change demonstrates a non-linear process and is related to other societal and biophysical disturbances through a series of transitions.

Land use transitions can be seen as primary forces driving the transformation and development of the rural–urban territorial system, and bringing about direct socio-economic and environmental effects on regional sustainability, e.g., resulting in farmland loss and soil degradation, affecting biodiversity and the ability of ecosystems to serve human needs, polluting the rural environment, influencing agricultural production and food security, and causing regional socio-economic and spatial restructuring. Land use transitions can be measured by changes in both the dominant morphology (e.g., quantity, structure, and spatial patterns) and the recessive morphology (e.g., quality, property rights, management mode, fixed input, productive ability, and function) of land use [1,2].

Land use transition is the change in land use morphology, which corresponds to the changes in regional socio-economic development. There is no doubt that rapid socio-economic development and urbanization processes will inevitably bring about drastic land use transitions. As China is emerging as a global economic superpower with a majority of 64% of the population representing the urban population, land use transitions during the rapid urbanization in China, with an urban–rural dualism, have received much more attention.

The aim of this Special Issue was to detect or examine the processes, patterns, and socio-economic and environmental effects of land use transitions in China, the mechanism of human–land interactions against the context of rapid urbanization and industrialization from a wide range of perspectives (i.e., geographical, social–political, ecological, etc.), and the provision of solutions for sustainable land use based on scientific findings.

The collection of peer-reviewed articles in Special Issue number twenty-six includes one review article and twenty-five research articles. The Special Issue is organized in the following way: after the first review paper serving as a background introduction, the following papers are presented under four major topics, such as (a) farmland use management, (b) rural restructuring and vitalization, (c) ecological and environmental effects, and (d) urban development.

1

In the review paper [3], the overall picture, development trends, key fields, and hot topics of land use transition research in the past two decades are systematically analyzed from a comprehensive perspective, with the aim to provide scientific support for sustainable land use and environmental management. This review incorporates two complementary parts: the systematic quantitative literature review (based on CiteSpace) and the traditional literature review. The results reveal that there are four key fields and hot topics in land use transition research, i.e., theories and hypotheses of land use transition: i. measuring land use transition; ii. the impacts of land use transition on "social-economic-ecological" systems; iii. drivers and regulation of land use transition. However, challenges remain; current land use transition research is still, to some extent, fragmented, and it should be enriched through an integration with land system science. The dominant morphology bias should be redressed by underlining the recessive morphology transition process. Accordingly, land use transition research needs to be further strengthened in the near future in the following aspects: a. carrying out land use transition research under the guidance of land system science; b. attaching importance to the transition of land tenure regimes; c. overcoming the challenges of detecting the recessive morphology of land use; d. linking local land use transition with globalization. Finally, it is clear that the scientific theory on land use transition lags behind the research practice. Despite considerable advances in land use transition research and related fields, an inclusive theory of land use transition or sets of theories has not emerged.

This progress and prospect thinking can be used to help to analyze the issues regarding land use transitions and urban–rural development in China and its eco-environmental effects that arise in the other papers in this Special Issue, which are introduced below under four broad topics.

## 2. Farmland Use Management

There are twelve papers which focus on farmland use transition and land use management in this Special Issue, from the perspectives of dominant morphology, recessive morphology, or both.

Lyu et al. [4] constructed a morphological evaluation index system to identify the characteristics of farmland use transition in the Sihong County of the Huang-Huai-Hai Plain, China. The dominant morphology in terms of area and landscape pattern and the recessive morphology focusing on function were considered in this work. The results show that the implementation of land consolidation projects increased the area and aggregation of farmland, while urbanization and road construction occupied and divided the farmland, leading to a reduction in the area and an increase in the degree of fragmentation. Urbanization increased the demand for agricultural products and the degree of large-scale agricultural production and had a positive impact on the crop production and eco-environmental function of farmland. The research suggested that the government should formulate efficient policies to curb farmland occupation for urban and traffic utilization.

Li et al. [5] and Yang et al. [6] both studied the multiple cropping index, an important recessive morphology of farmland use. Li et al. calculated the multiple cropping index of farmland in China using the S-G filtering method based on the remote sensing data of NDVI, and proposed an optimized regionalization scheme for the use of farmland. The findings reveal that the gap between the multiple cropping index and the potential multiple cropping index of farmland in China is increasingly widening from north to south. Accordingly, four types of areas were classified into key development areas, potential growth areas, moderate development areas, and restricted development areas. Finally, some suggestions such as rotation, fallow, fixed yields with water, and the offset of the balance index with the multiple cropping index were put forward based on different areas. Yang et al. collected MODIS remote sensing image data and land use classification data and conducted a remote sensing inversion on China's multiple cropping index to examine the spatio-temporal changes and factors influencing the multiple cropping index of farmland in China. The results show that natural conditions, the non-agricultural process, the cultivated land quality, and agricultural intensification demonstrated different degrees of impact on the multiple cropping index.

Lu et al. [7] explored the green transition of cultivated land use in the Yellow River Basin from the perspective of green utilization efficiency evaluation, which covers the comprehensive benefits of economy, ecology, and society, based on the requirements of ecological civilization and green development construction. The findings reveal that the green utilization efficiency of cultivated land in the Yellow River Basin presents a general distribution characteristic of spatial agglomeration.

Song and Zhang [8] studied the adjustment and optimization of the cultivated land use layout in the typical counties of the main grain production area in Northeast China, based on the planting suitability of the main food crops (rice, soybean, and maize). The findings show that by optimizing the layout of rice, soybean, and maize, the planting suitability level of the food crops and the concentration level of the spatial pattern of cultivated land use layout have been improved. The results may provide a scientific basis and guidance for adjusting the regional planting structure and solving the problem of food structural imbalance.

Zhou et al. [9] investigated the transitional characteristics and triggers of farmland use change through a linear regression analysis, and the internal mechanism of these transitions based on the Advocacy Coalition Framework. The main findings show that macroscopic economic and social changes determined the overall evolution of the farmland area, and there were two advocacy coalitions in the farmland transition policy subsystem— the farmland supplement and farmland consumption coalitions. Under the influence of macroscopic economic and social development, external events play a catalytic role in the transitions, and relatively stable parameters have an indirect but lasting effect in terms of the transition outcomes.

Guo and Wang [10] analyzed the stage characteristics of China's non-grain farming and investigated the spatial agglomeration of non-grain farming and its influencing factors from the perspective of spatial econometrics. The results show that the per capita disposable income of rural residents and the urbanization rate of local areas promoted the development of non-grain farming, while local per capita farmland, road density, and the functional orientation of the main grain-producing areas had a negative impact on non-grain farming. Ultimately, some targeted measures were proposed to promote China's agricultural development in the new era.

Wang et al. [11] studied the relationship between land use transitions and farm performance in China from the perspective of land fragmentation, using detailed household survey data at the crop level from ten provinces in China to construct four land fragmentation indicators and six farm performance indicators. The findings show that land fragmentation increased the input of production materials and labor costs, reduced the purchasing of mechanical services by farmers and the efficiency of ploughing, and may have increased technical efficiency. To improve farm performance, it is recommended that decision-makers speed up land transfer and land consolidation, stabilize land property rights, establish land-transfer intermediary organizations, and promote large-scale production.

Lou et al. [12] analyzed the recessive transition mechanism and the internal differences in arable land use modes of 31 provinces in mainland China by applying an evaluation model to the degree of coupling coordination between the input and output on arable land. The results show that the total amount and the amount per unit area of the input and output on arable land in China have presented different spatio-temporal trends, along with the mismatched movement of the spatial barycenter. The results of this study highlight the different recessive transition patterns of arable land use in different provinces of China, which points to the outlook for higher technical inputs, optimized planting structures, and the coordination of human–land relationships.

Chen et al. [13] investigated the impacts of the farmland transfer-in or transfer-out of different rural households on income structure in Heilongjiang province, the major cereal production area in China, based on the Propensity Score Matching model. The results show that the total income of all rural households transferring-in farmland increased significantly, while the income decreased after transferring-out farmland, and part-time households had the largest increase, followed by pure-agricultural households and part-time households. Accordingly, some suggestions were obtained.

Yang et al. [14] explored the characteristics of livelihood capital and the land transfer of farmers under different livelihood strategies and the effect of livelihood capital on land transfer in Sichuan Province. The findings show that pure farmers tended to shift other capital toward natural capital, so their livelihood capital total index value decreased. Part-time farmers had different shift characteristics but their livelihood capital total index value both increased first, and then decreased, and non-farmers tended to shift natural capital towards other livelihood capitals, so their livelihood capital total index value increased. Based on the above analysis, some policy implications were obtained.

Zheng and Zhang [15] analyzed the characteristics of farmers' defaults in different periods and locations based on court decisions regarding rural land mortgage defaults from 2014 to 2020. The empirical results reveal that the time and location of rural land mortgage default cases are widely distributed in China, especially in Heilongjiang Province. It was suggested that when making mortgage loan policies for rural land management rights, financial institutions should give farmers the most preferential treatment regarding the amount, term, and interest rate of loans. Farmers' social security should be improved, and agricultural insurance should be strengthened. Meanwhile, the credit review of small and short-term loan farmers should be heightened.

### 3. Rural Restructuring and Vitalization

Qu et al. [16] constructed a comprehensive research framework of rural residential land scale, structure, and function from the perspective of the combination of the macro and micro scales based on differences between the rural residential areas in the region and the village scale forms. Taking Shandong Province as an example and using model quantitative analysis and horizontal comparative analysis methods, this paper explored the process characteristics of rural residential land use scale transition and the corresponding stage differentiation law of spatial structure and system function. It pointed out that the transition of the rural residential area from the macro to the micro scale is also the process related to the realization of rural restructuring and rural revitalization.

Han et al. [17] proposed a hypothesis of rural-spatial restructuring based on the evaluation of ecological-production-living spaces in terms of the changes in the dominant and recessive morphologies of land use, and analyzed the changes in the dominant morphology of land use by identifying the distribution characteristics of the elements of ecological-production-living spaces and analyzing their structural changes, based on which the process of rural spatial restructuring was judged, and to lay a solid foundation for the next step of rural revitalization. The findings show that combining changes in the ecological-production-living spatial area and quantity as well as changes in functional suitability enables a better understanding of the impact of the national macro-policy shift regarding rural development.

Zhang et al. [18] analyzed the spatio-temporal processes and dilemmas involved in rural industrial land transition by constructing an analytical framework for rural industrial land transition based on spatial governance. The results show that the comprehensive governance of rural space under the analytical framework of "matter-ownership-organization" is an important starting point for analyzing the process of transition of rural industrial land. Rural spatial governance is conducive to promoting the transition of rural land use and the healthy development of rural space. Finally, the authors argued that the experience of semi-urbanized regions with rural revitalization is of vital significance for other regions.

Chai et al. [19] analyzed the changes in both the dominant and recessive morphology of land use in She Village located in the suburban areas of Nanjing by employing participatory rural appraisal, remote sensing, and geographic information systems. The findings show that the She Village witnessed three stages, including industrial development, ecological restoration, and service industry development, with more diversified management modes, multifunctional land use, and intensified land fragmentation. This research deepens the understanding of the development process of suburban villages and provides a reference for land policy making and planning in other similar villages.

Tang et al. [20] explored the significant role that village committees play in farmers' withdrawal from rural homesteads. The results show that farmers' withdrawal from rural homesteads was significantly affected by the triple roles of village committees, among which an information intermediary was the most effective role, followed by the trust builder, and then the coordinated manager. The authors argued that promising village committees should act as "all-round stewards" in the decision-making of rural households, which not only includes the transmission of information between those above and those below, but also includes the need to actively strive for farmers' trust by letting their voice be heard. Based on the empirical findings, this paper finally proposed some policy suggestions.

## 4. Ecological and Environmental Effects

There are two papers that deal with the ecological and environmental effects of land use transitions in the Loess Plateau of China. Feng and Li [21] established a quantitative evaluation model for establishing the ecological safety effects from the four aspects of dam safety, slope stability, efficient farmland, and effective management, and then scientifically measured the ecological safety effects of land use transitions promoted by land consolidation projects. Three small watersheds (Gutun, Yangjuangou, and Luoping) within the Gully Land Consolidation Project area were employed to verify the evaluation model for the ecological safety effects. The results show that the Gully Land Consolidation Project can effectively improve the ecological environment and promote the development of modern agriculture, but the ecological safety of gullies and slopes in some areas may also face a series of threats due to improper project management measures. Li et al. [22] analyzed land use change and the sense of place of farmers, and further explored the interaction between them in the Yangjuangou catchment of Liqu Town in Baota District, Shaanxi Province. The results indicate that the intensity of the sense of place of local farmers fluctuated downwards, and the decline in the intensity of place attachment and place dependence promoted the reduction in sloping fields, the growth of ecological land, and abandoned fields. This paper suggested that rural areas in the Loess Hilly and Gully Region should strengthen innovation in land use patterns and focus on sustaining farmers' livelihoods in order to promote the harmonious development of human–environmental relations.

Yin et al. [23] analyzed the dynamic patterns of land use transitions in the Yellow River Basin by using the geo-information Tupu method. The results indicate that the Tupu units of the land use transitions were mainly based on the mutual transformation of grassland and unused land, and cultivated land and forestland, which were widely distributed in the upper and middle reaches of the basin. These findings could have theoretical support and policy implications for land use planning and environmental services in the Yellow River Basin.

Chen et al. [24] investigated the habitat quality effect of land use transition and analyzed the cause and mechanism of such changes from an economic–social–ecological complex system perspective in the Henan Water Source area of the Middle Route of the South-to-North Water Transfer Project. The results of this study provided a basis for the improvement of habitat quality, ecosystem protection and restoration, land resource management, and related policies in the Henan Water Source area of the Middle Route of the South-to-North Water Transfer Project.

Pan et al. [25] evaluated the matching degree of water and land resources, and their respective matching degrees with the economic development in the Chengdu–Chongqing Economic Circle with the Gini coefficient method. The results showed that the water and land resources and the economic development of the Chengdu–Chongqing Economic Circle have a high matching degree, but the inner cities have a great difference. The development of the Chengdu–Chongqing Economic Circle needs to promote economic growth and technological progress, and at the same time optimize the use of water and land resources to reduce its constraints on economic growth. Finally, policy suggestions of matching water and land resources and economic growth in different regions were put forward.

Cui et al. [26] proposed a new spatially explicit evaluation framework of land use conflict that directly examines three aspects of conflict, namely, ecological and agricultural, agricultural and construction, and ecological and construction land conflicts based on ecological quality and agricultural suitability and evaluated the spatio-temporal dynamic pattern and driving factors of land use conflict in the Yangtze River Economic Belt of China. The findings show that total population, population density, per capita GDP, number of mobile phone users, and road density were strong drivers that influenced the land use conflict of territorial space. Multiple policy recommendations including improving territorial space planning and governance ability, and improving land use efficiency, were proposed to manage and resolve the land use conflict of territorial space.

## 5. Urban Development

Niu et al. [27] constructed a theoretical framework for the interactive relationship between urbanization and land use transition and measured the level of urbanization from the perspective of population urbanization, economic urbanization, and social urbanization, while also evaluating the level of land use morphologies from the perspective of the dominant and recessive morphologies of land use. The results show that the relationship between urbanization and land use transition is not a simple linear relationship but tends to be complex. The process of urbanization, and reasonable urbanization and land use morphologies will promote further benign coupling in the system.

Li et al. [28] revealed the mechanisms underlying the influence of urban land use transitions on the economic spatial spillovers of central cities to provide a reference for China to optimize the land space layout in cities and to promote their coordinated development. The results show that continuing to strengthen the intensive use of urban land, promoting the improvement of land marketization, and establishing and improving the coordination mechanism for the economic development of urban agglomerations will help to strengthen the economic spatial spillovers of central cities in urban agglomerations.

## 6. Concluding Comments

Research on land use transition has developed rapidly since Long introduced land use transition research into China in 2001 [2]. Undoubtedly, the launch of this special issue of *Land* including 26 papers accelerates the development of this process to some extent. This special issue has succeeded in compiling theoretical and empirical studies to highlight land use transitions in China. The papers make important conceptual–theoretical and empirical contributions to the growing literature on land use transitions.

The papers in this special issue focus on four major topics, i.e., farmland use management, rural restructuring and vitalization, ecological and environmental effects, and urban development. These topics are also important research themes that need to be strengthened in China during a rapid urbanization process. The recessive morphology concerning the quality, property rights, management mode, input-output, and function of land use obtains the attention it deserves in this special issue. Based on the research results, most papers aim to translate scientific findings on land systems into solutions for sustainable land use, which is what land system science advocates.

However, there is still much more room to improve for land use transition research in China, as far as the papers published in this special issue go. Although some policies have been suggested to tackle the corresponding regional land use or socio-economic development issues, the operationalization of these policies is not without flaws, due to the insufficient attention paid to the interrelations between the urban and the rural and the multi-dimensional driving forces of land use transitions.

On the one hand, China is a nation with strong rural roots. Despite rapid urbanization, nearly half of its population still lives in rural areas. Tremendous land use transitions have transformed the development in both rural and urban areas. Accompanying this rapid urban–rural transformation, various land use issues have occurred, giving rise to formulation of new policies directly affecting land use. Therefore, we should bear in mind that the volatility and complexity of land use transitions in China, doomed by the rampant urban–rural transformation and the special 'dual-track' structure of urban–rural development, will present ongoing challenges for further research on land use transitions and urban–rural integrated development, which also needs extensive disciplinary interaction.

On the other hand, the driving forces of land use transitions are not only confined to socio-economic disturbances and physical conditions in one region but are also affected by the displacement effect resulting from international trade [29]. Globalization and urbanization have been two important global trends since the middle of the 20th century. As a result of their interaction, a complex trade network has gradually formed between the urban and the rural as well as between different cities even different countries, thus forming an important tele-coupling context for land use transitions [2]. As such, we need to pay more attention to the distant drivers of land use transitions and link local land use transitions with globalization, which has received much less attention in this special issue.

**Author Contributions:** Conceptualization, H.L., X.K., S.H. and Y.L.; writing—original draft preparation, H.L.; writing—review and editing, X.K., S.H. and Y.L.; supervision, H.L.; project administration, H.L.; funding acquisition, H.L., X.K. and Y.L. All authors have read and agreed to the published version of the manuscript.

**Funding:** This research was supported by the National Natural Science Foundation of China (Grant Nos. 41731286, 41971216 and 41971220), and National Social Science Foundation of China (Grant Nos. 19ZDA096 and 21AZD039).

**Acknowledgments:** We would like to thank all of the contributors to this special issue for their hard work and commitment in producing this volume, and also give our sincere thanks to the editors of *Land*, for their always present support in this special issue and to the reviewers for their constructive engagement with the manuscripts.

**Conflicts of Interest:** The authors declare no conflict of interest.

# References

1. Long, H. *Land Use Transitions and Rural Restructuring in China*; Springer: Berlin/Heidelberg, Germany, 2020; p. 5.
2. Long, H.; Qu, Y.; Tu, S.; Zhang, Y.; Jiang, Y. Development of land use transitions research in China. *J. Geogr. Sci.* **2020**, *30*, 1195–1214. [CrossRef]
3. Long, H.; Zhang, Y.; Ma, L.; Tu, S. Land use transitions: Progress, challenges and prospects. *Land* **2021**, *10*, 903. [CrossRef]
4. Lyu, L.; Gao, Z.; Long, H.; Wang, X.; Fan, Y. Farmland use transition in a typical farming area: The case of Sihong County in the Huang-Huai-Hai Plain of China. *Land* **2021**, *10*, 347. [CrossRef]
5. Li, T.; Wang, Y.; Liu, C. Research on identification of multiple cropping index of farmland and regional optimization scheme in China based on NDVI data. *Land* **2021**, *10*, 861. [CrossRef]
6. Yang, R.; Luo, X.; Xu, Q.; Zhang, X.; Wu, J. Measuring the impact of the multiple cropping index of cultivated land during continuous and rapid rise of urbanization in China: A study from 2000 to 2015. *Land* **2021**, *10*, 491. [CrossRef]
7. Lu, X.; Qu, Y.; Sun, P.; Yu, W.; Peng, W. Green transition of cultivated land use in the Yellow River Basin: A perspective of green utilization efficiency evaluation. *Land* **2020**, *9*, 475. [CrossRef]
8. Song, G.; Zhang, H. Cultivated land use layout adjustment based on crop planting suitability: A case study of typical counties in Northeast China. *Land* **2021**, *10*, 107. [CrossRef]
9. Zhou, X.; Li, X.; Song, W.; Kong, X.; Lu, X. Farmland transitions in China: An advocacy coalition approach. *Land* **2021**, *10*, 122. [CrossRef]
10. Guo, Y.; Wang, J. Identifying the determinants of non-grain farming in China and its implications for agricultural development. *Land* **2021**, *10*, 902. [CrossRef]
11. Wang, S.; Li, D.; Li, T.; Liu, C. Land use transitions and farm performance in China: A perspective of land fragmentation. *Land* **2021**, *10*, 792. [CrossRef]
12. Lou, Y.; Yin, G.; Xin, Y.; Xie, S.; Li, G.; Liu, S.; Wang, X. Recessive transition mechanism of arable land use based on the perspective of coupling coordination of input–output: A case study of 31 provinces in China. *Land* **2021**, *10*, 41. [CrossRef]
13. Chen, L.; Chen, H.; Zou, C.; Liu, Y. The impact of farmland transfer on rural households' income structure in the context of household differentiation: A case study of Heilongjiang Province, China. *Land* **2021**, *10*, 362. [CrossRef]

14. Yang, H.; Huang, K.; Deng, X.; Xu, D. Livelihood capital and land transfer of different types of farmers: Evidence from panel data in Sichuan Province, China. *Land* **2021**, *10*, 532. [CrossRef]

15. Zheng, H.; Zhang, Z. Analyzing characteristics and implications of the mortgage default of agricultural land management rights in recent China based on 724 court decisions. *Land* **2021**, *10*, 729. [CrossRef]

16. Qu, Y.; Dong, X.; Zhan, L.; Si, H.; Ping, Z.; Zhu, W. Scale transition and structure–function synergy differentiation of rural residential land: A dimensionality reduction transmission process from macro to micro scale. *Land* **2021**, *10*, 647. [CrossRef]

17. Han, D.; Qiao, J.; Zhu, Q. Rural-spatial restructuring promoted by land-use transitions: A case study of Zhulin Town in Central China. *Land* **2021**, *10*, 234. [CrossRef]

18. Zhang, L.; Ge, D.; Sun, P.; Sun, D. The transition mechanism and revitalization path of rural industrial land from a spatial governance perspective: The case of Shunde District, China. *Land* **2021**, *10*, 746. [CrossRef]

19. Chai, Y.; Qiao, W.; Hu, Y.; He, T.; Jia, K.; Feng, T.; Wang, Y. Land-use transition of tourist villages in the metropolitan suburbs and its driving forces: A case study of She Village in Nanjing City, China. *Land* **2021**, *10*, 168. [CrossRef]

20. Tang, P.; Chen, J.; Gao, J.; Li, M.; Wang, J. What role(s) do village committees play in the withdrawal from rural homesteads? Evidence from Sichuan Province in Western China. *Land* **2020**, *9*, 477. [CrossRef]

21. Feng, W.; Li, Y. Measuring the ecological safety effects of land use transitions promoted by land consolidation projects: The case of Yan'an City on the Loess Plateau of China. *Land* **2021**, *10*, 783. [CrossRef]

22. Li, Y.; Li, Y.; Fang, B.; Qu, L.; Wang, C.; Li, W. Land use change and farmers' sense of place in typical catchment of the Loess hilly and gully region of China. *Land* **2021**, *10*, 810. [CrossRef]

23. Yin, D.; Li, X.; Li, G.; Zhang, J.; Yu, H. Spatio-temporal evolution of land use transition and its eco-environmental effects: A case study of the Yellow River Basin, China. *Land* **2020**, *9*, 514. [CrossRef]

24. Chen, M.; Bai, Z.; Wang, Q.; Shi, Z. Habitat quality effect and driving mechanism of land use transitions: A case study of Henan water source area of the middle route of the south-to-north water transfer project. *Land* **2021**, *10*, 796. [CrossRef]

25. Pan, Y.; Ma, L.; Tang, H.; Wu, Y.; Yang, Z. Land use transitions under rapid urbanization in Chengdu-Chongqing Region: A perspective of coupling water and land resources. *Land* **2021**, *10*, 812. [CrossRef]

26. Cui, J.; Kong, X.; Chen, J.; Sun, J.; Zhu, Y. Spatially explicit evaluation and driving factor identification of land use conflict in Yangtze River Economic Belt. *Land* **2021**, *10*, 43. [CrossRef]

27. Niu, B.; Ge, D.; Yan, R.; Ma, Y.; Sun, D.; Lu, M.; Lu, Y. The evolution of the interactive relationship between urbanization and land-use transition: A case study of the Yangtze River Delta. *Land* **2021**, *10*, 804. [CrossRef]

28. Li, H.; Chen, K.; Yan, L.; Zhu, Y.; Liao, L.; Chen, Y. Urban land use transitions and the economic spatial spillovers of central cities in China's urban agglomerations. *Land* **2021**, *10*, 644. [CrossRef]

29. Meyfroidt, P.; Lambin, E.F.; Erb, K.H.; Hertel, T.W. Globalization of land use: Distant drivers of land change and geographic displacement of land use. *Curr. Opin. Environ. Sustain.* **2013**, *5*, 438–444. [CrossRef]

*Review*

# Land Use Transitions: Progress, Challenges and Prospects

Hualou Long [1,2], Yingnan Zhang [3], Li Ma [4,*] and Shuangshuang Tu [2,5]

[1] School of Public Administration, Guangxi University, Nanning 530004, China; longhl@igsnrr.ac.cn
[2] Institute of Geographic Sciences and Natural Resources Research, Chinese Academy of Sciences, Beijing 100101, China; tuss@nnnu.edu.cn
[3] School of Public Affairs, Zhejiang University, Hangzhou 310058, China; zhangyingnan@zju.edu.cn
[4] School of Public Affairs, Chongqing University, Chongqing 400044, China
[5] Key Laboratory of Environment Change and Resources Use in Beibu Gulf, Nanning Normal University, The Ministry of Education, Nanning 530001, China
* Correspondence: mal.1991@cqu.edu.cn

**Abstract:** The study of land use transition has generally become an important breakthrough point to deeply understand the human-land interaction and reveal major socio-economic development issues and related environmental effects. Attempting to provide scientific support for sustainable land use and environmental management, this review systematically analyzes the overall picture, development trends, key fields and hot topics of land use transition research in the past two decades from a comprehensive perspective, which incorporates two complementary parts including the systematic quantitative literature review (based on CiteSpace) and the traditional literature review. The results reveal that: a. current research presents three characteristics, i.e., focusing on complex social issues, driven by realistic demand, and research branches becoming clearer and more systematic; b. there are four key fields and hot topics in land use transition research, i.e., i. theories and hypothesis of land use transition; ii. measuring land use transition; iii. the impacts of land use transition on "social-economic-ecological" system; iv. drivers and regulation of land use transition. However, challenges remain, current land use transition research is still to some extent fragmented, and it should be enriched by integrating with land system science. The dominant morphology biased should be redressed by underlining the recessive morphology transition process. Meanwhile, new techniques and methods are necessary to observe, track, monitor and model the recessive attributes. Finally, distant drivers of land use transition should not be ignored in this rapidly globalizing world.

**Keywords:** land use transition; land use morphology; land system science; literature review; CiteSpace; progress and prospects

**Citation:** Long, H.; Zhang, Y.; Ma, L.; Tu, S. Land Use Transitions: Progress, Challenges and Prospects. *Land* **2021**, *10*, 903. https://doi.org/10.3390/land10090903

Academic Editor: Hossein Azadi

Received: 31 July 2021
Accepted: 25 August 2021
Published: 27 August 2021



## 1. Introduction

Land is the spatial carrier of anthropogenic activities, the most basic production factor of socio-economic development, and the most fundamental survival resource for urban and rural residents. Since the end of the 20th century, increasing intensive land use activities have become an important factor affecting global sustainable growth. On the one hand, over-exploitation and uncontrolled utilization of land resources in areas with higher natural suitability has brought huge challenges to regional sustainability. On the other hand, farmland abandonment in marginal areas has brought about a greater threat to food security [1–3]. A series of problems such as increased pressure on agricultural land, soil pollution and decreased biodiversity caused by high-intensity land use have brought about many difficulties to the development, management and sustainable use of land resources, and also attracted wide attention [4]. Land use faces the challenge of how to address the relations between meeting human needs and maintaining the long-term ability of the biosphere to provide goods and services [5].

At present, the world is experiencing major changes, which are intertwined with epidemic such as the COVID-19. Climate change poses severe threats to human survival [6].

As the leading sources of greenhouse gas emissions, land use transition (LUT) has greatly challenged the functions of ecosystems, thus having an important impact on climate change [7]. How to take effective measures to deal with resource exhaustion and the impact of human activities on the environment, ensure food security and further understand the feedback relationship between the natural environment and human society, has become an important issue that needs to be solved urgently [8]. LUT research helps to provide comprehensive information for decision-makers in land use planning and environmental management, and has important practical significance for coordinating regional social, economic and ecological development goals. In recent years, the research projects and related papers concerning LUT have shown a rapid growth trend, but the comprehensive and systematic bibliometric analysis is still insufficient. Scholars' focus on LUT research is constantly changing and adjusting. Therefore, it is necessary to clarify the research focus of different periods and the network relations of the hot topics. Several questions should be answered:

(1)    What is the general trend of LUT research?
(2)    What are the distinguishing stage characteristics and hot topics of LUT research?
(3)    What are the major fields of LUT research?
(4)    What are the challenges and future directions of LUT research?

## 2. Data and Methods

The literature data in this paper comes from the core collection of the Web of Science database (http://apps.webofknowledge.com/, accessed on 20 March 2021). Web of Science is an important database for obtaining global academic information. It includes more than 13,000 authoritative and high-impact academic journals around the world, covering fields such as natural science, engineering technology, biomedical science, social science, art and humanities, with data dating back to 1900. Web of Science catalogs references cited in this paper. With a unique citation index, users can easily retrieve their citation and trace the origin and history of a research document by using an article, a patent number, a conference document, a journal or a book as the search term.

This paper analyzes the knowledge graph by CiteSpace. CiteSpace is a data mining and visualization analysis software jointly developed by Professor Chen Chaomei from the School of Information Science and Technology of Drexel University and WISE Laboratory of Dalian University of Technology. The software can dig the underlying information and intuitively present relevant information and the interrelationships between information entities through a visual knowledge map by extracting and analyzing the subject information such as keywords, topics, authors and institutions. This software also shows the development trend of a discipline or knowledge field in a certain period through the convergence of relevant information, and reveals the development status of scientific knowledge in this field. It is widely used in information science, economics, sociology and many other fields [9]. The search prerequisites of LUT research are set as follows: "TS = land use transition", with TS as the theme, time spans 1900–2021, the language is English, and the literature type is article. There are 8700 records were retrieved, and 8564 records remain after eliminating the literature that is not related to the research subject, the earliest year is 1987. Based on Citespace.5.6.R3, we set the parameters: the cutting time is set as 1a (year), the threshold positioning is Top 50; the node type determines the purpose of CiteSpace analysis, so we select keyword in node types. Co-occurrence analysis helps us understand the hot topics, topic distribution and subject arrangement [9]. Keyword co-occurrence analysis is an effective tool of analyzing the keywords provided by the authors in the data set. Relying on keyword co-occurrence analysis of 8564 records related to LUT, the literature was macroscopically visualized and the network map was obtained, and the research progress of LUT was discussed.

This review consists of two complementary parts: the systematic quantitative literature review (Section 3) and the traditional literature review (Section 4). Systematic quantitative literature review uses a large amount of literature analysis, to explore the critical path and knowledge inflection point of the evolution of the subject field, so as to help scholars quickly understand the relevant situation of the field. However, there are still some defects in this method. It is unable to review the previous studies in a more deeply way and clarify the research context of different branches. Therefore, based on the systematic quantitative literature review, this article further carries out traditional literature review in order to better understand the research of LUT.

## 3. Statistical Analysis of Literatures Concerning LUT Research

### 3.1. An Overview of LUT Research

The number and trends of published literatures concerning LUT research from 1987 to 2020 were analyzed (Figure 1). We found that the number of literatures in this field has shown a fluctuating upward trend, and the number of published papers showed a rapid upward trend after 2013. According to the number of annual publications, the research on LUT can be roughly divided into two stages: (1) Slow growth stage (1987–2006). Research on LUT has been developed from scratch, and some developed countries have begun to devote themselves to related research on forest transition. (2) Rapid development stage (2007–present). Research on LUT has gradually received attention, the number of papers related to the subject of LUT has increased rapidly, and scholars have carried out a series of researches from different disciplines and perspectives with a variety of methods and technical means.

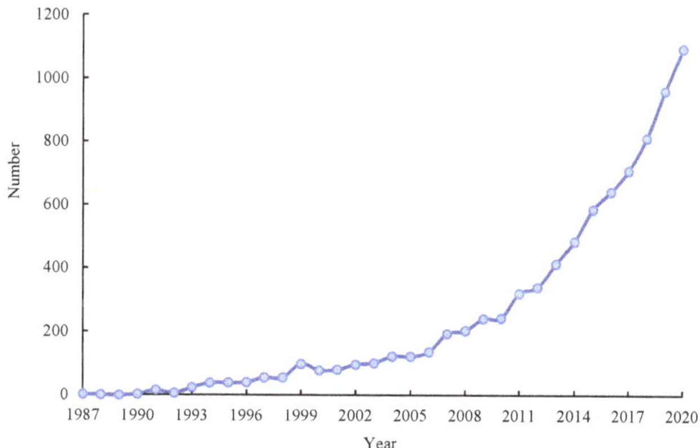


**Figure 1.** Number of literatures concerning LUT research from 1987 to 2020.

According to data from Web of Science, by the end of 2020, the top three countries with the number of publications on LUT research are USA (2982), China (1496) and Germany (844), followed by UK, Australia, Canada, Netherlands, France, Spain and Italy (Table 1). Research on LUT has attracted widespread attention in various fields. Statistical analysis shows that research results related to LUT have been published in more than 1600 SCI/SSCI indexed journals, covering multiple disciplines and fields such as geography, environmental science, ecology, sociology, economics and urban planning. The top 10 journals with publication volume are: *Land Use Policy, Sustainability, Remote Sensing of Environment, Remote Sensing, Science of the Total Environment, Land, Plos One, Applied Geography, Landscape and Urban Planning* and *Environmental Research Letters* (Figure 2).

**Table 1.** Major countries publishing articles concerning LUT.

| Rank | Country | Number of Articles | Centrality [a] |
|------|---------|--------------------|----------------|
| 1 | USA | 2982 | 0.32 |
| 2 | China | 1496 | 0.03 |
| 3 | Germany | 844 | 0.19 |
| 4 | UK | 727 | 0.14 |
| 5 | Australia | 563 | 0.16 |
| 6 | Canada | 515 | 0.17 |
| 7 | Netherlands | 454 | 0.1 |
| 8 | France | 435 | 0.14 |
| 9 | Spain | 358 | 0.08 |
| 10 | Italy | 350 | 0.05 |

Note: [a] Centrality is an indicator to measure the importance of nodes in the network [10]. The larger the value of centrality is, the more the number of publications cooperated with other countries.

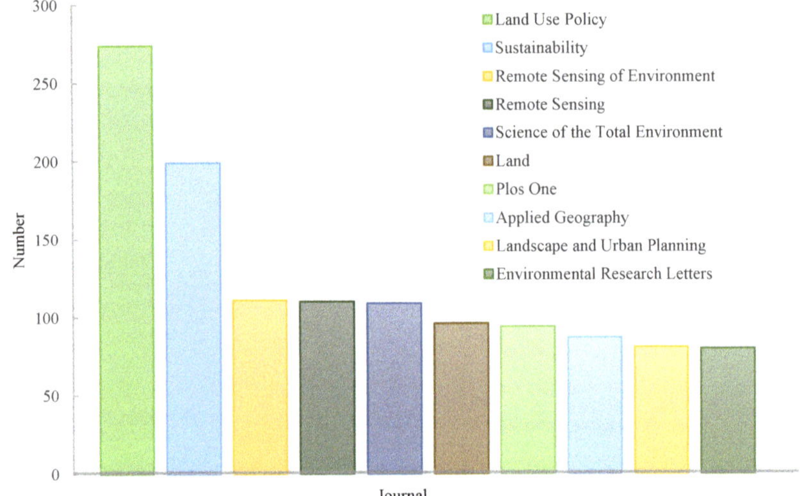


**Figure 2.** The top 10 journals with publication number concerning LUT during 1987–2020.

*3.2. Evolving Research Hot Topics*

### 3.2.1. Analysis of Keywords and Hot Topics Distribution

CiteSpace provides three visualizations methods: cluster view, timeline view and time-zone view. Among them, the timeline view focuses on delineating the relationship between clusters and the historical span of literature in a certain cluster. Based on CiteSpace, the keywords and hot topics related to LUT research since 2000 (few literatures on LUT previous to this) were analyzed. CiteSpace provides two indicators, modularity Q (Q) and weighted mean silhouette (S), based on the network structure and the definition of clustering. It can be used as a basis for us to judge the effect of atlas rendering. Generally speaking, Q value is generally within the interval of (0, 1), and Q > 0.3 means that the community structure divided is significant. Weighted mean silhouette means the homogeneity of the cluster. The higher the value is, the more consistent the members in the cluster will be. S > 0.7, means that the clustering is efficient and compelling [9]. The result showed the modularity Q and weighted mean silhouette of the cluster analysis are 0.6333 and 0.7154, respectively, indicating that the model clustering results are scientific and reasonable. Finally, the timeline map of LUT research from 2000 to 2020 was obtained (Figure 3). Related research hot topics can be roughly divided into 11 categories, i.e., LUT, land change, rural development, fallow management, circulation, shifting cultivation, change detection, habitat, land-use change, rural poverty alleviation and grassland traditional management. There are 10 keywords with a frequency of more than 30, i.e., dynamics, impact, China, deforestation, pattern, forest transition, cover change, urbanization, land use and model. Through the analysis of high-frequency words, it is found that the keywords of LUT research cover a wide range, and there are obvious differences in the research focus and hot topics at different stages. In general, it can be divided into the following three stages:

(1) Slow growth stage (2000–2007): This stage focuses on forest transition and land use change caused by large-scale deforestation due to population growth and agricultural expansion, as well as the impact of LUT on climate change, landscape, ecosystem, grassland management and agriculture policy.

(2) Fluctuant rising stage (2008–2012): At this stage, research on LUT has gradually attracted attention. The research focuses on land use change under the context of globalization, and the impact of farmland abandonment, grassland degradation and other factors on land use management and sustainable regional development.

(3) Rapid development stage (2013–present): Related research pays more attention to LUT and its resources and environmental effects in the process of globalization and rapid urbanization. Measuring methods and models of LUT have been explored extensively. At this stage, land abandonment and farmers' livelihood changes brought about by farmland and rural housing land transition have arisen the attention on the issues of ecosystem service changes.

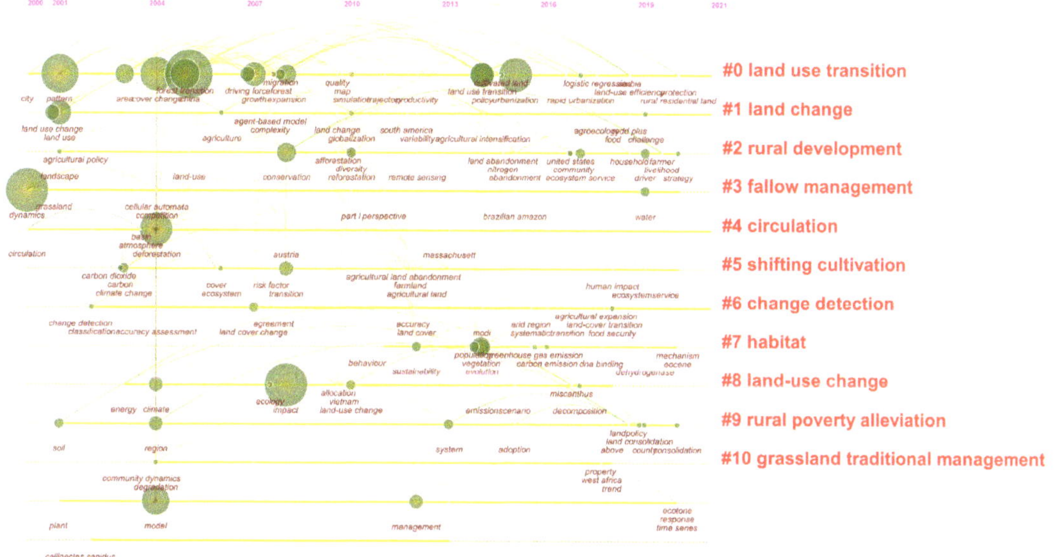



**Figure 3.** Timeline map of LUT research.

### 3.2.2. Burst Words Analysis

The keywords emergence degree can be used to explore the words with high frequency changes in a certain period of time from a large number of subject words, thereby reflecting the change of research hot topics during that period. Burst words represent the phenomenon that the keywords to be investigated transition in a short period of time. Burst words can detect words with a high frequency change rate in a certain period of time from a large number of subject words by investigating word frequency, emphasizing sudden change. Burst terms detection in CiteSpace was used to detect the emergent keywords in the LUT research from 2000 to 2020, and 25 emergent words were detected (Table 2). It can be seen from Table 2 that at different stages, there are obvious differences in research focus and hot topics areas. Before 2010, there were relatively few research on LUT, mostly focusing on the impact of grassland degradation and deforestation on ecosystems, as well as the spatio-temporal evolution characteristics of LUT and simulation studies. After 2010, the direction of LUT research has become more diverse, the frequency of hot topics has increased and more attentions have been paid to the research on complex issues caused by LUT. From 2010 to 2017, research topics such as land-change, transition-matrix, management, land-cover change and land-use change received more attentions. From 2017 to 2020, relevant research pays more attention to the impact of urbanization expansion and globalization on LUT. Among them, the spatio-temporal evolution of land-use change process, driving factors and its impacts on regional sustainability have become hot topics.

**Table 2.** Top 25 keywords with the strongest citation bursts during 2000–2020.

| Keywords | Year | Strength [a] | Begin | End | 2000–2020 [b] |
|---|---|---|---|---|---|
| vegetation | 2000 | 14.51 | 2000 | 2007 | |
| evolution | 2000 | 7.01 | 2000 | 2007 | |
| simulation | 2000 | 15.47 | 2000 | 2011 | |
| record | 2000 | 9.2 | 2000 | 2014 | |
| ecosystem | 2000 | 11.03 | 2002 | 2010 | |
| fire | 2000 | 8.45 | 2002 | 2011 | |
| pasture | 2000 | 7.41 | 2004 | 2011 | |
| grassland | 2000 | 8.67 | 2005 | 2009 | |
| deforestation | 2000 | 7.32 | 2007 | 2008 | |
| forest transition | 2000 | 7.17 | 2008 | 2013 | |
| land-change | 2000 | 15.71 | 2010 | 2010 | |
| transition-matrix | 2000 | 3.22 | 2010 | 2010 | |
| management | 2000 | 3.3 | 2012 | 2018 | |
| carbon stock | 2000 | 7.42 | 2014 | 2017 | |
| land cover change | 2000 | 3.44 | 2015 | 2016 | |
| land-use change | 2000 | 3.35 | 2015 | 2017 | |
| expansion | 2000 | 3.62 | 2017 | 2020 | |
| sustainable development | 2000 | 9.74 | 2018 | 2019 | |
| land use transition | 2000 | 4.11 | 2018 | 2020 | |
| life cycle assessment | 2000 | 7.05 | 2018 | 2020 | |
| urbanization | 2000 | 6.39 | 2018 | 2020 | |
| ecosystem service | 2000 | 11.36 | 2019 | 2020 | |
| renewable energy | 2000 | 9.2 | 2019 | 2020 | |
| politics | 2000 | 7.51 | 2019 | 2020 | |
| consolidation | 2000 | 7.45 | 2019 | 2020 | |

Note: [a] Strength is an indicator to measure the degree of a burst event. The larger the value is, the more active the keyword is in the research field. [b] The red line indicates the year with active burst words, and the green line indicates the year with inactive burst words.

### 3.2.3. Analysis of Institutional Cooperation Network

The institutional cooperation network map can show us how the various institutions are connected, as well as the contribution of each institution in the field of LUT research, which helps us identify researchers and institutions that deserve attention. Through analyzing the major research institutions and cooperation networks of LUT research, we found that LUT research has received extensive attentions in 88 countries and 420 research institutions all over the world (Figure 4). Universities and scientific research institutes have relatively close ties and cooperation. There are 47 institutions with more than 40 articles. The Chinese Academy of Sciences occupies a central position in the cooperation network in the field of LUT research, with University of Maryland, Beijing Normal University, University of Wisconsin, National Aeronautics and Space Administration, Colorado State University, Humboldt University, University of Copenhagen and Peking University, as the linkage of the network. In addition, Wageningen University, Michigan State University, University of Amsterdam, Arizona State University, Columbia University, Stanford University, Yale University and other research institutes have also published more fruitful works.

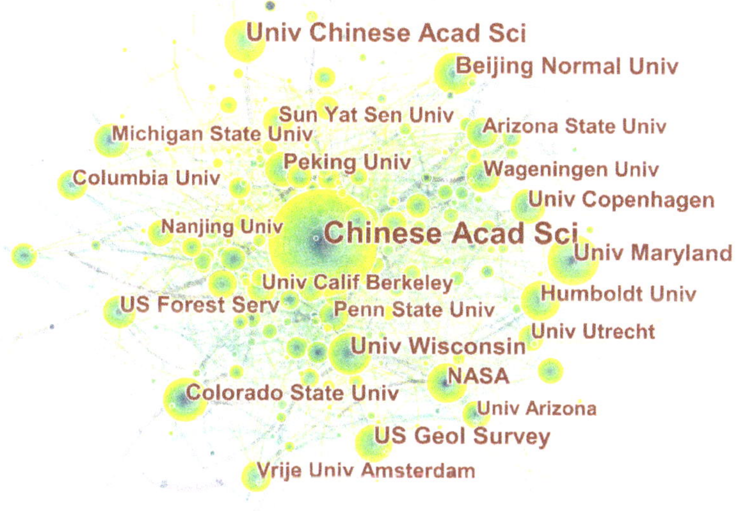


**Figure 4.** Institutional cooperation network map of LUT research.

## 4. Key Fields and Hot Topics of LUT Research

*4.1. Theories and Hypothesis of LUT*

Due to population growth, the global demand for food has accelerated the transformation of natural ecosystems into agricultural land. However, in some developed countries with diversified livelihood strategies, forest coverage has also increased. The latter trend is referred to as the forest transition, which is defined as the transition from net deforestation to net forest coverage increase [11]. In the early 1990s, Mather pioneered the forest transition hypothesis [12,13]. In 1995, Grainger proposed the concept of LUT from the perspective of land use morphology changes in forestry countries. He assumed that most forestry countries have to go through some stages of development: continuous deforestation and increased forest land until a new balance is reached between the forestry and agricultural sectors [14,15]. Forested land can even increase again due to self-regeneration and artificial afforestation. This turning point is what Mather calls forest transition, that is, at this point, the national forestry cover stops decreasing and starts to increase. In 2005, Foley constructed a stage model of LUT, suggesting that land may undergo five stages of transformation from natural ecosystems such as forests to territorial reclamation, subsistence agriculture and smallholder management, gradual intensification and intensive use [5]. Due to differences in the historical, social, economic conditions and ecological environment of different regions, the speed and stage of LUT are also different, and they are subject to socio-economic levels and national policies.

Human activities have modified the natural environment considerably. As the population grows, growing demand for food makes more land is needed to expand food production, which intensifies land-use and land-cover changes (LUCC) [16–18]. To gain better understanding of land-use and land-cover changes and of the physical and human driving forces behind these processes, LUCC project was cosponsored [19]. LUT is one of the manifestations of LUCC, and is also an important research content of GLP. Scholars have carried out research on the conceptual connotation, theoretical models, measurement methods, driving mechanism and environmental effects of LUT [20–24]. Long theorized land use transitions by developing and expanding the concept and connotations of land use morphology as dominant morphology and recessive morphology [23]. The dominant morphology refers to the land use structure of a certain region over a certain period of time, with features such as the quantity (area and proportion) and spatial pattern of land use types. While the recessive morphology includes the land use features in the aspects of quality (nutrient, pollution and degradation), property rights (state-owed and collective-owed), management mode (individual, joint-stock system and transfer and large-scale management), input (capital, technology and labor), output (yield, output value and input-output ratio) and function (production, living, ecology and culture). Accordingly, the concept of LUT was further developed as the changes in land use morphologies, including dominant morphology and recessive morphology, of a certain region over a certain period of time driven by socio-economic change and innovation, and it usually corresponds to the transformation of the socio-economic development stage [23,25]. Long put forward the theoretical model of regional land use transitions, i.e., as the socio-economic development, the competition/trade-off between different land use types presents a decreasing trend, and finally achieves a stable equilibria [25,26].

Based on the special socio-economic, socio-ecological and physical conditions, some scholars probed the research theoretical framework and hypotheses of LUT [27–32]. Qu and Long (2018), based on existing researches and the Environmental Kuznets Curve, put forward a theoretical hypothesis of the interactive mechanism among the land use transitions, the economic effect, the environmental effect and the land use management. Finding that there was a one-way Granger causality from urban construction land use transitions to economic development and environmental pollution, respectively, and no significant Granger causality was found from land use management to economic development or environmental pollution [33]. Some scholars supported that LUT refers to any change in land use systems from one state to another one, land use change is non-linear and different parts of the world are in different transition stages, depending on their history, social and economic conditions and ecological context [32,34].

### 4.2. Measuring LUT

The selection of land use morphology indicators and the measurement of its transition process are the premise and basis for analyzing the characteristics of LUT. The extension of land use morphology brings about opportunities and challenges as the qualitative aspect of land use transitions is reflected by the changes of recessive land use morphology, which is difficult to be measured or represented [35,36]. The research on the dominant morphology of land use is an important prerequisite for the recessive morphology research. The dominant and recessive morphologies are coupled to construct the characterization index of LUT, and various methods are used to quantify LUT. Comprehensive measurement helps to explore the characteristics and regularities of LUT from multiple perspectives and levels. Accordingly, Long put forward three innovative integrated approaches to study land use transitions: one is the multidisciplinary research framework for recessive LUT which involves disciplines including geography, management, economics and sociology [23]; another is the horizontal comparison research method with space to exchange for time [25]; the other is the transect research method based on the key gradient factor of regional socio-economic development [37]. Tsai used interactive LUT agent-based model by endogenizing the interactions of socio-ecological feedbacks and socio-economic factors in a generalizable model to simulate changes in land use caused by farmers' decision-making behaviors, and the recursive effects of land use changes on farmers' decision-making behaviors, and explored the conditions for forest transition in different scenarios [38]. Some scholars have used land satellite images and GIS to explore the trajectory of long-term series of forest cover changes, reveal the main driving paths of forest transition, and analyze the impact of forest transition on ecosystem products and services [39–41].

Through literature review, it is found that since the end of the 20th century, related research has shifted from focusing on single-dimensional LUT to multi-dimensional one [42–44]. At present, scholars are conducting research on the measurement, simulation, spatial differentiation characteristics and influence factors of LUT based on remote sensing data, national statistics data and survey data [45–49]. The measurement methods include classification and regression tree (CART) models, interactive land use transition agent-based model (ILUTABM), global land-use model (GLM), system of environmental-economic accounting (SEEA), center of gravity model, cold/hot spots analysis and other methods [22,50–58] (Table 3). At the same time, methods such as structural analysis, questionnaire interviews and the spatial econometric model have also been gradually applied to related researches. Abundant data sources and multiple models provide a variety of ideas for the measurement of LUT, and also provide scientific support for the research of LUT caused by socio-ecological feedback under the background of globalization. However, current researches focus on the measurement of the dominant morphology transformation of land use, while the measurement of the recessive morphology transformation of land use and its impact on "social-economic-ecological" still need to be further explored.

**Table 3.** Characterization and measurement methods of LUT.

| Data [a] | Methods | Object/Research Question | Reference |
|---|---|---|---|
| Remote sensing data | Classification and regression tree (CART) models | Land use transitions in unsustainable arid agro-ecosystems | Romo et al., 2014 [41]; Bonilla-Moheno and Aide, 2020 [50] |
| | Cellular automata models | Rules relate LUCC variables to the observed historical changes | Roodposhti et al., 2019 [51] |
| | Land-use transfer matrix | Regional land use type conversion | Liu and Long, 2016 [22]; Quintero-Gallego et al., 2018 [52] |
| | Interactive land use transition agent-based model (ILUTABM) | Simulates the land use changes resulting from farmers' decision | Tsai et al., 2019 [56] |
| Statistics data | Global land-use model (GLM); earth system models (ESMs) | Harmonization of land-use scenarios | Hurtt et al., 2020, 2011 [47,57] |
| | Transect research method | Rural housing land transition | Long et al., 2007 [37] |
| | Land use change (LUC) models, Dyna-CLUE model | Assessment Land use change modelling accuracy | Lü et al., 2020 [54] |
| | System of environmental-economic accounting (SEEA) | Land cover account | Wentland et al., 2020 [48]; Weber, 2007 [49] |
| Survey data | Ethnographic fieldwork | How customary land tenure systems mediate transformations of land use and livelihoods | Rignall and Kusunose, 2018 [58] |
| | Decoupling index model and balance index model | Coupling relationship of land use transition between cultivated land and rural residential land in China | Qu et al., 2019 [55] |

Note: [a] Measuring LUT is highly depended on the data sources, which is an important criterion and perspective for the classification of the techniques of measuring LUT. Therefore, we divided the measurement methods of LUT into three types based on data sources, i.e., remote sensing data, statistics data and survey data.

### 4.3. The Impacts of LUT on "Social-Economic-Ecological" System

#### 4.3.1. Impacts of LUT on Social Development

LUT is the result of the interaction between natural environmental conditions and socio-economic factors. Influential factors of LUT include endogenous socio-ecological forces and exogenous socio-economic factors [32]. On the one hand, various land use issues are related to the rapid urban-rural transformation development, which has significant impacts on land use policies [59–61]. On the other hand, socio-economic system and policies, especially those that related to land resources management, are important external factors that play a vital role in shaping land use morphology [62,63].

Considering the regions that LUT takes place, it can be divided into two counterparts-urban area and rural area. Urban LUT is a process of the expansion of construction land and reduction of cultivated land and forestland in the process of urbanization. Farmland transition and rural housing land transition are two crucial contents of rural LUT [64–66]. Against the context of globalization, marketization and urbanization, the growing foreign direct investment and tertiary industry accelerates the expansion of urban construction land, which encroaches on vast farmland and drives the changes of household livelihood and population flow, and, finally, induces the alteration of land use structure [67,68]. In view of rural regions, the variation of regional land use morphology is tightly associated with rural transformation development, and at the same time, is constrained by system vicissitude and national strategy [58,69]. The strategy to alleviate the pressure on land resources in some areas is to move production activities from one area to another [70], and it is not a sustainable way. Therefore, some scholars proposed sustainable land management scheme to assess the risk of land consolidation and agricultural development, reconcile environmental and agricultural policies, and to solve the problems of grassland abandonment and low land use efficiency [71,72].

#### 4.3.2. Impacts of LUT on Economic Growth

LUT is motivated by socio-economic changes. Due to the extensive exchange of energy, material, and information flows between the internal and external urban-rural territorial system, the main bodies of land use are more sensitive to the economic and social responses. Decisions relating to economic development demand often directly or indirectly change the supply of land services, thereby triggering the transformation of land use structure and functions [63]. With the increase of population, in order to meet people's various demands for land in production and living, productive land around the world has been extensively developed and converted [73]. Regarding competition for productive land, different scholars have different views, Malthusian believes that the stock of suitable land is finite, continuous development will lead to a shortage of productive land, which will have a negative impact on welfare. Ricardian reckoned that it becomes economically feasible to bring marginal land into use as prices of land-based commodities increase, but it comes at ever increasing economic, environmental and social costs. The economic impact of LUT is not directly proportional to the area loss, but is affected by the combined effects of soil capacity, dryland crop combination and local economic factors [74]. Due to the changes in socio-economic factors, such as the decline in soil quality, the increase in the opportunity cost of farming, the outmigration of rural labor, the adjustment of agricultural policies, and the reform of the land system, etc., land abandonment has become one of the important trends of global land use changes and it is crucial for agricultural production and landscape planning [1,52,75]. In response to the negative effects of LUT on rural economic development, Ojoyi pointed out that extra employment opportunities and livelihood support activities should be created to minimize dependence on natural resources [7]. Some scholars believe that through rural land use planning and advanced technologies, agglomerated economic production can be formed, which promotes the transition from the fragmented use of land under the subsistence agriculture model to the large-scale management under the intensive farm model, so as to reduce deforestation and relieve land pressure and improve land use efficiency [5,76–78].

Rural reform and development have always been the hot-spot issues of LUT, which function as the tool of regulating land use and promoting socio-economic development [33]. The internal driving force of LUT comes from the trade-offs and games between different stakeholders, which is manifested in the conflict of land use patterns. Driven by conflicts, the structure and function of land use are continuously adjusted to adapt to the new balance requirements, and, finally, LUT is realized through land services supply [4,79]. Farmland use is a complex process of rural agricultural economic reproduction and natural reproduction. The transition of cultivated land use has an important impact on the rural natural landscape and socio-economic development [53,80]. Especially in the context of ecological civilization construction and national food security, research on the mutual feedback mechanism of farmland transition and food security, and its impacts on farmers' livelihoods, rural industrial development and rural transformation development have received widespread attention [24,81].

### 4.3.3. Impacts of LUT on Ecosystem Services

At present, research on LUT and its environmental effects is mainly implemented by using GIS techniques, ecosystem service value assessment, landscape pattern index and ecological environment index, at the scales of regional, drainage basin, provincial, prefecture-level city, county level and township level. In the process of socio-economic development, the impacts of LUT on eco-environment have become one of the research priorities of global change research. The corresponding research contents range from atmospheric composition to terrestrial ecosystem [82–84], which generally can be divided into three aspects: (1) the impacts of LUT on atmospheric environment, water environment, soil environment, vegetation and biodiversity; (2) the impacts on overall ecosystem service; (3) the landscape ecological pattern responses, and the coupling relations between land use structure and land use multifunctionality [85–87].

Land development is revenue-oriented, the increase in human activities and commercial space is mainly at the expense of forest-covered ecosystems, farmland and pasture [19,88–90]. How to deal with the trade-offs between the value derived from new land uses and the cost of lost ecosystem services has become a very important proposition. The rapid transformation and fragmentation of land cover may lead to a series of problems such as biodiversity loss, land degradation, water quality decline, insufficient food supply, extinction of wildlife and environmental degradation [91–95]. Faced with the trade-off between environmental protection and food security, some scholars have proposed ecological plans for cropland reforestation and urban green projects through the production of commodities with high income and price elasticity to alleviate the pressure on the ecological environment caused by over-development of land resources [86,96]. In order to alleviate the pressure on grassland areas caused by the transformation of grassland to cultivated land, the EU sets minimum standards for the protection of the ratio of permanent grassland to protect the ecological value of grassland [97].

### 4.4. Drivers and Regulation of LUT

### 4.4.1. Research on the Driving Factors of LUT

In most cases, LUT is a random process [5]. Carrying out research on the driving factors of LUT will help scientifically regulate the quantity and quality of regional land resources, and is of great significance to regional land use planning, regional ecological environmental protection, mitigating global climate change and vegetation restoration strategies [7,98,99]. From the perspective of the land system, the driving factors of LUT can be divided into endogenous driving forces and external driving forces. The interaction and mutual influence of various factors have a comprehensive and complex impacts on urban-rural development and land use. On the one hand, with population growth, people's demand for productive land and residential land has increased. Urban land and agricultural land have largely replaced other land, and were limited by the location [100–102]. On the other hand, with socio-economic development, global power has become the main determinant of LUT. Facing the pressure of population growth and extreme poverty, national markets and policies created opportunities and constraints for new land uses [103–105].

In general, LUT is the result of the combined effect of endogenous socio-ecological feedback and exogenous socio-economic factors [106]. The intense flows of information, capital, commodities and people generated by the increasing interactions in this globalized world greatly influence the land use patterns, which highlights the forces of the remote markets [107]. The driving forces of LUT are related and heterogeneous in different time and space dimensions, and are affected by many complex factors such as nature, politics, economy, and culture [50,108]. Natural factors include natural disasters, endowment discrepancy and climate change; socio-economic factors incorporates globalization, urbanization, marketization, demand for agricultural products, agricultural production activities and population growth; political factors consist of national policies, land consolidation and land resource management systems.

### 4.4.2. Research on Optimal Regulation of LUT

At present, the root cause of many issues arise from LUT is the contradiction between socio-economic advancement and environmental protection, which results from the fact that, in most cases, economic growth is at the expense of environmental sustainability [77]. How to deal with the relationship between the social and economic benefits and resources and environmental benefits is the key to optimal regulation of LUT. Through literature review, it is found that the optimization and regulation of LUT is mainly realized through engineering and technological means, and policy and system innovation. The main cause of LUT lies in the fact that rural land has been intensively occupied by urban construction land. In terms of the regulation of LUT, it is necessary to change the way in which the external system of the rural area affects the internal system, promoting the free flow of urban and rural elements [109–111]. Land use planning and land consolidation are important engineering techniques to optimize and control the LUT. Rural land use planning is a way to ensure the best use of land. By evaluating and balancing the trade-offs between different social, economic and environmental goals, it discusses how to adjust the land use structure through spatial planning, so as to achieve the optimal land use status and promote the transformation of land use from single-function oriented to multifunctional land use [112–115]. As a policy tool to optimize the structure of land use and improve the efficiency of land use, land consolidation has the dual attributes of engineering projects and policy measures [72,116,117].

In response to the problems induced by LUT, relevant management departments have formulated a series of policy interventions to promote the sustainable use of land resources. Such as America's Endangered Species Act (ESA) and National Environmental Protection Act (NEPA), and land retirement programs and production subsidies, China's "1.8 billion mu red line" and "Grain for Green Policy", and Morocco's Customary land tenure [29–31]. France has adopted environmental policies aimed at the conservation of natural habitats and wildlife, and Cameroon enacted national forest law, which provided the legal basis for the implementation of a land use zoning [8,75,118,119]. It is possible to design alternative land-use management strategies to fight desertification processes [70]. Customary land tenure is essential for regulating land use and farmers' livelihoods and ensuring economic growth [69]. In addition, applications of dynamic land use classification have also been highly recognized. In order to facilitate the targeted implementation of land management strategy, some scholars divided territorial space into rural protection area, suburban coordination area, urban agglomeration area, restricted development area and conditional construction area, and propose corresponding management measures and policies according to the characteristics of each specific area to regulate land use activities and address the relationship between economic development and environmental protection [87,120].

## 5. Challenges and Prospects

The above analysis reveals that there still exists some drawbacks on LUT research that should be further improved. By reviewing the literature regarding the impacts of LUT on social and economic development, and the ecosystem services, we found that these literatures are quantitative biased, which mostly rely on the new approaches, especially the remote-sensing techniques. Researchers have intensively used the geographic information systems to map and quantify the impacts of LUT on ecosystem service values. However, this review of studies reveals a distinct paucity of the comprehensive research underlining both natural and human dimensions of land use activities. A complex systems approach can aid in organizing ideas regarding complex land use process relating to the corresponding policy/institution design, utilization behavior, socio-economic and environmental impacts. Therefore, it is necessary to examine the process and its consequences under the guidance of land system science. Besides, although land use change has been studied at a wide range of spatial and temporal scales, there is currently insufficient research into the LUT at the broadest-scale. Local or regional forces undoubtedly influence the LUT process, while the global driver cannot be ignored. Globalization has been an indispensable factor reshaping land use morphology, and land systems should be understood and modeled as open systems with large flows of goods, people and capital that connect local land use with global-scale factors [121].

### 5.1. LUT Research under the Guidance of Land System Science

Land system science provides a theoretical guidance for integrated LUT research. Current LUT research is to some extent fragmented, merely focusing on certain single land use type, e.g., farmland, forestland, rural housing land, etc. These studies have examined the process, patterns, mechanisms and impacts of land use transitions at the local and regional scale, and, have produced synthesized findings from individual case studies, as well as have generalized our understanding of LUT process [122]. However, only focusing on the one dimension or some key elements of land system cannot meet the demands of the research on LUT as land system not only represents the terrestrial components of the Earth system, but also encompasses all processes and activities related to the human use of land [123,124]. It is acknowledged that the architecture of land system is human–natural coupled, and requires to be studied from an integrated way. Land system research therefore has become an ideal tool to cope with the complexity of LUT. As a comprehensive concept, LUT is fully embodied in the trending variation of land use morphology, which is a so inclusive term that incorporates both dominant morphology (quantity involving area and proportion, spatial pattern of land use types, etc.) and recessive morphology (quality involving nutrient, pollution and degradation, and property rights involving state-owed, collective-owed, etc.). Thus, albeit LUT and land system are different concepts, they all attempt to provide a systematic understanding of land use. Land system science aims to improve the observation, monitoring, understanding, modelling and sustainability of land system and its changes [124,125]. LUT research should be proceeded within the research framework of land system science [126], and requires improved understanding and theorizing of the changes of land use morphologies as a highly dynamic and connected complex system transition process [127].

### 5.2. Attaching Importance to the Transition of Land Tenure Regime

Land tenure regime is one of the important factors affecting the recessive morphology of land use, and its variation and adjustment should be underlined as the existing LUT research is dominant morphology biased. Land system/polices/institutions are instruments of regulating land use activities, and plays a vital role in shaping land use morphology. In reality, LUT is the direct result of human decision-making at multiple scales, with far-reaching consequences for the land use morphology [128]. Policy/institution making is also a human-dominated process, which is complex and intricate. Therefore, to better understand land use transitions, it is necessary to scrutinize the relationships between land tenure regime evolution and the dynamic of land use morphology. Dominant morphology continues to be strongly shaped by policy/institution interventions. It can, therefore, be surmised that land tenure regime not only greatly influences the dominant morphology, but also functions as an ideal analytical lens to examine the regularities of LUT. For instance, China's rural land use system has been reformed and innovated towards an easy-to-transfer policy design [129]. While the scale of land management, input intensity, organization form and other corresponding attributes will change. A better insight into the recessive morphology evolution from the perspective of land tenure regime change is thereby required.

*5.3. Overcoming the Challenges of Detecting the Recessive Morphology of Land Use*

Innovating the technologies and methods of monitoring and modeling the recessive morphology of land use is needed to provide scientific underpinning for deepening LUT research. The key words and burst words analysis show that "land use change", "eocosystem service", "environmental impacts", etc., attracted more attentions in the past few decades. However, these studies mostly rely on the quantitative variation of land use, neglecting the human domain of land use activities. Understanding the consequences of LUT requires robust documentation on the characteristics of transition process. The observation and monitoring of land dominant morphology now mainly relying on remotely sensed data coupled with field observations and corroborating information describing the social, economic and physical dimensions of land use has achieved good detection results [130]. However, the attributes of the recessive morphology of land use encompass soil quality, property rights, management mode, etc., which are hidden, invisible, intricate and difficult to observe, monitor and quantify. Thus, applying state-of-the-art techniques and innovating new methods for understanding the socio-economic dimensions of LUT is of vital importance. For instance, the big data analysis technique is an effective tool of analyzing the land property information based on the land registration data, which can deal with huge volumes of data. Information technology may be an appropriate means of capturing the capital and information flow between urban and rural regions, and, of course, can be employed into analyzing land investment data. These approaches have been used successfully in different fields of LUT research. Yet it is not enough to reveal a full picture of the process of land use transitions [131]. Improved data, upgraded models and case studies in observation and estimation of LUT impacts, which depend only on exploring advanced techniques, are demanded for seeking a deeper understanding the transition of land recessive morphology.

*5.4. Linking Local LUT with Globalization*

The keywords (Table 2) and timeline map analysis (Figure 3) indicated that the impact globalization on LUT have been brought into focus. LUT research has a tradition of place-based studies, focusing on local/regional trending transition of the attributes of land use. As globalization proceeds, there are signs that distal interconnections have played an increasingly role in land use activities. Yet scant attention is given to the distant drivers of LUT. The various materials and non-material flows embodied in international trade and online activities generate direct and indirect changes on land use morphology and the affiliated impacts. In order to understand the consequences of international forces on local land use, approaches or methods from information geography are necessary to capture the visible or invisible information of land use. Causes of LUT are not confined to local factors, but incorporates distant influences, such as remote markets, diffusion of technologies and international political forces. Although short-term fluctuation and changes of land use morphology cannot be understood as LUT, the so-called "transition" stems from the accumulation of the progressive changes or refinement of land use morphology. The accumulation of these subtle and major changes will ultimately restructure local land use structure, and result in the transformation of land use functions. The relation between global land use changes and the emergence of new zoonotic diseases is still unclear. A quantitative analysis indicated that human encroachment into wildlife habitats may contribute to the emergence of zoonotic diseases [132]. However, there is not enough research examining its influence on the emergence of zoonotic disease. Understanding these emerging or hidden interactions and feedbacks between distant socio-economic activities and local land use poses theoretical and methodological challenges. The theoretical lenses through which the remote impacts on local LUT can be framed have been insufficiently explored. It is crucial to develop a new generation of multi-scale models and methods to couple local and global LUT processes. As the pandemic raged, interdisciplinary collaborations are urgently needed to advance knowledge on land use implications for zoonotic disease emergence [133].

## 6. Conclusions

LUT is a locally pervasive and globally significant social-ecological trend. The aims of this article are (1) to investigate the progress of LUT research based on both bibliometric and a systematic review of the literature; (2) to summarize key fields and research hot topics of LUT research; (3) to identify the challenges and suggest potential directions for future research.

We have demonstrated the following:

(1) The annual output of papers has exhibited a general upward trend during 1987–2020. This trend can be interpreted as an indication of the increasing importance of LUT in the research of land system science. Research networks and collaborations including both developing and developed countries have been established, which bring together researchers, practitioners and policymakers from multiple disciplines to work collaboratively on LUT.

(2) Research on LUT is characterized by focusing on complex social issues, driven by realistic demand, and research branch becoming clearer and more systematic. The key fields and hot topics of existing LUT research can be summarized into four aspects: i. theories and hypothesis of LUT; ii. measuring LUT; iii. the impacts of LUT on "social-economic-ecological" system; iv. drivers and regulation of LUT.

(3) The complexity of LUT research requires the diversity of disciplines, methodological approaches and research scales. It has become an interdisciplinary branch of sciences of geography, economics, land management, etc., integrating multiple methods including remote sensing, GIS and mathematical models. The research scale covers multiple levels such as township, region, country and global. Emerging factors continuously bring about both challenges and opportunities to LUT research. Globalization, information technology and other modern techniques complicate LUT process, the mechanism and potential pathways of LUT would be changed. Meanwhile, improvements in related technologies can particularly enhance observing, tracking, monitoring and modeling the recessive morphologies of LUT, thus deepening LUT research.

(4) Research on LUT has still many unresolved fundamental issues. LUT research is "science- and process-centred", theoretical discussions of LUT do not offer enough assistance for regulating and managing land use activities. A focus on local case studies based on contingent factors constrains the theoretical innovation of LUT research. LUT can be apprehended through theoretical generalizations that solves limitations of case studies. LUT theories could benefit from incorporating theories of land system to address the complex interactions, multi-causality and the contextual character of LUT process. Scientific theory on LUT lags behind the research practice. Despite considerable advances in LUT research and related fields, an inclusive theory of LUT or sets of theories have not emerged. Pursuit of the theoretical improvements should also enhance the connections between LUT and global environmental change, resilience and sustainability research, aiming at translating scientific findings on land system into solutions for sustainable land use.

**Author Contributions:** Conceptualization, H.L. and L.M.; methodology, H.L.; software, L.M.; validation, H.L., Y.Z. and S.T.; formal analysis, Y.Z.; investigation, S.T.; resources, H.L.; data curation, L.M.; writing—original draft preparation, H.L.; writing—review and editing, L.M.; visualization, Y.Z.; supervision, H.L.; project administration, H.L.; funding acquisition, H.L. All authors have read and agreed to the published version of the manuscript.

**Funding:** We gratefully acknowledge the support of the Key Program of National Natural Science Foundation of China (Grant No. 41731286).

**Institutional Review Board Statement:** Not applicable.

**Informed Consent Statement:** Not applicable.

**Data Availability Statement:** Not applicable.

**Conflicts of Interest:** The authors declare no conflict of interest.

## References


1. Díaz, G.I.; Nahuelhual, L.; Echeverría, C.; Marín, S. Drivers of land abandonment in Southern Chile and implications for landscape planning. *Landsc. Urban Plan.* **2011**, *99*, 207–217. [CrossRef]
2. Li, S.; Li, X. Global understanding of farmland abandonment: A review and prospects. *J. Geogr. Sci.* **2017**, *27*, 1123–1150. [CrossRef]
3. Xu, D.; Deng, X.; Guo, S.; Liu, S. Labor migration and farmland abandonment in rural China: Empirical results and policy implications. *J. Environ. Manag.* **2019**, *232*, 738–750. [CrossRef] [PubMed]
4. Long, H. Land use policy in China: Introduction. *Land Use Policy* **2014**, *40*, 1–5. [CrossRef]
5. Foley, J.A.; DeFries, R.; Asner, G.P.; Barford, C. Global consequences of land use. *Science* **2005**, *309*, 570–574. [CrossRef]
6. Quintero-Angel, M.; Coles, A.; Duque-Nivia, A.A. A historical perspective of landscape appropriation and land use transitions in the Colombian South Pacific. *Ecol. Econ.* **2021**, *181*, 106901. [CrossRef]
7. Ojoyi, M.M.; Mutanga, O.; Odindi, J.; Kahinda, J.M.; Abdel-Rahman, E.M. Implications of land use transitions on soil nitrogen in dynamic landscapes in Tanzania. *Land Use Policy* **2017**, *64*, 95–100. [CrossRef]
8. Swette, B.; Lambin, E.F. Institutional changes drive land use transitions on rangelands: The case of grazing on public lands in the American West. *Glob. Environ. Chang.* **2021**, *66*, 102220. [CrossRef]
9. Chen, C. CiteSpace II: Detecting and visualizing emerging trends and transient patterns in scientific literature. *J. Am. Soc. Inf. Sci. Technol.* **2006**, *57*, 359–377. [CrossRef]
10. Chen, C.; Ibekwe-Sanjuan, F.; Hou, J. The structure and dynamics of co-citation clusters: A multiple-perspective co-citation analysis. *J. Am. Soc. Inf. Sci. Technol.* **2014**, *61*, 1386–1409. [CrossRef]
11. Meyfroidt, P.; Lambin, E.F. The causes of the reforestation in Vietnam. *Land Use Policy* **2008**, *25*, 182–197. [CrossRef]
12. Mather, A.S. *Global Forest Resources*; Belhaven Press: London, UK, 1990.
13. Mather, A.S. The Forest Transition. *Area* **1992**, *24*, 367–379.
14. Grainger, A. National land use morphology: Patterns and possibilities. *Geography* **1995**, *80*, 235–245.
15. Grainger, A. The forest transition: An alternative approach. *Area* **1995**, *27*, 242–251.
16. De Sousa-Neto, E.R.; Gomes, L.; Nascimento, N.; Pacheco, F.; Ometto, J.P. *Land Use and Land Cover Transition in Brazil and Their Effects on Greenhouse Gas Emissions*; Elsevier: Amsterdam, The Netherlands, 2018; Chapter 20; pp. 309–321.
17. Verburg, P.H.; De Groot, W.T.; Veldkamp, A.J. Global environmental change and land use. In *Methodology for Multi-Scale Land-Use Change Modelling: Concepts and Challenges*; Springer: Amsterdam, The Netherlands, 2003; Part II.
18. Winkler, K.; Fuchs, R.; Rounsevell, M.; Herold, M. Global land use changes are four times greater than previously estimated. *Nat. Commun.* **2021**, *12*, 2501. [CrossRef] [PubMed]
19. Goldewijk, K.K. Estimating global land use change over the past 300 years: The HYDE Database. *Glob. Biogeochem. Cycles* **2001**, *15*, 417–433. [CrossRef]
20. Turner, B.L.; Lambin, E.F.; Reenberg, A. The emergence of land change science for global environmental change and sustainability. *Proc. Natl. Acad. Sci. USA* **2007**, *104*, 20666–20671. [CrossRef]
21. Yeo, I.Y.; Huang, C. Revisiting the forest transition theory with historical records and geospatial data: A case study from Mississippi (USA). *Land Use Policy* **2013**, *32*, 1–13. [CrossRef]
22. Liu, Y.; Long, H. Land use transitions and their dynamic mechanism: The case of the Huang-Huai-Hai Plain. *J. Geogr. Sci.* **2016**, *26*, 515–530. [CrossRef]
23. Long, H.; Qu, Y.; Tu, S.; Zhang, Y.; Jiang, Y. Development of land use transitions research in China. *J. Geogr. Sci.* **2020**, *30*, 1195–1214. [CrossRef]
24. Long, H.; Li, Y.; Liu, Y.; Woods, M.; Zou, J. Accelerated restructuring in rural China fueled by "increasing vs. decreasing balance" land-use policy for dealing with hollowed villages. *Land Use Policy* **2012**, *29*, 11–22. [CrossRef]
25. Long, H. *Land Use Transitions and Rural Restructuring in China*; Springer: Berlin/Heidelberg, Germany, 2020.

26. Long, H.; Qu, Y. Land use transitions and land management: A mutual feedback perspective. *Land Use Policy* **2018**, *74*, 111–120. [CrossRef]
27. Song, X. Discussion on land use transition research framework. *Acta Geogr. Sin.* **2017**, *72*, 471–487. (In Chinese)
28. Hu, S.; Tong, L.; Long, H. Land use transition potential and its assessment framework. *Geogr. Res.* **2019**, *38*, 1367–1377. (In Chinese)
29. Zhang, B.; Sun, P.; Jiang, G.; Zhang, R.; Gao, J. Rural land use transition of mountainous areas and policy implications for land consolidation in China. *J. Geogr. Sci.* **2019**, *29*, 1713–1730. [CrossRef]
30. Song, X.; Li, X. Theoretical explanation and case study of regional cultivated land use function transition. *Acta Geogr. Sin.* **2019**, *75*, 992–1010. (In Chinese)
31. Song, X.; Huang, Y.; Wu, Z.; Ouyang, Z. Does cultivated land function transition occur in China? *J. Geogr. Sci.* **2015**, *25*, 817–835. [CrossRef]
32. Lambin, E.F.; Meyfroidt, P. Land use transitions: Socio-ecological feedback versus socio-economic change. *Land Use Policy* **2010**, *27*, 108–118. [CrossRef]
33. Qu, Y.; Long, H. The economic and environmental effects of land use transitions under rapid urbanization and the implications for land use management. *Habitat Int.* **2018**, *82*, 113–121. [CrossRef]
34. Xu, J.; Sharma, R.; Fang, J.; Xu, Y. Critical linkages between land-use transition and human health in the Himalayan region. *Environ. Int.* **2008**, *34*, 239–247. [CrossRef]
35. Drummond, M.A.; Griffith, G.E.; Auch, R.F.; Stier, M.P.; Taylor, J.L.; Hester, D.J.; Riegle, J.L.; McBeth, J.L. Understanding recurrent land use processes and long-term transitions in the dynamic south-central United States, c. 1800 to 2006. *Land Use Policy* **2017**, *68*, 345–354. [CrossRef]
36. Mizutani, C. Construction of an analytical framework for polygon-based land use transition analyses. *Comput. Environ. Urban Syst.* **2012**, *36*, 270–280. [CrossRef]
37. Long, H.; Hellig, G.K.; Li, X.; Zhang, M. Socio-economic development and land-use change: Analysis of rural housing land transition in the Transect of the Yangtse River, China. *Land Use Policy* **2007**, *24*, 141–153. [CrossRef]
38. Tsai, Y.; Zia, A.; Koliba, C.; Bucini, G.; Guilbert, J.; Beckage, B. An interactive land use transition agent-based model (ILUTABM): Endogenizing human-environment interactions in the Western Missisquoi Watershed. *Land Use Policy* **2015**, *49*, 161–176. [CrossRef]
39. Carmona, A.; Nahuelhual, L. Combining land transitions and trajectories in assessing forest cover change. *Appl. Geogr.* **2012**, *32*, 904–915. [CrossRef]
40. Balthazar, V.; Vanacker, V.; Molina, A.; Lambin, E.F. Impacts of forest cover change on ecosystem services in high Andean mountains. *Ecol. Indic.* **2015**, *48*, 63–75. [CrossRef]
41. Romo-Leon, J.R.; van Leeuwen, W.J.D.; Castellanos-Villegas, A. Using remote sensing tools to assess land use transitions in unsustainable arid agro-ecosystems. *J. Arid. Environ.* **2014**, *106*, 27–35. [CrossRef]
42. Vereijken, P.H. Transition to multifunctional land use and agriculture. *Njas-Wagening. J. Life Sci.* **2003**, *50*, 171–179. [CrossRef]
43. Zhang, Y.; Long, H.; Ma, L.; Ge, D.; Tu, S.; Qu, Y. Farmland function evolution in the Huang-Huai-Hai Plain: Processes, patterns and mechanisms. *J. Geogr. Sci.* **2018**, *28*, 759–777. [CrossRef]
44. Bruzzone, L.; Cossu, R.; Vernazza, G. Detection of land-cover transitions by combining multidate classifiers. *Pattern Recognit. Lett.* **2004**, *25*, 1491–1500. [CrossRef]
45. Feng, Y.; Lei, Z.; Tong, X.; Gao, C.; Chen, S.; Wang, J.; Wang, S. Spatially-explicit modeling and intensity analysis of China's land use change 2000–2050. *J. Environ. Manag.* **2020**, *263*, 110407. [CrossRef]
46. Bruggeman, D.; Meyfroidt, P.; Lambin, E.F. Forest cover changes in Bhutan: Revisiting the forest transition. *Appl. Geogr.* **2016**, *67*, 49–66. [CrossRef]
47. Hurtt, G.C.; Chini, L.; Sahajpal, R.; Frolking, S.; Bodirsky, B.L.; Calvin, K.; Doelman, J.C.; Fisk, J.; Fujimori, S.; Goldewijk, K.K. Harmonization of global land use change and management for the period 850–2100 (LUH2) for CMIP6. *Geosci. Model Dev.* **2020**, *13*, 5425–5464. [CrossRef]
48. Wentland, S.A.; Ancona, Z.H.; Bagstad, K.J.; Boyd, J.; Hass, J.L.; Gindelsky, M.; Moulton, J.G. Accounting for land in the United States: Integrating physical land cover, land use, and monetary valuation. *Ecosyst. Serv.* **2020**, *46*, 101178. [CrossRef]
49. Weber, J. Implementation of land and ecosystem accounts at the European Environment Agency. *Ecol. Econ.* **2007**, *61*, 695–707. [CrossRef]
50. Bonilla-Moheno, M.; Aide, T.M. Beyond deforestation: Land cover transitions in Mexico. *Agric. Syst.* **2020**, *178*, 102734. [CrossRef]
51. Roodposhti, M.S.; Aryal, J.; Bryan, B.A. A novel algorithm for calculating transition potential in cellular automata models of land-use/cover change. *Environ. Model. Softw.* **2019**, *112*, 70–81. [CrossRef]
52. Quintero-Gallego, M.E.; Quintero-Angel, M.; Vila-Ortega, J.J. Exploring land use/land cover change and drivers in Andean mountains in Colombia: A case in rural Quindío. *Sci. Total Environ.* **2018**, *634*, 1288–1299. [CrossRef]
53. Ma, L.; Long, H.; Tu, S.; Zhang, Y.; Zheng, Y. Farmland transition in China and its policy implications. *Land Use Policy* **2020**, *92*, 104470. [CrossRef]
54. Lü, D.; Gao, G.; Lü, Y.; Ren, Y.; Fu, B. An effective accuracy assessment indicator for credible land use change modelling: Insights from hypothetical and real landscape analyses. *Ecol. Indic.* **2020**, *117*, 106552. [CrossRef]
55. Qu, Y.; Jiang, G.H.; Li, Z.; Tian, Y.; Wei, S. Understanding rural land use transition and regional consolidation implications in China. *Land Use Policy* **2019**, *82*, 742–753. [CrossRef]

56. Tsai, Y.H.; Stow, D.; An, L.; Chen, H.L.; Lewison, R.; Shi, L. Monitoring land-cover and land-use dynamics in Fanjingshan National Nature Reserve. *Appl. Geogr.* **2019**, *111*, 102077. [CrossRef]

57. Hurtt, G.C.; Chini, L.P.; Frolking, S.; Betts, R.A.; Feddema, J.; Fischer, G.; Fisk, J.P.; Hibbard, K.; Houghton, R.A.; Janetos, A. Harmonization of land-use scenarios for the period 1500–2100: 600 years of global gridded annual land-use transitions, wood harvest, and resulting secondary lands. *Clim. Chang.* **2011**, *109*, 117–161. [CrossRef]

58. Rignall, K.; Kusunose, Y. Governing livelihood and land use transitions: The role of customary tenure in southeastern Morocco. *Land Use Policy* **2018**, *78*, 91–103. [CrossRef]

59. Long, H.; Liu, Y.; Hou, X.; Li, T.; Li, Y. Effects of land use transitions due to rapid urbanization on ecosystem services: Implications for urban planning in the new developing area of China. *Habitat Int.* **2014**, *44*, 536–544. [CrossRef]

60. Shin, H.; Chae, S. Urbanisation and land use transition in a second-tier city: The emergence of small factories in Gimpo, South Korea. *Land Use Policy* **2018**, *77*, 534–541. [CrossRef]

61. Ge, D.; Wang, Z.; Tu, S.; Long, H.; Yan, H.; Sun, D.; Qiao, W. Coupling analysis of greenhouse-led farmland transition and rural transformation development in China's traditional farming area: A case of Qingzhou City. *Land Use Policy* **2019**, *86*, 113–125. [CrossRef]

62. Lambin, E.F.; Meyfroidt, P.; Rueda, X.; Blackman, A.; Börner, J.; Cerutti, P.O.; Dietsch, T.; Jungmann, L.; Lamarque, P.; Lister, J. Effectiveness and synergies of policy instruments for land use governance in tropical regions. *Glob. Environ. Chang.* **2014**, *28*, 129–140. [CrossRef]

63. Křeček, J.; Haigh, M. Land use policy in headwater catchments. *Land Use Policy* **2019**, *80*, 410–414. [CrossRef]

64. Long, H.; Ge, D.; Zhang, Y.; Tu, S.; Qu, Y.; Ma, L. Changing man-land interrelations in China's farming area under urbanization and its implications for food security. *J. Environ. Manag.* **2018**, *209*, 440–451. [CrossRef]

65. Chen, J.; Gao, J.; Chen, W. Urban land expansion and the transitional mechanisms in Nanjing, China. *Habitat Int.* **2016**, *53*, 274–283. [CrossRef]

66. Chen, K.; Long, H.; Liao, L.; Tu, S.; Li, T. Land use transitions and urban-rural integrated development: Theoretical framework and China's evidence. *Land Use Policy* **2020**, *92*, 104465. [CrossRef]

67. Nguyen, Q.; Kim, D. Reconsidering rural land use and livelihood transition under the pressure of urbanization in Vietnam: A case study of Hanoi. *Land Use Policy* **2020**, *99*, 104896. [CrossRef]

68. Zhu, C.; Zhang, X.; Wang, K.; Yuan, S.; Yang, L.; Skitmore, M. Urban–rural construction land transition and its coupling relationship with population flow in China's urban agglomeration region. *Cities* **2020**, *101*, 102701. [CrossRef]

69. Gao, J.; Jiang, W.; Chen, J.; Liu, Y. Housing-industry symbiosis in rural China: A multi-scalar analysis through the lens of land use. *Appl. Geogr.* **2020**, *124*, 102281. [CrossRef]

70. Martínez-Valderrama, J.; Ibáñez, J.; del Barrio, G.; Alcalá, F.J.; Sanjuán, M.E.; Ruiz, A.; Hirche, A.; Puidgefabregas, J. Doomed to collapse: Why Algerian steppe rangelands are overgrazed and some lessons to help land-use transitions. *Sci. Total Environ.* **2018**, *613–614*, 1489–1497. [CrossRef] [PubMed]

71. Hinojosa, L.; Lambin, E.F.; Mzoughi, N.; Napoléone, C. Constraints to farming in the Mediterranean Alps: Reconciling environmental and agricultural policies. *Land Use Policy* **2018**, *75*, 726–733. [CrossRef]

72. Djanibekov, U.; Finger, R. Agricultural risks and farm land consolidation process in transition countries: The case of cotton production in Uzbekistan. *Agric. Syst.* **2018**, *164*, 223–235. [CrossRef]

73. Lambin, E.F. Global land availability: Malthus versus Ricardo. *Glob. Food Secur.* **2012**, *1*, 83–87. [CrossRef]

74. Deines, J.M.; Schipanski, M.E.; Golden, B.; Zipper, S.C.; Nozari, S.; Rottler, C.; Guerrero, B.; Sharda, V. Transitions from irrigated to dryland agriculture in the Ogallala Aquifer: Land use suitability and regional economic impacts. *Agric. Water Manag.* **2020**, *233*, 106061. [CrossRef]

75. Hinojosa, L.; Tasser, E.; Rüdisser, J.; Leitinger, G.; Schermer, M.; Lambin, E.F.; Tappeiner, U. Geographical heterogeneity in mountain grasslands dynamics in the Austrian–Italian Tyrol region. *Appl. Geogr.* **2019**, *106*, 50–59. [CrossRef]

76. Garrett, R.D.; Lambin, E.F.; Naylor, R.L. Land institutions and supply chain configurations as determinants of soybean planted area and yields in Brazil. *Land Use Policy* **2013**, *31*, 385–396. [CrossRef]

77. Rudel, T.K.; Meyfroidt, P. Organizing anarchy: The food security–biodiversity–climate crisis and the genesis of rural land use planning in the developing world. *Land Use Policy* **2014**, *36*, 239–247. [CrossRef]

78. Bruggeman, D.; Meyfroidt, P.; Lambin, E.F. Impact of land-use zoning for forest protection and production on forest cover changes in Bhutan. *Appl. Geogr.* **2018**, *96*, 153–165. [CrossRef]

79. Bertoni, D.; Aletti, G.; Ferrandi, G.; Micheletti, A.; Cavicchioli, D.; Pretolani, R. Farmland use transitions after the CAP greening: A preliminary analysis using Markov Chains approach. *Land Use Policy* **2018**, *79*, 789–800. [CrossRef]

80. Ge, D.; Long, H.; Zhang, Y.; Ma, L.; Li, T. Farmland transition and its influences on grain production in China. *Land Use Policy* **2018**, *70*, 94–105. [CrossRef]

81. Skinner, M.W.; Kuhn, R.G.; Joseph, A.E. Agricultural land protection in China: A case study of local governance in Zhejiang Province. *Land Use Policy* **2001**, *18*, 329–340. [CrossRef]

82. Wu, C.; Chen, B.; Huang, X.; Dennis Wei, Y.H. Effect of land-use change and optimization on the ecosystem service values of Jiangsu province, China. *Ecol. Indic.* **2020**, *117*, 106507. [CrossRef]

83. Bettinger, P.; Merry, K. Land cover transitions in the United States South: 2007–2013. *Appl. Geogr.* **2019**, *105*, 102–110. [CrossRef]

84. Cuo, L.; Zhang, Y.; Gao, Y.; Hao, Z.; Cairang, L. The impacts of climate change and land cover/use transition on the hydrology in the upper Yellow River Basin, China. *J. Hydrol.* **2013**, *502*, 37–52. [CrossRef]
85. Pimm, S.L.; Raven, P. Biodiversity—Extinction by numbers. *Nature* **2000**, *403*, 843–845. [CrossRef] [PubMed]
86. Qiu, L.; Pan, Y.; Zhu, J.; Amable, G.S.; Xu, B. Integrated analysis of urbanization-triggered land use change trajectory and implications for ecological land management: A case study in Fuyang, China. *Sci. Total Environ.* **2019**, *660*, 209–217. [CrossRef] [PubMed]
87. Liang, X.; Jin, X.; Ren, J.; Gu, Z.; Zhou, Y. A research framework of land use transition in Suzhou City coupled with land use structure and landscape multifunctionality. *Sci. Total Environ.* **2020**, *737*, 139932. [CrossRef]
88. Izquierdo, A.E.; Grau, H.R. Agriculture adjustment, land-use transition and protected areas in Northwestern Argentina. *J. Environ. Manag.* **2009**, *90*, 858–865. [CrossRef] [PubMed]
89. Zhao, P.; Yang, H.; Kong, L.; Liu, Y.; Liu, D. Disintegration of metro and land development in transition China: A dynamic analysis in Beijing. *Transp. Res. Part A Policy Pract.* **2018**, *116*, 290–307. [CrossRef]
90. Odongo, V.O.; van Oel, P.R.; van der Tol, C.; Su, Z. Impact of land use and land cover transitions and climate on evapotranspiration in the Lake Naivasha Basin, Kenya. *Sci. Total Environ.* **2019**, *682*, 19–30. [CrossRef] [PubMed]
91. Fan, Z.; Li, J.; Yue, T. Land-cover changes of biome transition zones in Loess Plateau of China. *Ecol. Model.* **2013**, *252*, 129–140. [CrossRef]
92. Beardmore, L.; Heagney, E.; Sullivan, C.A. Complementary land use in the Richmond River catchment: Evaluating economic and environmental benefits. *Land Use Policy* **2019**, *87*, 104070. [CrossRef]
93. Liu, Y.; Long, H.; Li, T.; Tu, S. Land use transitions and their effects on water environment in Huang-Huai-Hai Plain, China. *Land Use Policy* **2015**, *47*, 293–301. [CrossRef]
94. Enaruvbe, G.O.; Keculah, K.M.; Atedhor, G.O.; Osewole, A.O. Armed conflict and mining induced land-use transition in northern Nimba County, Liberia. *Glob. Ecol. Conserv.* **2019**, *17*, e597. [CrossRef]
95. Pérez-Hugalde, C.; Romero-Calcerrada, R.; Delgado-Pérez, P.; Novillo, C.J. Understanding land cover change in a special protection area in central Spain through the enhanced land cover transition matrix and a related new approach. *J. Environ. Manag.* **2011**, *92*, 1128–1137. [CrossRef]
96. Meyfroidt, P. Trade-offs between environment and livelihoods: Bridging the global land use and food security discussions. *Glob. Food Secur.* **2018**, *16*, 9–16. [CrossRef]
97. Nitsch, H.; Osterburg, B.; Roggendorf, W.; Laggner, B. Cross compliance and the protection of grassland—Illustrative analyses of land use transitions between permanent grassland and arable land in German regions. *Land Use Policy* **2012**, *29*, 440–448. [CrossRef]
98. Arai, T.; Akiyama, T. Empirical analysis for estimating land use transition potential functions—Case in the Tokyo metropolitan region. *Comput. Environ. Urban Syst.* **2004**, *28*, 65–84. [CrossRef]
99. Nolte, C.; Gobbi, B.; le Polain De Waroux, Y.; Piquer-Rodríguez, M.; Butsic, V.; Lambin, E.F. Decentralized land use zoning reduces large-scale deforestation in a major agricultural frontier. *Ecol. Econ.* **2017**, *136*, 30–40. [CrossRef]
100. Braimoh, A.K. Random and systematic land-cover transitions in northern Ghana. *Agric. Ecosyst. Environ.* **2006**, *113*, 254–263. [CrossRef]
101. Li, T.; Long, H.; Liu, Y.; Tu, S. Multi-scale analysis of rural housing land transition under China's rapid urbanization: The case of Bohai Rim. *Habitat Int.* **2015**, *48*, 227–238. [CrossRef]
102. Nkeki, F.N.; Asikhia, M.O. Geographically weighted logistic regression approach to explore the spatial variability in travel behaviour and built environment interactions: Accounting simultaneously for demographic and socioeconomic characteristics. *Appl. Geogr.* **2019**, *108*, 47–63. [CrossRef]
103. Lambin, E.F.; Turner, B.L.; Geist, H.J.; Agbola, S.B.; Angelsen, A.; Bruce, J.W.; Coomes, O.T.; Dirzo, R.; Fischer, G.; Folke, C. The causes of land-use and land-cover change: Moving beyond the myths. *Glob. Environ. Chang.* **2001**, *11*, 261–269. [CrossRef]
104. Versace, V.L.; Ierodiaconou, D.; Stagnitti, F.; Hamilton, A.J. Appraisal of random and systematic land cover transitions for regional water balance and revegetation strategies. *Agric. Ecosyst. Environ.* **2008**, *123*, 328–336. [CrossRef]
105. Xu, M.; Zhang, Z. Spatial differentiation characteristics and driving mechanism of rural-industrial Land transition: A case study of Beijing-Tianjin-Hebei region, China. *Land Use Policy* **2021**, *102*, 105239. [CrossRef]
106. Tian, J.; Wang, B.; Zhang, C.; Li, W.; Wang, S. Mechanism of regional land use transition in underdeveloped areas of China: A case study of northeast China. *Land Use Policy* **2020**, *94*, 104538. [CrossRef]
107. Meyfroidt, P.; Lambin, E.F.; Erb, K.; Hertel, T.W. Globalization of land use: Distant drivers of land change and geographic displacement of land use. *Curr. Opin. Environ. Sustain.* **2013**, *5*, 438–444. [CrossRef]
108. Heilmayr, R.; Echeverría, C.; Fuentes, R.; Lambin, E.F. A plantation-dominated forest transition in Chile. *Appl. Geogr.* **2016**, *75*, 71–82. [CrossRef]
109. Johansen, P.H.; Ejrnæs, R.; Kronvang, B.; Olsen, J.V.; Præstholm, S.; Schou, J.S. Pursuing collective impact: A novel indicator-based approach to assessment of shared measurements when planning for multifunctional land consolidation. *Land Use Policy* **2018**, *73*, 102–114. [CrossRef]
110. Li, Y.; Li, Y.; Fan, P.; Long, H. Impacts of land consolidation on rural human–environment system in typical watershed of the Loess Plateau and implications for rural development policy. *Land Use Policy* **2019**, *86*, 339–350.

111. Aquilué, N.; De Cáceres, M.; Fortin, M.; Fall, A.; Brotons, L. A spatial allocation procedure to model land-use/land-cover changes: Accounting for occurrence and spread processes. *Ecol. Model.* **2017**, *344*, 73–86. [CrossRef]

112. Shen, Q.; Chen, Q.; Tang, B.; Yeung, S.; Hu, Y.; Cheung, G. A system dynamics model for the sustainable land use planning and development. *Habitat Int.* **2009**, *33*, 15–25. [CrossRef]

113. Kim, J.H. *Linking Land Use Planning and Regulation to Economic Development: A Literature Review*; SAGE Publications: Los Angeles, CA, USA, 2011; Volume 26, pp. 35–47.

114. Saunders, W.S.A.; Becker, J.S. A discussion of resilience and sustainability: Land use planning recovery from the Canterbury earthquake sequence, New Zealand. *Int. J. Disaster Risk Reduct.* **2015**, *14*, 73–81. [CrossRef]

115. Long, H.; Zhang, Y. Rural planning in China: Evolving theories, approaches, and trends. *Plan. Theory Pract.* **2020**, *21*, 782–786. [CrossRef]

116. Long, H.; Zhang, Y.; Tu, S. Rural vitalization in China: A perspective of land consolidation. *J. Geogr. Sci.* **2019**, *29*, 517–530. [CrossRef]

117. Janus, J.; Markuszewska, I. Forty years later: Assessment of the long-lasting effectiveness of land consolidation projects. *Land Use Policy* **2019**, *83*, 22–31. [CrossRef]

118. Sylvester, K.M.; Brown, D.G.; Deane, G.D.; Kornak, R.N. Land transitions in the American plains: Multilevel modeling of drivers of grassland conversion (1956–2006). *Agric. Ecosyst. Environ.* **2013**, *168*, 7–15. [CrossRef] [PubMed]

119. Bruggeman, D.; Meyfroidt, P.; Lambin, E.F. Production forests as a conservation tool: Effectiveness of Cameroon's land use zoning policy. *Land Use Policy* **2015**, *42*, 151–164. [CrossRef]

120. Wu, X.; Benjamin Zhan, F.; Zhang, K.; Deng, Q. Application of a two-step cluster analysis and the Apriori algorithm to classify the deformation states of two typical colluvial landslides in the Three Gorges, China. *Environ. Earth Sci.* **2016**, *75*, 146. [CrossRef]

121. Lambin, E.F.; Meyfroidt, P. Global land use change, economic globalization, and the looming land scarcity. *Proc. Natl. Acad. Sci. USA* **2011**, *108*, 3465–3472. [CrossRef] [PubMed]

122. Chowdhury, R.R.; Munroe, D.K.; de Bremond, A. Editorial overview: Seeking solutions to land challenges of the Anthropocene: A land systems science perspective. *Curr. Opin. Environ. Sustain.* **2019**, *38*, A1–A5. [CrossRef]

123. Aspinall, R.; Staiano, M. A conceptual model for land system dynamics as a coupled human–Environment system. *Land* **2017**, *6*, 81. [CrossRef]

124. Verburg, P.H.; Erb, K.H.; Mertz, O.; Espindola, G. Land system science: Between global challenges and local realities. *Curr. Opin. Environ. Sustain.* **2013**, *5*, 433–437. [CrossRef]

125. Robinson, T. Adaptation of a resilience program in NSW: The rural RAP. *Acta Neuropsychiatr.* **2006**, *18*, 306–307. [CrossRef]

126. Verburg, P.H.; Crossman, N.; Ellis, E.C.; Heinimann, A.; Hostert, P.; Mertz, O.; Nagendra, H.; Sikor, T.; Erb, K.H.; Golubiewski, N. Land system science and sustainable development of the earth system: A global land project perspective. *Anthropocene* **2015**, *12*, 29–41. [CrossRef]

127. Meyfroidt, P.; Chowdhury, R.R.; de Bremond, A.; Ellis, E.C.; Erb, K.H.; Filatova, T.; Garrett, R.D.; Grove, J.M.; Heinimann, A.; Kuemmerle, T. Middle-range theories of land system change. *Glob. Environ. Chang.* **2018**, *53*, 52–67. [CrossRef]

128. Garrett, R.D.; Lambin, E.F.; Naylor, R.L. The new economic geography of land use change: Supply chain configurations and land use in the Brazilian Amazon. *Land Use Policy* **2013**, *34*, 265–275. [CrossRef]

129. Zhu, F.; Zhang, F.; Ke, X. Rural industrial restructuring in China's metropolitan suburbs: Evidence from the land use transition of rural enterprises in suburban Beijing. *Land Use Policy* **2018**, *74*, 121–129. [CrossRef]

130. Lu Xiao Shi, Y.Y.; Chen, C.L.; Yu, M. Monitoring cropland transition and its impact on ecosystem services value in developed regions of China: A case study of Jiangsu Province. *Land Use Policy* **2017**, *69*, 25–40.

131. Chen, R.; Ye, C.; Cai, Y.; Xing, X.; Chen, Q. The impact of rural outmigration on land use transition in China: Past, present and trend. *Land Use Policy* **2014**, *40*, 101–110. [CrossRef]

132. Rulli, M.C.; DOdorico, P.; Galli, N.; Hayman, D.T.S. Land-use change and the livestock revolution increase the risk of zoonotic coronavirus transmission from rhinolophid bats. *Nat. Food* **2021**, *2*, 409–416. [CrossRef]

133. Plowright, R.K.; Reaser, J.K.; Locke, H.; Woodley, S.J.; Patz, J.A.; Becker, D.J.; Oppler, G.; Hudson, P.J.; Tabor, G.M. Land use-induced spillover: A call to action to safeguard environmental, animal, and human health. *Lancet Planet. Health* **2021**, *5*, e237–e245. [CrossRef]

*land*

*Article*

# Farmland Use Transition in a Typical Farming Area: The Case of Sihong County in the Huang-Huai-Hai Plain of China

Ligang Lyu [1,2,3], Zhoubing Gao [2], Hualou Long [1,4,*], Xiaorui Wang [5] and Yeting Fan [2,3]

[1] Institute of Geographic Sciences and Natural Resources Research, Chinese Academy of Sciences, Beijing 100101, China; liganglv@nufe.edu.cn
[2] School of Public Administration, Nanjing University of Finance & Economics, Nanjing 210023, China; shineover@foxmail.com (Z.G.); 9120191029@nufe.edu.cn (Y.F.)
[3] Key Laboratory of Coastal Zone Exploitation and Protection, Ministry of Natural Resources, Nanjing 210017, China
[4] School of Public Administration, Guangxi University, Nanning 530004, China
[5] Jiangsu Institute for Land Development and Consolidation, Nanjing 210017, China; njuwxr@163.com
* Correspondence: longhl@igsnrr.ac.cn

**Abstract:** An in-depth exploration of the dynamics and existing problems in farmland morphology is crucial to formulate targeted protection policies. In this study, we constructed a morphological evaluation index system to identify the characteristics of farmland use transition in Sihong County of the Huang-Huai-Hai Plain, China. The dominant morphology in terms of area and landscape pattern and the recessive morphology focusing on function were considered in this work. Based on this information, the driving factors of farmland use transition were quantitatively analyzed via the mixed regression model. The following major findings were determined: (1) The area showed a U-shaped change trend during 2009–2018. The patch density (PD) showed an upward trend, and the mean patch size (MPS) showed a downward trend, indicating that the degree of farmland fragmentation increased. The implementation of land consolidation projects increased the area and aggregation of farmland, while urbanization and road construction occupied and divided the farmland, leading to a reduction in area and increase in the degree of fragmentation. (2) The crop production, living security, and eco-environmental function of farmland showed a trend of first decreasing and then increasing. Urbanization increased the demand for agricultural products and the degree of large-scale agricultural production and had a positive impact on the crop production and eco-environmental function of farmland. Our research highlights that increasing farmland fragmentation should be addressed in the farming area. Therefore, the government should formulate efficient policies to curb farmland occupation for urban and traffic utilization.

**Keywords:** land use transition; farmland function; driving factors; Jiangsu Province; China



**Citation:** Lyu, L.; Gao, Z.; Long, H.; Wang, X.; Fan, Y. Farmland Use Transition in a Typical Farming Area: The Case of Sihong County in the Huang-Huai-Hai Plain of China. *Land* **2021**, *10*, 347. https://doi.org/10.3390/land10040347

Academic Editor: Stephen J. Leisz

Received: 16 February 2021
Accepted: 22 March 2021
Published: 31 March 2021



## 1. Introduction

As an important research element of the Global Land Project (GLP), land use transition has received extensive attention from scholars [1,2]. The concept of land use transition, which refers to the long-term and trend-based changes of regional land use morphology, was first proposed based on studies of forest transition [3–5]. Subsequently, research on forest transition as the foci of land use transition was developed, chiefly centering on theoretical progress and empirical studies [2,6,7] in European countries [5,8], as well as Asian [9] and American countries [10,11]. The land use morphology influenced by socio-economic development records the developmental process of the regional human–land system, which in turn affects socio-economic development. The bilateral interaction between these factors has contributed to land use transition [12,13]. Therefore, investigating land-use transition is of great practical value for exploring the methods for managing land resources and promoting the sustainable development of the regional social economy.

Understanding land use morphology is critical for interpreting land use transitions [14]. Land use morphology initially referred to the quantity and spatial structure of land-use types in a certain area within a specific period [3]. With in-depth research on land use transitions, this concept was further developed and expanded to include dominant (quantity and spatial patterns of land-use types) and recessive morphology (quality, property rights, management mode, input, output, etc.) [15]. Recessive morphology is a special type of morphology that relies on dominant morphology but can only be observed using analysis, detection, and investigation [2,15]. Thus, land use transition can be examined based on the changes in dominant land use morphology and recessive land use morphology. Existing studies focus on the theories and hypotheses of land use transition [16], rural housing land [17,18], urban and rural construction land [19], industrial land [20], and other single land-use types [7]; the eco-environmental effects of land use transition [21,22]; the driving mechanisms [23]; and other aspects [24]. Currently, there are many studies on land-use transition based on dominant morphology, but fewer from the perspective of recessive morphology, which is most closely related to land-use management [2].

Farmland is the most important and changeable land-use type in rural areas. Farmland use reflects the evolutionary dynamics of human–land relations in rural areas, as well as the current situation and problems in the development of agriculture and rural society. Therefore, changes in farmland morphology have a crucial impact on regional economic development, food security [25], and ecological security [26]. Recently, farmland use transition has also received significant attention [24]. These studies included the area, proportion, and spatial patterns of farmland [27,28]; farmland losses [29,30]; farmland landscape metrics [31]; farmland production functions [32]; farmland use intensity [33]; and farmland quality [34]. To date, most extant studies have analyzed the temporal and spatial characteristics of farmland use transition from the single perspective of dominant morphology or recessive morphology. However, few studies have comprehensively described the spatiotemporal patterns of farmland use transition combining both dominant and recessive morphology [24]. Moreover, several studies have used cross-section data of land use across different years. However, a lack of continuous data has restricted our understanding of the mechanisms and dynamics of farmland use transition.

The rapid urban–rural transition and development of the Huang-Huai-Hai Plain, whose land area and total population in agricultural areas accounted for 31.7% and 52.6% of China's plain agricultural areas, respectively, brought about an accelerated transition of land use [23]. Many studies have been conducted on the plain, mainly concentrating on the provincial [35] and cross-regional scales [36,37]. However, few studies used a typical county as the analysis object to carry out a long-term series of farmland use transition research. "County", the most basic unit of land management in China, is practically significant for the policy design of farmland protection to explore the characteristics of farmland use transition at the county level. This paper selected Sihong County, a typical region in the Huang-Huai-Hai Plain, as the study area. In recent years, the accelerated urbanization of Sihong led to increasing demand for various construction land, which will inevitably occupy farmland. Therefore, it is necessary to prevent the conversion of farmland into non-agricultural land and to ensure the sustainable use of land resources. This concern is the main problem facing the region at present, so the present study will comprehensively analyze the farmland use transition of Sihong from two complementary perspectives: (1) the spatial transition of farmland (the change in dominant morphology) and (2) the functional transition of farmland (the change in recessive morphology). By analyzing the spatial and functional evolution of the farmland in Sihong County, we further explore the driving factors behind this evolution. The present study will also have a broad implication for creating better land use policy design to optimize the allocation and regulation of regional land resources.

## 2. Materials and Methods

### 2.1. Study Area

Sihong County (33°08′–33°44′ N, 117°56′–118°46′ E) is located in the northwest area of Jiangsu Province, one of the typical agricultural regions in the Huang-Huai-Hai Plain in China (Figure 1). The terrain in this area is dominated by plains and hilly regions and covers about 2693.91 km², with an average elevation of 21.5 m. This area has an average annual temperature of 14.6 °C, and annual precipitation of 893.9 mm. In 2019, Sihong County contained 24 towns and 326 administrative villages, with a population of about 1.095 million, and the population density in this area was about 334 persons/km², with the rural population accounting for 43% of the total population. In 2019, the GDP per capita of Sihong County amounted to CNY 55,111, which was lower than that of China (CNY 70,892) during the same period, and the shares of primary industry, secondary industry, and tertiary industry in the GDP were 16.4%, 37.8%, and 46.1%, respectively. In 2019, the per capita net income of residents was CNY 23,750, which was lower than that of the nation (CNY 30,733) during the same period. This indicates that Sihong can be considered an underdeveloped area in China. Sihong has a long agricultural production history and abundant farmland resources. In 2018, there were 133,091.41 ha of farmland, comprising 49.41% of the total land area of Sihong.

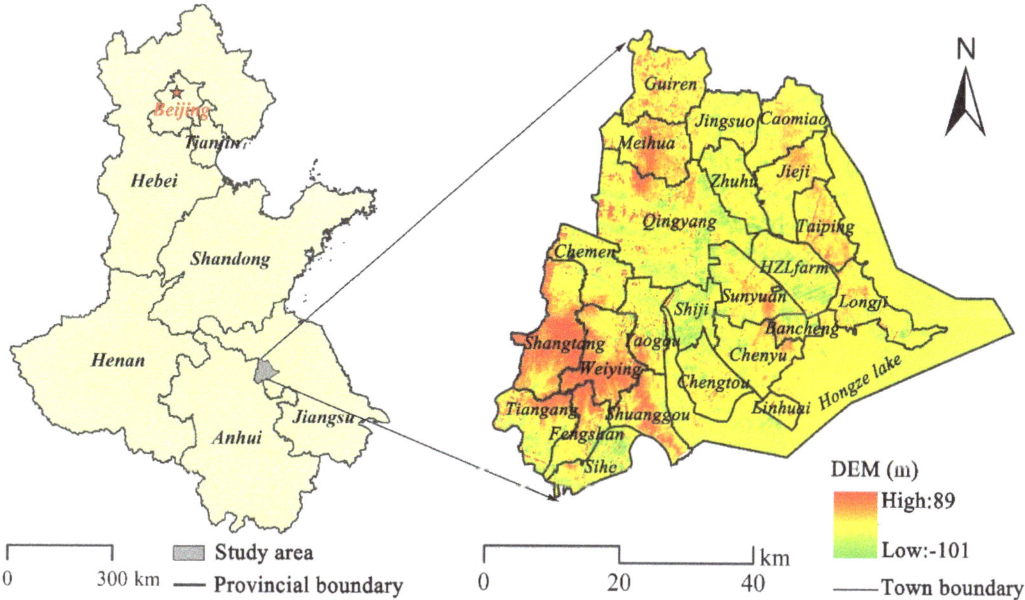


**Figure 1.** Location of Sihong County in the Huang-Huai-Hai Plain of China.

### 2.2. Data Sources

Due to the consistency and accuracy of land use data since the second national land survey in 2009 [38], our study selected the time span between 2009 and 2018 for farmland use transition analysis. The data were provided by the Department of Natural Resources of Jiangsu Province and adopted the standard land use classification system published in 2007 (GB/T 21010—2007), which consisted of 8 classes and 38 subclasses. Tailored to the needs of this study, land-use data were reclassified to 10 classes: farmland (FL), orchard land (OL), forest land (FRL), grassland (GL), urban land (UL), rural residential land (RL), mining land (ML), transportation land (TL), water body (WB), and other land (OTL). DEM data (spatial resolution: 30 m) for calculating the average elevation and slope of each town were downloaded from the geospatial data cloud website (http://www.gscloud.cn/. Accessed on 27 November 2020). Socio-economic data at the county and town level were mainly taken from the Suqian City Statistical Yearbook and Sihong County Statistical Yearbook. The quantitative data of land consolidation projects representing the intensity of land consolidation were provided by Jiangsu Institute for Land Development and Consolidation. The road data (including provincial road, national road, and highway) were extracted from the land-use data from 2009 to 2018 using ArcGIS 10.3.

### 2.3. Methods

#### 2.3.1. Measurement of Farmland Dominant Morphology

The assessment of farmland dominant morphologies was mainly based on the landscape pattern and quantity determined by the area of farmland. Based on previous landscape ecological studies [39–42], three indicators to measure farmland landscape patterns were adopted (Table 1): patch density (PD), mean patch size (MPS), and aggregation index (AI). These indicators were calculated at the class level by FRAGSTATS 4. Patch density (PD) is a measure of the fragmentation of farmland landscape patterns. Low PD values imply fewer patches and indicate farmland continuity, whereas higher values denote more patches, spatial dispersion, and discontinuity. The mean patch size (MPS) mainly describes the morphological changes of patches of farmland, where higher values indicate that the shapes of patches become more concentrated. The aggregation index (AI) reflects the degree of aggregation of patches of farmland. Low AI values indicate fewer aggregation levels of farmland, and vice versa.

**Table 1.** Description of the three indicators to measure farmland landscape patterns.

| Metrics | Formula | Description |
|---|---|---|
| Patch density (PD) | $PD = n/A$ (unit: N/ha) | $n$ = number of farmland patches; $A$ = total landscape area (ha); |
| Mean patch size (MPS) | $MPS = \sum_{i=1}^{n} a_i/n$ (unit: ha) | $a_i$ = area (ha) of farmland patch i; $n$ = number of farmland patches; |
| Aggregation index (AI) | $AI = \left[ \frac{g_{ii}}{\max \to g_{ii}} \right] \times 100$ (unit: Percent) | $g_{ii}$ = number of like adjacencies (joins) between pixels of farmland patches (class) i based on the single-count method. $\max \to g_{ii}$ = maximum number of like adjacencies (joins) between pixels of farmland patches (class) i based on the single-count method. |

#### 2.3.2. Measurement of Farmland Function Morphology

Recessive morphology was described through the farmland function. Based on the previous studies on the multi-functional value of farmland [37,43,44], an evaluation index system covering the crop production, living security, and eco-environmental functions of farmland was established (Table 2). Crop production function refers to the ability of farmland to produce grain, vegetables, melons and fruits, and other crops. The total crop-sown area of Sihong County in 2018 was 1.9849 million ha, and the sown area of grain, oilseeds, vegetables (including vegetable melon), melons (fruit melons), and cotton accounted for 88.44%, 1.50%, 6.43%, 2.26%, and 0.14% of the total sown area, respectively. Since the sown area of cotton was small and presented a decreasing trend, the grain, vegetable, fruit, and oilseed production of the farmland was taken into consideration to evaluate the crop production of the farmland. The living security function included food and employment security functions. The former was reflected by grain yield per unit and per capita farmland area, while the latter was reflected by the proportion of the employees engaging in the plantation industry. The eco-environmental function was greatly affected by the production activities of the farmland; therefore, this function was examined based on the negative effects of the production activities of farmland on the eco-environment. The use intensity of agricultural fertilizer, pesticides, and agricultural plastic film was taken into consideration when selecting indicators to evaluate the eco-environmental function of farmland. The eco-environmental function was expressed by reverse indexes.

**Table 2.** The evaluation indexes of farmland functions.

| Functional Classification | Index | Unit | Direction | Index Calculation Method | Weight |
|---|---|---|---|---|---|
| Crop production | Grain production | kg/hm$^2$ | + | Grain yield/farmland area | 0.483 |
| | Vegetable production | kg/hm$^2$ | + | Total yield of vegetables (including vegetable melons)/farmland area | 0.316 |
| | Melon and fruit production | kg/hm$^2$ | + | The total yield of melons (fruit melons)/farmland area | 0.104 |
| | Oilseed production | kg/hm$^2$ | + | Total oilseed production (peanut and rapeseed)/farmland area | 0.097 |
| Living security function | Per capita grain | kg/person | + | Grain yield/permanent resident population | 0.317 |
| | Per capita farmland area | hm$^2$/person | + | Farmland area/permanent resident population | 0.401 |
| | The proportion of employees in the plantation industry | — | + | Number of employees in plantation/number of rural employees | 0.282 |
| Eco-environmental function | Agricultural fertilizer use intensity | kg/hm$^2$ | − | Chemical fertilizer consumption/farmland area | 0.350 |
| | Pesticide use intensity | kg/hm$^2$ | − | Pesticide consumption/farmland area | 0.322 |
| | The intensity of agricultural plastic film use | kg/hm$^2$ | − | Plastic film consumption/farmland area | 0.328 |

Due to the different dimensions of the evaluation indicators, the maximum difference normalization method was employed to standardize the evaluation indicators in the first step:

$$f_i(k) = \frac{u_i(k) - \min u_i}{\max u_i - \min u_i} \text{ positive indexes Or}$$
$$f_i(k) = \frac{\max u_i - u_i(k)}{\max u_i - \min u_i} \text{ negative indexes. } (i = 1, 2 \dots, m; k = 1, 2 \dots,) \tag{1}$$

where $f_i(k)$ is the dimensionless value of the $i$th index in the $k$th year; $u_i(k)$ is the original value of the $i$th index in the $k$th year; $\max u_i$ and $\min u_i$ represent the maximum and minimum values of the $i$th indicator, respectively; M is the number of indicators; and n is the number of years. Then, each sub-function index was calculated using the formulas below [37]:

$$F(crop) = \sum_{i=1}^{n} W(crop)_i * f(crop)_i \tag{2}$$

$$\text{Or } F(living) = \sum_{i=1}^{n} W(living)_i * f(living)_i \tag{3}$$

$$\text{Or } F(ecol) = \sum_{i=1}^{n} W(ecol)_i * f(ecol)_i \tag{4}$$

where $F(crop)$, $F(living)$, and $F(ecol)$ denote the crop production, living security, and eco-environmental function indexes, respectively. Similarly, $W(crop)_i$, $W(living)_i$, and $W(ecol)_i$ represent the weights of index $i$ for each sub-function. This method combines entropy weighting and multiple correlation coefficient weighting to determine the weights of the indexes [37]. $f(crop)_i$, $f(living)_i$, and $f(ecol)_i$ represent the respective standardized index values. $F(crop)$, $F(living)$, and $F(ecol)$ range between 0 and 1, where the larger the value is, the higher the function indexes are.

### 2.3.3. Identifying Potential Important Driving Factors

Identifying major underlying factors of the farmland use transition was necessary for the rational use and management of farmland based on the comprehensive effects of the natural environment, social economy, land use policy, and other factors [35–37,44–46]. (1) The natural environment mainly included topography and climate. There was little difference in regional factors such as temperature and precipitation across the study area, while topography may have had a more prominent impact on the regional natural environment. Therefore, elevation (El) and slope (Slp) were selected to represent natural conditions. (2) In terms of the level of social and economic development, population density (Pd), urbanization of the population (Urp), farmers' net income (Fi), per capita GDP (Pgdp), and the total proportion of secondary and tertiary industry output value (Stp) were selected to represent social-economic conditions. (3) Since transportation infrastructure is an important spatial factor driving farmland use transition, road density (Rd) was selected as the measurement for this indicator. (4) The government implemented farmland protection policies through a land consolidation project and ultimately realized the control of farmland use transition directly or indirectly [35]. Therefore, the intensity of land consolidation (Lci, the quantitative data of land consolidation projects) was selected to represent land-use policy.

The influencing factors of farmland use transition were explored using the mixed regression model with data for the three periods of 2009, 2013, and 2018 at the town level. The mixed regression model was formulated as follows [46,47]:

$$Y_{mt} = \mu + \beta_{mt}X_{mt} + \varepsilon_{mt} \tag{5}$$

where $Y_{mt}$ is the dependent variable; $X_{mt}$ is the independent variable matrix; $\beta_{mt}$ is the regression coefficient; $\mu$ and $\varepsilon_{mt}$ are the intercept and error terms, respectively; $m$ is the town; and $t$ is the year. The model parameters were fitted using the ordinary least squares (OLS) method. To reduce the data fluctuation of variables, a logarithm transformation was carried out when the values of the variable were greater than 10 in the first step. Then, according to the test results of the variance inflation factor (VIF), the VIF of all variables was determined to be below 5, with no multicollinearity.

## 3. Results

### 3.1. Characteristics of Farmland Use Transition

3.1.1. Farmland Use Spatial Transition

The area of farmland experienced a continuous reduction and reached a recovery net growth during 2009–2018 in Sihong (Figure 2). The farmland area shrank from 132,470.26 ha in 2009 to 131,987.83 ha in 2013, with a decrease of 482.43 ha. However, the area increased to 133,091.41 ha in 2018. Farmland changed from continuous rapid consumption to low-speed consumption and ultimately achieved recovery with the development of economic and social development.

To analyze the internal conversion of farmland in Sihong county, three changing matrixes of farmland were utilized based on the three land use maps (Figure 3). From 2009 to 2013, the main decrease in farmland was mainly due to the occupation of urban land, rural residential land, and transportation land (Figure 4a). From 2013 to 2018, the restorative increase in farmland was caused mainly by the conversion of rural residential land and water bodies (Figure 4b). This occurred because, since 2012, Sihong has carried out the government-led land consolidation project of the "Million Hectares of Fertile Farmland", which converted rural residential land and water bodies into farmland and increased the farmland area. From 2009 to 2018, the main increase in farmland came from rural residential land and water bodies, with rates of 52.26% and 27.50%, respectively (Figure 4c).

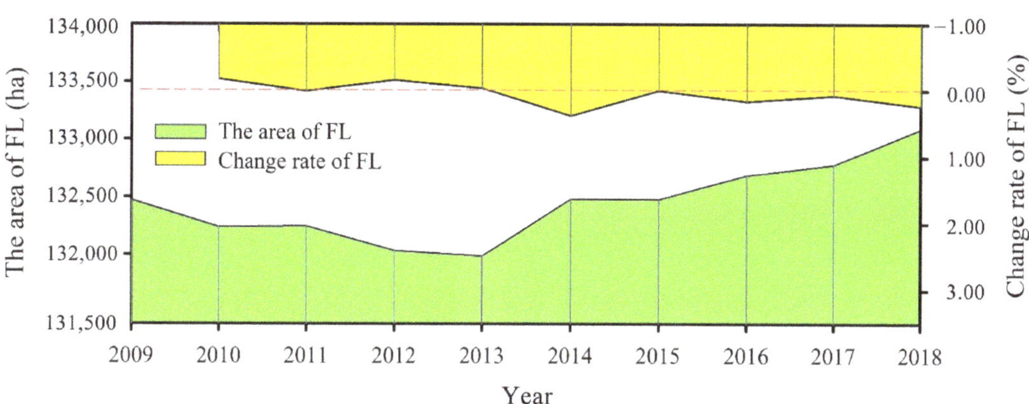

**Figure 2.** Changes in the area of farmland (FL) in Sihong from 2009 to 2018.

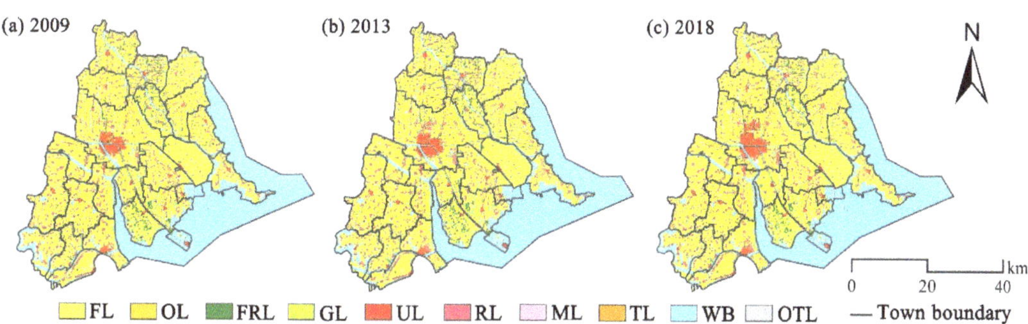



**Figure 3.** The land-use patterns of Sihong in 2009 (**a**), 2013 (**b**), and 2018 (**c**). Note: FL, farmland; OL, orchard land; FRL, forest land; GL, grassland; UL, urban land; RL, rural residential land; ML, mining land; TL, transportation land; WB, water body; OTL, other land.

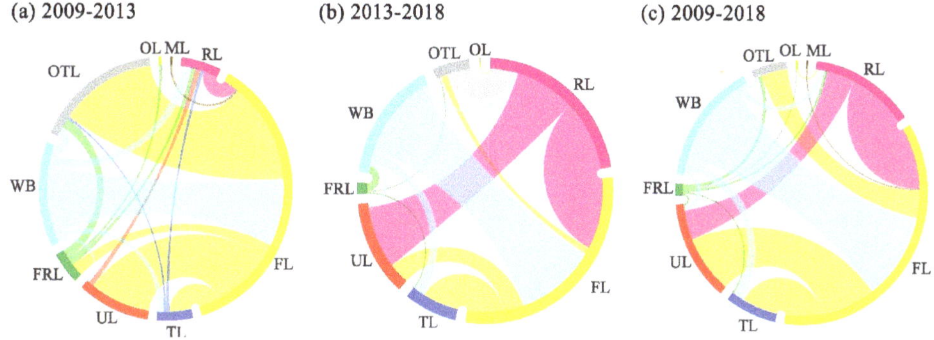



**Figure 4.** Land use conversion flows in Sihong during 2009–2013 (**a**), 2013–2018 (**b**), and 2009–2018 (**c**). The sizes of the lines are proportional in width to the contributions of each land-use type to the change.

The smallest increase in the PD of farmland was identified at the county scale in Figure 5a. The PD of farmland was 3.26 in 2009 and increased to 3.71 in 2018 (Figure 5a). By comparing the PD of farmland at the township scale in 2009, 2013, and 2018, similar changing characteristics to the county level were observed (Figure 6a–c). The MPS of farmland presented a general downward trend at the county scale and decreased from 16.71 ha in 2009 to 14.68 ha in 2018 (Figure 5b). The number of townships with low MPS values (1.77~9.05) increased from three in 2009 to six in 2018 (Figure 6d–f). This increase in PD and decrease in MPS suggests that the fragmentation degree of farmland increased. This mainly occurred because the farmland was invaded by urban land, villages, and transportation infrastructure construction and was spatially divided into fragmented patches, leading the fragmentation of the farmland to gradually increase. Generally speaking, land consolidation causes an increase in the MPS and decrease in the PD of farmland. However, in the current study area, each land consolidation project is small and dispersed and does not significantly increase the MPS of the farmland. Indeed, each project may have increased the farmland PD. The AI of farmland patches showed a falling and then rising trend (Figure 5c). The AI of farmland decreased from 96.67% in 2009 to 97.65% in 2013 and increased to 97.662% in 2018. This indicates that the degree of aggregation and connectivity possesses a transition characteristic of first decreasing and then increasing. This is mainly because Sihong has carried out the government-led land consolidation project of the "Million Hectares of Fertile Farmland" since 2012, which has improved the agglomeration and connectivity of the farmland landscape. The number of townships with high AI values (97.71–98.51%) decreased by 2 during 2009–2018, which indicates that the AI of farmland in some townships declined (Figure 6g–i).

### 3.1.2. Farmland Function Transition

At the county scale, the crop production function of farmland displayed a fall–rise trend during 2009–2018 (Figure 7a), and the index of the crop production function of farmland substantially rose from 0.210 in 2009 to 0.815 in 2018. In particular, the functions of grain and vegetable production of the farmland were substantially improved, while the fruit and oilseed production functions decreased substantially in the study area from 2009 to 2018 (Figure 7b). From the perspective of the township level, the crop production function of farmland in some townships also improved. The number of townships with high F (crop) index values (0.52–0.78) increased from four in 2009 to six in 2018 (Figure 8a–c).

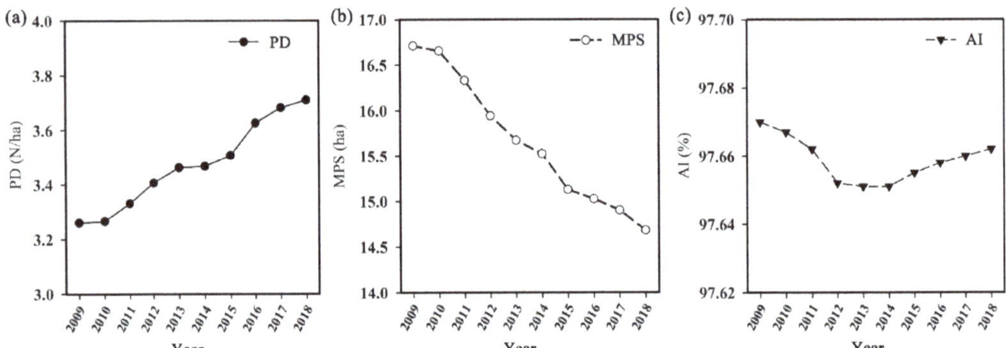

**Figure 5.** Changes in the PD (**a**), MPS (**b**), and AI (**c**) of farmland in Sihong from 2009 to 2018.

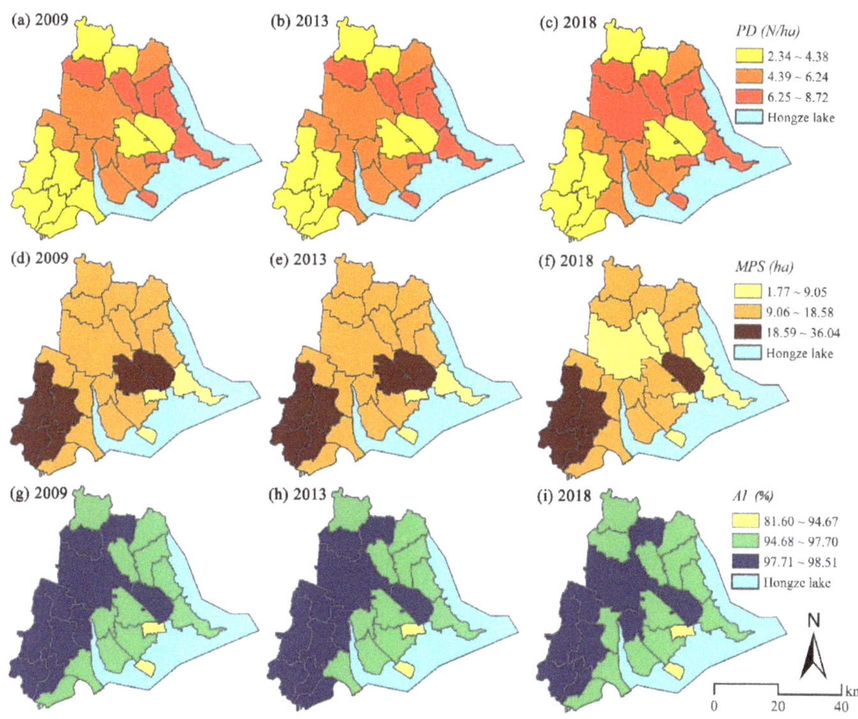

**Figure 6.** The spatial patterns of farmland for PD, MPS, and AI in Sihong at the town level in 2009, 2013, and 2018.

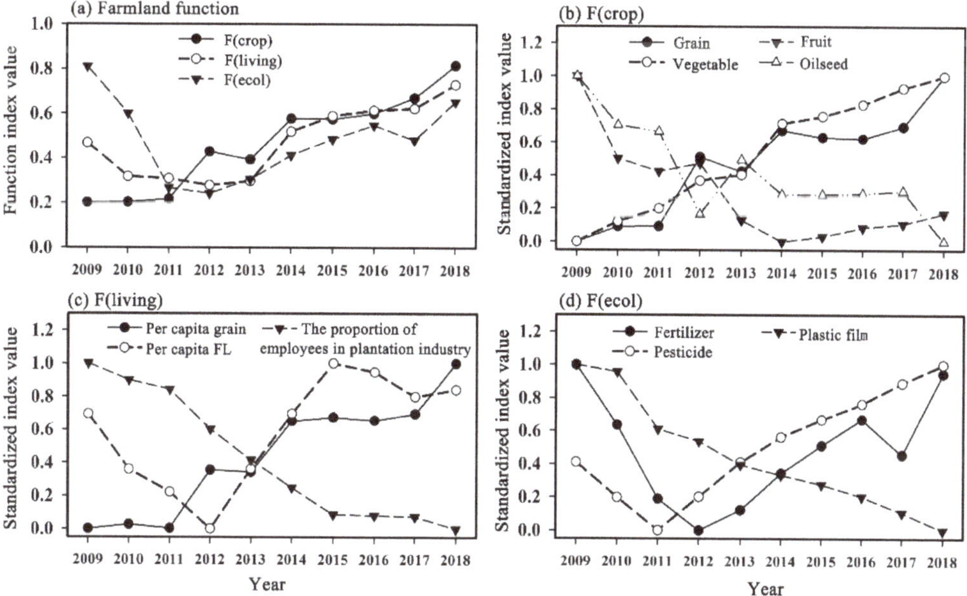

**Figure 7.** Changes in the farmland function in Sihong from 2009 to 2018.

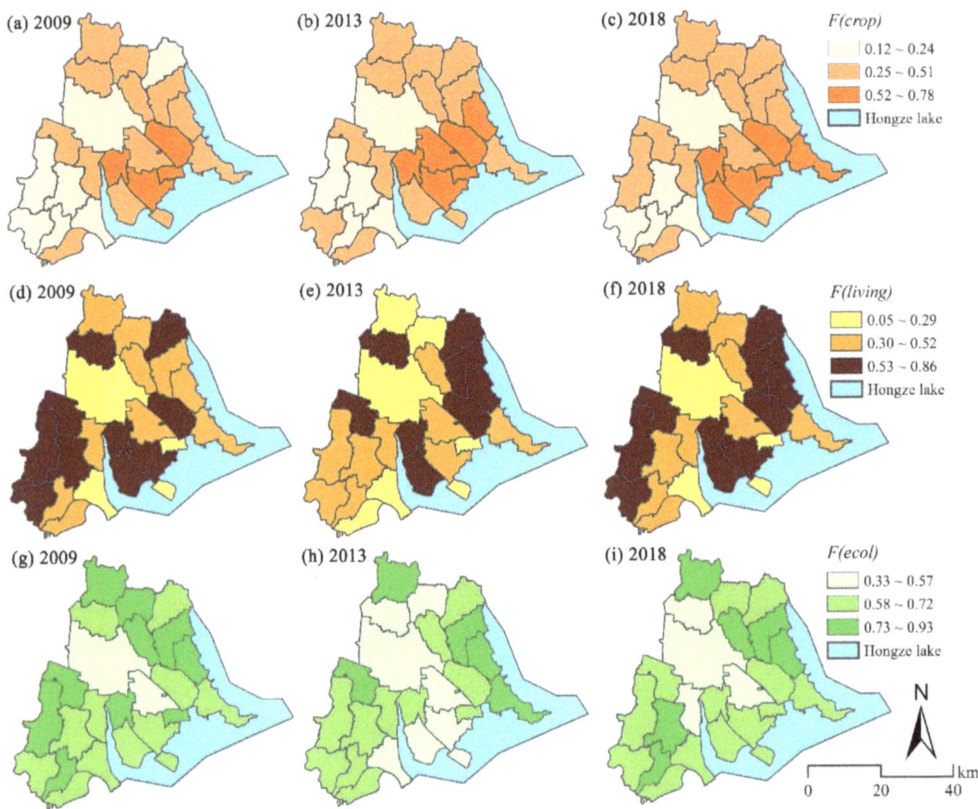



**Figure 8.** The spatial patterns of farmland use functions in Sihong at the town level in 2009, 2013, and 2018.

The living security production of farmland showed an initial falling trend and then rising trend during 2009–2018 (Figure 7a), and its index decreased from 0.468 in 2009 to 0.296 in 2013. However, this index increased to 0.729 in 2018 at the county scale. Specifically speaking, the indexes of grain yield per unit and per capita farmland area fluctuated upward, and the proportion of employees in the plantation decreased in the study area from 2009 to 2018. This shows that the function of farmland food security was enhanced, while the function of farmland employment security was significantly weakened (Figure 7c). The number of townships with low F (living) index values (0.05–0.29) increased from four in 2009 to seven in 2013, and these increased townships, including Guiren, Jinsuo, and Sihe, were restored in 2018 (Figure 8d–f). The low-value areas were mainly distributed in accordance with relatively high levels of urbanization.

The eco-environmental function of farmland showed an initial decrease and then increase trend during 2009–2018 (Figure 7a), and the index of the eco-environmental function of farmland decreased from 0.8103 in 2009 to 0.2409 in 2012; then, the index increased to 0.6530 in 2018, but it did not recover to the level in 2009. Specifically, the index of agricultural fertilizer intensity and pesticide intensity first decreased and then increased, indicating that the eco-environmental function of farmland increased in its fluctuation while the index of the use intensity of agricultural plastic film showed a downward trend (Figure 7d). The number of townships with high F(ecol) index values (0.73–0.93) decreased from 10 in 2009 to 5 in 2013, indicating that the eco-environmental function of farmland in some towns declined substantially. During 2013–2018, the number of high-value townships and medium-value townships also increased (Figure 8g–i).

### 3.2. Driving Factors of Farmland Use Transition

#### 3.2.1. Driving Factors of Farmland Use Spatial Transition

In the results of the regression analysis (Table 3), the standardized regression coefficient showed that the farmland use spatial transition at the town level is largely affected by natural environmental factors (e.g., elevation, slope), economic development factors (e.g., population density, urbanization), transportation infrastructure factors, and land use policy factors. In particular, the farmland area evolution had a positive correlation with the intensity of land consolidation ($p < 0.01$) and elevation ($p < 0.05$) and a negative correlation with road density ($p < 0.01$) and urbanization ($p < 0.05$). The PD of farmland was positively correlated with road density and urbanization ($p < 0.01$) but negatively influenced by elevation, slope, and population density ($p < 0.01$), as well as Stp ($p < 0.05$). The MPS of farmland was positively correlated with elevation, slope, and population density ($p < 0.01$). Additionally, the MPS was negatively correlated with urbanization and road density ($p < 0.01$) and farmers' net income ($p < 0.10$). The AI of farmland was positively correlated with slope and the intensity of land consolidation ($p < 0.01$), as well as Pgdp ($p < 0.10$), but was negatively correlated with road density and farmers' net income ($p < 0.01$).

**Table 3.** The regression analysis results for the morphology of farmland and the driving factors in Sihong in 2009, 2013, and 2018.

| Factors | Dominant Morphologies | | | | Functional Morphologies | | |
|---|---|---|---|---|---|---|---|
| | Area | PD | MPS | AI | F (crop) | F (living) | F (ecol) |
| *El* | 0.248 ** | −0.539 *** | 0.699 *** | 0.127 | −0.636 *** | 0.126 | −0.028 |
| *Slp* | 0.030 | −0.291 *** | 0.252 *** | 0.363 *** | −0.05 | −0.093 | 0.052 |
| ln*Pd* | 0.185 | −0.438 *** | 0.385 *** | 0.111 | −0.437 *** | −0.297 ** | 0.342 ** |
| *Urp* | −0.312 ** | 0.465 *** | −0.383 *** | −0.195 | 0.399 *** | −0.492 *** | 0.332 ** |
| ln*Pgdp* | −0.347 | 0.013 | 0.046 | 0.372 * | −0.088 | −0.325 | 0.863 *** |
| ln*Fi* | −0.103 | 0.214 | −0.299 * | −0.773 *** | 0.436 ** | 0.064 | −0.628 ** |
| *Stp* | 0.096 | −0.32 ** | 0.186 | 0.13 | −0.053 | 0.142 | −0.363 * |
| *Rd* | −0.425 *** | 0.414 *** | −0.588 *** | −0.758 *** | −0.146 | −0.148 | −0.562 *** |
| *Lci* | 0.929 *** | 0.071 | 0.169 | 0.604 *** | −0.247 | 0.196 | −0.241 |
| $R^2$ | 0.452 | 0.602 | 0.781 | 0.598 | 0.683 | 0.455 | 0.380 |
| Adjusted $R^2$ | 0.368 | 0.562 | 0.748 | 0.536 | 0.635 | 0.371 | 0.285 |
| *F* | 5.403 *** | 10.707 *** | 23.372 *** | 9.736 *** | 14.118 *** | 5.464 *** | 4.017 *** |

Note: * Significant at a 10% level; ** Significant at a 5% level; *** Significant at a 1% level. El, Elevation; Slp, Slope; Pd, Population density; Urp, Urbanization rate of population; Fi, Farmers' net income; Pgdp, Per capita GDP; Stp, The total proportion of secondary and tertiary industry output value; Rd, Road density; Lci, The quantitative data of land consolidation projects.

#### 3.2.2. Driving Factors of Farmland Function Transition

The function transition of farmland use at the town level was largely affected by economic development factors (e.g., population density, urbanization, Pgdp, farmers' net income) and transportation infrastructure factors (road density) (Table 3). Specifically, the crop production function of farmland was positively related to urbanization ($p < 0.01$) and farmers' net income ($p < 0.05$). The living security function evolution of farmland had a negative correlation with population density and urbanization ($p < 0.05$). The eco-environmental function of farmland was positively correlated with population density, urbanization, and Pgdp ($p < 0.01$). The eco-environmental function of farmland was negatively correlated with road density ($p < 0.01$), as well as farmers' net income ($p < 0.05$).

### 4. Discussion

Socio-economic factors were substantially correlated with farmland use transition individually (Table 3). Particularly, urbanization demonstrated the strongest correlation with almost all the indexes (excluding AI) of farmland morphology. Overall, urbanization caused a decrease in farmland area and MPS and an increase in the PD, mainly because part of the farmland was turned into urban land, which resulted in more fragmented farmland [48,49]. Urbanization was also substantially related to changes in the functional morphology of farmland. To be more specific, the increase in urbanization expanded market demand for local agricultural products and stimulated an increase in the supply capacity of agricultural products, which contributed to an improvement in the crop function of farmland [46]. The progress of the urbanization level prompted the labor force in the plantation industry to shift continuously to secondary and tertiary industries, which have absorbed a large number of rural laborers, making it possible to scale up farmland which came from the rural settlements reclaimed by the government, while the employment opportunities and social security functions of farmland have declined accordingly [50]. Urbanization also advanced agricultural production mechanization, industrialization, and modernization, which have reduced the use of chemical fertilizers and pesticides per unit area of farmland, thereby improving the eco-environmental functions [46,50,51]. However, agricultural production mechanization has also increased the amount of agricultural plastic film and reduced the eco-environmental functions of farmland. Moreover, the difficulties in plastic film decomposition will likely have a long-term impact on the eco-environmental function of farmland in the future [43,52].

Population density is another important socio-economic driving factor of farmland use transition [53,54]. Areas with high population density are generally economically developed areas where the contradiction between people and land is prominent [55] and land use is complex, leading to a high Pd and low MPS of the farmland. In these areas, agricultural employment opportunities are fewer, and the per capita farmland and food possession are too small. Population agglomeration also leads to an increase in demand for crop products and agricultural modernization, which together promote the crop production of farmland and reduce the use of chemical fertilizers and pesticides per unit area of farmland, thereby improving the eco-environmental function of the farmland. Moreover, the increase in farmers' income has reduced the MPS and AI of the farmland. This occurred because farmers converted their farmland into fishponds to achieve higher income, which resulted in more fragmented farmland. In addition, the increase in farmers' income also motivated rural laborers to engage in agricultural activities, which further promoted the crop production function of farmland [36].

The farmland protection policies represented by the intensity of land consolidation are also important driving factors of farmland use transition [56,57]. The implementation of consolidation projects has increased the area of farmland and the AI of farmland patches. In response to the large amount of farmland being occupied by construction land and non-agriculturalization in the process of urbanization and industrialization, this research has also demonstrated that the "Requisition–Compensation Balance of Farmland" policy and land consolidation project implemented by the Chinese government has succeeded in restoring the farmland area [58,59]. Sihong has carried out the government-led land reclamation project of the "Million Hectares of Fertile Farmland Reclamation Project" since 2012, which has improved the agglomeration and connectivity of the farmland landscape.

It was noted that the road network density had a significant impact on the regional farmland's morphological changes. The construction of road networks inevitably consumes and divides farmland, resulting in a decrease of the area of farmland [60], an increase of the density of farmland patches, and a reduction in the mean size and agglomeration of farmland patches. These factors caused an increase in the degree of fragmentation of the farmland. The increase in the degree of fragmentation of farmland constrained agricultural large-scale production and caused an increase in the use intensity of chemical fertilizers and pesticides per unit area of farmland, which eventually led to a decline in the eco-environmental function of the farmland.

Moreover, the regional natural environment had a certain impact on the farmland morphology [39,61], especially the dominant farmland morphologies. In areas with high altitudes and slopes, the possibility of farmland being occupied and divided by urban land was low, which led to a lower PD, higher MPS, and lower AI of farmland. However, due to poor natural conditions, the crop production function of farmland was low. In plain areas, due to the decline in the comparative benefit of agriculture and the increase in labor costs, farmers were more inclined to plant crops through mechanization, while in hilly areas, mechanization was difficult to achieve, so elevation had a negative effect on the crop production capacity of the farmland.

## 5. Conclusions and Policy Implications

### 5.1. Conclusions

This study examined the dominant controlling factors and characteristics of the farmland morphology variations in a typical agricultural area in China. From the perspective of the dominant morphological changes in farmland, the area of farmland showed a U-shaped change trend that first decreased and then increased during 2009–2018. However, the PD of the farmland patches showed an upward trend, and the MPS of farmland patches showed a downward trend, indicating that the degree of fragmentation of farmland increased. The spatial differences and dominant morphological changes of farmland were affected by several factors. Among them, the implementation of consolidation projects increased the area of farmland and the concentration of farmland patches, and the increase in urbanization and road density reduced the area and increased the degree of fragmentation of farmland. Elevation, slope, and population density also had a significant impact on the dominant form of cultivated land. Areas with a low altitude and slope and areas with a high population density had more complex land use conditions and a higher degree of cultivated land fragmentation. Regarding the recessive morphological changes of farmland (function morphology), the crop production, living security, and eco-environmental functions of farmland showed a falling and then rising trend. The spatial differences and changes in farmland functions were mainly affected by economic development factors and transportation factors. Urbanization had a positive impact on the crop production and eco-environmental functions of farmland, which reduced the living security function of the farmland. This suggests that urbanization increased the demand for agricultural products and the degree of large-scale agricultural production. These results could be useful for diagnosing the morphology of farmland in other agricultural areas, as morphology is very important for formulating reasonable farmland protection policies.

### 5.2. Implications for Land Use Policy

To meet the requirements of economic development, food security, and ecological environment security, regions should promptly adjust their land resource management policies and measures according to the current farmland morphology (including dominant and recessive morphology) and existing problems [15]. Our study demonstrated that the area of farmland has experienced a continuous reduction and reached a recovery net growth during 2009–2018 in the study area. This changing trend of farmland conformed to the "U-shaped" trend proposed by Song et al. [14] and Ge et al. [24]. The implementation of local land consolidation projects achieved an increase in the area and provided limited improvement in the function of the farmland. Therefore, it is necessary to scientifically formulate rural spatial planning; promote the implementation of comprehensive land consolidation projects throughout the region; and optimize agricultural, ecological, and construction space.

### 5.3. Limitation and Further Research

Since the land use data in this study were based on the Second National Land Survey led by the Chinese government (2009) and updated annual investigation (2010–2018), the data were mainly interpreted via high-resolution remote sensing images and field surveys and were obtained with high precision and temporal continuity. Although the Chinese government also conducted a land survey before 2009, it is difficult to effectively compare these surveys due to the inconsistency in land classification standards and accuracy with the second land survey made. Therefore, this study only conducted an analysis of the farmland use transition between 2009 and 2018, which was a relatively short research period. In future research, high-resolution remote sensing images should be adopted to interpret historical land use data and increase the study period to understand the underlying mechanisms and obtain the dynamics of farmland use transition.

The recessive morphology of farmland use is rich in connotations. The present study focuses on the farmland's functional morphology but does not cover the property rights, quality, and management modes of farmland. Thus, future research should be more systematic in analyzing the recessive morphology of farmland. In addition, although this study concentrated on farmland use transition at the microscopic scales of counties and towns, less attention has been paid at the village scale. The spatial and functional transition of farmland at the village scale are of significance for the formulation of policies to protect farmland and farmers' livelihoods. Therefore, more in-depth research on farmland use transition at the microscopic scale should be the focus of future research.

**Author Contributions:** Conceptualization, L.L. and H.L.; methodology, L.L.; software, Z.G.; validation, L.L., X.W. and Y.F.; formal analysis, L.L.; investigation, Z.G. and Y.F.; resources, X.W.; data curation, L.L.; writing—original draft preparation, L.L.; writing—review and editing, H.L.; visualization, L.L. and Z.G.; supervision, H.L.; project administration, L.L.; funding acquisition, H.L. and L.L. All authors have read and agreed to the published version of the manuscript.

**Funding:** This research was supported by the National Natural Science Foundation of China (Grant Nos. 41731286, 41801169 and 42001225), China Postdoctoral Science Foundation (2018M641456), the MOE (Ministry of Education in China) Project of Humanities and Social Sciences (18YJCZH120), the Natural Science Foundation of Jiangsu Province, China (BK20180819).

**Institutional Review Board Statement:** Not applicable.

**Informed Consent Statement:** Not applicable.

**Data Availability Statement:** Not applicable.

**Conflicts of Interest:** The authors declare no conflict of interest.

## References

1. Lambin, E.F.; Meyfroidt, P. Land use transition: Socio-ecological feedback versus socio-economic change. *Land Use Policy* **2010**, *27*, 108–118. [CrossRef]
2. Long, H.L.; Qu, Y. Land use transitions and land management: A mutual feedback perspective. *Land Use Policy* **2017**, *34*, 1607–1618. [CrossRef]
3. Grainger, A. The forest transition: An alternative approach. *Area* **1995**, *27*, 242–251.
4. Mather, A.S.; Fairbairn, J.; Needle, C.L. The course and drivers of the forest transition: The case of France. *J. Rural Stud.* **1999**, *15*, 65–90. [CrossRef]
5. Mather, A.S.; Needle, C.L. The forest transition: A theoretical basis. *Area* **1998**, *30*, 117–124. [CrossRef]
6. Barbier, E.B.; Burgess, J.C.; Grainger, A. The forest transition: Towards a more comprehensive theoretical framework. *Land Use Policy* **2010**, *27*, 98–107. [CrossRef]
7. Grainger, A. Difficulties in tracking the long-term global trend in tropical forestarea. *Proc. Natl. Acad. Sci. USA* **2008**, *105*, 818–823. [CrossRef]
8. Mather, A.S. Forest transition theory and the reforesting of Scotland. *Scott. Geogr. J.* **2004**, *120*, 83–98. [CrossRef]
9. Mather, A.S. Recent Asian forest transitions in relation to forest transition theory. *Int. For. Rev.* **2007**, *9*, 491–502.
10. Grau, H.R.; Aide, M. Globalization and land-use transitions in Latin America. *Ecol. Soc.* **2008**, *13*, 16. [CrossRef]
11. Yeo, I.-Y.; Huang, C. Revisiting the forest transition theory with historical records and geospatial data: A case study from Mississippi (USA). *Land Use Policy* **2013**, *32*, 1–13. [CrossRef]
12. Tuan, Y.F. Geography, phenomenology and the study of human nature. *Can. Geogr.* **1971**, *15*, 181–192. [CrossRef]
13. Long, H.L. *Land Use Transitions and Rural Restructuring in China*; Springer: Singapore, 2020.
14. Song, X.Q. Discussion on land use transition research framework. *Acta Geogr. Sin.* **2017**, *72*, 471–487.
15. Long, H.L.; Qu, Y.; Tu, S.S.; Zhang, Y.N.; Jiang, Y.F. Development of land use transitions research in China. *J. Geogr. Sci.* **2020**, *30*, 1195–1214. [CrossRef]
16. Li, X.B. Theoretical hypotheses about agricultural land use changes and the relevant propositions about environmental impacts. *Adv. Earth Sci.* **2008**, *23*, 1124–1129.
17. Long, H.L.; Heilig, G.K.; Li, X.B.; Zhang, M. Socio-economic development and land-use change: Analysis of rural housing land transition in the Transect of the Yangtse River, China. *Land Use Policy* **2007**, *24*, 141–153. [CrossRef]
18. Long, H.L.; Li, T.T. The coupling characteristics and mechanism of farmland and rural housing land transition in China. *J. Geogr. Sci.* **2012**, *22*, 548–562. [CrossRef]

19. Lyu, X.; Huang, X.J.; Zhang, Q.J. A literature review on urban-rural construction land transition. *City Plan Rev.* **2015**, *39*, 105–112.
20. Wang, T.; Kazak, J.; Han, Q.; de Vries, B.A. framework for path-dependent industrial land transition analysis using vector data. *Eur. Plan. Stud.* **2019**, *27*, 1391–1412. [CrossRef]
21. Amin, A.; Fazal, S.; Mujtaba, A.; Singh, S.K. Effects of land transformation on water quality of Dal Lake, Srinagar, India. *J. Indian Soc. Remote Sens.* **2014**, *42*, 119–128. [CrossRef]
22. Asabere, S.B.; Acheampong, R.A.; Ashiagbor, G.; Beckers, S.C.; Keck, M.; Erasmi, S.; Schanze, J.; Sauer, D. Urbanization, land use transformation and spatio-environmental impacts: Analyses of trends and implications in major metropolitan regions of Ghana. *Land Use Policy* **2020**, *96*, 104707. [CrossRef]
23. Liu, Y.Q.; Long, H.L. Land use transitions and their dynamic mechanism: The case of the Huang-Huai-Hai Plain. *J. Geogr. Sci.* **2016**, *26*, 515–530. [CrossRef]
24. Ge, D.Z.; Long, H.L.; Zhang, Y.N.; Ma, L.; Li, T.T. Farmland transition and its influences on grain production in China. *Land Use Policy* **2018**, *70*, 94–105. [CrossRef]
25. Ntihinyurwa, P.D.; de Vries, W.T. Farmland Fragmentation, Farmland Consolidation and Food Security: Relationships, Research Lapses and Future Perspectives. *Land* **2021**, *10*, 129. [CrossRef]
26. Willy, D.K.; Muyanga, M.; Jayne, T. Can economic and environmental benefits associated with agricultural intensification be sustained at high population densities? A farm level empirical analysis. *Land Use Policy* **2019**, *81*, 100–110. [CrossRef] [PubMed]
27. Qin, W.S.; Zhang, Y.F.; Li, G.D. Driving mechanism of cultivated land transition in Yantai Proper, Shandong Province, China. *Chin. Geogr. Sci.* **2015**, *25*, 337–349. [CrossRef]
28. Wang, Q.; Li, Y.B.; Luo, G.J. Spatiotemporal change characteristics and driving mechanism of slope cultivated land transition in karst trough valley area of Guizhou Province, China. *Environ. Earth Sci.* **2020**, *79*, 1–18. [CrossRef]
29. Islam, M.R.; Hassn, M.Z. Losses of agricultural land due to infrastructural development: A study on Rajshahi District. *Int. J. Sci. Eng. Res.* **2013**, *4*, 391–396.
30. Francis, C.A.; Hansen, T.E.; Fox, A.A.; Hesje, P.J.; Nelson, H.E.; Lawseth, A.E.; English, A. Farmland conversion to non-agricultural uses in the US and Canada: Current impacts and concerns for the future. *Int. J. Agric. Sustain.* **2012**, *10*, 8–24. [CrossRef]
31. Sun, B.; Zhou, Q.M. Expressing the spatio-temporal pattern of farmland change in arid lands using landscape metrics. *J. Arid Environ.* **2016**, *124*, 118–127. [CrossRef]
32. Tan, M. The transition of farmland production functions in metropolitan areas in China. *Sustainability* **2014**, *6*, 4028–4041. [CrossRef]
33. You, H.Y.; Hu, X.W.; Wu, Y.Z. Farmland use intensity changes in response to rural transition in Zhejiang province, China. *Land Use Policy* **2018**, *79*, 350–361. [CrossRef]
34. Slee, B.; Brown, I.; Donnelly, D.; Gordon, I.J.; Matthews, K.; Towers, W. The 'squeezed middle': Identifying and addressing conflicting demands on intermediate quality farmland in Scotland. *Land Use Policy* **2014**, *41*, 206–216. [CrossRef]
35. Shi, Y.Y.; Lyu, X.; Guo, G.C.; Gong, C. Temporal-spatial pattern and driving mechanism of cultivated land use transition based on GIS and spatial econometric model. *China Land Sci.* **2019**, *33*, 51–60.
36. Fu, H.; Liu, Y.J.; Sun, H.R.; Zhou, G.L. Spatiotemporal characteristics and dynamic mechanism of cultivated land use transition in the Beijing-Tianjin-Hebei region. *Prog. Geogr.* **2020**, *39*, 27–40.
37. Zhang, Y.N.; Long, H.L.; Ma, L.; Ge, D.Z.; Tu, S.S.; Qu, Y. Farmland function evolution in the Huang-Huai-Hai Plain: Processes, patterns and mechanisms. *J. Geogr. Sci.* **2018**, *28*, 759–777. [CrossRef]
38. Zhu, Z.Q.; Kong, X.S.; Li, Y.J. Identifying the Static and Dynamic Relationships Between Rural Population and Settlements in Jiangsu Province, China. *Chin. Geogr. Sci.* **2020**, *30*, 810–823. [CrossRef]
39. Jiang, P.H.; Li, M.C.; Lv, J.C. The causes of farmland landscape structural changes in different geographical environments. *Total Environ.* **2019**, *685*, 667–680. [CrossRef]
40. Zhang, Z.H.; Su, S.L.; Xiao, R.; Jiang, D.W.; Wu, J.P. Identifying determinants of urban growth from a multi-scale perspective: A case study of the urban agglomeration around Hangzhou Bay, China. *Appl. Geogr.* **2013**, *45*, 193–202. [CrossRef]
41. Xu, W.X.; Jin, X.; Liu, J.; Zhou, Y.K. Impact of cultivated land fragmentation on spatial heterogeneity of agricultural agglomeration in China. *J. Geogr. Sci.* **2020**, *30*, 1571–1589. [CrossRef]
42. He, S.W.; Yu, S.; Li, G.D.; Zhang, J.F. Exploring the influence of urban form on land-use efficiency from a spatiotemporal heterogeneity perspective: Evidence from 336 Chinese cities. *Land Use Policy* **2020**, *95*, 104576. [CrossRef]
43. Yang, X.; Tan, M.H. Changes and Relationships of Arable Land Functions in Beijing in Recent Years. *J. Nat. Resour.* **2014**, *29*, 733–743.
44. Song, X.Q.; Huang, Y.; Wu, Z.F.; Ouyang, Z. Does cultivated land function transition occur in China? *J. Geogr. Sci.* **2015**, *25*, 817–835. [CrossRef]
45. Song, X.Q.; Wu, Z.F.; Ouyang, Z. Route of cultivated land transition research. *Geogr. Res.* **2014**, *33*, 403–413.
46. Liu, J.Z.; Fang, Y.G.; Wang, R.R. Spatio-temporal evolution characteristics and driving mechanisms of agricultural multifunctions in Shandong province. *J. Nat. Resour.* **2020**, *35*, 2901–2915.
47. Wooldridge, J.M. *Introductory Econometrics: A Modern Approach*; Nelson Education: Toronto, ON, Canada, 2016.
48. Ntihinyurwa, P.D.; de Vries, W.T. Farmland fragmentation and defragmentation nexus: Scoping the causes, impacts, and the conditions determining its management decisions. *Ecol. Indic.* **2020**, *119*, 106828. [CrossRef]

49. Liu, X.L.; Wang, Y.; Li, Y.; Liu, F.; Shen, J.L.; Wang, J.; Xiao, R.L.; Wu, J.S. Changes in arable land in response to township urbanization in a Chinese low hilly region: Scale effects and spatial interactions. *Appl. Geogr.* **2017**, *88*, 24–37. [CrossRef]

50. Ma, L.; Long, H.L.; Tu, S.S.; Zhang, Y.N.; Zheng, Y.H. Farmland transition in China and its policy implications. *Land Use Policy* **2020**, *92*, 104470. [CrossRef]

51. Wang, X.; Shao, S.; Li, L. Agricultural inputs, urbanization, and urban-rural income disparity: Evidence from China. *China Econ. Rev.* **2019**, *55*, 67–84. [CrossRef]

52. Liu, E.K.; He, W.Q.; Yan, C.R. 'White revolution' to 'white pollution'—Agricultural plastic film mulch in China. *Environ. Res. Lett.* **2014**, *9*, 091001. [CrossRef]

53. Xu, Y.Q.; McNamara, P.; Wu, Y.F.; Dong, Y. An econometric analysis of changes in arable land utilization using multinomial logit model in Pinggu district, Beijing, China. *J. Environ. Manag.* **2013**, *128*, 324–334. [CrossRef] [PubMed]

54. Bucała-Hrabia, A. Long-term impact of socio-economic changes on agricultural land use in the Polish Carpathians. *Land Use Policy* **2017**, *64*, 391–404. [CrossRef]

55. Ann, T.W.; Wu, Y.Z.; Zheng, B.B.; Zhang, X.L.; Shen, L.Y. Identifying risk factors of urban-rural conflict in urbanization: A case of China. *Habitat Int.* **2014**, *44*, 177–185.

56. Long, H.L.; Zhang, Y.N.; Tu, S.S. Rural vitalization in China: A perspective of land consolidation. *J. Geogr. Sci.* **2019**, *29*, 517–530. [CrossRef]

57. Qu, Y.B.; Jiang, G.H.; Li, Z.T.; Tian, Y.Y.; Wei, S.W. Understanding rural land use transition and regional consolidation implications in China. *Land Use Policy* **2019**, *82*, 742–753. [CrossRef]

58. Liu, L.; Liu, Z.J.; Gong, J.Z.; Wang, L.; Hu, Y.M. Quantifying the amount, heterogeneity, and pattern of farmland: Implications for China's requisition-compensation balance of farmland policy. *Land Use Policy* **2019**, *81*, 256–266. [CrossRef]

59. Deng, Z.Q.; Zhao, Q.Y.; Bao, H.X. The Impact of Urbanization on Farmland Productivity: Implications for China's Requisition–Compensation Balance of Farmland Policy. *Land* **2020**, *9*, 311. [CrossRef]

60. Song, J.; Ye, J.T.; Zhu, E.Y.; Deng, J.S.; Wang, K. Analyzing the impact of highways associated with farmland loss under rapid urbanization. *ISPRS Int. J. Geoinf.* **2016**, *5*, 94. [CrossRef]

61. Fu, M.C.; Hu, Z.Q.; Wu, G.G.; Deng, J.S.; Wang, C. Analysis of evolutionary law rule of farmland landscape. *TXN. Chin. Soc. Agric. Eng.* **2005**, *6*, 54–58.

*Article*

# Research on Identification of Multiple Cropping Index of Farmland and Regional Optimization Scheme in China Based on NDVI Data

Tingting Li [1], Yanfei Wang [2,*], Changquan Liu [1] and Shuangshuang Tu [3,4]

[1]  Rural Development Institute, Chinese Academy of Social Sciences, Beijing 100732, China; litt@cass.org.cn (T.L.); liuchq@cass.org.cn (C.L.)
[2]  College of Land Science and Technology, China Agricultural University, Beijing 100193, China
[3]  Key Laboratory of Environment Change and Resources Use in Beibu Gulf, Nanning Normal University, The Ministry of Education, Nanning 530001, China; tuss@nnnu.edu.cn
[4]  Institute of Geographic Sciences and Natural Resources Research, Chinese Academy of Sciences, Beijing 100101, China
*  Correspondence: wangyf@cau.edu.cn

**Abstract:** The multiple cropping index of farmland is a significant characterization of land use intensity. Based on the NDVI data, this paper calculated the multiple cropping index of farmland in China using the S-G filtering method, and proposed an optimized regionalization scheme for the farmland use. The findings reveal that from 2000 to 2018, the multiple cropping index of farmland in China underwent the fluctuation of rising first, then falling and rising continuously, which was closely associated with the agricultural support policies enforced in China. Counties whose multiple cropping indexes decreased from 2009 to 2018 were mainly located in areas primarily producing grain, which exerted a greater influence on food security. The gap between the multiple cropping index and potential multiple cropping index of farmland is increasingly widening from north to south in China. Accordingly, four types of grain producing zones were delineated: key development zone, potential growth zone, appropriate development zone, and restricted development zone. Some suggestions, such as rotation, fallow, determination of yield by water and offsetting the quantity balance of farmland by increasing the multiple cropping index, are put forward based on different zones.

**Keywords:** NDVI; land use transition; multiple cropping index; farmland; regional optimization scheme



**Citation:** Li, T.; Wang, Y.; Liu, C.; Tu, S. Research on Identification of Multiple Cropping Index of Farmland and Regional Optimization Scheme in China Based on NDVI Data. *Land* **2021**, *10*, 861. https://doi.org/10.3390/land10080861

Academic Editor: Le Yu

Received: 1 July 2021
Accepted: 9 August 2021
Published: 16 August 2021

**Publisher's Note:** MDPI stays neutral with regard to jurisdictional claims in published maps and institutional affiliations.



## 1. Introduction

With global climate change, continuous population growth, and rapid urbanization, food security issues and policies remain a subject of concern to the international community. China feeds about 18% of the global population using 8% of its farmland [1]. The fundamental reality of more people and less farmland in China demonstrates that food security is crucial to the lasting political stability in China. With the impact of the epidemic and rising uncertainty in the international trade environment, the issues of ensuring baseline of food security and grasping the initiative in food security have become more prominent. As an important factor affecting food security, the change of farmland area has received more attention. Over the past 40 years of the reform and opening up in China, irreversible non-agricultural changes in a large amount of farmland have taken place according to the progress of fast industrialization and urbanization, which has led to the decrease of farmland area, and a threat to food security [2]. Strict observation of the red line of 1.2 million $km^2$ of farmland has become a national political task to ensure food security. However, with the continuous increase of total population and urbanization, it is very difficult to increase the food supply by increasing farmland area. On the contrary, it is more feasible to ensure national food security by improving the level of intensive use of existing farmland [3]. Multiple cropping is an important aspect of the intensive use of farmland [4]. From 1986 to 1995, the increased grain yields attributed to the increasing multiple crop index (MCI) of farmland, accounted for one third of the average annual total grain yields (429 billion kg) in China [5]. About 12% of global farmland applied multiple cropping in 2000. In addition, 34%, 13%, and 10% of rice, wheat, and maize crops, respectively, utilize multiple cropping, demonstrating the importance of such cropping systems for cereal production [6]. Moreover, compared with reclaimed farmland, existing farmland possesses better production conditions. Multiple cropping is, therefore, an effective way to increase the grain yields and ensure food security [3,7,8].

With the expansion of the connotation of food security, the objectives of researches on multiple cropping of farmland in different countries have also shown differences. In European and North American countries with a high level of economic and social development, the researches mainly focus on the impact of multiple cropping on pest control and soil improvement. The conclusions based on field experiments prove that increasing the level of MCI of farmland can increase diversity, thereby contributing to pest control and reducing herbicide intensity [9]. Exploring different multiple cropping modes can effectively enhance organic matter and microbial activities in the soil, thereby developing organic agriculture and obtaining a higher income [10]. In South America, Africa, and Asia, where the level of economic and social development is relatively low, the goal of related researches is mainly to increase grain yields [11–13]. In recent years, however, it has begun to shift to the direction of balanced nutrition [14]. It is worth noting that in Asia, human-land contradiction is very serious. The production system characterized by smallholder determines the necessity of increasing MCI to increase the grain yields and incomes [15]. Therefore, Asia has become a region of focus for researches on MCI [16,17]. In India, the zoning map of rice MCI was drawn and used to estimate the irrigation demand of different zones to provide a scientific basis for policy evaluation [18]. From north to south in China, great differences in terms of crop types and cropping systems are exhibited among eight temperature zones [19,20]. Influenced by agricultural production conditions and socio-economic development, the MCI in the major grain producing areas [21,22] and the rice-growing areas, where "double cropping to single cropping" is common [23,24], has noticeably declined in recent years.

Reasons for this are summarized into the following four aspects. Firstly, marginal incomes earned via a multi-cropping system decrease significantly as a result of the increasing production cost. MCI was changed from multiple cropping to single cropping to maximize the economic benefits [25]. Secondly, the labor marginal incomes from non-agricultural employment are much higher than those of agricultural production for Chinese farmers. Farmers are more inclined to transfer more labor time and production resources to part-time or non-agricultural production activities [26], thus resulting in seasonal or year-round abandonment of farmland especially in labor-intensive cash crops and regions closer to urban areas [22]. Thirdly, more farmers may face a poor harvest after using the "double cropping to single cropping" method, since those that plant double-cropping rice may be exposed to the intensive damages of insects, birds, and animals [27]. Fourthly, the adjustment of food policies will also cause changes of MCI through incentives and constraints on the planting behaviors of farmers. In a word, decreasing multi-cropping level and even abandoning cultivation, is a rational choice of farmers under low planting efficiency [28].

Existing researches mainly concern the influence of MCI on food security, and use MCI as the input variable to calculate variations in farmland area and grain yields. These prove that multi-cropping system can indeed increase the outputs of corn and rice [29], while decreasing the multi-cropping level can inhibit, and even decrease the growth of food output. This makes maintaining the self-sufficiency of cereal a challenge. However, some researches pointed out that the improvement of the multi-cropping system might influence the resource ecological environment. The practices in Pakistan prove that around 51% and 13% of water inefficiency are present under multiple and sole cropping systems, respectively [30]. The expansion of the multi-cropping system increased agricultural greenhouse gas emissions in the North Plain and neighboring regions in China [31], and the growth of the annual mean temperature, in return, can influence the growth of crops [32,33]. Evidently, pursuing high MCI blindly, and ignoring the water and temperature conditions would work against the increasing of grain yields and the sustainable development of the ecological environment [34].

The cropping system of China is not only experiencing a decline of MCI, but also facing the risk of spatial mismatch between cropping system and natural production conditions (including water, soil, gas, etc.). Firstly, there are abundant water and heat resources in South China. Historically, the food supply pattern entailed "transport from south regions to north regions", but now has changed to "transport from north regions to south regions", thus increasing consumption of farmland resources [35]. Secondly, the location of large and medium cities often highly overlaps with that of high-quality farmland [36]. A considerable amount of farmland with high-quality water and heat resources is occupied by urban construction sprawl, while the reclaimed farmland with poor production conditions is used to compensate for the loss of high-quality farmland with fertile soil and high MCI. The imbalance of the quality and production capacity of farmland has threatened China's food security [37]. In this regard, some studies have measured the potential multiple cropping index (PMCI) of farmland in China based on water and temperature conditions, which is the theoretical highest MCI of farmland based on the natural environment conditions [3]. Based on PMCI, some researches inferred the most sown area [38] and grain yields [1] under the optimal cropping system.

Scientific analysis of the relationship between multi-cropping system and potential multi-cropping system is conducive to deepening our understanding of farmland use and the scientific exploration of the potential of farmland, as well as providing references and supports for the implementation of a strategy that "stores foods in farmland". Based on the normalized difference vegetation index (NDVI), this paper will analyze the spatio-temporal characteristics of the MCI of farmland in China from 2000 to 2018. The distortion of water-land resources will be judged by the gap of MCI and PMCI. Finally, suggestions will be put forward to give full play to the production potential of high-quality farmland so as to achieve a win-win situation for food security and ecological security. Compared with existing researches on MCI of farmland [16–24], one of the innovations of this study is the problem of increasingly serious farmland abandonment introduced into the study of multi-cropping system. We will further divide the decline of MCI into "seasonal" abandonment and year-round abandonment, so as to respond to attentions on abandonment of farmland in China's farmland protection system. Another innovation was the delineation of four types of grain producing zones, namely key development zone, potential growth zone, appropriate development zone, and restricted development zone, and the provision of references for optimization of food production layout and benefit compensation mechanism design.

## 2. Materials and Methods

### 2.1. Materials

NDVI, also called the standardized vegetation index, is a comprehensive reflection of vegetation type, coverage form, and growth conditions in unit pixel. The value of NDVI is determined by the vegetation coverage and leaf area index (LAI). The physical growth processes of crop sowing, seedling, heading, maturing, and harvest in a year reflect fluctuations of NDVI with time, and peaks correspond to the time phases when the biomass of crop populations is the largest. According to this principle, the MCI of farmland is gained by extracting the peaks number of NDVI in one year. NDVI ranges between minus 1 and 1. Specifically, a negative value represents that a surface is covered by cloud, water, or snow; 0 represents rocks or naked soils; a positive value indicates vegetation coverage, which increases with the increase coverage [39]. In this study, the monthly (January to December) NDVI sequence from 2000 to 2018 is generated by the maximum value combination based on continuous time series of SPOT/VEG satellite remote sensing data. The spatial resolution of NDVI was 1 km × 1 km.

The spatial distribution data of potential multi-cropping system in China is estimated by the Global Agro-Ecological Zones (GAEZ) model developed by the FAO and IIASA together based on data of DEM, soil, farmland, and meteorological. On this basis, the ideal cropping system can be realized for the farmland. The potential multi-cropping system data includes single cropping, double cropping, and triple cropping in a year, with a spatial resolution of 10 km × 10 km.

In this study, the farmland grid data in five phases (2000, 2005, 2010, 2015, and 2018) were used to restrict the identification range of cropping system in farmland and eliminate interferences of other land use types. The spatial resolution of it is 1 km × 1 km. The number of farmland grids has been decreasing continuously since 2000, and it experienced a sharp reduction from 2005 to 2010 and since 2015 (Figure 1).

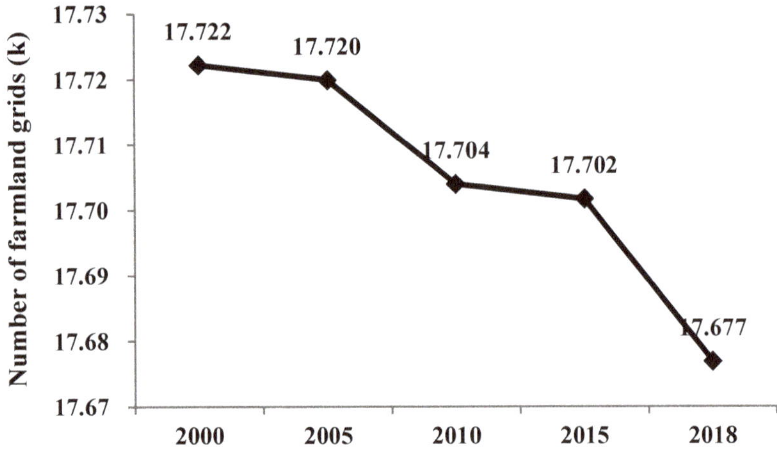


Figure 1. Variations of the number of farmland grids in China (2000–2018).

The above three types of data are provided by the Data Registration and Publishing System of Data Center for Resources and Environmental Sciences, Chinese Academy of Sciences (https://www.resdc.cn/data.aspx?DATAID=254, accessed on 12 October 2019). They cover 31 provinces in China, except Taiwan, Hong Kong, and Macau, as shown in Figure 2. In addition, it is necessary to introduce the locations of several important agricultural areas used in the paper. Huang-Huai-Hai Plain is composed of Hebei, Beijing, Tianjin, most of Henan, and northern Anhui and Jiangsu. It is the most important grain producing area in China due to its balanced rain and heat and flat terrain. The Loess Plateau includes Shanxi, Ningxia, north of the Qinling Mountains in Shaanxi, and southeastern Gansu. The terrain of this area is complex and diverse, and the ecological environment is fragile.

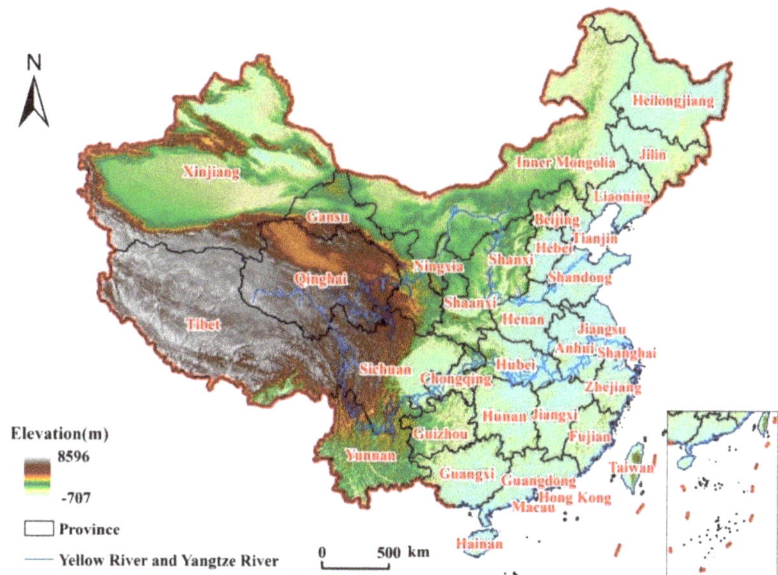


**Figure 2.** Location of the study area.

## 2.2. Methods

Step 1: Extract monthly NDVI sequence of farmland

This study concerns the MCI of farmland. Hence, the NDVI dataset of farmland was extracted only to eliminate interferences of other land use types. Firstly, the grids with farmland attribute in the dataset of national land use were clipped to build up a mask of farmland. Secondly, the mask of farmland was overlapped and spatial registration with NDVI data from January to December using ArcGIS software, thus extracting the NDVI sequences of farmland.

Step 2: Reconstruct NDVI sequence through the Savitzky-Golay (S-G) filter

The NDVI sequence of farmland had noise caused by atmospheric interferences or other reasons, thus making the NDVI value lower than the true value. Hence, the S-G filter was used for further smoothing and denoising of the NDVI sequence. As a result, high-quality NDVI sequence, which represented the growth trend of crops, was gained. The S-G filter is a convolutional smoothing approach based on the least square method [40], and it performs the polynomial least square fitting to the adjacent values in a local window. The S-G filter needs two parameters, which are the width of the smoothing window (m) and the degree of the polynomial (d). It requires that the m is shorter than the length of the NDVI sequence and is an odd number, and d is less than m. The larger the m and the smaller the d, the smoother the filtering result, but it also possibly eliminate more real details. Attributes of each farmland grid were the NDVI values from January to December and the sequence length was 12. According to the principle of parameter determination, three filtering windows of (m = 3, d = 2), (m = 5, d = 3) and (m = 5, d = 4) were chosen.

Step 3: Combine the original curve and fitting curve of NDVI

This step was used to maintain high values and decrease abnormal low values. When the original value of the NDVI was higher than the fitting value obtained in Step 2, the original NDVI value was retained. Otherwise, when the original NDVI value was smaller than the fitting value, it was replaced by the fitting value, and the NDVI curve was rebuilt.

Step 4: Calculate the fitting effect coefficient

This step was used to judge the fitting effect between the newly NDVI sequence obtained in Step 3 and the original NDVI sequences. The smaller coefficient indicates the better fitting effect. The calculation formula of the fitting effect coefficient is as follows:

$$fc = \sum_{i=1}^{n} \left( \left| N_i^1 - N_i^0 \right| \times W_i \right) \tag{1}$$

$$W_i = \begin{cases} 1, N_i^1 < N_i^0 \\ 1 - \frac{d_i}{d_{max}}, N_i^1 > N_i^0 \end{cases} \tag{2}$$

where $fc$ is the fitting effect coefficient of S-G filter; n is the length of NDVI sequence (which is 12); $i$ is the serial number of elements in the NDVI sequence; $N_i^1$ and $N_i^0$ are the fitting value and original value of the NDVI of the element $i$, respectively; $W_i$ is the weight of element $i$; $d_i$ is the absolute residual error between the original value and fitting value of the NDVI of element $i$; and $d_{max}$ is the maximum value in $d_i$. The $fc$ of the three filtering windows in Step 2 was compared and that of the (m = 5, d = 4) was the smallest. The coefficient of (m = 5, d = 4) was 0.054, which was 0.014 and 0.013 lower than that of (m = 3, d = 2) and (m = 5, d = 3), respectively. The fitting accuracy of filtering windows of (m = 5, d = 4) increased by about 20% (Table 1). Therefore, the fitting values of the original NDVI data were performed by filtering windows (m = 5, d = 4).

**Table 1.** Fitting effect coefficient (fc) of different filtering windows.

| Filtering Windows | (m = 5, d = 3) | (m = 3, d = 2) | (m = 5, d = 4) |
|---|---|---|---|
| fc | 0.067 | 0.068 | 0.054 |

Step 5: Peak recognition

The first-order differential method was used to recognize the peaks and valleys of the NDVI sequence. In the NDVI sequence of three successive months that first rises and then falls, the middle value was recognized as peak. On the contrary, in the NDVI sequence of three successive months that first falls and then rises, the middle value was recognized as valley. Meanwhile, a statistical analysis on the peak number of each grid was carried out to represent the MCI in each farmland grid.

Step 6: Eliminate interference peaks

In order to reduce the error of the MCI, this paper set up some criteria to remove interference peaks : ① The NDVI value of each peak was higher than 0.4, which was the empirical value of relevant researches [41]; ② to remove interference peaks, the occurrence time of real peaks was limited from April to October; and ③ the NDVI difference between peak and its adjacent two valleys cannot be smaller than 20% of the difference between the maximum and minimum in the NDVI sequences of 12 months. When one peak could not meet the above three criteria at the same time, it was regarded as an interference peak and deleted.

## 3. Results

### 3.1. Changes in the Level of Multiple Cropping Index of Farmland in China

#### 3.1.1. Stage Characteristics

The variation trend of the MCI of farmland gained from the NDVI was similar to that of MCI calculated by statistical data (national farmland area divided by total sowing area in the same year). From 2000 to 2018, the above two types of MCI of farmland in China underwent the fluctuation of rising first, then falling and rising continuously (Figure 3). In 2018, the MCI extracted by the NDVI was 124.9%, which is only slightly different when compared with the 126.2% calculated using statistical data. In summary, the identification method of MCI of farmland based on NDVI data is feasible and reliable.

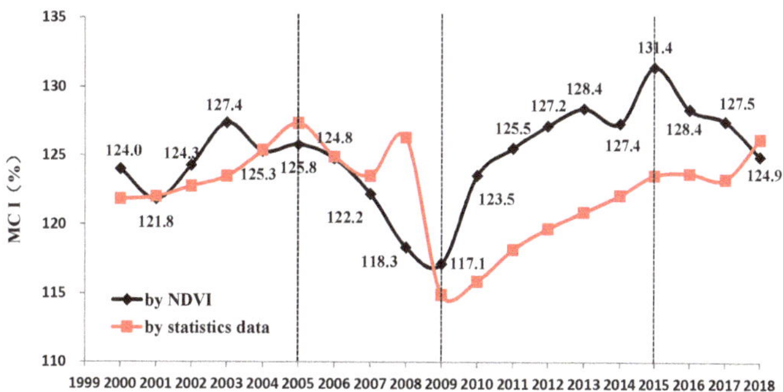


**Figure 3.** Stage characteristics of MCI of farmland in China (2000–2018).

Since the 21st century, the mean MCI of farmland in China was 125.4%, with a maximum and minimum of 131.4% and 117.1%, respectively (Figure 3). The fluctuation amplitude reached 14.3%, accounting for 11.4% of the mean. The MCI of farmland in China varies greatly from year to year, and it is necessary to analyze the characteristics of different stages. In general, it experienced four stages, which agree with the change in agricultural support policies enforced in China. From 2000 to 2005, the MCI of farmland remained stable and slowly grew from 124.0% to 125.8% (Figure 3). In the same period, the total grain yield in China in 2005 was 4.7% higher than that in 2000, which was due to the slight increase of MCI to a certain extent.

From 2005 to 2009, the MCI of farmland in China dropped sharply by 8.7% from 125.8% to 117.1% (Figure 3). During this stage, several food support policies, such as the lowest purchase price of rice and wheat and the temporary storage system of corn, were launched successively. Due to the low purchase price in the early stage of policy implementation and the hysteretic effect of incentive to farmers after the policy implementation, the MCI was not increased in this stage. Moreover, food support policies led to a rise in the prices of agricultural production materials and a rapid increase in production costs, thus further weakening the marginal benefits of agriculture. For farmers with more non-agricultural employment opportunities, seasonal or annual abandonment of farmland was the rational selection to achieve the maximum benefits. Hence, the MCI of farmland dropped dramatically.

From 2009 to 2015, the MCI of farmland increased continuously by 14.3% from 117.1% to 131.4% (Figure 3), which had something to do with the increasing supports for food production by the Chinese government in this period. With the annual growth of the lowest purchase price, farmers' enthusiasm in grain production was improved significantly, and they were able to gain more benefits by expanding the sowing area. In the same period, the grain yield in China in 2015 was higher by 18.2% compared with that in 2009. The increase of the MCI played an important role in the growth of grain yield.

From 2015 to 2018, the MCI of farmland began to decline again. With the inversion of domestic and foreign grain price, the problems of "high yield, high import and high inventory" became increasingly prominent. In 2016, China made a considerable adjustment to agricultural support policies, i.e., China canceled the temporary storage system of corn which had been implemented for eight years, and reduced the lowest purchase price of early indica rice. In 2017, the lowest purchase prices of all kinds of rice were declined as well. In 2018, the lowest purchase prices of both rice and wheat further decreased. In this context, the MCI of farmland decreased by 6.5% from 131.4% to 124.9% (Figure 3), which reflected the sharp reduction of farmer's enthusiasm for grain production.

### 3.1.2. Deconstruction of MCI Based on the Microscopic Perspective of Land Use Transition

Dynamic changes in the MCI of farmland are an important characterization of land use transition [42]. From the microscopic perspective, dynamic changes of the MCI reflect the production behavior response of farmers, who are independent "rational economists", to farmland use intensity under the principle of optimal allocation of production factors and maximum benefits. The proportion of farmland with double or triple cropping in total farmland area was calculated to determine the rate of multiple cropping. In general, the farmland in China was dominated by single cropping and the proportion had remained above 60% for a long period. The rate of multi-cropping (MCI > 1) of farmland in China experienced a sharp reduction in 2009 and a stable growth in 2015. In 2009, about 1/5 of farmland was engaged in a multi-cropping system, which increased to 1/3 in 2015. This also indicated that about 13% of farmland changed from a single cropping system into a multi-cropping system from 2009 to 2015. Due to the adjustment of agricultural support policies, the rate of multi-cropping of farmland decreased to 26.4% in 2018 (Figure 4).

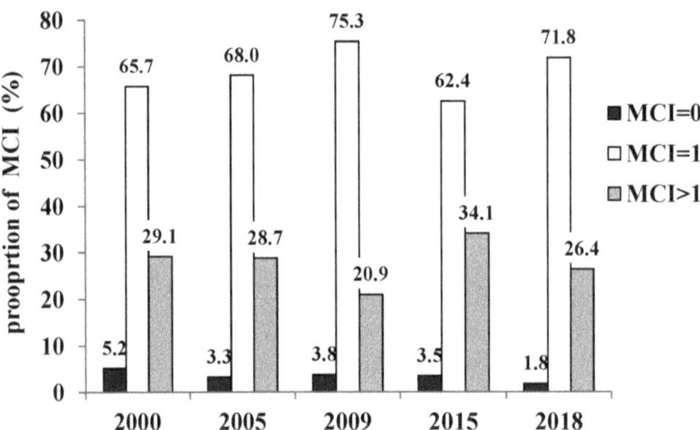


**Figure 4.** The proportion of different MCI of farmland in China.

To pursue maximum marginal benefits of agricultural production, farmers will adjust the production behaviors, i.e., choosing annual or seasonal abandonment of farmland or increasing the MCI. Variations in the planting behaviors of farmers can be further understood by analyzing the transfer matrix among different MCIs. From 2000 to 2018, production behaviors of farmers were mainly manifested via the following features (Table 2). Firstly, the seasonal abandonment of farmland by "transforming double cropping to single cropping" was the most universal. In the double cropping region in 2000, 52.0% of farmland changed from double cropping to single cropping, and only 46.4% continued to be double cropping. Secondly, the seasonal abandonment of farmland occurred widely among the triple cropping system. Only 1.6% of farmland was maintained as triple cropping system, while 59.4% and 37.4% of farmland changed to a single cropping system and double cropping system, respectively. On the contrary, 0.4% of farmland with single cropping and 0.3% of farmland with double cropping were adjusted to the triple cropping. This also indicated that the farmland with triple cropping experienced great spatial transfer. Thirdly, the "transformation from single cropping system to double cropping system" compensated for the decreased area of farmland with double cropping. About 17.8% of single cropping farmland was transformed into double cropping farmland. Moreover, since the proportion of single cropping farmland accounted for as high as 65.3% in 2000, the above transfer could compensate for the reduced double cropping farmland as much as possible. Fourthly, the massive reclamation of abandoned farmland was another important feature in this stage. Approximately 96.4% of abandoned farmland was all reclaimed. Due to the improvement of mechanization and the popularization of agricultural socialized services, the time cost of part-time farmers or migrant farmers to grow grain was reduced, which helped to slow down the occurrence of seasonal or year-round abandonment. During this period, the proportion of abandoned farmland fell by 3.5%.

**Table 2.** Transfer matrix among different MCIs (%) (2000–2018).

| 2000 \ 2018 | Abandoned Farmland | Single Cropping | Double Cropping | Triple Cropping | Total in 2000 |
|---|---|---|---|---|---|
| Abandoned farmland | 3.6 | 74.8 | 21.6 | 0.0 | 5.3 |
| Single cropping | 1.9 | 79.9 | 17.8 | 0.4 | 65.3 |
| Double cropping | 1.3 | 52.0 | 46.4 | 0.3 | 29.4 |
| Triple cropping | 1.6 | 59.4 | 37.4 | 1.6 | 0.1 |
| Total in 2018 | 1.8 | 71.8 | 26.0 | 0.4 | 100.0 |

### 3.1.3. Deconstruction of MCI Based on the Macroscopic Perspective of Land Use Transition

Influences on the production behaviors of farming households on the multi-cropping system of farmland have attracted considerable attention. However, variations of the MCI caused by land use transition based on macroscopic perspective are often ignored [43]. Since the 21st century, China's urbanization has been rapid. The urbanization rate increased from 36.22% to 59.58% from 2000 to 2018. The growth of the permanent population in urban areas led to continuous expansion of urban space, and a great deal of farmland was occupied by construction. Under the constraints of the farmland occupancy-compensation balance system, the contradiction between protecting farmland and guaranteeing construction land was alleviated by actively supplementing farmland through land consolidation and reclamation. However, quantity balance between occupation and supplementation of farmland was a hard requirement, and the attention paid to the improvement of quality was inadequate. Shoddy farmland for quality farmland was relatively universal, which generally influences the improvement of the MCI of farmland.

By comparing the average MCI of exited farmland, increased farmland, and unchanged farmland, it can be judged how the land use transition characterized by land use type change affects the multi-cropping system of farmland. It was found that there were three common laws in the four periods of 2000–2005, 2005–2009, 2009–2015, and 2015–2018 (Table 3). Firstly, the MCI of exited farmland was higher than that of unchanged farmland in the same period (1.26 > 1.24; 1.27 > 1.26; 1.32 > 1.30; and 1.34 > 1.31). This is because the exited farmland occupied by construction usually located in suburbs, and is high-level farmland with good irrigation, high soil fertility, flat terrain, and convenient traffic conditions, where farmers could increase the MCI of the farmland through intensive utilization. Secondly, the MCI of increased farmland was lower than that of unchanged farmland in the same period (1.04 < 1.26; 1.22 < 1.30; 1.06 < 1.32; and 1.25 < 1.26). In order to maintain the balance of farmland quantity, increased farmland is mainly reclaimed in areas with poor cultivated conditions. Due to the poor location and cultivated conditions, the increased farmland is prone to be non-agricultural, non-grain, and even abandoned directly. Thirdly, the MCI of exited farmland is higher than that of increased farmland (1.26 > 1.04; 1.27 > 1.22; 1.32 > 1.06; and 1.34 > 1.25). This again reflects the fact of the exiting of high-quality farmland and the compensation of low-quality farmland. It can be seen that the implementation of the policy of quantity balance between exited farmland and increased farmland was difficult to offset the degeneration of quality, and to achieve a balance of production capacity of farmland. Fourthly, the MCI of unchanged farmland showed a continuous upward trend over time. Evidently, stable expectations of farmers were conducive to the increase of the MCI.

**Table 3.** Comparative analysis of average MCI of exited, increased, and unchanged farmland.

| Year | 2000–2005 | | 2005–2009 | | 2009–2015 | | 2015–2018 | |
|---|---|---|---|---|---|---|---|---|
| | 2000 | 2005 | 2005 | 2009 | 2009 | 2015 | 2015 | 2018 |
| Decreased farmland | 1.26 | - | 1.27 | - | 1.32 | - | 1.34 | - |
| New increased farmland | - | 1.04 | - | 1.22 | - | 1.06 | - | 1.25 |
| Unchanged farmland | 1.24 | 1.26 | 1.26 | 1.30 | 1.30 | 1.32 | 1.31 | 1.26 |

### 3.2. Spatial Variation Characteristics of MCI of Farmland in China

### 3.2.1. Grouping Structure of MCI of Farmland by County

The MCI of farmland in each county was obtained by summarizing and calculating the MCI of all grids within its range. As shown in Table 4, the MCIs of farmland of most counties in China were mainly distributed between 80–160%. Such counties accounted for 84.2% and 86.9% of all in 2000 and 2018, respectively. From 2000 to 2018, counties with decreasing MCI accounted for 33.5%, while counties with increasing MCI accounted for 62.8%. Based on the grouping structure of MCIs, the proportion of counties in group of 80–100% and 100–120% decreased from 2000 to 2018, while the proportion of counties in group of 120–140% and 140–160% rose. The phenomenon of the increase of counties with higher MCIs led to the overall upward trend of MCI of farmland in China. In addition, it was worth noting that the proportion of counties in group of >160% declined to some degree. These counties were generally the dominant production areas, where natural conditions such as water, soil, light, and heat were more suitable for food production. The decline of MCIs in these areas had prominently adverse effects on food production.

**Table 4.** Grouping structure of MCIs of farmland in all counties.

| Proportion (%)<br>Group of MCI (%) | 2000 | 2005 | 2009 | 2015 | 2018 |
|---|---|---|---|---|---|
| 0 | 2.4 | 2.3 | 3.3 | 2.5 | 2.7 |
| 0–80 | 0.6 | 0.4 | 6.4 | 0.4 | 0.1 |
| 80–100 | 19.5 | 19.5 | 32.5 | 19.0 | 18.8 |
| 100–120 | 28.4 | 20.7 | 31.1 | 7.3 | 11.6 |
| 120–140 | 23.9 | 29.8 | 8.3 | 17.9 | 39.4 |
| 140–160 | 12.4 | 13.3 | 4.7 | 37.2 | 17.1 |
| >160 | 12.8 | 13.8 | 13.6 | 15.7 | 10.3 |

### 3.2.2. Spatio-Temporal Pattern of MCI of Farmland by County

The MCI of farmland is determined jointly by physical geographical environment and human economic activities. Overall, the MCI of farmland in China took on a general pattern of higher values in the south and lower values in the north (Figure 5). The areas where the MCI was lower than 100% were chiefly distributed in the northeast, northwest, and northern China. Restricted by the natural environment such as low temperature and less precipitation, the MCIs of farmland in these areas were dominated by single cropping. However, the spatial distribution of MCI does not entirely coincide with the spatial pattern of temperature and precipitation in China, which demonstrates that the MCI of farmland is also influenced by agricultural farming conditions and socio-economic conditions. In particular, the areas with MCI between 150% and 200% were not only distributed in the southern areas of Guangdong and Guangxi, but also extensively distributed in northern Jiangsu, Henan, northern Anhui, southern Shaanxi, and southern Gansu. These agricultural areas possessed the advantages of flat terrain and good farming conditions, which means it was easy to make use of large-scale machinery for farming and was favorable for saving labor input in agricultural production. In addition, the plains in the middle and lower reaches of the Yangtze River provide favorable light and heat conditions and have developed economy. So the MCI of farmland had maintained at around 140%. The hilly areas along the southeast coast, Sichuan Basin, Loess Plateau, Yunnan-Guizhou Plateau, and the hilly areas along the south of the Yangtze River are characterized by undulating terrain, limited farming conditions, and difficulty in using large-scale machinery. As a result, the MCIs of those areas were between 100% and 130%. From the perspective of administrative areas, the provinces and cities with the highest MCI of farmland in China were concentrated in Jiangsu, Guangdong, Henan, Guangxi and Anhui, while the provinces and cities with lowest MCI were concentrated in Liaoning, Xinjiang, Beijing, Jilin, Inner Mongolia, and other places.

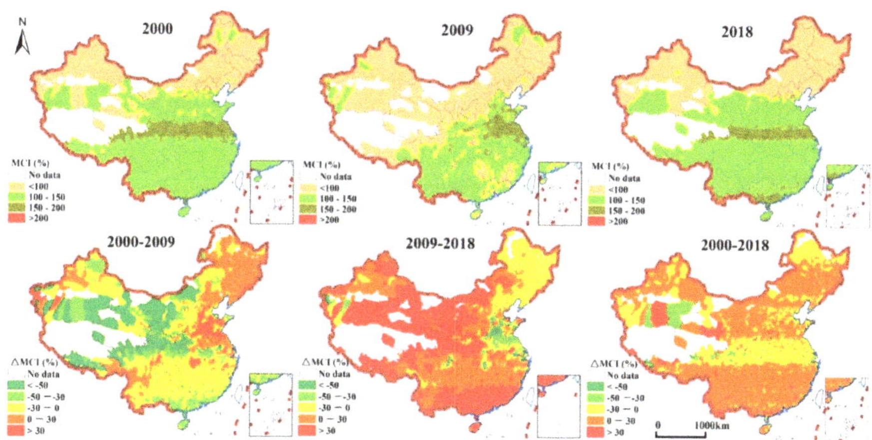

**Figure 5.** Characteristics of the MCI of farmland in Chinese counties from 2000 to 2018.

Comparing MCI in different periods, the counties with decreasing MCIs from 2000 to 2009 were chiefly concentrated in western China (such as Xinjiang, Gansu, Qinghai, and Sichuan) and in southern China (such as Guangdong, Jiangxi, Hunan, and Fujian). During this period, China enforced the national strategy of developing the western area, which launched large-scale construction of infrastructure and key industries in the western areas. The strategy expedited the course of transferring agricultural population to cities and towns, which resulted in the reduction of investment in agriculture and the decline of the MCI. From 2009 to 2018, the counties with declining MCIs were chiefly distributed in the Huang-Huai-Hai Plain and the lower reaches of the Yangtze River, which were the major grain producing areas. These areas are economically developed, which makes it easier for farmers to move to cities and decrease labor input in agricultural production. Although the enforcement of the policy of the lowest purchase price of grain aroused farmers' enthusiasm for growing grain and facilitated the growth of MCI of each province in the short term, while the MCI in developed areas declined first after the lowest purchase price lowered further. It was confirmed in relevant studies that there appear to be labor-saving planting methods, such as planting trees by shrinking the grain-planting area in Hebei Province, and changing two-season rice into one-season rice in Hubei Province, etc. Generally, there existed both counties with rising and falling MCI in China. But the principal areas with falling MCIs from 2009 to 2018 were concentrated in two major grain producing areas, which exert a more adverse influence on the food security of the country.

### 3.3. Gaps between MCI and PMCI of Farmland and Regional Optimization Scheme

Agricultural production is particularly dependent upon the endowment of natural resources. Theoretically, if the agricultural production entirely coincides with the resource endowment, it will help to boost the efficiency of agricultural production. Although agricultural technology can avoid some natural conditions to a certain extent, it is still hard to form structural changes. Thus, it is essential to carry out a comparative analysis of the MCI and PMCI to survey the degree to which the major grain production areas and the dominant production areas spatially match. To avoid the influence of inter-annual fluctuation, this paper made a comparative analysis of the average MCI from 2015 to 2018 and PMCI.

The findings reveal that farmland was overused in northern China and underutilized in southern China (Figure 6). From the north to the south in China, the gap between MCI and PMCI increasingly widened. In the northern China (such as Xinjiang, Inner Mongolia, Heilongjiang and Jilin), where single cropping system dominated, the gap between the MCI of farmland and the PMCI was kept within 10%, which means agricultural production was effectively matched with regional water and heat resources. It is worth noting that the ecotone between agriculture and animal husbandry in the north and south, which covers the water conservation areas of Beijing-Tianjin-Hebei, the soil erosion areas of Loess hills and gullies, and the fragile ecological environment areas of Qinghai-Tibet Plateau, exhibits the phenomenon of over-utilization of farmland. Some counties employ the agricultural production mode of triple cropping in two years or double cropping in one year, which is higher than the upper limit of single cropping in one year determined by local water and heat conditions and exceeds the carrying capacity of local resources and the environment. Meanwhile, there is a narrow strip along the Bohai Rim, the southern part of Huang-Huai-Hai region, the Qinling Mountains, Yunnan and South China, where the MCI of farmland increased by 10–100%. There are abundant water and heat conditions in the middle and lower reaches of the Yangtze River, south of the Yangtze River, and Sichuan-Guizhou area, and most areas are suitable for double cropping or even triple cropping in a year. However, the actual MCIs of farmland in those regions were not more than 150%, and even only single cropping per year.

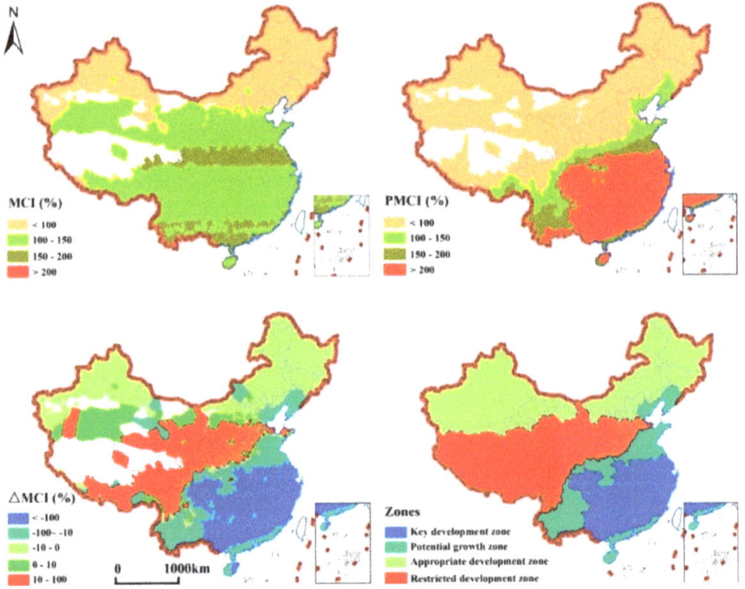


**Figure 6.** Gaps between MCI and PMCI of farmland in China and the division of types.

Based on the gap between the MCI and PMCI in counties, China was classified into four types of grain producing zones, such as key development zone, potential growth zone, appropriate development zone, and restricted development zone (Figure 6), which serve as a basis for the layout optimization of grain production and the design of the benefit compensation mechanism. The key development zone is intended to raise the MCI of farmland by 100% and make full use of abundant water and heat resources in southeast China. The potential growth zone aims to raise the MCI of farmland by 50%. The appropriate development zone continues to keep the current cropping system of farmland, and chiefly produces one-season crops. The restricted development zone is laid out to restrict the use of farmland appropriately, employs reasonable fallow and rotation measures to evade over-exploitation of groundwater resources and soil erosion, and protects the fragile local ecological environment.

## 4. Discussion and Conclusions

Based on the NDVI data, this paper calculated the MCI of farmland in China from 2000 to 2018, and explored the spatio-temporal characteristics of MCI. In addition, the spatial optimization scheme of farmland was put forward according to the gap between MCI and PMCI. The conclusions are drawn as below: from 2000 to 2018, China's MCI of farmland underwent the fluctuation of rising first, then falling, before rising continuously. These fluctuations were closely associated with the agricultural support policies enforced in different stages and farmers' reduction in the intensity and utilization of farmland, owing to the low income earned from growing grain. The areas with high MCIs in China were situated in the major grain producing areas, such as Huang-Huai-Hai plain and the southern areas including Guangdong and Guangxi provinces. The proportion of counties with declining MCIs from 2009 to 2018 was lower than that from 2000 to 2009, but the counties with declining MCIs in the later stage were chiefly situated in the major grain producing areas, which exert adverse influence on food security. Compared with the PMCI, the utilization intensity of farmland in northern China was high, while most areas in southern China boasted great potential to increase the MCI. Four different regions and relative optimizing countermeasures were proposed.

In consideration of China's basic national conditions, i.e., more people and less farmland, small-scale production restricts the increase of the income from agricultural production. Despite that, the overall MCI in China has shown a rising trend since 2000, with still more than 30% of the counties having experienced a downturn. Firstly, China's floating population was 376 million in 2020, and most of them were rural migrants. Along with the acceleration of China's industrialization and urbanization, a large number of laborers will still be transferred to cities and towns, and the input of agricultural labor will continue to be decreased. Farmland in Chinese agricultural areas takes the form of collective property rights, and some migrant farmers choose to make use of farmland in an extensive manner for fear of the loss of their rights and interests arising from land transfer. Secondly, government departments have not yet controlled the abandonment of farmland or the change of two seasons to one season according to legal instruments. Following the present development trend, the MCI of some counties will take a downturn trend in the future. It is necessary to attach enormous importance to the phenomenon of decreasing the MCI in principal grain producing areas to avoid food security problems as a result of a large-scale occurrence. Thirdly, non-agricultural farmland is inevitable due to rapid urbanization. In the pursuit of the balance of arable land, low quality "new" arable land reduces the multiple cropping of arable land, which leads to ecological problems.

There is an interactive relationship between the MCI of farmland and the natural environment. In the long term, the high level of climatic variability affects the MCI by affecting the PMCI. With the improvement of farmland irrigation facilities, the water supply limitations noticeably decreased under the irrigated scenario. The growth of the annual mean temperature was identified as the main reason underlying the increase of the PMCI. Furthermore, the area found to be suitable only for single cropping farming decreased, while the area suitable for triple cropping farming increased significantly from the 1960s to the 2000s. The magnitude of change of the PMCI showed a pattern of increase both from northern China to southern China and from western China to eastern China. However, the fluctuations of the MCI calculated in this paper are mainly related to agricultural policies and farmers' decision-making, and were not always increasing. Furthermore, the spatial pattern of the MCI does not change with longitude and latitude, and the proportion of single cropping farmland increased significantly. Clearly, the spatio-temporal characteristics of the MCI and PMCI variability are not consistent. It can be seen that from a short-term perspective, the MCI variability is mainly affected by farmers' planting behavior, whose goal is to maximize economic benefits, even at the expense of ecological resources and the environment. Among them, the impact on water has received widespread attention.

Thus far, there is no conclusive conclusion on whether increasing MCI can improve the efficiency of water utilization. Studies in Pakistan [30] and Brazil [44] proved that multi-cropping system improved water use efficiency. This situation is likely due to the fact that crop species selection is determined, in large part, based on farmers' financial interests, but not necessarily on which crop is the most suitable. However, the spatial mismatch between the MCI and PMCI in China is bound to increase the overall water consumption. Southern China possesses favorable water and heat conditions, but low efficiency in terms of the utilization of farmland, while northern China possesses high intensity utilization of farmland. In particular, groundwater has been over-exploited in the Huang-Huai-Hai plain for the development of irrigation agriculture. As a result, serious underground funnels appeared in this area.

Some suggestions were or will be put forward to solve the problem of increased water consumption. Firstly, the Chinese government has begun to enforce measures to close pumping wells in some areas of Hebei, and decrease over-exploitation of groundwater by way of cropping rotation. Secondly, following the concept of the green development of agriculture, China's agricultural production should optimize the pattern of utilization of farmland in the future, and reinforce the determination of yield by water in areas with high utilization intensity. Meanwhile, the government should increase the utilization intensity of farmland in key development zones. In southern China, it is necessary to issue agricultural support policies, such as comprehensive production subsidies, agricultural socialization services, and financial loans, etc., to encourage and support the adjustment of planting structure or land transfer, and promote the shift from single cropping to multiple cropping, or the rotation of food crops and cash crops. Thirdly, the distortion in the implementation of policy of "pothook of city construction land increase and rural residential land decrease" has caused a decline in quality and implicit decrease in quantity of farmland, respectively. To solve this problem, policies can be adjusted to establish a supplementary mechanism based on farmland productivity. The increase in the MCI of farmland can be used to offset the amount of farmland balance index.

In this paper, by comparing the MCI of farmland calculated based on NDVI data with statistics, we proved that the MCI of farmland, gained by ways of S-G filtering, peak extraction, and peak elimination, is credible and endowed with a reference and popularization value. Grain production in China is mainly carried out by smallholders, and the farmland vegetation is complex, diverse, and irregularly distributes on the surface, particularly in mountainous and hilly areas where there are few large-scale agricultural crops of the same type. Since remote sensing images are made up of regular grids of equal size and influenced by mixed pixels, there will inevitably be certain errors in our results. In future research, perfecting the selection of extraction methods and making use of higher resolution data, in order to raise the accuracy of the MCI of farmland, should be attempted.

**Author Contributions:** T.L. conceived and designed the structure of this paper, processed the data, and wrote part of the paper. Y.W. wrote part of the literature review, results, and discussion section. C.L. participated in the framework discussion and data processing. S.T. made suggestions for the revision of the paper. All authors have read and agreed to the published version of the manuscript.

**Funding:** This research was funded by the National Natural Science Foundation of China (grant nos. 41901216, 41731286, 41971216, 41701200, 72003194), the Key Program of National Social Science Foundation of China (grant no. 21AZD039), the Bagui Scholars Program of Guangxi Zhuang Autonomous Region, and Program of Science and Technology Plan of Guangxi Zhuang Autonomous Region (grant no. AD19110158).

**Institutional Review Board Statement:** Not applicable.

**Informed Consent Statement:** Not applicable.

**Data Availability Statement:** The data used in the manuscript is provided by the Data Registration and Publishing System of Data Center for Resources and Environmental Sciences, Chinese Academy of Sciences (https://www.resdc.cn/data.aspx?DATAID=254, accessed on 1 June 2021).

**Conflicts of Interest:** The authors declare no conflict of interest.

# References


1.	Pittelkow, C.M.; Liang, X.; Linquist, B.A.; Groenigen, K.; Kessel, C.V. Productivity limits and potentials of the principles of conservation agriculture. *Nature* **2015**, *517*, 365–368. [CrossRef]
2.	He, C.; Liu, Z.; Min, X.; Ma, Q.; Dou, Y. Urban expansion brought stress to food security in China: Evidence from decreased cropland net primary productivity. *Sci. Total Environ.* **2017**, *576*, 660–670. [CrossRef]
3.	Wu, W.; Yu, Q.; You, L.; You, L.; Chen, K.; Tang, H.; Liu, J. Global cropping intensity gaps: Increasing food production without cropland expansion. *Land Use Policy* **2018**, *76*, 515–525. [CrossRef]
4.	Long, H.L.; Qu, Y.; Tu, S.S.; Zhang, Y.N.; Jiang, Y.F. Development of land use transitions research in China. *J. Geogr. Sci.* **2020**, *30*, 1195–1214. [CrossRef]
5.	Wang, H.G. *70 Years of Farming System in China*; Agricultural Press: Beijing, China, 2005; p. 123.
6.	Katharina, W.H.; Jan, P.D.; Felix, T.P.; Stefan, S.; Philip, K.T.; Alberte, B.; Mario, H. Multiple cropping systems of the world and the potential for increasing cropping intensity. *Glob. Environ. Chang.* **2020**, *64*, 102131.
7.	Zuo, L.J.; Wang, X.; Liu, F.; Yi, L. Spatial exploration of multiple cropping efficiency in China-based on time series remote sensing data and econometric model. *J. Integr. Agric.* **2013**, *12*, 903–913. [CrossRef]
8.	Yu, Q.Y.; Wu, W.B.; You, L.Z.; Zhu, T.J.; Vliet, J.V.; Verburg, P.H.; Liu, Z.H.; Li, Z.G.; Yang, P.; Zhou, Q.B.; et al. Assessing the harvested area gap in China. *Agric. Syst.* **2017**, *153*, 212–220. [CrossRef]
9.	Strehlow, B.; Mol, F.D.; Gerowitt, B. Herbicide intensity depends on cropping system and weed control target: Unraveling the effects in field experiments. *Crop Prot.* **2020**, *129*, 105011. [CrossRef]
10.	Virginia, S.N.; Zornoza, R.; Faz, N.; Fernández, J.A. Comparison of soil organic carbon pools, microbial activity and crop yield and quality in two vegetable multiple cropping systems under Mediterranean conditions. *Sci. Hortic.* **2020**, *261*, 109025.
11.	Hampf, A.C.; Tommaso, S.; Michael, B.M.; Kawohl, T.; Kilian, M.; Nendel, C. Future yields of double-cropping systems in the Southern Amazon, Brazil, under climate change and technological development. *Agric. Syst.* **2020**, *177*, 102707. [CrossRef]
12.	Eliakira, K.N.; Frederick, B.; Patrick, A.N. Productivity of intercropping with maize and common bean over five cropping seasons on smallholder farms of Tanzania. *Eur. J. Agron.* **2020**, *113*, 125964.
13.	Xie, H.L.; Liu, G.Y. Spatiotemporal differences and influencing factors of multiple cropping index in China during 1998–2012. *J. Geogr. Sci.* **2015**, *25*, 1283–1297. [CrossRef]
14.	Pinki, M.; Ruth, D.F.; Jessica, C.; Nicole, F.; Arif Md, A.; Aurelie, H.; Shauna, D.; Jessica, F. Multiple cropping alone does not improve year-round food security among smallholders in rural India. *Environ. Res. Lett.* **2021**, *16*, 065017.
15.	Devendra, C.; Thomas, D. Smallholder farming systems in Asia. *Agric. Syst.* **2002**, *71*, 17–25. [CrossRef]
16.	Zuo, L.J.; Zhang, Z.X.; Dong, T.T.; Wang, X. Progress in the research on the multiple cropping index. *J. Nat. Resour.* **2009**, *24*, 553–560.
17.	Gray, J.; Friedl, M.; Frolking, S.; Ramankutty, N.; Gumma, M.K. Mapping Asian cropping intensity with MODIS. *IEEE J. Sel. Top. Appl. Earth Obs. Remote Sens.* **2014**, *7*, 3373–3379. [CrossRef]
18.	Frolking, S.; Jagadeesh, B.Y.; Douglas, E. New district-level maps of rice cropping in India: A foundation for scientific input into policy assessment. *Field Crop. Res.* **2006**, *98*, 164–177. [CrossRef]
19.	Yan, H.M.; Liu, J.Y.; Cao, M.K. Remotely sensed multiple cropping index variations in China during 1981–2000. *Acta Geogr. Sin.* **2005**, *60*, 559–566.
20.	Ding, M.J.; Chen, J.; Xin, L.J.; Lin, L.H.; Li, X.B. Spatial and temporal variations of multiple cropping index in China based on SPOT-NDVI during 1999–2013. *Acta Geogr. Sin.* **2015**, *70*, 1080–1090.
21.	Yan, H.M.; Xiao, X.M.; Huang, H.Q. Satellite observed crop calendar and its spatio-temporal characteristics in multiple cropping area of Huang-Huai-Hai Plain. *Acta Ecol. Sin.* **2010**, *30*, 2416–2423.
22.	Shen, J.; Chang, Q.R.; Li, F.L.; Qin, Z.F.; Xie, B.N. Dynamic monitoring of cropping index in Guanzhong area using remote sensing in 2000–2013. *Trans. Chin. Soc. Agric. Mach.* **2016**, *47*, 280–287.
23.	Li, W.Y.; Jiang, L.G.; Li, P. The spatio-temporal pattern of rice cropping systems in the polder area of Poyang lake during 2000–2010. *Resour. Sci.* **2014**, *36*, 809–816.
24.	Xu, X.B.; Yang, G.S. Spatial and temporal changes of multiple cropping index in 1995–2010 in Taihu Lake basin, China. *Trans. Chin. Soc. Agric. Eng.* **2013**, *29*, 148–155.
25.	Zhang, C.J.; He, H.M.; Mokhtar, A. The impact of climate change and human activity on spatio-temporal patterns of multiple cropping index in South West China. *Sustainability* **2019**, *11*, 5308. [CrossRef]
26.	Jiang, M.; Li, X.B.; Xin, L.J.; Tan, M.H. Paddy rice multiple cropping index changes in Southern China: Impacts on national grain production capacity and policy implications. *J. Geogr. Sci.* **2019**, *29*, 1773–1787. [CrossRef]
27.	Cohn, A.S.; VanWey, L.K.; Spera, S.A.; Mustard, J.F. Cropping frequency and area response to climate variability can exceed yield response. *Nat. Clim. Chang.* **2016**, *6*, 601–604. [CrossRef]
28.	Li, S.F.; Li, X.B.; Xin, L.J.; Tan, M.H.; Wang, X.; Wang, R.J.; Jiang, M.; Wang, Y.H. Extent and distribution of cropland abandonment in Chinese mountainous areas. *Resour. Sci.* **2017**, *39*, 1801–1811.
29.	Arianti, F.D.; Minarsih, S.; Nurwahyuni, E. Optimizing dry land through the implementation of maize and rice multiple cropping system in Pemalang Regency. In *IOP Conference Series: Earth and Environmental Science, Proceedings of the 1st International Conference on Sustainable Tropical Land Management, Bogor, Indonesia, 16–18 September 2020*; IOP Publishing Ltd.: Bristol, UK, 2021; Volume 648, p. 012070.

30. Hina, F.; Lal, K.A.; Sehrish, H. Comparative water efficiency analysis of sole and multiple cropping systems under tunnel farming in Punjab-Pakistan. *J. Water Resour. Prot.* **2020**, *12*, 455–471.

31. Zhang, X.Y. Multiple cropping system expansion: Increasing agricultural green house gas emissions in the North China Plain and neighboring regions. *Sustainability* **2019**, *11*, 3941. [CrossRef]

32. Zuo, L.J.; Wang, X.; Zhang, Z.X.; Zhao, X.L.; Liu, F.; Yi, L.; Liu, B. Developing grain production policy in terms of multiple cropping systems in China. *Land Use Policy* **2014**, *40*, 140–146. [CrossRef]

33. Kang, Q.L.; Li, C.L.; Zhang, Y.H. The cropping index change and impact factor analysis in Jiangsu Province between 2001 and 2010. *J. Cap. Norm. Univ. Nat. Sci. Ed.* **2017**, *38*, 86–94.

34. Araya, A.; Prasad, P.V.V.; Ciampitti, I.A.; Jha, P.K. Using crop simulation model to evaluate influence of water management practices and multiple cropping systems on crop yields: A case study for Ethiopian highlands. *Field Crop. Res.* **2021**, *260*, 108004. [CrossRef]

35. Ben, P.Q.; Wu, S.H.; Li, X.T.; Zhou, S.L. China's inter-provincial grain trade and its virtual cultivated land flow simulation. *Geogr. Res.* **2016**, *35*, 1447–1456.

36. Long, H.L.; Zhang, Y.N. Rural planning in China: Evolving theories, approaches, and trends. *Plan. Theory Pract.* **2020**, *21*, 782–786.

37. Lin, G.C.S.; Ho, S.P.S. China land: Insights from the 1996 Land Survey. *Land Use Policy* **2003**, *20*, 87–107. [CrossRef]

38. Li, T.T.; Long, H.L.; Zhang, Y.N.; Ge, D.Z.; Li, Y.R. Analysis of the spatial mismatch of grain production and farmland resources in China based on the potential crop rotation system. *Land Use Policy* **2017**, *60*, 26–36. [CrossRef]

39. Zhu, X.L.; Li, Q.; Sheng, M.G.; Chen, J.; Wu, J. A methodology for multiple cropping index extraction based on NDVI time-series. *J. Nat. Resour.* **2008**, *23*, 534–544.

40. Savitzky, A.; Golay, M. Smoothing and differentiation of data by simplified least squares procedures. *Anal. Chem.* **1964**, *36*, 1627–1639. [CrossRef]

41. Fan, J.L.; Wu, B.F. A methodology for retrieving cropping index from NDVI profile. *J. Remote Sens.* **2004**, *8*, 628–636.

42. Long, H.L. *Land Use Transitions and Rural Restructuring in China*; Springer Nature: Singapore, 2020.

43. Long, H.L.; Qu, Y. Land use transitions and land management: A mutual feedback perspective. *Land Use Policy* **2018**, *74*, 111–120. [CrossRef]

44. Langeveld, J.W.A.; Dixon, J.; Van, K.H.; Quist-Wessel, P.M.F. Analyzing the effect of biofuel expansion on land use in major producing countries: Evidence of increased multiple cropping. *Biofuels Bioprod. Biorefin.* **2014**, *8*, 49–58. [CrossRef]

*Article*

# Measuring the Impact of the Multiple Cropping Index of Cultivated Land during Continuous and Rapid Rise of Urbanization in China: A Study from 2000 to 2015

Ren Yang [1], Xiuli Luo [1], Qian Xu [2,*], Xin Zhang [1] and Jiapei Wu [1]

[1] School of Geography and Planning, Sun Yat-sen University, Guangzhou 510275, China; yangren666@mail.sysu.edu.cn (R.Y.); luoxli6@mail2.sysu.edu.cn (X.L.); zhangx697@mail2.sysu.edu.cn (X.Z.); wujp23@mail2.sysu.edu.cn (J.W.)

[2] School of Public Administration, Guangdong University of Finance & Economics, Guangzhou 510320, China

* Correspondence: xuqian@gdufe.edu.cn

**Abstract:** With the continuous and rapid rise of urbanization in China, land use transition research has been carried out extensively. Multiple cropping is the content of land use recessive morphology research, and it is also a common agricultural system in China. Accordingly, further research on multiple cropping index (MCI) can enrich the land use transition research and help to evaluate China's food security. In order to examine the spatiotemporal changes and factors influencing the MCI of cultivated land in China, we collected MODIS remote sensing image data and land use classification data and conducted a remote sensing inversion on China's MCI from 2000, 2005, 2010, and 2015. The spatial distributions and evolution processes of the MCI were explored through spatial mapping, statistical analysis, and processing with the Geographic Information System; moreover, the influencing factors of MCI were explored quantitatively with principal component regression. The results were as follows: (1) at the provincial scale, the average MCI across Guangdong, Guangxi, Hainan, Henan, Anhui, and Jiangsu was high; meanwhile, the average MCI across Heilongjiang, Inner Mongolia, Ningxia, and Qinghai was low. Between 2000 and 2015, the number of provinces with low MCI increased gradually, and the average MCI decreased greatly in the southern provinces. (2) At the county scale, the Taihang Mountains, Qinling Mountains, and Hengduan Mountains formed the boundary of China's single cropping and multiple cropping indices. Dynamic changes in China's MCI were obvious, and the number of counties with MCI change values lower than 0 increased gradually. Last, (3) natural conditions, nonagricultural process, cultivated land quality, and agricultural intensification demonstrated different degrees of impact on the MCI; in particular, the influence of nonagricultural industries, pesticides, and agricultural plastic film on the MCI proved especially important. Future research should strengthen the existing work on related transformations in farmers' livelihoods, especially in terms of the return of rural labor force, the body of agricultural production, agricultural ecological issues, and the balance between increased crop production and reduced environmental pollution. In addition, agricultural policy design should pay more attention to cultivated land quality, the farmer who cultivates the land, and the multiple cropping potential of cultivated land.

**Keywords:** multiple cropping; cultivated land; land use transition; food security; influencing factors; China

**Citation:** Yang, R.; Luo, X.; Xu, Q.; Zhang, X.; Wu, J. Measuring the Impact of the Multiple Cropping Index of Cultivated Land during Continuous and Rapid Rise of Urbanization in China: A Study from 2000 to 2015. *Land* **2021**, *10*, 491. https://doi.org/10.3390/land10050491

Academic Editor: Hossein Azadi

Received: 2 April 2021
Accepted: 27 April 2021
Published: 6 May 2021

**Publisher's Note:** MDPI stays neutral with regard to jurisdictional claims in published maps and institutional affiliations.

## 1. Introduction

In 2015, the United Nations adopted the 17 Sustainable Development Goals, one of which is to eradicate hunger by 2030. However, the global demand for food production is continuing to rise due to population growth, diet changes, and increasing biofuel use [1]. Given the projected demand of our current course, global crop production must double by 2050 [2,3]; however, this is unlikely to happen, because climate change threatens farm



*Land* **2021**, *10*, 491. https://doi.org/10.3390/land10050491     https://www.mdpi.com/journal/land

yields [4,5]. Indeed, several studies have shown that global grain production faces slow growth, stagnancy, or decline [3,6,7]. More specifically, four key global crops—maize, rice, wheat, and soybeans—demonstrate insufficient yield trends for doubling the global crop production by 2050 [4]. We need to ensure that the world's agricultural systems can produce enough food to feed the world's growing population; however, because our current agricultural systems are destroying the world's land, water, biodiversity, and climate, we must at once make our agriculture systems sustainable land use systems to reduce the adverse ecological impacts [8,9]. Future increases in food production should come from intensifying our use of existing land by halting agricultural expansion, improving cropping efficiency [10,11], closing "yield gaps" on underperforming lands, reducing waste, and shifting diets [8].

Asia's population is growing at a rate of 56 million people per year, putting increasing pressure on food demand and land use [12]. Recent studies suggested that urban expansion will result in a 1.8–2.4% loss of global croplands by 2030, with about 80% of the loss taking place in Asia and Africa [13]. China is the most populous country in Asia, with 19% of the world's population and only 8% of the world's croplands; notably, this asymmetry may affect China's—and, more broadly, the world's—food security [14,15]. Between the 1950s and 1990s, China's urbanization caused the rapid transition of cultivated land use, many cultivated lands transformed into nonagricultural uses, and rural areas served urban development [16,17]; accordingly, millions of rural agricultural laborers flowed into cities and nonagricultural industries. As a result, the proportion of China's rural population decreased from 82.08% in 1978 to 42.7% in 2016, and the proportion China's economy comprising the agricultural output decreased from 50.5% in the early 1950s to 8.6% in 2016 [18]. On the one hand, China's loss of its rural population led to the aging and hollowing of its rural areas [19,20]. On the other hand, it led to the serious abandonment of cultivated land, the weakened function of agricultural production, and the weakened status of farmers. For example, two million hectares of agricultural land fall out of production each year in China [14]. Despite rural depopulation, the land area of rural settlements has not decreased correspondingly. On the contrary, rural settlement areas nearly tripled from 1967 to 2008, and most of these increased areas occupied agricultural land, exacerbating the destruction of agricultural land [21]. In addition, China's agriculture has been characterized by "small-scale farming", "traditional farming methods", and "self-sufficiency" for a long time; therefore, low agricultural labor productivity has become a key weakness of agricultural competitiveness and sustainable development in China [22,23]. China already has the strictest cultivated land protection system, including land use planning; a basic cultivated land protection system; a land use regulation system; and a balanced system for the requisition, compensation, and consolidation of cultivated land and land consolidation. Although these policies have made remarkable achievements, they have not been able to address the shrinking of the country's cultivated land area and the deteriorating quality of regionally cultivated land. Today, the problem of the continuous nonagricultural, nongrain, and extensive use of cultivated land is worsening in China [16,24].

Some scholars claimed that increasing the farmland multiple cropping index (MCI) was the simplest method to enhance tillage [25–27]. Here, it is helpful to note that the cropping index refers to the number of times a crop is planted in a year, and a high MCI is related to an increase of the sown area, thus boosting food output [28]. Therefore, global crop production may be enhanced by agricultural intensification through multiple cropping to expand acreage without increasing the area of cultivated land [1].

Scholars from different countries have researched MCI in different regions of the world and focused on various aspects of the topic, including the evolution of Brazilian soybean double-cropping systems [29]; remote sensory monitoring of African cropping systems [30]; the relationship between crop rotation systems, water distribution, the political system, and the rational behavior of farmers in Egypt [31]; chemical weed control on crop yields in the Czech Republic [32]; and the effects of crop rotation on agrobiodiversity in Vanuatu [33]. Some scholars found that crop rotation and multiple cropping were used as methods to change the physical properties of soil. For example, legumes can enhance the soil in a multiple crop lands by enhancing its organic carbon, fertility, aggregate stability, and vegetable yields under semi-arid conditions [34–36]. Meanwhile, scholarships have also been done to develop the first spatially explicit measure of the cropping intensity gap and, later, to uncover the differences between the potential and actual cropping intensity gaps; ultimately, these studies found that Latin America had a tremendous potential to expand its grain harvest by eliminating the cropping intensity gap, followed by Asia [1].

The multiple cropping system is a common farming practice in Asia, and agricultural production in Asia mainly emphasizes intensive multiple cropping production of rice and wheat [37]. Some researchers found that, in agricultural areas in Asia with large-scale irrigation systems, the land was mainly cultivated under the double-cropping system [38]. The current research on Asia focuses on India, Vietnam, and China. For example, scholarship done in India mostly used remote sensing satellite data to obtain crop rotation maps [39–41]; analyze the area and spatial distribution of multiple cropping crops [42]; and evaluate the efficiency and sustainability of India's cropping system by the MCI, area diversity index, and cultivated land utilization index [40]. In Vietnam, a time series of MODIS data was used to monitor the rice cropping intensity of the Mekong Delta [12]. For instance, based on MODIS time series imagery and field interviews, some studies explored the relationship between seasonal changes in the river and the temporal–spatial distribution of cropping systems and rice phenology in the Mekong Delta; they found that the change of the water environment was closely related to the changes of rice cropping systems [43,44].

Benefiting from a monsoon climate, China has one of the highest MCI in the world, with nearly 57% of its land cultivated using multiple cropping [45,46]. On the one hand, MCI is an important measure for evaluating China's food security, as its increase is essential to meet the growing demand for food [47,48]. On the other hand, MCI is an important evaluation index of land use transition research [49]. MCI researches the changes of recessive morphologies of land use transition and reflects the land management model by measuring the intensity level of the cultivated land use [50,51]. Therefore, it is of great practical significance to research MCI and its influencing factors to support national policy formulation in China [28,52,53]. In recent years, in order to evaluate the current situation of China's multiple cropping system, domestic scholars have actively researched the MCI, using different geographical scales. On the national scale, scholars have used various methods to calculate the actual MCI, the potential MCI, multiple cropping efficiency, and other indicators, such as the econometrics model [54], stochastic frontier analysis [28,17], Theil index [53], and continuous wavelet transform [55]; moreover, they have analyzed the spatiotemporal distribution of China's multiple cropping system and its regional differences. Xie and Liu discovered that China's MCI increased year after year from 1998 to 2012 [53]; meanwhile, Qiu et al. found that China's cropping intensity increased remarkably from 1982 to 1999 but declined slightly from 2001 to 2013 [55]. On the regional and provincial scales, the multiple cropping systems in some major agricultural production areas have been studied. For instance, Peng et al. found that Zhejiang Province's MCI decreased from 2001 to 2003 before increasing in 2004 and that the MCI in the southwest was higher than in the northeast [56]. Additionally, Zhang et al. extracted double-cropping systems in Northern China by using a Fourier analysis from the time series MODIS data [57]. Li et al. combined the results of field surveys and Landsat data and found that rice cropping systems in the Poyang Lake region showed an increasing trend from 2004 to 2010 [58], with an increased rate of about 20.2%. Feng et al. studied the effects of the planting patterns on the growth, yield, and economic benefits of cotton in a wheat–cotton double-cropping system versus monoculture cotton [59].

Previous research on the MCI mainly focused on a single scale in China, and it seldom considered the whole country through a multi-scale perspective. It focused on how to improve the technology and technical aspects to extract MCI with remote sensing data, as well as the spatiotemporal distribution characteristics; however, there is insufficient research on the regional differences and driving forces. Cultivated land is an important land type in the research of rural land use transition, and it is closely related to human production and life. Cultivated land use reflects the evolution of human–environment relationship in rural areas and also reflects the current situation and problems of China's agriculture and rural society. Therefore, it is necessary to have a comprehensive understanding of the national MCI. Based on the limitations of the previous research, this study sought to answer the following questions: What are the characteristics of the dynamic evolution and spatiotemporal distribution of MCI? Where are these characteristics distributed at the provincial and county scales? What accounts for the differences in the spatiotemporal distributions?

## 2. Research Design

### 2.1. Research Structure

This research assumed "phenomenon–problem–pattern–process–mechanism" as the main logical structure, and its contents mainly included the research foundation, scientific problems, data processing, spatiotemporal analysis, and influencing factor analysis; accordingly, the detailed research process for each part was designed as follows (Figure 1).

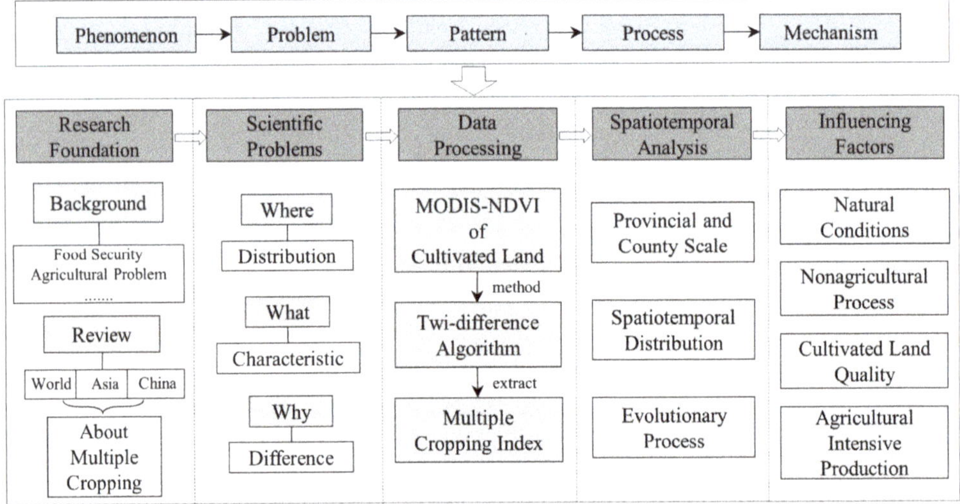


**Figure 1.** Framework for examining the spatiotemporal characteristics and influencing factors of the MCI.

This paper takes the human–land relationship as its core principle and built a research framework concerning the influencing factors starting from four aspects: natural conditions, nonagricultural process, cultivated land quality, and agricultural intensive production (Figure 2).

First, the natural conditions of the agricultural climate conditions, topography, landform, and soil have deeply determined the cropping system of traditional agriculture. Second, the urban–rural transition has dramatically changed the interactions between urban and rural production; notably, following the acceleration of rural economic development, significant changes took place in rural areas—for example, rural labor forces, industrial structures, and employment structures began to move away from agriculture. Third, cultivated land quality affects the increase of MCI and is at once affected by human activities and natural disasters. Human activities include the use of agricultural fertilizers, pesticides, and agricultural films. Finally, agricultural modernization and the agricultural production efficiency directly impact intensive agricultural production, thereby changing the MCI. Agricultural modernization is influenced by factors such as traffic, irrigation, the conditions of agricultural machinery, and investment level. As the main body of agricultural production, farmers' input and enthusiasm for farming determine the level of agricultural production efficiency, which is also reflected in per capita cultivated land area and grain per labor. Moreover, because farmers are rational, economic individuals, their enthusiasm is directly affected by their agricultural income, which also influences the MCI. Based on the above analysis and the availability of data, 19 indexes were selected to investigate the factors impacting the MCI (Table 1).

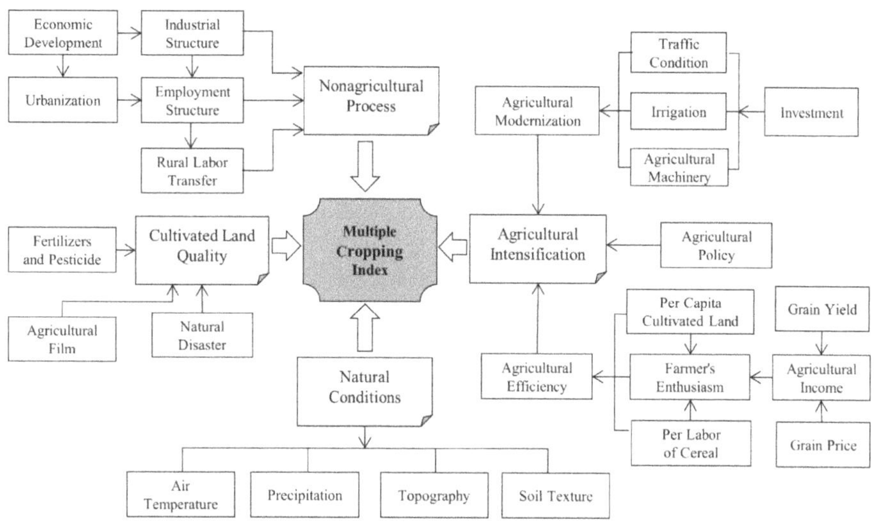


**Figure 2.** The research framework concerning the influencing factors.

**Table 1.** The influencing factors index of the MCI in China.

| Types | Indexes | | Calculation Method and Data Description |
|---|---|---|---|
| Nonagricultural process | Nonagricultural population | $x1$ | Nonagricultural population/Total population |
| | Nonagricultural industry | $x2$ | GDP of secondary and tertiary industry/GDP |
| | Urbanization rate | $x3$ | China Statistical Yearbook |
| | Per capita GDP | $x4$ | China Statistical Yearbook |
| Cultivated land quality | Density of agricultural fertilizer | $x5$ | Agricultural fertilizer/Agricultural acreage |
| | Density of pesticide | $x6$ | Pesticide/Agricultural acreage |
| | Density of agricultural plastic film | $x7$ | Agricultural plastic film/Agricultural acreage |
| | Natural disaster | $x8$ | Natural disaster/Crop sown area |
| Agricultural efficiency | Cultivated area per capita | $x9$ | Cultivated land area/Rural population |
| | Grain yield per unit area | $x10$ | Grain total yield/Cultivated land area |
| | Grain yield per labor force | $x11$ | Grain total yield/Agricultural population |
| | Farmers' net income per capita | $x12$ | China Rural Statistical Yearbook |
| Agricultural modernization | Irrigation rate | $x13$ | Irrigated area of cultivated land/Cultivated land area |
| | Agricultural machinery per unit area | $x14$ | Agricultural machinery/Agricultural land area |
| | Investment conditions | $x15$ | Total investment in fixed assets/Administrative area of land |
| Natural conditions | Average annual precipitation | $x16$ | Resource and Environment Data Cloud Platform |
| | Average annual temperature | $x17$ | Resource and Environment Data Cloud Platform |
| | Soil Texture | $x18$ | Resource and Environment Data Cloud Platform |
| | Relief degree of land surface | $x19$ | Global Change Data & Discovery |

### 2.2. Methods and Procedures

#### 2.2.1. Extraction Methods of the MCI

The MCI refers to the times of sequential crop planting in the same cultivated land in one year, reflecting the utility degree of arable land to be used at a certain time. It is desirable to extract the MCI and its spatial distribution information by remote sensing image data. In this study, the arable land coverage regions of the MODIS-NDVI remote sensing image data were extracted from the land use classification data for 2000, 2005, 2010, and 2015. The extraction methods of the MCI were as follows [56,60].

As shown in Figure 3, a coordinate system was established, with time as an x-coordinate and normalized difference vegetation index (NDVI) as a y-coordinate, showing a periodic dynamic of crops in the sowing, germinating, earring, and harvesting stages. Specifically, Figure 3a shows crops' performances under the single-cropping system (one-peak curve). Figure 3b shows crops' performances under the double-cropping system (two-peak curve). Figure 3c illustrates crops' performances under the triple-cropping system (triple-peak curve). Last, Figure 3d concerns bare land or abandoned land, characterized by insignificant peaks and low NDVI. Hence, the number of NDVI peaks within a year can be understood as the MCI of the area.

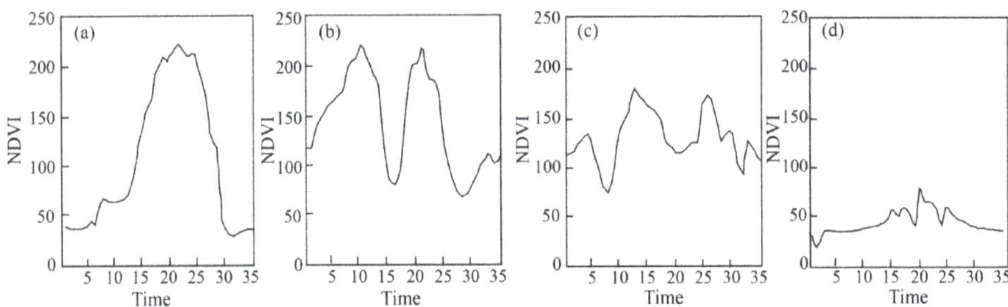

**Figure 3.** The NDVI time series curve smoothing by HANTS. Note: time as an x-coordinate and NDVI as a y-coordinate, without dimensionality. (**a**) Crops' performances under the single-cropping system (one-peak curve). (**b**) Crops' performances under the double-cropping system (two-peak curve). (**c**) Crops' performances under the triple-cropping system (triple-peak curve). (**d**) Crops' performances under bare land or abandoned land, characterized by insignificant peaks and low NDVI.

Considering the relationship between the NDVI time series curve and the MCI, it can be concluded that the calculation of the MCI is equal to the extraction process of the peak frequency of the NDVI time series. The extraction formula is as follows:

$$F_i = \frac{S_{sumapex_i}}{S_{sumpixel_i}} \tag{1}$$

where $F_i$ represents the frequency of NDVI time series peaks of the administrative units providing the MCI. $S_{sumapex_i}$ is the total number of peaks formed within a year on all pixel values. $S_{sumpixel_i}$ refers to the total number of pixels in the NDVI curves. As for the frequency of the peaks, it was extracted using the Twi-difference algorithm, for the NDVI time series of each pixel can be considered as a sequence of the discrete points. Specifically, the differences of the neighboring NDVI values should be computed with Formula (2) to conclude $S_1$ (Sequence 1), and the plus or minus would be determined by Formula (3). If the value is positive, 1 is recorded; alternatively, $-1$ is recorded to yield $S_2$ (Sequence 2). Finally, the differences of $S_{2i}$ and $S_{2i+1}$ can be calculated using Equation $S_3$ (Sequence 3).

$$S_{1_i} = NDVI_{i-1} - NDVI_i \tag{2}$$

$$S_2 = \begin{cases} 1 & S_{1_i} > 0 \\ -1 & S_{1_i} < 0 \end{cases} \tag{3}$$

$$S_{3_i} = S_{2_i} - S_{2_{i+1}} \tag{4}$$

In all the formulas above, $i$ represents the $i$th pixels in different sequences. It is concluded that the peak of crop growth occurred in $S_3$, where the pixel value was $-2$, and the pre and post values were both 0. MODIS-NDVI data from 16 consecutive days in 2000, 2005, 2010, and 2015, temporally and spatially based on the spatial vector data of the land use type, was analyzed using MATLAB and ArcGIS; the NDVI peaks' frequency of each raster cell is shown in Figure 4—the inversion of the MCI was accomplished. Next, the provincial and county administrative units were used to carry on the statistics related to the MCI and the MCI means of Chinese counties and provinces were presented using the ArcGIS10.2 platform to prepare them for the subsequent analysis.

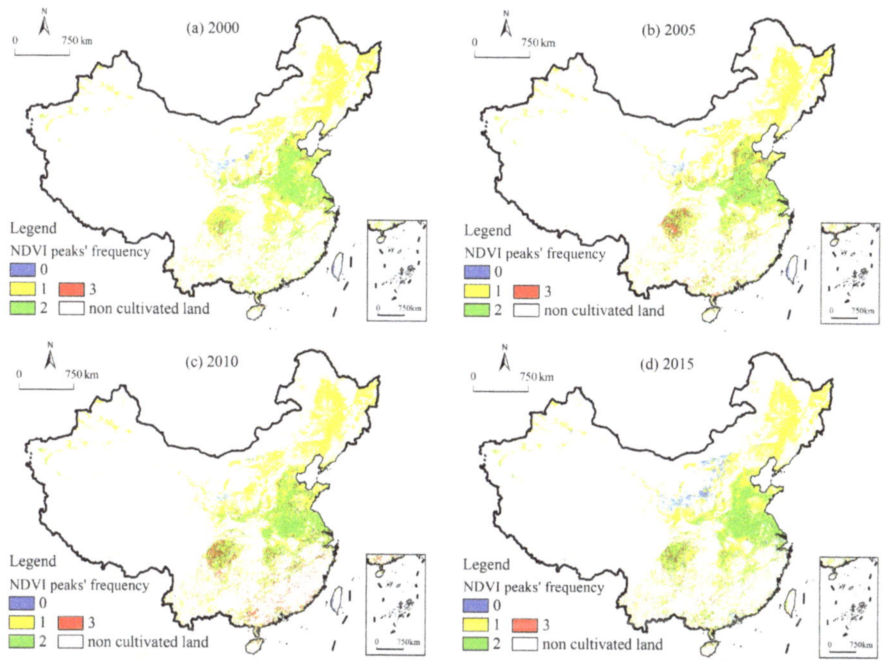



**Figure 4.** Spatial distribution of the NDVI peak frequency of each raster cell in different years in China. (**a**) NDVI peak frequency in 2000. (**b**) NDVI peak frequency in 2005. (**c**) NDVI peak frequency in 2010. (**d**) NDVI peak frequency in 2015.

### 2.2.2. Principal Component Regression

Pearson first proposed the principal component analysis method [61], which was popularized and developed by Hotelling [62]. Later, Massy proposed principal component regression based on the idea of the principal component analysis [63]. The core idea of principal component regression is to transform multiple independent variables into a few principal components through dimension reduction to eliminate collinearity among original independent variables; without changing the original independent variables' interpretation of the dependent variables, the transformed principal component is used to replace the original independent variable for the regression analysis.

The years 2000 and 2015 were selected to explore the factors influencing the MCI from 2000 to 2015, and the research sample was taken from 31 provinces of mainland China. Due to the large number of indicators but small sample sizes, there was severe multicollinearity among the independent variables; thus, the principal component regression was used to increase the reliability of the results. The MCI was the dependent variable, and the $x1$–$x19$ indicators in Table 1 were the independent variables. To make the indicators comparable, the deflator index of each province was used to eliminate the interference of the price factors for the economic indicators, such as per capita GDP, total investment in fixed assets, and farmers' net income per capita. The processed data were analyzed by principal component regression using SPSS18.0 (https://www.ibm.com/analytics/spss-statistics-software) (accessed on 21 October 2019).

First, the factor analysis tool of SPSS18.0 was used for the principal component analysis, and Kaiser–Meyer–Olkin (KMO) and Bartlett's tests of sphericity were used as the criteria for the applicability judgment. The results showed that the KMO in 2000 and 2015 were 0.656 and 0.646, and both met the test criteria. Bartlett's test of sphericity met the significance level. Hence, the samples selected in this research were suitable for the principal component analysis, because both conditions met the applicability requirements. Next, the principal components were determined by the basic principle that the cumulative variance rate was more than 85% and the eigenvalue was close to 1. As shown in Table 2, five principal components were confirmed in 2000 and 2015, respectively. The cumulative variance rate for 2015 was 84.10%, close to 85%, and the eigenvalue of the sixth principal component differed significantly from 1; therefore, five principal components were retained in 2015. There was no collinearity between the five new independent variables.

**Table 2.** Principal component eigenvalues and total variances for 2000 and 2015.

| Component | 2000 | | 2015 | |
| --- | --- | --- | --- | --- |
| | Eigenvalue | % of Cumulative Variance | Eigenvalue | % of Cumulative Variance |
| 1 | 7.84 | 41.25 | 7.57 | 39.82 |
| 2 | 5.04 | 67.78 | 4.00 | 60.86 |
| 3 | 1.61 | 76.26 | 1.96 | 71.19 |
| 4 | 1.24 | 82.81 | 1.54 | 79.32 |
| 5 | 0.93 | 87.69 | 0.91 | 84.10 |

Second, by multiplying the factor score and the square root of the eigenvalue resulting from the principal component analysis, the principal component score was obtained as the new independent variable score. Next, the dependent variable—namely, the MCI—was standardized by the z-score method, and the standardized MCI and five principal component scores were analyzed by linear regression. The results are shown in Table 3.

**Table 3.** Principal component regression results in 2000 and 2015.

| Type | 2000 | | 2015 | |
| --- | --- | --- | --- | --- |
| | Unstandardized Coefficients (*B*) | Sig. | Unstandardized Coefficients (*B*) | Sig. |
| (Constant) | 0.00 | 1.00 | 0.00 | 1.00 |
| Principal Component 1 | 0.24 ** | 0.00 | 0.19 ** | 0.00 |
| Principal Component 2 | −0.24 ** | 0.00 | −0.34 ** | 0.00 |
| Principal Component 3 | −0.01 | 0.92 | 0.15 ** | 0.03 |
| Principal Component 4 | −0.14 | 0.12 | 0.07 | 0.37 |
| Principal Component 5 | 0.12 | 0.26 | 0.02 | 0.85 |

(Note: ** represents sig. $< 0.05$).

The results show that Principal Component 1 ($F_1$) and Principal Component 2 ($F_2$) reached significant levels (sig. $< 0.05$) in 2000, and Principal Component 1 ($F_1'$), Principal Component 2 ($F_2'$), and Principal Component 3 ($F_3'$) reached significant levels (sig. $< 0.05$) in 2015. Thus, we put the unstandardized coefficients (*B*) into the formulas $F_{2000} = B_1 F_1 + B_2 F_2$ and $F_{2015} = B_1' F_1' + B_2' F_2' + B_3' F_3'$. Next, we divided the component matrix in tSPSS18.0 (principal component analysis) by the square root of the eigenvalue to get the eigenvector ($a_i$) of each principal component before replacing $a_i$ in the principal component formula $F_k = a_{1k} Z_1 + a_{2k} Z_2 + \ldots + a_{nk} Z_n$ ($Z$ is the standard matrix of the independent variable). Finally, by substituting $F_k$ into formulas $F_{2000}$ and $F_{2015}$, the normalized independent variable $X$ and dependent variable $Y$ were reduced to the original data $y$ and $x$, and the final regression formulas $y_{2000}$ and $y_{2015}$ were obtained.

*2.3. Data Sources*

First, MODIS-NDVI remote sensing image data and vector data in 2000, 2005, 2010, and 2015 on the spatiotemporal distributions of the agricultural lands were provided by the Earth System Scientific Data Sharing Platform, Chinese Academy of Science (data source of Figure 4). Second, the indexes of the influencing factors included information on the economic, social, and natural conditions. The economic and social data, such as population, gross domestic product, cultivated land area, and grain output, were derived from the China Statistical Yearbooks (2001 and 2016), China Rural Statistical Yearbooks (2001 and 2016), China Population & Employment Statistics Yearbooks (2001 and 2015), and other statistical yearbooks for the provinces (data sources of Table 1 and Figure 5). In particular, the data for China's agricultural population and nonagricultural population were only available until the end of 2014 (China Population & Employment Statistics Yearbook 2015); therefore, the data for 2015 were replaced by those from the end of 2014. Data on China's cultivated land area were obtained from China's Land and Resources Bulletin (data source of Table 1 and Figure 5). Third, the natural condition data, such as soil texture, annual average temperature, and annual average precipitation, were obtained from the Resource and Environment Data Cloud Platform, Chinese Academy of Science (data source of Table 1). Data sources for the soil texture include the percentages of sand, silt, and clay. According to the China soil texture classification (1985) [64], the soil texture data were divided into three types: clay, loam, and sand. They were marked as 1, 2, and 3, respectively, in the regression analysis. Besides, the relief degree of the land surface was derived from the research data of You et al. [65], who shared it in the Global Change Research Data Publishing & Repository (data source of Table 1). Finally, the deflation index used to process the economic data, the calculation process involved in the GDP, and the GDP index were taken from the China Statistical Yearbook (data source of Table 1).

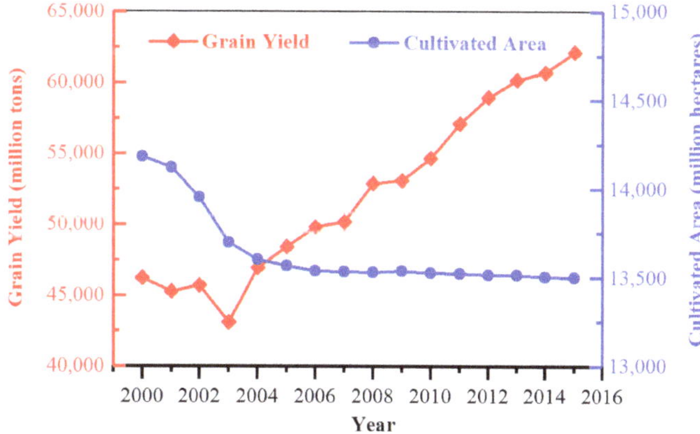


**Figure 5.** China's cultivated land areas and grain yields, 2000–2015.

In 2000, 2005, 2010, and 2015, mainland China did not suffer from severe agricultural disasters; therefore, it is reasonable to choose these four years as representatives to study the MCI from 2000 to 2015 without much contingency. In addition, due to the limited availability of the data, this research only studied mainland China and did not involve Hong Kong, Macao, or Taiwan.

## 3. Results

### 3.1. Cultivated Land Area and Grain Yield

The cultivated land area in China from 2000 to 2008 was reported by the first national land survey and the cultivated land area from 2009 to 2015 by the second national land survey. However, the criteria and techniques used in the two surveys were different, and the values between them could not be simply compared. Therefore, referring to the research of Chen et al. [66], this paper adjusted the cultivated land area from 2000 to 2008 to the cultivated land area based on the second survey, as shown in Figure 5.

In general, China's cultivated land areas showed a downward trend from 2000 to 2015, from 141.83 million hectares in 2000 to 135.00 million hectares in 2015, with an average annual decline of 0.46 million hectares. Although China's cultivated land area had been declining over the past 15 years, it did not break the "red line of 1.80 billion mu of cultivated land in China". The changes in grain yields fluctuated. Overall, the grain yield showed an upward trend, rising from 462.18 million tons in 2000 to 621.44 million tons in 2015, with an average annual growth of 10.62 million tons. Notably, the grain yields from 2000 to 2001 and from 2002 to 2003 showed a declining trend, and the grain yield in 2003 was the lowest in the past 15 years; meanwhile, the grain yields from 2003 to 2015 showed an upward trend with a relatively fast rate of increase and an average annual increase of 15.90 million tons.

### 3.2. Spatiotemporal Change of MCI at the Provincial Scale

Figure 6 shows the cultivated land MCI of 31 provinces in China in 2000, 2005, 2010, and 2015. The research results showed that the MCI varied dramatically in different provinces. Notably, the MCI was high in the coastal provinces of Southern China. For example, Guangdong and Guangxi had the highest average MCI (190%) from 2000 to 2015, followed by Hainan (187%). In 2005 and 2010, the MCI in Guangdong exceeded 200%, and Fujian, Guangxi, and Jiangxi had more than 200% MCI in 2010; meanwhile, the MCI of Hainan, Guangxi, Henan, Anhui, and Jiangsu were high and stable above 160% from 2000 to 2015. Most of these areas with high MCI are located in the south of China, which is characterized by a subtropical monsoon climate and good water and heat conditions conducive to the multi-cropping of the cultivated land. Moreover, Henan, Jiangsu, and Guangdong are among China's major agricultural provinces and, thus, areas where the development of agriculture is particularly supported. Along these lines, their level of agricultural modernization is high, which makes their MCI higher and the rate of change relatively stable. Most the provinces with smaller MCI are located in the northeast and northwest. From 2000 to 2015, the MCI of the provinces of Heilongjiang, Inner Mongolia, Ningxia, and Qinghai were lower than 100%, and the number of provinces with MCI lower than 100% increased year after year, from five provinces in 2000 to 11 provinces in 2015. The Shaanxi, Shanxi, and Gansu Provinces were newly added to this list in 2015.

This paper divided the 2000–2015 interval into three stages to analyze the changes in the MCI of the cultivated land (Figure 7). It was concluded that the number of provinces with a declining MCI increased with time and that the declining range of the MCI increased. An MCI of 61.29% of the provinces showed an increasing trend from 2000 to 2005; in Sichuan, Guangdong, and Guangxi, the MCI increased by 50.38%, 25.52%, and 23.59%, respectively. The most rapid decline was in the economically developed city of Beijing, which decreased by 30.59%. Generally, an MCI of 51.61% of the provinces showed a trend of growth from 2005 to 2010. Meanwhile, compared with the previous stage, the number of provinces with a declining MCI increased. Shandong and Hebei had the fastest decline in their MCI at 25.03% and 23.12%, respectively. The southern provinces, mainly Chongqing, Fujian, Jiangxi, Zhejiang, Guangdong, and Hunan, had rapid growth rates of the MCI, with Fujian showing the fastest growth rate at 75.58%. The change in the MCI from 2010 to 2015, as shown by the green line in Figure 7, was almost lower than 0. In fact, the MCI in 93.55% of the provinces showed a downward trend, while the MCI of Shandong and Hebei showed a small increase, with 8.34% and 0.54%, respectively. In the southern provinces, the MCI decreased by a large margin, and the five provinces of Fujian, Guangdong, Jiangxi, Zhejiang, and Shanghai showed decreases of more than 50%. From the above results, the dynamic changes of the MCI of the southern provinces were obvious. Before 2010, China's MCI was in a growth trend, but this was replaced by a downtrend and substantial decline after 2010. This trend is not conducive to China's food security, because most of the double-cropping and multi-cropping rice areas in China are located in its southern regions.

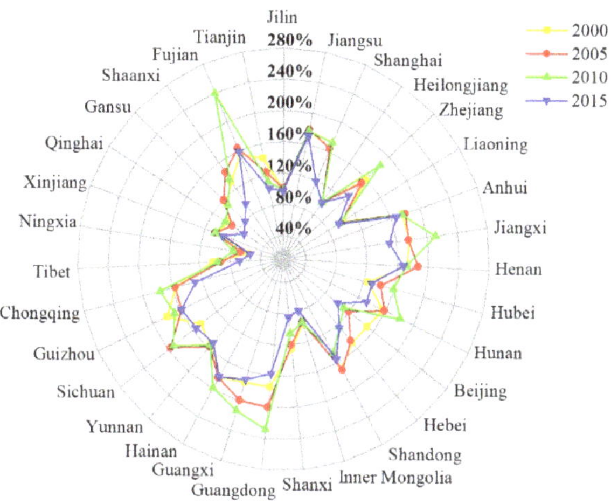

**Figure 6.** The value of the MCI in Chinese provinces, 2000–2015.

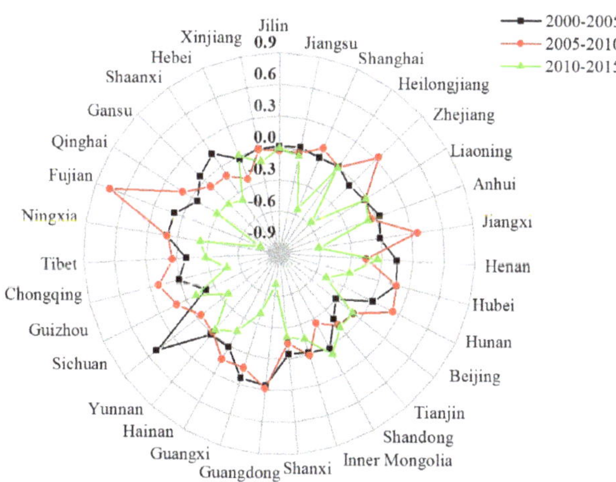

**Figure 7.** The changes in the values of the MCI in Chinese provinces from 2000 to 2005, 2005 to 2010, and 2010 to 2015.

## 3.3. Spatiotemporal Change of MCI at the County Scale

Figure 8 shows the spatial distribution of China's MCI from 2000 to 2015, and Figure 9 illustrates the statistics of the MCI of counties across different delimited ranges. China's MCI was roughly divided into a single-cropping system and a multiple-cropping system by the "Taihang Mountains–Qinling Mountains–Hengduan Mountains" boundary from 2000 to 2015 (Figure 8). The MCI in the north of the boundary was less than 100% and more than 100% in the south.

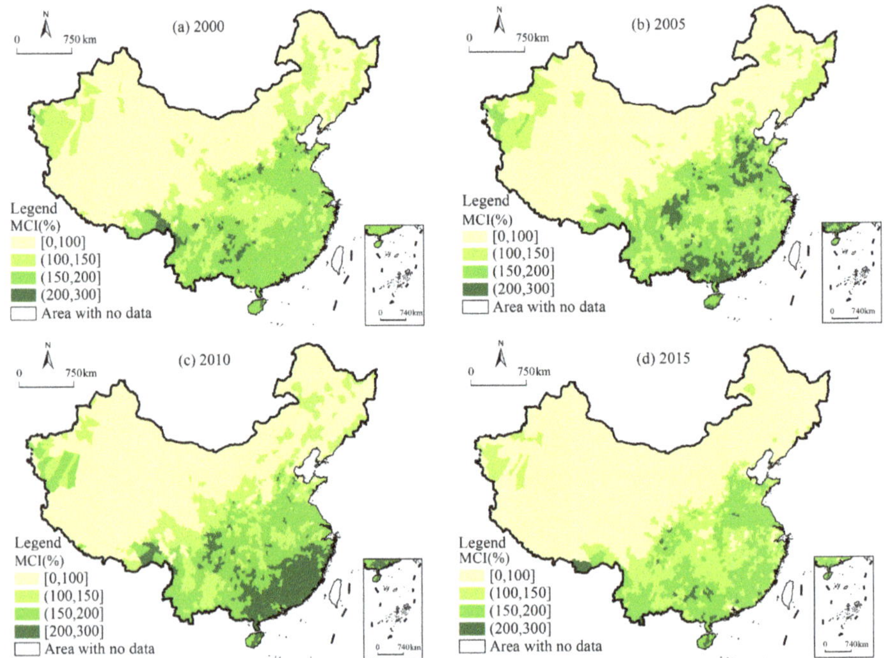

**Figure 8.** The spatial distributions of the MCI in counties in different years. (**a**) The MCI in counties in 2000. (**b**) The MCI in counties in 2005. (**c**) The MCI in counties in 2010. (**d**) The MCI in counties in 2015.

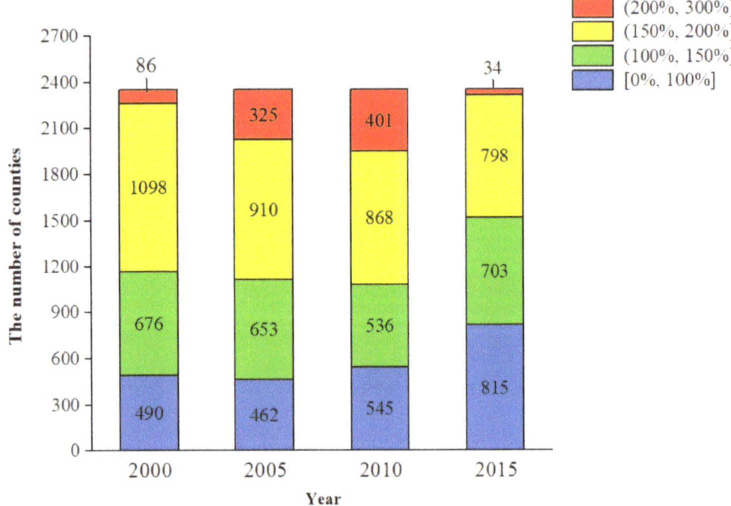

**Figure 9.** The number of counties demonstrating each MCI range.

In 2000, the average value of China's MCI was 143.16%. The number of counties with MCI in the 150–200% range was the largest, with a total of 1098 units, accounting for 46.72% of China's counties (Figure 9); notably, these counties were primarily located in the East China region, South China region, and Yunnan Province (Figure 8a). Meanwhile, the number of counties with MCI in the 100–150% range was 676, accounting for 28.77%

of China's counties; these counties were mainly distributed in the provinces of the mid-Yangtze River, Northeast China, and Southwest Xinjiang. The number of counties with MCI in the 0–100% range was 490, accounting for 20.85% of China's counties; these counties were mainly distributed in the north of the boundary of "Taihang Mountains–Qinling Mountains–Hengduan Mountains". Only 3.66% of the counties demonstrated an MCI greater than 200%, and the spatial distribution was scattered.

In 2005, the average value of China's MCI was 148.66%, an increase from 2000. In particular, the number of counties with MCI in the 200–300% range increased to 325, accounting for 13.83% of China's counties (Figure 9); these counties were mainly distributed in the North China region, South China region, and Chengdu Plain (Figure 8b). Counties with MCI in the 150–200% range still had the highest proportion but decreased to 38.72% of China's counties; these counties were primarily distributed in the North China Plain and to the south of the Yangtze River. The spatial distribution with MCI lower than 150% was similar to that observed in 2000.

In 2010, the average value of China's MCI was 152.03%, an increase from 2005. The number of counties with MCI in the 200–300% range continuously increased to 401, accounting for 17.06% of China's counties (Figure 9); these counties were mainly distributed in the hilly region of Southeast China and Chengdu Plain (Figure 8c). The number of counties with MCI in the 100–200% range accounted for 66.51%, which were mainly distributed in the south of the boundary of "Taihang Mountains–Qinling Mountains–Hengduan Mountains" and some areas in the southwest of Xinjiang Province and Northeast China. The areas where the MCI was lower than 100% were mainly distributed in the north of the boundary of "Taihang Mountains–Qinling Mountains–Hengduan Mountains".

In 2015, the average value of China's MCI was 125.95%. Compared to 2010, the average value of the MCI decreased by 26.08% and 5.22% annually. The spatial distribution of MCI in the 200%–300% range decreased sharply to 34 counties (Figure 9) that were scattered across the southern region (Figure 8d), accounting for 1.45%. The number of counties with MCI in the 0–100% range increased sharply, mainly distributed in the north of the boundary of "Taihang Mountains–Qinling Mountains–Hengduan Mountains". However, in some counties south of the boundary, the MCI was also lower than 100%, especially in the economically developed Pearl River Delta and Yangtze-Huaihe regions, and the MCI of many counties in Hubei Province were also lower than 100%. The number of counties with MCI in the 100–150% range increased significantly to 703, accounting for 29.91% of China's counties; these counties were mainly distributed in the provinces of the middle and lower reaches of the Yangtze River and Yunnan. The number of counties with MCI in the 150–200% range decreased to 798, accounting for 33.96% of China's counties; these counties were mainly distributed in Eastern China, Southern China, and southwest of Yunnan Province.

Algebraic operations were carried out for the four years between 2000 and 2015, and the variations in the MCI values for the four periods—2000–2005, 2005–2010, 2010–2015, and 2000–2015—were obtained (Figure 10).

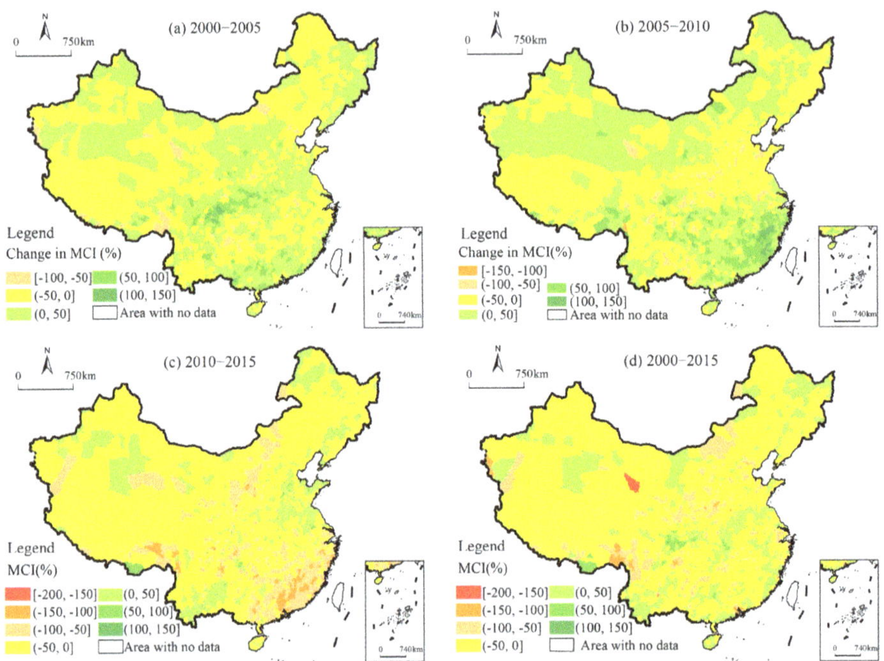

**Figure 10.** Spatial distributions of the MCI value changes across Chinese counties. (**a**) The MCI value changes from 2000 to 2005. (**b**) The MCI value changes from 2005 to 2010. (**c**) The MCI value changes from 2010 to 2015. (**d**) The MCI value changes from 2000 to 2015.

From 2000 to 2005, 56.17% of China's counties demonstrate variations in the MCI higher than 0; these counties were mainly distributed across the central provinces, Xinjiang, three northeastern provinces, and the coastal provinces of Guangdong and Guangxi (Figure 10a). The rates of change in the MCI values of most counties were under 50%, while changes greater than 50% were evident in only 6.17% of the counties—these counties were distributed across the areas of the Chengdu Plain, Guangxi, and Guangdong. Meanwhile, 43.83% of the counties showed a change in the MCI values lower than 0; these counties were widely distributed across the northeast region, northwest region, Tibet, Yunnan, and Guizhou. Only 2.26% of the counties demonstrated a variation in value in the range between −200% and −50%, and their distribution was relatively scattered.

From 2005 to 2010, 49.23% of the counties exhibited a variation in the MCI value greater than 0, with a decrease of 6.94% compared with the previous stage; the counties were mainly distributed across the southeastern provinces, northwest, and northeast regions (Figure 10b). The number of counties in which the MCI variation values were greater than 50% increased, accounting for 9.79% of the counties; these counties were mainly distributed in Guangdong, Fujian, Jiangxi, and Zhejiang. In contrast, 50.77% of the counties demonstrated a change in MCI values lower than 0; this was higher than the percentage of counties with a change higher than 0; notably, these counties were mainly distributed across the north of the Qinling–Huaihe River Line, the south of the Inner Mongolia Autonomous Region, and—widely—in Tibet. Moreover, the number of counties with a change in MCI values in the range between −200% and −50% increased, accounting for 4.81%, and the distribution remained scattered.

From 2010 to 2015, the proportion of counties with a change in MCI values higher than 0 dropped sharply to 20.47%, and the proportion of counties with changes in the MCI lower than 0 was almost four times that of the proportion of counties with changes in the MCI higher than 0 (Figure 10c). In terms of spatial distribution, the MCI changes across the cultivated land in China were generally lower than 0. More precisely, the proportion of counties with variations in the MCI in the range between −200% and −50% increased sharply to 23.53%, and these counties were mainly distributed across China's southeast coastal provinces. However, only the Bohai Rim region, Heilongjiang, Yunnan, and Xinjiang showed MCI changes higher than 0, of which only 0.68% of the county MCI were higher than 50%.

Overall, from 2000 to 2015, MCI changes greater than 0 were evident in 24.55% of the Chinese counties, and the number of counties with MCI changes lower than 0 was more than three times that of the number of counties with values higher than 0 (Figure 10d). During these 15 years, China's MCI showed a significant downward trend—in particular, the MCI declined greatly in the three provincial border regions of Yunnan, Sichuan, and Tibet; some provinces in the upper and middle reaches of the Yellow River basin; the Beijing–Tianjin–Hebei region; the Yangtze River Delta; and the Pearl River Delta, with variation values in the range between −200% and −50%. Conversely, the districts of counties with MCI higher than 0 were scattered, mainly across Northeast China, the Chengdu Plain, Jianghan Plain, Henan Province, Xinjiang Province, the Western Inner Mongolia Autonomous Region, and the southeast coastal provinces. The change value was only 1.53% above 50%, which was distributed only in Southern Tibet, the Chengdu Plain, Jianghan Plain, and other regions.

### 3.4. Influencing Factors of MCI

The final regression formula is $y = \sum_{i=1}^{n} b_i x_i + c$ ($n = 19$), $c$ is the constant, and $b_i$ is the coefficient of the influencing factor ($x_i$). The results of $b_i$ and $c$ are shown in Table 4. Specifically, the size of $b_i$ reflects the degree of influence of this factor on the MCI, and the positive and negative values of $b_i$ reflect the effect of this factor on the MCI.

**Table 4.** Principal component regression results of the influencing factors for the MCI.

| Type | | 2000 | | 2015 | |
| --- | --- | --- | --- | --- | --- |
| Index | $x$ | Coefficient: $b$ | Constant: $c$ | Coefficient: $b$ | Constant: $c$ |
| Non-agricultural population | $x1$ | −0.0025 | | −0.0019 | |
| Non-agricultural industry | $x2$ | −0.0543 | | −0.6093 | |
| Urbanization rate | $x3$ | −0.0007 | | −0.0011 | |
| Per capita GDP | $x4$ | $5.5 \times 10^{-7}$ | | $-8.2 \times 10^{-7}$ | |
| Density of agricultural fertilizer | $x5$ | 0.3645 | | 0.2152 | |
| Density of pesticide | $x6$ | 9.7628 | | 3.7945 | |
| Density of agricultural plastic film | $x7$ | 1.4343 | | −0.4621 | |
| Natural disaster | $x8$ | −0.5069 | | −0.4206 | |
| Cultivated area per capita | $x9$ | −1.0038 | | −0.0876 | |
| Grain yield per unit area | $x10$ | 0.0354 | 0.1599 | 0.0233 | 1.7466 |
| Grain yield per labor force | $x11$ | −0.1731 | | −0.0140 | |
| Farmers' net income per capita | $x12$ | $1.7 \times 10^{-5}$ | | $-4.9 \times 10^{-7}$ | |
| Irrigation rate | $x13$ | 0.1615 | | 0.0351 | |
| Agricultural machinery per unit area | $x14$ | 0.0047 | | 0.0044 | |
| Investment conditions | $x15$ | $3.0 \times 10^{-5}$ | | $-7.0 \times 10^{-6}$ | |
| Average annual precipitation | $x16$ | $9.2 \times 10^{-6}$ | | $1.8 \times 10^{-5}$ | |
| Average annual temperature | $x17$ | 0.0014 | | 0.0009 | |
| Soil texture | $x18$ | −0.0725 | | −0.0841 | |
| Relief degree of land surface | $x19$ | −0.0460 | | −0.0265 | |

Table 4 shows the relationship between the MCI and its impact factors. As for the nonagricultural process factors, in 2000 and 2015, the regression results showed that the influence of the four factors of the nonagricultural process on the MCI was almost negative. Especially for the nonagricultural industry, its influence coefficient changed from −0.0543 in 2000 to −0.6093 in 2015, demonstrating an obvious negative effect on the MCI. As for cultivated land quality, in 2000, all the other impact factors were positively correlated with the MCI except natural disasters. However, in 2015, the effect of agricultural plastic film on the MCI became negative, and the density of pesticides showed a significant downward trend, from 9.7628 to 3.7945. As for the agriculturally intensive production factors, the impact coefficient of the grain yield per unit area, irrigation rate, and agricultural machinery per unit area were all positive, whereas the remaining factors were almost negative, but the negative effect of cultivated area per capita and grain yield per labor force on the MCI significantly declined by 2015. As for the qualities of the natural conditions, the results showed that the average annual precipitation and average annual temperature positively impacted the MCI; meanwhile, the soil texture and the relief degree of land surface negatively affected the MCI.

## 4. Discussion

### 4.1. Effects of Natural Conditions on MCI

Regarding natural conditions, hydrothermal conditions are the basic influencing factors for the multiple croppings of cultivated land, and the soil texture and the relief degree of the land surface also have important effects on the multiple croppings of cultivated land. Studies have shown that the realization of a multiple-cropping system depends largely on temperature and precipitation—sufficient accumulated temperature and rainfall are necessary to realize multiple croppings [48,67–69]. Affected by climate change, the northern limits of multiple-cropping systems have moved northward, and the projected area of cultivated land for multiple croppings may significantly expand during the 21st century in China [67]. Notably, the soil texture is closely related to soil aeration and water and fertilizer conservation. Poor soil texture inhibits the implementation of multiple croppings. Studies have confirmed that soil fertility is a key factor affecting the crop yield in the cropping system—notably, high soil fertility can increase the crop yield in a multiple-cropping system [59], but poor soil texture is the main reason a multiple-cropping system becomes ineffective [47]. Moreover, the land's relief degree is generally associated with multiple croppings: the higher the topography of the cultivated land, the more difficult it is to cultivate and achieve multiple-cropping systems [70]. Crops are suitable for cultivation in areas with relatively flat topography; high topographies can be hot spots for geological disasters, thus hindering the cultivation of crops. Additionally, areas with high topography generally have low levels of agricultural modernization, including the serious abandonment of cultivated land, which can dampen the MCI. However, technological developments can weaken the negative effect of topography on the MCI.

### 4.2. Adverse Effects of Nonagricultural Process on MCI

Nonagricultural processes in rural areas have an inhibitory effect on the increase of the MCI, especially the nonagricultural effects of the industry and population, which are the most direct factors that lead to the weakening of agricultural production subjects. Seeking higher economic benefits, the labor force engaged in agricultural production of the nonagricultural industry, with a large number of farmers working in cities and settling down as the main force to promote the process of urbanization [17,71,72]. Given the stable transformations in peasants' livelihoods, the rural labor force keeps reducing, and the problems of land abandonment and non-grain conversion become prominent [17,73], and the nonagricultural use of cultivated land has become one of the most typical trends of land use transitions in China [50,74], producing a negative impact on the MCI of cultivated land. In recent years, the Chinese government has focused on deploying major strategies to support the development of agriculture and rural areas, resulting in the adjustment of the agricultural structure, changing the morphology of cultivated land use, promoting the diversification of rural regional functions, and gradually diversifying the rural types. The rural development does not accomplish only an agricultural function; the secondary and tertiary industries rise gradually. The flow and agglomeration of various factors offer possibilities for the diversification of the rural industry. To some extent, this process weakens the status of agriculture, makes the labor force originally left in the countryside turn to a new type of nonagricultural industry and then encourages the emergence of the nonagricultural industry and nonagricultural employment [70,75]. To prevent the lack of the main body of agricultural production, the study on the transformation of farmers' livelihoods and the multiple-cropping potential of cultivated land should be further strengthened, focusing on the identification of effective measures to encourage the return of rural labor force through micro-case studies to stabilize the main body of agricultural production.

*4.3. Effect of Intensive Agricultural Production on the MCI*

Intensive agricultural production promotes the MCI by improving the agricultural production efficiency—taking the grain yield as a measure of the agricultural economic benefits and agricultural modernization as a measure of the agricultural production efficiency, these items have a positive impact on multiple croppings. The grain yield per unit area directly affects the economic income of farmers growing grain, and the high agricultural income achieved by a high grain output can encourage farmers to engage in multiple croppings. However, improved multiple croppings of the cultivated land are limited by water resources; therefore, adequate irrigation is necessary to improve the MCI of the cultivated land significantly [29,54]. Here, it is helpful to note that the agricultural machinery represents the level of agricultural modernization. The higher the productivity and efficiency of the cultivated land with superior agricultural machinery conditions, the more inclined farmers are to carry out multiple croppings [55,76]. Our research found that farmers with more cultivated areas per capita may not be accordingly more enthusiastic about multiple croppings—this may be related to the small scales, insufficient production input, and weak labor force of some farming operations. In China, fertile lands in many mountainous areas have been abandoned, but the Chinese government is currently stepping up efforts to reform the land market, such as developing the rural land rental market to facilitate larger-scale farming operations and reduce the area of abandoned cultivated land [77,78].

*4.4. The Effect of Cultivated Land Quality on the MCI*

Cultivated land quality affects the morphology of cultivated land use, determines whether the land can be cultivated sustainable, and guarantees the long-term implementation of a multiple-cropping system. Natural disasters are a direct factor affecting the growth of multiple croppings. On the one hand, natural disasters objectively block the normal growth period of crops. On the other hand, natural disasters destroy the balance of soil and water, changing the spatial structure of the cultivated land, causing soil erosion and soil nutrient loss, resulting in a decline in the cultivated land quality, which will continue to affect the planting of the next crop. Multiple cropping requires high soil nutrient contents; therefore, large amounts of pesticides and fertilizers are required. The input of pesticides and fertilizers can improve the grain yield, ensure the production efficiency of multiple croppings, and make farmers willing to retain multiple-cropping behaviors. Notably, agricultural plastic film plays a very important role in the moisture and heat preservation of crops. The appropriate use of agricultural film can promote crop growth and is conducive to the multiple croppings of cultivated lands. However, the overuse and improper treatment of pesticides, fertilizers, and the film changes the soil's properties, causes agricultural nonpoint source pollution, and contributes to a number of environmental problems [79,80]. The continuation of this unscientific use will make it difficult to maintain multiple-cropping systems. What is important to take away here is that pesticides, fertilizers, and the film will continue to play vital roles in China's multiple-cropping system, which, as noted above, is of great significance to food security. Therefore, in the future, we need to balance the relationship between increasing the crop yields and reducing the environmental pollution while at once strengthening research on green agricultural systems, biodegradable film promotion, pesticide and fertilizer use efficiency, and alternative chemical fertilizers.

### 4.5. China's Cultivated Land Protection Policy and the MCI

This research showed that cultivated lands in China are facing the double pressure of area reduction and MCI reduction. Some studies support the idea that agricultural policies can stimulate cultivation, thus stabilizing the MCI [55,70]. China has been implementing the strictest cultivated land protection policies for many years, such as the policies of "taking grain as the highest priority", "red line of 1.8 billion mu of cultivated land in China", "basic cultivated land protection system", and "cultivated land requisition-compensation balance" [81]. Although the "red line of 1.8 billion mu of cultivated land" in China has not been broken, merely focusing on maintaining the total area of cultivated land has not prevented the loss and abandonment of a large portion of high-quality cultivated land. Many studies have shown that a large amount of high-quality cultivated land had been lost in the process of urbanization. Land development, land reclamation, land consolidation, and other measures lead to the occupation of high-quality cultivated land by low-quality cultivated land, and the existing policies for cultivated land have not managed to guarantee quality and quantity, resulting in the imbalance of cultivated land quality across the whole country [81–83]. How do the agricultural policies really guarantee food security? In addition to quantity, more attention should be paid to cultivated land quality, the farmer who cultivates the land, and the multiple-cropping potential of the cultivated land. It is important to tap the multiple-cropping potential of cultivated lands and make good use of the advantages of the regional MCI and incorporate the increased sowing area of the MCI into the assessment system of local officials to stimulate the enthusiasm of local governments to ensure food security.

### 4.6. Research Limitations

In this research, MODIS-NDVI was used as the data source to extract the MCI of the cultivated lands with the Twi-difference algorithm. Compared with the statistical data, remote sensing data can eliminate human disturbance and are more objective. However, the extraction of the MCI from remote sensing data may be affected by mixed pixels, which results in uncertainty. Due to the lack of field investigation in this paper, the accuracy of the extracted results cannot be verified. Further research will need to address this issue. As for the analysis of the factors and mechanisms influencing the MCI, given the limited availability of the data, this study selected the provincial scale for exploratory research. Although most of the selected indicators were based on rural statistical indicators, the provincial scale was too large, and the research results were uncertain. For example, the influence of important factors such as temperature and precipitation was positive, but the coefficient was close to 0, which can only reflect the general situation at the macro level. In the future, the scale of the research should be reduced to improve the accuracy of the results.

## 5. Conclusions

Using MODIS data, the spatial distribution and dynamic changes of the MCI in China in 2000, 2005, 2010, and 2015 were presented at different scales. Meanwhile, principal component regression was used to study the influencing factors. Our results suggest that the number of provinces with lower MCI increased gradually, and the dynamic changes in the MCI in the southern provinces were obvious at the provincial scale, especially in the MCI of the coastal provinces such as Fujian, Guangdong, Jiangxi, Zhejiang, and Shanghai, which declined significantly from 2000 to 2015. The mean MCI in Guangdong, Guangxi, Hainan, Henan, Anhui, and Jiangsu was high, whereas that of Heilongjiang, Inner Mongolia, Ningxia, and Qinghai was low. However, at the county scale, we found that the spatial distribution of the MCI differed from that at the provincial scale. The single-cropping index and MCI of China from 2000 to 2015 were roughly bounded by the "Taihang Mountains–Qinling Mountains–Hengduan Mountains"—the MCI was lower than 100% at the north of the boundary and higher than 100% at the south of the boundary. From 2000 to 2015, the dynamic changes of China's MCI were obvious, and the proportion of the MCI with change values lower than 0 continued to rise. Notably, from 2005 to 2010 and 2010 to 2015, the MCI changed dramatically, mainly in the southeastern coastal areas; more specifically, the MCI rose in the former stage and dropped sharply in the latter stage.

We also examined and discussed the factors affecting the MCI. First, natural conditions are important factors in agriculture, among which hydrothermal conditions determine the upper limit of the MCI, and soil texture and the relief degree of the land surface inhibit the increase of the MCI. Second, nonagricultural processes adversely affect the MCI, especially the growth of the nonagricultural population and nonagricultural industry, which weaken the identity of farmers as the main body of agriculture and the rural agricultural production function. Third, the factors of intensive agricultural production, such as grain yield per unit area, irrigation rate, and agricultural machinery condition, promote the MCI by improving the agricultural production efficiency. Fourth, the cultivated land quality determines whether the land can be cultivated sustainably. For example, natural disasters cause water and soil imbalances. Meanwhile, pesticides, fertilizers, and a film can improve the grain yield and ensure the production efficiency of multiple croppings; however, their unscientific use will make farmlands unsustainable.

Given the continuous, rapid growth of nonagricultural processes, more research should be done on the transformations occurring in farmers' livelihoods, especially on the return of the rural labor force and the main body of agricultural production. Moving forward, it will be necessary for scholars and practitioners alike to balance the increases in crop production with the reductions in environmental pollution to grapple the increasingly severe agricultural ecological problems faced by China. Agricultural policies are of great significance to agricultural development; accordingly, moving forward, the policies should not only focus on the quantity of agricultural output but, also, on advancing the cultivated land quality, the farmer who cultivates the land, and the multiple-cropping potential of the cultivated land.

**Author Contributions:** Conceptualization, R.Y.; methodology, X.L. and J.W.; software, X.L. and J.W.; validation, R.Y. and X.L.; formal analysis, X.L.; investigation, X.L. and X.Z.; resources, Q.X. and R.Y.; data curation, Q.X.; writing—original draft preparation, X.L.; writing—review and editing, R.Y., X.L. and X.Z.; visualization, X.L. and X.Z.; supervision, Q.X. and R.Y.; project administration, Q.X.; funding acquisition, Q.X. and R.Y. All authors have read and agreed to the published version of the manuscript.

**Funding:** This research was financially supported by the National Natural Science Foundation of China (No. 41871177 and No. 41801088) and Natural Science Foundation of Guangdong Province(2018A0303130097).

**Institutional Review Board Statement:** Not applicable.

**Informed Consent Statement:** Not applicable.

**Data Availability Statement:** The data presented is primarily reflected in the article, more details are available on request from the corresponding author.

**Conflicts of Interest:** The authors declare no conflict of interest.

# References


1. Wu, W.B.; Yu, Q.Y.; You, L.Z.; Chen, K.; Tang, H.J.; Liu, J.G. Global cropping intensity gaps: Increasing food production without cropland expansion. *Land Use Policy* **2018**, *76*, 515–525. [CrossRef]
2. Tilman, D.; Balzer, C.; Hill, J.; Befort, B.L. Global food demand and the sustainable intensification of agriculture. *Proc. Natl. Acad. Sci. USA* **2011**, *108*, 20260–20264. [CrossRef] [PubMed]
3. Ray, D.K.; Ramankutty, N.; Mueller, N.D.; West, P.C.; Foley, J.A. Recent patterns of crop yield growth and stagnation. *Nat. Commun.* **2012**, *3*, 1293. [CrossRef]
4. Ray, D.K.; Mueller, N.D.; West, P.C.; Foley, J.A. Yield trends are insufficient to double global crop production by 2050. *PLoS ONE* **2013**, *8*, e66428. [CrossRef] [PubMed]
5. Pugh, T.A.M.; Müller, C.; Elliott, J.; Deryng, D.; Folberth, C.; Olin, S.; Schmid, E.; Arneth, A. Climate analogues suggest limited potential for intensification of production on current croplands under climate change. *Nat. Commun.* **2016**, *7*, 12608. [CrossRef] [PubMed]

6. Brisson, N.; Gate, P.; Gouache, D.; Charmet, G.; Oury, F.X.; Huard, F. Why are wheat yields stagnating in Europe? A comprehensive data analysis for France. *Field Crop. Res.* **2010**, *119*, 201–212. [CrossRef]

7. Finger, R. Evidence of slowing yield growth–the example of Swiss cereal yields. *Food Policy* **2010**, *35*, 175–182. [CrossRef]

8. Foley, J.A.; Ramankutty, N.; Brauman, K.A.; Cassidy, E.S.; Gerber, J.S.; Johnston, M.; Mueller, N.D.; O'connell, C.; Ray, D.K.; West, P.C.; et al. Solutions for a cultivated planet. *Nature* **2011**, *478*, 337–342. [CrossRef] [PubMed]

9. Wu, W.B.; Yu, Q.Y.; Peter, V.H.; You, L.Z.; Yang, P.; Tang, H.J. How could agricultural land systems contribute to raise food production under global change? *J. Integr. Agric.* **2014**, *13*, 1432–1442. [CrossRef]

10. Keys, E.; McConnell, W.J. Global change and the intensification of agriculture in the tropics. *Glob. Environ. Chang.* **2005**, *15*, 320–337. [CrossRef]

11. Ali, A.M.S. Population pressure, agricultural intensification and changes in rural systems in Bangladesh. *Geoforum* **2007**, *38*, 720–738. [CrossRef]

12. Chen, C.F.; Son, N.T.; Chang, L.Y. Monitoring of rice cropping intensity in the upper Mekong Delta, Vietnam using time-series MODIS data. *Adv. Space Res.* **2012**, *49*, 292–301. [CrossRef]

13. D'Amour, C.B.; Reitsma, F.; Baiocchi, G.; Barthel, S.; Güneralp, B.; Erb, K.H.; Haberl, H.; Creutzig, F.; Seto, K.C. Future urban land expansion and implications for global croplands. *Proc. Natl. Acad. Sci. USA* **2017**, *114*, 8939–8944. [CrossRef]

14. Wang, Q.B.; Halbrendt, C.; Johnson, S.R. Grain production and environmental management in China's fertilizer economy. *J. Environ. Manag.* **1996**, *47*, 283–296. [CrossRef]

15. FAOSTAT. Statistical Database of the Food and Agricultural Organization of the United Nation. 2012. Available online: http://faostat3.fao.org/ (accessed on 21 October 2019).

16. Liu, Y.S.; Qiao, L.Y. Innovating system and policy of arable land conservation under the new-type urbanization in China. *Econ. Geogr.* **2014**, *34*, 1–6. (In Chinese) [CrossRef]

17. Liu, Y.S.; Li, Y.H. Revitalize the world's countryside. *Nature* **2017**, *548*, 275–277. [CrossRef] [PubMed]

18. Ge, D.Z.; Long, H.L.; Zhang, Y.N.; Tu, S.S. Analysis of the coupled relationship between grain yields and agricultural labor changes in China. *J. Geogr. Sci.* **2018**, *28*, 93–108. [CrossRef]

19. Yang, R.; Liu, Y.S.; Chen, Y.F. Comprehensive measure and partition of rural hollowing in China. *Geogr. Res.* **2012**, *31*, 1697–1706. (In Chinese)

20. Liu, Y.S.; Yang, R.; Li, Y.H. Potential of land consolidation of hollowed villages under different urbanization scenarios in China. *J. Geogr. Sci.* **2013**, *23*, 503–512. [CrossRef]

21. Liu, Y.S.; Yang, R.; Long, H.L.; Gao, J.Y.; Wang, J. Implications of land-use change in rural China: A case study of Yucheng, Shandong province. *Land Use Policy* **2014**, *40*, 111–118. [CrossRef]

22. Chen, Y.F.; Liu, Y.S.; Xu, K.S. Characteristics and mechanism of agricultural transformation in typical rural areas of eastern China: A case study of Yucheng City, Shandong Province. *Chin. Geogr. Sci.* **2010**, *20*, 545–553. [CrossRef]

23. Chen, Y.F.; Li, X.D.; Liu, Y. Increasing China's agricultural labor productivity: Comparison and policy implications from major agrarian countries. *J. Resour. Ecol.* **2018**, *9*, 575–585. [CrossRef]

24. Liu, D.; Gong, Q.W.; Yang, W.J. The evolution of farmland protection policy and optimization path from 1978 to 2018. *Chin. Rural Econ.* **2018**, *408*, 39–53. (In Chinese)

25. Hayami, Y.; Ruttan, V.W. *Agricultural Development: An International Perspective*; The Johns Hopkins Press: Baltimore, MD, USA; London, UK, 1971.

26. Turner, B.L., II; Hanham, R.Q.; Portararo, A.V. Population pressure and agricultural intensity. *Ann. Assoc. Am. Geogr.* **1977**, *67*, 384–396. [CrossRef]

27. Turner, B.L., II; Doolittle, W.E. The concept and measure of agricultural intensity. *Prof. Geogr.* **1978**, *30*, 297–301. [CrossRef]

28. Verburg, P.H.; Chen, Y.Q.; Veldkampa, T. Spatial explorations of land use change and grain production in China. *Agric. Ecosyst. Environ.* **2000**, *82*, 333–354. [CrossRef]

29. Abrahão, G.M.; Costa, M.H. Evolution of rain and photoperiod limitations on the soybean growing season in Brazil: The rise (and possible fall) of double-cropping systems. *Agric. For. Meteorol.* **2018**, *256–257*, 32–45. [CrossRef]

30. Jönsson, P.; Eklundh, L. TIMESAT-a program for analyzing time-series of satellite sensor data. *Comput. Geosci.* **2004**, *30*, 833–845. [CrossRef]

31. Kato, H.; Kimura, R.; Elbeih, S.F.; Iwasaki, E.; Zaghloul, E.A. Land use change and crop rotation analysis of a government well district in Rashda village-Dakhla Oasis, Egypt based on satellite data. *Egypt. J. Remote Sens. Space Sci.* **2012**, *15*, 185–195. [CrossRef]

32. Mayerová, M.; Madaras, M.; Soukup, J. Effect of chemical weed control on crop yields in different crop rotations in a long-term field trial. *Crop Prot.* **2018**, *114*, 215–222. [CrossRef]

33. Blanco, J.; Pascal, L.; Ramon, L.; Vandenbroucke, H.; Carrière, S.M. Agrobiodiversity performance in contrasting island environments: The case of shifting cultivation in Vanuatu, Pacific. *Agric. Ecosyst. Environ.* **2013**, *174*, 28–39. [CrossRef]

34. Yusuf, A.A.; Abaidoo, R.C.; Iwuafor, E.N.O.; Olufajo, O.O.; Sanginga, N. Rotation effects of grain legumes and fallow on maize yield, microbial biomass and chemical properties of an Alfisol in the Nigerian savanna. *Agric. Ecosyst. Environ.* **2009**, *129*, 325–331. [CrossRef]

35. Luce, M.S.; Grant, C.A.; Zebarth, B.J.; Ziadi, N.; O'Donovan, J.T.; Blackshaw, R.E.; Harker, K.N.; Johnson, E.N.; Gan, Y.; Lafond, G.P.; et al. Legumes can reduce economic optimum nitrogen rates and increase yields in a wheat-canola cropping sequence in western Canada. *Field Crop. Res.* **2015**, *179*, 12–25. [CrossRef]

36. Sánchez-Navarro, V.; Zornoza, R.; Faz, Á.; Fernández, J.A. Comparing legumes for use in multiple cropping to enhance soil organic carbon, soil fertility, aggregates stability and vegetables yields under semi-arid conditions. *Sci. Hortic.* **2019**, *246*, 835–841. [CrossRef]

37. Devendra, C.; Thomas, D. Smallholder farming systems in Asia. *Agric. Syst.* **2002**, *71*, 17–25. [CrossRef]

38. Canisius, F.; Turral, H.; Molden, D. Fourier analysis of historical NOAA time series data to estimate bimodal agriculture. *Int. J. Remote Sensin* **2007**, *28*, 5503–5522. [CrossRef]

39. Panigrahy, S.; Sharma, S.A. Mapping of crop rotation using multidate Indian Remote Sensing Satellite digital data. *Isprs J. Photogramm. Remote Sens.* **1997**, *52*, 85–91. [CrossRef]

40. Panigrahy, S.; Manjunath, K.R.; Ray, S.S. Deriving cropping system performance indices using remote sensing data and GIS. *Int. J. Remote Sens.* **2005**, *26*, 2595–2606. [CrossRef]

41. Frolking, S.; Yeluripati, J.B.; Douglas, E. New district-level maps of rice cropping in India: A foundation for scientific input into policy assessment. *Field Crop. Res.* **2006**, *98*, 164–177. [CrossRef]

42. Biradar, C.M.; Xiao, X.M. Quantifying the area and spatial distribution of double-and triple-cropping croplands in India with multi-temporal MODIS imagery in 2005. *Int. J. Remote Sens.* **2011**, *32*, 367–386. [CrossRef]

43. Sakamoto, T.; Nguyen, N.V.; Ohno, H.; Ishitsuka, N.; Yokozawa, M. Spatio–temporal distribution of rice phenology and cropping systems in the Mekong Delta with special reference to the seasonal water flow of the Mekong and Bassac rivers. *Remote Sens. Environ.* **2006**, *100*, 1–16. [CrossRef]

44. Sakamoto, T.; Phung, C.V.; Kotera, A.; Nguyen, K.D.; Yokozawa, M. Analysis of rapid expansion of inland aquaculture and triple rice-cropping areas in a coastal area of the Vietnamese Mekong Delta using MODIS time-series imagery. *Landsc. Urban Plan.* **2009**, *92*, 34–46. [CrossRef]

45. Ding, M.J.; Chen, Q.; Xin, L.J.; Li, L.H.; Li, X.B. Spatial and temporal variations of multiple cropping index in China based on SPOT-NDVI during 1999–2013. *Acta Geogr. Sin.* **2015**, *70*, 1080–1090. (In Chinese)

46. Yang, X.L.; Chen, Y.Q.; Pacenka, S.; Gao, W.S.; Ma, L.; Wang, G.Y.; Yan, P.; Sui, P.; Steenhuis, T.S. Effect of diversified crop rotations on groundwater levels and crop water productivity in the North China Plain. *J. Hydrol.* **2015**, *522*, 428–438. [CrossRef]

47. Zuo, L.J.; Wang, X.; Liu, F.; Yi, L. Spatial exploration of multiple cropping efficiency in China based on time series remote sensing data and econometric model. *J. Integr. Agric.* **2013**, *12*, 903–913. [CrossRef]

48. Gao, J.Q.; Yang, X.G.; Zheng, B.Y.; Liu, Z.J.; Zhao, J.; Sun, S.; Li, K.N.; Dong, C.Y. Effects of climate change on the extension of the potential double cropping region and crop water requirements in Northern China. *Agric. For. Meteorol.* **2019**, *268*, 146–155. [CrossRef]

49. Ma, L.; Long, H.L.; Tu, S.S.; Zhang, Y.N.; Zheng, Y.H. Farmland transition in China and its policy implications. *Land Use Policy* **2020**, *92*, 104470. [CrossRef]

50. Long, H.L. *Land Use Transitions and Rural Restructuring in China*; Springer: Singapore, 2020.

51. Long, H.L.; Qu, Y.; Tu, S.S.; Zhang, Y.N.; Jiang, Y.F. Development of land use transitions research in China. *J. Geogr. Sci.* **2020**, *30*, 1195–1214. [CrossRef]

52. Lin, G.C.S.; Ho, S.P.S. China's land resources and land-use change: Insights from the 1996 land survey. *Land Use Policy* **2003**, *20*, 87–107. [CrossRef]

53. Xie, H.L.; Liu, G.Y. Spatiotemporal differences and influencing factors of multiple cropping index in China during 1998–2012. *J. Geogr. Sci.* **2015**, *25*, 1283–1297. [CrossRef]

54. Zuo, L.J.; Wang, X.; Zhang, Z.X.; Zhao, X.L.; Liu, F.; Yi, L.; Liu, B. Developing grain production policy in terms of multiple cropping systems in China. *Land Use Policy* **2014**, *40*, 140–146. [CrossRef]

55. Qiu, B.W.; Lu, D.F.; Tang, Z.H.; Song, D.J.; Zeng, Y.H.; Wang, Z.Z.; Chen, C.C.; Chen, N.; Huang, H.Y.; Xu, W.M. Mapping cropping intensity trends in China during 1982–2013. *Appl. Geogr.* **2017**, *79*, 212–222. [CrossRef]

56. Peng, D.; Huang, J.F.; Jin, H.M. Monitoring the sequential cropping index of arable land in Zhejiang province of China using MODIS-NDVI. *Agric. Sci. China* **2007**, *6*, 208–213. [CrossRef]

57. Zhang, M.W.; Zhou, Q.B.; Chen, Z.X.; Liu, J.; Zhou, Y.; Cai, C.F. Crop discrimination in Northern China with double cropping systems using Fourier analysis of time-series MODIS data. *Int. J. Appl. Earth Obs. Geoinf.* **2008**, *10*, 476–485. [CrossRef]

58. Li, P.; Feng, Z.M.; Jiang, L.G.; Liu, Y.J.; Xiao, X.G. Changes in rice cropping systems in the Poyang Lake Region, China during 2004–2010. *J. Geogr. Sci.* **2012**, *22*, 653–668. [CrossRef]

59. Feng, L.; Wang, G.P.; Han, Y.C.; Li, Y.B.; Zhu, Y.; Zhou, Z.G.; Cao, W.X. Effects of planting pattern on growth and yield and economic benefits of cotton in a wheat-cotton double cropping system versus monoculture cotton. *Field Crop. Res.* **2017**, *213*, 100–108. [CrossRef]

60. Yang, R.; Liu, Y.S.; Chen, Y.F.; Li, T.T. The remote sensing inversion for spatial and temporal changes of multiple cropping index and detection for influencing factors around Bohai Rim in China. *Sci. Geogr. Sin.* **2013**, *33*, 588–593. (In Chinese) [CrossRef]

61. Pearson, K. LIII. On lines and planes of closest fit to systems of points in space. *Lond. Edinb. Dublin Philos. Mag. J. Sci.* **1901**, *2*, 559–572. [CrossRef]

62. Hotelling, H. Analysis of a complex of statistical variables into principal components. *J. Educ. Psychol.* **1933**, *24*, 417–441. [CrossRef]

63. Massy, W.F. Principal components regression in exploratory statistical research. *J. Am. Stat. Assoc.* **1965**, *60*, 234–256. [CrossRef]

64. Wu, K.N.; Zhao, R. Soil texture classification and its application in China. *Acta Pedol. Sin.* **2019**, *56*, 227–241. (In Chinese) [CrossRef]

65. You, Z.; Feng, Z.M.; Yang, Y.Z. Relief degree of land surface dataset of China (1 km). *Dig. J. Global Change Data Repos.* **2018**. [CrossRef]
66. Chen, Y.J.; Yi, X.Y.; Fang, L.N.; Yang, R.Z. Analysis of cultivated land and grain production potential in China. *Sci. Agric. Sin.* **2016**, *49*, 1117–1131. (In Chinese) [CrossRef]
67. Yang, X.G.; Chen, F.; Lin, X.M.; Liu, Z.J.; Zhang, H.L.; Zhao, J.; Li, K.N.; Ye, Q.; Li, Y.; Lv, S.; et al. Potential benefits of climate change for crop productivity in China. *Agric. For. Meteorol.* **2015**, *208*, 76–84. [CrossRef]
68. Pires, G.F.; Abrahão, G.M.; Brumatti, L.M.; Oliveira, L.J.; Costa, M.H.; Liddicoat, S.; Liddicoat, S.; Kato, E.; Ladle, R.J. Increased climate risk in Brazilian double cropping agriculture systems: Implications for land use in Northern Brazil. *Agric. For. Meteorol.* **2016**, *228–229*, 286–298. [CrossRef]
69. Hampf, A.C.; Stella, T.; Berg-Mohnicke, M.; Kawohl, T.; Kilian, M.; Nendel, C. Future yields of double-cropping systems in the Southern Amazon, Brazil, under climate change and technological development. *Agric. Syst.* **2020**, *177*, 102707. [CrossRef]
70. Ding, M.J.; Chen, Q.; Xiao, X.M.; Xin, L.J.; Zhang, G.L.; Li, L.H. Variation in cropping intensity in Northern China from 1982 to 2012 based on GIMMS-NDVI data. *Sustainability* **2016**, *8*, 1123. [CrossRef]
71. Van den Berg, M.M.; Hengsdijk, H.; Wolf, J.; Van Ittersum, M.K.; Wang, G.H.; Roetter, R.P. The impact of increasing farm size and mechanization on rural income and rice production in Zhejiang province, China. *Agric. Syst.* **2007**, *94*, 841–850. [CrossRef]
72. Hong, Y.; Berentsen, P.; Heerink, N.; Shi, M.J.; van der Werf, W. The future of intercropping under growing resource scarcity and declining grain prices—A model analysis based on a case study in Northwest China. *Agric. Syst.* **2019**, *176*, 102661. [CrossRef]
73. Wang, R.J.; Li, X.B.; Tan, M.H.; Xin, L.J.; Wang, X.; Wang, Y.H.; Jiang, M. Inter-provincial differences in rice multi-cropping changes in main double-cropping rice area in China: Evidence from provinces and households. *Chin. Geogr. Sci.* **2019**, *29*, 127–138. [CrossRef]
74. Chen, K.Q.; Long, H.L.; Liao, L.W.; Tu, S.S.; Li, T.T. Land use transitions and urban-rural integrated development: Theoretical framework and China's evidence. *Land Use Policy* **2020**, *92*, 104465. [CrossRef]
75. Feike, T.; Doluschitz, R.; Chen, Q.; Graeff-Hönninger, S.; Claupein, W. How to overcome the slow death of intercropping in the North China Plain. *Sustainability* **2012**, *4*, 2550–2565. [CrossRef]
76. Van Loon, J.; Woltering, L.; Krupnik, T.J.; Baudron, F.; Boa, M.; Govaerts, B. Scaling agricultural mechanization services in smallholder farming systems: Case studies from sub-Saharan Africa, South Asia, and Latin America. *Agric. Syst.* **2020**, *180*, 102792. [CrossRef] [PubMed]
77. Tian, Q.; Holland, J.H.; Brown, D.G. Social and economic impacts of subsidy policies on rural development in the Poyang Lake Region, China: Insights from an agent-based model. *Agric. Syst.* **2016**, *148*, 12–27. [CrossRef]
78. Min, S.; Waibel, H.; Huang, J.K. Smallholder participation in the land rental market in a mountainous region of Southern China: Impact of population aging, land tenure security and ethnicity. *Land Use Policy* **2017**, *68*, 625–637. [CrossRef]
79. Li, Y.J.; Kahrl, F.; Pan, J.J.; Roland-Holst, D.; Su, Y.F.; Wilkes, A.; Xu, J.C. Fertilizer use patterns in Yunnan Province, China: Implications for agricultural and environmental policy. *Agric. Syst.* **2012**, *110*, 78–89. [CrossRef]
80. Ren, C.C.; Liu, S.; van Grinsven, H.; Reis, S.; Jin, S.Q.; Liu, H.B.; Gu, B.J. The impact of farm size on agricultural sustainability. *J. Clean. Prod.* **2019**, *220*, 357–367. [CrossRef]
81. Jin, X.B.; Zhang, Z.H.; Wu, X.W.; Xiang, X.M.; Sun, W.; Bai, Q.; Zhou, Y.K. Co-ordination of land exploitation, exploitable farmland reserves and national planning in China. *Land Use Policy* **2016**, *57*, 682–693. [CrossRef]
82. Song, W.; Pijanowski, B.C. The effects of China's cultivated land balance program on potential land productivity at a national scale. *Appl. Geogr.* **2014**, *46*, 158–170. [CrossRef]
83. Song, W.; Liu, M.L. Farmland conversion decreases regional and national land quality in China. *Land Degrad. Dev.* **2017**, *28*, 459–471. [CrossRef]

*Article*

# Green Transition of Cultivated Land Use in the Yellow River Basin: A Perspective of Green Utilization Efficiency Evaluation

Xiao Lu [1,2,*], Yi Qu [2], Piling Sun [2], Wei Yu [2] and Wenlong Peng [2]

[1] School of Humanities and Law, Northeastern University, Shenyang 110169, China
[2] School of Geography and Tourism, Qufu Normal University, Rizhao 276826, China; quyi9611@163.com (Y.Q.); spling86@qfnu.edu.cn (P.S.); yuwei@qfnu.edu.cn (W.Y.); pwl9506@163.com (W.P.)
* Correspondence: lvxiao@mail.neu.edu.cn; Tel.: +86-187-6931-9508

Received: 1 November 2020; Accepted: 24 November 2020; Published: 27 November 2020

**Abstract:** Exploring the green transition of cultivated land use from the perspective of green utilization efficiency evaluation has become an important content of deepening the study of cultivated land use transition, which is of great significance to promote food security and ecological civilization construction. At present, there are few studies on the green utilization efficiency of cultivated land (GUECL), which covers the comprehensive benefits of economy, ecology and society, combined with the requirements of ecological civilization and green development. Taking 65 cities (regions and autonomous prefectures) of the Yellow River Basin as the basic evaluation unit, the GUECL of the Yellow River Basin is evaluated with a Super-SBM model. In general, the GUECL of the Yellow River Basin was not high at four time points of 2000, 2006, 2012 and 2018, which presents a trend of "rising first and then falling". Analyzing its temporal and spatial evolution pattern, the GUECL in the upper, middle and lower reaches presented an order of the upper reaches area > the lower reaches area > the middle reaches area; and the spatial variation trend showed a decrease from west to east, and a U-shaped change in the south-north direction. Using spatial correlation analysis, except for the year 2000, the GUECL in the Yellow River Basin presents a general distribution characteristic of spatial agglomeration, which is positively correlated in 2006, 2012 and 2018. The change of spatio-temporal pattern is the result of internal and external factors. The former mainly displays in the main characteristics of farmers, family characteristics and farmers' cognition, while the latter is reflected in natural, social and policy factors.

**Keywords:** land use transition; green utilization efficiency of cultivated land; spatial and temporal pattern; the Yellow River Basin

---

## 1. Introduction

As the basis of human survival and development, agriculture feeds more than 7 billion people in the world [1]. With the increase of the global population and the demand for food, the food production system has gained more attention from scholars [2,3]. Maximizing food production and minimizing the use of critical resources is a vital challenge for global sustainable development [4]. To solve the problem of food safety, an important way is to explore the "double high" agricultural path, that is, resource utilization efficiency can be improved by increasing crop output per unit area [5]. As an important material condition to ensure the national food safety and ecological security, stabilize the economic and social order and promote the coordinated development of urban and rural areas [6], cultivated land is of great importance in sustainable and efficient utilization, which requires more attention in developing countries and in some resource-poor countries.

In the past few decades, China's agriculture has developed rapidly. However, with the comprehensive advancement of urbanization and industrialization processes in China [7,8], a large amount of cultivated land has been occupied by nonagricultural construction, causing a rapid decrease of cultivated land. As we all know, the total amount of cultivated land is limited and difficult to increase rapidly, and once converted to nonagricultural construction land, it will be hard to be reclaimed. In addition, China faces a complex and urgent transition situation of cultivated land use, and the utilization and protection of cultivated land is facing great challenges due to the phenomena of farmland abandonment [9], and low efficiency and extensive utilization [10], as well as the problems of excessive intensification of cultivated land utilization and nonpoint-source pollution aggravation. In recent years, China has continuously promoted the reform of the rural land system, improved the land use management policy system, explored the law and path of cultivated land use transition, and made efforts to guarantee the long-term food safety through the strict protection and efficient use of cultivated land [11]. Currently, facing the strategic demands of ecological civilization construction and national food safety guarantee, China is promoting the transition of cultivated land utilization to green and efficiency under the guidance of ecological environmental friendliness and sustainable utilization of resources, so as to realize the trinity pattern of "quantity, quality and ecology" of cultivated land utilization and protection in the new era.

In recent years, the study of land use transition has been widely concerned as a new approach to the comprehensive study of national/regional land use/cover change [12]. Land use transition refers to the transition process of regional land use morphology corresponding to the transition of economic and social development stage in a period of time with the change and innovation of economic and social development stage. Among them, land use morphology includes dominant morphology (quantity, structure and spatial pattern) and recessive morphology (quality, property right, management mode, input, output and function, etc.) [13]. As the main type of land use in the process of rapid development of regional social economy, cultivated land is more frequently converted with other land use types, and its transition process has become an important content and extension direction in the field of land use transition [14]. Many scholars have carried out a series of studies on the spatial–temporal characteristics [15] and driving mechanism [16] of cultivated land use transition, but the research focuses on the transition of dominant morphology of cultivated land use, while less attention is paid to the transition of recessive morphology. Compared with the dominant morphology, the transition of the recessive morphology of cultivated land use is more easily affected by the structure of property rights, the scale of operation, the efficiency of land use and the evolution of multifunction. It pays more attention to the natural and social attributes of cultivated land, and is more closely related to the sustainable development of agriculture and the green use of cultivated land. Therefore, it has become a new research focus to change the utilization morphology of economic output to green and sustainable utilization. As the main behavior of laborers and the main embodiment of the relationship between human and land in the process of cultivated land utilization, the input-output in the process of cultivated land use directly affects the efficiency of cultivated land use. Exploring the cultivated land green utilization efficiency from the perspective of input-output recessive morphology has become an important content of deepening the research on cultivated land use transition. Therefore, constructing an evaluation and analysis framework for the green utilization efficiency of cultivated land (GUECL) based on the concept of green development, and selecting of typical regions to analyze their spatial–temporal evolution pattern and influence mechanism. It plays an important supporting role in promoting the green transition of cultivated land use and promoting the coordination between ecological civilization and food safety.

The Yellow River Basin is an important ecological barrier in China [17], and also a key area for ensuring national food safety. In 2018, the total area of cultivated land in Yellow River Basin accounted for 34.89% of the total cultivated land area in China, and the grain output accounted for 35.37% of the total grain output in China. The green utilization of cultivated land and the high-quality agricultural development in the Yellow River Basin are directly related to the national food safety, ecological security and long-term stability. In September 2019, President Xi Jinping held a forum on ecological protection and high-quality development of the Yellow River Basin in Zhengzhou and delivered an important speech. The ecological protection and high-quality development in the Yellow River Basin rose to a major national strategy. Therefore, the research on the green transition of cultivated land use in the Yellow River Basin has become a basic subject to respond to the national strategic requirements and support the ecological protection and high-quality development of the Yellow River Basin.

The process of urbanization and industrialization in developed countries is relatively early. For the study of land use efficiency, researchers pay more attention to urban land use efficiency [18,19]. With the expansion of urban scale, economic, environmental and ecological issues have become prominent. Scholars have begun to link the concept of land use efficiency with the long-term sustainability of development from an ecological and socio-economic perspective [20]. There are relatively few studies on agricultural land use efficiency [21], especially the research on utilization efficiency of cultivated land. There is no unified evaluation standard, and the selected indices include yield ratio [22] and land equivalent ratio [23]. In the research methods, some scholars use Data Envelopment Analysis (DEA) [24,25] and the mathematical process of linear programming to evaluate the relative efficiency of decision-making units (DMU) scientifically and systematically; and some scholars use Stochastic Frontier Analysis (SFA) [26,27] methods considering the high volatility of agricultural output. Regarding the influencing factors of agricultural land use efficiency, a study on the eco-efficiency of cultivated land samples in the rural region of Le Marche (Italy), which found that most arable farms exhibit a modest level of eco-efficiency in relation to the use of fertilizers and pesticides, while farms are more eco-efficient if they are led by young farmers and participate to agri-environmental schemes [28]. A study on rice farms in the southwest of Niger shows that factors, such as farm size, experience in rice farming and land ownership had a direct impact on technical efficiency [29]. Land use rights also affected the decisions made by farmers to invest in land and to improve efficiency [30]. In addition, intercropping hybrid poplar and switchgrass can improve land use efficiency [31]. Regarding the green utilization of cultivated land, the EU Common Agricultural Policy has established legal and institutional requirements for cultivated land protection, emphasizing the greening of farming methods and the precise use of cultivated land. In order to protect the ecological environment, Germany promotes green farming in accordance with the EU policy framework, maintaining green areas and planting intercropping crops to achieve sustainable use of arable land resources, which initially shows the effect of green subsidies [32]. In order to promote the highly intensive and ecological development of agriculture, the Dutch government has introduced and implemented a strict ecological environment protection system [33]. In Europe, the formation of the common agricultural policy has a greater role in promoting the green development of European agriculture.

In recent years, the exploration of utilization efficiency of cultivated land in China has increased. In the existing study, the construction of evaluation index system, the selection of evaluation methods and the influencing factors of utilization efficiency of cultivated land have received extensive attention. Most of the previous studies selected input indices from the three aspects of land, capital and labor [34]; and took total agricultural output value and total grain output as output indices [10]. Considering the economic and social benefits of cultivated land resources, some scholars selected per capita net income as output index [35]. The utilization of cultivated land creates the desirable output for the economy and society, while it produces the environmental pollution at the same time, that is, the undesirable output. Therefore, environmental pollution should also be included in the rating index system [36]. In terms of the selection of research methods, since the DEA model was introduced into China in the 1990s, more and more studies have been conducted on efficiency measurement with this method, mainly including BC2 model, super-efficiency slacks-based measure (Super-SBM) model and epsilon-based measure (EBM) model in the classic DEA model [34,37,38]. In addition, some scholars use the stochastic frontier production function and nonradial directional distance function to measure utilization efficiency of cultivated land [39,40]. On the whole, there are many methods to choose from that are relevant for different research purposes. In the aspect of influencing factors of utilization efficiency of cultivated land, both the main characteristics of cultivated land utilization (e.g., age of agricultural labor force and farmer differentiation [41], and the characteristics of cultivated land utilization and resource endowment, such as the multiple cropping index of cultivated land, quality of cultivated land, farmland right verification and farmland transfer [42], have been paid more attention. In addition, considering that as the main body of cultivated land production and management, local concepts or parochialism will be formed on the land emotionally and psychologically, and the land values of farmers will also be studied as an influential factor of the utilization efficiency of cultivated land [43].

To sum up, the existing research on utilization efficiency of cultivated land has made a lot of achievements, but there are many differences in the data source, index system, model selection and other aspects. Especially in the construction of index system, there is still plenty of scope for improvement. Most of the existing index systems explore the utilization efficiency of cultivated land from the perspective of economic benefits of input-output, while the social and ecological benefits are ignored. Some studies [36] have considered ecological benefit indices such as nonpoint-source pollution and carbon emissions, but have not paid attention to social benefit indices. On the whole, there are few studies on the GUECL which cover the comprehensive benefits of economy, ecology and society, combined with the requirements of ecological civilization and green development. In addition, as far as the research unit is concerned, the existing research focuses on the province, city and county as the research object, and the research on watershed scale and cross-administrative area is relatively few.

The research purposes of this paper are as follows: (1) under the background of ecological civilization construction in China, a theoretical analysis framework of green utilization of cultivated land is constructed, and an evaluation index system of GUECL is constructed by comprehensively considering economic, social and ecological benefits; (2) taking the Yellow River Basin, the latest national strategic region in China, as the research object, the constructed index system and Super-SBM model are used to calculate the GUECL in the Yellow River Basin, and the spatial–temporal evolution pattern and driving factors are analyzed; (3) trying to put forward some countermeasures and suggestions to promote the green transition of cultivated land use in the Yellow River Basin in order to provide support for the ecological protection and high-quality development of the Yellow River Basin.

## 2. Materials and Methods

### 2.1. Study Area

Originating from the BaYanKaLa mountains in Qinghai Province, China, the Yellow River flows through 9 provinces (autonomous regions), including Qinghai, Sichuan, Gansu, Ningxia, Inner Mongolia, Shaanxi, Shanxi, Henan and Shandong, a total of 65 cities (districts, autonomous prefectures) (Figures 1 and 2). It flows into the Bohai Sea in Kenli District, Dongying city, Shandong Province, with a basin area of 795,000 km$^2$ (including an area of 42,000 km$^2$ in the internal flow area). The topography of the Yellow River Basin is high in the west and low in the east, with great differences in geomorphic features. The western part is composed of a series of high mountains and developed glacier geomorphic features; the middle part is loess geomorphic features with serious soil and water loss; the eastern part mainly consists of the alluvial plain of the Yellow River. The basin is more complexly affected by the atmospheric circulation and monsoon circulation, so the climate difference is significant. There are various soil types in the Yellow River Basin, mainly including meadow soil, tidal soil, chestnut soil, soft soil and brown soil. It is rich in natural resources, which occupies an important position in China with great development potential.

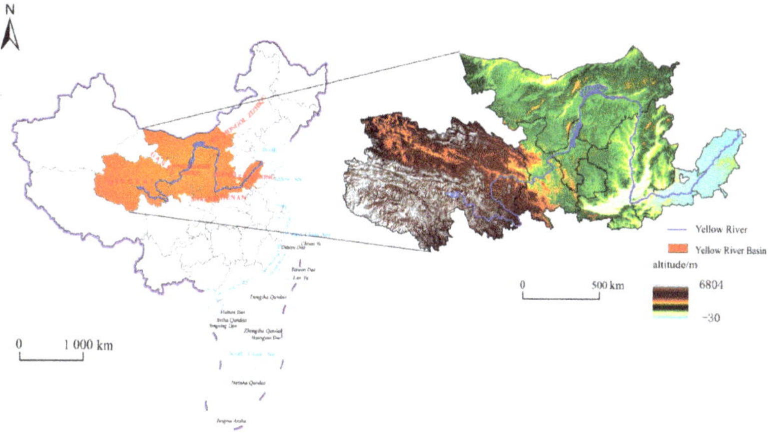

**Figure 1.** Location map of the Yellow River Basin.

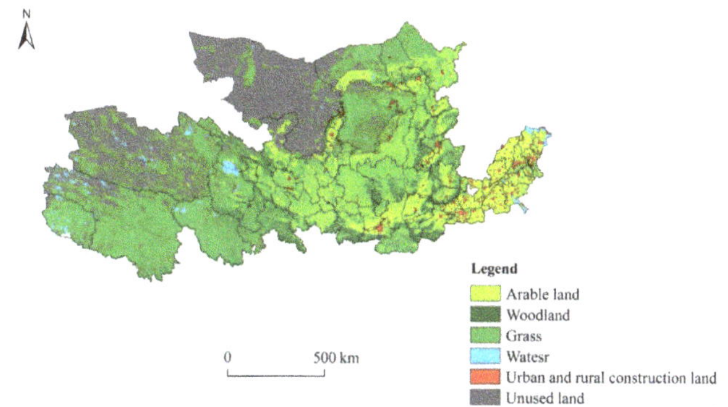


**Figure 2.** Land use map of the Yellow River Basin in 2018.

In 2018, the GDP of the 9 provinces and autonomous regions through which the Yellow River flows was CNY 23.86 trillion, accounting for about 26.50% of the national total. The total population was 420 million, accounting for about 30% of the national total. The cultivated land area was 47.0316 million hm$^2$, and the per capita cultivated land in these regions was 0.11 hm$^2$, 1.15 times of the national per capita cultivated land. The Yellow River Basin has been an agricultural economic development area in China for a long time. The upper Ningxia–Inner Mongolia Hetao Plain, the middle Fen-Wei basin and the lower areas along Yellow River, with rich soil and high agricultural production level, are the three major agricultural production bases in the Yellow River Basin [44]. The land use of the basin is mainly agricultural land, and the regional economy is dominated by agriculture and animal husbandry. The main crops are wheat, corn, millet, potato, cotton, oil, etc., especially wheat and cotton, which occupy an important position in China.

Refer to the Yellow River Volume and Yellow River Yearbook published since 1949 and the Yellow River Basin Flood Control Plan approved by the State Council in 2008, "from Hekou Town, Tuoketuo County, Inner Mongolia, to Taohuayu, Xingyang City, Henan Province, is the middle reaches of the Yellow River, from the Taohuayu to the estuary are classified as the lower reaches of the Yellow River". Taking into account the influence of administrative divisions, the regions and cities involved in Qinghai, Gansu, Ningxia, and Inner Mongolia in the Yellow River Basin are classified as the upper reaches, and the relevant cities in Shanxi and Shaanxi are classified as the middle reaches, the cities involved in Henan and Shandong are classified as lower reaches.

*2.2. Analysis Framework and Research Method*

2.2.1. Analysis Framework and Indicator System

Since the 18th National Congress of the Communist Party of China, green development has become an important part of China's new development concept. Under the background of fully implementing green development and promoting agricultural modernization, China's agriculture has accelerated green transition. For a long time, high-intensity land use and excessive input of chemicals have threatened the ecosystem and environment. The contradiction among economy, resources and environment has made people realize to promote agriculture green development and sustainable development. The agriculture green development is the continuation and deepening of the agriculture sustainable development. Compared with the agriculture sustainable development, it has more specific objectives to achieve, which pays more attention to resource conservation, environmental friendliness, ecological conservation and product quality, and pursues the win-win of ecology, economy and society. As the material basis of agricultural production, cultivated land plays an important role in ensuring the national food safety. Therefore, the green transition of cultivated land utilization has become an important link in promoting the agriculture green development.

At present, there is no clear definition of green utilization of cultivated land. Generally speaking, it is to change the traditional way of cultivated land utilization based on the concept of green development. In this process, ecological agriculture theory, sustainable development theory, circular economy theory, etc., except for the green development theory, all provide reference ideas for the definition of green utilization of cultivated land and have become the important theoretical basis for understanding the green utilization of cultivated land. Generally speaking, under the guidance of ecological agriculture theory, the ecological utilization of cultivated land is to combine traditional technology with modern technology, to optimize the allocation of elements input, and to achieve the unity of economic, ecological and social benefits. The connotation of sustainable utilization of cultivated land under the guidance of sustainable development theory mainly includes two aspects: making use of cultivated land, creating wealth and promoting economic development; and improving ecological environment to meet the needs of human survival from the perspective of taking into account the interests of future generations. The concept of circular economy produced in the evolution process of the contradiction between man and nature, is to implement the management and regulation of "reducing, reusing and recycling" of the resource flow mode in social production and reproduction activities, which is a new economic development model with high ecological efficiency [45]. Agricultural circular economy builds agricultural development on the harmonious coexistence with the environment. It is to reduce the consumption of natural resources as much as possible in the process of agricultural production, especially to control the input amount of nonrenewable resources, pay attention to ecological protection and take into account both the economic benefits of agricultural production and the benefits of ecological environment. Based on the above understanding, the green development concept of harmonious coexistence between person and nature as well as sustainable development is introduced into the process of cultivated land utilization, and the primary understanding of green utilization of cultivated land is formed.

Based on this, we believe that the green utilization of cultivated land can be regarded as one of the forms of cultivated land use transition. Compared to the cultivated land use of the space morphology the trend of change and the evolution, the green utilization of cultivated land is more focused on the cultivated land use in the process of input and output, mode of operation, efficiency, benefit, change in the morphology of cultivated land quality and other functions or attributes change. It places emphasis more on the ecological output or social effect of land use activities, which plays an important role to maintain the ecosystem service function, which is closer to the goal and concept of transition of recessive morphology of cultivated land use. The connotation of green utilization of cultivated land includes three aspects: (1) in terms of economy, at a certain level of inputs, more economic benefits can be obtained as far as possible; (2) in terms of society, we will ensure national food safety and social stability, improve people's living standards, and pay attention to the impact of the use of cultivated land on food safety and farmers' lives; (3) in terms of ecology, in the process of cultivated land utilization, we can reduce the damage to the environment and ecosystem to the greatest extent. Based on this, from the perspective of input-output, the GUECL is an efficiency measurement concept that takes the output of economic and social dimensions as the desirable output and the environmental pollution as the undesirable output, and its goal is to promote the maximization of economic and social output and the minimization of environmental pollution through scientific evaluation (Figure 3). It is also similar to "the optimal green efficiency of arable land use" defined by Xie et al. [40], but it increases the desirable output of social dimension and enriches the connotation of GUECL.

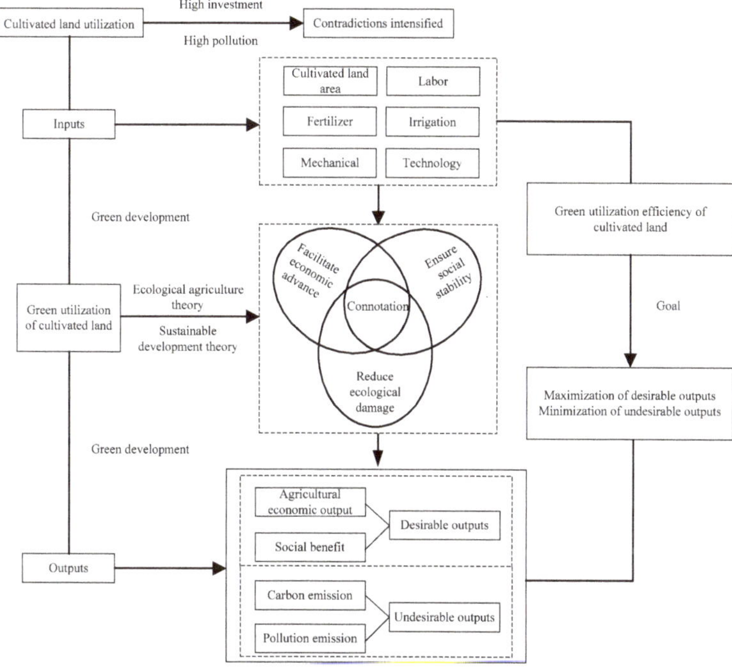



**Figure 3.** Theoretical analysis framework of the green utilization efficiency of cultivated land (GUECL).

The utilization of cultivated land is a complicated process. Utilization efficiency of cultivated land reflects the rationality of the allocation of various resources invested in cultivated land utilization in the process of agricultural production, and shows the degree of realization of cultivated land resource value output in agricultural production. Therefore, utilization efficiency of cultivated land is an important index to evaluate the level and degree of cultivated land utilization [46]. As one of the agricultural input factors, cultivated land must be combined with labor, agricultural machinery, fertilizer, pesticide and other input factors to promote production [22]. Referring to the existing research [10,36], we selected cultivated land area, labor, fertilizer, irrigation and machinery as input indices. We also added technology into the input indices, considering that technology is also the key factor in promoting green development. Therefore the above 6 indices represented the input variables of green utilization of cultivated land. The utilization of cultivated land has the dual functions of supplying agricultural products and releasing a large number of carbon emissions, which can exert a negative impact on regional ecological environment to a certain extent. Therefore, if the negative impact of cultivated utilization is not taken into account in the selection of output indices, the research results will inevitably be biased [47]. In view of the threat to the cultivated land ecosystem caused by cultivated land pollution resulted from fertilizer, pesticide and plastic film residues as well as carbon emissions in the process of cultivated land utilization, combined with the existing research [10,36], two indices of pollution emissions and carbon emissions were selected to represent the undesirable outputs in the process of green utilization of cultivated land, and the agricultural output value was selected as the desirable economic output. Finally, considering the importance of food safety to social stability, we chose the food safety coefficient as the desirable social output. In conclusion, the evaluation index system of GUECL from the input-output perspective was constructed (Table 1).

**Table 1.** Evaluation index system of GECLU.

| Primary Indices | Secondary Indices | Variable | Remarks |
| --- | --- | --- | --- |
| Inputs | Cultivated land area | Total sown area of crops/thousand hectare | |
| | Labor | Agricultural employees/10,000 people | Referring to relevant literature [36], agricultural employees are equal to the number of employees of agriculture forestry animal husbandry and fishery multiplied by the proportion of agricultural output value in the total output value of agriculture forestry animal husbandry and fishery |
| | Fertilizer | Fertilizer application amount (converted to net)/10,000 tons | |
| | Irrigation | Effective irrigation area/thousand hectares | |
| | Mechanical | Total agricultural machinery power/ten thousand kw | |
| | Technology | Agricultural technician/one person | |
| Desirable Outputs | Agricultural economic output | Total agricultural output /CNY 100 million | Unified conversion to 2000, eliminate the impact of price factors |
| | Social benefit | Food safety coefficient | Per capita grain output/400 kg |
| Undesirable Outputs | Carbon emission | Carbon emission of cultivated land use/ten thousand tons | Refer to literature [36] |
| | Pollution emission | Chemical fertilizer pollution/10,000 tons Pesticide pollution/10,000 tons<br><br>Agricultural film pollution/10,000 tons | Referring to relevant literature [48], the amount of chemical fertilizer pollution is equal to the amount of fertilizer application multiplied by the pollution rate of chemical fertilizer, and the pollution rate of chemical fertilizer is calculated as 65%.The calculation method of pesticide pollution is similar to that of chemical fertilizer, and the pesticide pollution rate is calculated as 50%. The amount of agricultural film pollution refers to the residual amount of agricultural film, and the residual rate is calculated as 10% |

### 2.2.2. Super-SBM Model for Evaluating GUECL

Data Envelopment Analysis (DEA) is a nonparametric method that uses mathematical tools to evaluate the effectiveness of the production frontier of an economic system. After continuous improvement, domestic and foreign scholars have summarized and explored a variety of evaluation models. The relaxation-based nonradial model (SBM) was first proposed by Tone [49]. It can be used to evaluate the efficiency of multiple inputs and multiple outputs. However, multiple decision-making units may be fully effective at the same time. In view of this, Tone [50] expanded the model and the Super-SBM model was further proposed, which combined the advantages of the DEA model and the SBM model, and effectively solved the defects of the previous model. The model is constructed as:

$$\rho = \min \frac{\frac{1}{m} \sum_{i=1}^{m} \frac{\bar{x}_i}{x_{i0}}}{\frac{1}{s_1+s_2}\left(\sum_{r=1}^{s_1} \frac{\bar{y}_r^g}{y_{r0}^g} + \sum_{j=1}^{s_2} \frac{\bar{y}_j^b}{y_{j0}^b}\right)}$$

$$
s.t. \begin{cases} x_0 = X\lambda + S^-, y_0^g = Y^g\lambda - S^g, y_0^b = Y^b\lambda + S^b \\ \overline{x} \geq \sum_{j=1,\neq 0}^{n} \lambda_j x_j, \; \overline{y}^g \leq \sum_{j=1,\neq 0}^{n} \lambda_j y_j^g, \; \overline{y}^b \leq \sum_{j=1,\neq 0}^{n} \lambda_j y_j^b \\ \overline{x} \geq x_0, \overline{y}^g \leq y_0^g, \overline{y}^b \leq y_0^b \\ \sum_{j=1,\neq 0}^{n} \lambda_j = 1, S^- \geq 0, S^g \geq 0, S^b \geq 0, \overline{y}^g \geq 0, \lambda \geq 0 \end{cases}
\tag{1}
$$

where $\rho$ is the value of ecological efficiency; $x$, $y^g$ and $y^b$ represent input, desirable outputs and undesirable outputs, respectively; $m$, $s_1$ and $s_2$ represent the number of indicators for inputs, desirable outputs and undesirable outputs; $S$, $S^g$ and $S^b$ are slacks of input, desirable outputs and undesirable outputs, $\lambda$ is the weight vector. Specifically, DMU is relatively efficient if $\rho \geq 1$ and DMU is relatively inefficient if $\rho < 1$ [51].

### 2.3. Data Sources and Processing Software

Considering the availability of data, the integrity of administrative boundaries and other factors, this paper took 65 cities (regions and autonomous prefectures) flowing through the Yellow River as the research objects, and selected four time points of 2000, 2006, 2012 and 2018 for research. The data for the indicators in this paper were obtained from the relevant years of China Statistical Yearbook, China Rural Statistical Yearbook, China City Statistical Yearbook, China Statistical Yearbook for Regional Economy, Statistical Yearbook of each province and the Statistical Communique on National Economic and Social Development. According to the needs of the research content and methods, the technical tools we used include ArcGIS, MaxDEA and GeoDa software.

## 3. Results

### 3.1. Spatial-Temporal Variation of GUECL in the Yellow River Basin

Using MaxDEA software, we calculated the GUECL in 65 areas of the Yellow River Basin, as summarized in Table 2.

According to the average value of GUECL of 65 evaluation units in the Yellow River Basin in 2000, 2006, 2012 and 2018 (Table 2), the overall GUECL in the Yellow River Basin is not high, generally presents a trend of "rising first and then falling", indicating that the green utilization of cultivated land in the Yellow River Basin has made some progress during 2000–2012, but the utilization efficiency has decreased in recent years, and there is still a large space for green utilization of cultivated land.

According to the principle of Super-SBM model, when the GUECL value is greater than or equal to 1, the DMU is relatively effective; when the GUECL value is less than 1, the DMU is relatively invalid [51]. As can be seen from Table 2, there were 20 evaluation units with GUECL value greater than or equal to 1 in the Yellow River Basin in 2000, accounting for about 31% of the 65 evaluation units, of which 10 were relatively effective in the upper reaches, 4 in the middle reaches and 6 in the lower reaches, which indicates that about 69% of the evaluation units were in relatively ineffective green utilization state in 2000. Compared to 2000, in 2006 and 2012, the number of relatively effective rating units increased to 21 and 26, respectively, accounting for 32% and 40%. In 2018, there were 24 relatively effective evaluation units for GUECL in the Yellow River Basin, accounting for about 37%, and among them, the upper, middle and lower reaches are 14, 6 and 4, respectively. The GUECL of middle and lower reaches has fewer relatively effective evaluation units.

**Table 2.** GUECL in 65 cities (regions, autonomous prefectures) in the Yellow River Basin.

| | DMU | 2000 | 2006 | 2012 | 2018 | | DMU | 2000 | 2006 | 2012 | 2018 |
|---|---|---|---|---|---|---|---|---|---|---|---|
| | Xining | 0.12 | 0.18 | 0.32 | 0.39 | | Weinan | 0.27 | 0.28 | 0.36 | 0.33 |
| | Tibetan Autonomous Prefecture of Golog | 1.50 | 1.85 | 1.52 | 1.85 | | Shangluo | 1.01 | 1.01 | 1.06 | 1.05 |
| | Yushu Tibetan Autonomous Prefecture | 1.10 | 1.13 | 1.16 | 1.25 | | Tongchuan | 0.29 | 0.44 | 1.09 | 1.14 |
| | Haixi Mongolian and Tibetan Autonomous Prefecture | 0.46 | 1.11 | 1.18 | 1.10 | Middle reaches area | Xianyang | 1.03 | 1.04 | 1.11 | 1.01 |
| | Tibetan Autonomous Prefecture of Hainan | 0.43 | 1.69 | 1.07 | 1.15 | | Xinzhou | 0.13 | 0.22 | 0.33 | 0.32 |
| | Tibetan Autonomous Prefecture of Huangnan | 1.08 | 1.19 | 1.30 | 1.34 | | Yuncheng | 0.16 | 0.29 | 0.50 | 0.52 |
| Upper reaches area | Haidong | 0.12 | 0.21 | 0.48 | 0.64 | | Shuozhou | 0.52 | 0.42 | 0.62 | 0.73 |
| | Tibetan Autonomous Prefecture of Haibei | 0.26 | 0.44 | 0.44 | 1.07 | | Changzhi | 0.32 | 0.44 | 0.39 | 0.34 |
| | Pingliang | 0.26 | 1.00 | 0.71 | 1.01 | | Jincheng | 0.27 | 0.41 | 0.41 | 0.47 |
| | Tianshui | 1.04 | 0.39 | 1.04 | 0.52 | | Linfen | 0.13 | 0.25 | 0.39 | 0.34 |
| | Dingxi | 0.27 | 0.30 | 0.49 | 0.32 | | Taiyuan | 0.27 | 0.38 | 0.38 | 0.44 |
| | Baiyin | 0.61 | 0.36 | 0.37 | 0.37 | | Lvliang | 0.06 | 0.19 | 0.30 | 0.22 |
| | Linxia Hui Autonomous Prefecture | 0.59 | 0.33 | 0.48 | 0.32 | | Jinzhoong | 0.08 | 0.33 | 0.41 | 0.45 |
| | Wuwei | 0.41 | 0.43 | 0.49 | 0.41 | | Jiyuan | 1.12 | 1.27 | 1.19 | 1.13 |
| | Lanzhou | 1.13 | 0.22 | 0.26 | 0.21 | | Luoyang | 0.15 | 0.58 | 0.69 | 0.46 |
| | Qingyang | 1.05 | 0.46 | 1.06 | 0.37 | | Zhengzhou | 0.13 | 0.32 | 0.32 | 0.25 |
| | Gannan Tibetan Autonomous Prefecture | 0.68 | 0.46 | 1.05 | 1.07 | | Anyang | 0.25 | 0.59 | 0.57 | 0.40 |
| | Shizuishan | 0.33 | 1.04 | 1.08 | 1.04 | | Jiaozuo | 1.01 | 1.02 | 0.76 | 1.02 |
| | Zhongwei | 1.03 | 0.46 | 1.01 | 1.39 | | Xinxiang | 0.25 | 0.43 | 0.45 | 0.32 |
| | Wuzhong | 0.33 | 0.50 | 0.59 | 0.45 | Lower reaches area | Kaifeng | 1.04 | 1.10 | 1.08 | 1.02 |
| | Yinchuan | 0.24 | 1.03 | 1.00 | 1.02 | | Puyang | 0.56 | 0.68 | 0.65 | 0.50 |
| | Guyuan | 0.09 | 0.32 | 1.01 | 1.05 | | Sanmenxia | 1.03 | 1.09 | 1.09 | 1.10 |
| | Wuhai | 1.32 | 1.14 | 1.15 | 1.08 | | Jinan | 1.08 | 1.03 | 1.05 | 0.38 |
| | Ordos | 0.28 | 1.14 | 0.54 | 0.42 | | Liaocheng | 0.19 | 0.34 | 0.44 | 0.34 |
| | Ulanqab | 1.01 | 1.12 | 0.40 | 0.35 | | Dezhou | 0.37 | 0.37 | 0.47 | 0.35 |
| Middle reaches area | Alxa League | 1.08 | 1.29 | 1.20 | 1.17 | | Jining | 0.27 | 0.35 | 0.47 | 0.34 |
| | Bayan Nur | 0.21 | 1.06 | 0.52 | 0.33 | | Zibo | 0.49 | 0.53 | 1.01 | 0.53 |
| | Baotou | 0.17 | 1.03 | 1.04 | 0.40 | | Taian | 1.02 | 0.55 | 0.76 | 0.49 |
| | Hohhot | 0.15 | 0.73 | 0.43 | 0.28 | | Dongying | 0.42 | 0.49 | 0.46 | 0.39 |
| | Yulin | 0.32 | 0.44 | 1.06 | 1.26 | | Binzhou | 0.49 | 0.49 | 0.59 | 0.39 |
| | Yan'an | 1.01 | 0.62 | 1.09 | 1.08 | | Laiwu | 0.41 | 0.49 | 1.01 | 0.79 |
| | Baoji | 0.21 | 0.35 | 0.41 | 0.34 | | Heze | 0.20 | 0.28 | 0.31 | 0.28 |
| | Xi'an | 1.05 | 0.38 | 0.57 | 1.05 | | Average | 0.54 | 0.65 | 0.73 | 0.68 |

GUECL in the upper, middle and lower reaches also shows large spatial–temporal differences. From the perspective of time change trend (Figure 4), the efficiency value in the middle reaches showed an upward trend, and the efficiency value in the upper and lower reaches both showed a first upward trend and then a downward trend during 2000–2018. In 2000, 2006 and 2012, the GUECL value in the middle reaches was lower than that in the lower reaches, while in 2018, it exceeded that in the lower reaches. In recent years, China's western development strategy and the rising strategy in central region have been continuously promoted, and the economic development in the middle and upper reaches of the Yellow River Basin has been improved. In addition, the Land Management Law strictly protects cultivated land. Central Document No. 1 concerns the issue of "agriculture, rural areas and farmers". The state has continuously promoted the green development of agriculture, introduced the green cropping system in rural areas, implemented cropland rotation, fallow and other planting methods, implemented the policy of returning farmland to forests and grass, and abolish the agricultural tax, etc., which, to a certain extent, will affect the GUECL. From 2000 to 2018, the average GUECL values in the upper, middle and lower reaches of the Yellow River Basin were 0.74, 0.53 and 0.61, respectively. In comparison, the upstream GUECL is significantly higher, which may be related to the natural environment and socio-economic conditions. Most of the cultivated land in the upper reaches of Qinghai Province is concentrated in the Yellow River Basin. Xining and Haidong are the agricultural production intensive areas of Qinghai Province. Ningxia has paid more attention to green investment in recent years. Inner Mongolia also has good resources and environment conditions, and the development momentum of its primary industry is promising; while Shanxi Province in the middle reaches is one of the key provinces of soil and water loss in the country, and the Shanxi section of the Yellow River Basin is the most serious area of soil and water loss in Shanxi Province, which greatly affects the agriculture green development and green utilization of cultivated land. In addition, in order to pursue the highest yield and profit in agricultural production, there are often over-exploitation and extensive utilization of cultivated land [52]. Henan Province and Shandong Province in the lower reaches are mainly plains, which are suitable for agricultural production. The economic development of the areas along the Yellow River is generally higher than that of the areas not along the Yellow River. However, the rapid development of Central Plains City Cluster and Shandong Peninsula City Cluster make a large amount of cultivated land occupied, at the same time, a large number of pollutants are discharged and the heavy population pressure also makes the GUECL relatively low.

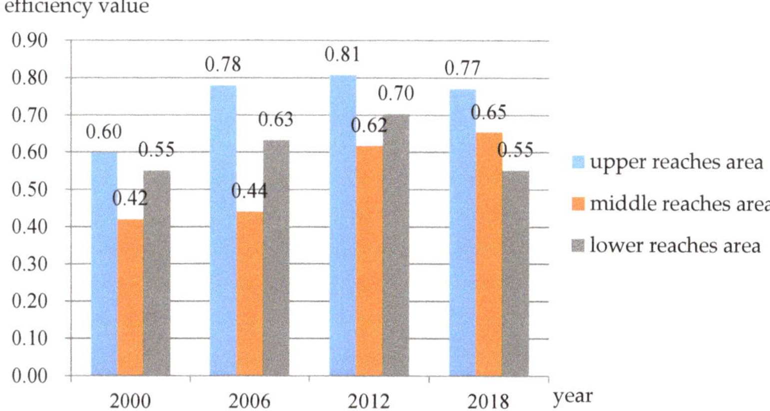


**Figure 4.** Changes of GUECL in the upper, middle and lower reaches of the Yellow River Basin.

From the level of prefecture and city, the GUECL in Tibetan Autonomous Prefecture of Golog was the highest from 2000 to 2018, which was 1.67. It's mainly due to the fact that the prefecture has actively explored a new way of high-quality development oriented by ecological priority and green development in recent years, so as to maximize benefits through optimal allocation of resources. The GUECL in Lvliang City is the lowest, only 0.19, which may be related to its natural environment. Located on the Loess Plateau, the city is characterized by complex landforms, severe erosion, shallow tillage soil, and low and concentrated rainfall. It is a typical region of Shanxi Province with drought of 9 years out of 10 [52]. The second reason may be extensive cultivation and low cultivation management technology, which leads to low GUECL.

*3.2. GUECL Spatial Pattern and Its Changes in the Yellow River Basin*

In order to understand the spatial distribution of GUECL in the Yellow River Basin more intuitively, the GUECL values of 65 evaluation units are divided into five grades according to the natural break jenks method by using ArcGIS (Figure 5). As can be seen from Figure 5, in 2000, the areas with high GUECL were mostly distributed in the upper and middle reaches, with the value of 0.69–1.50. The efficiency values of some areas in Henan Province and Shandong Province in the lower reaches were also high, while areas with low efficiency values were mainly located in Shanxi Province, with the efficiency value of 0.06–0.19. In 2006, the number of areas with high GUECL decreased, mainly distributed in the upper reaches of Qinghai Province. The efficiency values of Hainan Tibetan Autonomous Prefecture and Tibetan Autonomous Prefecture of Golog were the highest, ranging from 1.30 to 1.85, and the GUECL value in some areas of Gansu Province and Shanxi Province were relatively low, ranging from 0.18 to 0.36. In 2012, areas with high GUECL were mainly distributed in the upstream of Qinghai Province and Inner Mongolia Autonomous Region; the efficiency values of the middle reaches of Shaanxi Province and the lower reaches of Shandong Province were also relatively high; Tibetan Autonomous Prefecture of Golog, Huangnan Tibetan Autonomous Prefecture, Alxa League, Jiyuan City, Haixi Mongolian Autonomous Prefecture, Yushu Tibetan Autonomous Prefecture and Wuhai City had the highest efficiency values, which ranged from 1.12 to 1.52; while areas with low GUECL were mostly located in Shanxi Province, ranging from 0.26 to 0.41. In 2018, areas with high GUECL were mainly distributed in Huangnan Tibetan Autonomous Prefecture and Tibetan Autonomous Prefecture of Golog of upper Qinghai Province and Zhongwei City of Ningxia Hui Autonomous Region, while the GUECL value was still low in parts of Gansu, Inner Mongolia, Shanxi and Shandong provinces.

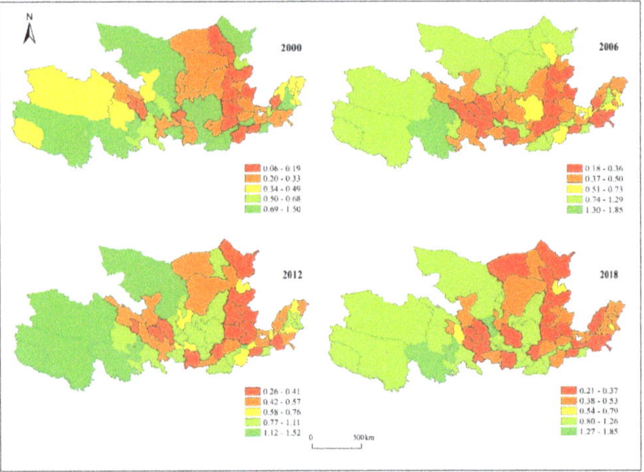

**Figure 5.** Spatial pattern of GUECL in the Yellow River Basin.

To reveal the overall trend of the spatial pattern change of the GUECL in the Yellow River Basin, the trend analysis tool of the statistical analysis module of ArcGIS software was used to generate the trend chart of GUECL in the Yellow River Basin in 2000, 2006, 2012 and 2018. As can be seen from Figure 6, the spatial projections of GUECL in 2000, 2006, 2012 and 2018 were relatively similar, showing a downward trend in the east-west direction and a U-shaped change trend in the south-north direction. With different changes, it indicated that the regional difference in the south-north direction occupied a dominant position. In 2000, the change of the south-north direction was relatively smooth, and in 2006, compared with 2000, the change of the trend line increased, indicating an increase in the difference. In 2012 and 2018, the spatial projection of GUECL increased significantly from west to east, while the south-north U-shaped trend decreased. This indicated an increase in the east-west spatial difference while the south-north difference existed.

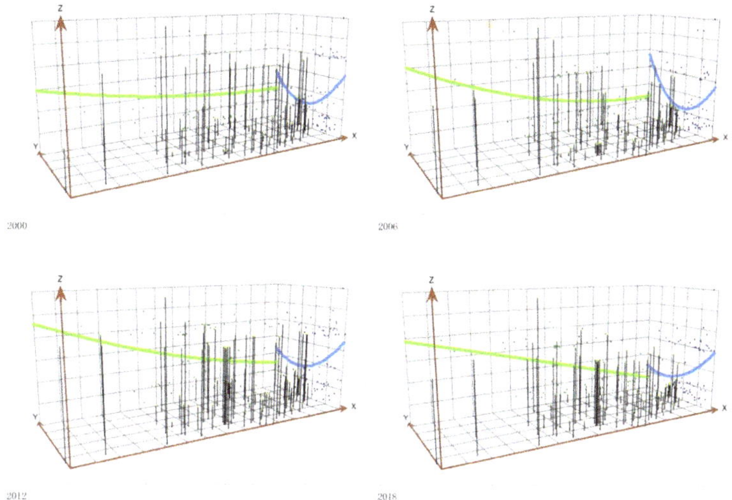


**Figure 6.** Trends in GUECL in the Yellow River Basin.

*3.3. GUECL Spatial Correlation Analysis*

### 3.3.1. GUECL Global Spatial Autocorrelation Analysis

The global autocorrelation Moran's I index of GUECL at four evaluation points (Table 3) was calculated by the GeoDa software, and 999 substitutions were selected. Except for 2000, the Moran's I index of GUECL in the Yellow River Basin was positive, p value was far less than 0.05, and z value was more than 1.96, showing that under the 5% significant level, the GUECL in the Yellow River basin as a whole had obvious spatial agglomeration and distribution characteristics in 2006, 2012 and 2018. In general, Moran's I index decreased first and then increased, from 0.2025 in 2006 to 0.1601 in 2012, and increased to 0.2226 in 2018. The spatial agglomeration and distribution characteristics were obvious.

**Table 3.** GUECL Moran's I value and significance test results in the Yellow River Basin.

| Year | I | z-Value | p-Value |
|------|------|------|------|
| 2000 | −0.0086 | −0.1176 | 0.4250 |
| 2006 | 0.2025 | 3.7902 | 0.0020 |
| 2012 | 0.1601 | 3.2397 | 0.0040 |
| 2018 | 0.2226 | 4.3117 | 0.0010 |

### 3.3.2. GUECL Local Spatial Autocorrelation Analysis

In order to further analyze the spatial differences of GUECL in the Yellow River Basin, a local spatial autocorrelation analysis was conducted to obtain the local indicators of spatial association (LISA) agglomeration diagram of GUECL in the relevant years (Figure 7).

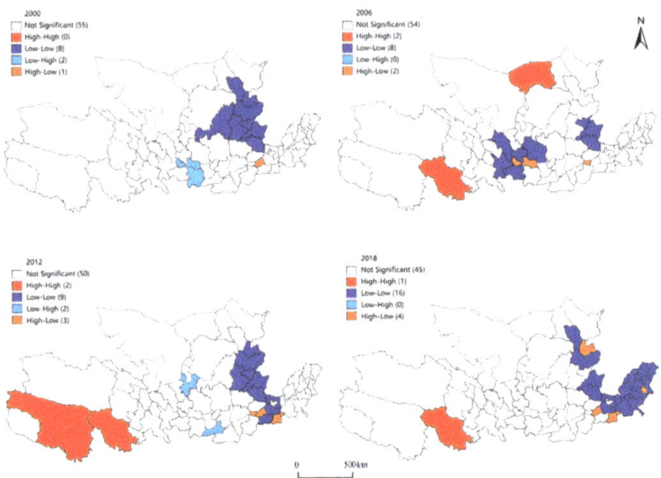

**Figure 7.** Local indicators of spatial association (LISA) aggregation map of GUECL in the Yellow River Basin.

As can be seen from Figure 7, in 2000, 2006, 2012 and 2018, most evaluation units in the Yellow River Basin had no obvious agglomeration characteristics. The number of evaluation units presenting low-low type agglomeration was the largest and increased significantly with time, namely 8, 9 and 16, respectively, indicating that GUECL clustering in the Yellow River basin was mainly of low-low type agglomeration and the regional agglomeration phenomenon increased gradually. In 2000, the low-low type agglomeration areas were mainly distributed in the middle reaches, such as Shuozhou, Xinzhou, Lvliang, Taiyuan, Jinzhong, Changzhi and Yulin in Shaanxi Province. In 2006, the low-low type agglomeration shifted to Tianshui, Dingxi, Baiyin, Guyuan and Qingyang of Gansu Province in the upper reaches of the Yellow River Basin, which may be related to the resource endowment of Gansu Province. In 2012, low-low type agglomeration was again distributed in Shuozhou, Xinzhou, Lvliang, Taiyuan, Jinzhong and Changzhi in Shanxi Province; Yulin in Shaanxi Province withdrew from this type of area, and Anyang, Xinxiang and Zhengzhou in Henan Province entered this type agglomeration. In 2018, low-low type agglomeration moved to the lower reaches of the Yellow River, and all areas flowing through the Yellow River in Shandong Province except Laiwu City were low-low type agglomeration. This might be due to the rapid economic development in the lower reaches of the Yellow River Basin, but the impact on the environment was neglected, and the excessive application of pesticides and fertilizers brought more unexpected output. The increase of low-low type agglomeration indicates that the low efficiency of GUECL may have an infectious effect. The number of evaluation units of high-low type agglomeration is also increasing, but its spatial distribution is more scattered. The GUECL of Jiaozuo, Jiyuan and Kaifeng in Henan Province is higher, which gradually forms a gap with the surrounding areas. The quantity fluctuation of high-high type agglomeration mainly happens in the upper reaches, and the quantity of evaluation units of low-high type agglomeration is relatively small. Pingliang in Gansu Province has changed from low-high type agglomeration in 2000 to high-low type agglomeration in 2006, which indicates that it is improving its own GUECL.

## 4. Discussion

### 4.1. Driving Factors of GUECL Pattern Change in the Yellow River Basin

The concept of green development has not been determined in detail since it was put forward. However, international organizations and scholars from various countries have reached a basic consensus when defining green development: compared with the traditional extensive growth mode, green development is a sustainable development path to realize human society and nature on the basis of saving resources and protecting regional ecological environment, and advocates that the harmony between human and nature's harmonious coexistence will also be the main direction of future economic transformation of countries and regions. The concept of green development is not only an evaluation and analysis of the current development of the green economy but also a scientific approach to improving the efficiency of green development based on understanding the existing green development level [53]. This paper studies the green utilization efficiency of cultivated land. The green utilization of cultivated land is to implement the concept of green development, pay more attention to the negative effect of "unexpected" output on production, strengthen the rational allocation of input factors in the process of cultivated land utilization and develop a green and sustainable cultivated land utilization mode with low chemical fertilizer and pesticide inputs. Based on the measurement of GUECL in 65 cities (regions, autonomous prefectures) through which the Yellow River flows, this paper attempts to analyze the driving mechanism of GUECL's spatial and temporal pattern differences, so as to provide reference for guiding farmers to make rational and efficient use of cultivated land, realizing green utilization of cultivated land in the Yellow River Basin and promoting agriculture green development. At present, many scholars have carried out extensive and in-depth research on utilization efficiency of cultivated land, which has played a good role in promoting the efficient use of cultivated land in China, but the study of GUECL in the Yellow River Basin has not been reported yet. From the comprehensive evaluation results and the actual development of the Yellow River Basin, the spatial and temporal pattern and its change of GUECL are the result of the comprehensive effect of internal factors (main characteristics of farmers, family characteristics, etc.) and external factors (nature, economy, policy, etc.).

### 4.1.1. Internal Factors

The influence of the main characteristics of farmers on GUECL is mainly reflected in the age and education level of farmers. Generally speaking, farmers with older age and a lower education level are less physically able to engage in farming, less able to accept new things and technologies [54], and less able to grasp green production knowledge and skills such as the scientific use of fertilizers and pesticides, so GUECL also decreases. However, other studies have shown that in villages with higher nonagricultural employment level, although the physical strength of elderly farmers is somewhat reduced, they have rich farming experience and better understanding of how inputs can increase outputs, so they can make more effective use of cultivated land than the young labor force [29]. At present, farmers in the middle and lower reaches of the Yellow River Basin have a higher education level than those in the upper reaches, and a stronger ability to accept new things and bear risks. Therefore, the spatial and temporal pattern of GUECL is different to a certain extent.

With regard to the family characteristics of farmers, some studies believe that the higher the proportion of agricultural income in the total household income, the higher the utilization efficiency of cultivated land [41]. Other studies have shown that families mainly engaged in planting industry, which took land as the source of family income and basic living security. They are highly dependent on land and have high expectations. Therefore, a lot of means of production such as chemical fertilizers, pesticides and agricultural film will be invested in the limited land in order to obtain a greater income [55]. However, excessive application of chemical fertilizers and pesticides will lead to agricultural non-point source pollution and affect the green utilization of cultivated land. In the upper reaches of the Yellow River Basin, agricultural income accounts for a large proportion of the total income of farm households, and the family's livelihood is highly dependent on cultivated land resources. The secondary and tertiary industries are developed in the lower reaches developed areas of the Yellow River Basin. There are fewer people in the household working in agriculture and it is no longer the primary source of livelihood. For farmers mainly engaged in agricultural production activities, they may choose intensive cultivation and invest a large amount of chemicals in pursuit of greater benefits, ignoring the effective green utilization of cultivated land. In order to become a new type of agricultural operator, smallholder families must change their original farming concepts and methods, and move towards large-scale, regional and standardized farming, which also lays a foundation for the green utilization of cultivated land to some extent.

Farmers' cognition is also an important factor affecting GUECL. As for the cognition of the current ecological environment, some farmers have a vague environmental consciousness, ignoring the harm of their behavior to the environment, and blindly pursue output, which leads to the appearance of agricultural nonpoint-source pollution. Moreover, some farmers are aware of the harm of chemical fertilizers and pesticides, which are rarely taken into account due to the short-term quick effect, more incomes from increased crop yields than the expenditures on chemical fertilizers and pesticides [55], their weak awareness of using organic fertilizers and biological pesticides, and the external characteristics of chemical fertilizer and pesticide pollution, resulting in the increasing agricultural nonpoint-source pollution, and the severe challenges faced by the green utilization of cultivated land. If farmers are able to recognize that ecological environmental problems caused by agricultural production and understand its harm, it is possible to change extensive farming practices and adopt more green and sustainable farming practices. In addition, farmers' cognition of relevant policies will also have an impact on farming behavior. Laws and regulations can guide and bind their farming behavior, but if farmers do not understand the relevant systems and policies, their awareness of the green utilization of cultivated land will not be strong, thus affecting the GUECL.

### 4.1.2. External Factors

First, natural factors. Generally speaking, the plain area has flat terrain, superior natural conditions such as light, water and heat, and high grain productivity [10]. Compared with other mountainous and plateau areas, the Ningxia Plain, Hetao Plain and North China Plain through which the Yellow River flows have superior natural conditions, which are easy to carry out agricultural cultivation, good cultivated land quality and high multiple cropping index, so the conditions for green utilization of cultivated land are relatively well. However, the desertification and salinization of cultivated land in Ningxia, Inner Mongolia, Shaanxi and other regions will seriously affect the quality of cultivated land and the improvement of grain yield per unit area in the basin, while in areas with serious soil erosion, the arid soil will also affect agricultural production [40]. In addition, water resources will also affect the GUECL. The yield of agricultural products depends on the availability of sufficient irrigation water, and the quality of agricultural products depends on the quality of agricultural irrigation water. One study shows that, the agriculture, industry and urban residential areas accounted for influence of 40%, 26%, and 16% on water amount reduce, respectively [56]. In fact, it is short of water resources in the Yellow River Basin. With the rapid development of economy, agricultural water can not be guaranteed, and is occupied during the water shortage period, resulting in crop yield reduction. Compared with other regions, the lower reaches of the Yellow River is an "above ground river" with superior self-flow irrigation conditions. Climate change is obvious in the Yellow River Basin. The west of Lanzhou belongs to the Tibetan Plateau monsoon region, and the rest areas are temperate and subtropical monsoon areas. The annual precipitation of the Yellow River Basin decreases gradually from southeast to northwest. It is rainy in the southeast and arid in the northwest, and the precipitation distribution is very uneven, which also affects GUECL and its spatio-temporal differences to a certain extent.

Second, economic factors. With the improvement of the economic level, more funds may be invested in the utilization of cultivated land, and the infrastructure of cultivated land utilization will be further improved, which will provide good production conditions for the green utilization of cultivated land. In addition, the input in machinery and technology will be increased, too. Therefore, mechanical input will affect GUECL. The development of agricultural modernization can not be separated from scale operation. Land circulation and land consolidation can make the cultivated land centralized and connected to facilitate the use of machinery, and the reduction in production costs, as well as the improvement of scale management level and GUECL. The total power of agricultural machinery per unit area of cultivated land can reflect agricultural mechanization level of a region. However, it is not that the more agricultural machinery per unit area of cultivated land has, the higher the utilization efficiency of cultivated land. Studies have shown that the per capita cultivated land in Poyang Lake Ecological Economic Zone is insufficient, and the plots are fragmented, which makes it difficult to carry out large-scale operation. Therefore, they are negatively correlated to some extent [57]. For the continuous progress of agricultural technology, it is possible to transform the natural environmental conditions and promote the maximum utilization of cultivated land resources. Agricultural technicians can also help farmers to improve their farming methods, and develop new technologies to prevent farmers from blindly using cultivated land, as well as ensure the yield and quality of agricultural products [58]. While improving infrastructure, high input will inevitably lead to an increase of the undesirable output in the utilization of cultivated land [36], which will affect the green utilization of cultivated land. In addition, with the rapid development of economy and the continuous improvement of urbanization level, a large number of farmers turn to cities and towns to engage in secondary and tertiary industries, which results in the abandonment of farmland, and to a certain extent limits the improvement of GUECL.

Third, policy factors. The introduction of the national food security policy has greatly increased China's grain supply, but also led to the occurrence of agricultural pollution, such as the excessive application of chemical fertilizers and pesticides by farmers simply to increase grain output. In recent years, the Central Document No.1 and the Central Rural Work Conference have all made arrangements centering on the agriculture green development. Under the background of ecological civilization construction, farmers pay more attention to the protection of the environment in the process of cultivated land utilization. As agriculture, farmers and rural issues continue to receive attention, a series of agricultural subsidies and protection policies have been introduced in response to the actual conditions in various regions [57], which not only reduces the agricultural input cost of farmers, but also provides a policy guarantee for farmers to change traditional farming methods and adopt new agricultural technologies, so as to promote the green utilization of cultivated land and improve GUECL significantly. Compared with the middle and lower reaches of the Yellow River, the economic development level of Qinghai Province in the upper reaches of the Yellow River is relatively backward. In recent years, with the support of national funds and policies, agriculture in some areas of the upper and middle reaches of the Yellow River has also developed rapidly.

### 4.2. Policy Suggestions for Improving GUECL in the Yellow River Basin

In order to improve GUECL and promote the ecological protection and high-quality development in the Yellow River Basin, the following suggestions are put forward: (1) We will work out a scientific national land spatial plan for the Yellow River Basin. With the six concepts of "innovation, coordination, green, openness, sharing and security", we actively explore and carry out the research and compilation practice of national land spatial plan. Considering the needs of agricultural production and ecological protection, we will coordinate the relationship between national land spatial development and protection in combination with the actual situation in the upper, middle and lower reaches of the Yellow River Basin and ecologically important area in the upper reaches, and promote the restoration and construction of cultivated land ecological protection. Soil erosion in the middle reaches is serious, where soil and water conservation and pollution control should be the focus. The pollution of the cultivated land in the lower reaches is very serious, where ecological protection is worthy of attention. In order to improve the carrying capacity of national land space through efficient use of resources, we will carry out comprehensive national land spatial improvement and ecological restoration work, and build a green ecological barrier to realize the high-quality green utilization of cultivated land. (2) We will accelerate the construction of a three-in-one pattern of quantity, quality and ecology for the utilization and protection of cultivated land at the basin level. With the implementation of the strictest cultivated land protection system, we will firmly hold the red line of 120 million hectares of cultivated land to realize the stability of the quantity of cultivated land resources. As the basis of agricultural production activities, the stability of cultivated land resources is directly related to the country's food security and social stability. We will improve the cultivated land protection and compensation mechanism, and control the cultivated land occupied for construction, as well as ensure that the quality of cultivated land does not decline to solve practical problems in the protection of cultivated land. We will carry out ecological land renovation, promote the mechanization and large-scale operation of cultivated land while protecting cultivated land resources, and guide the green transition of cultivated land utilization with the concept of green development. We will improve the cultivated land rehabilitation system and the ecological compensation mechanism to ensure national food security. We will vigorously promote the technology of soil formula fertilization to improve the utilization rate of chemical fertilizers and reduce non-point source pollution caused by fertilization. Finally, we will achieve the purpose of green utilization of cultivated land to improve the efficiency of utilization by improving the quality and productivity of cultivated land as well as the ecological environment. (3) We will promote agriculture green development. At present, ecological civilization and green development have risen to national strategy, and agriculture green development has become the main direction to promote the structural reform on the supply side of agriculture. Facing many challenges, it is necessary to accelerate institutional innovation and promote agriculture green development. In view of the upper reaches of the Yellow River Basin where the ecological status is important but the ecological environment is fragile, green agriculture should be developed, industrial development and ecological protection should be promoted as a whole, and ecological environmental advantages should be transformed into industrial advantages. We will improve the innovation drive and incentive constraint mechanism in the agriculture green development, standardize the production behavior of agricultural means of production, solve pollution caused by input of means of production in the production of agricultural products from the source and promote agriculture green development through scientific and technological innovation.

*4.3. Deficiency*

In view of the author's research level and data availability, there are still some deficiencies in this paper. First, the selection of indicators may not be comprehensive enough. The utilization of cultivated land is a long-term and complex process involving all aspects, so the index system may not fully reflect the connotation of the green utilization of cultivated land. Secondly, it is not deep and comprehensive to explore the reasons for the spatial and temporal pattern changes of GUECL in the Yellow River Basin with a preliminary analysis from the qualitative point of view. In the next step, more in-depth studies can be carried out from the perspective of quantitative and qualitative combination to obtain more practical results.

## 5. Conclusions

Based on Super-SBM model, this paper calculated the GUECL of 65 evaluation units in the Yellow River Basin in 2000, 2006, 2012 and 2018 with the consideration of undesirable outputs. Based on this, the change characteristics of GUECL spatial and temporal pattern and its influencing factors were preliminarily analyzed. The following conclusions were drawn:

(1)  On the whole, GUECL in the Yellow River Basin is not high. At four time points in 2000, 2006, 2012 and 2018, GUECL in the Yellow River Basin generally presents a trend of "rising first and then falling". The GUECL presents an order of the upper reaches > the lower reaches > the middle reaches. Compared with the upper and lower reaches, the GUECL in the middle reaches has a better upward trend. The spatial variation trend shows a decrease from west to east, and a U-shaped change in the south-north direction.

(2)  Except for the year 2000, GUECL in the Yellow River Basin was positively correlated in 2006, 2012 and 2018, showing spatial agglomeration and distribution characteristics in the overall situation. The local spatial autocorrelation is mainly low-low type agglomeration, and the regional agglomeration phenomenon has gradually strengthened. The number of evaluation

units of high-low type agglomeration increased, and there was a fluctuation of high-high type agglomeration and low-high type agglomeration changes.

(3) Factors that influence the GUECL spatial and temporal pattern changes in the Yellow River Basin can be divided into internal factors and external factors. The former mainly includes the main characteristics of farmers, family characteristics and farmers' cognition, while the latter is reflected in natural, social and policy factors. Internal and external factors have a comprehensive effect on the GUECL in the Yellow River Basin.

**Author Contributions:** Conceptualization, X.L.; methodology, Y.Q.; software, P.S.; formal analysis, W.P.; data curation, W.Y. and W.P.; writing-original draft preparation, X.L. and Y.Q.; writing-review and editing, X.L.; visualization, Y.Q. and P.S.; validation, P.S.; project administration, W.Y.; and funding acquisition, X.L. All authors have read and agreed to the published version of the manuscript.

**Funding:** This research was funded by the National Natural Science Foundation of China (41671176, 42071226), LiaoNing Revitalization Talents Program (XLYC1807060) and Youth Innovation and Technology Plan of Shandong Colleges and Universities (2020RWG010).

**Conflicts of Interest:** We declare that we have no conflict of interest, there is no professional or other personal interest of any nature or kind in any product, service and/or company that could be construed as influencing the position presented in, or the review of, the manuscript entitled, "Exploring the spatiotemporal differentiation of green utilization efficiency of cultivated land in the Yellow River Basin, China". The funders had no role in the design of the study; in the collection, analyses, or interpretation of data; in the writing of the manuscript, or in the decision to publish the results.

## References

1.	Clark, M.; Tilman, D. Comparative analysis of environmental impacts of agricultural production systems, agricultural input efficiency, and food choice. *Environ. Res. Lett.* **2017**, *12*, 064016. [CrossRef]

2.	Allouche, J. The sustainability and resilience of global water and food systems: Political analysis of the interplay between security, resource scarcity, political systems and global trade. *Food Policy* **2011**, *36* (Suppl. S1), s3–s8. [CrossRef]

3.	Tiwari, P.C.; Joshi, B. Natural and socio-economic factors affecting food security in the Himalayas. *Food Secur.* **2012**, *4*, 195–207. [CrossRef]

4.	Sarkar, D.; Kar, S.K.; Chattopadhyay, A.; Shikha; Rakshit, A.; Tripathi, V.K.; Dubey, P.K.; Abhilash, P.C. Low input sustainable agriculture: A viable climate-smart option for boosting food production in a warming world. *Ecol. Indic.* **2020**, *8*, 106412. [CrossRef]

5.	Shen, J.B.; Cui, Z.L.; Miao, Y.X.; Mi, G.H.; Zhang, H.Y.; Fan, M.S.; Zhang, C.C.; Jiang, R.F.; Zhang, W.F.; Li, H.G.; et al. Transforming agriculture in China: From solely high yield to both high yield and high resource use efficiency. *Glob. Food Secur.* **2013**, *2*, 1–8. [CrossRef]

6.	Liu, Y.S.; Wang, J.Y.; Long, H.L. Analysis of arable land loss and its impact on rural sustainability in Southern Jiangsu Province of China. *J. Environ. Manag.* **2010**, *91*, 646–653. [CrossRef]

7.	Zhang, Y.S.; Lu, X.; Liu, B.Y.; Wu, D.T. Impacts of urbanization and associated factors on ecosystem services in the Beijing-Tianjin-Hebei urban agglomeration, China: Implications for Land Use Policy. *Sustainability* **2018**, *10*, 4334. [CrossRef]

8.	Lu, X.; Peng, W.L.; Huang, X.J.; Fu, Q.Q.; Zhang, Q.J. Homestead Management in China from the "Separation of Two Rights" to the "Separation of Three Rights": Visualization and analysis of hot topics and trends by mapping knowledge domains of academic papers in China National Knowledge Infrastructure (CNKI). *Land Use Policy* **2020**, *97*, 104670. [CrossRef]

9.	Xie, H.L.; Wang, P.; Yao, G.R. Exploring the dynamic mechanisms of farmland abandonment based on a spatially explicit economic model for environmental sustainability. *Sustainability* **2014**, *6*, 1260–1282. [CrossRef]

10.	Zhang, L.X.; Zhu, D.L.; Xie, B.P.; Du, T.; Wang, X. Spatiotemporal pattern evolvement and driving factors of cultivated land utilization efficiency of the major grain producing area in China. *Resour. Sci.* **2017**, *39*, 608–619. (In Chinese)

11.	Yang, C.H.; Wu, L.; Lin, H.L. Analysis of total-factor cultivated land efficiency in China's agriculture. *Agric. Econ. (AGRICECON)* **2010**, *56*, 231–242. [CrossRef]

12. Chen, R.S.; Ye, C.; Cai, Y.L.; Xing, X.S.; Chen, Q. The impact of rural out-migration on land use transition in China: Past, present and trend. *Land Use Policy* **2014**, *40*, 101–110. [CrossRef]

13. Long, H.L.; Qu, Y. Land use transitions and land management: A mutual feedback perspective. *Land Use Policy* **2018**, *74*, 111–120. [CrossRef]

14. Ge, D.Z.; Long, H.L.; Zhang, Y.N.; Ma, L.; Li, T.T. Farmland transition and its influences on grain production in China. *Land Use Policy* **2018**, *70*, 94–105. [CrossRef]

15. Lu, X.; Shi, Y.Y.; Chen, C.L.; Yu, M. Monitoring cropland transition and its impact on ecosystem services value in developed regions of China: A case study of Jiangsu Province. *Land Use Policy* **2017**, *69*, 25–40. [CrossRef]

16. Gong, Y.; Li, J.; Li, Y. Spatiotemporal characteristics and driving mechanisms of arable land in the Beijing-Tianjin-Hebei region during 1990–2015. *Socio-Econ. Plan. Sci.* **2019**, *70*, 100720. [CrossRef]

17. Zhang, Y.S.; Lu, X.; Liu, B.Y.; Wu, D.T.; Fu, G.; Zhao, Y.T.; Sun, P.L. Spatial relationships between ecosystem services and socioecological drivers across a large-scale region: A case study in the Yellow River Basin. *Sci. Total Environ.* **2020**. [CrossRef]

18. Zitti, M.; Ferrara, C.; Perini, L.; Carlucci, M.; Salvati, L. Long-term urban growth and land use efficiency in southern Europe: Implications for sustainable land management. *Sustainability* **2015**, *7*, 3359–3385. [CrossRef]

19. Masini, E.; Tomao, A.; Barbati, A.; Corona, P.; Serra, P.; Salvati, L. Urban growth, land-use efficiency and local socioeconomic context: A comparative analysis of 417 metropolitan regions in Europe. *Environ. Manag.* **2019**, *63*, 322–337. [CrossRef]

20. Pili, S.; Grigoriadis, E.; Carlucci, M.; Clemente, M.; Salvati, L. Towards sustainable growth? A multi-criteria assessment of (changing) urban forms. *Ecol. Indic.* **2017**, *76*, 71–80. [CrossRef]

21. Van Zanten, H.H.E.; Mollenhorst, H.; Klootwijk, C.W.; van Middelaar, C.E.; de Boer, I.J.M. Global food supply: Land use efficiency of livestock systems. *Int. J. Life Cycle Assess.* **2016**, *21*, 747–758. [CrossRef]

22. Seufert, V.; Ramankutty, N.; Foley, J.A. Comparing the yields of organic and conventional agriculture. *Nature* **2012**, *485*, 229–232. [CrossRef] [PubMed]

23. Agegnehu, G.; Ghizaw, A.; Sinebo, W. Yield performance and land-use efficiency of barley and faba bean mixed cropping in Ethiopian highlands. *Eur. J. Agron.* **2006**, *25*, 202–207. [CrossRef]

24. Umetsu, C.; Lekprichakul, T.; Chakravorty, U. Efficiency and technical change in the Philippine rice sector: A malmquist total factor productivity analysis. *Am. J. Agric. Econ.* **2003**, *85*, 943–963. [CrossRef]

25. Skevas, T.; Stefanou, S.E.; Lansink, A.O. Pesticide use, environmental spillovers and efficiency: A DEA risk-adjusted efficiency approach applied to Dutch arable farming. *Eur. J. Oper. Res.* **2014**, *237*, 658–664. [CrossRef]

26. Ferreira, M.D.P.; Féres, J.G. Farm size and Land use efficiency in the Brazilian Amazon. *Land Use Policy* **2020**, *99*, 104901. [CrossRef]

27. Barath, L.; Ferto, I. Heterogeneous technology, scale of land use and technical efficiency: The case of Hungarian crop farms. *Land Use Policy* **2015**, *42*, 141–150. [CrossRef]

28. Bonfiglio, A.; Arzeni, A.; Bodini, A. Assessing eco-efficiency of arable farms in rural areas. *Agric. Syst.* **2017**, *151*, 114–125. [CrossRef]

29. Oumarou, B.; Zhou, H.Q.; Muhammad, A.R.; Sidra, G. Analysis on technical efficiency of rice farms and its influencing factorsin south-western of Niger. *J. Northeast Agric. Univ.* **2016**, *23*, 67–77.

30. Akram, M.W.; Akram, N.; Wang, H.S.; Andleeb, S.; Rehman, K.U.; Kashif, U.; Mehmood, A. Impact of land use rights on the investment and efficiency of organic farming. *Sustainability* **2019**, *11*, 7148. [CrossRef]

31. Emi, K.; Steven, C.F.; Harold, P.C.; Brian, J.S.; Austin, H.; Jeffrey, S.; Stephen, O.G.; William, J.J. Effect of inter cropping hybrid poplar and switch grass on biomass yield, forage quality, and land use efficiency for bioenergy production. *Biomass Bioenergy* **2018**, *111*, 31–38.

32. Lin, H.C.; Hülsbergen, K.J. A new method for analyzing agricultural land-use efficiency, and its application in organic and conventional farming systems in southern Germany. *Eur. J. Agron.* **2017**, *83*, 15–27. [CrossRef]

33. Buitenhuis, Y.; Candel, J.J.L.; Termeer, C.J.A.M.; Feindt, P. Does the Common Agricultural Policy enhance farming systems' resilience? Applying the Resilience Assessment Tool (ResAT) to a farming system case study in the Netherlands. *J. Rural Stud.* **2020**. [CrossRef]

34. Kuang, B.; Lu, X.; Zhou, M.; Chen, D.L. Provincial cultivated land use efficiency in China: Empirical analysis based on the SBM-DEA model with carbon emissions considered. *Technol. Forecast. Soc. Chang.* **2020**, *151*, 119874. [CrossRef]

35. Li, Q.; Gu, G.F.; Yao, L.; Wang, J.K.; Chen, Y.Y. The changes of cultivated land use efficiency in Heilongjiang Land Reclamation Area. *J. Arid Land Resour. Environ.* **2016**, *30*, 30–35. (In Chinese)
36. Feng, Y.G.; Peng, J.; Deng, Z.B.; Wang, J. Spatial-temporal variation of cultivated land's utilization efficiency in China based on the dual perspective of non-point source pollution and carbon emission. *China Popul. Resour. Environ.* **2015**, *25*, 18–25. (In Chinese)
37. Han, H.; Zhang, X. Exploring environmental efficiency and total factor productivity of cultivated land use in China. *Total Environ.* **2020**, *726*, 138434. [CrossRef]
38. Xie, H.L.; Zhang, Y.W.; Yongrok, C. Measuring the cultivated land use efficiency of the main grain-producing areas in China under the constraints of carbon emissions and agricultural nonpoint source pollution. *Sustainability* **2018**, *10*, 1932. [CrossRef]
39. Wang, L.J.; Li, H. Cultivated land use efficiency and the regional characteristics of its influencing factors in China: By using a panel data of 281 prefectural cities and the stochastic frontier production function. *Geogr. Res.* **2014**, *33*, 1995–2004. (In Chinese)
40. Xie, H.L.; Chen, Q.R.; Wang, W.; He, Y.F. Analyzing the green efficiency of arable land use in China. *Technol. Forecast. Soc. Chang.* **2018**, *133*, 15–28. [CrossRef]
41. Xu, H.Z.; Guo, Y.Y.; Wu, G.C. The effect of farmer differentiation on the efficiency of cultivated land use. *Chin. Rural Econ.* **2012**, *31*, 39–47. (In Chinese)
42. Yang, J.; Li, Z. Transfer-in and technical efficiency of farmland use in the context of households labor division: Case of rural households in Ganfu Plain area. *China Land Sci.* **2015**, *29*, 50–57. (In Chinese)
43. Zhang, Y.J.; Chen, Y.; Liu, Y.; Lu, Z. The influence of farmers' land values on the cultivated land utilization efficiency. *J. Arid Land Resour. Environ.* **2017**, *31*, 19–25. (In Chinese)
44. Chen, Y.J.; Wu, K.; Lu, B.; Yuan, Z.; Xu, Y.X. The present situation and the agricultural structure adjustment of agricultural production in the Huanghe River Basin. *Prog. Geogr.* **2005**, *24*, 106–113. (In Chinese)
45. Henry, M.; Schraven, D.; Bocken, N.; Frenken, K.; Hekkert, M.; Kirchherr, J. The battle of the buzzwords: A comparative review of the circular economy and the sharing economy concepts. *Environ. Innov. Soc. Transit.* **2021**, *38*, 1–21. [CrossRef]
46. Han, H.B.; Zhang, X.Y. Static and dynamic cultivated land use efficiency in China: Aminimum distance to strong efficient frontier approach. *J. Clean. Prod.* **2020**, *246*, 119002. [CrossRef]
47. Lu, X.H.; Kuang, B.; Li, J.; Han, J.; Zhang, Z. Dynamic evolution of regional discrepancies in carbon emissions from agricultural land utilization: Evidence from Chinese provincial data. *Sustainability* **2018**, *10*, 552. [CrossRef]
48. Wu, X.Q.; Wang, Y.P.; He, L.M.; Lu, G.F. Agricultural eco-efficiency evaluation based on AHP and DEA model: A case of Wuxi city. *Resour. Environ. Yangtze Basin* **2012**, *21*, 714–719. (In Chinese)
49. Tone, K.A. Slacks-based measure of efficiency in data envelopment analysis. *Eur. J. Oper. Res.* **2011**, *130*, 498–509. [CrossRef]
50. Tone, K.A. Slacks-based measure of super-efficiency in data envelopment analysis. *Eur. J. Oper. Res.* **2012**, *143*, 32–41. [CrossRef]
51. Zhao, P.J.; Zeng, L.E.; Lu, H.Y.; Zhou, Y.; Hu, H.Y.; Wei, X.Y. Green economic efficiency and its influencing factors in China from 2008 to 2017: Based on the super-SBM model with undesirable outputs and spatial Dubin model. *Total Environ.* **2020**, *741*, 140026. [CrossRef] [PubMed]
52. Liu, J.M.; Ma, J.G. Problems and solutions of sloping farmland in the Yellow River Valley of Shanxi Province. *China Land Sci.* **2003**, *2*, 59–64. (In Chinese)
53. Wu, J.; Lu, W.; Li, M. A DEA-based improvement of China's green development from the perspective of resource reallocation. *Total Environ.* **2020**, *717*, 137106. [CrossRef] [PubMed]
54. Khai, H.V.; Yabe, M. Technical efficiency analysis of rice production in Vietnam. *J. ISSAAS* **2011**, *17*, 135–146.
55. Wang, Y.; Zhu, Y.C.; Zhang, S.X.; Wang, Y.Q. What could promote farmers to replace chemical fertilizers with organic fertilizers? *J. Clean. Prod.* **2018**, *199*, 882–890. [CrossRef]
56. Chen, Y.P.; Fu, B.J.; Zhao, Y.; Wang, K.B.; Zhao, M.M.; Ma, J.F.; Wu, J.H.; Xu, C.; Liu, W.G.; Wang, H. Sustainable development in the Yellow River Basin: Issues and strategies. *J. Clean. Prod.* **2020**, *263*, 121223. [CrossRef]
57. Xie, H.L.; Zhang, D.B.; Wang, W.; Xie, X.; Wu, Q.; Cheng, L.J. Analysis of cultivated land use efficiency and its influencing factors for Poyang Lake Ecological Economic Zone. *Res. Soil Water Conserv.* **2016**, *23*, 214–221. (In Chinese)

58.    Souza, G.; Gomes, E.G. Improving agricultural economic efficiency in Brazil. *Int. Trans. Oper. Res.* **2015**, *22*, 329–337. [CrossRef]

**Publisher's Note:** MDPI stays neutral with regard to jurisdictional claims in published maps and institutional affiliations.

*Article*

# Cultivated Land Use Layout Adjustment Based on Crop Planting Suitability: A Case Study of Typical Counties in Northeast China

**Ge Song * and Hongmei Zhang**

School of Humanities and Law, Northeastern University, Shenyang 110169, China; 1710017@stu.neu.edu.cn
* Correspondence: songge@wfxy.neu.edu.cn; Tel.: +86-159-4038-0677

**Citation:** Song, G.; Zhang, H.
Cultivated Land Use Layout
Adjustment Based on Crop Planting
Suitability: A Case Study of Typical
Counties in Northeast China. *Land*
**2021**, *10*, 107. https://doi.org/
10.3390/land10020107

Academic Editors: Hualou Long,
Xiangbin Kong, Shougeng Hu and
Yurui Li

Received: 23 December 2020
Accepted: 21 January 2021
Published: 23 January 2021

**Publisher's Note:** MDPI stays neutral
with regard to jurisdictional claims in
published maps and institutional affiliations.

**Abstract:** Cultivated land use layout adjustment (CLULA) based on crop planting suitability is the refinement and deepening of land use transformation, which is of great significance for optimizing the allocation of cultivated land resources and ensuring food security. At present, people rarely consider the land suitability of crops when using cultivated land, resulting in an imbalance between crop distribution and resource conditions such as water, heat, and soil, and adversely affects the ecological security and utilization efficiency of cultivated land. To alleviate China's grain planting structural imbalance and efficiency loss, this paper based on the planting suitability of main food crops (rice, soybean, and maize) to adjust and optimize the cultivated land use layout (CLUL) in the typical counties of the main grain production area in Northeast China, using the agent-based model for optimal land allocation (AgentLA) and GIS technology. Findings from the study show that: (1) The planting suitability of rice, soybean, and maize in the region is obviously different. Among them, the suitability level of soybean and maize is high, and that of rice is low. The current CLUL of the food crops needs to be further optimized and adjusted. (2) By optimizing the layout of rice, soybean, and maize, the planting suitability level of the food crops and the concentration level of the CLUL spatial pattern have been improved. (3) The plan for CLULA is formulated: The study area is divided into rice stable production area, maize-soybean rotation area, maize dominant area, and soybean dominant area, and town or village is identified as the implementation unit of CLULA. The plan for CLULA will be conducive to the concentrated farming of food crops according to the suitable natural conditions and management level. The research realized the optimization of spatial structure and cultivated land use patterns of different food crops integrating farming with protecting land. The significance of the study is that it provides a scientific basis and guidance for adjusting the regional planting structure and solving the problem of food structural imbalance.

**Keywords:** land use transition; cultivated land use layout adjustment (CLULA); suitability; food crops; planting structure; Northeast China

## 1. Introduction

Land use transition is the result of the comprehensive effect of regional own conditions and external environmental factors in the process of economic and social development, and it also brings direct socio-economic and environmental impacts on regional sustainability [1]. As a new approach to the comprehensive study of national/regional land use/cover change [2], land use transition refers to the changes in regional land use morphology corresponding to the transition of the socio-economic development stage over a certain period of time, including the dominant morphological transition and the recessive morphological transition [3]. The dominant morphology is presented through the land use features of quantity, structure, and spatial pattern, and the recessive morphology is presented through the land use features of quality, property rights, management mode, input, output, and function [4]. Cultivated land resources are an important part of land resources, and the core element contributing to food security and socio-economic development [5], and cultivated land use transition is an important manifestation of land use transition. The research on cultivated land use transition not only includes the exploration of regional cultivated land use morphology, transition pattern, and transition mechanism at different stages, but also needs to answer the question of what kind of cultivated land use transition should be adopted, and what measures should be taken to ensure its implementation [6]. Cultivated land use layout adjustment (CLULA) is an important part of cultivated land use transition. Based on grasping the regional cultivated land spatial distribution, planting structure characteristics (cultivated land use dominant morphology), and the spatial distribution of cultivated land use natural and socio-economic elements (cultivated land use recessive morphology) under a certain socio-economic development stage, CLULA is to reasonably match various cultivated land use spatial elements guided by certain goals and practical needs, so as to realize cultivated land use layout optimization and clear direction of cultivated land use transition. CLULA enriches the elements and connotation of cultivated land use transition, and further answers the questions of "what kind of transition" and "how to transform".

In 2019, China's urbanization reaching above 60%, which means that China has entered the mid-late stage of urbanization, and it is an important stage for digesting the accumulated contradictions during the rapid expansion of urbanization [7]. Accelerated urbanization and the subsequent increase in human activities are triggering land use transitions in China [4]. China Statistical Yearbook data shows that from 2006 to 2015, China's corn area continued to increase, with an average annual increase of over 1 million hectares, while soybean area decreased by more than 400,000 hectares annually, resulting in an imbalance in crop planting structure. From 2004 to 2015, despite the twelfth consecutive increase in China's total grain volume, the grain market showed a simultaneous increase in imports, production, and stocks. The phased supply of corn exceeds demand, while the gap between supply and demand of soybeans has expanded year by year, which has intensified the urgency of grain planting structure adjustment and cultivated land use transition. In China, the food security problem has changed from insufficient quantity to structural contradiction, which has created new food security problems. From 2015 to 2018, the Chinese government successively issued policies to take advantage of the main grain-producing area and adjust the planting structure.

In recent years, the quantitative structure of food crops in China has been appropriately adjusted under policy guidance, but the spatial distribution of regional food crops lacks scientific basis. Under the current rural land system in China, cultivated land use layout (CLUL) is increasingly affected by factors such as agricultural market price, farmers' planting preferences, and agricultural techniques, while the climate, terrain, soil conditions [8], and spatial suitability for crop cultivating are not considered adequately. The current CLUL, especially the food crops layout, has a problem of unbalanced allocation of land, water, and heat resources [9], resulting in loss of efficiency and land degradation [10], threatening the sustainable use of cultivated land [11,12]. The spatial planting structure of regional food crops needs to be further optimized and adjusted. Land suitability analysis for crops is a prerequisite to achieve optimum utilization of cultivated land resources [13,14]. It is necessary to comprehensively consider the spatial matching relationship between the natural conditions, human factors of cultivated land utilization, and actual requirements of crops to explore the planting suitability and production potential of different crops [15,16]. Cultivated land use layout adjustment (CLULA) based on crop planting suitability has become the essential path to guarantee food supply without degradation of cultivated land.

There are abundant research results on the adjustment and optimization of land use layout, which have gone through a research process from qualitative analysis to quantitative calculation, and quantitative structural configuration to spatial distribution. Most of the research is based on the consideration of land quality, ecology, function, food production, and economic benefit to optimize the land use layout [17–21], which can be divided into two aspects: Quantity optimization and space optimization of land use.

Socio-economic factors are important factors considered in land use quantity optimization, which are mainly reflected in the guiding role of regional grain production [22], input–output [23], and net income [24] in the optimization. Land evaluation is often used as one of the key processes for land use space optimization [25], including land suitability evaluation, land quality evaluation, land ecological evaluation [26], etc. Factors such as climate, topography, soil, hydrology, land management conditions, as well as socio-economic aspects, are often selected [27,28], and a wide range of methods have been developed, including Multi-criterion evaluation (MCE) [29–31], analytic hierarchy process (AHP) [32,33], Storie index (SI) [34], rule-based classification method [35,36], etc., for land evaluation. In terms of research methods, the related research has used multi-objective programming model (MOP) [37], cellular automata (CA) [38], comparative advantage index [39], agent-based model (ABM) [40], and multiple optimization algorithms [41–43] to optimize land use layout in terms of quantitative structure, spatial pattern, and benefit optimization. However, the planting suitability of food crops and the spatial structure relationship between crops, as well as the reasonable matching of natural and human factors of cultivated land use for different food crops have not received sufficient attention. CLULA based on the planting suitability of food crops is an extension and refinement of the research on cultivated land use and protection issues, taking the research deep into the essence of cultivated land use, which needs to be further explored. Thus, this paper takes the typical area of the main grain production area in Northeast China as the study area, based on the planting suitability of the main grain crops: Rice, soybean, and maize in the region, comprehensively considering the environmental conditions and requirements of crop planting such as climate, terrain, soil, and management, using AgentLA model and GIS methods to match various cultivated land use factors in space, and achieve the layout optimization and spatial relationship coordination of food crops. This research will provide a scientific basis for guiding the regional planting structure adjustment, alleviating the problem of food structural contradictions, at the same time clearing the way for the practice of regional cultivated land use transition.

## 2. Materials and Methods

### 2.1. Description of the Study Area

Keshan County, Yi'an County, and Baiquan County are the three adjacent counties situated in Heilongjiang Province, the main grain production area in Northeast China, and it is an important commodity grain base and soybean production base. It is also located in the black soil belt in the northern part of the Songnen Plain of China (Figure 1), in which the planting system is bringing in one harvest a year. The region has a temperate continental monsoon climate and receives annual precipitation of approximately 500 mm. Plains and hills are the main geomorphic types of the region. The soil types are mainly black soil and chernozem.

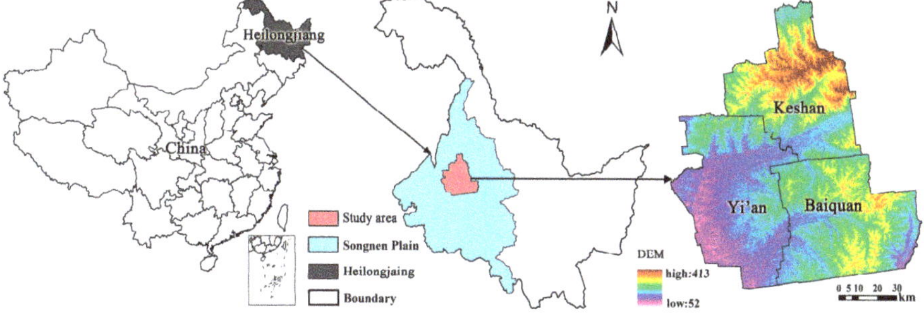


**Figure 1.** Location of the study area.

According to the remote sensing images interpretation results in the study, the total area of the study area is about 10,460 km$^2$, in which 8307.43 km$^2$ (79.42%) is covered by cultivated land in 2018. The main crops are maize, soybean, and rice, which have a total area of 8000.70 km$^2$, accounting for 96.31% of the cultivated land in the region. The ratio of the planted area of rice, soybean, and maize is about 1:8.71:8.64. The rice is concentrated in the north of central Yi'an County and northern Keshan County, the maize is mainly distributed in Yi'an County and southern Baiquan County, and the soybean is mostly distributed in the north and east of Keshan County and Baiquan County. According to the Qiqihar Economic Statistical Yearbook (1989−2019), from 2000 to 2015, the maize area increased by 308,152 hectares in the region, an increase rate of 353.17%, while the area of soybeans decreased significantly, especially from 2009 to 2015, when its area decreased by 175,363 hectares. In order to alleviate the imbalance of crop planting structure, in 2016, the Chinese government released the policies of "Guiding Opinions on the Adjustment of Maize Structure in the 'Sickle Bend' Region" and "Pilot Scheme on Exploring the Implementation of Cultivated Land Fallow and Rotation System". Keshan County, Yi'an County, and Baiquan County in the study area were listed as key pilots for cultivated land fallow and rotation in 2017 and 2018, which is very typical.

### 2.2. Data Sources and Processing

In this paper, meteorology, topography, soil, land use data, and remote sensing image data are available for consideration. (1) Meteorological data are obtained from the monthly data set of China's ground cumulative value (1981–2010) of National Meteorological Administration of China (http://data.cma.cn) and the Chinese meteorological background data set of the Resource and Environmental Science Data Center of Chinese Academy of Sciences (http://www.resdc.cn), including accumulated temperature, monthly average temperature, and precipitation during the crop growth period. The meteorological data are the cumulative annual averages of meteorological stations in the study area from 1981 to 2010. ArcGIS was used to make universal Kriging interpolation analysis on the temperature, precipitation, and accumulated temperature of the meteorological stations, and the spherical function was selected to obtain meteorological spatial raster data in the region, with a spatial resolution of 100 m. (2) Topographic data includes Digital Elevation Model (DEM), slope, and landform type. DEM and slope are the 30 m resolution data in 2009, obtained from the ASTER Global Digital Elevation Model (ASTER GDEM) provided by Geospatial Data Cloud site, Computer Network Information Center, Chinese Academy of Sciences (http://www.gscloud.cn). The landform types data comes from the 2009 Atlas of Landforms of the People's Republic of China (1: 1 million), which is raster data with a resolution of 1 km. (3) Soil data comes from the 1:1 million soil type vector map and the dataset of main traits of cultivated land quality in Heilongjiang Province, China in 2010, including soil texture, soil thickness, organic matter, available potassium, available phosphorus, and soil pH. (4) Land use data were spatial vector data extracted from the 2018 China land change survey database, including farmland ditches, water surfaces, shelter forests, residential areas, and farmland plots, etc. By calculating the Euclidean distances from each farmland plot to farmland ditches, water surfaces, and residential areas with ArcGIS, the spatial raster data of the drainage capacity, irrigation potential, and farming convenience were obtained. Besides, the kernel density analysis tool in ArcGIS was used to calculate the spatial density of the shelter forests, obtain the spatial raster data of the farmland shelterbelt density, and the calculation geometry tools were used to obtain the perimeter and area of each farmland plot. The field shape regularity of farmland plots was calculated by the formula of "4*(area/perimeter) ^1/2". (5) Landsat8 (OLI) satellite remote sensing image data are obtained from the USGS EROS Data Center. Combined with the characteristics of crop phenology, remote sensing image quality, and cloudiness in the region, multi-period remote sensing images were selected. Through remote sensing image processing such as radiometric calibration, atmospheric correction, geometric correction, and band fusion (6, 5, 2 bands), obtained the raster image map for crop interpretation, with a spatial resolution of 30 m. Combining the remote sensing image map and the 2018 cultivated land vector diagram in the study area extracted from the annual change database of the second national land survey in China, the spatial distribution vector map of the main crops was got through manual visual interpretation based on the image characteristics of different crops such as color and texture. To calculate and analyze various types of spatial data uniformly, ArcGIS was used to convert all raster data into vector data.

### 2.3. Land Suitability Evaluation of Crops

Land suitability evaluation is the process of estimating the land performance for alternative kinds of use [15,44,45], and its basic features are the comparison of the requirements of land use with the resources offered by the land [46]. Land suitability analysis for crops is a function of crop requirement and land characteristics reflected in final decisions [15], which can be identified as a multi-criteria evaluation (MCE) approach [35]. Drawing on the Food and Agricultural Organization (FAO) framework for land suitability [47–50], agricultural land suitability evaluation (ALSE) process [28], and other research results [5,34], land suitability evaluation for crops could mainly include (1) structuring the MCE model by identification of the environmental requirements of crops; (2) selecting standardization functions and determining the quantitative relationship between each considered environmental factor and the requirement of the target crop; (3) calculating the suitability score for a single factor for each evaluation unit; and (4) combining of the scores from all the considered factors [51,52]. Given this, the specific evaluation methods of this research are as follows:

### 2.3.1. Evaluation Indicators and Criteria

Sixteen indicators from the aspects of climate, topography, soil, and land use management were selected in the evaluation (Table 1). Climate, topography, and soil indicators characterize the natural conditions of crop planting, which are the prerequisite factors affecting crop suitability. Management indicators reflect the utilization conditions of crop planting, adding human factors based on natural conditions. The irrigation potential means the distance from the land to canals and dams. The drainage capacity represents the distance from the land to the ditches, and the farming convenience indicates the distance from the land to the settlement. The closer the distance, the greater the irrigation potential, drainage capacity, and farming convenience of the land. The field shape regularity reflects the difficulty of mechanized farming, and the shelterbelt density reflects the wind-proof and sand-fixing conditions of the land. Land management factors are also important factors affecting crop planting suitability and can be improved through engineering measures.

The indicators including restrictive, and non-restrictive indicators. The restrictive indicators have specific criteria for different crops with the most suitable range, maximum threshold, and minimum threshold, while the non-restrictive indicators have no specific suitability standards, which need to be further defined. The FAO guidelines [47,53], regional land evaluation materials [54,55], and relevant research results of the study area [56] were referenced in determining the evaluation criteria. For restrictive indicators, such as accumulated temperature, precipitation, slope, soil thickness, etc., the crop growth model [57,58] and membership function method were used to quantify the relationship between environmental conditions and crop requirements, and then calculate the crop planting suitability for each restrictive indicator. For non-restrictive indicators, such as available potassium, available phosphorus, farming convenience, shelterbelt density, etc., combined with the expert scoring of the relevant land evaluation results in the study area [54,55], the statistics-based classification method was used to define the evaluation criteria, and the scores of non-restrictive indicators were obtained. The quantified suitability results of each indicator are converted to 0–100 points, which is used to calculate the comprehensive score of suitability for the next step.

### 2.3.2. Evaluation Indicators Weight

The analytic hierarchy process (AHP) is an important tool for system analysis developed after mechanism analysis and statistical analysis. In this study, the AHP was used to determine the weight of the indicators. The specific steps mainly include (1) build a hierarchical structure model: The decision goals, decision criteria, and decision objects are layered according to their mutual relationship, and a hierarchical structure diagram is drawn, which is divided into the target layer, the middle layer, and the lowest layer. The target layer is the problem to be solved. In this study, it refers to the land suitability evaluation of crops. The middle layer refers to the criteria for decision-making, including climate suitability, topography suitability, soil suitability, and management suitability. The lowest layer refers to alternatives in decision-making, that is, evaluation indicators; (2) construct a judgment matrix: Based on the hierarchical structure model, the judgment matrix is determined by comparing the importance of indicators to the target. The 1–9 scale method was used to compare the importance of indicators in pairs and rank them according to their importance; (3) hierarchical sorting and consistency check: Hierarchical sorting is to calculate the weight of factors (indicators) at the same level based on their importance. The sum-product method is used to calculate the maximum eigenvalue $\lambda_{max}$ of the judgment matrix and the corresponding eigenvector $W_i$. The random consistency index *RI* and the consistency index *CI* are introduced to check the consistency of the hierarchical sorting results, and the random consistency index can be obtained in the table of *RI* standard value.

$$CI = \frac{\lambda_{\max} - n}{n - 1} \tag{1}$$

$$CR = \frac{CI}{RI} \tag{2}$$

where $\lambda_{max}$ is the maximum eigenvalue, $n$ is the order of the judgment matrix. *CI* is the consistency index, and the smaller the index value, the more consistent it is. *CR* is the consistency ratio, and the larger the value, the more inconsistent it is. When $CR < 0.1$, the consistency test is passed.

The consistency test results show that the target layer and factor layer of the land suitability evaluation for the three main crops all pass the consistency test (Table 2).

### 2.3.3. Comprehensive Score of Land Suitability

On the basis of determining evaluation indicators, criteria, and weights, the multi-index comprehensive score calculation method was used to assess the land suitability score of the crops. Based on the Cannikin Law, when a score of a certain indicator is "0", it will become a "shortboard" factor for suitability results, and the suitability scores of other indicators will have no effect on the comprehensive score. Therefore, when the score of an indicator of the evaluated unit is "0", the comprehensive suitability score of the unit is "0", regardless of the suitability of other indicators. When none of the indicator scores of the evaluated unit is "0", the comprehensive score will be calculated by weighting and summing the scores of all indicators. It can be expressed as the following formula:

$$S = 0, \text{ (When the score of a restrictive indicator of a unit is "0")} \tag{3}$$

$$S = \sum_{i=1}^{n} s_i * w_i, \text{ (When none of the restrictive indicator scores of a unit is "0")} \tag{4}$$

where $S$ is the comprehensive score of the evaluated unit, $n$ is the number of evaluation indicators, $s_i$ and $w_i$ are the score and weight of the $i$-th evaluation indicator, respectively.

**Table 1.** Land suitability evaluation index system and quantitative standard for crops.

| Factors | Indicators | Soybean | Maize | Rice |
|---|---|---|---|---|
| | | Criteria and Weights (*a-b, c, d, w*) | | |
| Climatic factors | Daily mean temperature during the growth period* (°C) | 20−23, 15, 25, 0.0941 | 20−23, 16, 28, 0.0797 | 22−24, 16, 28, 0.0678 |
| | Precipitation during the growth period* (mm) | 450−500, 270, 680, 0.1078 | 450−500, 220, 630, 0.1448 | -, -, -, 0.1231 |
| | ≥10 °C accumulated temperature* (°C) | >2500, 2000, -, 0.2467 | >2800, 2300, -, 0.2630 | >2600, 2300, -, 0.2237 |
| Terrain factors | Slope* (%) | 0-8, -, 30, 0.0419 | 0-8, -, 30, 0.0295 | 0-2, -, 8, 0.1827 |
| | landform type* | -, -, -, 0.0839 | -, -, -, 0.0886 | -, -, -, 0.0609 |
| Soil factors | Soil texture* | -, -, -, 0.0613 | -, -, -, 0.0727 | -, -, -, 0.0121 |
| | Soil thickness * (cm) | >50, -, -, 0.0568 | > 50, -, -, 0.0340 | >100, -, -, 0.0383 |
| | Organic matter (g/kg) | -, -, -, 0.0305 | -, -, -, 0.0340 | -, -, -, 0.0121 |
| | Available potassium (mg/kg) | -, -, -, 0.0146 | -, -, -, 0.0139 | -, -, -, 0.0050 |
| | Available phosphorus (mg/kg) | -, -, -, 0.0146 | -, -, -, 0.0139 | -, -, -, 0.0050 |
| | pH* | 6.0−6.5, 5.2, 7.5, 0.1101 | 5.0−7.0, 5.2, 8.0, 0.1076 | 5.5−6.0, 5.2, 8.2, 0.0258 |
| Management factors | Irrigation potential | -, -, -, 0.0386 | -, -, -, 0.0259 | -, -, -, 0.0841 |
| | Drainage capacity | -, -, -, 0.0551 | -, -, -, 0.0445 | -, -, -, 0.0841 |
| | Farming convenience | -, -, -, 0.0092 | -, -, -, 0.0075 | -, -, -, 0.0146 |
| | Field shape regularity | -, -, -, 0.0183 | -, -, -, 0.0201 | -, -, -, 0.0319 |
| | Shelterbelt density | -, -, -, 0.0165 | -, -, -, 0.0201 | -, -, -, 0.0290 |

Note: The suitability reference values of the indicators are listed in the order of *a-b, c, d*, and *w*, where *a-b* is the most suitable range of the indicator, *c* and *d* are the lowest and highest thresholds of the indicator, and *w* is the indicator weight. Indicators with "*" are restrictive indicators, and those without "*" are non-restrictive indicators. "-" means that the indicator has no suitable range or thresholds.

**Table 2.** Consistency test result of evaluation indicators weight.

| Target Layer | Consistency Ratio | Factor Layer | Consistency Ratio |
|---|---|---|---|
| Planting Suitability of Rice | 0.0304 | Climatic factors | 0.0088 |
| | | Terrain factors | 0.0000 |
| | | Soil factors | 0.0299 |
| | | Management factors | 0.0214 |
| Planting Suitability of Maize | 0.0579 | Climatic factors | 0.0088 |
| | | Terrain factors | 0.0000 |
| | | Soil factors | 0.0348 |
| | | Management factors | 0.0308 |
| Planting Suitability of Soybean | 0.0172 | Climatic factors | 0.0176 |
| | | Terrain factors | 0.0000 |
| | | Soil factors | 0.0594 |
| | | Management factors | 0.0292 |

## 2.4. Cultivated Land Use Layout Adjustment Model

CLULA in this paper includes crop layout optimization and spatial relationship co-ordination. The crop planting suitability is the main basis for this study to optimize the crop layout. To facilitate the large-scale management of cultivated land, the cultivated land concentration is also an important factor to be considered. The agent-based model for optimal land allocation (AgentLA) can well take into account the land suitability and space compactness [59], which can be used for crop layout optimization. Based on the optimized crop layout, the planting characteristics, farming system, food needs, and planning orientation are take into consideration to clarify the priority order of crops occupying the same space. ArcGIS spatial analysis tools are used to achieve spatial relationship coordination between crops, and then the plan for cultivated land use layout adjustment is determined (Figure 2).

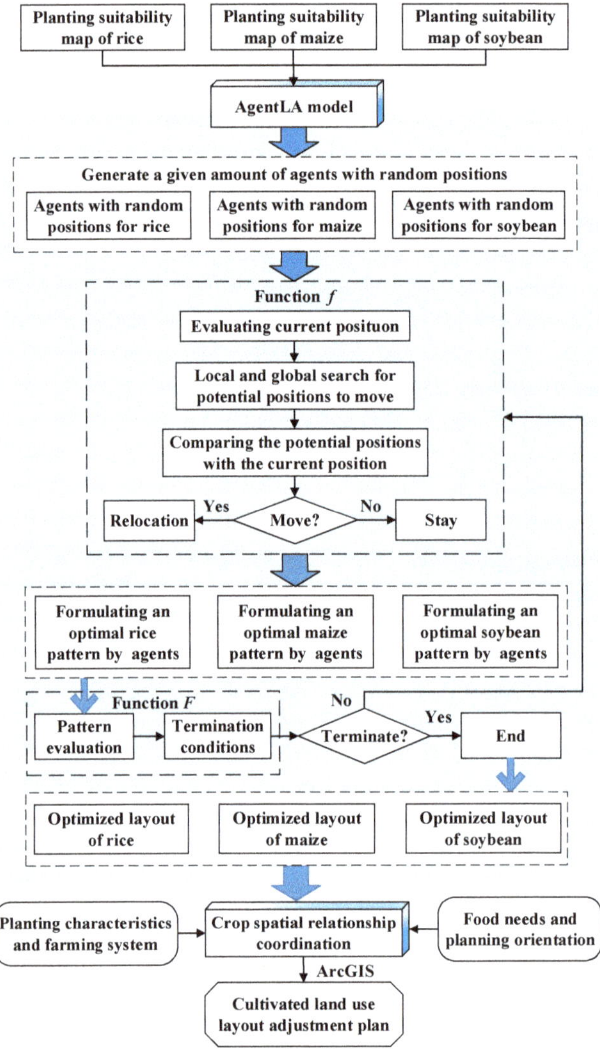



**Figure 2.** Procedure of cultivated land use layout adjustment.

In the AgentLA model, an agent is used to identify a single unit of the given crop type to be allocated, and each agent occupies a spatial grid. In this study, the scale of each spatial grid is set to 100 m × 100 m, which can distinguish the spatial distribution of different crop types. All the agents are different, but have the same criteria for determining a move to a better location. Initially, the model generates a spatial agent group with random positions according to the spatial scale of the study area. The population size of agents is equal to the given amount of land use to be located. Each agent will use a fitness function *f* to assess a potential site for relocation. An agent will move to a better location based on the assessment of the site by collecting local and global information. After all agents complete the decision, another function *F* is used to evaluate the optimization degree and feasibility of the whole simulated pattern [60]. By repeatedly iterating until the value change of function *F* tends to be stable and less than the preset threshold, the model stops iterating and outputs the final optimization result.

A fitness function $f$ is defined to measure whether a position is worth occupying by an agent, and it includes two objectives: Suitability and space compactness. The two objectives have different weights, which directly affect the fitness function value and the final optimized layout. The formulas are as following:

$$f = w_v v + w_c c \tag{5}$$

$$c = \frac{\sum\limits_{i \in \Omega} x_i \exp(-d_i/\gamma)}{\sum\limits_{i \in \Omega} \exp(-d_i/\gamma)} \tag{6}$$

where $f$ is the fitness function, $v$ is the crop suitability value, $c$ is the spatial compactness, and $w_v$ and $w_c$ are their weights, respectively ($w_v + w_c = 1$), $x_i$ is a binary variable which equals 1 if cell $i$ is occupied and 0 otherwise. $\Omega$ represents the Moore neighborhood of the central agent. $d_i$ is the Euclidean distance from cell $i$ to the focal agent. $\gamma$ is a compensation parameter that ranges from 1 to 10.

A function $F$ is equivalent to the evaluation function of the optimization result. It considers both suitability and spatial efficiency and considers the two factors to be equally important. The $F$ value is between 0 and 1. The higher the $F$ value, the better the optimization effect. $F$ function is defined as follows:

$$F = SV \text{ - } SL \tag{7}$$

$$SV = \sum_{i=1}^{n} v_i / V_{MaxSum} \tag{8}$$

$$SL = \frac{L_{Sum} - L_{MinSum}}{L_{ManSum} - L_{MinSum}} \tag{9}$$

where $n$ is the number of agents, $v_i$ is the suitability value of the $i$-th cell, $V_{MaxSum}$ is the suitability value of the most suitable cell, $SV$ is used to measure how well the objective of suitability is achieved, $SL$ is a spatial morphological feature, which measures the dispersion of the simulated pattern, $L_{Sum}$ represents the sum of the perimeter of the current pattern, $L_{MinSum}$ is the sum of the perimeter when the pattern is assumed to be the most compact, $L_{MaxSum}$ is the sum of the perimeter when the assumption is that all agents exist in isolation and are not adjacent to each other. The wellness of the resulted pattern increases when $SV$ is higher and $SL$ is smaller.

## 3. Results and Discussion

### 3.1. Planting Suitability of Food Crops

There are obvious differences in the planting suitability of rice, maize, and soybean in the study area. The natural breaks (syn. Jenks) method is used to classify the suitability, and to facilitate the comparison of suitability between crops, the classification results of the three crops are integrated into five classes: Highly suitable (S1), suitable (S2), moderately suitable (S3), marginally suitable (S4), and unsuitable (N). The results of crop planting suitability classification and distribution are shown in Table 3 and Figure 3.

### 3.1.1. Planting Suitability of Rice

The suitability level of rice is the lowest among the three main corps. It is indicated that 52.66 km², 286.37 km², 475.09 km², and 694.89 km² of cultivated land are located in S1, S2, S3, and S4 suitability classes respectively, accounting for only 18.16% of the cultivated land in the region. Nearly 6798.41 km² (81.84%) of cultivated land are not suitable (N) for rice, which covers most parts of the study area (Table 3). The main constraint for rice planting is the slope and geomorphological types in the region. Low slope and plain areas are suitable for cultivating rice. Rice cultivation has high requirements for terrain factors. The best slope is 0–2°, once the slope is more than 8°, it is not suitable for rice cultivation.

The average slope of most cultivated land in the study area is greater than 8°, and the landform type is mainly low elevation platform, which is not conducive to planting rice. Considering the suitability map (Figure 3a), the highly suitable and suitable (S1-S2) areas for cultivating rice are concentrated in the north of central Yi'an County and the south of Keshan County. The moderately suitable and marginally suitable (S3-S4) areas are mainly located in the southwest of Yi'an County and north of Keshan County. Baiquan County is less suitable for rice cultivation among the three counties, and the area suitable for rice cultivation is small and scattered.

### 3.1.2. Planting Suitability of Maize

The suitability results for maize showed that almost all cultivated land in the study area is suitable for cultivating maize. The suitable (S2) and moderately suitable (S3) areas for cultivating rice is 6636.15 km$^2$ (79.88%), which accounts for the highest proportion of cultivated land area in the region. Only 1.19% (99.10 km$^2$) of the cultivated land were classified as highly suitable (S1), and the marginally suitable (S4) area is 1572.16 km$^2$, accounting for 18.93% of the cultivated land area in the region (Table 3). The land suitability map for maize (Figure 3b) showed that the best areas for cultivating maize (S1-S2) are concentrated in the north of Yi'an County and southwest of Keshan County, and scattered in the south of Yi'an County and Baiquan County. Areas that are moderately suitable (S3) are mainly distributed around the suitable (S2) area, and the two occupy a similar proportion of cultivated land. The marginally suitable (S4) area is mainly located in the northern of Keshan County, south of central Yi'an County, and southwest of Baiquan County. Keshan County is less suitable for maize cultivation among the three counties, and nearly half of the area is marginally suitable (S4) area. According to the distribution characteristics of land suitability of maize, the main limiting factors for maize planting are climate and soil. Compared with rice and soybeans, maize cultivation has higher requirements for mean daily temperature and accumulated temperature.

### 3.1.3. Planting Suitability of Soybean

The suitability level of soybean is the highest among the three main corps. It was shown that, while 1051.28 km$^2$ (12.65%) of the cultivated land is unsuitable (N) for soybean cultivation, 1146.73 km$^2$ and 4671.95 km$^2$ of cultivated land are located in S1 and S2 suitability classes respectively, which covered 70.04% of the cultivated land. S3 and S4 suitability classes consisted of a 62.13 km$^2$ and a 1375.34 km$^2$ area, respectively, which only covered about 0.75% and 16.56% of the cultivated land area in the region (Table 3). As can be seen from Figure 3c, the unsuitable (N) area is concentrated in the southwest of the study area, which is mainly affected by the limiting factor of soil pH. Compared with rice and maize, soybean has higher requirements for soil pH. When the pH is greater than 7.5, it is not conducive to soybean cultivation. The moderately suitable (S3) and marginally suitable (S4) areas are located mainly in the northeast and south of the region, and most of them are adjacent to the unsuitable area. The highly suitable (S1) and suitable (S2) areas are located over large areas of the study area such as the north, northwest, southeast, and central parts.

Comparing the current distribution and the planting suitability maps of crops in the study area (Figure 3), it is found that the rice is mainly concentrated in the central area of Yi'an County, while the planting suitability of rice in some areas of central Yi'an County is low. The southern area of Keshan County is very suitable for rice cultivation, but it is not used effectively. The current area of maize in the study area is very large, accounting for a relatively high proportion in three counties. However, the planting suitability level of maize in Keshan County is low, indicating that the area of maize in this area needs to be reduced. Furthermore, it can be seen that the current soybean distribution has not fully covered the high suitability area. For example, the planting suitability of soybean is high in the north and northeast of Yi'an County, but fewer soybeans are planted in this area. The above analysis shows that the current crop layout needs to be further optimized. The results show that the current distribution of crops in the study area needs to be further optimized and adjusted.

**Table 3.** Results of planting suitability classification of rice, maize, and soybean.

| Suitability | Classification | N | S4 | S3 | S2 | S1 |
|---|---|---|---|---|---|---|
| | score | 0 | (0,75] | (75,80] | (80,85] | (85,100] |
| Rice | Area (km$^2$) | 6798.41 | 694.89 | 475.09 | 286.37 | 52.66 |
| | Proportion (%) | 81.84 | 8.36 | 5.72 | 3.45 | 0.63 |
| Soybean | Area (km$^2$) | 1051.28 | 62.13 | 1375.34 | 4671.95 | 1146.73 |
| | Proportion (%) | 12.65 | 0.75 | 16.56 | 56.24 | 13.80 |
| Maize | Area (km$^2$) | 0.02 | 1572.16 | 3317.53 | 3318.62 | 99.10 |
| | Proportion (%) | 0.00 | 18.93 | 39.93 | 39.95 | 1.19 |

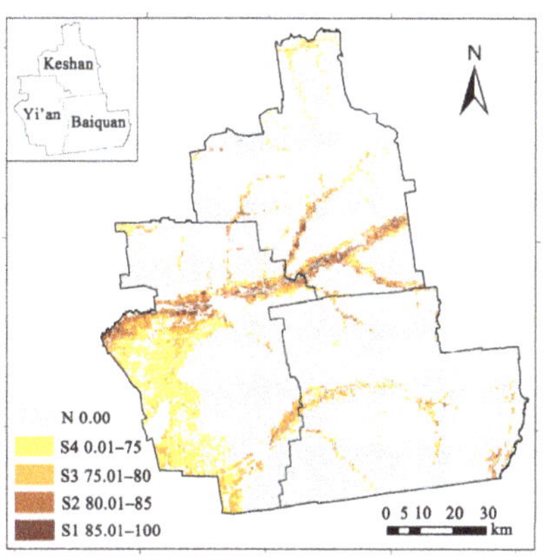


(a) Planting suitability map of rice.

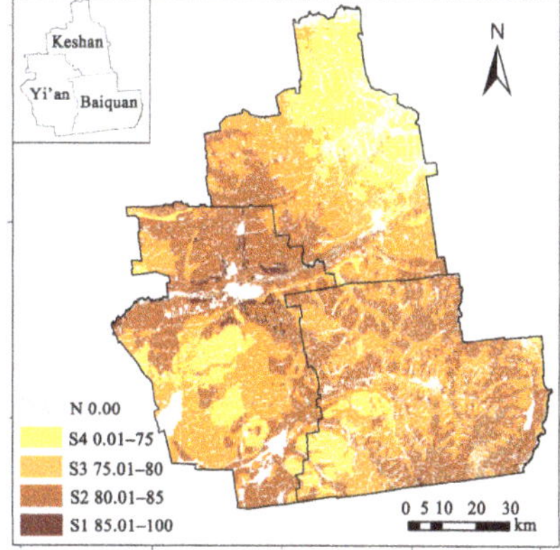


(b) Planting suitability map of maize.

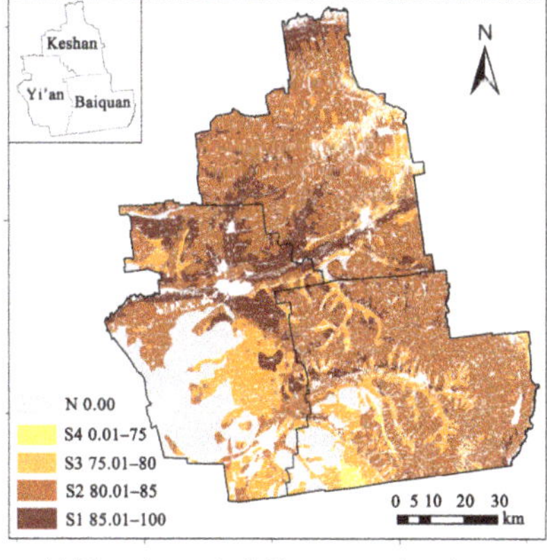


(c) Planting suitability map of soybean.

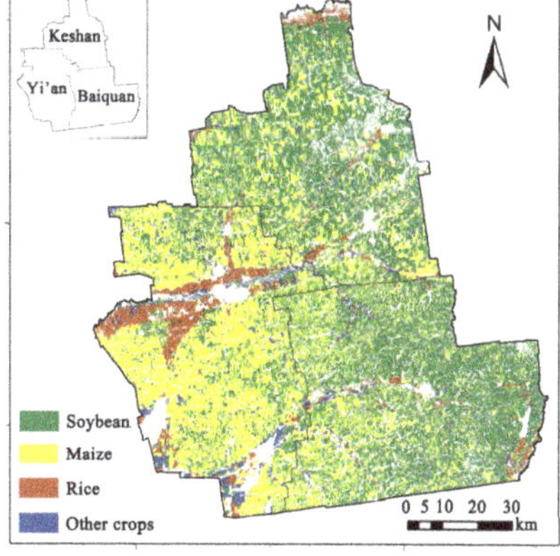


(d) Current distribution of main food crops.

**Figure 3.** The current distribution (**a**) and the planting suitability maps (**b–d**) of main food crops in the study area.

### 3.2. Optimized Layout of Food Crops

To improve the suitability level of crops and facilitate the large-scale management of cultivated land use, this study takes into account the planting suitability of crops and the concentration of cultivated land to optimize the layout of the main food crops in the study area. Before the AgentLA model runs, the agent number to be optimized for the target crop needs to be set. Under the influence of agricultural policies in recent years, the food crop quantitative structure in the study area has been adjusted to some degree, so this research only optimizes the spatial structure of the food crops. It is assumed that the quantity structure of the three main food crops is the same as the current year, in 2018. According to the results of the remote sensing interpretation of the current crop distribution in this study, taking 100 m × 100 m grids as the cell, the agent number to be optimized for rice, maize, and soybean are set to 43,650, 379,668, and 376,897, respectively. Through multiple experiments, it is found that when the weight $w_v$ value is 0.7, the fitness function value $F$ of the three crops are the largest. $F$ values of these three outcomes are 0.87, 0.92, and 0.92, respectively, which is a relatively optimal result. Therefore, the suitability weight and the space efficiency weight are set to 0.7 and 0.3, respectively. Rice, soybean, and maize went through 134, 125, and 127 iterations, respectively, and the fitness function values tended to be stable. The iterations were stopped, and the results of cultivated land layout optimization for food crops were output. The optimized layout of rice is mainly distributed in the north of central Yi'an County and the south of Keshan County. The optimized layout of soybean is concentrated in Keshan County and the north of Yi'an County, and the optimized layout of maize is mainly distributed in Yi'an County and the west of Keshan County. For Baiquan county, the proportion of maize and soybean optimized layout is also large, but the distribution is relatively scattered (Figure 4).

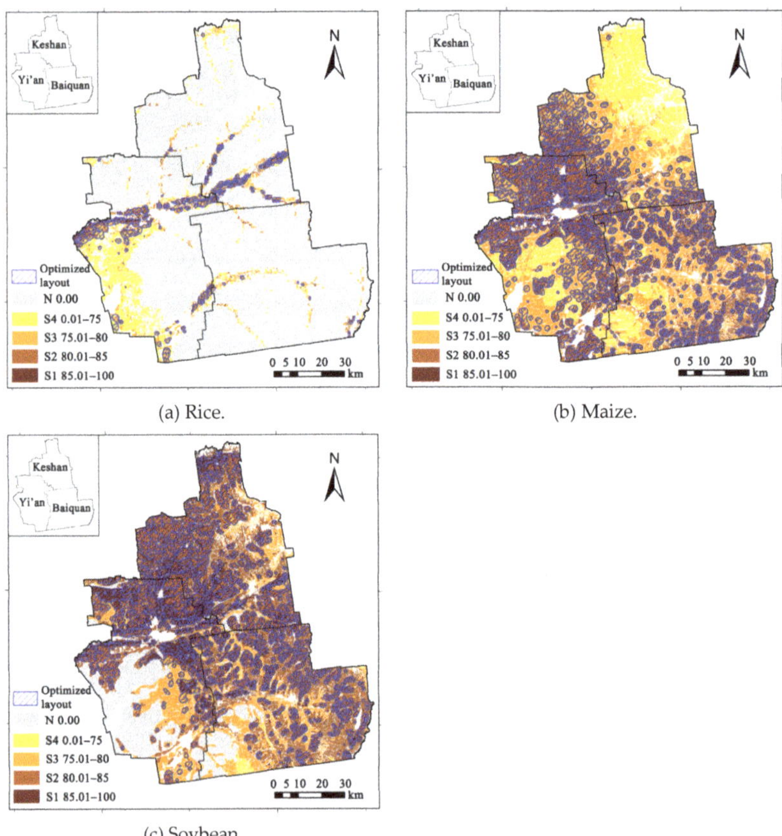

(a) Rice.

(b) Maize.

(c) Soybean.

**Figure 4.** Optimized layout of rice (**a**), maize (**b**), and soybean (**c**) in the study area.

To further verify the feasibility of the optimization results, the current layout of the three crops were evaluated using the *F* function to compare with the optimization results. Through calculation, the evaluation values *F* of the current layout of rice, soybean, and maize were 0.59, 0.89, and 0.87, respectively, which were lower than the optimized value. The results show that the land suitability level of the three food crops and the concentration level of the spatial pattern of CLUL have been improved. The optimization results effectively take into account the space suitability and space efficiency of the crops and realize the optimal allocation of the cultivated land use space of food crops. Among them, the optimization effect of rice was the best, indicating that the current rice distribution is not reasonable, and the land production potential can be further tapped. This is mainly because the planting suitability level of the current rice layout is low, and the improvement space is large.

### 3.3. Adjustment Plan of CLUL

Food crops and cultivated land have a unique correspondence in space. The spatial combination of multiple crops forms the CLUL. On the same land space, there are both competition and cooperation between crops. How to coordinate the spatial relationship between crops is a key issue to be considered in CLULA. This study combines the planting characteristics, farming system, and policy orientation of the food crops in the region, and considers factors such as food demand, cultivated land management, and cultivated land ecological security to coordinate the spatial layout relationship of crops, and then to delineate the adjustment areas of CLUL and determine the adjustment unit.

### 3.3.1. Distribution of CLULA Areas

The remote sensing interpretation results of the current crop distribution in this study show that, in 2018, the planting ratio of rice, soybean, and maize in the region was about 1:8.71:8.64. In 2016, the Ministry of Agriculture of China issued the "National Plant Restructuring Plan (2016−2020)" and put forward requirements for the crop planting structure in Northeast China, that is, by stabilizing the rice area, reducing the maize area, expanding soybeans, miscellaneous grains, potatoes, and forage crops, and constructing a reasonable rotation system to adjust the crop planting structure. To protect paddy field resources in the study area and stabilize rice production, the first step is to determine a rice stable production area based on the agricultural policy guidance and the optimized layout of food crops. In this study, the rice optimized layout (Figure 4a) was directly identified as the rice stable production area. To further adjust the cultivated land layout of other crops, we used the toolbox of ArcGIS to subtract the stable rice production area from the optimized layout of maize (Figure 4b) and soybeans (Figure 4c), and extract the optimized layout of maize and soybeans that do not overlap with the stable rice production area, which can be defined as area M and area N. The continuous cultivation of the same crop for many years will cause problems such as soil environmental damage and the decline of cultivated land fertility [61]. To ensure the ecological security and the sustainable use of cultivated land, priority should be given to determine the crop rotation area to reduce the ecological cost of continuous farming. Among the major crops in the region, maize and soybeans have similar climate, humidity, and temperature requirements. Soybean can absorb and fix nitrogen in the air, which can effectively improve soil fertility. Maize is a nitrogen-loving crop. Soybean and maize rotation planting will have a good effect on the storage of soil nutrients. Therefore, the second step was to determine the maize–soybean rotation area. The ArcGIS toolbox was used to take the intersection of area M and area N, and get the maize–soybean rotation area. The final step was to determine maize and soybean layout adjustment areas. The ArcGIS toolbox was used to subtract the maize–soybean rotation area from the area M and area N respectively, to obtain the maize dominant area and the soybean dominant area, which can guide the maize and soybean planting decisions such as reduction and expansion.

According to the statistics of the research result (Figure 5a), the total area of the stable rice production area is about 436.50 km$^2$, accounting for 5.25% of the total cultivated land area of the region, which is mainly distributed in Keshan County and Yi'an County. The overall shape of the stable rice production area is like a stripe running through the east–west direction, concentrated in the central region and scattered in the southern part of the study area. The maize–soybean rotation area accounts for 30.32% of the total cultivated land in the region, with a total area of about 2519.65 km$^2$, concentrated in the west of Keshan County and the north of Yi'an County, and scattered in the south of Yi'an County and Baiquan County. The area of the maize dominant area and the soybean dominant area are similar, about 950.40 km$^2$ and 967.20 km$^2$, respectively, accounting for 11.44% and 11.64% of the total area of cultivated land in the region, but the spatial distributions of the two areas are quite different. The soybean dominant area is mainly distributed in the northern part of the region, and Keshan County occupies a relatively high proportion. The maize dominant area is scattered in the southern part of the region, and Yi'an County and Baiquan County have a relatively high proportion (Figure 5b).

The results of CLULA show that the planting suitability and spatial agglomeration of the three main crops have been improved, which can be verified by the $F$ value mentioned above. It can be estimated that after adjustment, the $F$ value of corn and soybeans has increased by about 5%, and that of rice has increased by 47%. The improvement of planting suitability can effectively increase the potential for food production, and the centralized crop layout can increase the scale benefits of farmland. Besides, corn and soybean rotation can effectively improve the fertility of farmland and ensure sustainable land use.

### 3.3.2. Implementation Unit of CLULA

Based on the adjustment area for the utilization of cultivated land, this study determined the adjustment unit according to the distribution characteristics of the adjustment areas. There are obvious differences in the distribution of different adjustment areas. Among them, the soybean dominant area, and maize–soybean rotation area in the north and west of Keshan County are concentrated, while the soybean dominant area, rice stable production area, and maize–soybean rotation area in the east and south of the county are scattered. In the northern, central, and eastern parts of Yi'an County, the distribution of maize–soybean rotation area and rice stable production area are concentrated, while the distribution of maize dominant area and maize-soybean rotation area in the southeast of the county are scattered. In contrast, the distribution of the four adjustment areas in Baiquan County is scattered. To ensure the efficiency and effectiveness of the implementation of CLULA, considering the distribution characteristics of the adjustment areas and the scale of the internal administrative units of each county, the adjustment units are divided into towns and villages. The areas where the adjustment areas are concentrated are adjusted in units of towns. For example, town A in the north of Keshan County is identified as an adjustment unit of soybean dominant area, and town B in the west of Keshan County is identified as an adjustment unit of maize–soybean rotation area. The areas where the adjustment areas are scattered are adjusted in units of villages. For example, village C in the west of Yi'an County is identified as an adjustment unit of rice stable production area, and village D in the south of Baiquan County is identified as an adjustment unit of maize dominant area (Figure 5c). Town or village should be selected as the implementation unit for CLULA according to local conditions to effectively implement the adjustment plan.

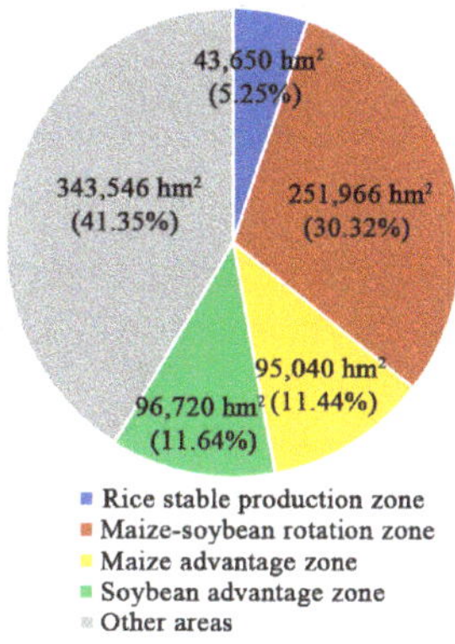


(a) Statistics of cultivated land use layout adjustment (CLULA) areas.

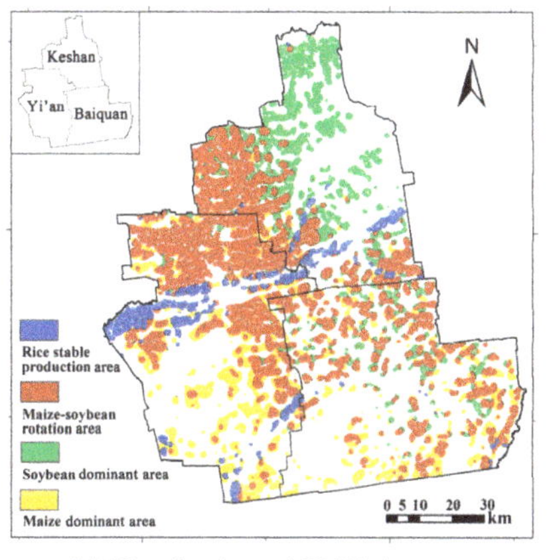


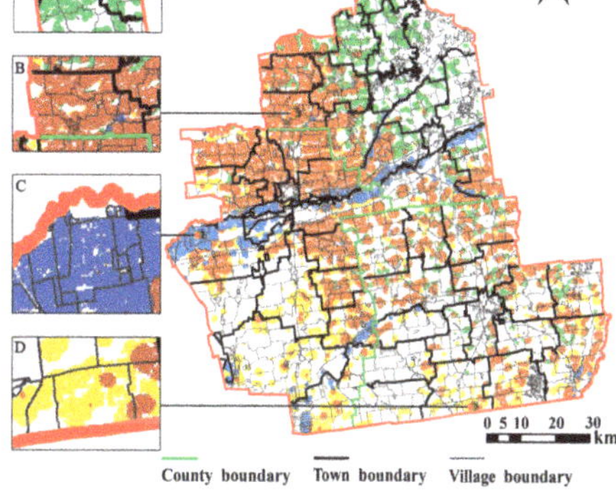


(b) Distribution of CLULA areas.

(c) Implementation unit of CLULA.

**Figure 5.** Statistics (**a**), distribution (**b**), and implementation unit (**c**) of CLULA.

The study effectively coordinated the spatial relationship between different food crops, and obtained an adjustment plan for CLUL integrating farming with protecting land, which provides a scientific basis for the spatial implementation of CLULA, and enables the adjustment plan within the scope of town or village to be carried out quickly. The research results also provide a reference for CLULA in other similar areas. While delineating the adjustment areas and determining the implementation unit, the guidance of farmers in different CLULA areas should be strengthened, and it is an important way to effectively implement the CLULA. For the rice, stable production area, soybean dominant area, and maize dominant area, the cultivation of dominant crops should be guided. Special attention should be paid to expanding soybean in soybean dominant areas and reducing maize in non-dominant maize areas, to alleviate the current structural contradiction between grain supply and demand. It should be noted that soybean continuous cropping for 2 years will lead to a yield decline. Therefore, soybean–wheat, soybean–tuber, and other crop rotation models should be supported in the soybean dominant area, and the selection of specific crop rotation models also needs to consider the suitability of other crops. Considering the differences in production costs and prices of different crops, the CLULA will lead to changes in the economic interests of farmers, and planting income is an important factor affecting farmers' planting decisions. It is necessary for the government to improve the price policy of agricultural products and guide farmers to adjust the planting structure according to market demand. For the maize–soybean rotation area, combined with the soil, climate, and other natural conditions in the region, the "maize–soybean" annual rotation or "maize–maize–soybean" rotation pattern should be adopted. The specific pattern can be selected based on market prices and the actual demand, which not only guarantees economic benefits, but is also conducive to land recuperation. Additionally, it is advisable to carry out pilots for cultivated land rotation in units of towns or villages, and gradually expand the scope of cultivated land rotation by summing up the experience of the pilots. In addition, advancing agricultural technological innovation and improving the level of agricultural machinery are also crucial to promote CLULA.

CLULA is a complicated systematic project, which not only involves the optimal allocation of spatial factors for cultivated land use, but also involves the specific implementation process, including coordinating the interests of the participants, constructing a scientific adjustment model, formulating safeguard measures, and designing policy systems, etc. This study mainly guided the precise implementation of CLULA from the spatial level. How to establish a scientific guidance mechanism to ensure the effective implementation of CLULA needs further discussion. Furthermore, under the guidance of the agricultural planting structure adjustment policy in China, the imbalance of the grain planting structure in the Northeast region has been alleviated to some extent, but the scientific nature of the quantitative structure remains to be discussed in depth. To further ensure the scientificity and rationality of the adjustment of the planting structure in terms of quantity and space allocation, it is necessary to simultaneously optimize the quantitative structure and spatial structure of crops while taking into consideration the natural and management conditions of crops, as well as the needs of socio-economic development at the current stage. Cultivated land use layout research that takes into account the optimization of quantitative structure and spatial structure will become the focus of the next step of exploration.

## 4. Conclusions

To alleviate the current planting structural imbalance and the grain structural contradiction in China, this paper is based on the planting suitability of food crops, taking into account the crop suitability level and the CLUL concentration level, to optimize the spatial layout of the main food crops in the study area. By coordinating the spatial relationship of different food crops for cultivated land use, the adjustment plan for CLUL is determined.

The results from the present study show that, affected by factors such as terrain and climate, there are obvious differences in the suitability level of rice, soybean, and maize planting in the study area. The current layout of the three crops needs to be further optimized and adjusted, and the crop planting suitability provides an important basis for the optimization of crop layout. Selecting areas with high suitability for crop planting will be conducive to enhancing the production potential and efficiency of cultivated land.

Compared with the current distribution of the crop, the optimized crops layout cover areas with high land suitability, and the spatial pattern of CLUL is more concentrated and contiguous. The optimized layout of rice is concentrated in the north of central Yi'an County and the south of Keshan County, the optimized layout of soybean is concentrated in the north of Yi'an County and the north and west of Keshan County, and the optimized layout of maize is concentrated in the north of Yi'an County and the west of Keshan County. The optimized results are conducive to the implementation of large-scale crop structure adjustment and optimization, and effectively promote the large-scale management of cultivated land.

The adjustment plan for CLUL fully considers the characteristics of food crop planting, planning policy orientation, and cultivated land ecological security factors. Four adjustment areas for CLUL were delineated: Rice stable production area, maize-soybean rotation area, maize dominant area, and soybean dominant area. Town or village was identified as the implementation unit for CLULA in the study area. The research results achieved a reasonable spatial matching of multiple cultivated land use factors such as climate, terrain, soil, and management of different crops, which will be conducive to the sustainable use of cultivated land that adapts to the laws of natural ecology and coordinates multiple development goals. Furthermore, enhancing regional guidance to farmers and improving agricultural product price policies are the key paths to ensure the effective implementation of CLULA.

This study provides useful guidance that can be used to make fine management for regional crop planting structure adjustment and cultivated land layout optimization. Additionally, the study on CLULA is used to refine and deepen land use transition on the basis of grasping the dominant morphology and recessive morphology characteristics of cultivated land use. Through the guidance and control of land use morphology by means of CLULA, land use transformation will achieve the expected goal, and then realize the sustainable and efficient use of urban and rural land resources.

**Author Contributions:** Conceptualization, G.S.; methodology, H.Z.; software, H.Z.; validation, G.S.; formal analysis, G.S. and H.Z.; resources, G.S. and H.Z.; data curation, H.Z.; writing—original draft preparation, H.Z.; writing—review and editing, G.S. and H.Z.; visualization, H.Z.; supervision, G.S.; project administration, G.S.; funding acquisition, G.S. All authors have read and agreed to the published version of the manuscript.

**Funding:** This study was funded by the National Natural Science Foundation of China (grant numbers 41971247, 41571165), and National Social Science Fund Major Project of China (grant numbers 19ZDA096).

**Acknowledgments:** The authors gratefully acknowledge the editors and anonymous reviewers for their constructive comments on our manuscript. We also thank the data support from the "National Meteorological Administration of China (http://data.cma.cn)", "Data Center for Resources and Environmental Sciences, Chinese Academy of Sciences (RESDC) (http://www.resdc.cn)", and "Geospatial Data Cloud site, Computer Network Information Center, Chinese Academy of Sciences (http://www.gscloud.cn)".

**Conflicts of Interest:** The authors declare no conflict of interest.

# References

1. Grainger, A. National land use morphology: Patterns and possibilities. *Geography* **1995**, *80*, 235–245.
2. Chen, R.S.; Ye, C.; Cai, Y.L.; Xing, X.S.; Chen, Q. The impact of rural out-migration on land use transition in China: Past, present and trend. *Land Use Policy* **2014**, *40*, 101–110. [CrossRef]
3. Long, H.; Qu, Y. Land use transitions and land management: A mutual feedback perspective. *Land Use Policy* **2018**, *74*, 111–120. [CrossRef]
4. Long, H.; Qu, Y.; Tu, S.; Zhang, Y.; Jiang, Y. Development of land use transitions research in China. *J. Geogr. Sci.* **2020**, *30*, 1195–1214. [CrossRef]
5. Ge, D.; Long, H.; Zhang, Y.; Ma, L.; Li, T. Farmland transition and its influences on grain production in China. *Land Use Policy* **2017**, *70*, 94–105. [CrossRef]
6. Qu, Y.; Long, H. A framework of multi-disciplinary comprehensive research on recessive farmland transition in China. *Acta Geogr. Sin.* **2018**, *73*, 1226–1241.
7. Wang, K.; Lin, C.; Wu, C. Trends and planning choices after china's urbanization rate reaching above 60%. *City Plan. Rev.* **2020**, *44*, 9–17.
8. Jin, T. The adjustment of China's grain cropping structure and its effect on the consumption of water and land resources. *J. Nat. Resour.* **2019**, *34*, 14–25.
9. Zhang, Y.; Lei, G.; Zhang, H.; Lin, J. Spatiotemporal dynamics of land and water resources matching of cultivated land use based on micro scale in Naoli River Basin. *Trans. Chin. Soc. Agric. Eng.* **2019**, *35*, 185–194.
10. Kazemi, H.; Akinci, H. A land use suitability model for rainfed farming by Multi-criteria Decision-making Analysis (MCDA) and Geographic Information System (GIS). *Ecol. Eng.* **2018**, *116*, 1–6. [CrossRef]
11. Yin, D.; Li, X.; Li, G.; Zhang, J.; Yu, H. Spatio-Temporal Evolution of Land Use Transition and Its Eco-Environmental Effects: A Case Study of the Yellow River Basin, China. *Land* **2020**, *9*, 514. [CrossRef]
12. Song, G.; Wang, Y.; Lei, G. Effect Mechanism Research of Influential Factors of Cultivated Land Use System Security of Black Soil Region in Songnen High Plain: A Case Study of Bayan County in Heilongjiang Province. *J. Nat. Resour.* **2014**, *2*, 13–26.
13. Akpoti, K.; Kabo-Bah, A.T.; Zwart, S.J. Agricultural land suitability analysis: State-of-the-art and outlooks for integration of climate change analysis. *Agric. Syst.* **2019**, *173*, 172–208. [CrossRef]

14. Kihoro, J.; Bosco, N.J.; Murage, H. Suitability analysis for rice growing sites using a multicriteria evaluation and gis approach in great Mwea region, Kenya. *Springerplus* **2013**, *2*, 1–9. [CrossRef]

15. Prakash, T.N. *Land Suitability Analysis for Agricultural Crops: A Fuzzy Multicriteria Decision Making Approach*; Science in Geoinformatics: Enschede, The Netherlands, 2003; pp. 6–13.

16. Amiri, F.; Shariff, A.R.B.M. Application of geographic information systems in land use suitability evaluation for beekeeping: A case study of Vahregan watershed (Iran). *Afr. J. Agric. Res.* **2012**, *7*, 89–97.

17. Shu, H.; Xiong, P. Reallocation planning of urban industrial land for structure optimization and emission reduction: A practical analysis of urban agglomeration in China's Yangtze River Delta. *Land Use Policy* **2019**, *81*, 604–623. [CrossRef]

18. Xu, W.; Zeng, Z.; Xu, Z.; Li, X.; Chen, X.; Li, X.; Xiao, R.; Liang, J.; Chen, G.; Lin, A.; et al. Public health benefits of optimizing urban industrial land layout—The case of Changsha, China. *Environ. Pollut.* **2020**, *263*, 114388. [CrossRef]

19. Zeng, L.; Li, J.; Zhou, Z.; Yu, Y. Optimizing land use patterns for the grain for Green Project based on the efficiency of ecosystem services under different objectives. *Ecol. Indic.* **2020**, *114*, 106347. [CrossRef]

20. Wang, Y.; Song, G.; Zhang, H. The Research of Land Use Pattern Optimization on Country Level in Heilongjiang Province. *Econ. Geogr.* **2016**, *36*, 147–151.

21. Lu, M.; Wei, L.; Ge, D.; Sun, D.; Zhang, Z.; Lu, Y. Spatial optimization of rural settlements based on the perspective of appropriateness–domination: A case of Xinyi City. *Habitat Int.* **2020**, *98*, 102148. [CrossRef]

22. Liao, F.H.; Gordon, B.; DePhelps, C.; Saul, D.; Fan, C.; Feng, W. A Land-Based and Spatial Assessment of Local Food Capacity in Northern Idaho, USA. *Land* **2019**, *8*, 121. [CrossRef]

23. Lou, Y.; Yin, G.; Xin, Y.; Xie, S.; Li, G.; Liu, S.; Wang, X. Recessive Transition Mechanism of Arable Land Use Based on the Perspective of Coupling Coordination of Input–Output: A Case Study of 31 Provinces in China. *Land* **2021**, *10*, 41. [CrossRef]

24. Das, B.; Singh, A.; Panda, S.N.; Yasudal, H. Optimal land and water resources allocation policies for sustainable irrigated agriculture. *Land Use Policy* **2015**, *42*, 527–537. [CrossRef]

25. Yu, J.; Chen, Y.; Wu, J. Computers, Environment and Urban Systems Modeling and implementation of classification rule discovery by ant colony optimisation for spatial land-use suitability assessment. *Comput. Environ. Urban Syst.* **2011**, *35*, 308–319. [CrossRef]

26. Wu, X.; Wang, S.; Fu, B.; Liu, Y.; Zhu, Y. Land use optimization based on ecosystem service assessment: A case study in the Yanhe watershed. *Land Use Policy* **2018**, *72*, 303–312. [CrossRef]

27. Bojorquez-Tapia, L.A.; Diaz-Mondragon, S.; Ezcurra, E. GIS-based approach for participatory decision making and land suitability assessment. *Int. J. Geogr. Inf. Sci.* **2001**, *15*, 129–151. [CrossRef]

28. Elsheikh, R.; Mohamed Shariff, A.R.B.; Amiri, F.; Ahmad, N.B.; Balasundram, S.K.; Soom, M.A.M. Agriculture Land Suitability Evaluator (ALSE): A decision and planning support tool for tropical and subtropical crops. *Comput. Electron. Agric.* **2013**, *93*, 98–110. [CrossRef]

29. Malczewski, J. Ordered weighted averaging with fuzzy quantifiers: GIS-based multicriteria evaluation for land-use suitability analysis. *Int. J. Appl. Earth Obs. Geoinf.* **2006**, *8*, 270–277. [CrossRef]

30. Nguyen, T.T.; Verdoodt, A.; Van, Y.; Delbecque, T.; Tran, N.; Ranst, E. Design of a GIS and multi-criteria based land evaluation procedure for sustainable land-use planning at the regional level. *Agric. Ecosyst. Environ.* **2015**, *200*, 1–11. [CrossRef]

31. Zolekar, R.B.; Bhagat, V.S. Multi-criteria land suitability analysis for agriculture in hilly zone:Remote sensing and GIS approach. *Comput. Electron. Agric.* **2015**, *118*, 300–321. [CrossRef]

32. Cay, T.; Uyan, M. Evaluation of reallocation criteria in land consolidation studies using the Analytic Hierarchy Process (AHP). *Land Use Policy* **2013**, *30*, 541–548. [CrossRef]

33. Jayanthi, M.; Thirumurthy, S.; Samynathan, M.; Manimaran, K.; Duraisamy, M.; Muralidhar, M. Assessment of land and water ecosystems capability to support aquaculture expansion in climate-vulnerable regions using analytical hierarchy process based geospatial analysis. *J. Environ. Manag.* **2020**, *270*, 110952. [CrossRef]

34. Vasu, D.; Srivastava, R.; Patil, N.G.; Tiwary, P.; Chandran, P.; Kumar Singh, S. A comparative assessment of land suitability evaluation methods for agricultural land use planning at village level. *Land Use Policy* **2018**, *79*, 146–163. [CrossRef]

35. Reshmidevi, T.V.; Eldho, T.I.; Jana, R. A GIS-integrated fuzzy rule-based inference system for land suitability evaluation in agricultural watersheds. *Agric. Syst.* **2009**, *101*, 101–109. [CrossRef]

36. Yu, J.; Chen, Y.; Wu, J.; Khan, S. Cellular automata-based spatial multi-criteria land suitability simulation for irrigated agriculture. *Int. J. Geogr. Inf. Sci.* **2011**, *25*, 131–148. [CrossRef]

37. Chang, Y.C.; Ko, T.T. An interactive dynamic multi-objective programming model to support better land use planning. *Land Use Policy* **2014**, *36*, 13–22. [CrossRef]

38. Guzman, L.A.; Escobar, F.; Pea, J.; Cardona, R. A cellular automata-based land-use model as an integrated spatial decision support system for urban planning in developing cities: The case of the bogotá region. *Land Use Policy* **2020**, *92*, 104445. [CrossRef]

39. Liu, Y.; Wang, S.; Chen, B. Optimization of national food production layout based on comparative advantage index. *Energy Procedia* **2019**, *158*, 3846–3852. [CrossRef]

40. Li, F.; Li, Z.; Chen, H.; Chen, Z.; Li, M. An agent-based learning-embedded model (ABM-learning) for urban land use planning: A case study of residential land growth simulation in Shenzhen, China. *Land Use Policy* **2020**, *95*, 104620. [CrossRef]

41. Tongur, V.; Ertunc, E.; Uyan, M. Use of the Migrating Birds Optimization (MBO) Algorithm in solving land distribution problem. *Land Use Policy* **2020**, *94*, 104550. [CrossRef]

42. Uyan, M.; Tongur, V.; Ertunc, E. Comparison of different optimization based land reallocation models. *Comput. Electron. Agric.* **2020**, *173*, 105449. [CrossRef]

43. Wang, Q.; Liu, R.; Men, C.; Guo, L. Application of genetic algorithm to land use optimization for non-point source pollution control based on CLUE-S and SWAT. *J. Hydrol.* **2018**, *560*, 86–96. [CrossRef]

44. He, Y.; Yao, Y.; Chen, Y.; Ongaro, L. Regional Land Suitability Assessment for Tree Crops Using Remote Sensing and GIS. In Proceedings of the 2011 International Conference on Computer Distributed Control and Intelligent Environmental Monitoring (CDCIEM) IEEE, Changsha, China, 19–20 February 2011; pp. 354–363.

45. Mu, Y. *Developing a Suitability Index for Residential Land Use: A Case Study in Dianchi Drainage Area*; University of Waterloo: Waterloo, ON, Canada, 2006.

46. AbdelRahman, M.A.E.; Natarajan, A.; Hegde, R. Assessment of land suitability and capability by integrating remote sensing and GIS for agriculture in Chamarajanagar district, Karnataka, India. *Egypt. J. Remote Sens. Space Sci.* **2016**, *19*, 125–141. [CrossRef]

47. FAO. A framework for land evaluation. *FAO Soil Bull.* **1976**, *32*, 27–36.

48. FAO. Guidelines: Land evaluation for rainfed agriculture. *FAO Soil Bull.* **1983**, *52*, 237.

49. FAO. Guidelines: Land evaluation for irrigated agriculture. *FAO Soil Bull.* **1985**, *55*, 231.

50. FAO. Land evaluation: Towards a revised framework. *Electron. Publ. Policy Support Branch.* **2007**, *124*, 107.

51. Liu, Y.; Jiao, L.; Liu, Y.; He, J. A self-adapting fuzzy inference system for the evaluation of agricultural land. *Environ. Model. Softw.* **2013**, *40*, 226–234. [CrossRef]

52. Ferretti, V.; Montibeller, G. Key challenges and meta-choices in designing and applying multi-criteria spatial decision support systems. *Decis. Support Syst.* **2016**, *84*, 41–52. [CrossRef]

53. Fischer, G.; Shah, M.; van Velthuizen, H. Agro-ecological zones assessments. In *Land Use, Land Cover and Soil Sciences*; EOLSS Publishers: Oxford, UK, 2006.

54. Ren, Y.; Huang, J.; Xue, Y. *Thematic Evaluation of Cultivated Land Quality in Maize Superiority Areas in Northeast China*; China Agriculture Press: Beijing, China, 2017.

55. National Agro-Tech Extension and Service Center. *Thematic Evaluation of Cultivated Land Quality in Rice Superiority Areas in Northeast China*; China Agriculture Press: Beijing, China, 2017.

56. Liu, D. Study on Evaluation Index System of Climate Suitability of Main Crops. *Heilongjiang Agric. Sci.* **1991**, *4*, 9–14.

57. Sun, Y.; Shen, S. Research Progress in Application of Crop Growth Models. *Chin. J. Agrometeorol.* **2019**, *40*, 444–459.

58. Hou, Y.; Zhang, Y.; Wang, L.; Lv, H.; Song, Y. Climate Suitability Model for Spring Maize in Northeast China. *Chin. J. Agrometeorol.* **2013**, *24*, 3207–3212.

59. Chen, Y.; Li, X.; Liu, X.; Li, S. Coupling Geosimulation and Optimization (GeoSOS) for Zoning and Alerting of Agricultural Conservation Areas. *Acta Geogr. Sin.* **2010**, *65*, 1137–1145.

60. Chen, Y.; Li, X.; Liu, X.; Liu, Y. An agent-based model for optimal land allocation (AgentLA) with a contiguity constraint. *Int. J. Geogr. Inf. Sci.* **2010**, *24*, 1269–1288. [CrossRef]

61. Wezel, A.; Casagrande, M.; Celette, F.; Vian, J.F.; Ferrer, A.; Peigné, J. Agroecological practices for sustainable agriculture. A review. *Agron. Sustain. Dev.* **2014**, *34*, 1–20.

*Article*

# Farmland Transitions in China: An Advocacy Coalition Approach

Xiaoping Zhou [1], Xiaotian Li [1], Wei Song [2,*], Xiangbin Kong [3] and Xiao Lu [4]

1   School of Government, Beijing Normal University, Beijing 100875, China; zhouxp@bnu.edu.cn (X.Z.); lixt@mail.bnu.edu.cn (X.L.)
2   Key Laboratory of Land Surface Pattern and Simulation, Institute of Geographic Sciences and Natural Resources Research, Chinese Academy of Sciences, Beijing 100101, China
3   College of Land Science and Technology, China Agricultural University, Beijing 100083, China; kxb@cau.edu.cn
4   School of Humanities and Law, Northeastern University, Shenyang 110169, China; lvxiao@mail.neu.edu.cn
*   Correspondence: songw@igsnrr.ac.cn

**Abstract:** In recent decades, global social and economic development has resulted in substantial land-use transitions. This was first observed with respect to losses of forested land, attracting worldwide concern. Forest transitions have an important impact on global ecology, whilst farmland transitions are key in terms of global food security. However, research into farmland transitions is lacking, particularly with respect to mechanistic analysis. Using data on China's farmland areas between 1950 and 2017, we investigated the transitional characteristics, and triggers, of farmland change through linear regression analysis. Furthermore, based on the Advocacy Coalition Framework, we reveal the internal mechanism of these transitions. Our main findings are as follows: (1) between 1950 and 2017, China's farmland area exhibited significant growth, and there were two transitions, namely in 1984 and 2004; (2) macroscopic economic and social changes determine the overall evolution of the farmland area; (3) there were two advocacy coalitions in the farmland transition policy subsystem—the farmland supplement and farmland consumption coalitions; (4) under the influence of macroscopic economic and social development, external events play a catalytic role in the transitions, and relatively stable parameters have an indirect but lasting effect in terms of transition outcomes.

**Keywords:** land change science (LCS); farmland transitions; advocacy coalition; China

**Citation:** Zhou, X.; Li, X.; Song, W.; Kong, X.; Lu, X. Farmland Transitions in China: An Advocacy Coalition Approach. *Land* **2021**, *10*, 122. https://doi.org/10.3390/land10020122

Academic Editor: Stephen J. Leisz
Received: 9 November 2020
Accepted: 21 January 2021
Published: 27 January 2021

**Publisher's Note:** MDPI stays neutral with regard to jurisdictional claims in published maps and institutional affiliations.

## 1. Introduction

Land change science (LCS) is an interdisciplinary research approach that integrates social science, environmental and geographical information, and remote sensing science [1]. Recently, LCS has become an important part of global environmental change and sustainable development sciences [2]. Recently, scholars have started to focus on trends in land-use changes [3], that is, on land-use transformations. Land-use transitions have become complex processes involving multiple land-use types [4–6]. Among them, forest land transformations have attracted worldwide concern [7]. However, studies on farmland transitions are scarce, although farmland is the basis of food production. With rapid global industrialization and urbanization, large amounts of farmland have been permanently converted into non-agricultural areas [8,9], and the world's per capita farmland area declined from 0.41 ha. in 1960 to 0.21 ha. in 2019 [10]. For the vast number of developing countries that face the pressure of food security and the need for economic development, there is a sharp conflict between the protection and loss of farmland [11,12]. In this sense, farmland transitions in developing countries should become a research hotspot.

Numerous studies have explored the concept of land-use transition. In 1987, Walker [7] first proposed the term "land-use transition" when studying deforestation in developing countries; however, at that time, the focus was on describing the process of discarded logging land being reclaimed as agricultural land. In the early 1990s, Mather [13] studied the dynamics of forest areas in developed countries and proposed a forest transition hypothesis. Further, he predicted the spatiotemporal dynamics of forest land by invoking the forest transition curve [14]. Based on the theory of forest land transitions, Grainger [15] explained land-use transitions from the perspective of national land-use morphology. After long-term exploration by the academic community, a land-use transition was defined as a change in the land-use system from one state to another [3]. Since then, others have improved this concept, believing that land-use transitions refer to long-term land-use trends in a certain area in the context of socio-economic development [16], accounting for farmland, forest land, urban land, and homesteads [17]. Research on land-use transformations approaches the topic from one of two perspectives: trends in single-type land-use patterns and overall land-use patterns in a region [18,19].

In the study of land-use transitions, forest land and farmland are the two most typical land-use types, and their primary transitional characteristics have been widely discussed. Regarding forest land transitions, Rudel's [20] analysis of cross-country data from five consecutive world forest resource surveys concluded that in many countries, forest cover has undergone a trend reversal from deforestation during economic poverty to reforestation during economic development. Studies on forest land transitions in Vietnam [21] and Germany [22] have verified this conclusion. Similarly, Culas [23] states that forest land transitions exhibit different trends in different countries and regions; the author used the Environmental Kuznets Curve (EKC) to analyze panel data spanning 43 countries from 1970 to 1994. Generally, the EKC is applied to express the relationship between environmental quality and per capita income. In this article, the horizontal axis is "per capita income", and the vertical axis is "deforestation". The results showed that the inverted U-type EKC is suitable for Latin America and Africa, whereas a U-shaped function is suitable for Asia. In comparison to forest land transitions, most scholars researching farmland transitions have neglected to focus on changes in trends and the reasons underlying those changes [24].

Changes in farmland areas in different countries have been explored, for example, using case-studies in northern Ghana where they have steadily expanded for 31 years [25], while farmlands in the plateau region of northern Argentina have decreased despite increases in other farmland areas in other regions [26]. Ge et al. [27] constructed a theoretical model to explain the temporal transition of farmlands in China by using per capita farmland as the measurement index and analyzing the transition between 1990 and 2010. Using the year 2000 as the turning point, the results showed that 71% of China's farmland areas have experienced a steady transition from gradual decline in farmland per capita to gradual growth. However, the short time-span of this sample does not cover important Chinese policies such as "The Great Leap Forward" (1958) or the economic reform and opening-up (1978); therefore, they cannot fully reflect the overall position of China's farmland transitions.

The driving forces and mechanisms behind land-use transitions reveal important foci which should be emphasized to, and by, policy-makers and other decision-makers. For example, factors such as the natural environment [28], population changes [29], economic development [30], and energy changes [31] are important driving forces behind land-use transitions. Regarding farmland transitions, it has been stated that natural factors account for the basic elements of the transition itself, and the influences of elevation and slope are particularly significant [27]. Socio-economic factors are also the focus of research on driving forces and include population, GDP, fixed asset investment, and per capita disposable income [32]. According to one study, immigration plays an important role in land-use transitions because it leads to urban expansion and farmland occupation. For example, in the United States, urban expansion not only directly converted farmland to urban use, but also left it idle due to the spillover effects of urbanization [33]. At the same time, political and policy factors in the process of land use transitions play an important role and can therefore not be ignored. After Spain joined the European Economic Community in

1986, the agricultural land area in that country decreased from 45 to 38% [34]. The driving forces mentioned above, especially regarding policies, need to be taken into account when analyzing the farmland transition in China; policy factors play a crucial role in the transition under the strict planning and management characteristics of land-use in China.

In the 1980s, American political scientist Paul Sabatier proposed the Advocacy Coalition Framework (ACF). This is a systematic analytical tool that covers numerous aspects such as natural resources, socio-economic structures, political environment, emergencies, actors, and values. The ACF uses the policy belief system as the standard with which to integrate different policy actors and to analyze how the coalitions of these actors affect policy outputs [35]. Scholars have leveraged this framework in different country contexts and have applied different policy issues to test it. For example, forest certification is a process that uses market mechanisms to ensure that timber harvesting sources are legal, thereby promoting sustainable forest management. When analyzing forest certification issues in Indonesia, Canada, and Switzerland, Elliott and Schlaepfer [36] asserted that the ACF is an effective analytical tool that allows an understanding of multiple participants and involves a policy-learning process. Leifeld [37] traced the change from a single hegemonic advocacy coalition to a stable coalition by analyzing changes in the alliance belief system in the German pension security policy subsystem and corrected the deficiencies of the policy innovation change concept in the ACF. These studies have helped to verify the rationality and science-based approach of the ACF and have also led to revision and refinement of the framework itself to render it more sophisticated and applicable.

Several practices suggest that the ACF is suitable for analyzing governance issues with serious value differences. Land is a scarce resource, and when coupled with its status as a comprehensive space carrier, the use process is full of competition over interests and value uncertainties. In this context, some scholars have introduced the ACF into land-use research. For example, Heinmiller and Pirak [38] studied the urban changes of the Greater Golden Horseshoe region in Canada based on the ACF. The research showed that three coalitions (the agricultural, environmentalist, and developer coalitions) formed a policy subsystem around their basic and unique core beliefs of the policy. Similar to urban changes, the transition of farmland areas is also a complex land-use problem. From a global perspective, regardless of whether a country's land areas are publicly or privately owned, changes in farmland areas are, to varying degrees, affected by policies. As a country that is typically characterized by public land ownership, the spatiotemporal dynamics of China's farmlands reflect stronger policy measures. Farmland transitions can be seen as a product of land-use policy changes, and the ACF explains the processes of these changes from the perspective of the learning and interactions of policy advocates' coalitions. In this sense, the ACF may be an effective framework to study the mechanism underlying farmland transitions.

Overall, although forest land transitions have been widely studied, available research and exploration of farmland, as a hybrid artificial-natural land system, is insufficient. In particular, research on farmland use has not adequately considered trends in farmland transitions, and long-term studies on farmland transitional characteristics are lacking. In addition, regarding the underlying mechanisms, natural resources, social economic structure, and the political environment, and other factors, also need to be considered.

Against this background, to fill these knowledge gaps, the objectives of this study are as follows: (1) to reveal the inflection point of the farmland transitions and the change in trend characteristics at different stages over a long-term scale by analyzing farmland data for China from 1950 to 2017; (2) to analyze the policy actors, coalition composition, and belief systems present in farmland transition contexts based on the perspective of the ACF and build a theoretical framework that systematically explains the mechanisms of the transitions; (3) to analyze the impacts of coalition changes and external events at different transition stages on farmland transitions and reveal the mechanism underlying farmland transitions in China.

## 2. Materials and Methods

### 2.1. Overview

This article analyzes farmland transitions from the perspective of quantity changes. Because a "farmland transition" refers to long-term changes in farmland area, a linear regression method was used to visualize changes in farmland area in China from 1950 to 2017. Changes in land-use patterns tend to exhibit substantive volatility, and transition means that trend changes have occurred. We determined these points from images of farmland area change.

### 2.2. Advocacy Coalitions Affecting Farmland Transitions

The ACF identifies policy actors in a policy subsystem that share a particular set of beliefs and take action based on their shared beliefs [35]. The framework can be demarcated into three components: relatively stable parameters, external events, and policy subsystems (Figure 1). Among them, relatively stable parameters, such as natural resources and social structures, do not change considerably over a certain period and are therefore not considered herein. This study primarily aims to analyze the operational mechanism of the policy subsystem and the role of external events in farmland transitions in China. In the policy subsystem, the participants involved were first identified, and subsequently, the coalitions that these participants formed were determined. The policy belief structures of the coalitions were then analyzed, and the policy outputs formed by the interactions among these coalitions were identified. Finally, these policy outputs formed the farmland transitions.

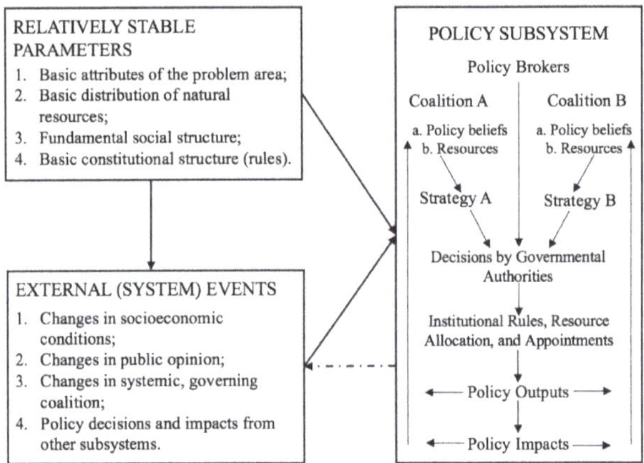


**Figure 1.** The Advocacy Coalition Framework, revised according to Weible et al. [39].

According to the ACF, members of the same coalition have no disputes over long-term coordinated actions [35]. Therefore, when identifying members, a coalition is often divided according to the different attitudes held by members towards a certain coordinated action. In developed countries, due to the high availability of recorded texts concerning the processes of policy agendas, formal minutes are a favored source of data, e.g., from parliaments, hearings, and legislative committees. Within these records, one can recognize which social groups are involved in the policy agenda and their policy claims. Therefore, by identifying these policy claims, members with similar claims can be classified as being in the same advocacy coalition [40,41].

However, in China, due to the difficulty in obtaining texts on the processes of policy agendas, it was necessary to make certain adjustments to the method of identifying alliance members. First, as farmland transitions reflect trend changes, it was possible to list the activities that led to increases (Table 1: A1–A4) and decreases in farmland area (Table 1: B1–B4) from the land activity codebook in Table 1. Second, the text data of various plans that had direct and significant impacts on farmlands were collected and analyzed according to the codebook (Table 1). Subsequently, information on the various land activities involved in the text was extracted. The organizations responsible for managing such land activities may be the standard from which the organizations' interests were affected; therefore, these participants were identified in the policy subsystem of the transition. Finally, according to their different policy opinions on different land-use activities, policy participants were divided into different advocacy coalitions.

**Table 1.** Land activity codebook.

| | Land Activity | Definition |
|---|---|---|
| Farmland Area Increase Activity | A1. Land development | Activities that bring unused land into a cultivable status through engineering, biological, or multi-disciplinary measures |
| | A2. Land arrangement | Adjusting the layout of land-use through the integration of farmland and centralized merging of settlements |
| | A3. Land reclamation | Taking remediation measures vis-a-vis land destroyed by mines to make it available. Some reclaimed land is used as farmland |
| | A4. Positive agricultural restructuring | Changing the agricultural production structure according to changes in the market demand for agricultural products, resulting in an increase in farmland areas |
| Farmland Area Decrease Activity | B1. Construction | Non-agricultural construction spanning industry, mining, transportation, and real estate occupy farmland, resulting in a decrease in farmland areas |
| | B2. Damaged by disaster | Farmland that has been washed away, burned, or buried due to various disasters and that cannot be restored in the short term |
| | B3. Withdrawing from farmland for ecological reasons | Due to ecological needs, the state plans a stepwise conversion of farmland to forests, grasslands, and lakes |
| | B4. Negative agricultural restructuring | Changing the agricultural production structure according to changes in the market demand structure for agricultural products, resulting in a decrease in farmland areas |

In the ACF, the beliefs of the policy actors drive their actions, and they want these beliefs to be reflected in policy outputs. In a coalition, if the coalition members have no disputes over long-term coordinated actions, they must have the same belief system, which acts as the bond and core driving force behind the formation of the policy actors. This belief system plays a linking role and can be divided into three levels: deep core, policy core, and secondary beliefs (Figure 2). The deep core beliefs at the innermost core are abstract and include long-term judgments on axiomatic issues and attitudes towards nature, and they remain virtually unchanged [35]. The core policy beliefs are located in the middle layer and apply the deep core beliefs to specific policy subsystems, creating the basis for uniting different policy actors to form an advocacy coalition [35]. The outermost secondary beliefs are instrumental decisions formed by different coalitions based on experiences and necessary information searches to realize the core policy beliefs of the coalitions [35,42]. If a coalition wants to occupy an advantageous position in a policy agenda, it needs to continuously expand the influence of its belief system. In the farmland transition policy subsystem, by identifying the differences in the belief systems of the different coalitions, the coalition's influence on land policy outputs and farmland transitions can be further analyzed.

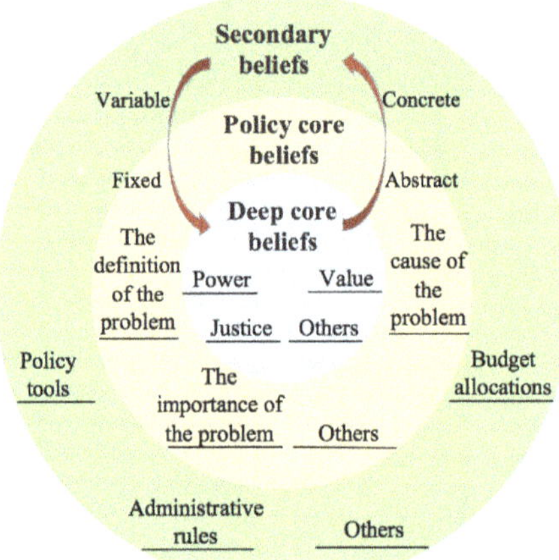



**Figure 2.** Structure of policy belief systems, adapted from Sabatier [35].

Generally, major policy changes are affected by external events in the policy's subsystem. Such events include changes in the socio-economic environment, changes in public opinion, changes in the ruling coalition's system, policy decisions, and influences from other subsystems. Policy changes will affect the spatiotemporal dynamics of farmland areas, thus affecting farmland transitions. Therefore, when studying this transition, it is crucial to also pay attention to various external events that can affect policy changes.

*2.3. Data Sources*

Due to the lack of accurate farmland survey data in the People's Republic of China (PRC) during the 1950s–1990s, farmland data from 1950 to 1995 were based on the work of Feng et al. [43]. Those authors analyzed, inverted, and reconstructed China's farmland data in stages. Using the farmland survey data obtained in 1953 as the basis for determining agricultural tax as a reference point, the verification results suggest that the data from 1949 to 1952 have a good connection. China's statistical work was conducted from 1953 to 1960, and the statistical results were appropriately consistent with the net reduction in farmland areas. Overall, the data from 1949 to 1960 can directly reflect the amount of farmland in China. The relevant data sources from 1986 to 1995 are diverse and vary greatly; they include data on the spatiotemporal dynamics of farmland published by the Ministry of Land and Resources (MLR), which is the most reliable source. Therefore, based on the results of the detailed national land survey in 1996, Feng et al. inverted the MLR data from 1986 to 1995, on a yearly basis. The statistical data from 1960 to 1985 were inconsistent with the actual changes in farmland trends; therefore, farmland areas were fit and reconstructed based on grain output data. The premise for adopting this method is that China's grain data are authentic at this stage, and it has been verified that farmland area and grain output were strongly correlated during this period. Considering that the "household contract responsibility system", initiated in 1978, was a watershed moment in China's agricultural development, Feng et al. simulated the grain-farmland area relationship between 1961 and 1978 and between 1978 and 1985, using data from 1949 to 1960 and 1986 to 1995, respectively. The farmland area data after reconstruction from 1950 to 1995 were well connected between the different periods and comparison with other reference data also showed good consistency. Therefore, the results of Feng et al. were used as the basis of the data analysis in this article. Farmland data between 1996 and 2008 were obtained from the "China Statistical Yearbook" [44]. It should be noted that a national land survey was conducted in China from 2007 to 2009, and the results of this survey were adopted in the farmland area of 2009. Changes in survey techniques and methods led to a significant "jump" in the amount of farmland between 2008 and 2009 (Figure 3), and therefore, statistics after 2009 cannot reliably reflect the changing trend in farmland area between 2008 and 2009 because the presence of one or more substantive artefacts cannot be ruled out. To improve the practicability of the data, 2009–2017 farmland data refer to the results of farmland data reconstruction by Wang et al. [45]. Based on original statistical data, the authors used the ARIMA model (Autoregressive Integrated Moving Average model)[1] to predict and extrapolate the amount of farmland before and after 2009 and to correct the data.

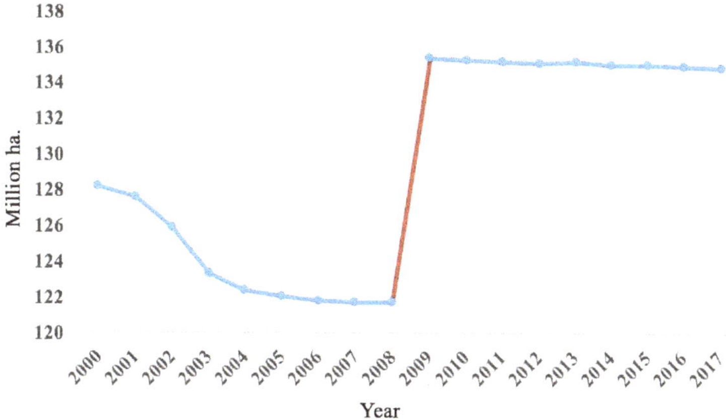


**Figure 3.** Changes in China's farmland area between 1996 and 2017, including the adjustments that resulted from the 2007–2009 national land survey.

When analyzing the policy subsystem of the farmland transitions, the text data included the "Outline of the National Land Use Master Plan" [46,47] (1997–2010 and 2006–2020: two periods in total), the "National Economic and Social Development Master Plan" [48] (1953–1957 and 1958–1962: once every 5 years after 1966, comprising a total of 13 periods), the "National Land Improvement Plan" [49] (2011–2015 and 2016–2020: two periods in total), and other important planning texts that had a direct impact on land-use.

## 3. Results and Discussion

### 3.1. China's Farmland Transitions

Regression analysis of China's farmland area (Figure 4) from 1950 to 2017 reveals alternating trends between increases and decreases. However, overall, after nearly 70 years of development, China's farmland areas have generally increased.

Throughout the study period, there were two inflection points on the curve of farmland area change, i.e., in 1984 and 2004. From 1950 onward, the farmland area curve showed a moderate and stable upward trend, but the total amount was far below the average value in the sample. This trend was interrupted after 1957, and from 1958 to 1960, the farmland area in China experienced a short but drastic period of decline, with large farmland losses. Of the entire study period, 1960 was the year with the lowest farmland area in China. However, from 1961 to 1984, farmland area increased rapidly, reaching a peak in 1984 before starting to decline once again. After 2005, when the farmland area gradually approached the "arable land minimum"[2], the downward trend weakened, and the farmland area remained reasonably stable.

---

[1]  A method for converting non-stationary time series to stationary time series by differential processing.

[2]  "Arable land minimum" refers to the minimum area of the arable land that should be protected. China sets this value at 120 million hectares.

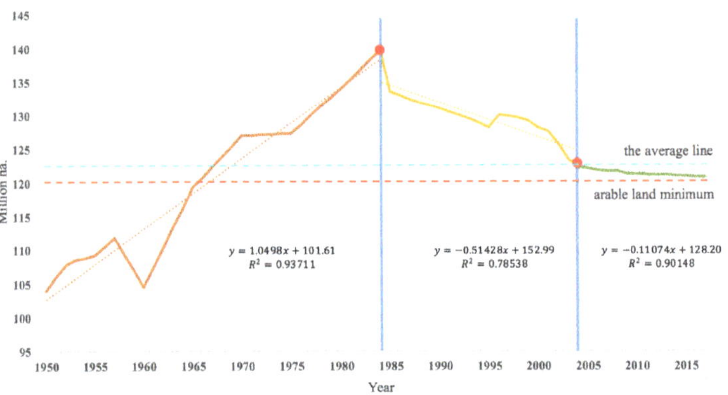


**Figure 4.** Regression analysis of farmland area changes in China between 1950 and 2017.

Evidently, farmland area experienced two transitions in terms of quantity. The first is from a fast growth period to a rapid descent period in 1957, and the second is the transition to a stable period in 2004.

### 3.2. Advocacy Coalitions Affecting Farmland Transitions

There are two advocacy coalitions in the process of policy changes that affect farmland transitions: the farmland supplement coalition and the farmland consumption coalition (Table 2). The farmland supplement coalition supports land activities related to the expansion of farmland areas, such as land development, arrangement, and reclamation. The primary policy actors include the Ministry of Land and Resources of the PRC, the Ministry of Agriculture of the PRC, the State Administration of Grain, and think tanks that study farmland protection. The first three of these policy actors are responsible for the protection of farmland and food security throughout the country and represent the backbone of the farmland supplement coalition. Land planning experts and policy researchers are collectively referred to as think tanks. They use their expertise to provide policy advice and suggestions on farmland protection, farmland reclamation, and the improvement of farmland quality. The farmland consumption coalition primarily involves policy actors involved in land development, land reclamation, and farmland occupation. The farmland occupation activities primarily include transportation, water conservancy, urban construction activities, real estate development, rural housing construction activities, and energy extraction activities.

Additionally, there are two other important policy actors in the farmland transition policy subsystem. Their positions and attitudes are relatively contradictory in the policy actions related to farmland, and they do not belong to any coalition but are defined as "policy brokers". The National Development and Reform Commission department guides overall reform of the economic system and provides macro-control. Furthermore, it is responsible for managing the State Administration of Grain in the farmland supplement coalition and the National Energy Administration in the farmland consumption coalition. To meet economic and agricultural production goals, it is necessary to take a macro-perspective which, in turn, means impacting on various land-use arrangements. Under China's administrative system, local governments are simultaneously responsible for various domains such as economic development, farmland protection, or withdrawing from farmland for ecological reasons; they act as brokers for all parties in the policy subsystem of farmland transitions.

**Table 2.** Advocacy coalitions in the farmland transition policy subsystem.

| | Increase in Farmland Activity | | | | | Decrease in Farmland Activity | | |
|---|---|---|---|---|---|---|---|---|
| | Land Development | Land Arrangement | Land Reclamation | Positive Agricultural Restructuring | Construction | Damaged by Disaster | Withdrawing from Farmland for Ecological Reasons | Negative Agricultural Restructuring |
| Farmland supplement coalition | | | | | | | | |
| Ministry of Land and Resources | Y | Y | Y | | | | | |
| Ministry of Agriculture | Y | Y | Y | Y | | | | N |
| State Administration of Grain | Y | Y | Y | Y | | | | N |
| Think tanks | Y | Y | Y | | | N | Y | |
| Farmland consumption coalition | | | | | | | | |
| Rural collective economic organizations | | | | N | Y | | | Y |
| National Energy Administration | N | | N | | Y | | | |
| Ministry of Transport | N | | N | | Y | | | |
| Ministry of Housing and Rural Development | N | | N | | Y | | | |
| State Forestry Administration | | | | N | | | Y | Y |
| Ministry of Water Resources | N | | N | | Y | | Y | |
| Industrial and mining enterprises | N | | N | | Y | | | |
| Real estate development enterprises | N | | N | | Y | | | |
| Policy brokers | | | | | | | | |
| National Development and Reform Commission | Y | Y | Y | | Y | | Y | |
| Local government | Y | Y | Y | | Y | | Y | |

Note: Y = yes (supports this type of land activity); N = no (does not support this type of land activity).

Table 3 shows that the farmland supplement and farmland consumption coalitions have opposite belief systems. In terms of deep core beliefs, the key to the conflict between the two coalitions is whether, under the existing external environment, they should continue to pay attention to protecting the right to survival or give priority to the right to development. The farmland supplement coalition believes that "food for the program" is still a prudent mantra and farmland areas should be increased to maintain the fundamental bottom line of national food security, while the farmland consumption coalition is more in agreement with "taking economic construction as the center" and believes that sufficient land support should be provided for economic and social construction. The core policy beliefs are a representation of the deep core beliefs, and the contradiction between the two coalitions' beliefs revolves around the question of the type of land that should be focused on for protection. The farmland supplement coalition focuses on farmland protection and farmland area supplementation, making this coalition's line-up stable for a long period of time. However, the actors in the farmland consumption coalition are involved in all aspects of construction (including energy, transportation, and real estate), and they will only join the coalition when they determine that it is beneficial for them to do so. Against this background, changes in the coalition's membership occur relatively frequently. Such changes result in the two coalitions having a disparate number of policy participants, yet they are virtually equal in their overall strength and thus maintain the possibility of having a balanced competition of interests. From the perspective of their basic policy mechanisms, the mandatory planning force of the farmland supplement coalition is stronger than the market-based means of the farmland consumption coalition, and therefore, changes in farmland area exhibit an overall upward trend. Secondary beliefs are instrumental decisions that are made to achieve the core policy beliefs and necessary information searches under special circumstances. In terms of instrumental decisions, the two coalitions have further refined their basic policy mechanisms and selected different policy tools. Regarding information searches, in recent years, China has paid increased attention to "eco-civilization" [50]. In this context, the two coalitions have formed different perceptions of the ecological function of farmland. Since the 1990s, the theory and practice of good governance has flourished [51], emphasizing the diversification of management methods. In this context, the two coalitions hold different positions regarding the direction of system reform.

**Table 3.** The advocacy coalitions' belief systems on farmland transitions.

| Levels and Types of Beliefs | Farmland Supplement Coalition | Farmland Consumption Coalition |
|---|---|---|
| | Deep core beliefs | |
| Value priority | Right to survival. | Right to development. |
| Basic standards of distributive justice | Food for the program[3]. | Taking economic construction as the center[4]. |
| | Policy core beliefs | |
| The definition of the problem | Farmland is the basic resource on which mankind depends and is the fundamental guarantee of national food security. | Development and construction are the most important ways to improve national economic strength. It is possible to reduce farmland by developing agricultural technology, thereby increasing the output per unit area. |
| The importance of the problem | China's population continues to grow and the demand for grain is large, but the area for farmland is decreasing. Therefore, farmland protection is urgent. | The rapid development of China's economy requires adequate land security for urban construction, infrastructure construction, industrial and mining construction, among others. |
| The cause of the problem | Construction encroachment and withdrawing from farmland for ecological reasons lead to farmland area reduction. | Some scattered farmland areas have impeded urban and industrial development. |

**Table 3.** *Cont.*

| Levels and Types of Beliefs | Farmland Supplement Coalition | Farmland Consumption Coalition |
| --- | --- | --- |
| Basic policy mechanism | Defining "arable land minimum" and adopting mandatory planning and legislative means to effectively protect and supplement farmland. | Market methods should be used to coordinate various land-use relations, and the use and protection of farmland should be incorporated into the market mechanism. |
| | Secondary beliefs | |
| Policy tools | Approval system for the conversion of farmland, a farmland acquisition–compensation balance system, and so on. | Farmland occupation tax, development rights transfer mechanism, farmland index transaction, among others. |
| Adequacy of ecological protection | Farmland has ecological service functions, and the farmland ecosystem meets the needs of environmental protection. | The farmland ecosystem is unstable, and withdrawing from farmland cultivation for ecological reasons meets environmental protection requirements. |
| The direction of system reform | Farmland protection needs to be gradually legislated and institutionalized. To adapt to the new situation, it is necessary to constantly improve the farmland protection system. | At present, farmland protection in China is too dependent on compulsory institutional measures and should be appropriately adjusted to reduce the control of areas by improving the availability of space for economic development. |

### 3.3. The Farmland Transition Mechanism Based on the ACF

Farmland transitions reflect coalitions' interactions with farmland use and protection against the background of national macroeconomic and social development. Economic and social transitions at the national macro-level determine the overall direction of farmland area changes. Both the farmland supplement coalition and the farmland consumption coalition launched their competitive interests over the policy agenda related to land use. Under the combined influences of external events and relatively stable parameters, a series of policy outputs were formed. These policy outputs determine changes in farmland area to a certain extent. Combined with the change of macroeconomic policy environment, this paper analyzes the reasons underlying the farmland transition, based on the ACF, and explores the important policy outputs concerning farmland use in China over the last 70 years (Figure 5 and Appendix A).

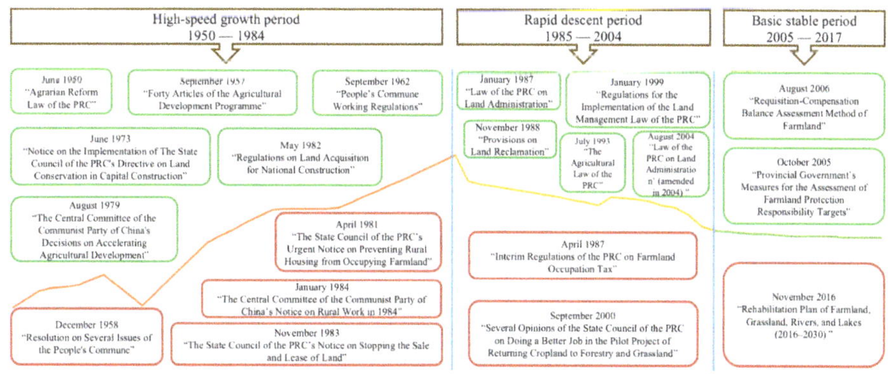


**Figure 5.** Changes in China's farmland use policies; further details can be found in the Appendix A.

---

<sub>3</sub> This was a slogan put forward by Chinese government in the 1950s to address the issue of insufficient grain supply; it called for concentrated efforts to develop grain cultivation and ensure food production.

<sub>4</sub> This was a priority for China to promote economic and social development in the 1980s, calling for a concentrated effort to develop social productive forces and facilitate national industrialization.

### 3.3.1. Macro-Economic and Social Dynamics Determine Farmland Transitions

First, changes in demographic policy influenced farmland transitions in China to a large degree. From the 1950s to the early 1980s, China's population experienced rapid growth due to the concept of "many children, many blessings". The Chinese government encouraged population growth at this stage. However, the limited natural resources could not sustain the growing population, resulting in poverty and hunger. For this reason, farmland increased rapidly to enhance grain production. However, out of fear of excessive population growth, the Chinese government shifted from encouraging population growth to slowing down population growth in 1982. This change also slowed down farmland expansion.

Second, changes in the household registration system influenced farmland transitions in China. The "Regulations of the People's Republic of China on Household Registration", issued in 1958, contained a strict urban-rural dual household registration system. Until the 1980s, this system severely restricted migration within China, especially rural-to-urban migration. Therefore, the growth of the rural labor force at this stage was positively correlated with the growth of farmland area. After 1984, China began to implement the resident identity card system, gradually relaxing household registration control. The rural labor force was strongly attracted to industries and cities. In the context of relaxing household registration control, a large part of the rural population migrated to cities, which resulted in a decrease in farmland area.

Last, economic restructuring also influences the overall evolution of farmland area. During the research period, China underwent a transition from a planned economy to a market economy. During the period of the former (from the 1950s to 1980s), China's economic and social development depended on the substantial development of agriculture, characterized by the expansion of farmland areas to increase agricultural production. Since the 1980s, China has gradually been establishing a market economy system. The Chinese government has vigorously promoted industrialization and urbanization, which has led to an increase in the demand for construction land. Thus, large areas of farmland were converted into built-up land.

### 3.3.2. Changes in the Coalition Power Balance Influence Farmland Transitions

There was a period of rapid farmland growth between 1950 and 1984. Based on the policy outputs of this period (Figure 6), it is evident that for a country that was just emerging from the Second Sino-Japanese War (1931–1945) and the Chinese People's War of Liberation (1946–1949) and was in despair, food was the cornerstone of people's livelihoods, alongside national stability. As such, with the support of all salient actors, restorative growth of farmland was possible. However, the contradiction between the limited land resources and the expanding population began to be obvious. The reserve land could not meet the growing needs of farmland and construction land, and the opposition between the two coalitions was formed. After a period of development, the core policy beliefs of the "food for the program" of the farmland supplement coalition (proposed in 1958) were relatively stable. From the perspective of policy outputs, the coalition gradually developed and matured in minor respects, such as in terms of issues surrounding wasteland and lake reclamation. During this period, the farmland consumption coalition initially formed the core policy belief of "taking economic construction as the center" (proposed in 1978). The main actors in the coalition were rural collective economic organizations. The consumption of farmland was reflected in rural housing construction and the development of social enterprises. Nevertheless, these areas only occupied a small part of the farmland area. Overall, the strength of the farmland supplement coalition at this stage was higher than that of the farmland consumption coalition. Thus, the farmland area continued to grow rapidly. Although there was a brief period with a sharp decline in farmland between 1958 and 1960 due to poor judgements by top-level system designers, the overall upward trend again resumed in the 1960s.

From 1985 to 2004, farmland areas experienced a period of long-term steady descent. During this stage, China's economy expanded rapidly, accompanied by increased urbanization and a higher demand for construction land. Under the double promotion of expansion (demand for urban construction and expansion of rural autonomy), all types of illegal farmland occupation were frequent. During this period, although there were several policies related to farmland protection, they showed obvious problem orientations and post-recovery characteristics. With the participation of policy actors spanning industry, mining, and transportation, the power of the farmland supplement coalition greatly increased and occupied a dominant position in the policy agenda.

After 2005, the declining trend in farmland area was curbed, especially with the development of the Ministry of Land and Resources of the PRC, which was established in 1998 and became the backbone of the farmland supplement coalition. In terms of policy output, rationalization and institutionalization of management also occurred and the introduction of a series of systems such as the dynamic equilibrium of the total cultivated areas, acquisition–compensation balance, and a farmland protection-targeted responsibility system, enabled China's farmland protection measures to gradually form a systematic and comprehensive policy network. This led to the narrowing of the power gap between the two major coalitions. In the context of a more modernized and urban economy, the share of the agricultural sector in GDP and the share of the rural income structure are currently both decreasing [52], making it hard to recognize a new transition from declining to increasing farmland area. However, China's farmland area is expected to remain stable for a long time to come, as the power of the coalitions becomes more balanced.

### 3.3.3. Catalytic Action of External Events

During policy changes, external events act as catalysts to enhance or weaken the power of the alliance and often lead to major policy changes, directly affecting the policy subsystem in a short period of time. On the one hand, the mechanism by which external events play a catalytic role directly affects existing coalitions and changes their right to speak in the policy arena. On the other hand, it influences the strength of coalitions by changing the composition of those coalitions.

During the "Great Leap Forward"[5] between 1958 and 1960, iron and steel smelting was promoted for building large railways. Capital construction investments expanded rapidly, agricultural production stagnated, a large amount of farmland was wasted, and the power of the farmland supplement coalition was weakened. This led directly to short-term drastic reductions in farmland areas from 1957 to 1960.

Economic reform and opening-up began in 1978 and this had a profound impact on all aspects of Chinese society. At the beginning of this movement, with increasing per capita incomes, China's market for agricultural products expanded. Further, the "household contract responsibility system" enhanced the enthusiasm for production and improved the supply capacity of agricultural products [53]. Under the stimulation of both supply and demand, the cultivated land area continued to grow. However, with the deepening of the reform and opening-up, the rapid rise of the economy kicked various construction projects in China into full swing, and the threat to various types of farmland was high. At the same time, the increase in grain yield and grain imports temporarily suppressed the urgency to protect farmland. As a result, the farmland consumption coalition gradually came to dominate the policy subsystem, leading to another interruption to the upward trend and ushering in the first sudden change in farmland area in 1984.

In the 1990s, economic globalization became an important characteristic of the global economy. International competition and free trade accelerated the flow of factors and the international division of labor, which had far-reaching impacts on the economic systems of all countries in the world [54]. In the tide of globalization, China became the "workshop of

---

[5] This was a nationwide social production campaign in China between 1958 and 1960 that set a series of unrealistic economic tasks and targets, such as catching up with and surpassing the UK in the production of major industrial products over 15 years.

the world", taking full advantage of its cheap labor force. In this context, the development of processing and manufacturing industries accelerated construction on occupied farmland. The large number of farmers pouring into the cities to work also led to the partial abandonment of farmland. These are all important factors that contributed to the steady decline in farmland area in China since 1984. After China's accession to the WTO (World Trade Organization) in 2001, massive imports of agricultural products had a significant impact on the country's traditional natural economy [55], making the period of 2001–2004 a small climax of the decline in cultivated land area. However, at this stage, there were also some important external events affecting the farmland supplement coalition. In 1994, Lester R. Brown, director of the World Watch Institute, published a report entitled "Who will feed China? Wake-up call for a small planet" [56], which caused controversy about China's food security. As a result, China paid more attention to its domestic food security and farmland protection, and the strength of the farmland supplement coalition was enhanced. Under the influence of this event, in 1995, there was a small short-term increase in China's farmland area, albeit in the context of an overall downward trend. In 1998, a catastrophic flood occurred in the Yangtze River Basin. Upon investigation, it was found that the large amount of reclamation land had reduced the storage capacity of the rivers and lakes, which played an important role in the disaster. This also sounded the alarm for the "land reclamation" method of supplementing farmland areas, and the approach of "withdrawing from farmland for ecological reasons" started to play an important role in the policy subsystem. As a result of these external events, the farmland transition entered a new stage in 2005.

### 3.3.4. Relatively Stable Parameters Have Indirect but Long-Lasting Effects on Farmland Transitions

Generally, relatively stable parameters do not directly affect the interaction of coalition members, but objectively determine the resource constraints and the probability of policy changes. As shown in Figure 5, the overall trend of farmland change does not always correspond to policy outputs, especially in the period 1985–2004, when the policy outputs of the farmland supplement coalition were considerably higher than those of the farmland consumption coalition, but the farmland area declined for a long time. This was largely due to changes in the relatively stable parameters.

The transition from agricultural rural societies to industrial urban societies is the economic and social development path followed by most countries in the world. During the period from 1950 to 2017 (especially after 1978), China also experienced such structural changes [57], among which the urban and rural structure had the largest impacts on farmland changes. During the study period, China's urbanization rate increased from 11.18% in 1950 to 59.52% in 2017, with an increase of about 770 million permanent urban residents. Numerous rural people migrated to the cities, which resulted in farmland abandonment. This largely explains the steady decline in farmland since 1984.

In the process of social structure change, people's understanding of farmland also changed, which means that the basic attributes of the problem domain changed. In a subsistence-based agricultural economy, increasing farmland is seen as the only source of livelihood, especially for farmers, whose most important means of production is farmland. With the advent of industrialization and urbanization, the economic value of land and labor for cultivation was much lower than that of development and construction. As a result, awareness of the importance of farmland has become diversified, and some farmland protection policies have even run counter to the wishes of farmers. This is one of the reasons why the spatiotemporal dynamics of farmland are not consistent with policy outputs.

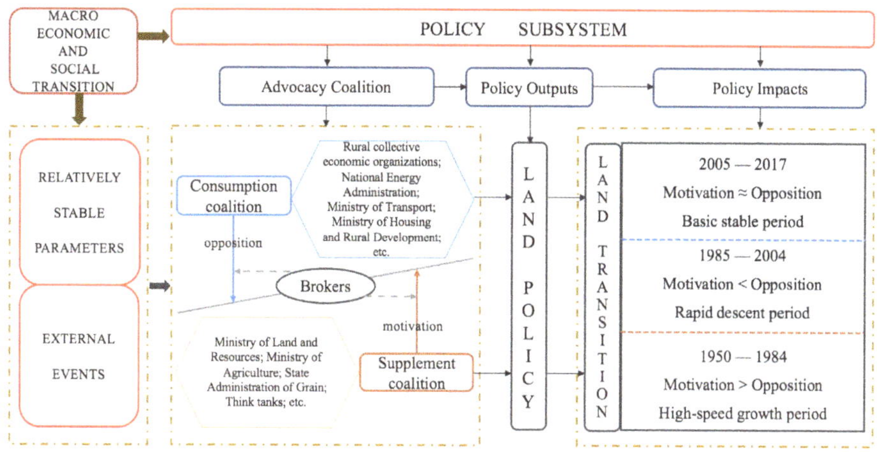


**Figure 6.** China's farmland transformation mechanism from the perspective of the Advocacy Coalition Framework.

## 4. Conclusions

Transitions emphasize changes in trends. Therefore, when studying farmland transitions, it is critical to determine inflection points. We used a linear regression method to analyze changes in China's farmland area from 1950 to 2017. We searched for, and sought to understand the reasons for, abrupt transition points. Based on our results, there are two inflection points (1984 and 2004) during the study period which means that there have been two transitions in China's farmland area between 1950 and 2017. From 1950 to 1984, China was in a period of sustained fast growth, with the total farmland area increasing rapidly and continuously. In the period from 1985 to 2004, there was a sharp decline in farmland area. After 2005, the range of change of farmland area became smaller and entered a reasonably stable period.

Farmland, as a type of national land arrangement, is most significantly affected by macro-level economic and social transitions. Since 1950, changes in China's macroeconomic policies have led to demographic transitions in the country, a change in the household registration management system and the reform of the economic system, all of which have ultimately determined the overall manner and direction of farmland area change.

We applied the ACF to study the mechanism of farmland transition. We highlight two advocacy coalitions in China—farmland supplement and farmland consumption—that held opposing belief systems. Of the two, the former was the driving force for the growth of farmland areas, whereas the latter represented the opposition. In this policy subsystem, the policy brokers consisted of local governments and the National Development and Reform Commission, acting simultaneously on the two coalitions and playing a coordinating role. Competition of interests and learning among the coalitions promoted the introduction of various land-use policies. In the policy output process, external events acted as catalysts. The implementation of various land policy outputs had an impact on the spatiotemporal dynamics of farmland areas, which, in turn, affected the transition. In this context, and from the perspective of the framework of advocacy coalitions, the transition of farmland areas reflects, to some extent, policy influence. During farmland transition, external events act as catalysts, whereas relatively stable parameters have indirect but lasting effects once they change.

We predict that, with the change from extensive to intensive land use, the farmland supplement coalition will no longer rely on the increase in farmland area to achieve agricultural development. In addition, the farmland consumption coalition will also tend to increase the efficiency of land use instead of occupying more farmland. This way, the conflict between the two coalitions is expected to decrease. The two alliances may move from confrontation to collaboration, with the aim to achieve growth of both agriculture and industry, based on new technologies.

Although this paper focuses on the farmland transition in China, this is also a worldwide problem, especially in developing countries. By advocating the framework of coalitions, a more comprehensive analytical tool is established. According to the identification of policy coalitions, the analysis of belief systems, and the analysis of external factors, it is helpful to understand the reasons for transitions and to predict future trends.

However, it should be noted that the research framework developed and applied herein depends on certain assumptions: the existence of a single stable government, public land ownership, and land use which is strictly planned and managed by the government. As such, applications in other country contexts need to ensure that the underlying framework is modified accordingly to ensure that it is fit for purpose.

**Author Contributions:** Conceptualization, X.Z. and X.L. (Xiaotian Li); methodology, X.Z., W.S., and X.L. (Xiaotian Li); formal analysis, X.L. (Xiao Lu), X.L. (Xiaotian Li) and X.K.; writing—original draft preparation, X.Z., X.L. (Xiaotian Li); writing—review and editing, W.S.; funding acquisition, X.K. and W.S. All authors have read and agreed to the published version of the manuscript.

**Funding:** This research was funded by the National Social Science Fund of China (Grant No. 19ZDA096) and The Second Tibetan Plateau Scientific Expedition and Research (Grant No. 2019QZKK0603).

**Institutional Review Board Statement:** Not applicable.

**Informed Consent Statement:** Not applicable.

**Data Availability Statement:** No new data were created or analyzed in this study. Data sharing is not applicable to this article.

**Conflicts of Interest:** The authors declare no conflict of interest. The funders played no role in the design of the study, in the collection, analyses, or interpretation of data, in the writing of the manuscript, or in the decision to publish the results.

# Appendix A

**Table A1.** Farmland use polies in China over time.

| Period | Policy Text | Policy Outputs |
| --- | --- | --- |
| | | Fast growth period (1950–1984) |
| June 1950 | "Agrarian Reform Law of the PRC" | The implementation of farmers' land ownership, the liberation of rural productivity, and the vigorous development of agricultural production led to a rapid increase in the demand for farmland nationwide. |
| September 1957 | "Forty Articles of the Agricultural Development Program" (revised draft) | This was the program's document for the agricultural "Great Leap Forward" and raised the requirements of agriculture, forestry, animal husbandry, by-products, and fishery production in a short period of time; it also raised excessively high production requirements, leading to the transformation of lakes into fields during its peak period. |
| December 1958 | "Resolution on Several Issues of the People's Commune" | This document proposed a "three-three" farming system, which required reducing the planting area of crops by using one-third of the total farmland for crops, one-third for trees and grass, and one-third for leisure. |
| September 1962 | "People's Commune Working Regulations" (draft amendment) | This document allowed the production team to reclaim wasteland, manage barren hills, and make full use of all possible resources within the scope of the team. |
| June 1973 | "Notice on the Implementation of The State Council of the PRC's Directive on Land Conservation in Capital Construction" | This required construction land to be used only after strict examination and approval to save land and avoid industrial land occupying farmland. |
| August 1979 | "The Central Committee of the Communist Party of China's Decisions on Accelerating Agricultural Development" | The document proposed to implement the "food for the program" policy, develop agricultural production, systematically reclaim wasteland, and convert lakes into fields. |
| April 1981 | "The State Council of the PRC's Urgent Notice on Preventing Rural Housing from Occupying Farmland" | This stated that due to occupied farmland being used for building houses and establishing enterprises in rural areas, extensive publicity and education should be carried out, the layout should be reasonably planned, arable land should not be occupied (as much as possible), building materials should be reformed, and damage to farmland should be reduced. |
| May 1982 | "Regulations on Land Acquisition for National Construction" | The document stipulated that any wasteland that could be used should not occupy farmland and stipulated the land compensation fee and resettlement compensation standard for the requisitioned farmland. During the resettlement process, wasteland should be properly developed to expand the farmland area. |
| November 1983 | "The State Council of the PRC's Notice on Stopping the Sale and Lease of Land" | The notice required that rural organizations must be firmly prevented from occupying farmland through privately negotiated conditions. They frequently occupied farmland and vegetable fields by renting, buying, and selling houses or by occupying the land by means of "jointly building houses", "jointly setting up factories", and "jointly building warehouses". This required that existing farmland must be protected. |
| January 1984 | "The Central Committee of the Communist Party of China's Notice on Rural Work in 1984" | This stated that the land contract period should be more than 15 years and should prohibit the conversion of contracted land to non-agricultural land, accelerate the development of mountainous areas, water areas, and grasslands, and develop rural transportation, post and telecommunications, and rural industry. |

**Table A1.** *Cont.*

| Period | Policy Text | Policy Outputs |
| --- | --- | --- |
| | | Rapid descent period (1985–2004) |
| January 1987 | "Law of the PRC on Land Administration" | This implemented a land-use control system, compiled an overall land-use plan, and stipulated land use as divided into agricultural land, construction land, and unused land. It strictly restricted the conversion of agricultural land into construction land, controlled the total amount of construction land, and implemented special protection for farmland. |
| April 1987 | "Interim Regulations of the PRC on Farmland Occupation Tax" | Units and individuals occupying farmland to build houses or engage in other non-agricultural construction were regarded as tax obligors of the farmland occupation tax. |
| November 1988 | "Provisions on Land Reclamation" | This stated that enterprises and individuals who cause land damage due to production and construction activities would carry out land reclamation in accordance with regulations and pay compensation for loss of farmland. If the reclaimed land was used for agricultural, forestry, animal husbandry, or fishery production, the relevant state regulations would receive agricultural tax relief. |
| July 1993 | "The Agricultural Law of the PRC" | This stipulated that governments both at and above the county level should delineate basic farmland protection areas in accordance with regulations and implement special protection for farmland in the basic farmland protection areas. |
| January 1999 | "Regulations for the Implementation of the Land Management Law of the PRC" | This implemented a dynamic balance system of total farmland and adopted a series of administrative, economic, and legal measures to ensure that the total area of existing farmland in China could only increase, not decrease, within a certain period, with the aim of gradually improving the quality of farmland. |
| September 2000 | "Several Opinions of the State Council of the PRC on Doing a Better Job in the Pilot Project of Returning Cropland to Forests and Grassland" | This implemented a system of withdrawing from farmland for ecological reasons. The Ministry of Finance, the Ministry of Water Resources, the Ministry of Agriculture, and other departments were jointly responsible for returning farmland to forests and grasslands. |
| August 2004 | "Law of the PRC on Land Administration" (amended in 2004) | This implemented an approval system for the conversion of farmland. If the land occupied by construction involved the conversion of farmland, then approval procedures for this conversion should be followed. |
| | | Basic stable period (2005–2017) |
| October 2005 | "Provincial Government's Measures for the Assessment of Farmland Protection Responsibility Targets" | This implemented a target responsibility system for farmland protection. The Ministry of Land and Resources, the Ministry of Agriculture, and other relevant departments should propose assessment indicators for the amount of farmland and the basic farmland protection area of the provincial government. |
| August 2006 | "Requisition-Compensation Balance Assessment Method of Farmland" | This implemented a system of farmland acquisition–compensation balance. The Ministry of Land and Resources' management department conducted assessments based on construction land projects and determined the quantity, quality, and capital of the farmland to be added according to the supplemented farmland plan. |
| November 2016 | "Rehabilitation Plan of Farmland, Grassland, Rivers, and Lakes" (2016–2030) | This implemented a rehabilitation system, strictly observed the "arable land minimum", established reasonable rotation and fallow systems, improved the quality of farmland, and ensured the safety of the soil environment. Land that was not suitable for farming needed to be withdrawn for ecological reasons. |

## References

1. Turner, B.L., II; Lambin, E.F.; Reenberg, A. The emergence of land change science for global environmental change and sustainability. *Proc. Nat. Acad. Sci. USA* **2008**, *104*, 20666–20671. [CrossRef] [PubMed]
2. Rindfuss, R.R.; Walsh, S.J.; Turner, B.L.; Fox, J.; Mishra, V. Developing a science of land change: Challenges and methodological issues. *Proc. Nat. Acad. Sci. USA* **2004**, *101*, 13976–13981. [CrossRef] [PubMed]
3. Lambin, E.F.; Meyfroidt, P. Land use transitions: Socio-ecological feedback versus socio-economic change. *Land Use Policy* **2010**, *27*, 108–118. [CrossRef]
4. Angel, S.; Parent, J.; Civco, D.L.; Blei, A.; Potere, D. The dimensions of global urban expansion: Estimates and projections for all countries, 2000–2050. *Prog. Plan.* **2011**, *75*, 53–107. [CrossRef]
5. Belay, K.T.; Van Rompaey, A.; Poesen, J.; van Bruyssel, S.; Deckers, J.; Amare, K. Spatial analysis of land cover changes in Eastern Tigray (Ethiopia) from 1965 to 2007: Are there signs of a forest transition? *Land Degrad. Dev.* **2015**, *26*, 680–689. [CrossRef]
6. Ge, D.; Long, H. Coupling relationship between land use transitions and grain yield in Huang-Huai-Hai Plain, China. *J. Agric. Res. Environ.* **2017**, *34*, 319–327. [CrossRef]
7. Walker, R.T. Land use transition and deforestation in developing countries. *Geog. Anal.* **1987**, *19*, 18–30. [CrossRef]
8. Valerià, P.; Mckenzie, F.H. Peri-urban farmland conservation and development of alternative food networks: Insights from a case-study area in metropolitan Barcelona (Catalonia, Spain). *Land Use Policy* **2013**, *30*, 94–105. [CrossRef]
9. Chanel, O.; Delattre, L.; Napoléone, C. Determinants of local public policies for farmland preservation and urban expansion: A French illustration. *Land Econ.* **2014**, *90*, 411–433. [CrossRef]
10. Masini, E.; Tomao, A.; Barbati, A.; Corona, P.; Serra, P.; Salvati, L. Urban growth, land-use efficiency and local socioeconomic context: A comparative analysis of 417 metropolitan regions in Europe. *Environ. Manag.* **2018**, *63*, 322–337. [CrossRef]
11. Liu, Y.; Fang, F.; Li, Y. Key issues of land use in China and implications for policy making. *Land Use Policy* **2014**, *40*, 6–12. [CrossRef]
12. Liu, Y. Introduction to land use and rural sustainability in China. *Land Use Policy* **2018**, *74*, 1–4. [CrossRef]
13. Mather, A.S. The forest transition. *Area* **1992**, *24*, 367–379. [CrossRef]
14. Mather, A.S.; Needle, C.L. The forest transition: A theoretical basis. *Area* **1998**, *30*, 117–124. [CrossRef]
15. Grainger, A. National land use morphology: Patterns and possibilities. *Geography* **1995**, *80*, 235–245. [CrossRef]
16. Song, X. Discussion on land use transition research framework. *Acta Geogr. Sin.* **2017**, *72*, 471–487. [CrossRef]
17. Li, T.; Long, H.; Liu, Y.; Tu, S. Multi-scale analysis of rural housing land transition under China's rapid urbanization: The case of Bohai Rim. *Habitat Int.* **2015**, *48*, 227–238. [CrossRef]
18. Liu, Y.; Long, H.; Li, T.; Tu, S. Land use transitions and their effects on water environment in Huang-Huai-Hai Plain, China. *Land Use Policy.* **2015**, *47*, 293–301. [CrossRef]
19. Liu, Y.; Yang, R.; Long, H.; Gao, J.; Wang, J. Implications of land-use change in rural China: A case study of Yucheng, Shandong province. *Land Use Policy.* **2014**, *40*, 111–118. [CrossRef]
20. Rudel, T.K. Is there a forest transition? Deforestation, reforestation, and development. *Rural Soc.* **1998**, *63*, 533–552. [CrossRef]
21. Lambin, E.F.; Meyfroidt, P. Forest transition in Vietnam and its environmental impacts. *Glob. Chang. Biol.* **2008**, *14*, 1319–1336. [CrossRef]
22. Plieninger, T.; Schleyer, C.; Mantel, M.; Hostert, P. Is there a forest transition outside forests? Trajectories of farm trees and effects on ecosystem services in an agricultural landscape in Eastern Germany. *Land Use Policy* **2012**, *29*, 233–243. [CrossRef]
23. Culas, R.J. REDD and forest transition: Tunnelling through the environmental Kuznets curve. *Ecol. Econ.* **2012**, *79*, 44–51. [CrossRef]
24. D'Amour, C.B.; Reitsma, F.; Baiocchi, G.; Barthel, S.; Güneralp, B.; Erb, K.-H.; Haberl, H.; Creutzig, F.; Seto, K.C. Future urban land expansion and implications for global croplands. *Proc. Nat. Acad. Sci. USA* **2016**, *114*, 8939–8944. [CrossRef]
25. Kikuko, S.; Braimoh, A.K.; Ram, A.; Saito, O. Land transition and intensity analysis of cropland expansion in Northern Ghana. *J. Environ. Manag.* **2018**, *62*, 892–905. [CrossRef]
26. Izquierdo, A.E.; Grau, H.R. Agriculture adjustment, land-use transition, and protected areas in north-western Argentina. *J. Environ. Manag.* **2009**, *90*, 858–865. [CrossRef]
27. Ge, D.; Long, H.; Zhang, Y.; Ma, L.; Li, T. Farmland transition and its influences on grain production in China. *Land Use Polocy* **2017**, *70*, 94–105. [CrossRef]
28. Jiang, Y.; Liu, J.; Cui, Q.; An, X.; Wu, C. Land use/land cover change and driving force analysis in Xishuangbanna Region in 1986–2008. *Front. Earth Sci.* **2011**, *5*, 288–293. [CrossRef]
29. Milestad, R.; Svenfelt, A.; Dreborg, K.-H. Developing integrated explorative and normative scenarios: The case of future land use in a climate-neutral Sweden. *Futures* **2014**, *60*, 59–71. [CrossRef]
30. Chen, J.; Liu, Z. Socio-economic model of Jiangsu Province based on panel data. *Asian J. Agric. Res.* **2009**, *1*, 53–55. [CrossRef]
31. Zhang, L.; Lu, D.S.; Li, Q.Z.; Lu, S. Impacts of socioeconomic factors on cropland transition and its adaptation in Beijing, China. *Environ. Earth Sci.* **2018**, *77*, 575. [CrossRef]
32. Berry, D.; Plaut, T. Retaining agricultural activities under urban pressures: A review of land use conflicts and policies. *Policy Sci.* **1978**, *9*, 153–178. [CrossRef]
33. Daniels, T.L.; Nelson, A.C. Policies to preserve prime farmland in the USA: A comment; Reply to comment: The analytical basis for an effective farmland landscape preservation scheme in the USA. *J. Rural Stud.* **1990**, *6*, 331–336. [CrossRef]

34. Corbelle-Rico, E.; Butsic, V.; Enríquez-García, M.J.; Radeloff, V.C. Technology or policy? Drivers of land cover change in northwestern Spain before and after the accession to European Economic Community. *Land Use Policy* **2015**, *45*, 18–25. [CrossRef]
35. Sabatier, P.A. An advocacy coalition framework of policy change and the role of policy-oriented learning therein. *Policy Sci.* **1988**, *21*, 129–168. [CrossRef]
36. Elliott, C.; Schlaepfer, R. Understanding forest certification using the Advocacy Coalition Framework. *Policy Econ.* **2001**, *2*, 257–266. [CrossRef]
37. Leifeld, P. Reconceptualizing major policy change in the Advocacy Coalition Framework: A discourse network analysis of German pension politics. *Policy Stud.* **2013**, *41*, 169–198. [CrossRef]
38. Heinmiller, B.T.; Pirak, K. Advocacy coalitions in Ontario land use policy development. *Rev. Policy Res.* **2017**, *34*, 168–185. [CrossRef]
39. Weible, C.M.; Sabatier, P.A.; Mcqueen, K. Themes and variations: Taking stock of the Advocacy Coalition Framework. *J. Policy Stud.* **2010**, *37*, 121–140. [CrossRef]
40. Burnett, M.; Davis, C. Getting out the cut: Politics and national forest timber harvests, 1960–1995. *Admin. Soc.* **2002**, *34*, 202–228. [CrossRef]
41. Nicolle, S.; Leroy, M. Advocacy coalitions and protected areas creation process: Case study in the Amazon. *J. Environ. Manag.* **2017**, *198*, 99–109. [CrossRef] [PubMed]
42. Mockshell, J.; Birner, R. Donors and domestic policy makers: Two worlds in agricultural policymaking? *Food Policy* **2015**, *55*, 1–14. [CrossRef]
43. Feng, Z.; Liu, B.; Yang, Y. A study of the changing trend of Chinese cultivated land amount and data reconstructing: 1949–2003. *J. Nat. Res.* **2005**, *20*, 37–45. [CrossRef]
44. China Statistics Press. China Statistical Yearbook, 1990–2017. Available online: http://navi.cnki.net/KNavi/YearbookDetail?pcode=CYFD&pykm=YINFN&bh (accessed on 5 November 2020).
45. Wang, J.; Li, X. Research on the change trend of farmland quantity in China for recent 20 years and its driving factors. *Chin. J. Agric. Resour. Reg. Plan.* **2019**, *40*, 171–176.
46. Ministry of Land and Resources of China (MLRC), 1997–2010. Available online: http://china.findlaw.cn/fangdichan/tudiguanli/ztgh/22225.html (accessed on 5 November 2020).
47. Ministry of Land and Resources of China (MLRC), 2006–2020. Available online: http://www.gov.cn/zxft/ft149/content_1144625.htm (accessed on 5 November 2020).
48. National Development and Reform Commission, 2016–2020. Available online: http://www.gov.cn/xinwen/2016-03/17/content_5054992.htm (accessed on 5 November 2020).
49. Ministry of Land and Resources of China, National Development and Reform Commission (MLRC and NDRC), 2016–2020. Available online: https://www.ndrc.gov.cn/fggz/fzzlgh/gjjzxgh/201705/t20170517_1196769.html (accessed on 5 November 2020).
50. Jong, M.D. From eco-civilization to city branding: A neo-marxist perspective of sustainable urbanization in China. *Sustainability* **2019**, *11*, 5608. [CrossRef]
51. Weiss, T.G. Governance, good governance and global governance: Conceptual and actual challenges. *Third World Q.* **2000**, *21*, 795–814. [CrossRef]
52. Winters, P.; Essam, T.; Zezza, A. Patterns of rural development: A cross-country comparison using microeconomic data. *J. Agric. Econ.* **2010**, *61*, 628–651. [CrossRef]
53. Gong, B. Agricultural reforms and production in China: Changes in provincial production function and productivity in 1978–2015. *J. Dev. Econ.* **2018**, *132*, 18–31. [CrossRef]
54. Neumann, P. Globalization and its impact on the forms of international economic governance/pavel neumann. *Enterp. Soc.* **2011**, *12*, 115–124. [CrossRef]
55. Yu, W.; Frandsen, S.E. China's WTO commitments in agriculture and impacts of potential OECD Agricultural Trade Liberalizations. *Asian Econ. J.* **2005**, *19*, 1–28. [CrossRef]
56. Brown, L.R. *Who Will Feed China? Wake-Up Call for a Small Planet*; Earthscan Publications: London, UK, 1995.
57. Chen, R.; Ye, C.; Cai, Y. The impact of rural out-migration on land use transition in China: Past, present and trend. *Land Use Policy.* **2014**, *40*, 101–110. [CrossRef]

*Article*

# Identifying the Determinants of Nongrain Farming in China and Its Implications for Agricultural Development

**Yuanzhi Guo** [1,2,*] **and Jieyong Wang** [1,2]

1   Institute of Geographic Sciences and Natural Resources Research, Chinese Academy of Sciences, Beijing 100101, China; wjy@igsnrr.ac.cn
2   Key Laboratory of Regional Sustainable Development Modeling, Chinese Academy of Sciences, Beijing 100101, China
*   Correspondence: guoyz.16b@igsnrr.ac.cn

**Abstract:** Promoted by rapid industrialization and urbanization, the structure and spatial pattern of farming in China has changed greatly, and nongrain farming (NGF) has become more common. However, excessive NGF in some areas is not conducive to sustainable agricultural development and threatens China's food security. In this study, we briefly analyze the stage characteristics of NGF in China and investigate the spatial agglomeration of NGF and its influencing factors from the perspective of spatial econometrics. The results showed that the average annual growth rate of NGF in China from 1985 to 2019 was 0.64%, and there was a growing positive spatial correlation between NGF in each province. Spatial Durbin model (SDM) estimation showed that both the per capita disposable income of local rural residents and the local urbanization rate promoted the development of NGF, while local per capita farmland, road density, and the functional orientation of the main grain-producing areas had a negative impact on NGF. The per capita disposable income of rural households and urbanization rate in neighboring areas had a promoting effect on the development of NGF, while road density in neighboring areas was negatively correlated with NGF. Ultimately, some targeted measures are proposed to promote China's agricultural development in the new era.

**Keywords:** nongrain farming; spatial correlation; spatial Durbin model; food security; agricultural supply-side structural reform

**Citation:** Guo, Y.; Wang, J. Identifying the Determinants of Nongrain Farming in China and Its Implications for Agricultural Development. *Land* **2021**, *10*, 902. https://doi.org/10.3390/land10090902

Academic Editor: Le Yu

Received: 19 July 2021
Accepted: 23 August 2021
Published: 27 August 2021

**Publisher's Note:** MDPI stays neutral with regard to jurisdictional claims in published maps and institutional affiliations.

## 1. Introduction

Agriculture is the source of food and clothing, the foundation of human survival, and the most basic material for production department in the national economy [1,2]. If the national economy is a building, agriculture is the cornerstone of that building. This status is determined by its own nature and will not change with the development of society and the economy or a decline in its representation in GDP [3]. As an important part of agriculture, farming refers to the social production sector that uses the biological functions of plants to obtain products such as grain, non-staple foods, feed and industrial raw materials through artificial cultivation. Since the transition from a hunting-and-gathering society to an agricultural society, the primary source of nutrients for human survival and development, whether plant or animal, has been farming [4–6]. Therefore, farming is a top priority in social and economic development [7], and due to its unique and important position in agricultural production, a reasonable farming structure is essential to promoting the sustainable development of agriculture.

China has been a large agricultural country since ancient times, and farming plays a significant and unique role in national economic and social development [8]. Since the reform and opening up in 1978, rapid industrialization and urbanization have promoted remarkable changes in China's farming structure [9–11], and the outstanding performance is the continuous decrease in grain crops and the rapid increase in nongrain crops such as vegetable and oil crops [12,13]. However, as the most populous country in the world,

145

China's per capita grain output has long been hovering right at the standard line of food safety (400 kg/person) set by the FAO for a long time [14–16], which suggests a grim outlook for national food security. In addition, unreasonable nongrain farming (NGF) in some areas has exacerbated this structural imbalance between the supply and demand of agricultural products, which increases the challenges to food security and is not conducive to agricultural development [11,17]. Against this background, NGF has not only become a focus of public attention and government work but also an important object of research by scholars.

In terms of research content, existing studies on China's NGF have mainly focused on analyzing the impacts of industrialization and urbanization [10,11,18,19], economic growth [11,20], the income of rural residents [21,22], and other variables on the development of NGF and have investigated how to protect cultivated land and ensure national food security in a context of growing NGF [11,20,23,24]. With the advancement of land system reform and the improvement of the rural land market, increasing attention has also been given to the role of NGF in the process of land circulation, focusing on investigating the causes of NGF and proposing the appropriate regulatory path to guide agricultural development [19,20,25]. In general, these studies are mainly conducted from a cost-benefit perspective and the factors affecting cost and benefit [26]. Regarding the research methods, most are based on case studies or micro studies at the level of rural households, and analysis of the regional patterns and influencing factors of China's NGF at different spatial scales has been overlooked. Thus, the formulation of regional agricultural development policies lacks sufficient scientific support, which is not conducive to realizing the goal of the modernization of agriculture and the countryside.

With the rapid development of society and the economy, peoples' living standards have greatly improved. As a result, peoples' demand for agricultural products is no longer limited to simply quantity, and increasing attention is being given to the quality and diversification of agricultural products [27]. Corresponding with this change, the Chinese government has actively promoted agricultural supply-side structural reform to ensure that the supply of agricultural products meets the needs of consumers, forming an agricultural product supply system with strong guarantees and reasonable structure [28]. However, agricultural supply-side structural reform in some areas is simply regarded as reducing grain production and increasing nongrain production, which restricts the sustainable development of agriculture. Therefore, a scientific understanding of NGF has become one of the key issues in ensuring national food security and deepening agricultural supply-side structural reform in China. Using a dataset of China's social and economic development during the period of 1985–2019, this study analyzes the spatial-temporal pattern of China's NGF at the national and provincial levels and employs a spatial panel data model to investigate the factors influencing regional differentiation in NGF. These findings not only will deepen understanding of NGF but also merit particular attention from policy makers aiming to ensure national food security and promote agricultural high-quality development in China.

## 2. Materials and Methodology

### 2.1. Methods

#### 2.1.1. Measurement of NGF

Farming refers to the cultivation of farm crops including cereals, beans, tubers, cotton, oil-bearing crops, sugar crops, fiber crops, tobacco, vegetables, orchards, fruits, nuts, beverage and spice crops, medicinal herbs, and others [12]. In China, cereals, beans and tubers are classified as grain crops, and the rest are classified as nongrain crops. Here, NGF refers to the production of nongrain crops in farming, and we use the proportion of sown area of nongrain crops to the total sown area to quantify NGF. Thus, China's NGF can be calculated as follows:

$$NGF = 1 - \frac{Area_{grain}}{Area_{total}} = 1 - \frac{Area_{cereals} + Area_{beans} + Area_{tubers}}{Area_{total}} \tag{1}$$

where $Area_{grain}$ is the sown area of grain crops, including cereals, beans and tubers; and $Area_{total}$ denotes the total sown area of crops, which refers to the area of all land sown or transplanted with crops that are harvested within the calendar year. All crops harvested within the year are counted as sown area, regardless of being sown in this year or the previous year, and crops sown this year but will be harvested in the coming year are excluded. In general, the value of NGF ranges from 0 to 1.

### 2.1.2. Spatial Autocorrelation Analysis

According to Tobler's first law of geography, there are agglomeration, random, regular, and other relationships between geographical items or their attributes in spatial distributions [29]. Spatial autocorrelation analysis aims to reveal this potential dependency, and the commonly used indicators include Moran's *I* statistic [30], Getis G [31], and Geary's C ratio [32]. Drawing lessons from related studies [33–35], this study employs Moran's *I* statistic, which includes global and local Moran's *I*, to measure the spatial autocorrelation between China's NGF in each province.

The global Moran's *I* measures the relationship between the attribute values of adjacent spatial objects. A positive value indicates a positive correlation in the distribution of NGF, a negative value indicates a negative correlation, and zero indicates no spatial correlation. According to existing research, the formula for calculating Moran's *I* is as follows [30]:

$$I = \left( n \sum_{i=1}^{n} \sum_{j=1}^{n} w_{ij} (y_i - \overline{y})(y_j - \overline{y}) \right) / \left( S^2 \sum_{i=1}^{n} \sum_{j=1}^{n} w_{ij} \right) \tag{2}$$

$$S^2 = \frac{1}{n} \sum_{i=1}^{n} (y_i - \overline{y})^2 \,, \ \overline{y} = \frac{1}{n} \sum_{i=1}^{n} y_i \tag{3}$$

where $y_i$ is the NGF of province $i$, $n$ is the number of provinces, and $w_{ij}$ is a component of the spatial weight matrix $W$ determined by the principle of geographic adjacency; it equals 1 if the two provinces are adjacent, and otherwise equals 0. The value of the global Moran's *I* ranges from $-1$ to 1, where a larger value indicates higher spatial correlation, and a lower value indicates the opposite.

To diagnose the outliers, local Moran's *I*, which is also known as the local indicator of spatial association (*LISA*), is used to measure the correlation degree of NGF between province $i$ and its neighboring provinces and to identify the characteristics of the spatial spillover of NGF. Essentially, the local Moran's *I* decomposes the global Moran's *I* into a local scale and can be calculated as follows [36]:

$$I_i = z_i \sum_{j=1}^{n} w_{ij} z_j \tag{4}$$

where $z_i$ and $z_j$ are the normalized NGFs of provinces $i$ and $j$, respectively, and $w_{ij}$ is the spatial weight matrix of row standardization.

### 2.1.3. Spatial Panel Model

The quantitative inspection and estimation methods for general panel models are mature and have been widely used in empirical research [37,38]. However, due to the spatial correlation between geographic things [29], problems such as the deviation of test statistics between levels and inconsistent parameter estimation will be caused if the general panel model is used to study the related scientific problems [39]. Thus, spatial effects should be taken into consideration when investigating the mechanisms of China's NGF; that is, it is necessary to use the spatial panel model to explore the factors influencing regional inequality in China's NGF. In general, the spatial panel model includes spatial autoregression model (SAR) and spatial error model (SEM), where the former indicates that there is a spatial lag term in the explained variable of the model, and the latter indicates that the error terms of the model are spatially correlated. Additionally, Lesage and Pace [40] proposed the spatial Durbin model (SDM), which contains the spatial lag terms of dependent and independent variables and represents a more generalized form of SAR and SEM, to analyze the mechanism of spatial heterogeneity.

In general, the SAR can be calculated as follows:

$$y_{it} = \delta \sum_{j=1}^{N} w_{ij} y_{jt} + \alpha + \beta X_{it} + \mu_i + \lambda_t + \varepsilon_{it} \tag{5}$$

where $y_{it}$ is the dependent variable of province $i$ at time $t$; $\sum_j w_{ij} y_{jt}$ denotes the interaction between $y_{it}$ and $y_{jt}$ of adjacent province $j$; $w_{ij}$ is a component of the $N \times N$ dimensional nonnegative spatial weight matrix $W$ determined by the features of sample provinces; $\delta$ is an endogenous parameter that reflects the spatial interaction between dependent variables; $X_{it}$ is a $1 \times K$ dimensional exogenous variable; $\beta$ is the corresponding $K \times 1$ dimensional coefficient vector; $\mu_i$ and $\lambda_t$ are the spatial and temporal specific effects, respectively; and $\varepsilon_{it}$ is a random error term.

The basic form of the SEM is as follows:

$$y_{it} = \alpha + \beta X_{it} + \mu_i + \lambda_t + \varphi_{it}, \ \varphi_{it} = \rho \sum_{j=1}^{N} w_{ij} \varphi_{jt} + \varepsilon_{it} \tag{6}$$

where $\varphi_{it}$ is a spatial autocorrelation error term, $\sum_j w_{ij} y_{jt}$ denotes the influence of the error term of adjacent province $j$ on province $i$, and $\rho$ is the spatial autocorrelation coefficient between error terms.

The formula for the SDM is as follows:

$$y_{it} = \delta \sum_{j=1}^{N} w_{ij} y_{jt} + \alpha + \beta X_{it} + \sum_{j=1}^{N} w_{ij} X_{jt} \gamma + \mu_i + \lambda_t + \varepsilon_{it} \tag{7}$$

where $X_{jt}$ is a $1 \times K$ dimensional exogenous variable with spatial lag, and $\gamma$ denotes the corresponding $K \times 1$ dimensional parameter vector. Under certain conditions, the SDM can be simplified into SAR and SEM.

### 2.2. Variable Selection

Although there are many studies on China's NGF, they have mainly focused on the nongrain of cultivated land. Here, we briefly review research on the nongrain of cultivated land and then select influencing factors of NGF. Based on a rural household survey, Chen et al. [41] discussed the causes of farmers' willingness to grow grain from the perspectives of individual and family characteristics as well as grain price. Additionally, Jin [42] analyzed the factors influencing farmers' willingness to produce grain considering aspects such as market factors, natural conditions and preferential policies. Zhang and Jiang [22] studied the differences of nongrain in land transferred by different types of farmers and investigated the contributing factors from regional economic development, agricultural production conditions and other aspects. Zhao et al. [19] explored the mechanisms of China's NGF based on an analysis of urbanization rate, per capita disposable income of rural households, proportion of nonagricultural industry in GDP, per household cultivated land, number of agricultural employees and other variables. Su et al. [11] analyzed the driving forces of different nongrain production types by using multinomial logistic regression modeling with geophysical, proximate, neighborhood and policy variables.

Reviewing the existing studies, it is obvious that the driving factors of China's NGF can be divided into four types: natural conditions, which mainly refer to the conditions of farmland, such as quantity and fertility; macro socioeconomics, including industrialization, urbanization, and regional economic development; individual characteristics, including labor capacity, age, etc.; and policy systems, such as land transfer and agricultural subsidies. This study focuses on exploring the spatial-temporal pattern of provincial NGF in China and its influencing factors. Therefore, the micro factors of rural households are not included in the follow-up analysis.

Given the analysis above, the following seven factors are selected to investigate the mechanisms of the regional imbalance of China's NGF (Table 1): ① Per capita farmland. Agriculture has a significant effect on scale economies, and an increase in the agricultural production scale is conducive to improving agricultural total factor productivity and decreasing agricultural production costs [43]; ② urbanization rate, which can reflect the degree to which rural people are transferring to urban areas and the potential market scale of regional agricultural products; ③ per capita GDP, which is an indicator reflecting regional economic development. In general, the higher the level of the regional economy is, the higher people's living standards. As a result, their needs are increasingly diversified and advanced, which promotes the diversification of agricultural production [44]; ④ per capita disposable income of rural households. An increase in farmers' incomes is conducive to improving their production conditions, thus affecting their agricultural production decisions; additionally, farmers gain a stronger ability to meet their various needs; ⑤ road density, which reflects regional traffic situation and has an important impact on the sale of agricultural products; ⑥ rural population aging. Agriculture is a labor-intensive industry, and the aging of rural population directly reduces working populations engaged in agricultural production; and ⑦ function orientation of main grain-producing areas. In the main grain-producing areas, the development of NGF is strictly restricted, and the government has issued a series of policies and measures to encourage grain production. According to the "opinions on the reform and improvement of policies and measures for comprehensive agricultural development" issued in December 2003, Hebei, Inner Mongolia, Liaoning, Jilin, Heilongjiang, Jiangsu, Henan, Shandong, Hubei, Hunan, Jiangxi, Anhui, and Sichuan are identified as the main grain-producing areas in China.

**Table 1.** The selection of influencing factors behind NGF.

| Variables | Description |
| --- | --- |
| 1. Per capita farmland (*PFARM*) | Farmland area per person employed in primary industry |
| 2. Urbanization rate (*UR*) | Proportion of urban resident population in total resident population |
| 3. Per capita GDP (*PGDP*) | Calculated according to the caliber of resident population |
| 4. Per capita disposable income of rural households (*PCDIR*) | Excluding migrant workers, but including college students who are supported by the family |
| 5. Road density (*RDEN*) | Excluding urban streets, dead end highways, streets built for agricultural production and inside factories |
| 6. Rural population aging (*AGING*) | Proportion of rural population aged 65 and above in total rural population |
| 7. Function orientation of main grain-producing areas (*FUNO*) | If it is the main grain-producing area, FUNO is as-signed "1", otherwise it is 0. |

*2.3. Materials*

This study makes full use of data on the sown areas of farm crops in China at different spatial scales. Sown areas of farm crops come from the China Statistical Yearbook. Administrative divisions and digital elevation models are downloaded from the National Geomatics Center of China and WebGIS, respectively. Data on *PFARM* come from the China Statistical Yearbook on Environment and China Statistical Yearbook on Land and Resources. *UR, PGDP, PCDIR* and *RDEN* are from the China Statistical Yearbook and the provincial Statistical Yearbooks. *AGING* and *EDU* are derived from the China Population Statistical Yearbook and the China Population and Employment Statistical Yearbook. The missing data are replaced by the data of adjacent years or supplemented through the method of trend extrapolation. According to the research design, Hong Kong, Macao and Taiwan are excluded. Thus, a total of 31 provincial administrative units were obtained.

Because the data of some independent variables before 1990 are difficult to obtain, a dataset from 1990 to 2019 is used for the spatial panel analysis.

## 3. Results Analysis

### 3.1. Spatiotemporal Pattern of NGF in China

### 3.1.1. Historical Evolution of NGF in China

Formula (1) is used to calculate China's NGF at the national level (Figure 1). From 1985 to 2019, China's total sown area of farm crops, sown area of grain crops, sown area of nongrain crops and NGF achieved different degrees of growth, with average annual growth rates of 0.43%, 0.19%, 1.07% and 0.64%, respectively. Specifically, the total sown area of farm crops showed a fluctuating rising trend; the sown area of grain crops remained relatively stable before 1999, then declined until 2003, and then increased steadily until shifting to another downward trend after 2016; the sown area of nongrain crops increased steadily before 2003, then had a downward trend until 2006, and remained stable during the period of 2006–2016 before shifting to a rising trend after 2016; the evolutionary trend of NGF was similar to that of the sown area of nongrain crops, but the change was more obvious.

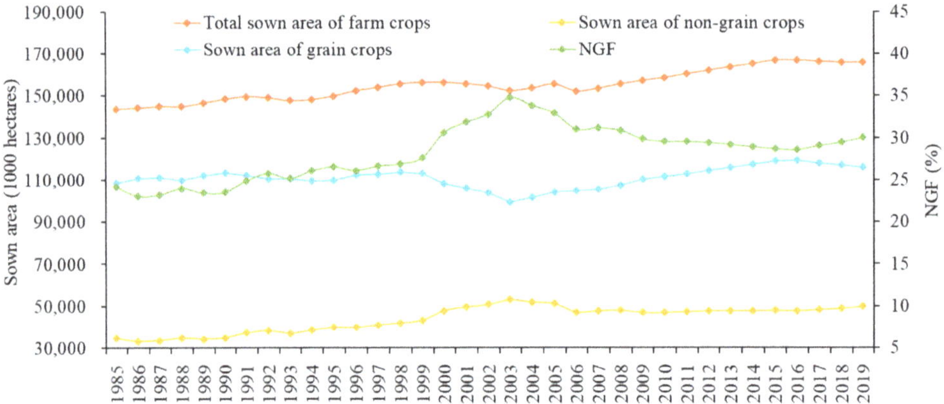


**Figure 1.** The evolution of NGF and sown areas of farm crops in China from 1985 to 2019.

With the continuous advancement of the reform and opening up, the increasing population in China requires the expansion of the sown areas of farm crops to meet people's needs. In addition, the diversified demand caused by improvements in living standards requires the diversification of agricultural production, which increases the sown area of nongrain crops and their proportion in total farming. By the end of the 20th century, China actively promoted the adjustment of its agricultural structure to cope with the impact of WTO accession on agricultural production. As a result, the sown area of grain crops decreased rapidly while that of nongrain crops increased rapidly. Correspondingly, NGF increased rapidly from 1999 to 2003, seriously threatening national food security. Focusing on the serious decline in grain production and the transformation of its domestic economy, China established the guiding principle of industry supporting agriculture and cities supporting villages in 2004 and issued a series of policies to support and benefit agriculture, including the abolition of agricultural taxes, grain subsidies, comprehensive subsidies for agricultural means of production, a minimum purchase price system for grain, etc. These preferential policies have greatly improved farmers' enthusiasm for growing grain, increased the sown area of grain crops, and caused NGF to decline. With improvement in the guaranteed ability to meet national food security needs, NGF began to rise again in 2016 under the influence of market supply and demand.

### 3.1.2. Regional Pattern of NGF in China

Employing formula (1), NGF at the provincial level from 1985 to 2019 is also calculated, and the results are divided into three grades: low-value area (0 < NGF ≤ 30%), mid-value area (30% < NGF ≤ 50%), and high-value area (50% < NGF ≤ 70%). According to previous analysis on the evolution of NGF, the starting and ending years as well as the two transitional years, 2003 and 2016, were selected to reveal the spatial pattern of NGF in China (Figure 2). In 1985, the NGF of most provinces was less than 30%, and only five provinces belonged to the mid-value area, including Shanghai, Jiangxi, Hubei, Hunan and Xinjiang (Figure 2a). In 2003, some provinces had entered the high NGF value group, specifically, Beijing, Shanghai, Hubei and Xinjiang, while the spatial scope of the low-value areas was significantly reduced from 26 provinces to 10 provinces, including Shanxi, Inner Mongolia, Liaoning, Jilin, Heilongjiang, Chongqing, Yunnan, Tibet, Shaanxi, and Ningxia (Figure 2b). In 2016, Beijing and Hubei dropped out the ranks of the high-value areas and became mid-value areas, while Guangxi and Hainan joined the ranks of high-value areas; the provinces with a low-value of NGF were Tianjin, Hebei, Shanxi, Inner Mongolia, Liaoning, Jilin, Heilongjiang, Jiangsu, Anhui, Henan, Tibet and Shanxi (Figure 2c). In 2019, Zhejiang, Guangdong and Guizhou were added to the ranks of high-value areas on the basis of 2016; areas with low-value NGF were concentrated in Northeast and North China, including Tianjin, Hebei, Shanxi, Inner Mongolia, Liaoning, Jilin, Heilongjiang, Jiangsu, Anhui, Shandong, Henan, and Shaanxi (Figure 2d).

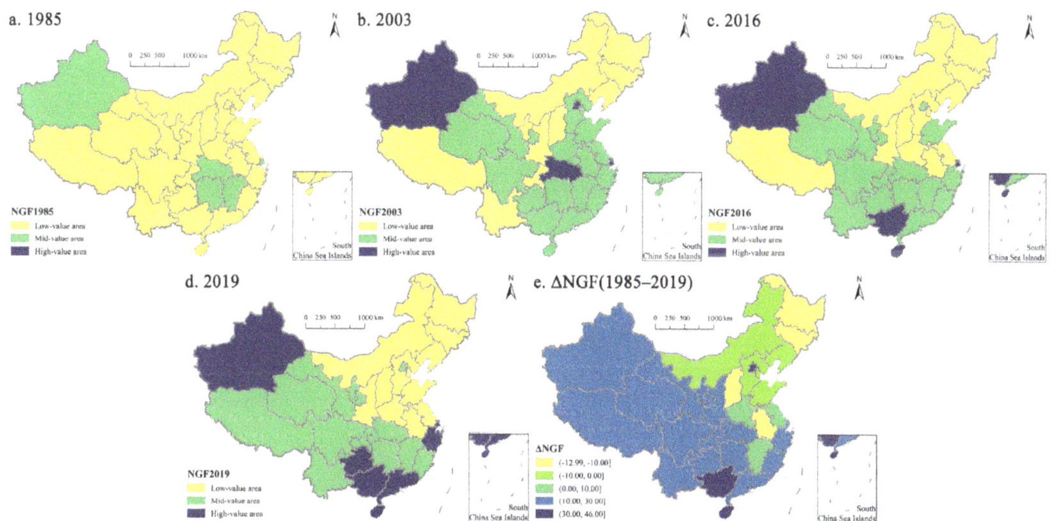


**Figure 2.** Spatial patterns of NGF and its changes in China from 1985 to 2019 (**a–e**).

In terms of the changes in NGF from 1985 to 2019, there were only nine provinces with declining NGF, mainly distributed in North and Northeast China, with this decline averaging −7.63%. Among them, the provinces with an increase of NGF below −10% were Heilongjiang (−12.99%), Shanxi (−11.91%), Jilin (−11.49%) and Anhui (−10.92%). By contrast, there were 22 provinces with a rising NGF, and the average value was 20.12%. Specifically, the provinces with an increase of NGF greater than 30% were Hainan (45.06%), Guangxi (31.72%) and Beijing (30.29%); and three provinces had an increase in NGF less than 10%, that was, Jiangxi (0.99%), Jiangsu (2.86%) and Henan (4.32%) (Figure 2e).

### 3.1.3. Spatial Autocorrelation of NGF in China

Exploratory spatial data analysis (ESDA) is used to explore the spatial correlation of NGF at the provincial level in China. The results show that the global Moran's *I* of NGF during the period of 1985–2019 was greater than zero at a 10% significance level, which indicates that there is a significant positive correlation between NGF in different provinces. In other words, the spatial distribution of NGF is not random but indicates that provinces with similar levels of NGF tend to cluster in a specific geographic space. From the perspective of evolutionary trends, the global Moran's *I* index has a fluctuating rising trend, from 0.052 in 1985 to 0.434 in 2019, which indicates that the degree of spatial agglomeration among NGFs at the provincial level was gradually increasing. Specifically, it had a trend of first increasing from 1985 to 1996, then decreasing until 2000, then increasing again, and then remaining relatively stable during the period of 2009–2019 (Figure 3).

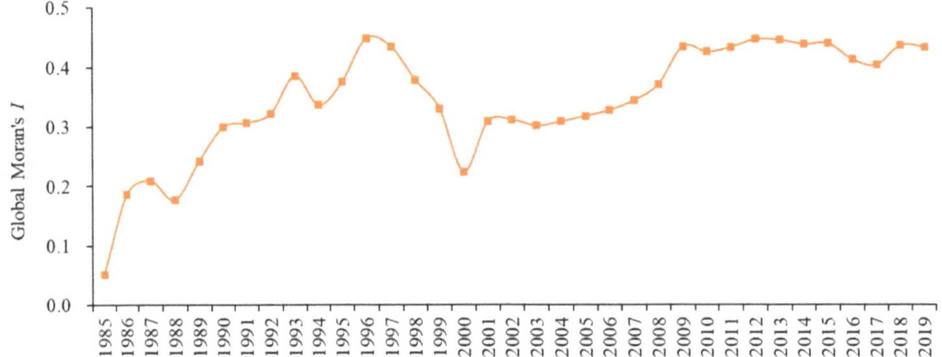

**Figure 3.** Changes in the global Moran's *I* of NGF in China from 1985 to 2019.

To investigate the spatial agglomeration characteristics of NGF, the local Moran's *I* is calculated, and the Moran scatter plots in 1985, 2003, 2016 and 2019 are selected to analyze the spatial pattern of NGF for different local areas (Figure 4). In 1985, 2003, 2016 and 2019, there were 9, 11, 12 and 13 provinces in the first quadrant and 10, 12, 10 and 11 provinces in the third quadrant, respectively. Therefore, the number of provinces with positive spatial correlations accounted for 61.29%, 74.19%, 70.97%, and 77.42% of the 31 provinces in 1985, 2003, 2016, and 2019, respectively, indicating that the local spatial agglomeration of NGF in China was mainly characterized by high-high and low-low clusters. Specifically, in 1985, provinces featuring a high-high cluster included Shanghai, Jiangsu, Zhejiang, Anhui, Jiangxi, Shandong, Hubei, Hunan and Guangdong, and those featuring a low-low cluster included Shanxi, Liaoning, Jilin, Heilongjiang, Sichuan, Yunnan, Shaanxi, Gansu, Qinghai and Ningxia (Figure 4a). In 2003, the provinces featuring a high-high cluster included Beijing, Tianjin, Shanghai, Jiangsu, Zhejiang, Fujian, Jiangxi, Hunan, Guangdong, Guangxi and Hainan, and those featuring a low-low cluster included Shanxi, Inner Mongolia, Liaoning, Jilin, Heilongjiang, Henan, Chongqing, Sichuan, Guizhou, Yunnan, Shaanxi and Ningxia (Figure 4b). In 2016, the provinces featuring a high-high cluster included Shanghai, Zhejiang, Fujian, Hunan, Guangdong, Guangxi, Hainan, Chongqing, Guizhou, Yunnan, Qinghai, and Xinjiang, and those featuring a low-low cluster included Tianjin, Hebei, Shanxi, Inner Mongolia, Liaoning, Jilin, Heilongjiang, Shandong, Henan, and Shaanxi (Figure 4c). In 2019, the provinces featuring a high-high cluster included Shanghai, Zhejiang, Fujian, Hunan, Guangdong, Guangxi, Hainan, Chongqing, Sichuan, Guizhou, Yunnan, Qinghai, and Xinjiang, and those featuring a low-low cluster included Tianjin, Hebei, Shanxi, Inner Mongolia, Liaoning, Jilin, Heilongjiang, Anhui, Shandong, Henan, and Shaanxi (Figure 4d).

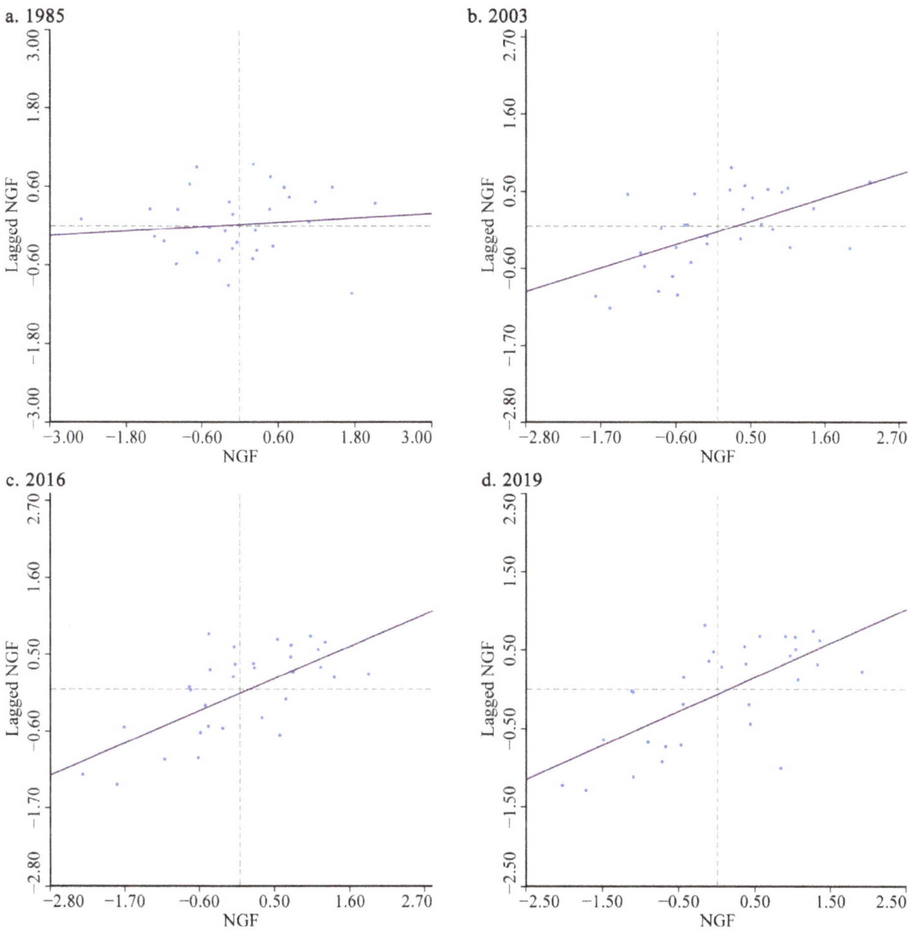

**Figure 4.** Moran scatter plot of NGF in China in 1985, 2003, 2016, and 2019 (**a–d**).

From the evolution of the different types of spatial agglomeration, it can be seen that the provinces featuring a high-high cluster of NGF became increasingly concentrated in the south, while those featuring a low-low cluster were increasingly concentrated in northeast and north China, which indirectly reflected the northward trend of the center of gravity of China's grain production from 1985 to 2019 [14].

### 3.2. Factors Influencing NGF in China

According to the previous analysis of the spatial agglomeration of NGF, there are obvious spatial correlation characteristics in China's provincial NGF. Therefore, it is necessary to establish a spatial econometric model to analyze its influencing factors. The results of the Lagrange multiplier (*LM*) test show that all statistics reach a 1% significance level (Table 2), which also indicates that the spatial panel is better than the nonspatial panel, and there are spatial error effects and spatial lag effects. Therefore, we reject the mixed multiple regression model. In addition, the *LR* test and *Wald* test are used to analyze the original hypothesis that the SDM is simplified into SAR and SEM: $H_0^1$: $\gamma = 0$ and $H_0^2$: $\gamma + \delta\beta = 0$. The results of the *LR* test and *Wald* test show that the SDM should be employed for the empirical analysis of influencing factors (Table 3). The results of the SDM estimation show that model (2) has the largest R-square of (Table 4). Therefore, the SDM with time fixed effects should be chosen to explore the influencing factors of NGF at the provincial level in China.

**Table 2.** Results of the Lagrange multiplier (*LM*) test.

| Test | Spatial Error | | Spatial Lag | |
|---|---|---|---|---|
| | Statistic | *p*-Value | Statistic | *p*-Value |
| Moran's $I$ | 12.794 | 0.000 | | |
| Lagrange multiplier | 157.831 | 0.000 | 270.165 | 0.000 |
| Robust Lagrange multiplier | 53.748 | 0.000 | 166.081 | 0.000 |

**Table 3.** Results of the likelihood ratio (*LR*) and *Wald* test.

| *LR* Test | | | *Wald* Test | | |
|---|---|---|---|---|---|
| | LR chi$^2$(7) | Prob > chi$^2$ | | chi$^2$(6) | Prob > chi$^2$ |
| Comparison of SDM and SAR | 245.50 | 0.0000 | The first test method | 12.44 | 0.0528 |
| Comparison of SDM and SER | 211.44 | 0.0000 | The second test method | 16.51 | 0.0113 |

**Table 4.** SDM estimation of the influencing factors of NGF in China.

| | Model (1) | Model (2) | Model (3) | Model (4) |
|---|---|---|---|---|
| | NGF | NGF | NGF | NGF |
| *PFARM* | −0.713 *** | −0.458 *** | −0.195 *** | −0.112 ** |
| | (0.0408) | (0.0521) | (0.0587) | (0.0560) |
| *UR* | −0.0999 *** | 0.0812 ** | 0.114 *** | 0.116 *** |
| | (0.0271) | (0.0367) | (0.0322) | (0.0307) |
| ln*PGDP* | 3.974 ** | −2.139 | 9.565 *** | 7.026 *** |
| | (1.659) | (1.911) | (1.238) | (1.453) |
| ln*PCDIR* | 2.184 | 12.55 *** | −8.411 *** | −5.157 ** |
| | (2.085) | (2.414) | (2.196) | (2.321) |
| *RDEN* | −1.965 | −3.634 *** | −2.077 ** | −3.620 *** |
| | (1.235) | (1.133) | (0.845) | (0.882) |
| *AGING* | −0.211 | −0.109 | −0.123 | −0.333 ** |
| | (0.183) | (0.175) | (0.133) | (0.131) |
| *FUNO* | −7.250 *** | −9.150 *** | | |
| | (0.608) | (0.705) | | |
| *Constant* | −7.831 * | | | |
| | (4.129) | | | |
| *PFARM* × *W* | | −0.0542 | −0.0136 | 0.107 |
| | | (0.133) | (0.135) | (0.132) |
| *UR* × *W* | | 0.628 *** | 0.0303 | 0.0176 |
| | | (0.0665) | (0.0664) | (0.0672) |
| ln*PGDP* × *W* | | −3.024 | 1.227 | −4.672 |
| | | (4.218) | (1.558) | (3.065) |
| ln*PCDIR* × *W* | | 39.96 *** | 1.348 | 10.28 * |
| | | (4.701) | (2.523) | (5.474) |
| *RDEN* × *W* | | −8.208 *** | −11.33 *** | −15.20 *** |
| | | (2.456) | (1.473) | (1.902) |
| *AGING* × *W* | | −0.215 | 0.0986 | −0.672 *** |
| | | (0.355) | (0.203) | (0.244) |
| *FUNO* × *W* | | −2.369 | | |
| | | (1.707) | | |
| *Rho* | | −0.266 *** | 0.206 *** | 0.0224 |
| | | (0.0452) | (0.0401) | (0.0444) |
| *Observations* | 930 | 930 | 930 | 930 |
| *R-square* | 0.371 | 0.217 | 0.154 | 0.0966 |

Note: Standard errors in parentheses; * $p < 0.1$, ** $p < 0.05$, *** $p < 0.01$.

According to the model estimation results, regional NGF is not only affected by local factors, such as per capita disposable income and national policy orientation, but also closely related to variables in neighboring areas, such as urban development, road density and other factors. In terms of the influencing degree, the impact of local variables on the development of NGF ranked from highest to lowest is *PCDIR, FUNC, RDEN, PFARM,* and *UR*, while the same ranking of the impact from variables in neighboring areas is *PCDIR, RDEN,* and *UR.*

Specifically: ① local *PFARM* has a significant negative effect on NGF, while that in neighboring areas has no significant effect. The fact that a large population with relatively little land causes grain production to have unique significance in regional agriculture, the increase in *PFARM* helps to achieve scale operations, reduce agricultural production costs, improve agricultural production efficiency, boost the development of regional grain production, and ultimately reduce the proportion of nongrain crops in farming. This finding is consistent with the conclusions of existing studies [19,45]; ② *UR*s in both local and neighboring areas have a positive effect on NGF, but the degree of their influence is smallest when compared with that of other variables. The development of urbanization consumes much farmland and compresses the space of grain production. On the other hand, demands for various means of production and living caused by the agglomeration of population, industry and other factors promote the development of NGF. In addition, against the background of a rural population flowing to urban areas in China, the relatively low income from grain production causes a large number of transferred lands to be used for nongrain crop production [11,46]; ③ *PCDIR*s in both local and neighboring areas have a significant positive effect on the development of NGF. The increase in farmers' income improves their lives, and their consumption habits and diet structure have also changed correspondingly, gradually shifting from a grain-based diet to a diverse assortment of grain, fruits, vegetables, etc. In addition, the diversified demand for consumption generated by the improvement in living standards of rural residents in surrounding areas also promotes the diversification of regional farming; ④ *RDEN*s in both local and neighboring areas have a significant negative effect on the development of NGF. Due to the regional differences in grain production in China, transportation plays an important role in grain circulation. Good internal and external traffic conditions help to promote grain circulation, ensure the income from grain production and reduce the proportion of nongrain crops in farming; and ⑤ local *FUNC* has a significant negative effect on the development of NGF, while that in neighboring areas has no significant effect. The main grain-producing areas are designated by the central government to ensure food security. To ensure the implementation of the national strategy, the government has issued corresponding preferential policies, thus limiting the development of NGF in these areas. However, these policies are exclusive; therefore, whether the neighboring areas are major grain-producing areas has no impact on the local NGF.

## 4. Discussion and Policy Implications

### 4.1. Scientific Understanding of China's NGF

Although NGF has challenged the national economy and social development, we should treat this phenomenon objectively. On the one hand, China is a country with a vast territory, and the regional differences in natural conditions such as water and soil require different regions to choose appropriate crops according to local conditions to meet people's needs, which determines the diversity of agricultural production. This kind of diversity is not only manifested in the diversity of crop species and germplasm resources, but also reflected by many varieties and varying performances of the same crop. In addition, the diversification of agricultural planting can improve productivity and resource utilization efficiency by optimizing access to biomass and water resources, thus strengthening the capacity, vitality and competitiveness of agroecosystems [47]. On the other hand, NGF is an inevitable outcome of social and economic development [48]. In the initial stage of agricultural development, people were mainly engaged in grain production, such as millet and rice; with the development of productivity and the enhancement of interregional links, the types of grain crops in specific areas are becoming increasingly diversified, and vegetable crops, feed crops, cash crops and other crops have gradually become important objects of labor. As a result, the diversification of farming has been greatly improved.

Currently, the completion of the goal of building a well-off society in an all-round way indicates that China's social and economic development has entered a new stage. Accordingly, the principal social contradiction has evolved into the contradiction between people's ever-growing needs for a better life and unbalanced and inadequate development. Against this background, people's demand for agricultural products has transformed from a focus on quantity to one on quality, which indicates higher requirements for the variety and quality of food [49,50]. Compared with the past, when people simply wanted to have sufficient food to eat, they are now more concerned about eating well and eating healthily [27]. At the end of 2015, agricultural supply-side structural reform became an important component of the transformation and upgrading of China's agriculture in the new era, with the aim of improving the quality and efficiency of the agricultural supply system and realizing the transformation of agricultural product supply and demand from simply being a matter of maintaining a low-level total balance to needing to supply high-quality structural coordination [51]. After a period of development, the structure of agricultural production in China continues to be optimized, and the regional pattern tends to be reasonable, which provides not only a guarantee for meeting people's multilevel and diversified needs but also solid support for stabilizing the overall situations of economic and social development. However, in this process, some areas are experiencing excessive NGF, mainly including the nongrain of food production and the nongrain of nonfood production. Additionally, these phenomena show a trend of accelerated development [52] and have brought great challenges to China's fragile grain security.

### 4.2. Policy Implications for China's Agricultural Development

Since 2004, grain production in China has had bumper harvests annually, and by the end of 2019, the per capita output of grain had reached 475 kg [12]. However, there are still some serious problems behind this achievement, such as the imbalance between supply and demand, increasing resource use and environmental pressures, which have meant that the balance between grain supply and demand has been tenuous for a long time [53]. Focusing on the transition in the principal social contradiction, the transformation and upgrading of the dietary structure of urban and rural residents in China has become a general trend, and agricultural supply-side structural reform has become the key to agricultural and rural development. In this context, it is critical that the relationship between national grain security and agricultural supply-side structural reform should be scientifically addressed to promote the sustainable development of agriculture, increase farmers' income, and ultimately realize the goal of rural revitalization.

First, a system for protecting farmland must be strictly implemented. The government needs to strictly maintain the red line for farmland, especially basic farmland, strengthen land use control and law enforcement supervision, comprehensively implement strategic tasks for ensuring quantity and improving the quality of farmland, and seriously investigate and punish behaviors that occupy and indiscriminately abuse farmland, thus protecting and optimizing grain production capacity. Second, more attention should be given to developing moderate-scale operations. By adhering to the fundamental rural management system, all regions should vigorously develop new-type agricultural operation subjects and service subjects, and accelerate the development of various forms of moderate-scale operation through the circulation of management rights, joint-stock cooperation, land trusteeship and other methods, thus improving the efficiency of agricultural production. Third, the agricultural subsidy system needs to be improved. The government should further improve current agricultural subsidy policies, focus on the actual operators rather than the owners of contracting rights, and issue preferential policies for major grain-producing areas, entities with moderate-scale operations and green ecological agriculture. Fourth, the regional agricultural pattern should be continuously optimized. Based on the national main functional area planning and regional layout planning of superior agricultural products, the government should scientifically and reasonably designate grain production functional areas, such as rice, wheat and corn, and production protection areas for important agricultural products, such as soybean, cotton and rapeseed. On this basis, incentive mechanisms and support policies can be established and improved, thus constantly implementing the responsibilities of construction and management entities. Fifth, the main grain-producing areas need to be further deepened and refined. In some main grain-producing areas, especially economically developed provinces, the grain production function of many counties has been seriously degraded, while there are some counties with large grain output in non-main grain-producing areas, but they lack corresponding policy support. Therefore, delimitation of the main grain-producing areas should be based on the county as the spatial unit to improve policy pertinence. Through multipoint efforts and strategies, an agricultural product supply system with reasonable structure and strong guarantees can be established, and agricultural production will better meet people's demands for variety and quality, enhancing the driving force and ability of sustainable agricultural development.

### 4.3. Limitations and Future Developments

Currently, rural China has entered a new historical stage, and the principal contradiction faced by agricultural development has changed from insufficient quantity to structural disequilibrium [54,55]. In terms of the contents, structural disequilibrium is mainly reflected on the supply side, which includes not only an imbalance in the proportion of agriculture, forestry, animal husbandry and fishery, but also disharmony in the internal structure of planting, forestry, animal husbandry and fishery production. In this study, we only focus on farming in the narrow sense. Therefore, more attention should be given to strengthening research on agriculture in the broad sense to provide support for the modernization of agriculture and the countryside in the new era. At the spatial scale, this study analyzes the regional pattern of China's NGF with the provincial administrative region as the spatial unit, which makes the granularity of the spatial analysis slightly coarse. In future research, it is necessary to strengthen research at the county and village scales, thus better revealing the spatial characteristics and internal mechanisms of NGF at the meso and micro levels and enhancing the practical significance of the research. Furthermore, the specific types of NGF are worthy of further study, thus providing scientific support for constructing a modern agricultural industrial system, production system and management system. In addition, NGF in rural China is mainly characterized by the nongrain of cultivated land [56,57] Therefore, strengthening research on the nongrain of cultivated land is necessary to better guide the implementation of national strategies, such as cultivated land protection and food security.

## 5. Conclusions

Due to the low comparative benefits of grain crops, farmers' enthusiasm for grain production is declining, and the phenomena of extensive grain production and farmland abandonment are common in rural China [58,59]. Excessive NGF in some areas is not conducive to the high-quality development of agriculture in the new era. To reverse this situation, the central government strictly requires that the amount of farmland should not be reduced. However, compared with a one-sided focus on cultivated land area, sown area, which can reflect the planting situation of different crops, deserves more attention. This study used data on the sown areas of farm crops to analyze the spatial and temporal patterns of NGF in China, explored its influencing factors, and proposed countermeasures to promote sustainable agricultural development.

The results showed that the evolution of China's NGF from 1985 to 2019 had an overall upward trend, where performance first rose, then decreased, and then increased. During the research period, there was a significant positive spatial correlation between NGFs in different provinces, and this feature had a gradually increasing trend. In terms of the spatial pattern, provinces featuring high-high clusters of NGF were increasingly concentrated in South China, and those featuring low-low clusters were increasingly concentrated in North and Northeast China. The SDM estimation showed that local *PCDIR* and *UR* promoted the development of NGF, while local *PFARM, RDEN,* and *FUNO* had a negative impact. In terms of variables in neighboring areas, *PCDIR* and *UR* were positively correlated with the development of NGF, while *RDEN* was negatively correlated with the development of NGF. These findings can serve as a scientific basis for the policy-making of China's high-quality development of agriculture and rural revitalization in the new era, and we hope that this paper will encourage the development of similar studies.

**Author Contributions:** Conceptualization, Y.G. and J.W.; methodology, Y.G. and J.W.; software, Y.G.; validation, Y.G. and J.W.; formal analysis, Y.G.; investigation, Y.G.; resources, Y.G.; data curation, Y.G.; writing—original draft preparation, Y.G.; writing—review and editing, Y.G. and J.W.; visualization, Y.G.; supervision, Y.G. and J.W.; project administration, Y.G.; funding acquisition, Y.G. All authors have read and agreed to the published version of the manuscript.

**Funding:** This research was supported by the National Natural Science Foundation of China (Grant No. 42001203) and the Strategic Priority Research Program of the Chinese Academy of Sciences (Grant No. XDA23070300).

**Institutional Review Board Statement:** Not applicable.

**Informed Consent Statement:** Not applicable.

**Data Availability Statement:** The associated dataset of the study is available upon request to the corresponding author.

**Acknowledgments:** The authors would like to thank the anonymous reviewers for their comments and suggestions which contributed to the further improvement of this paper.

**Conflicts of Interest:** The authors declare that they have no conflict of interest.

## References

1. Han, S.; Pan, W.; Zhou, Y.; Liu, Z. Construct the prediction model for China agricultural output value based on the optimization neural network of fruit fly optimization algorithm. *Future Gener. Comput. Syst.* **2018**, *86*, 663–669. [CrossRef]
2. Gepts, P.; Papa, R. *Evolution During Domestication*; ELS Language Centers: Princeton, NJ, USA, 2002.
3. Johnston, B.F.; Mellor, J.W. The role of agriculture in economic development. *Am. Econ. Rev.* **1961**, *4*, 566–593.
4. Zhao, R.; Wang, E.; Zhang, X.; Liu, J.; Zhou, S.; Li, G.; Han, M. *Human Geography*; Higher Education Press: Beijing, China, 2010.
5. Cajete, G. *A People's Ecology: Explorations in Sustainable Living*; Book Marketing Group: Santa Fe, NM, USA, 1999.
6. Welch, R.M.; Graham, R.D. Breeding for micronutrients in staple food crops from a human nutrition perspective. *J. Exp. Bot.* **2004**, *396*, 353–364. [CrossRef]
7. Ayanlade, A.; Radeny, M. COVID-19 and food security in Sub-Saharan Africa: Implications of lockdown during agricultural planting seasons. *Npj Sci. Food* **2020**, *4*, 13. [CrossRef] [PubMed]
8. Guo, Y.; Liu, Y. The process of rural development and paths for rural revitalization in China. *Acta Geogr. Sin.* **2021**, *6*, 1408–1421.
9. Chen, J. Rapid urbanization in China: A real challenge to soil protection and food security. *Catena* **2007**, *69*, 1–15. [CrossRef]
10. Long, H.; Tang, G.; Li, X.; Heilig, G.K. Socio-economic driving forces of land-use change in Kunshan, the Yangtze River Delta economic area of China. *J. Environ. Manag.* **2007**, *3*, 351–364. [CrossRef]
11. Su, Y.; Qian, K.; Lin, L.; Wang, K.; Guan, T.; Gan, M. Identifying the driving forces of non-grain production expansion in rural China and its implications for policies on cultivated land protection. *Land Use Policy* **2020**, *92*, 104435. [CrossRef]
12. National Bureau of Statistics of China (NBS). *China Statistics Yearbook*; China Statistics Press: Beijing, China, 2020.
13. Johnson, D.G. Does China have a grain problem? *China Econ. Rev.* **1994**, *5*, 1–14. [CrossRef]
14. Wang, J.; Zhang, Z.; Liu, Y. Spatial shifts in grain production increases in China and implications for food security. *Land Use Policy* **2018**, *74*, 204–213. [CrossRef]
15. Wang, Y. The challenges and strategies of food security under rapid urbanization in China. *Sustainability* **2019**, *2*, 542. [CrossRef]
16. Yin, P.; Fang, X.; Tian, Q.; Ma, Y. The changing regional distribution of grain production in China in the 21st century. *J. Geogr. Sci.* **2006**, *4*, 396–404. [CrossRef]
17. Zhang, T. Why are Chinese small farmers planting more grain crops? —Review of the Logic of Grain Planting (2018). *China Agric. Econ. Rev.* **2019**, *1*, 173–175. [CrossRef]
18. Hou, M.; Deng, Y.; Yao, S. Spatial Agglomeration pattern and driving factors of grain production in China since the reform and opening up. *Land* **2021**, *1*, 10. [CrossRef]

19. Zhao, X.; Zheng, Y.; Huang, X.; Kwan, M.P.; Zhao, Y. The effect of urbanization and farmland transfer on the spatial patterns of non-grain farmland in China. *Sustainability* **2017**, *8*, 1438. [CrossRef]
20. Qiu, T.; Choy, S.T.B.; Li, S.; He, Q.; Luo, B. Does land renting-in reduce grain production? Evidence from rural China. *Land Use Policy* **2020**, *90*, 104311. [CrossRef]
21. Zhao, X.; Li, Y.; Zheng, Y.; Jin, Z.; Zhang, X. Impact of industrial structure and farmers' income structure on the non-grain conversion of arable land. *Sci. Technol. Manag. Land Resour.* **2019**, *5*, 66–77.
22. Zhang, O.; Jiang, C. Analysis on differences of 'non-grain' of different types farmers in transfer-in farmland. *J. Financ. Trade Res.* **2016**, *4*, 24–31.
23. Pan, J.; Chen, Y.; Zhang, Y.; Chen, M.; Fennell, S.; Luan, B.; Wang, F.; Meng, D.; Liu, Y.; Jiao, L.; et al. Spatial-temporal dynamics of grain yield and the potential driving factors at the county level in China. *J. Clean. Prod.* **2020**, *255*, 120312. [CrossRef]
24. Zhong, F.; Zhu, J. Food security in China from a global perspective. *Choices* **2017**, *32*, 1–5.
25. Liu, Y.; Wang, C.; Tang, Z.; Nan, Z. Will farmland transfer reduce grain acreage? Evidence from Gansu province, China. *China Agric. Econ. Rev.* **2018**, *2*, 277–292. [CrossRef]
26. Wu, S.; Yu, H.; Chu, Y. Limitations and breakthroughs in the study of "non-grain" in cultivated land under agricultural scale management. *J. Northwest A&F Univ.* **2019**, *3*, 142–151.
27. Huang, Y.; Tian, X. Food accessibility, diversity of agricultural production and dietary pattern in rural China. *Food Policy* **2019**, *84*, 92–102. [CrossRef]
28. Kong, X. The basic connotation and policy suggestions on the structural reform of agricultural supply side. *Reform* **2016**, *2*, 104–115.
29. Tobler, W.R. A computer movie simulating urban growth in the Detroit Region. *Econ. Geogr.* **1970**, *46* (Suppl. 1), 234–240. [CrossRef]
30. Moran, P.A. Notes on continuous stochastic phenomena. *Biometrika* **1950**, *37*, 17–23. [CrossRef]
31. Getis, A.; Ord, J.K. The Analysis of Spatial Association by Use of Distance Statistics. *Geogr. Anal.* **1992**, *3*, 89–206. [CrossRef]
32. Geary, R.C. The contiguity ratio and statistical mapping. *Inc. Stat.* **1954**, *5*, 115–146. [CrossRef]
33. Wang, J.; Liao, Y.; Liu, X. *Spatial Data Analysis Tutorial*; Science Press: Beijing, China, 2010.
34. Liu, Y.; Liu, J.; Zhou, Y. Spatio-temporal patterns of rural poverty in China and targeted poverty alleviation strategies. *J. Rural Stud.* **2017**, *52*, 66–75. [CrossRef]
35. Prasannakumar, V.; Vijith, H.; Charutha, R.; Geetha, N. Spatio-temporal clustering of road accidents: GIS based analysis and assessment. *Procedia-Soc. Behav. Sci.* **2011**, *21*, 317–325. [CrossRef]
36. Anselin, L. Local indicators of spatial association—LISA. *Geogr. Anal.* **1995**, *2*, 93–115. [CrossRef]
37. Eilat, Y.; Einav, L. Determinants of international tourism: A three-dimensional panel data analysis. *Appl. Econ.* **2004**, *12*, 1315–1327. [CrossRef]
38. Hsiao, C. Panel data analysis—Advantages and challenges. *Test* **2007**, *16*, 1–22. [CrossRef]
39. Wang, Z.; Cui, B.; Zhang, Y. *Spatial Econometrics: Modern Models and Methods*; Peking University Press: Beijing, China, 2017.
40. LeSage, J.; Pace, R.K. *Introduction to Spatial Econometrics*; Chapman and Hall/CRC: New York, NY, USA, 2009.
41. Chen, S.; Zeng, F.; Liu, H.; Xu, H. Factors affecting farmers' willingness to grow grain in key grain-producing areas: Based on the investigation of 475 rural households of Hunan Province. *J. Hunan Agric. Univ.* **2012**, *5*, 29–35.
42. Jin, T. Analysis on Farmers' willingness to grow grain and its influencing factors in major grain producing areas. *Stat. Decis.* **2013**, *17*, 91–95.
43. Paul, C.; Nehring, R.; Banker, D.; Somwaru, A. Scale economies and efficiency in US agriculture: Are traditional farms history? *J. Prod. Anal.* **2004**, *3*, 185–205. [CrossRef]
44. Max-Neef, M. Development and human needs. In *Development Ethics*; Routledge: London, UK, 2017; pp. 169–186.
45. Wu, S. A study of the regional characteristics of China's grain production and its cause of formation: An empirical analysis since the market-oriented reform. *Econ. Res. J.* **2000**, *10*, 38–45.
46. Liu, Y.; Fang, F.; Li, Y. Key issues of land use in China and implications for policy making. *Land Use Policy* **2014**, *40*, 6–12. [CrossRef]
47. FAO. *The 10 Elements of Agroecology: Guiding the Transition to Sustainable Food and Agricultural Systems*; Food and Agriculture Organization of the United Nations: Rome, Italy, 2018.
48. Long, H.; Tu, S.; Ge, D.; Li, T.; Liu, Y. The allocation and management of critical resources in rural China under restructuring: Problems and prospects. *J. Rural Stud.* **2016**, *47*, 392–412. [CrossRef]
49. Luo, Z.; Rao, H.; Huang, L.; Liu, M.; Chen, T.; Yao, M.; Li, J. Effect of Cu stress on minerals in rice by analyzing husk based on laser-induced breakdown spectroscopy. *Appl. Phys. B* **2020**, *126*, 8. [CrossRef]
50. Zhou, Y. Environmental planning and design of pollution-free pig farms in rural areas. *Rev. Científica Fac. Cienc. Vet.* **2019**, *5*, 1345–1354.
51. Fang, F. Seeking the theoretical origins of supply-side structural reform. *Soc. Sci. China* **2018**, *4*, 37–52.
52. Kong, X. Problems, causes and countermeasures of "non grain" of cultivated land. *China Land* **2020**, *11*, 17–19.
53. Shang, Q. Understanding of tight grain balance should be persisted for a long time. *China Grain Econ.* **2014**, *4*, 34–37.
54. Guo, Y.; Liu, Y. Poverty alleviation through land assetization and its implications for rural revitalization in China. *Land Use Policy* **2021**, *105*, 105418. [CrossRef]

55.  Han, H.; Wu, S. Structural change and its impact on the energy intensity of agricultural sector in China. *Sustainability* **2018**, *12*, 4591. [CrossRef]
56.  Wang, Y.; Chen, Y.; Yi, X.; Xiao, B. The non-grain problem in the process of land transfer and the countermeasures. *Chin. J. Agric. Resour. Reg. Plan.* **2011**, *4*, 13–16.
57.  Yang, R.; Chen, Y.; Yi, X.; Fang, L. The causes and countermeasures on the excessive "non-food" tendency in the process of land transfer. *Chin. J. Agric. Resour. Reg. Plan.* **2012**, *3*, 14–17.
58.  Li, T.; Long, H.; Zhang, Y.; Tu, S.; Ge, D.; Li, Y.; Hu, B. Analysis of the spatial mismatch of grain production and farmland resources in China based on the potential crop rotation system. *Land Use Policy* **2017**, *60*, 26–36. [CrossRef]
59.  Qian, W.; Wang, D.; Zheng, L. The impact of migration on agricultural restructuring: Evidence from Jiangxi Province in China. *J. Rural Stud.* **2016**, *47*, 542–551. [CrossRef]

*Article*

# Land Use Transitions and Farm Performance in China: A Perspective of Land Fragmentation

Shukun Wang, Dengwang Li *, Tingting Li and Changquan Liu

Rural Development Institute, Chinese Academy of Social Sciences, Beijing 100732, China; wangshukun@cass.org.cn (S.W.); litt@cass.org.cn (T.L.); liuchq@cass.org.cn (C.L.)
* Correspondence: lidw@cass.org.cn

**Abstract:** Land fragmentation (LF) is widespread worldwide and affects farmers' decision-making and, thus, farm performance. We used detailed household survey data at the crop level from ten provinces in China to construct four LF indicators and six farm performance indicators. We ran a set of regression models using OLS methods to analyse the relationship between LF and farm performance. The results showed that (1) LF increased the input of production material and labour costs; (2) LF reduced farmers' purchasing of mechanical services and the efficiency of ploughing; and (3) LF may increase technical efficiency (this result, however, was not sufficiently robust and had no effect on yield). Generally speaking, LF was negatively related to farm performance. To improve farm performance, it is recommended that decision-makers speed up land transfer and land consolidation, stabilise land property rights, establish land-transfer intermediary organisations and promote large-scale production.

**Keywords:** land fragmentation; farm performance; land transfer; China

**Citation:** Wang, S.; Li, D.; Li, T.; Liu, C. Land Use Transitions and Farm Performance in China: A Perspective of Land Fragmentation. *Land* **2021**, *10*, 792. https://doi.org/10.3390/land10080792

Academic Editors: Hualou Long, Xiangbin Kong, Shougeng Hu, Yurui Li and Elisa Marraccini

Received: 8 May 2021
Accepted: 24 July 2021
Published: 28 July 2021

**Publisher's Note:** MDPI stays neutral with regard to jurisdictional claims in published maps and institutional affiliations.

## 1. Introduction

Over the past four decades, China's agriculture productivity has improved significantly [1], with a real growth rate of 6.1% in the Gross Value of Agricultural Output (GVAO) per year throughout 1978–2015 [2]. With the economic development of urban and rural areas, China's land use has also undergone a transition. Agricultural arable land from land fragmentation to large-scale management is regarded as an essential transition path. Since the late 1980s, the Chinese Government began consciously encouraging land consolidation and promoting land transfer between rural households, as many rural labourers began to enter the cities for employment [3–5]. With the government's firm policy of support, the land rental market has experienced rapid development. According to Ministry of Agriculture data, the total rental area of arable land in China was 0.6 million hectares in 1994 and exceeded 35.9 million hectares in 2018, with an average increase of 18.6%. According to the third national agricultural census results, the proportion of large-scale farming (more than 3.33 hectares in China's southern provinces and more than 6.67 hectares in its northern provinces) across the total arable land area reached 28.6% in 2016.

In recent years, China's land transfer rate has slowed [6]. One of the main reasons for this is severe land fragmentation (LF). LF is a typical characteristic of China's traditional agriculture. The status quo of LF in China can be traced back to the very beginning of rural reform in the late 1970s when the Household Responsibility System (HRS) had just been introduced. The HRS reallocated collective agricultural land to individual rural households equally, giving them relative autonomy over land-use decisions and crop selection. Due to differences in fertility and topography, the arable land was divided into different grades, and each household was assigned several plots of different land quality [7,8]. In some areas, farmers obtained as many as dozens of plots. Although this land allocation method guaranteed fairness and justice among rural households, it was extremely inconvenient in terms of agricultural production.

161

Generally speaking, LF affects farmers' production decisions and, thus, their farm performance. Most scholars and policymakers believe LF decreases farm performance due to an inability to achieve an economy of scale. Additionally, LF simultaneously increases the area taken up by ridges, ditches, and roads between plots, wasting arable land resources. As the plots are scattered, the irrigation and mechanical efficiency of the cultivated land is significantly reduced [9,10]. When farmers want to operate on a large scale, large numbers of plots need to be transferred, in turn increasing land transaction costs [11]. Further, some people propose that land fragmentation increases the input cost of fertilisers and pesticides [12,13], reducing the rural labour emigration [14] while also reducing yields and household income [15–18]. Due to a series of negative effects induced by LF, many scholars believe that land transfer [19] and consolidation [20] should be accelerated to promote large-scale land production.

Some scholars believe that LF has a positive impact on farm performance. Those who support this view believe farmers can adjust their planting structure according to the characteristics of different plots to alleviate the lack of seasonal supply of rural labour, especially in traditional agriculture, which can profit from intensive cultivation [21,22]. According to different plots, LF allows farmers to allocate production factors to increase land productivity and food diversification, acceptability, accessibility, and sovereignty at the local level [23,24]. At the same time, a variety of crops can be planted to reduce market risks and farmers' exposure to weather variability [25,26]. LF can also speed up the transfer of rural labour out of agriculture [16].

Even if the same indicators are selected, research findings differ across countries. For example, Tan et al. [27] adopted China's farm household survey data and found that the number of plots had a positive impact on technical efficiency. In contrast, Rahman and Rahman [17] used the same indicators and methods on farmers' survey data in Bangladesh and found that LF decreased technical efficiency. The research mentioned above focused primarily on a single farm performance indicator, and few scholars have systematically analysed multiple indexes of LF on farm performance. Only Latruffe and Piet [28] used data from Brittany, France, to analyse the impact of LF on 15 farm performance indicators. Comparatively, China has more people, less land per capita than France, and land fragmentation has become more serious. Several studies have investigated the impact of LF on agriculture production in China, including its effect on technical efficiency [27], product costs [29], and returns relative to scale [30,31]. However, these studies have two main deficiencies: (1) the study sites were concentrated in one area, and (2) the sample sizes were small. Few studies have comprehensively examined the relationship between LF and multiple farm performance indicators in China.

This paper used crop-level data from ten provinces in China to determine the relationship between LF and multiple production indicators to identify the advantages and/or disadvantages of LF on farm performance. To achieve this goal, we selected four LF indicators and six farm performance indicators, giving a total of 24 regression models, using the unique large-scale survey data in China. With this, we empirically analysed the relationship between LF and farm performance. The findings of the current research address a gap in the existing literature and have essential reference significance for land policy recommendations.

The remainder of this paper is structured as follows. Section 2 provides a literature review. Section 3 outlines the methods used in the study, including data collection, indicator selection, model specification, and variable selection. Section 4 reports the empirical results and discusses the data. Section 5 presented the robustness test. Finally, Section 6 presents the study's conclusions and policy implications.

## 2. Land Reform and Literature Review

### 2.1. An Overview of Land Reform in China

In China, the land is owned by the State or village collectives. According to China's constitution and Land Administration Law, land in urban areas is owned by the State, while the State expropriates rural land except for that owned by the collectives, which means that the village or group effectively owns it. Before the Household Responsibility System (HRS) was introduced, rural land was owned by collectives and jointly utilised by their members—that is, agricultural operations were organised at the collective level. This, however, was widely regarded as inefficient [31].

The HRS reform began in the late 1970s, assigned collective agricultural land to individual rural households in an equalitarian way with contracts of up to 15 years, and entitled them to relative autonomy over land-use decisions and crop selection. Due to China's limited land resources, land quality varies greatly, even within the same village. To ensure an even distribution, the land is generally divided into three grades (i.e., good, medium, and poor), and these three land types are equally distributed according to the number of people in each household. Although this method of allocating land accords fairness and justice to farmers, it results in a farmer owning multiple plots, with the land parcels often far apart. These strategies led to land fragmentation among Chinese farmers.

For a long time following the introduction of HRS, although the land-use right of rural land belonged to farmers, the circulation of the right among farmers was not legitimate. Facing a growing voice of legalising the land rental market in the context of rapid urbanisation and with rural labourers flooding cities to work in China, the government revised the Rural Land Contract Law. Under the usufruct right, the renting out and mortgaging of rural land is now permitted. Nowadays, as the rural labour force in China continues to decrease and agricultural mechanisation increases, the Chinese Government is encouraging farmers to transfer land and expand their scale of planting.

### 2.2. Literature Review

To conduct an extensive literature search, a keyword search was initially conducted on specific literature databases such as Google Scholar, Web of Science, the Web of Knowledge, Research in Agricultural & Applied Economics (AgEcon-Search) and others. Keywords such as "Land fragmentation", "LF", "agriculture production" (and combinations of these) were used. The indicators of LF and farm performance, and the relationship between them, are presented in Table 1.

One stream of previous research has discussed the concept and measurement methods of LF, building the foundation for further research. Two broad viewpoints can be distinguished concerning LF measurement, namely single indicator methods and comprehensive index methods. The former treats LF as one or several indicators, such as plot number, the average area of each plot [21,32–34], the ratio of plot number to farm size, and the time spent for all parcels [35]. The comprehensive index methods combine all single indicators to generate a comprehensive index. The three most commonly used indicators include Januszewski's Index, Simpson's Index, and Igbozurike's Index [36–38].

Another consideration that previous studies have investigated is the effect of LF on agriculture production and farm performance. The vast majority of studies suggest that LF is not conducive to farm performance. One factor is the impact of land fragmentation on production costs. Kawasaki [12] used data from rice farmers in Japan to calculate costs using the stochastic frontier method based on a C-D production function and translog production function, revealing that the impact of land plot number on cost was negative. Specifically, the decrease in agricultural productivity led to the transfer of labour to non-agricultural sectors. The second is the impact of land fragmentation on income. Based on household survey data from Vietnam, Tran et al. [18] used the Simpson's Index to measure land fragmentation, concluding that land fragmentation significantly reduced household income after overcoming endogeneity. The other factor is the effect on productivity. Based on household survey data, both Looga et al. [37] and Rahman et al. [17] concluded that LF decreases yield. Most of the abovementioned studies focus on analysing a single production performance measure, and few scholars have used the same set of data to study the impact of LF on multiple production performance indicators.

Arguably, the current controversy involves the impact of LF on technical efficiency. This area of the current research mainly involves the use of the stochastic frontier method or the DEA method to first calculate the technical efficiency and then evaluate the impact of LF on technical efficiency through the use of the OLS method. Most scholars believe that land fragmentation has a negative impact on productivity. For example, Latruffe et al. [28] used French farm household survey data and adopted the OLS method to find that land fragmentation decreased total factor productivity and technical efficiency. However, some scholars have found a positive impact. Tan et al. [27] believe that LF enables farmers to more effectively allocate input elements and improve technical efficiency. Ciaian et al. [34] believe that in areas where rural labour is surplus, land fragmentation can make full use of agricultural labour and improve technical efficiency.

**Table 1.** Literature review on LF, farm performance, and their relationship.

| Author and Year | LF Indictor | Farm Performance Indictor | Country | Relationship |
|---|---|---|---|---|
| **Panel A: LF has a "bad" relationship with performance** | | | | |
| Jabarin et al. (1994) [39] | Whether farmers located in production region | Production cost | Jordan | + |
| Kawasaki et al. (2010) [12] | The number of plots per farm household | Production cost | Japan | + |
| Tan et al. (2010) [27] | Average distance of the plots to the homestead | Technical efficiency | China | - |
| Austin et al. (2012) [40] | Januszewski Index | Value of farm output to the value of inputs per hectare | Nigeria | - |
| Jia et al. (2014) [41] | The number of plots per farm household | Marginal productivity of labour | China | - |
| Latruffe et al. (2014) [28] | The number of plots per farm household | Technical efficiency | France | - |
| Alemu et al. (2017) [42] | Average distance of parcels from homestead; cultivated parcel number | Yields | Ethiopia | - |
| Tran et al. (2019) [18] | Simpson's Index | Per capita household income | Vietnam | - |
| Lu et al. (2019) [15] | The number of plots per farm household | Marginal productivity of labour | China | - |
| Lu, Xie et al. (2018) [31] | Simpson's Index | Scale elasticity | China | - |
| Tan (2005) [43] | Average distance of the plots to the homestead | Production cost | China | + |
| He (2014) [29] | Simpson's Index | Production cost | China | + |
| Wan et al. (2001) [30] | The number of plots per farm household | Scale elasticity | China | - |
| **Panel B: LF has a "good" relationship with performance** | | | | |
| Tan et al. (2010) [27] | The number of plots, average plot size | Technical efficiency | China | +/+ |
| Ciaian et al. (2018) [34] | The number of plots per farm household | Technical efficiency | Albania | + |
| Kadigi et al. (2017) [44] | The size and number of parcels; average distance to parcels; the Simpson Index | Yield | Tanzania | + |
| Veljanoska (2018) [45] | Number of plots; number of plots with different soil texture; number of plots with different slope; Simpson's Index | Adapting to climate change | Uganda | + |
| Knippenberg et al. (2020) [22] | Number of plots; Simpson's Index | Food security | Ethiopia | + |
| Looga et al. (2018) [37] | Schmook Index; Januszewski Index; average size of the parcel; number of parcels; area-weighted mean size of the parcels; total area of landholding | Productivity per working hour and net value added per working hour | Estonia | U-shape |

Source: The conclusions outlined above were derived from the author's review of the literature presented in the current study.

## 3. Methodology

### 3.1. Data

The data used in this study were collected from a rural household survey organised by the Rural Development Institute of the Chinese Academy of Social Sciences in 2020. The total sample size of the survey was 3833 farmer households from ten major agricultural production provinces in China. The surveying involved stratified random sampling in selecting the sample provinces, counties, villages, and rural households to be used in the study. First, ten provinces, including Heilongjiang province in the northeast, Zhejiang, Shandong, and Guangdong in the east, Anhui and Henan in the central region, and Guizhou, Sichuan, Shaanxi, and Ningxia in the west, were selected based on the level of provincial economic development. Second, within each province, the survey categorised all counties into five groups (i.e., high, relatively high, middle, relatively low, and low) according to their per capita GDP, and then randomly selected one county from each group. Next, the survey randomly selected three towns from each county and two villages from each town. Finally, 12–14 rural households in each village were randomly selected. Of the total 3833 sampled households, 64% were engaged in agricultural production. Of the households engaged in agricultural production, 30.90% and 41.67% planted one and two kinds of crops, respectively. The distribution structure of the survey area is shown in Figure 1.

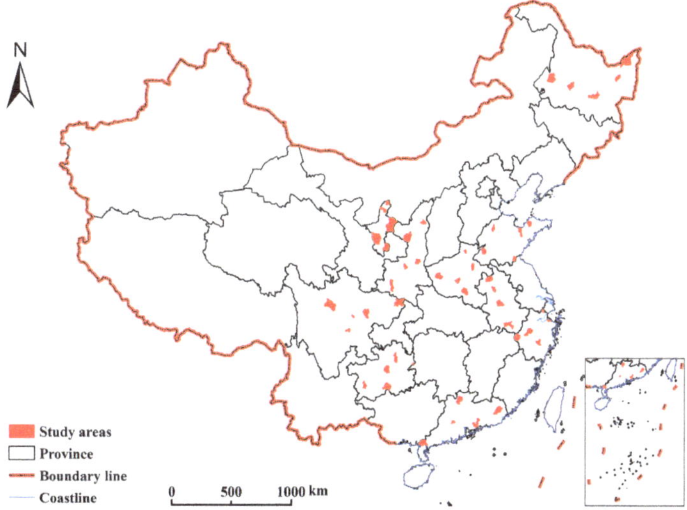


**Figure 1.** Distribution of household survey sites.

The surveys collected information at the village, household, and crop levels. At the household level, the data contain the following information: (1) detailed information about household agricultural production, including the output and input of each crop; (2) plot-level information, including the number of plots, the area of the largest three plots, and the average distance of the largest three plots from the household residence; (3) household characteristics, such as family size, the structure of the family labour force, and family members' age and education level. The survey data contains general information on the village population and geographical information about land at the village level.

In China, it is common for farmers to produce various crops within a year for two main reasons; the first is diversified planting. Specifically, to avoid natural and market risks, farmers may plant multiple crops simultaneously. The second is, planting two or three seasons worth of crops in a year. This situation is more common in North and South China, compared to Western and Northeastern region. For example, corn is planted after wheat is harvested in North China, and rapeseed is planted after the rice is harvested in the country's southern provinces. In the study sample, 69% of farmers planted more than one crop in one year. Therefore, we selected crop-level data for empirical analysis. In the survey, we asked for information on common crops with a total area of more than 0.5 mu per household. These mainly included wheat, rice, corn, soybeans, rapeseed and silage maize.

*3.2. Measuring Farm Performance and Land Fragmentation*

3.2.1. Farm Performance

A series of indicators were selected to reflect farm performance, and these were divided into three categories. The first category is cost-benefits for farmers, and two indicators were included: (1) materials cost usage (i.e., the total of material costs per mu of pesticides, fertilisers, seeds, and irrigation for each crop); and (2) labour cost usage (i.e., the total of the labour costs and the opportunity cost of their own labour—here, the average price of hired labour in the village was used to estimate the opportunity cost of their own labour).

The second category is the use of mechanisation by farmers, in which two indicators were used: (1) the total input of mechanised services purchased in different production stages of each crop, including ploughing, sowing, spraying, fertilising, irrigation, harvesting, and other production processes; and (2) the efficiency of mechanical ploughing for each crop. During the survey, if the farmer used mechanical ploughing, we asked farmers to answer the scale of ploughed land per hour for different crops, and we used this indicator to measure the efficiency of ploughed land.

The third category is productivity, in which two indicators of yield and technical efficiency were used. Yield constitutes the average output per mu of different crops (as answered by the farmers) and technical efficiency. Technical efficiency constitutes the effectiveness with which a given set of inputs were used to produce an output. In general, technical efficiency can be calculated in two ways: (1) using the parametric stochastic frontier analysis (SFA) approach, or (2) using non-parametric data envelopment analysis (DEA). Based on the research by Ciaian et al. [34] and Belotti et al. [46], the SFA approach was adopted for the current study. The model is described in detail in Section 3.3.

### 3.2.2. Land Fragmentation

From Table 1, the current research indictors of LF are the Simpson's Index [18,22,29], Januszewski Index [37,40], number of plots [12,28,41], average plot size [37,44], and average distance from plot to home [10,27,42] (among others). The Simpson and Januszewski indexes are considered ideal indicators because they simultaneously reflect the number of plots, plot size, and total scale, which can be defined as $SI = 1 - \sum_{a=1}^{n} a_i^2 / \sum_{a=1}^{n} a_i$ and $JI = \sqrt{\sum_{a=1}^{n} a_i} / \sum_{a=1}^{n} \sqrt{a_i}$, where $a_i$ is the size of the plot $i$. At the same time, because the calculation of the abovementioned two indexes requires the size of each piece of land, it is difficult to obtain data from some countries where the land is fragmented. Most scholars use multiple indicators to measure LF, including the number of plots, the average plot size, and the average distance from the plot to home (Table 1). We investigated the number of plots on each crop, the sown area of each crop, the area of the largest three plots, and the average distance of the largest three plots to home on a farmer's planting of more than 0.5 mu of cereals crops. We selected three indicators to represent LF at the crop level from different farmers. The number of plots constitutes the first LF indicator used at the crop level; the second indicator is the proportion of the area of the largest three plots to the total area; the third indicator is the average distance from the largest three plots to farmers' homes, and the fourth indicator is the average plot size. The largest three pieces of land can be used to measure the concentration of farmers' farmland. In the absence of the size and distance of each piece, we believe that the proportion of the largest three pieces of land to the total area and the average distance from the largest three pieces of land can also be used to measure the fragmentation of farmers' land.

### 3.3. Empirical Specification

The model specification is composed of two main parts. The first part calculates the indicators of farm performance, and the second calculates the impact of LF on farm performance. The survey of farmers directly supplied some indicators of farm performance—for example, yields and ploughing efficiency—while some indicators were obtained through simple calculations—for example, the material cost is the sum of chemical fertilisers, pesticides and seed costs. After obtaining the above indicators, the OLS model was used to measure the impact of LF on a series of indicators representing farm performance. We used four LF indicators and six indicators of farm performance, giving 24 regression models. The relevant model is detailed below.

(1) OLS estimation model

$$Y_{hik} = \beta_0 + \beta_1 LF_{hif} + \beta_2 C_{hi} + \beta_3 F_h + \beta_4 H_h + I_i + D_d + \varepsilon_{hi} \tag{1}$$

Based on findings from previous studies [22,28], we used the OLS model to estimate a series of farm performance indicators in addition to technical efficiency (see model [1]). $Y_{hik}$ represents the $k$-th farm performance indicator of the $h$-th farmer, planting the $i$-th crop (where $i$ represents the six crops of wheat, corn, rice, soybean, rapeseed, and silage corn), and $LF_{hif}$ constitutes the key variable in the model, which represented the $f$-th LF index of the $h$-th farmer planting the $i$-th crop. Further, $C_{hi}$ represents the control variables at the crop level, such as disaster damage, irrigation conditions. and sown area; $F_h$ represents the control variables at the household level, which represents the family size and whether there were non-agricultural workers in the $h$-th farmer's household; and $H_h$ represents the control variables at the household-head level, including the gender of the household head, education level, and whether he or she was a village cadre ('Village cadre' mainly refers to the management personnel in the village, which mainly includes the village leader, village director, and accountant). $I_i$ and $D_d$ denote crop and regional control variables (including the eastern, central, western, and northeastern regions of China), respectively. Finally, $\varepsilon_{hi}$ represents the error term.

(2) Technical efficiency estimation model

A stochastic production frontier (SFA) model was used to estimate the effect of LF on technical efficiency. Many researchers have often incorporated exogenous effects using a two-step approach. The first step is to use the production function to estimate technical inefficiency, and the second is to use the regression method to estimate the impact of some factors on technical inefficiency. As pointed out by Wang and Schmidt [46] and Belotti et al. [47], this approach leads to severely biased results; thus, we focused only on model extensions based on simultaneous estimation. The model settings are shown in Formulas (2)–(4):

$$\ln Y_{hi} = \beta_0 + \sum_{j=1}^{4} \beta_j \ln X_{hij} + \frac{1}{2} \sum_{j=1}^{4} \sum_{k=1}^{4} \beta_{hjk} \ln X_{hij} \ln X_{hik} + v_{hi} - \mu_{hi} \tag{2}$$

$$\mu_{hi} = N^+(\mu_{hi}, \sigma_u^2) \tag{3}$$

$$\mu_{hi} = z'_{hi} \varphi \tag{4}$$

In the formulas outlined above, subscript $h$ represents the household, subscript $i$ represents different crops, and subscript $j$ represents different input variables. Therefore, ln $Y_{hi}$ in Equation (2) is the logarithm of yield on crop $i$ of farmer $h$, including wheat rice, maise, soybean, rapeseed and silage maise. $X_{hij}$ represents the material costs (e.g., fertilisers, seeds, and irrigation), labour costs, mechanical service costs, and land costs for the $i$th crop of farmer $h$. The term $u_{hi}$ specified in Equation (2) is the log difference between the maximum and actual output; and $u_{hi}$ is a realisation from an exponential distribution variable; $z_i$ is a vector of the exogenous variables; and $\varphi$ is the vector of unknown parameters to be estimated (the so-called inefficiency effects). In model (4), the $z_{hi}$ represents different LF indicators. STATA16 was used to estimate the models given above.

Notably, the abovementioned models may have endogenous problems—for example, the impact of LF on yield. Farmers with severe LF may have low agricultural production technology and management capabilities, leading to biased estimation results. Generally speaking, instrumental variables are used to solve the problem. For example, Wang et al. [10] used geographic dummy variables as instrumental variables for LF to study the impact of LF on income. The current study selected six indicators to measure farm performance. It was difficult in the current study to address the endogeneity of each indicator, as this cannot easily be done within the scope of a single study. Research by Latruffe et al. [28] and Ciaian et al. [34] predominantly used the OLS method to explore the correlation between China's LF and farm performance from multiple dimensions. Although causality cannot be inferred in the current study, we believe our findings have important reference significance for academic research and policymaking.

*3.4. Description of Variables*

Table 2 is a descriptive statistical analysis of all variables, predominantly reporting the mean, standard deviation, minimum, and maximum values. The table shows that the average number of plots at the crop level was 3.6, and the number of plots varied greatly, with a maximum value of 21 and a minimum value of 1. The average proportion of the largest three plots of land to the total sown area was 0.85, which is relatively concentrated. The largest three land plots were 0.95 km away from the farmers' homes, which was generally close to home and ranged between 0 and 9 km. Those farmers land that was 0 km from the household were mainly farming arable land next to their yard. The average plot size was 3.87 mu. In the following model, the four indicators representing LF are expressed by LF1, LF2, LF3 and LF4, respectively.

In terms of the farm performance indicators, the average material cost input of chemical fertilisers, seeds, pesticides, and irrigation was found to be 277.16 yuan per mu. The average yield of six crops was 417 kg per mu. Labour input costs mainly included the sum of farmers' labour opportunity costs and employment labour costs, with an average of 259.59 yuan per mu. The average purchase of mechanised services for the whole crop growth process was found to be 97.67 yuan per mu, including tilling land, fertilisation, sowing, harvesting, and other links. This index fluctuated greatly, and the minimum value of 0 indicates that manual or animal labour was used during crop production. The efficiency of mechanised ploughing was 3.27 mu per hour, which varied greatly, with the ploughed area ranging from 0.2 mu to 17.3 mu per hour. We defined the total income minus the material cost input as the profitability indicator, and the average profitability of the crop was found to be 476.85 yuan per mu.

Various control variables were chosen based on the household decision-making model, and findings from previous studies were used in all 18 regressions. The average resident population per household was 3.28, the average sowing area per household was 18.43 mu, 66% of the crops were not affected by natural disasters, and 53% of the farmers did not take out agricultural insurance. Household heads were predominantly male, and their education level was mainly at the primary and junior high school level, accounting for 79% of the total. Village cadres were 18% of household heads, and only 5% of the people were engaged in non-agricultural work. In the sample, the percentages of wheat, rice, corn, soybean, rapeseed, and silage corn were 23%, 19%, 47%, 5%, 3%, and 4%, respectively.

A specific correlation between farm size and productivity was found. Most studies have concluded there is an inverse relationship between farm size and agriculture productivity [48–50], even though several scholars have reported a U-shaped relationship [51]. Under this assumption, if there is a correlation between land fragmentation and farm size, a confounding effect in which farmers are allocating inputs due to farm size rather than LF can be considered. Figure 2 shows the data depicts a positive correlation between farm size and LF. Therefore, it was necessary to control the variable of farm size in the regression model, and this estimated result was considered together with the LF variable.

**Table 2.** Descriptive statistics and variable definitions.

| Variable | Definition or Unit | Mean | Std. Dev. | Min. | Max. |
|---|---|---|---|---|---|
| Land Fragmentation Indicator | | | | | |
| LF1 | Number of plots | 3.63 | 3.06 | 1 | 21 |
| LF2 | The proportion of the largest three plots of land to the total crop area. | 0.85 | 0.23 | 0 | 1 |
| LF3 | The average distance of the three largest plots from home (km) | 0.95 | 0.93 | 0 | 9 |
| LF4 | Average plot size | 3.87 | 7.24 | 0.03 | 78 |
| Farm Performance Indicator | | | | | |
| Material cost | Included fertiliser, seed, pesticide, and irrigation cost (yuan/mu) | 277.16 | 151.46 | 14.1 | 2660 |
| Yield | Kg/mu | 417 | 182.16 | 2 | 1000 |
| Labour cost | Including the cost of employment and the opportunity cost of own labour (Yuan/mu) | 259.59 | 341.15 | 2 | 2480 |
| Purchase machinery service | Including ploughing land, fertilisation, sowing, harvest, and other processes (Yuan/mu) | 97.67 | 90.69 | 0 | 520 |
| Tilling efficiency | Mechanical arable area per hour (Mu/hour) | 3.27 | 3.35 | 0.2 | 17.3 |
| Profitability | Revenue minus material costs without farm subsidies (Yuan/mu) | 476.85 | 383.31 | -657.5 | 1832 |
| Control Variables | | | | | |
| Head of household characteristics | | | | | |
| Gender | 1 = female; 0 = male | 0.04 | 0.2 | 0 | 1 |
| Education level | | | | | |
| illiteracy | 1 = yes; 0 = other | 0.07 | 0.26 | 0 | 1 |
| <=Primary school | 1 = yes; 0 = other | 0.3 | 0.46 | 0 | 1 |
| Junior high school | 1 = yes; 0 = other | 0.49 | 0.5 | 0 | 1 |
| Senior high school | 1 = yes; 0 = other | 0.12 | 0.33 | 0 | 1 |
| >=College | 1 = yes; 0 = other | 0.11 | 0.11 | 0 | 1 |
| Village cadre | 1 = yes; 0 = other | 0.18 | 0.38 | 0 | 1 |
| Non-agricultural employment | 1 = yes; 0 = other | 0.05 | 0.22 | 0 | 1 |
| Household characteristics | | | | | |
| Household size | Total number of permanent family members | 3.28 | 1.45 | 1 | 10 |
| Crop characteristics | | | | | |
| Farm size | Mu | 18.43 | 73.22 | 0.1 | 2200 |
| Disaster | 1 = The crop has not suffered natural disasters | 0.66 | 0.47 | 0 | 1 |
| Insurance | 1 = The crop is insured | 0.53 | 0.5 | 0 | 1 |
| Land rent | Yuan/mu | 516.2 | 319.5 | 30 | 2342 |
| Crop | | | | | |
| wheat | 1 = yes; 0 = other | 0.23 | 0.42 | 0 | 1 |
| rice | 1 = yes; 0 = other | 0.19 | 0.4 | 0 | 1 |
| maize | 1 = yes; 0 = other | 0.47 | 0.5 | 0 | 1 |
| soybean | 1 = yes; 0 = other | 0.05 | 0.21 | 0 | 1 |
| rapeseed | 1 = yes; 0 = other | 0.03 | 0.18 | 0 | 1 |
| silage maize | 1 = yes; 0 = other | 0.04 | 0.19 | 0 | 1 |

Note: Data from author survey; 1 Mu = 1/15 Ha; Yuan is the currency of China (1 USD = 6.90 Yuan in 2019).

In the literature, many studies have reported measurement errors on farm size based on the use of self-reported data [52,53]. However, this is not the case in China, since household farms hold better knowledge of the land areas in operation. The Ministry of Agriculture and Rural Affairs of China began conducting a large-scale land titling pilot in 2009. According to the investigation, code of practice for the right to rural contractual management (which the government developed), farmland details, including location, area, and owner of use rights, are being investigated in this pilot with the help of GPS/GNSS and drones. The results of the land titling pilot will officially be announced in the village for several days, and the farmers must confirm the outcomes. Then, the government will grant land certificates to the farmers. By the end of 2020, approximately 96% of the land in China was contractual. Therefore, we believe that the sample of farmers in the current study self-reported the details of their farmland with few errors. While collecting the data from the farmers in the sample, their land certificates and land rental contracts were also inspected.

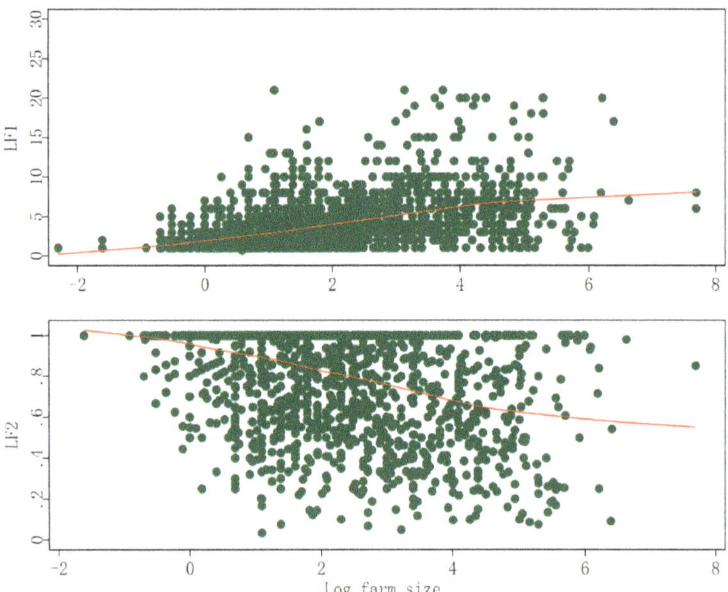

**Figure 2.** Relationship between farm size and land fragmentation (Bandwidth = 0.8).

## 4. Results and Discussion

### *4.1. The Results of LF on the Costs and Benefits*

#### 4.1.1. The Influence of LF on Material Cost

The natural logarithm of the per mu material cost was used as the dependent variable to reduce the skewness distributions within the data. In Table 3, models (1)–(4) represent the relationships between the number of plots, the proportion of the largest three plots, the average distance of the largest three plots to home, average plot size and farmers' material costs at the crop level. The results generated from models (1)–(3) indicate that LF was associated with increased material cost, but the results from model (3) were slightly different. In model (1), after controlling for other variables, the number of plots was significantly positively related to material cost. The coefficient of the LF1 variable was 0.0192, and this was statistically significant at the level of 1%. As the number of plots increased, the material cost per mu increased by 1.92%. Based on the average material cost of each crop of 277, each additional piece of a plot increased the average cost per mu by 5.32 Yuan. In model (2), it can be concluded that the proportion of the largest three plots of land and material costs were negatively correlated and statistically significant. In Model (3), LF was measured using the average distance of the largest three plots from home, showing that the further the average distance from home, the lower the material cost. When plots are further from home, farmers may reduce their material inputs, such as their use of pesticides and chemical fertilisers, due to the high costs associated with transportation and production. Although model (4) was not significant, it revealed that the larger the average plot was, the less material cost was input. The estimated result of model 3 diverged from Tan's [44] findings that the average distance from land to home revealed a positive relationship with material cost. However, Tan [44] used the average distance from all plots to a residence, and the survey data was collected from the Jiangxi Province, where land fragmentation is more severe.

Table 3. Results of LF on production material cost.

| | Model (1) | Model (2) | Model (3) | Model (4) |
|---|---|---|---|---|
| LF1 | 0.0192 *** | | | |
| | −5.057 | | | |
| LF2 | | −0.0888 * | | |
| | | (−1.646) | | |
| LF3 | | | −0.0224 ** | |
| | | | (−1.975) | |
| LF4 | | | | −0.0006 |
| | | | | (−0.388) |
| Log farm size | −0.0796 *** | −0.0576 *** | −0.0465 *** | −0.0541 *** |
| | (−6.561) | (−5.174) | (−4.643) | (−4.905) |
| Head of household characteristics | | | | |
| Gender | 0.0149 | 0.0134 | −0.0017 | −0.0021 |
| | −0.289 | −0.264 | (−0.034) | (−0.041) |
| Education level | | | | |
| <=Primary school | −0.0773 ** | −0.0764 * | −0.0627 | −0.0755 * |
| | (−1.973) | (−1.920) | (−1.642) | (−1.881) |
| Junior high school | −0.0318 | −0.0251 | −0.0213 | −0.0280 |
| | (−0.853) | (−0.665) | (−0.590) | (−0.733) |
| Senior high school | −0.0617 | −0.0625 | −0.0573 | −0.0612 |
| | (−1.383) | (−1.376) | (−1.295) | (−1.344) |
| >=College | 0.0317 | 0.0364 | 0.0479 | 0.0356 |
| | −0.265 | −0.304 | −0.404 | −0.297 |
| Village cadre | −0.0082 | −0.0113 | −0.0078 | −0.0755 * |
| | (−0.337) | (−0.460) | (−0.315) | (−1.881) |
| Non−agricultural employment | 0.0268 | 0.0385 | 0.0517 | 0.0379 |
| | −0.611 | −0.87 | −1.153 | −0.862 |
| Family characteristics | | | | |
| Family members | 0.0107 | 0.0112 * | 0.0138 ** | 0.0128 * |
| | −1.592 | −1.654 | −2.018 | −1.886 |
| Disaster | −0.0528 ** | −0.0510 ** | −0.0481 ** | −0.0523 ** |
| | (−2.431) | (−2.344) | (−2.223) | (−2.402) |
| Insurance | −0.0252 | −0.0346 * | −0.0279 | −0.0357 * |
| | (−1.268) | (−1.734) | (−1.400) | (−1.782) |
| Crop controls | yes | yes | yes | yes |
| Region controls | yes | yes | yes | yes |
| Constant | 5.5009 *** | 5.5803 *** | 5.4828 *** | 5.5027 *** |
| | −61.06 | −51.496 | −61.425 | −61.402 |
| Observations | 2413 | 2405 | 2386 | 2427 |
| R-squared | 0.19 | 0.184 | 0.178 | 0.179 |

Note: Robust *t*-statistics in parentheses; *** significance at 1%; ** significance at 5%; * significance at 10%.

### 4.1.2. The Influence of LF on Labour Input

Similarly, we used the logarithm of labour input as a dependent variable. Table 4 shows the results of the effects of LF on the labour cost at the crop level. After controlling for other variables, through models (1) to (4), we can see that LF and labour input were positively related. However, model (4) did not reveal an obvious significant relationship. From model (1), each increase in the number of plots at the crop level increased the average labour force cost per mu by 5.46%, and it was highly significant at the 1% level. Based on the average values of the sample, a saving of 14.17 Yuan for each additional plot could be made. The results of model (2) showed that when the proportion of the largest three plots increased by 1%, the labour input decreased by 0.54 Yuan and was highly significant at the 1% level. The results of model (3) revealed that for every average increase of one km of the largest three plots, the labour input increased by 4.3%, or 11.16 Yuan, according to the average value of the sample. In the survey, farmers were asked about the household labour input in each production stage, such as cultivated land, sowing, spraying insecticide, and harvesting, and then the opportunity cost based on the village-level labour employment price was calculated. This was then added to the hired labour input. Household labour input included preparation time and actual working time, with preparation time including the time taken to travel between the residence and the plot. It was determined that the further the farmer was from their residence, the more labour input cost he/she incurred. This finding is consistent with He's [29] findings derived from the use of the Simpson's Index.

**Table 4.** Results of LF on labour cost.

| | Model (1) | Model (2) | Model (3) | Model (4) |
|---|---|---|---|---|
| LF1 | 0.0546 *** | | | |
| | −6.264 | | | |
| LF2 | | −0.5421 *** | | |
| | | (−5.269) | | |
| LF3 | | | 0.0433 * | |
| | | | −1.902 | |
| LF4 | | | | −0.0002 |
| | | | | (−0.048) |
| Log farm size | −0.4360 *** | −0.3769 *** | −0.3324 *** | −0.3351 *** |
| | (−0.026) | (−0.023) | (−0.021) | (−15.029) |
| Head of household characteristics controls | yes | yes | yes | yes |
| Household characteristics controls | yes | yes | yes | yes |
| Crop controls | yes | yes | yes | yes |
| Region controls | yes | yes | yes | yes |
| Constant | 4.3286 *** | 4.7434 *** | 4.0605 *** | 4.1170 *** |
| | −22.715 | −21.233 | −21.14 | −21.926 |
| Observations | 2523 | 2509 | 2489 | 2541 |
| R-squared | 0.517 | 0.51 | 0.503 | 0.506 |

Note: Robust *t*-statistics in parentheses; *** significance at 1%; ** significance at 5%; * significance at 10%.

The analysis presented above shows that LF can significantly increase material and labour costs. Although the total cost of the reduction is small, it is meaningful in terms of proportion. Taking wheat production as an example, if material capital, labour cost, and land lease cost (including the opportunity of owning land) are deducted, the income per mu is only 51.6 Yuan (excluding subsidies). According to our regression results, under the condition of controlling the total sown area, each additional piece of land can increase the material cost for farmers by 5.32 Yuan and the labour cost by 14.17 Yuan, which accounts for 37.78% of the maize income. Additionally, from an environmental perspective, LF is not conducive to reducing the input of material capital such as pesticides and fertilisers.

### 4.2. The Results of LF on the Use of Mechanisation

The data in Table 5 show the correlation between LF and the farmers' purchase of machinery services. According to the estimated results of model (1), the cost of purchasing mechanised services for farmers with more plots was lower and highly significant at the 1% level, with each increase in the number of plots reducing the cost of purchasing services by an average of 10.33%. Model (2) showed that the higher the proportional area of the largest three plots, the higher the cost of mechanisation, which was significant at the 1% level. Models (3) and (4) also validated this conclusion: the further the average distance of the largest three plots, the lower the cost of purchasing machinery services for farmers. Additionally, the larger the plot, the higher the input for purchasing machinery.

**Table 5.** Results of LF on purchase machinery service.

|  | Model (1) | Model (2) | Model (3) | Model (4) |
|---|---|---|---|---|
| LF1 | −0.1033 *** | | | |
| | (−6.072) | | | |
| LF2 | | 1.1295 *** | | |
| | | −5.625 | | |
| LF3 | | | −0.0979 ** | |
| | | | (−2.232) | |
| LF4 | | | | 0.0107 * |
| | | | | −1.697 |
| Log farm size | 0.4200 *** | 0.3265 *** | 0.2189 *** | 0.2815 *** |
| | (−0.052) | (−0.050) | (−0.042) | −6.133 |
| Head of household characteristics | yes | yes | yes | yes |
| Household characteristics | yes | yes | yes | yes |
| Crop controls | yes | yes | yes | yes |
| Region controls | yes | yes | yes | yes |
| Constant | 3.7600 *** | 2.7223 *** | 4.2169 *** | 4.1684 *** |
| | −11.639 | −6.517 | −12.86 | −13.186 |
| Observations | 2635 | 2623 | 2597 | 2648 |
| R-squared | 0.288 | 0.275 | 0.268 | 0.271 |

Note: Robust *t*-statistics in parentheses; *** significance at 1%; ** significance at 5%; * significance at 10%. For the few farmers who did not purchase machinery services, we added one and then logarithm to this variable.

The area of (mechanized) ploughed land per hour of different crops was used as an index to measure mechanical efficiency. The estimated results of the relationship between LF and mechanical efficiency are shown in Table 6. Both models (1) and (2) show that LF significantly reduced ploughed land efficiency, and this was highly significant at the 1% level. For example, in model (1), it was found that when other variables remain unchanged, each additional piece of land for farmers reduced the area of ploughed land by 0.03 mu per hour. The results from model (3) showed some differences—specifically, the further farmers' average distance from home and higher cultivated land efficiency.

**Table 6.** Results of LF on ploughing efficiency.

|  | Model (1) | Model (2) | Model (3) | Model (4) |
|---|---|---|---|---|
| LF1 | −0.0259 *** | | | |
| | (−5.103) | | | |
| LF2 | | 0.2011 *** | | |
| | | (2.968) | | |
| LF3 | | | 0.0313 ** | |
| | | | (2.059) | |
| LF4 | | | | 0.0079 *** |
| | | | | (3.313) |
| Log farm size | 0.1617 *** | 0.1422 *** | 0.1143 *** | 0.0952 *** |
| | (−0.015) | (−0.014) | (−0.013) | (6.892) |
| Head of household characteristics controls | yes | yes | yes | yes |
| Household characteristics controls | yes | yes | yes | yes |
| Crop controls | yes | yes | yes | yes |
| Region controls | yes | yes | yes | yes |
| Constant | 1.6080 *** | 1.3979 *** | 1.6328 *** | 1.3662 *** |
| | (15.661) | (10.614) | (15.455) | (11.734) |
| Observations | 1997 | 1979 | 1974 | 1979 |
| R-squared | 0.283 | 0.276 | 0.262 | 0.272 |

Note: Robust *t*-statistics in parentheses; *** significance at 1%; ** significance at 5%; * significance at 10%.

### 4.3. The Results of LF on Productivity

The yield and technical efficiency were selected to measure crop productivity. Data in Tables 7 and 8 reported the impact of LF on both, respectively. Unlike the above model, the production function model was used to analyse the relationship between LF and yield. Therefore, it was necessary to add logarithmic material capital and labour cost to the control variables. From Table 7, it can be seen that the number of plots, the proportion of the largest three plots and the average plot size had no significant effect on crop yield, while the average distance to a residence significantly reduced crop yield. The yield decreased by 0.03 kg for every 1 km increase in the average distance from home. Additionally, it can be seen that the relationship between LF and yield was not significant.

**Table 7.** Analysis of the effect of LF on yield.

|  | Model (1) | Model (2) | Model (3) | Model (4) |
|---|---|---|---|---|
| LF1 | 0.0034 |  |  |  |
|  | (0.667) |  |  |  |
| LF2 |  | 0.0328 |  |  |
|  |  | (0.511) |  |  |
| LF3 |  |  | −0.0316 ** |  |
|  |  |  | (−2.430) |  |
| LF4 |  |  |  | −0.0020 |
|  |  |  |  | (−0.884) |
| Log(farm size) | 0.0021 | 0.0158 | 0.0168 | 0.0160 |
|  | (0.149) | (1.206) | (1.495) | (1.317) |
| Log(labour cost) | 0.0016 | 0.0026 | 0.0040 | 0.0030 |
|  | (0.207) | (0.207) | (0.318) | (0.238) |
| Log(material capita) | 0.1792 *** | 0.1771 *** | 0.1760 *** | 0.1761 *** |
|  | (6.070) | (6.010) | (5.926) | (6.005) |
| Head of household characteristics | yes | yes | yes | yes |
| Household characteristics | yes | yes | yes | yes |
| Crop controls | yes | yes | yes | yes |
| Region controls | yes | yes | yes | yes |
| Constant | 5.0393 *** | 4.9577 *** | 5.0343 *** | 5.0407 *** |
|  | (24.726) | (22.524) | (25.586) | (25.163) |
| Observations | 2175 | 2167 | 2156 | 2190 |
| R-squared | 0.374 | 0.375 | 0.366 | 0.373 |

Note: Robust *t*-statistics in parentheses; *** significance at 1% ** significance at 5%; * significance at 10%.

**Table 8.** Analysis of the effect of LF on technical inefficiency.

|  | Model (1) | Model (2) | Model (3) | Model (4) |
|---|---|---|---|---|
| Frontier |  |  |  |  |
| lnX1 | 0.3970 ** | 0.2450 | 0.4003 * | 0.3686 * |
|  | (1.967) | (1.233) | (1.946) | (1.781) |
| lnX2 | 0.0687 | 0.1233 | 0.0560 | 0.0726 |
|  | (0.782) | (1.465) | (0.636) | (0.818) |
| lnX3 | 0.4846 *** | 0.5304 *** | 0.5493 *** | 0.5422 *** |
|  | (3.035) | (3.571) | (3.420) | (3.340) |
| lnX4 | 0.0475 | 0.0656 | 0.0466 | 0.0314 |
|  | (0.840) | (1.223) | (0.818) | (0.541) |
| lnX1 * lnX1 | 0.0017 | 0.0142 | −0.0006 | 0.0025 |
|  | (0.093) | (0.803) | (−0.030) | (0.131) |
| lnX2 * lnX2 | −0.0060 | −0.0033 | −0.0032 | −0.0040 |
|  | (−1.533) | (−0.910) | (−0.816) | (−1.010) |
| lnX3 * lnX3 | 0.0018 | −0.0082 | −0.0041 | −0.0053 |
|  | (0.154) | (−0.757) | (−0.335) | (−0.434) |

**Table 8.** *Cont.*

|  | Model (1) | Model (2) | Model (3) | Model (4) |
|---|---|---|---|---|
| lnX4 * lnX4 | −0.0020 | −0.0013 | −0.0015 | −0.0006 |
|  | (−0.480) | (−0.340) | (−0.354) | (−0.142) |
| lnX1 * lnX2 | 0.0170 | 0.0040 | 0.0175 | 0.0141 |
|  | (1.250) | (0.310) | (1.283) | (1.012) |
| lnX1 * lnX3 | −0.0614 *** | −0.0474 ** | −0.0574 ** | −0.0551 ** |
|  | (−2.739) | (−2.247) | (−2.527) | (−2.415) |
| lnX1 * lnX4 | −0.0086 | −0.0130 * | −0.0089 | −0.0086 |
|  | (−1.051) | (−1.676) | (−1.079) | (−1.011) |
| lnX2 * lnX3 | −0.0216 ** | −0.0238 *** | −0.0240 ** | −0.0232 ** |
|  | (−2.155) | (−2.578) | (−2.372) | (−2.267) |
| lnX2 * lnX4 | 0.0057 | 0.0062 * | 0.0071 * | 0.0082 ** |
|  | (1.435) | (1.718) | (1.804) | (1.990) |
| lnX3 * lnX4 | 0.0006 | −0.0001 | −0.0010 | −0.0005 |
|  | (0.094) | (−0.023) | (−0.150) | (−0.078) |
| Head of household characteristics controls | yes | yes | yes | yes |
| Household characteristics controls | yes | yes | yes | yes |
| Crop controls | yes | yes | yes | yes |
| Region controls | yes | yes | yes | yes |
| Inefficiency term |  |  |  |  |
| LF1 | −0.0787 *** |  |  |  |
|  | (−4.2404) |  |  |  |
| LF2 |  | −15.8755 |  |  |
|  |  | (−1.464) |  |  |
| LF3 |  |  | 0.0153 |  |
|  |  |  | (0.409) |  |
| LF4 |  |  |  | −0.0092 * |
|  |  |  |  | (−1.702) |
| Log farm size | −0.0384 | −1.8754 | −0.1634 *** | −0.1192 *** |
|  | (−1.039) | (−1.385) | (−4.635) | (−3.327) |
| Usigma | −0.4200 *** | 2.0118 *** | −0.4854 *** | −0.4733 *** |
|  | (−6.484) | (3.133) | (−7.897) | (−8.100) |
| Vsigma | −3.5335 *** | −3.3879 *** | −3.5686 *** | −3.5692 *** |
|  | (−38.966) | (−44.515) | (−38.940) | (−38.665) |
| Constant | 3.1460 *** | 3.2421 *** | 2.9362 *** | 3.0536 *** |
|  | (3.615) | (3.831) | (3.354) | (3.460) |
| Observations | 2111 | 2106 | 2091 | 2078 |

Note: z-statistics in parentheses; *** significance at 1%; ** significance at 5%; * significance at 10%; lnX1, lnX2, lnX3, and lnX4 represent the natural logarithm of the material cost, labour cost, land rent, and purchase machinery service variables.

Table 8 shows the relationship between LF and technical inefficiency. Interestingly, when other variables remained unchanged, the number of land plots (model [1]) was negatively correlated with technical inefficiency. Models (2)–(4) showed that there was no significant relationship between LF and technical inefficiency. The results of the above four models revealed that the impact of LF on technical inefficiency was not robust. Using survey data from rice farmers in China, Tan et al. [27] showed that the number of land plots had a positive impact on technical efficiency, which is consistent with the results of the current study. Those who have obtained a positive relationship believe that the production of multiple plots of land can reduce risk, with the variation effect exceeding the management effect [23–26]. There is extensive heterogeneity in the different plots in China and some differences in land type, insect pests and irrigation facilities. Under the condition of controlling for the total sown area, farmers with more plots can make full use of the characteristics of different plots to allocate production factors to achieve higher technical efficiency. Based on the above analysis, it is important to be cautious about the relationship between LF and technical inefficiency. Additionally, we used the C-D production function to estimate the impact of LF on technical inefficiency (see Table A1 in Appendix A), and the estimated results showed little difference. The LR test showed that the model specification of the translog production function was superior to the C-D production function.

## 5. Robustness

### 5.1. Valuation of Household Labour

When analysing the relationship between LF and labour costs, the average price of hired labour in the village was used to calculate farmers' labour input costs, which implicitly assumes that all labourers have the opportunity to be hired as agricultural labourers in their communities. However, some agricultural labourers may not fully enter the non-agricultural market in rural China due to ageing or a lack of technology. They are likely to put the surplus labour into their own agricultural production, inferring that their actual labour price is lower than the market labour value. Subsequently, to validate the stability of the results, half of the average price of the non-agricultural labour market price was used to calculate the farmers' labour opportunity cost. The estimated results are shown in Table 9. When comparing the estimation results with those presented in Table 4, the estimation coefficient of LF changed very little, indicating that the impact of fragmentation on labour input was robust.

**Table 9.** Analysis of the effect of LF on labour cost (the labour cost of farmers' labour was calculated at one-half the prevailing wage).

| | Model (1) | Model (2) | Model (3) | Model (4) |
|---|---|---|---|---|
| LF1 | 0.0517 *** | | | |
| | (5.937) | | | |
| LF2 | | −0.5274 *** | | |
| | | (−5.101) | | |
| LF3 | | | 0.0417 * | |
| | | | (1.847) | |
| LF4 | | | | 0.0024 |
| | | | | (0.527) |
| Log farm size | −0.4134 *** | −0.3548 *** | −0.3127 *** | −0.3226 *** |
| | (−15.430) | (−14.881) | (−14.871) | (−14.389) |
| Head of household characteristics controls | yes | yes | yes | yes |
| Household characteristics controls | yes | yes | yes | yes |
| Crop controls | yes | yes | yes | yes |
| Region controls | yes | yes | yes | yes |
| Constant | 3.6462 *** | 4.0447 *** | 3.3836 *** | 3.4301 *** |
| | (19.367) | (18.172) | (17.855) | (18.535) |
| Observations | 2523 | 2509 | 2489 | 2541 |
| R-squared | 0.515 | 0.507 | 0.501 | 0.505 |

Note: Robust *t*-statistics in parentheses; *** significance at 1%; **significance at 5%; * significance at 10%.

### 5.2. Ploughing Methods

When analysing the relationship between LF and farmland efficiency, the speed at which a farmer can plough their land was used as an index. However, the ploughing methods used by farmers may differ. For example, some use cattle to plough their land, and some use mechanical ploughing. If the ploughing method is related to LF and the scale of cultivated land, the estimated result is biased. Consequently, this complication was considered using the following means.

The data presented in Table 10 show that the farmers in the sample had a very high degree of mechanisation of ploughing rapeseed, which showed the lowest ploughing mechanisation, reaching 93.49%. Wheat and soybean fields, on the other hand, essentially reached 100%. Therefore, we did not have to consider any complications caused by mechanical or cattle ploughing. However, even if farmers used machinery to plough the land, the different mechanical horsepower used by different farmers may also have caused biases in the estimation results. Unfortunately, we did not investigate the horsepower of the machinery used by farmers when ploughing the land. Generally speaking, if a farmer purchases machinery services, the machinery horsepower will exceed the farmers' machinery horsepower. Therefore, we used whether a farmer purchased machinery services as a proxy variable of the mechanical horsepower and controlled for it in the model. The estimated results are shown in Table 11. A comparison of the data presented in Tables 8 and 11 shows that the results changed very little, further validating their robustness.

**Table 10.** Percentage of different crops ploughed by machine.

| Crops | Ploughing (%) |
|---|---|
| Maize | 99.44 |
| Rice | 97.42 |
| Maize | 96.13 |
| Soybean | 99.43 |
| Rapeseed | 93.49 |

**Table 11.** Analysis of the effect of LF on tilling efficiency (controlling for whether to purchase mechanical service variables).

| | Model (1) | Model (2) | Model (3) | Model (4) |
|---|---|---|---|---|
| LF1 | −0.0242 *** | | | |
| | (−4.800) | | | |
| LF2 | | 0.1789 *** | | |
| | | (2.634) | | |
| LF3 | | | 0.0346 ** | |
| | | | (2.277) | |
| LF4 | | | | 0.0082 *** |
| | | | | (3.340) |
| Log farm size | 0.1565 *** | 0.1382 *** | 0.1118 *** | 0.0859 *** |
| | (10.809) | (9.876) | (8.868) | (6.350) |
| Purchase of mechanical services | 0.1147 *** | 0.1126 *** | 0.1432 *** | 0.1556 *** |
| | (2.756) | (2.647) | (3.259) | (3.672) |
| Head of household characteristics | yes | yes | yes | yes |
| Family characteristics | yes | yes | yes | yes |
| Crop controls | yes | yes | yes | yes |
| Region controls | yes | yes | yes | yes |
| Constant | 1.6080 *** | 1.3979 *** | 1.6328 *** | 1.4982 *** |
| | (15.661) | (10.614) | (15.455) | (13.633) |
| Observations | 1997 | 1979 | 1974 | 1996 |
| R-squared | 0.283 | 0.276 | 0.262 | 0.266 |

Note: Robust $t$-statistics in parentheses; ***significance at 1%; **significance at 5%; *significance at 10%.

### 5.3. Land Fragmentation Indicator Check

The LF indicators of LF2 and LF3 were based on the three largest plots. Sometimes the farmers had less than three plots, and their actual plot number may have been between 1 and 21. To check the robustness of the estimated results, we selected samples with more than three plots for regression. From Table 12, it can be seen that except for the effect of LF2 on ploughing efficiency and the effect of LF3 on ploughing efficiency and technical inefficiency, the other estimates underwent little change, further validating the robustness of the results.

**Table 12.** Robust check of LF2 and LF3 on farm performance.

|  | Material Cost | Labour Cost | Machinery Service | Ploughing Efficiency | Yield | Technical Inefficiency |
|---|---|---|---|---|---|---|
| | Panel A: Relationship between LF2 and farm performance. | | | | | |
| LF2 | −0.1369 * | −0.4705 *** | 1.4493 *** | 0.1272 | 0.0413 | −1.5010 *** |
| | (−1.803) | (−3.191) | (4.796) | (1.201) | (0.471) | (−8.635) |
| Log farm size | −0.0748 *** | −0.4361 *** | 0.4539 *** | 0.1914 *** | 0.0161 | −0.2356 *** |
| | (−3.988) | (−10.924) | (5.012) | (7.260) | (0.873) | (−5.766) |
| | Panel B: Relationship between LF3 and farm performance. | | | | | |
| LF3 | −0.0405 ** | 0.0262 * | −0.0082 ** | −0.0011 | −0.0256 * | −0.0231 |
| | (−2.342) | −1.813 | (−2.139) | (−0.049) | (−1.949) | (−0.331) |
| Log farm size | −0.0660 *** | −0.3970 *** | 0.3037 *** | 0.1760 *** | 0.0145 | −0.4873 *** |
| | (−3.712) | (−10.959) | (3.751) | (7.333) | (0.796) | (−12.781) |

Note: Robust *t*-statistics in parentheses; *** significance at 1%; ** significance at 5%; * significance at 10%; the sample only retained farmers with more than three plots, and the control variables were the same as before (they are omitted here).

### 5.4. Outlier Value check

Many outliers were present in the farmers' survey data. For example, the yield per mu ranged from 2.0 kg to 1000 kg per mu. Although we took into account the natural logarithm of the relevant variables in the model, we were concerned the results would be biased due to outlier values. Therefore, we checked the robustness in two ways, which involved dropping and winsorising 5% of samples for dependent variables on both sides, respectively. Tables 13 and 14 show the impact of LF on labour cost and yield. Compared with Tables 4 and 7, the estimated results revealed little change, indicating that the previous results are robust.

**Table 13.** Robustness test of LF on labour cost.

| Model (1) | Model (2) | Model (3) | Model (4) |
|---|---|---|---|
| Panel A: Five percent of the samples were dropped from the dependent variable on both sides. | | | |
| 0.0381 *** | −0.4629 *** | 0.0225 | 0.0036 |
| (4.723) | (−4.888) | (1.097) | (0.874) |
| Panel B: Five percent of the samples were winsorised at 5% on both sides. | | | |
| 0.0515 *** | −0.5394 *** | 0.0392 * | 0.0017 |
| (6.220) | (−5.527) | (1.840) | (0.438) |

Note: Panel A and pane B represents two methods used to check the outliers of the dependent variables; the control variables in the above model are the same as those in Table 4 (they are omitted here); models (1)–(4) represent the regression of LF1–LF4 on labour cost, respectively; *** significance at 1%; ** significance at 5%; * significance at 10%.

**Table 14.** Robustness test of LF on yield.

| Model (1) | Model (2) | Model (3) | Model (4) |
|---|---|---|---|
| Panel A: Five percent of the samples were dropped from the dependent variable on both sides. | | | |
| 0.0034 | 0.0328 | −0.0316 ** | −0.0020 |
| (0.667) | (0.511) | (−2.430) | (−0.884) |
| Panel B: Five percent of the samples were winsorised on both sides. | | | |
| 0.0028 | 0.0266 | −0.0232 ** | −0.0019 |
| (0.740) | (0.543) | (−2.149) | (−1.231) |

Note: Panel A and pane B represent two methods used to check the outliers of the dependent variables; the control variables in the above model are the same as those in Table 7 (they are omitted here); models (1)–(4) represent the regression of LF1–LF4 on yield, respectively; *** significance at 1%; ** significance at 5%; * significance at 10%.

## 6. Conclusions

Based on the unique crop-level survey data from ten provinces in China, the current study investigated the correlation between four indicators representing land fragmentation (LF) and six indicators representing farm performance. The OLS model was used to empirically analyse the relationship between LF and farm performance. To the authors' knowledge, this study is the first to investigate the relationship between LF and multiple production indicators and the first to comprehensively compare and analyse the "good" and "bad" effects of LF on farm performance. The results revealed that while controlling for other variables, LF increased the input of farmers' material cost and increased the cost of labour input. Due to the fragmented land in China, it is difficult for farmers to implement large-scale production. Some new equipment and new technologies that save labour and material costs are also challenging to implement in this context. For example, drone spraying can save pesticides and labour and improve insecticidal efficiency. However, the large-scale production of one crop is difficult to implement in areas in China where the land is severely fragmented. In terms of mechanisation, the fragmentation of land was found to reduce the purchase of machinery services by farmers. Accordingly, the larger the scale of the land, the easier it is for farmers to purchase supporting mechanical services from the market. For small-scale land, it was found that farmers tended to use their own machinery or their own labour for agricultural production. Additionally, it was identified that the more severe the LF, the lower the efficiency of mechanical ploughing. It was also found that LF did not affect crop yields in terms of productivity but had a positive effect on technical efficiency, although these results were not sufficiently robust.

The current study found that LF correlated differently with farm performance and, overall, the adverse effects dominated. These results confirm that incentives to encourage land-leasing and consolidation are justified to improve farm performance from a policy perspective. Simultaneously, in areas with severe LF, the development of small agricultural machinery should be encouraged to meet the needs of small or impoverished farmers. As China's economy develops, LF to large-scale production will be the trend dominating land-use transitions. However, integrating land should be undertaken cautiously. Large-scale production was negatively correlated with some farm performance, such as technical efficiency. Therefore, increasing the scale of land should improve farmers' management level and prevent the reduction of technical efficiency.

**Author Contributions:** S.W. conceived and designed the structure of this study and cleaned the data, ran models, and wrote part of the paper. D.L. wrote the introduction and literature review. T.L. and C.L. participated in the paper discussion and revision. All authors have read and agreed to the published version of the manuscript.

**Funding:** This research was funded by the Youth Program of National Natural Science Foundation of China (Grant No. 72003194, 41901216), the China Agriculture Research System of MOF and MARA, the Youth Program of National Social Science Fund of China (Grant No. 18CJY032), the Program of Comprehensive Survey on Rural Revitalization and China Rural Survey Database funded by Chinese Academy of Social Sciences (Grant No. GQDC2020017), the Bagui Scholars Program of Guangxi Zhuang Autonomous Region, and the Program of Science and Technology Plan of Guangxi Zhuang Autonomous Region (Grant No. AD19110158).

**Institutional Review Board Statement:** Not applicable.

**Informed Consent Statement:** Not applicable.

**Data Availability Statement:** As the data gathered came from field surveys conducted by the research team, it is not publicly available. The data used in this paper are available upon request from the corresponding author.

**Conflicts of Interest:** The authors declare no conflict of interest.

## Appendix A

**Table A1.** Analysis of the effect of LF on technical inefficiency using the C-D production function.

|  | Model (1) | Model (2) | Model (3) | Model (4) |
|---|---|---|---|---|
| Frontier |  |  |  |  |
| lnX1 | 0.0834 *** | 0.0817 *** | 0.0860 *** | 0.0849 *** |
|  | (4.998) | (3.137) | (5.104) | (4.976) |
| lnX2 | 0.0025 | −0.0295 ** | 0.0056 | 0.0075 |
|  | (0.316) | (−2.378) | (0.708) | (0.938) |
| lnX3 | 0.0659 *** | 0.0463 ** | 0.0711 *** | 0.0678 *** |
|  | (5.268) | (2.194) | (5.603) | (5.231) |
| lnX4 | 0.0213 *** | 0.0140 ** | 0.0195 *** | 0.0183 *** |
|  | (4.579) | (2.255) | (4.184) | (3.827) |
| Head of household characteristics controls | yes | yes | yes | yes |
| Household characteristics controls | yes | yes | yes | yes |
| Crop controls | yes | yes | yes | yes |
| Region controls | yes | yes | yes | yes |
| Inefficiency term |  |  |  |  |
| LF1 | −0.0841 *** |  |  |  |
|  | (−4.166) |  |  |  |
| LF2 |  | −15.8495 |  |  |
|  |  | (−1.469) |  |  |
| LF3 |  |  | 0.0133 |  |
|  |  |  | (0.352) |  |
| LF4 |  |  |  | −0.0089 |
|  |  |  |  | (−1.625) |
| Log farm size | −0.0440 | −1.9071 | −0.1686 *** | −0.1280 *** |
|  | (−1.187) | (−1.396) | (−4.803) | (−3.554) |
| Usigma | −0.4021 *** | 2.0207 *** | −0.4688 *** | −0.4560 *** |
|  | (−6.285) | (3.162) | (−7.729) | (−7.905) |
| Vsigma | −3.5100 *** | −3.3786 *** | −3.5465 *** | −3.5445 *** |
|  | (−40.193) | (−45.391) | (−39.997) | (−39.720) |
| Constant | 5.3674 *** | 5.4390 *** | 5.2966 *** | 5.3290 *** |
|  | (41.269) | (44.813) | (40.434) | (40.429) |
| Observations | 2111 | 2106 | 2091 | 2078 |
| **LR test**: Assumption of C-D production function nested in the translog production function. |  |  |  |  |
| LR chi2 (10) | 27.07 | 25.64 | 26.00 | 25.18 |
| Prob > chi2 | 0.0026 | 0.0043 | 0.0037 | 0.0050 |

Note: z-statistics in parentheses; *** significance at 1%; ** significance at 5%; * significance at 10%; lnX1, lnX2, lnX3, and lnX4 represent the natural logarithm of the material cost, labour cost, land rent, and purchase machinery service variables.

## References


1. Huang, J.; Yang, G. Understanding recent challenges and new food policy in China. *Glob. Food Secur.* **2017**, *12*, 119–126. [CrossRef]
2. Gong, B. Agricultural reforms and production in China: Changes in provincial production function and productivity in 1978–2015. *J. Dev. Econ.* **2018**, *132*, 18–31. [CrossRef]
3. Zhao, Y. Leaving the Countryside: Rural-to-Urban Migration Decisions in China. *Am. Econ. Rev.* **1999**, *89*, 281–286. [CrossRef]
4. Long, H. Land consolidation: An indispensable way of spatial restructuring in rural China. *J. Geogr. Sci.* **2014**, *24*, 211–225. [CrossRef]
5. Xu, D.; Yong, Z.; Deng, X.; Zhuang, L.; Qing, C. Rural-Urban Migration and its Effect on Land Transfer in Rural China. *Land* **2020**, *9*, 81. [CrossRef]
6. Huang, J.; Ding, J. Institutional innovation and policy support to facilitate small-scale farming transformation in China. *Agric. Econ.* **2016**, *47*, 227–237. [CrossRef]
7. Lin, J.Y. Rural factors markets in China after the household responsibility system reform. In *Chinese Economic Policy: Economic reform at Midstream*; Reynolds, B., Ed.; Paragon House: New York, NY, USA, 1989; pp. 487–509.
8. Wang, H.; Tong, J.; Su, F.; Wei, G.; Ran, T. To reallocate or not: Reconsidering the dilemma in China's agricultural land tenure policy. *Land Use Policy* **2011**, *28*, 805–814. [CrossRef]

9.  Manjunatha, A.; Anik, A.R.; Speelman, S.; Nuppenau, E. Impact of land fragmentation, farm size, land ownership and crop diversity on profit and efficiency of irrigated farms in India. *Land Use Policy* **2013**, *31*, 397–405. [CrossRef]

10. Wang, X.; Yamauchi, F.; Huang, J.; Rozelle, S. What constrains mechanization in Chinese agriculture? Role of farm size and fragmentation. *China Econ. Rev.* **2020**, *62*, 101221. [CrossRef]

11. Deininger, K.; Savastano, S.; Carletto, C. Land fragmentation, cropland abandonment, and land market operation in Albania. *World Dev.* **2012**, *40*, 2108–2122. [CrossRef]

12. Kawasaki, K. The costs and benefits of land fragmentation of rice farms in Japan. *Aust. J. Agric. Resour. Econ.* **2010**, *54*, 509–526. [CrossRef]

13. Deininger, K.; Monchuk, D.; Nagarajan, H.K.; Singh, S.K. Does Land Fragmentation Increase the Cost of Cultivation? Evidence from India. *J. Dev. Stud.* **2016**, *53*, 82–98. [CrossRef]

14. Yan, X.; Bauer, S.; Huo, X. Farm size, land reallocation, and labour migration in rural China. *Popul. Space Place* **2014**, *20*, 303–315. [CrossRef]

15. Lu, H.; Xie, H.; Yao, G. Impact of land fragmentation on marginal productivity of agricultural labor and non-agricultural labor supply: A case study of Jiangsu, China. *Habitat Int.* **2019**, *83*, 65–72. [CrossRef]

16. Van den Berg, M.M.; Hengsdijk, H.; Wolf, J.; Van Ittersum, M.K.; Wang, G.; Roetter, R.P. The impact of increasing farm size and mechanization on rural income and rice production in Zhejiang province, China. *Agric. Syst.* **2007**, *94*, 841–850. [CrossRef]

17. Rahman, S.; Rahman, M. Impact of land fragmentation and resource ownership on productivity and efficiency: The case of rice producers in Bangladesh. *Land Use Policy* **2009**, *26*, 95–103. [CrossRef]

18. Tran, T.Q.; Van Vu, H. Land fragmentation and household income: First evidence from rural Vietnam. *Land Use Policy* **2019**, *89*, 104247. [CrossRef]

19. Kan, K. Creating land markets for rural revitalization: Land transfer, property rights and gentrification in China. *J. Rural. Stud.* **2021**, *81*, 68–77. [CrossRef]

20. Li, Y.; Wu, W.; Liu, Y. Land consolidation for rural sustainability in China: Practical reflections and policy implications. *Land Use Policy* **2018**, *74*, 137–141. [CrossRef]

21. Fenoaltea, S. Risk, transaction costs, and the origin of medieval agriculture. *Explor. Econ. Hist.* **1976**, *13*, 129–151. [CrossRef]

22. Knippenberg, E.; Jolliffe, D.; Hoddinott, J. Land Fragmentation and Food Insecurity in Ethiopia. *Am. J. Agric. Econ.* **2020**, *102*, 1557–1577. [CrossRef]

23. Bentley, J.W. Economic and ecological approaches to land fragmentation: In defense of a much-maligned phenomenon. *Annu. Rev. Anthropol.* **1987**, *16*, 31–67. [CrossRef]

24. Ntihinyurwa, P.D.; de Vries, W.T. Farmland fragmentation, farmland consolidation, and food security: Relationships, research lapses and future perspectives. *Land* **2021**, *10*, 129. [CrossRef]

25. Benin, S.; Smale, M.; Pender, J.; Gebremedhin, B.; Ehui, S. The economic determinants of cereal crop diversity on farms in the Ethiopian highlands. *Agric. Econ.* **2004**, *31*, 197–208. [CrossRef]

26. Ntihinyurwa, P.D.; de Vries, W.T.; Chigbu, U.E.; Dukwiyimpuhwe, P.A. The positive impacts of farmland fragmentation in Rwanda. *Land Use Policy* **2019**, *81*, 565–581. [CrossRef]

27. Tan, S.; Heerink, N.; Kuyvenhoven, A.; Qu, F. Impact of land fragmentation on rice producers' technical efficiency in South-East China. *NJAS Wagening. J. Life Sci.* **2010**, *57*, 117–123. [CrossRef]

28. Latruffe, L.; Piet, L. Does land fragmentation affect farm performance? A case study from Brittany, France. *Agric. Syst.* **2014**, *129*, 68–80. [CrossRef]

29. He, M. An Analysis of the Impact of Land Fragmentation on Agricultural Production Cost: Evidence from Farmers in Gansu Province, P.R. China. s.n. 2014. Available online: http://edepot.wur.nl/313215 (accessed on 5 April 2021).

30. Wan, G.H.; Cheng, E. Effects of land fragmentation and returns to scale in the Chinese farming sector. *Appl. Econ.* **2001**, *33*, 183–194. [CrossRef]

31. Lu, H.; Xie, H.; He, Y.; Wu, Z.; Zhang, X. Assessing the impacts of land fragmentation and plot size on yields and costs: A translog production model and cost function approach. *Agric. Syst.* **2018**, *161*, 81–88. [CrossRef]

32. Binns, B.O. *The Consolidation of Fragmented Agricultural Holdings*; Food and Agriculture Organization, United Nations: Washington, DC, USA, 1950.

33. Jha, R.; Nagarajan, H.K.; Prasanna, S. *Land Fragmentation and Its Implications for Productivity: Evidence from Southern India*; Australia South Asia Research Centre (ASARC) Working Paper 2005/01; ASARC: Canberra, Australia, 2005.

34. Pavel, C.; Miroslava, R.; Guri, F.; Zhllima, E.; Shahu, E. The impact of crop rotation and land fragmentation on farm productivity in Albania. *Stud. Agric. Econ.* **2018**, *120*, 116–125. [CrossRef]

35. Dovring, F.; Dovring, K. *Land and Labor in Europe in 1900–1950*, 2nd ed.; Springer: The Hague, The Netherlands, 1960; pp. 42–51.

36. Blare, B.; Hazell, P.; Place, F.; John, Q. The economics of farm fragmentation: Evidence from Ghana and Rwanda. *World Bank Econ. Rev.* **1992**, *6*, 233–254. [CrossRef]

37. Looga, J.; Jürgenson, E.; Sikk, K.; Matveev, E.; Maasikamäe, S. Land fragmentation and other determinants of agricultural farm productivity: The case of Estonia. *Land Use Policy* **2018**, *79*, 285–292. [CrossRef]

38. Wang, Y.; Li, X.; Lu, D.; Yan, J. Evaluating the impact of land fragmentation on the cost of agricultural operation in the southwest mountainous areas of China. *Land Use Policy* **2020**, *99*, 105099. [CrossRef]

39. Jabarin, A.S.; Epplin, F.M. Impacts of land fragmentation on the cost of producing wheat in the rain-fed region of northern Jordan. *Agric. Econ.* **1994**, *11*, 191–196. [CrossRef]

40. Austin, O.C.; Ulunma, A.C.; Sulaiman, J. Exploring the Link between Land Fragmentation and Agricultural Productivity. *Int. J. Stat. Appl.* **2012**, *2*, 30–34. [CrossRef]

41. Jia, L.; Petrick, M. How does land fragmentation affect off-farm labor supply: Panel data evidence from China. *Agric. Econ.* **2014**, *45*, 369–380. [CrossRef]

42. Alemu, G.T.; Ayele, Z.B.; Berhanu, A.A. Effects of Land Fragmentation on Productivity in Northwestern Ethiopia. *Adv. Agric.* **2017**, *2017*, 1–9. [CrossRef]

43. Tan, S. Land Fragmentation and Rice Production: A Case Study of Small Farms in Jiangxi Province, P.R. China. Ph.D. Thesis, Wageningen University and Research, Wageningen, The Netherlands, 2005.

44. Kadigi, R.M.; Kashaigili, J.J.; Sirima, A.; Kamau, F.; Sikira, A.; Mbungu, W. Land fragmentation, agricultural productivity and implications for agricultural investments in the Southern Agricultural Growth Corridor of Tanzania (SAGCOT) region. *Tanzan. J. Dev. Agric. Econ.* **2017**, *9*, 26–36.

45. Veljanoska, S. Can land fragmentation reduce the exposure of rural households to weather variability? *Ecol. Econ.* **2018**, *154*, 42–51. [CrossRef]

46. Wang, H.J.; Schmidt, P. One-step and two-step estimation of the effects of exogenous variables on technical efficiency levels. *J. Product. Anal.* **2002**, *1*, 129–144. [CrossRef]

47. Belotti, F.; Daidone, S.; Ilardi, G.; Atella, V. Stochastic Frontier Analysis using Stata. *Stata J. Promot. Commun. Stat.* **2013**, *13*, 719–758. [CrossRef]

48. Deolalikar, A.B. The Inverse Relationship between Productivity and Farm Size: A Test Using Regional Data from India. *Am. J. Agric. Econ.* **1981**, *63*, 275–279. [CrossRef]

49. Gautam, M.; Ahmed, M. Too Small to Be Beautiful? The Farm Size and Productivity Relationship in Bangladesh. *Food Policy* **2018**, *84*, 165–175. [CrossRef]

50. Wineman, A.; Jayne, T.S. Factor Market Activity and the Inverse Farm Size-Productivity Relationship in Tanzania. *J. Dev. Stud.* **2020**, *57*, 443–464. [CrossRef]

51. Helfand, S.M.; Levine, E.S. Farm Size and the Determinants of Productive Efficiency in the Brazilian Center-West. *Agric. Econ.* **2004**, *31*, 241–249. [CrossRef]

52. Carletto, C.; Gourlay, S.; Winters, P. From guesstimates to GPStimates: Land area measurement and implications for agricultural analysis. *J. Afr. Econ.* **2015**, *24*, 593–628. [CrossRef]

53. Desiere, S.; Jolliffe, D. Land productivity and plot size: Is measurement error driving the inverse relationship? *J. Dev. Econ.* **2018**, *130*, 84–98. [CrossRef]

*Article*

# Recessive Transition Mechanism of Arable Land Use Based on the Perspective of Coupling Coordination of Input–Output: A Case Study of 31 Provinces in China

Yi Lou, Guanyi Yin *, Yue Xin, Shuai Xie, Guanghao Li, Shuang Liu and Xiaoming Wang

College of Geography and Environment, Shandong Normal University, Jinan 250000, China; 2019020678@stu.sdnu.edu.cn (Y.L.); 2018020745@stu.sdnu.edu.cn (Y.X.); 2020020735@stu.sdnu.edu.cn (S.X.); 201914010224@stu.sdnu.edu.cn (G.L.); 201814010119@stu.sdnu.edu.cn (S.L.); 204020@sdnu.edu.cn (X.W.)
* Correspondence: 616071@sdnu.edu.cn

**Abstract:** In the rapid process of urbanization in China, arable land resources are faced with dual challenges in terms of quantity and quality. Starting with the change in the coupling coordination relationship between the input and output on arable land, this study applies an evaluation model of the degree of coupling coordination between the input and output (D_CCIO) on arable land and deeply analyzes the recessive transition mechanism and internal differences in arable land use modes in 31 provinces on mainland China. The results show that the total amount and the amount per unit area of the input and output on arable land in China have presented different spatio-temporal trends, along with the mismatched movement of the spatial barycenter. Although the D_CCIO on arable land increases slowly as a whole, 31 provinces show different recessive transition mechanisms of arable land use, which is hidden in the internal changes in the input–output structure. The results of this study highlight the different recessive transition patterns of arable land use in different provinces of China, which points to the outlook for higher technical input, optimized planting structure, and the coordination of human-land relationships.

**Keywords:** land use transition; arable land use; input–output; spatio-temporal variation; movement of spatial barycenter; optimization of arable land use

**Citation:** Lou, Y.; Yin, G.; Xin, Y.; Xie, S.; Li, G.; Liu, S.; Wang, X. Recessive Transition Mechanism of Arable Land Use Based on the Perspective of Coupling Coordination of Input–Output: A Case Study of 31 Provinces in China. *Land* **2021**, *10*, 41. https://doi.org/10.3390/land10010041

Received: 3 December 2020
Accepted: 30 December 2020
Published: 5 January 2021



## 1. Introduction

Historically, the world has not witnessed such rapid urbanization as that which has taken place over the past several decades in China [1]. Measured as the proportion of permanent urban population in the total population, the urbanization of China has risen from 17.9% in 1978 (the year of the reform and opening-up in China) to 59.85% in 2018. Along with the urbanization process, China's economic structure is also undergoing rapid transition. The proportion of China's nonagricultural economy (the ratio of the GDP of the secondary and tertiary industries in the total national GDP) increased from 72% in 1978 to 93% in 2018, showing a significant shift in economic development focus towards nonagricultural industries. In this vast wave, the use of arable land is facing various challenges. On the one hand, accompanied by a decrease in the rural population (from 790.14 million in 1978 to 56.401 million in 2018), the urban population increased from 172.45 million in 1978 to 831.37 million in 2018 [2]. By 2018, more than 200 million rural residents had migrated to urban areas, indicating a large transition of farmers' livelihood from traditional agricultural production to employment in secondary and tertiary industries in cities [3,4]. On the other hand, along with a rapid expansion of construction land area (from 5845 $hm^2$ in 1978 to 56,075.9 $hm^2$ in 2018, with an annual growth rate of 21.5%), the urban built-up area in China increased synchronously from 6720.5 to 58,455.7 $hm^2$, with an annual increase rate of 19.2%. During the same period, the annual growth rate of the urbanization rate in China was 1.04%. This striking development pattern indicated that the "urbanization of land" is much faster than the "urbanization of population," which has posed high pressure on the loss of arable land resources [5]. From 2004 and 2018, 10,693.20 hm$^2$ of arable land was converted into construction land in China. It is undeniable that the development of urbanization in China emphasizes the limitation of arable land resources and stimulates arable land intensity. However, it also brings more dual challenges of both the agricultural labor force and arable land resources [6].

In recent years, the question of how to effectively protect arable land and ensure food security has been a hot topic for policymakers and researchers. Since 1998, the Chinese government has implemented a series of arable land protection policies. In China's land use constraint system, users are required to use the land strictly in accordance with the prescribed land uses and emphasize that the transition of arable land to construction land must be examined and approved by the higher government. In the balanced system of requisition–compensation for arable land, local governments are required to complement the same area of high-quality arable land if construction occupies arable land. In the stripping and reuse system of tillage soil, users of construction land who occupied high-quality arable land are required to strip the topsoil of arable land and move the fertile soil to other arable lands for further reuse. In the compensation system of arable land protection, the rural collective economic organization and the farmers who implement farmland protection can obtain an annual subsidy of 6.3–20 RMB/hm$^2$, which varies in different regions. Every year, the No.1 document of the State Council of China restates the red line of $1.2 \times 10^8$ hm$^2$ of arable land, highlights the stipulation of arable land rotation and land fallow, emphasizes the implementation of a permanent basic farmland protection system, and keeps improving the balance system of requisition–compensation of arable land. A series of policies have curbed the rapid decline in arable land area and has realized the coordination and unity of protecting arable land resources to a certain extent [7,8].

However, the demand for land continued to increase with the rapid urbanization process, which caused negative results in the amount and quality of arable land. In the practice of balancing the system of requisition–compensation of arable land, the supplemented arable land tends to be insufficient and low-quality. In some local areas, high-quality arable land is used for construction, and low-quality land is used for supplementing arable land, which gives great uncertainty to the overall efficiency of arable land use [9,10]. In terms of arable land quality, scholars also point out various quality problems in arable land use. The decrease in nutrient content, loss of cultivated layer soil, nonpoint agricultural pollution caused by excess pesticide and fertilizer use are frequently discussed in arable land use [11–15]. From the perspective of the internal complexity of arable land use, the fluctuating multiple cropping index, increasing nongrain planting structure, and arable land abandonment problems have had intricate far-reaching influences on arable land use [16,17]. This phenomenon formed an arable land use mode with the characteristic of "heavy use and light maintenance," which will inevitably lead to the continuous decline of arable land use efficiency in China. For fear of further deterioration, wide observations of arable land use change, and strong measures of arable land use optimization are urgently needed.

Faced with the above problems, the importance and necessity of land use transition have been highlighted. With socioeconomic development, Long et al. proposed that land use transitions will appear in conflicts and transitions among different land use types, which will result in a new balance along with periodic change [18]. Among the various transitions of land use, Chen highlighted that the most typical transition in China is the transition from arable land to construction land, which needs to be optimized to better promote integrated urban-rural development [19]. Lyu et al. believes that sustainable of arable land use requires rational of input structure, improvement of output and greener ecological environment [20]. In this perspective, the change in input–output structure will profoundly regulate sustainable intensified arable land use. More specifically, the arable land use should not blindly increase output by increasing the input, but reasonably coordinate the spatio-temporal relation of input-output matters most. An atypical characteristic of the input–output change on arable land is the coupling relationship between the specific input system (combining fertilization, sowing, pesticide, and mechanical power) and the output of arable land. This kind of internal change in the coupling relationship between input and output on arable land is vital but invisible, it may constitute the recessive characteristic of arable land use transition. What kind of coupling relationship between the input and output on arable land is presenting? How does the arable land use transition change in different areas? Is there any space for arable land use transition? These questions are becoming meaningful perspectives for arable land use improvement. To our knowledge, previous studies mostly focus on the discussion of arable land quality and quantities and the evaluation of the overall land use efficiency and risk [21–24]. Less attention has been given to the arable land use transformation mechanism from the perspective of input–output change. As a result, the above questions have not yet been fully answered.

To fulfill the research gaps, this study aims to further explore the law of the recessive transition mechanism of arable land use based on the input-output perspective. By evaluating the degree of coupling coordination of input–output on arable land (D_CCIO) in 31 provinces of mainland China, this study raises the following scientific questions:

1. In the rapid urbanization wave of China, what are the spatiotemporal changes in the input–output of arable land?
2. Is the input and output of arable land highly coupled and sustainable?
3. What are the characteristics of arable land use transition among different regions, and what is the enlightenment for the policy of optimizing arable land use mode?

Based on the above research objectives, this paper is divided into the following parts: Section 2 introduces the research area, methods, and data; Section 3 introduces the research results; Section 4 discusses the enlightenment of the research results in depth, and Section 5 summarizes the main conclusions of this paper.

## 2. Research Area and Methods

### 2.1. Research Area and Data Source

This paper selects 31 provincial administrative units of mainland China as the research area (Figure 1). In the research process, the basic data involved in the calculation of each indicator are obtained from the China rural statistical yearbook (2009, 2019) and China population and employment statistical yearbook (2009, 2019). The spatial boundaries of administrative areas are obtained from the Resource and Environment Science and Data Center of the Institute of Geographic Sciences and Natural Resources Research, Chinese Academy of Sciences.

**Figure 1.** The 31 provinces of China in this study.

*2.2. Indicator System of Input–Output Analysis on Arable Land*

This study establishes an evaluating indicator system of the coupling coordination for input–output on arable land. Indicators involved in the evaluation were chosen based on the principles of comprehensiveness, independence, and accessibility. First of all, since 2015, the Ministry of Agriculture in China began to advocate zero growth of the use of fertilizers and pesticides. Because fertilizers and pesticides are essential elements in the growth of crops, therefore, the rational use of them is of vital importance for improving arable land use efficiency [25–27]. Secondly, mulching film can improve the utilization efficiency of arable land and water, which is widely used in areas with water shortages and low temperatures. However, it is important to deeper analyze the change of mulching film use, because the residues are difficult to degrade and will destroy soil pore structure [28]. Finally, as complementary roles of planting, sowing, and harvesting, observing the change of labor force and agro-machinery is necessary for analyzing arable land use [29,30].

Considering the vast territory of China, different climatic and geographical conditions nourish different local crops, so it is necessary to select a variety of crops in the output indicators. Because the total output of grain, oil crops, cotton, sugar crops, tobacco, vegetables, and fruits accounted for more than 95% of the country's crop output, the yields of the above seven types of crops were chosen as the output indicators (Table 1) [21,31–33].

**Table 1.** Indicators of the subsystem of input–output on arable land.

| Subsystem | Content | | Subsystem | Content | |
|---|---|---|---|---|---|
| | Nitrogen fertilizer (kg) | I1 | | Grain (kg) | O1 |
| | Phosphorus fertilizer (kg) | I2 | | Oil crop (kg) | O2 |
| | Potash fertilizer (kg) | I3 | | Cotton (kg) | O3 |
| Input of arable land (I) | Compound fertilizer (kg) | I4 | Output of arable land (O) | Sugar crop (kg) | O4 |
| | Mulching film (kg) | I5 | | Tobacco (kg) | O5 |
| | Mechanical power (kg) | I6 | | Fruit (kg) | O6 |
| | Pesticide (kg) | I7 | | Vegetable (kg) | O7 |
| | Labor (person) | I8 | | | |

*2.3. Evaluation of the Degree of Coupling Coordination Between Input and Output on Arable Land (D_CCIO)*

2.3.1. Quantification of the Input–Output Subsystems on Arable Land

Based on thermodynamic principles, the entropy method has been widely used as an objective method in engineering, social and economic fields. This method uses the information entropy to calculate the entropy value according to the variation degree of each indicator and can effectively solve the problem of information overlap between multiple index variables [34]. After determining the weight of each indicator using the entropy method, we can calculate the indexes of the input–output subsystem using the following process.

(1) Standardization of indicators:

To eliminate the influence of data dimensions and units on the evaluation result, the original data are usually transformed into dimensionless data. The specific formula is as follows:

$$U_{ij} = \frac{P_{ij} - min P_{ij}}{max P_{ij} - min P_{ij}} \tag{1}$$

where $U_{ij}$ represents the standardized value for the $j$-th indicator of the $i$-th item and $P_{ij}$ is the value for the $j$-th indicator of the $i$-th item.

(2) Calculate the intermediate parameters ($M$):

$$M = \frac{P_{ij}}{\sum_{i=1}^{n} P_{ij}} \tag{2}$$

(3) Calculate the entropy ($e_j$) of the $j$-th indicator:

$$e_j = \frac{-\left(\sum_{i=1}^{n} M_{ij} \times \ln M_{ij}\right)}{\ln n} \tag{3}$$

(4) Calculate the utility value of each index ($d_j$):

$$d_j = 1 - e_j \tag{4}$$

(5) Calculate the index weight ($w_j$):

$$w_j = \frac{d_j}{\sum_{j=1}^{n} d_j} \tag{5}$$

(6) Calculation of each subsystem score:

$$I = \sum_{j=1}^{n} w_j M_{ij} \quad O = \sum_{j=1}^{n} w_j M_{ij} \tag{6}$$

where $I$ and $O$ indicate the input index and output index on arable land, respectively.

2.3.2. Evaluating the Degree of Coupling Coordination of Input and Output (D_CCIO)

To study the degree of coupling coordination for the input–output on arable land, a coupling coordination evaluation model was built as follows [35]:

$$C = 2\sqrt{\frac{I \times O}{(I+O)^2}} \tag{7}$$

where $C$ is the degree of coupling, indicating the interaction intensity of the two-word systems, and $C \in [0, 1]$. The greater the coupling degree is, the stronger the interaction between the two subsystems, and vice versa.

The coupling model is limited in that it cannot reflect the development level of two systems, and "false" high coupling results may appear in two low-level systems. To avoid this problem, the coupling coordination degree model was used to accurately evaluate the coupling coordination relationship between the input and output of arable land at the provincial level.

$$T = \alpha I + \beta O \tag{8}$$

$$D\_CCIO = \sqrt{C \times T} \tag{9}$$

where $T$ is the comprehensive coordinating index of the input and output of arable land, which reflects the effect or contribution of the integrated synergy of the input and output of arable land. Both $\alpha$ and $\beta$ are weights to be determined. The input system and output system of arable land are equally important and reference previous achievements $\alpha = \beta = 0.5$ [34,36]. D_CCIO is the degree of coupling coordination between input and output on arable land, abbreviated as $D\_CCIO$, $D\_CCIO \in [0, 1]$. By referring to relevant references and combining them with the actual situation of the research area, $D\_CCIO$ is divided into sections (Table 2), and quantitative judgment and analysis are made according to the criteria given herein [37].

**Table 2.** Classifications of degree of coupling coordination between input and output on arable land.

| Categories | Development Modes between Sub Systems of Input and Output | Grades of D_CCIO | Classes |
|---|---|---|---|
| Balanced arable land use | I > O: Balanced arable land use with output lagged<br>I = O: Balanced arable land use with input and output synchronized<br>I < O: Balanced arable land use with input lagged | 0.91~1.00<br>0.81~0.90<br>0.71~0.80<br>0.61~0.70 | Extremely balanced arable land use<br>Seriously balanced arable land use<br>Moderately balanced arable land use<br>Slightly balanced arable land use |
| Transitional arable land use | I > O: Transitional arable land use with output lagged<br>I = O: Transitional arable land use with input and output synchronized<br>I < O: Transitional arable land use with input lagged | 0.51~0.60<br>0.41~0.50 | Barely balanced development<br>Barely unbalanced development |
| Unbalanced arable land use | I > O: Unbalanced arable land use with output lagged<br>I = O: Unbalanced arable land use with input and output synchronized<br>I < O: Unbalanced arable land use with input lagged | 0.31~0.40<br>0.2~0.3<br>0.1~0.2<br>0~0.1 | Slightly unbalanced arable land use<br>Moderately unbalanced arable land use<br>Seriously unbalanced arable land use<br>Extremely unbalanced arable land use |

*2.4. Standardized Deviation Ellipse of the Input and Output on Arable Land*

The analysis of the standard deviation ellipse (SDE) was first proposed by Lefever in 1926 and has been used as a feasible tool to describe the directivity of the spatial distribution [38,39]. The spatial and temporal evolution of geographical elements in spatial distribution range, direction, and shape can be described through the variation of parameters such as the barycenter, the size of the major and minor axes, and the standard difference of the major and minor axes of the ellipse to comprehensively reveal the spatial and temporal evolution characteristics and process of regional development from multiple perspectives [40]. The main parameters of the SDE are calculated as follows:

$$\overline{X} = \frac{\sum_{i=1}^{n} x_i}{n}, \ \overline{Y} = \frac{\sum_{i=1}^{n} y_i}{n} \tag{10}$$

$$SDE_x = \sqrt{\frac{\sum_{i=1}^{n} (x_i - \overline{X})^2}{n}}, \ SDE_y = \sqrt{\frac{\sum_{i=1}^{n} (y_i - \overline{Y})^2}{n}} \tag{11}$$

where $x_i$ and $y_i$ are the coordinates for feature $i$, $(\overline{X}, \overline{Y})$ represents the coordinates of the spatial barycenter for the features, and n is equal to the total number of features. $SDE_x$ and $SDE_y$ represent the major and minor axes of the ellipse. By using the SDE method in ArcGIS 10.2, the directivity of the spatial distribution of input and output on arable land can be visualized. The evolution process and law of the spatial distribution of the input and output factors of arable land in the province of China were obtained.

*2.5. Analysis of the Spatial Barycenter of Input and Output on Arable Land*

Barycenter modeling, which has been extensively utilized in the fields of urban planning, economic geography, and land use science, is a preferred modeling approach that traces the spatial movement direction of barycenters for targeted objects. Moreover, movement direction and distance to the center of gravity can reflect changes in quantity and changing trends of the targeted object over time [41]. Differing from the qualitative description of the spatial change in arable land, the law of barycenter migration can represent the whole dynamic evolution process of element distribution. The equation of the movement distance for the barycenter can be expressed as follows:

$$Distance = \sqrt{(X_{t2} - X_{t1})^2 + (Y_{t2} - Y_{t1})^2} \tag{12}$$

where *Distance* is the movement distance of barycenter(km), and $X_{t1}$, $X_{t2}$, and $Y_{t1}$, $Y_{t2}$ are coordinates of spatial barycenter of different inputs and outputs for the years *t1* and *t2*.

## 3. Result Analysis

### 3.1. The Spatiotemporal Change in the Input of Production Material on Arable Land

3.1.1. Provincial Input of Production Material

In 2008–2018, three kinds of production material inputs on arable land experienced a significant decrease. The input of nitrogen fertilizer, pesticide, and labor on arable land decreased by $2.37 \times 10^9$ kg, $1.69 \times 10^8$ kg, and $4.80 \times 10^7$ person, respectively. The reduction rate of the above three inputs on arable land fell by more than 10.0%. In contrast, other inputs on arable land showed a rising trend. The most significant one is the decline in compound fertilizer input, which increased from $1.61 \times 1010$ to $2.27 \times 1010$ kg in 2008–2018 (with an increase rate of 40.9%). In addition, the input of plastic film increased from $1.11 \times 10^9$ to $1.40 \times 10^9$ kg, and the input of mechanical power increased from $8.22 \times 10^8$ to $1.00 \times 10^9$ kw, both of which increased by more than 20%. Comparatively, the increase rate of phosphate fertilizer and potash fertilizer was not significant (<10%).

Figures 2 and 3 represent the input factors on arable land in 2008 and 2018, respectively. The subfigures A-H illustrate the input of nitrogen fertilizer, phosphate fertilizer, potash fertilizer, compound fertilizer, pesticide, mulching film, mechanical power and labor, respectively. The production material input on arable land in the Huang-Huai-Hai Plain area is relatively prominent. Various kinds of input on arable land in this area accounted for over 28% of the national total. In particular, mechanical power accounted for as high as 46.97% in 2008. The standardized ellipse shows that the directional distribution of various inputs is of two types: northeast-southwest directional distribution (Figure 2A–H except Figure 2F) and northwest-southeast directional distribution (Figures 2F and 3F). Except for the input of plastic film, various inputs showed directivity along the northeast-southwest direction, which is roughly consistent with the distribution of China's major grain-producing areas. The high-value area of various inputs also coincides with the distribution of the main grain-producing areas, indicating that the production material input is much higher in these areas. Conversely, the plastic film shows a distribution directivity along the northwest–southeast, and the high-value areas are mainly concentrated in Xinjiang, Gansu, Sichuan, and Yunnan provinces. In addition, the proportion of provinces with the top 2 highest inputs of the mulching film increases from 30% to 38%, showing an increasing degree of spatial aggregation for mulching film input.

In terms of specific elements, the polarization characteristic of nitrogen fertilizer input is prominent. Although the total amount of nitrogen fertilizer across the country showed a decreasing trend, the number of provinces with the highest top 2 inputs and the minimum input of nitrogen fertilizer increased by eight and four, respectively. This indicated that the max-min range among different polar means is prominent. Conversely, the total amount of mechanical power across the country increased, but provinces with the highest input and the lowest input both decreased. Similarly, the spatial distribution of pesticide input (Figures 2E and 3E) maintained a "U"-shaped pattern, but the number of provinces with the highest input decreased significantly.

### 3.1.2. Input of Production Material on Per Unit Area

Considering that different arable land areas exist among different provinces, this study also uses the input of production material per unit area to reflect the "intensity" of input. Except for phosphate fertilizer, the change trend of input per unit area and total input of all production materials was consistent. The input of nitrogen fertilizer per unit area decreased from 203.85 to 163.31 $kg/hm^2$. The reduction rates of pesticide and labor force input per unit area were also high (the former 20.05%, the latter 18.57%). Unexpectedly, the increase in the total input of phosphate fertilizer was accompanied by a decrease in input per unit area (from 61.65 to 52.61 $kg/hm^2$). Conversely, the national average input of compound fertilizer per unit area has increased significantly from 139.20 to 171 $kg/hm^2$ (increase rate of 22.85%). It is worth noting that although the increment of potash fertilizer is very low (from 0.23 to 0.28 $kg/hm^2$), its rate of increase is the highest (23.01%). Comparatively, the growth rate of plastic film and mechanical power input per unit area was low (<10%).

Figures 4 and 5 represent the input factors on per unit area of arable land in 2008 and 2018, respectively. The subfigures A–H illustrate the input of nitrogen fertilizer, phosphate fertilizer, potash fertilizer, compound fertilizer, pesticide, mulching film, mechanical power and labor on per unit area of arable land, respectively. The spatial distribution of production material inputs per unit area is significantly different from that of total inputs. Unlike the northeast–southwest directivity of the standardized ellipse (Figures 2 and 3), the input of nitrogen fertilizer, phosphorus fertilizer, mechanical power, and labor force in the unit area did not show obvious spatial directivity. A possible reason is that the input of pesticides and fertilizers per unit area in the main grain-producing areas is no longer significantly higher than that in the nonmain grain-producing areas (the former 88.30 $kg/hm^2$, the latter 77.68 $kg/hm^2$).

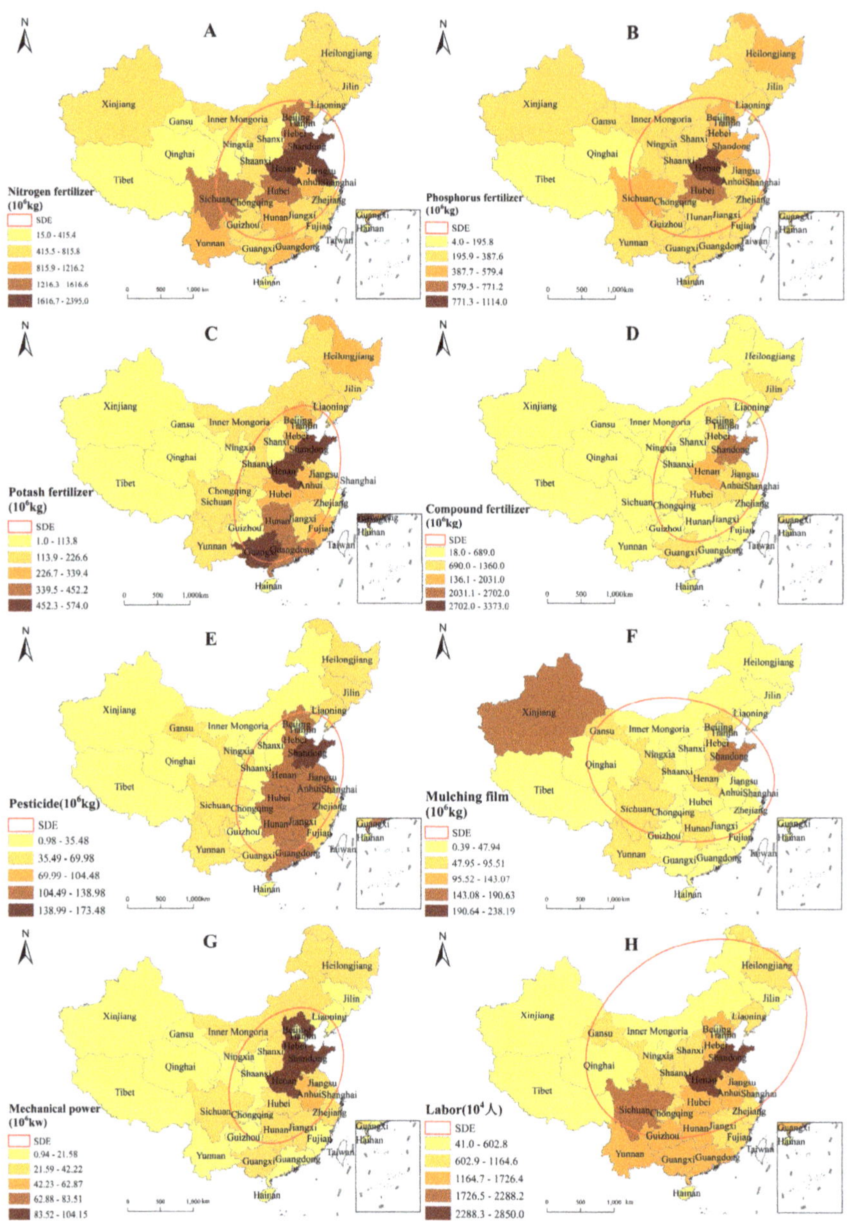

**Figure 2.** Input of production material on arable land in 2008.

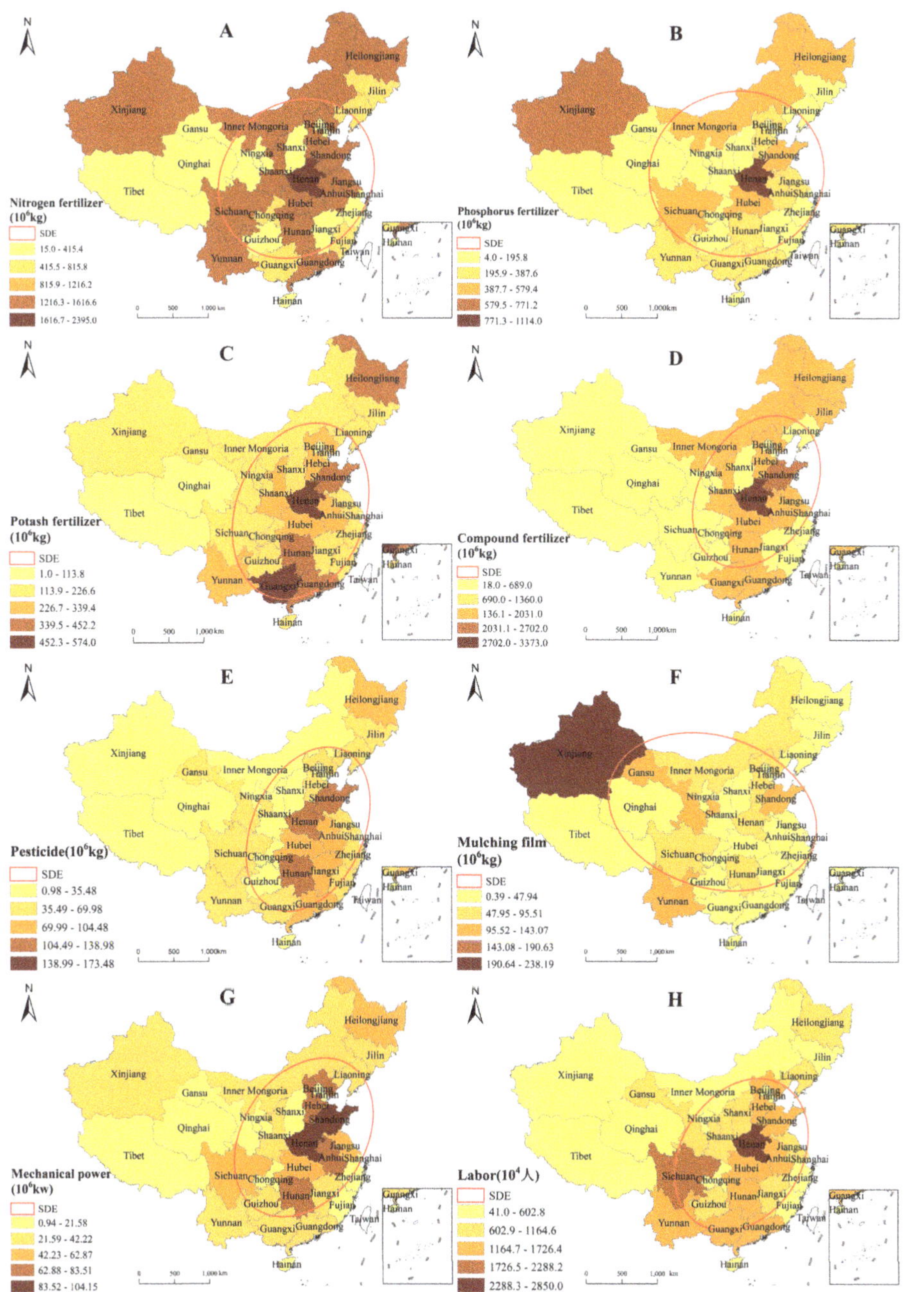

**Figure 3.** Input of production material on arable land in 2018.

In terms of specific inputs, in contrast to the distribution of total input (Figures 2 and 3), the inputs of nitrogen fertilizer and pesticide per unit area are increasingly concentrated in five provinces along the southeast coast (including Jiangsu, Zhejiang, Shanghai, Min, and Guangdong). In addition, both phosphate fertilizer and the plastic film showed an

SDE with an obvious northwest orientation. Represented by the high value of phosphate fertilizer and plastic film in Xinjiang Province, the trend was further increased.

*3.2. The Spatiotemporal Change in the Output on Arable Land*

### 3.2.1. Provincial Output of Crop Yield

In 2008–2018, the yield of grain, vegetable, fruit, and oil crops increased, in which the most obvious crops were grain and vegetable. As pillar crops, the grain yield and vegetable yield increased from $5.29 \times 10^{11}$ and $5.92 \times 10^{11}$ kg to $6.58 \times 10^{11}$ and $7.03 \times 10^{11}$ kg, respectively (the annual growth rates were as high as 2.44% and 1.87%, respectively). Conversely, the total output of cotton, sugar crops, and tobacco showed a downward trend, of which sugar crops decreased the most (from $1.34 \times 10^{11}$ to $1.19 \times 10^{11}$ kg, with an average annual decline rate of 1.11%). Tobacco showed the fastest decline rate, with the yield decreasing by $5.97 \times 10^8$ kg, with an average annual decline rate as high as 2.10%.

Figures 6 and 7 represent the output on arable land in 2008 and 2018, respectively. The subfigures A-G illustrate the output of grain, vegetable, cotton, oil crop, sugar crop, fruit and tobacco, respectively. From the perspective of the distribution pattern, each crop has obvious clustering directivity. Except for tobacco, sugar, and cotton, the yields of other crops on the Huang-Huai-Hai Plain accounted for more than 35%. Compared with other areas, the Huang-Huai-Hai Plain maintained a very high level of output. Specifically, the proportion of fruit yield in this region to the whole country was up to 50.44%~52.59% (Figure 6F). Among them, the output of grain crops (Figure 6A) in the main producing areas was as high as $3.99 \times 10^{11}$ kg, accounting for 75.50% of the national total, which increased to 78.74% in 2018, occupying an absolute advantage.

In addition, the high-value area of grain yield shifted to the north. With Qinling Mountain and the Huaihe River as the dividing line, the ratio of grain yield between North China and South China was 54:46 in 2008, which changed to 59:41 in 2018, forming a pattern of high north–low south grain production. The SDE of tobacco (Figures 6G and 7G) showed an obvious southwest directivity. The output of tobacco in Yunnan and Guizhou provinces accounted for 44.44% of the total tobacco output in the country, which increased to 48.91% in 2018. In contrast, the SDE of cotton (Figure 6C) shows northwest directivity. In 2008, there were two cotton agglomeration areas in Xinjiang and Huang-Huai-Hai Plain, with the output accounting for 40.39% and 42.75% of the country, respectively. By 2018, the proportion of the cotton output of Xinjiang Province to the country increased to 83.75%, indicating an increasing spatial polarization phenomenon.

### 3.2.2. Output of Crop Yield Per Unit Area

The crop yield per unit area of most crops (except sugar crops) showed the same changing trend as that of total crop output. Four crops continued to increase in both total yield and yield per unit area, including grain, vegetable, fruit and oil crops. Although the total amount of grain crops increased the most, the yield per unit area of grain did not increase significantly (from 4990.59 to 5580.04 kg/hm$^2$ in 2008–2018, with a growth rate of 11.81%). Oil yield per unit area increased the least (only increased by 385.61 kg/hm$^2$), yet it showed the highest growth rate (17.62%). It is worth noting that although the total output of sugar decreased, the yield per unit area of sugar increased significantly (from 40,646.39 to 43,487.71 kg/hm$^2$ in 2008–2018), showing the highest productivity per unit area. The changes in other crops were smaller (growth rate <10%).

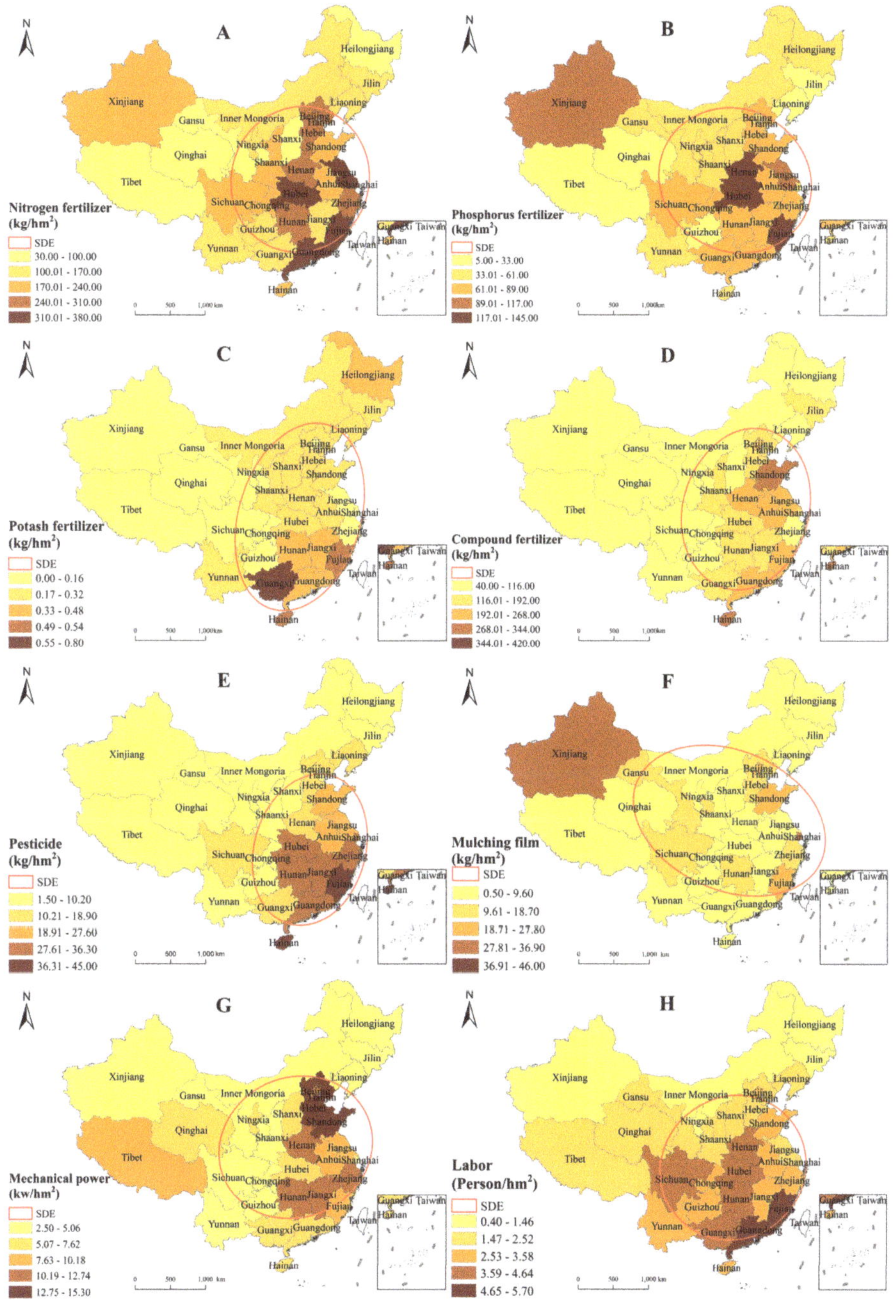



**Figure 4.** Input of production material on per unit area of arable land in 2008.

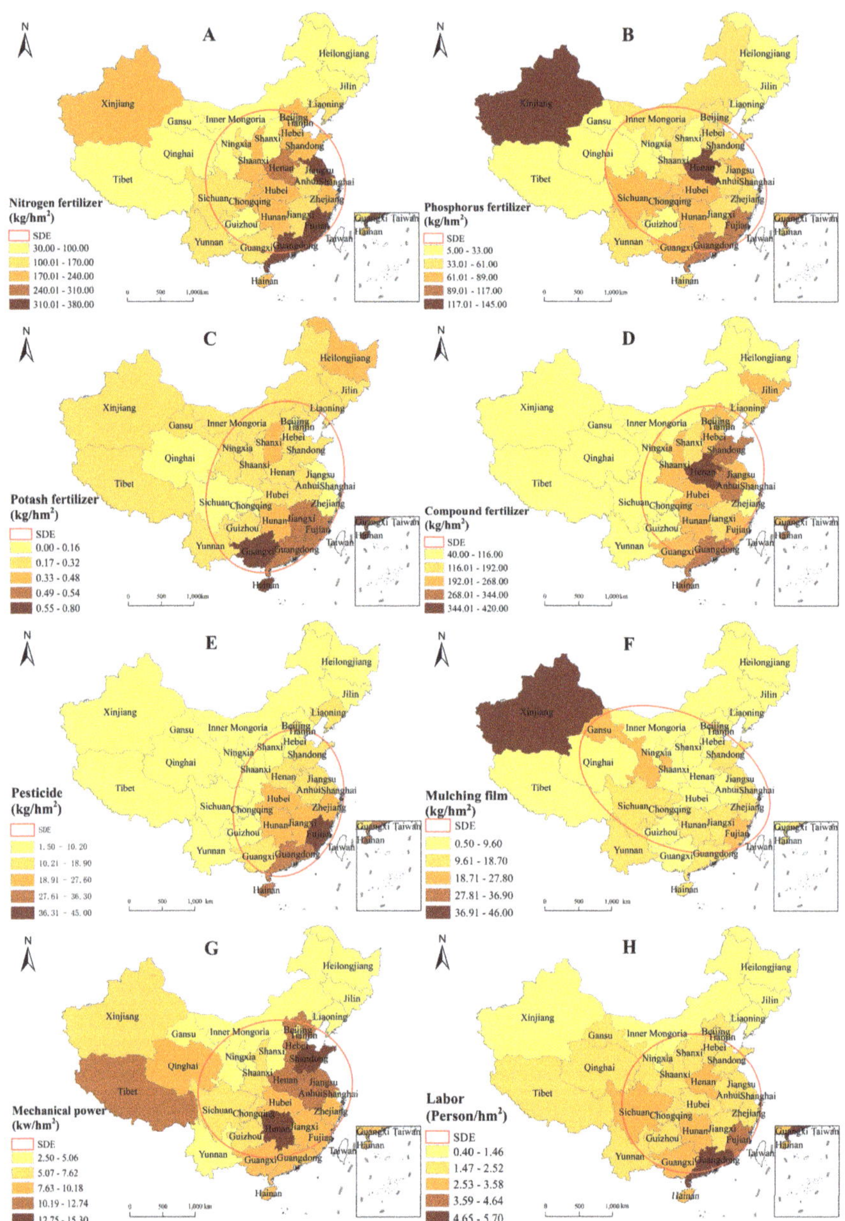

**Figure 5.** Input of production materials on per unit area of arable land in 2018.

Figures 8 and 9 represent the output on per unit area of arable land in 2008 and 2018, respectively. The subfigures A-G illustrate the output of grain, vegetable, cotton, oil crop, sugar crop, fruit and tobacco on per unit area, respectively. Interestingly, the spatial distribution of output per unit area and total output in each province are inconsistent. Taking grain as an example, although the grain yield of most major producing areas was higher than that of nonmajor grain producing areas (the former was 5464.21 kg/hm$^2$, while the latter was 4720.75 kg/hm$^2$), there are contrasting examples. The total grain yield of Heilongjiang in the main grain-producing areas ranked 1st in the country, but its grain yield per unit was low (rank 24th in 2018). Conversely, the total grain outputs of Tibet, Guangdong, Fujian, and Zhejiang (members of nonmajor grain-producing areas) are low, yet their grain yields per unit area are relatively high. The distribution patterns of vegetables and fruits were similar: the high-value regions were concentrated in northern China. Even so, the yield per unit area of fruit still gradually formed a new high-value region in northeast China by 2018.

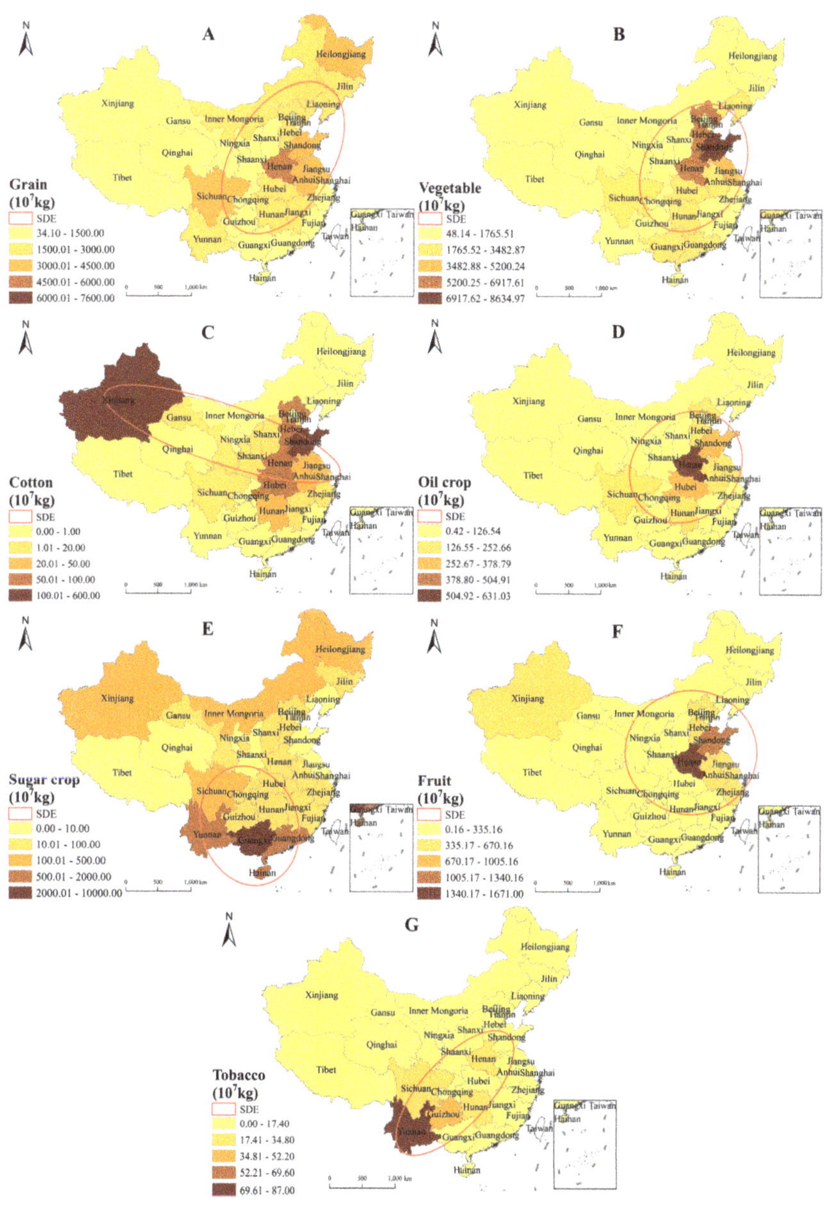

**Figure 6.** Output of crop yields on arable land in 2008.

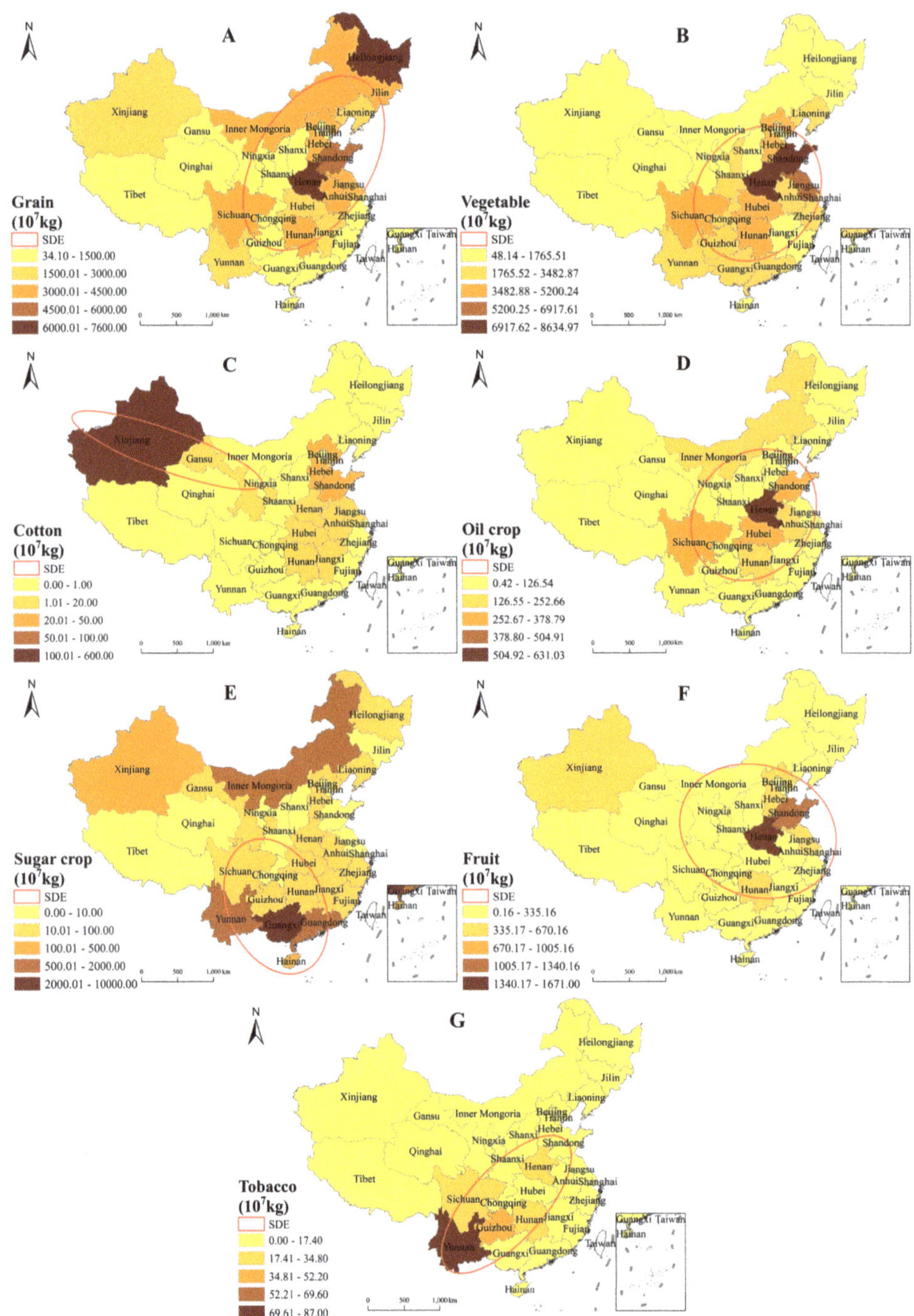



**Figure 7.** Output of crop yields on arable land in 2018.

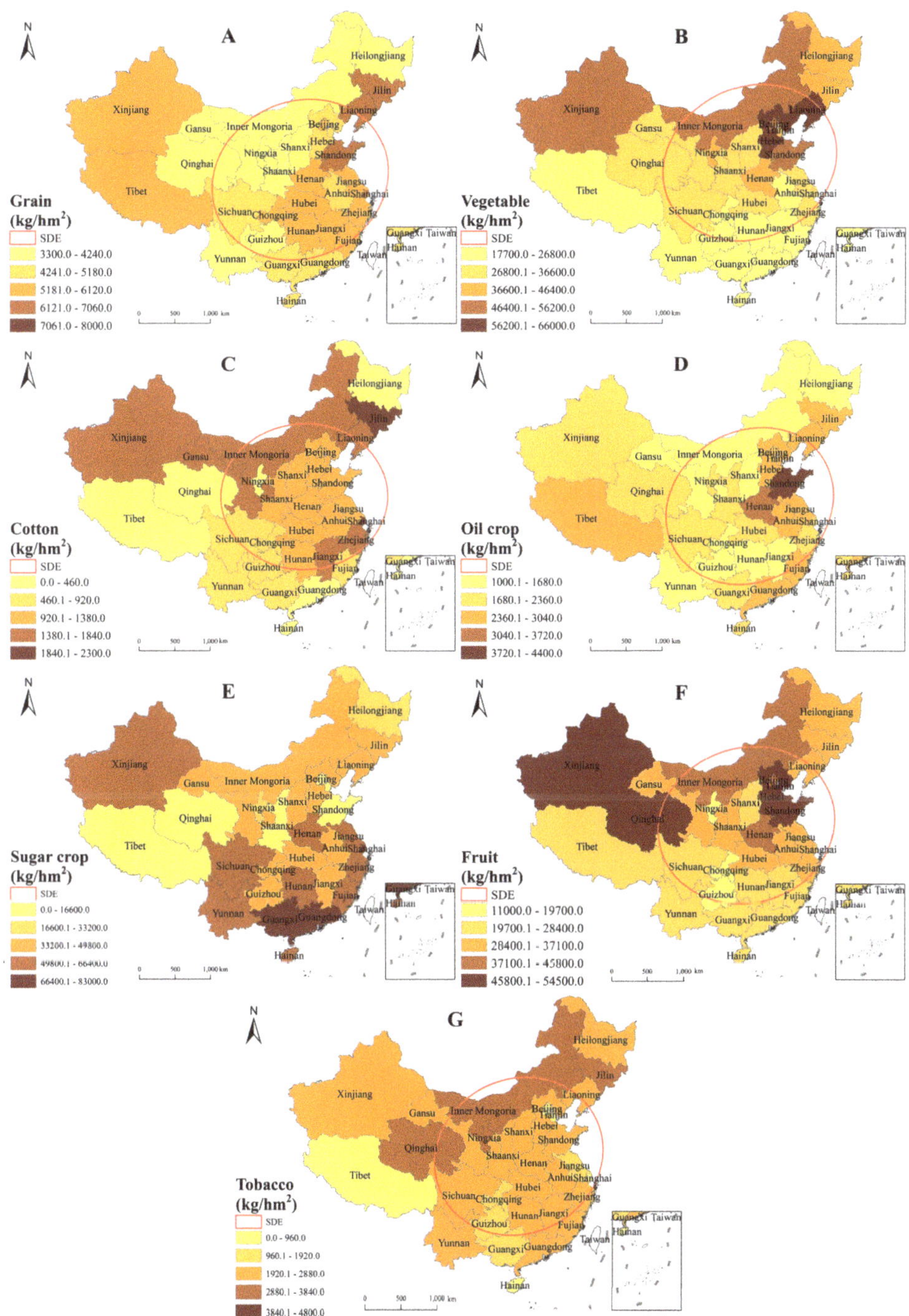


**Figure 8.** Output on per unit area of arable land in 2008.

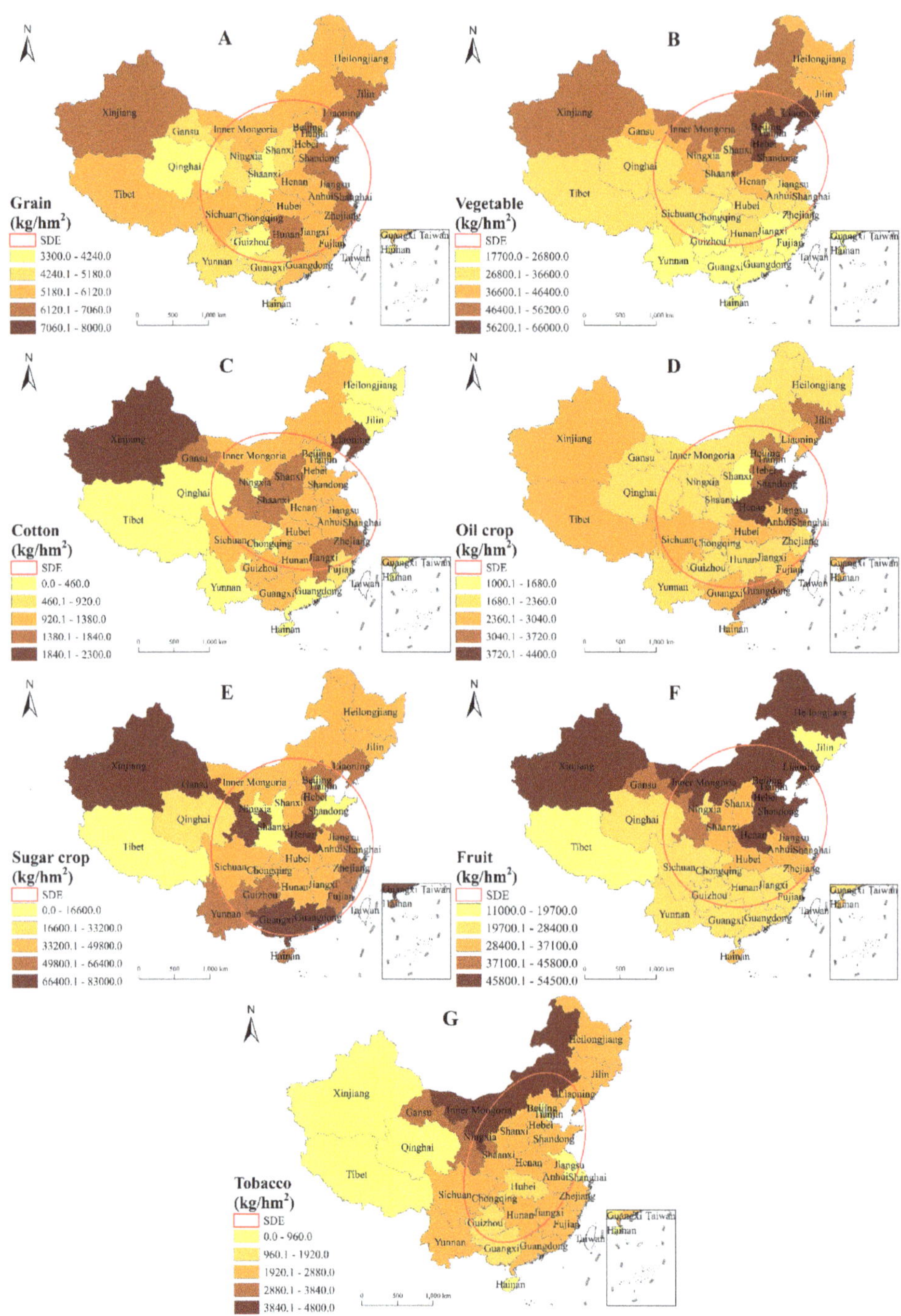



**Figure 9.** Output on per unit area of arable land in 2018.

*3.3. The Movement of Spatial Barycenter for the Input and Output on Arable Land*

The movement of the spatial barycenter for the input on arable land is shown in Figure 10, in which subfigure a and c illustrate the movement of barycenters of different kinds of inputs, while b and d illustrate the movement of barycenters of different inputs on per unit area. The spatial barycenters of all kinds of inputs were located in the central region of China (Henan, Hubei, and Shaanxi) and moved towards the west, which showed obvious consistency. In terms of the total amount of input (the red arrows in Figure 10), the spatial barycenter migrated towards the northwest. However, in terms of the input per unit area (the blue arrows in Figure 10), except for potash fertilizer and compound fertilizer, the spatial barycenter of all inputs moved towards the southwest.

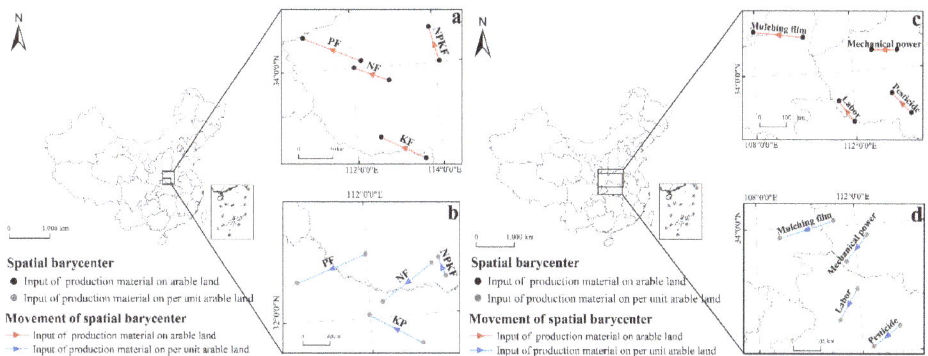


**Figure 10.** The movement of the spatial barycenter for the input on arable land (NF refers to nitrogen fertilizer; PF refers to phosphorus fertilizer; KF refers to potash fertilizer; NPKF refers to compound fertilizer).

In terms of movement distance, the spatial barycenter of nitrogen fertilizer moved 77.52 km westward and 25.63 km northward. The spatial barycenter of phosphate fertilizer moved 126.07 km westward and 46.99 km towards northward. The spatial barycenter of potash fertilizer moved 98.05 km westward and 43.54 km northward. Compound fertilizer moved 23.74 km westward and 71.62 km northward, respectively. Mulching film moved 176.95 km westward and 16.11 km northward. Pesticide moved 34.59 km westward and 39.08 km northward. Mechanical power moved 41.77 km westward and 1.09 km northward. Labor moved 25.66 km westward and 22.66 km northward. In total, the furthest movement westward is mulching film, the furthest movement northward is phosphorus fertilizer, and the furthest movement southward is mechanical power. Among all kinds of production inputs, the spatial barycenter of the plastic film showed the longest migration distance (177.68 km) and the fastest migration speed, and the linear migration distance was as high as 177.68 km. Pesticides showed the shortest movement distance (only 34.24 km) and the slowest moving speed.

The movement of the spatial barycenter for the outputs on arable land is shown in Figure 11, in which subfigure a and c illustrate the movement of barycenters of different kinds of outputs, while b and d illustrate the movement of barycenters of different outputs on per unit area. From the perspective of total output (the red arrow in Figure 11), the barycenter of grain moved towards the northeast, cotton and sugar crops moved towards the northwest, and the other crops all moved towards the southwest. In terms of yield per unit area (the blue arrow in Figure 11), only the fruits, vegetables, and sugar crops moved in the same direction as their total yield. The barycenter of both total output and yield per unit area of fruits and vegetables moved towards the southwest. Conversely, the barycenter of both total output and yield per unit area of sugar crops shifted towards the northwest.

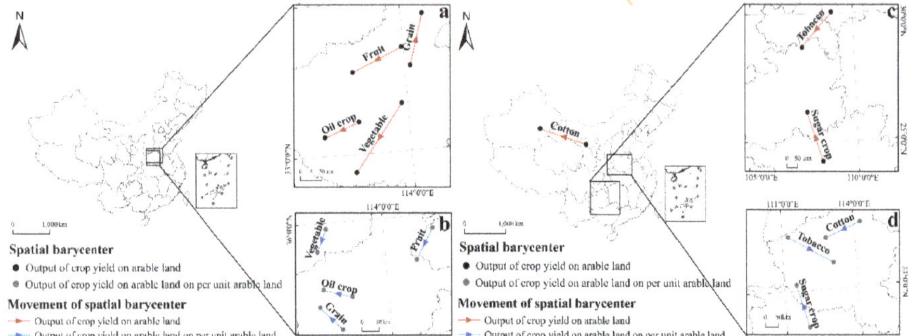

**Figure 11.** The movement of the spatial barycenter for the outputs on arable land.

In terms of movement distance, the spatial barycenter of grain moved 27.75 km eastward and 118.25 km northward. Vegetables migrated 103.49 km westward and 156.79 km southward. Fruits moved 111.48 km westward and 57.87 km southward, and oil crops moved 77.07 km westward and 35.96 km southward. Tobacco moved by 80.90 km westward and 101.64 km southward. Sugar crops moved 28.61 km westward and 37.24 km northward. Cotton moved 1112.35 km westward and 358.34 km northward. In total, in the east–west and north–south directions, the spatial barycenter of cotton moved by the longest distance (1168.64 km, with a speed of 116.8 km/year). Conversely, sugar crops moved by the shortest distance (46.96 km, with a speed of 4.70 km/year).

### 3.4. Degree of Coupling Coordination for the Input–Output on Arable Land

The degree of coupled coordination of input–output (D_CCIO) in 2008 and 2018 was illustrated in Figure 12a,b, respectively. The D_CCIO on arable land was low but increased gradually (average D_CCIO increased from 0.50 to 0.52 in 2008–2018). In terms of the grades of D_CCIO, the number of high-grade provinces increases while the number of low-grade provinces decreases.

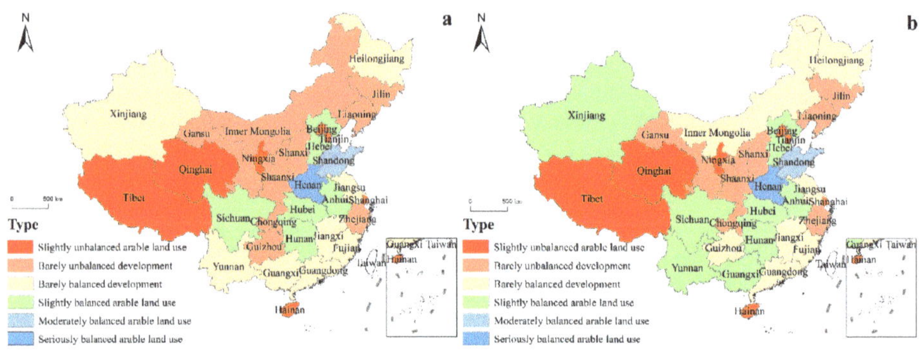

**Figure 12.** The change of D_CCIO among different provinces in 2008–2018.

Specifically, Henan and Shandong were the only provinces that showed seriously balanced and excited balanced grades. More provinces showed slightly balanced grades, including Hebei, Anhui, Hubei, Hunan, and Sichuan in 2008, with three newly added members (Guangxi, Yunnan, and Xinjiang) in 2018.

The number of provinces with barely balanced D_CCIO decreased. In 2008, Xinjiang, Guizhou, Yunnan, Guangxi, Heilongjiang, Jiangsu, Jiangxi, Fujian, Guangdong, and Guizhou were excluded by 2018. The number of provinces with barely unbalanced D_CCIO decreased, including Inner Mongolia, Jilin, Liaoning, Gansu, Shaanxi, Shanxi, Chongqing, Guizhou, and Zhejiang, among which Inner Mongolia and Guizhou were excluded. The provinces with the lowest D_CCIO remained unchanged, including Tibet, Qinghai, Ningxia, Beijing, Tianjin, Shanghai, and Hainan. By comparing the input index and output index, except Yunnan, Guizhou, and Shanghai, the output of all provinces lags behind the input.

In terms of the change in D_CCIO (Figure 13a, a represents the change of input-output, and b represents the change of D_CCIO), only 5 provinces experienced changes in the D_CCIO grades, including Inner Mongolia, Xinjiang, Yunnan, Guizhou, and Guangxi. In detail, by comparing the change of input and output, this study found five types of changing patterns in different provinces (Figure 13b): (1) the output grew faster than the grow of input, (2) the input index grew faster than the grow of output, (3) the input declined faster than the decline of output, (4) the input decreased whereas the output increased, and (5) the input increased whereas the output decreased. The 1st and 4th changing patterns were concentrated in South China, whereas the 2nd changing pattern was concentrated in Northwest China, and the 3rd pattern was concentrated in the North China Plain.

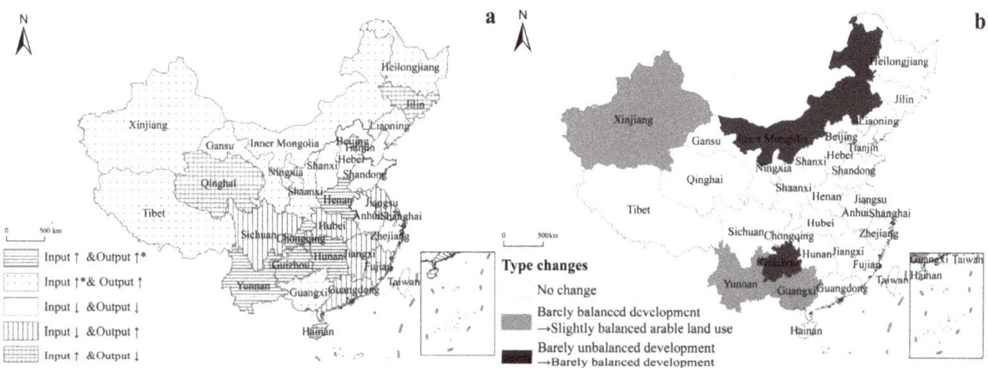


**Figure 13.** Change of the input–output and the coupling coordination type (the * in Figure 12a indicates the one which changed more. For instance, input ↑ and output ↑ * means the output increased more than the increase of input).

The research results found 13 "excellent provinces" that have good farmland utilization patterns (marked with * in Table 3). First, Henan and Shandong continued to show the highest D_CCIO from 2008–2018, which played a leading role in high-efficiency arable land use. Second, Yunnan, Guizhou, Guangxi, Xinjiang, and Inner Mongolia showed a significant increasing trend of D_CCIO, which indicated that the arable land use pattern was improving towards higher coordination among input and output. Third, Sichuan, Hubei, Jiangsu, Jiangxi, Fujian, and Southern Guangdong provinces showed an "input drops and output increases" trend, which proved a changing trend of arable land use towards higher productivity with lower input. These "excellent provinces" illustrated reasonable arable land use patterns towards better coordination between input and output, which was beneficial for the long-term use of scarce arable land.

**Table 3.** Different changing mode of degree of coupling coordination between the input and output (D_CCIO) among the 31 provinces.

| Level of D_CCIO | Change of Input (I) and Output (O) | Characteristics | Provinces |
|---|---|---|---|
| Seriously balanced | Both I and O increased | $\Delta I < \Delta O$ | Henan * |
| Moderately balanced | Both I and O decreased | $\Delta I > \Delta O$ | Shandong * |
| Slightly balanced | Both I and O increased | $\Delta I < \Delta O$<br>$\Delta I > \Delta O$ | Yunnan *, Hunan<br>Xinjiang *, Guangxi * |
| | Both I and O decreased | $\Delta I > \Delta O$<br>$\Delta I = \Delta O$ | Hebei<br>Anhui |
| | I and O changed reversely | I decreased and O increased | Sichuan *, Hubei * |
| Barely balanced | Both I and O increased | $\Delta I < \Delta O$<br>$\Delta I > \Delta O$ | Guizhou *<br>Inner Mongolia *, Heilongjiang |
| | I and O changed reversely | I decreased and O increased | Jiangsu *, Jiangxi, Fujian, Guangdong |
| Barely unbalanced | Both I and O increased | $\Delta I < \Delta O$<br>$\Delta I > \Delta O$ | Chongqing<br>Gansu, Shaanxi, Liaoning ×<br>Shanxi, Zhejiang |
| | Both I and O decreased | $\Delta I > \Delta O$ | |
| | I and O changed reversely | I increased and O decreased | Jilin × |
| Slightly unbalanced | Both I and O increased<br>Both I and O decreased | $\Delta I > \Delta O$<br>$\Delta I > \Delta O$ | Tibet, Ningxia<br>Beijing, Tianjin, Shanghai |
| | I and O changed reversely | I increased and O decreased | Qinghai, Hainan |

Note: * refers to the increase in the efficiency of the use of arable land and × refers to input increased and output decreased.

However, some of these "excellent provinces" have objective limits on arable land use. For instance, Yunnan, Guizhou, and Fujian are restricted by local mountainous terrain. Although plenty of rain and light exists in Sichuan, mountainous areas concentrated in western Sichuan bring a mountainous plateau climate, which causes insufficient heat for arable land use (annual temperature is only 4~12 °C). Moreover, as one of the most economically developed areas in China, Guangdong shares a low agricultural contribution rate (agricultural output accounted for only 3.94% in the province by 2018). Except for Henan, Shandong, Inner Mongolia, and Xinjiang, these "excellent provinces" are distributed in southern China. This forms a similar distribution with the current economic pattern (higher nonagricultural economic output in southern China and lower in northern China), which may bring increasingly higher pressure for agricultural development in these areas. In the future, the rapid nonagricultural economic development in the "excellent provinces" may affect the arable land amount and quality and further disturb the input–output coordination on arable land.

In those provinces with the lowest D_CCIO, the lack of arable land resources is the main limitation, as megacities in China, Shanghai, Beijing and Tianjin had low arable land areas (lower than $4.37 \times 10^5$ hm$^2$), which ranked last, second last and third last among the studied provinces (Table 3). Located in the western high-altitude zone, Tibet and Qinghai lack ideal water and heat resources and maintain low cultivated land area ($<5.91 \times 10^5$ hm$^2$), which ranks 4th and 5th from the bottom, respectively. Hainan Province is also limited by the amount of arable land resources ($<7.22 \times 10^5$ hm$^2$), which ranked 6th from the bottom. Located on the Loess Plateau, Ningxia faces the challenge of the natural environment caused by water resource shortages and soil erosion (cultivated land area $<12.9 \times 10^5$ hm$^2$, the 7th from the bottom). These regions all showed a characteristic of a large change in input and a small change in output. Surprisingly, this study found that two major national agricultural regions showed significantly weaker land use:

Jilin and Liaoning Provinces (marked with $^\times$ in Table 3). As an important agricultural region in northeastern China, Jilin and Liaoning enjoyed a deal area of arable land (the former $4.972 \times 10^6$ hm$^2$, ranked 5th, the latter $6.987 \times 10^6$ hm$^2$, ranked 13th), whereas they showed lower D_CCIO and a less ideal change trend (input increase but output decrease in Jilin, output growth slower than the growth of input in Liaoning). The "output lags behind input" phenomenon in important agricultural production provinces may go against the overall arable land use efficiency.

## 4. Discussion

### 4.1. Differences in the Recessive Transfortion of Arable Land Use Patterns

Facing the dynamic changes of society–economy–ecology in rapidly urbanizing areas, land use transition often becomes a common practice. Previous studies have pointed out two mechanisms of land use transition: explicit transition and recessive transition [42]. The explicit land use transition is mainly manifested as regional construction land expansion and arable land contraction in rapid urbanization process [43,44]. The recessive transition of arable land use may be more invisible and complex [45], such as the recessive transition of the input structure on arable land [46]. By comparison, this study confirms the existence of recessive transition of arable land use from the perspective of input–output change. Because the 31 provinces belong to different physical geographic regions of China (Figure 1), the details of recessive transition of arable land use in this study can be discussed as follows.

First, located in North China District, the Hebei, Shandong, Henan showed a "high input and high output" characteristic, accompanied by internal structural transition of input. Most inputs in this region showed a downward trend (except compound fertilizer and mechanical power). Especially, the decrease in nitrogen fertilizer, pesticide, and labor input was more than 1.25 times the national average level, in which the decline rate of nitrogen fertilizer was close to twice the national average. However, the increase in mechanical power was only 20.56% of the national average. Although these areas enjoy unique natural advantage of flat terrain and rich farmland resources, affected by the traditional small farmer production mode, arable land patches are mostly fragmented and large-scale mechanized farming is inadequate [47,48]. In addition, the groundwater funnel caused by excessive groundwater exploitation for irrigation will also force these areas to seek water-saving arable land use patterns [49]. Therefore, the transition mechanism of arable land use in this area can be summarized as "increasing compound fertilizer and machinery power input, and decreasing the overall input, accompanied by incomplete mechanization".

In contrast, though Jilin, Liaoning, and Heilongjiang Province are located in North-East District, one of the most important agricultural production areas in China, showed low D_CCIO. Enjoying the fertile black soil and large-scale management of arable land, the grain sown area in these areas have been proved to increase continuously in recent years [50]. However, the growth rate of input was high (except labor force, all input factors increased by more than 20%, far exceeding the national average level), with a lag in the output on arable land. One of the reasons for this low D_CCIO in this area is the lag of yield per unit area. Specifically, the high total output but low output per unit area in this area have highlighted the lag of productivity per unit area, which is also a manifestation of the recessive transition of arable land use. Therefore, the recessive transition mechanism of arable land use in northeast China can be summarized as "increasing overall inputs and sparing labor force on the premise of expanding sown area, accompanied by lagging productivity per unit area".

Located in Central China, Hunan, Hubei, Anhui, Jiangxi, Jiangsu, Zhejiang and Fujian Province enjoy ideal climate condition (average annual precipitation is 1300–1900 mm, average annual temperature is 16~19 °C), meanwhile, fertile irrigation from the Yangtze River gives this area natural advantages of planting. However, the D_CCIO of this area is not the highest. The main reason is the difference of terrain and economic development mode. Specifically, abundant arable land on the plain in Hunan and Hubei give them the chance to increase the ago-mechanical input significantly (with a growth rate of 58%). With mountainous and hilly areas accounting for 80%, Fujian and Jiangxi lack the advantage of expanding to the farming scale, which makes an overall decline in input and output growth. Therefore, the recessive transition mechanism of arable land use in this area can be summarized as "polarization of mechanization level and planting scale caused by terrain difference".

Though located in the North-West District with severe water shortage and diurnal temperature variation, Xinjiang showed a high D_CCIO in a typical oasis agricultural mode. A closer look reveals that the total inputs in this region are not high, but the use of mulching film and phosphate fertilizer is very prominent, ranking 1st and 2nd among the 31 provinces, respectively. The main reason comes from the predominant crop (the total yield, per unit yield and sown area of cotton in Xinjiang ranks 1st in China). As two highly needed factors during the growth of cotton [51,52], mulching film and phosphorus fertilizer become the most significant input on arable land. Similarly, located in South China District, Yunnan enjoys subtropical monsoon climate with sufficient water-heat resources and fertile red soil, which makes ideal living environment for tobacco and sugar crops [53]. Potash fertilizer, the indispensable factor for tobacco and sugar crops, showed a growth rate with 7.5 times that of the average level in China. Superior crop in Xinjiang and Yunnan shows a similar influence on the input–output pattern on arable land. As a result, the transition mechanism of arable land use in these two regions can be summarized as "specializing plantation of predominant crop, with priority of resource input and production output."

Located in the Qinghai-Tibet District with an average altitude above 4000 m, Qinghai and Tibet have a typical plateau mountain climate. Water, heat and arable land are scarce in this area, which limited the arable land use and make the D_CCIO extremely low. Though far lower than the national level, most inputs kept increasing. However, natural constraints hinder the increase of output on arable land. Under such objective constraints, the transition mechanism of arable land use in this region can be summarized as "retreat from agricultural development and weaken the use of arable land".

To sum up, significant different transition mechanisms of arable land use among different provinces are hidden in the internal structural change of input–output on arable land [20]. The constraints of arable land use lay in the deficient agricultural mechanization input caused by the smallholder management mode, the lagging productivity per unit area, and the polarized input–output pattern caused by dominant crop planting. To improve the efficiency and rationality of arable land use, the basic premise of "taking measures according to local conditions" is needed. Characteristics of regional geographical conditions and planting structures should be taken into deeper consideration, which will be a necessary mean for optimized transition of arable land use.

### 4.2. Plenty of Room for Optimizing the Planting System on Arable Land

Based on microscale experimental research, previous studies have emphasized the importance of optimizing planting systems, which may contribute to the efficiency of comprehensive resource use [54,55]. In contrast, based on a geographical spatiotemporal perspective, this study proves that there is plenty of room for optimizing the planting system. The spatial distribution of total output and output per unit area differs greatly, which may be a breakthrough point for the optimization of planting structure. In addition, the movement of spatial barycenters of the total output and output per unit area did not match well, which also confirms the necessity of planting structure optimization. The movement of the spatial barycenter of the total grain yield towards the northeast and the grain yield per unit area towards the northwest are the east-west opposite. The movement of the spatial barycenter of the total oil crop yield towards the southwest and its yield per unit area towards the northwest are south–north opposite. The total yield of tobacco towards the southwest and the yield per unit towards the southeast are the east–west opposite. Similarly, the total yield of cotton towards the northwest and the yield per unit area towards the southwest are south-north opposite.

Theoretically, the yield per unit area could better reflect the productivity of arable land. The "opposite movement" of the total yield and yield per unit area is probably a result of the unreasonable allocation of cultivated land scale and planting structure. Taking Heilongjiang and Xinjiang as examples, Heilongjiang's total grain output ranks 1st in China, while its per unit area yield ranks 24th. The total grain output of Xinjiang ranks 15th in China, while its per unit area yield ranks 2nd. This large contrast may be caused by regional resource endowment differences and agricultural scales. Although the base of arable land area in Xinjiang is small (5148.1 hm$^2$), the inland rivers supplied by melting water from snow in this area provide up to 94.86% irrigation. Although Heilongjiang has as much as 15,845.9 hm$^2$ of arable land, its effective irrigation rate is only 38.62%. In addition, the long sunshine time, high photosynthetic efficiency, and sparse population make Xinjiang a flexible place for arable land use. "Plant crops if the land is good, and graze if the land is bad," which makes the average yield per unit area in Xinjiang enjoy certain advantages.

The mismatch between the total yield and yield per unit indicated that the increase in crop yield might rely heavily on the increase in sown area rather than land productivity, which emphasizes the importance of optimizing the planting structure. Previous research also shows that there is no significant scale benefit in arable land, i.e., expansion of arable land or sown area does not necessarily lead to an increase in crop yield [56]. From this perspective, this study found break-through points for the places where total yield and per unit yield were uncoupled. The grain of Xinjiang, the vegetables of Inner Mongolia, Xinjiang, Liaoning and Hebei, the cotton of Jilin, the oil crops of Hebei, Guangdong and Jilin, the fruits of Inner Mongolia, Heilongjiang, Liaoning and Hebei, and the tobacco of Inner Mongolia and Gansu all show typical uncoupling characteristics of "high yield per unit area but low total yield." In response, based on local arable land resources, gradually transforming the planting structure to these special local crops may become an important means for the transition and optimization of arable land use.

### 4.3. Policy Implications of Arable Land Use Transition

To better improve national arable land use on a macroscale, the results pointed to some possible policy-making points. First, in view of the recessive transition of the input–output structure of arable land use, optimizing the mode of arable land utilization by taking measures according to local conditions should be taken seriously. Henan and Shandong should strengthen the protection of arable land amount and give full play to agro-mechanization. For those with low input and low output levels, such as Tibet, Ningxia, Beijing, Tianjin, Shanghai, Qinghai, and Hainan, which have disadvantages in arable land scale, they should appropriately reduce the intensity of agricultural production and take regional economic transition as the main direction. For those with high input-low output levels, such as Inner Mongolia, Heilongjiang, Gansu, Shaanxi, and Liaoning, they should focus on enriching the soil fertility, improving the irrigation rate, and reasonable fallowing.

Second, long-term arable land use planning should focus on long-term observations of productivity among different regions. The grain of Xinjiang, the vegetables in Inner Mongolia, Xinjiang, Liaoning and Hebei, the cotton in Jilin, etc., all show a decoupling phenomenon of high yield per unit and low total yield, which highlights the necessity of long-term monitoring on a macro scale. Dynamically allocating agricultural production tasks on the basis of long-term observations can help to maximize the productivity of arable land and give full play to regional advantages.

Finally, the policy emphasis on the transition of arable land use lies in the coordinated relationship between humans and land. China's rapid urbanization caused more than 200 million farmers to leave arable land and seek jobs in cities in 1978–2018, which may further affect national arable land use efficiency [2–4,57]. Scholars insist that arable land use should focus on saving arable land and shifting from labor-intensive land to technology-intensive land [18]. However, in areas with relatively abundant arable land resources, such as Northeast China and Xinjiang, there is still much potential for the large-scale management of arable land. In these areas, either bringing labor back to arable land or letting in capital and markets, the formation of specialized farm operations, and widely improving the technical advantages of the use of arable land are suitable paths for the region. In contrast, in the North China Plain with many small farmers, the policy should focus on the promotion of arable land circulation. It should provide professional farmers with flexibility to expand the arable land area and give stable contract rights to those farmers who rent out their arable land. On the other hand, in the hilly mountainous areas of Southwest China, policy-making should focus on preventing arable land abandonment, prohibiting the occupation of flat farmland, and supplementing sloping arable land to increase farmers' enthusiasm for arable land use [58].

## 5. Conclusions

By analyzing the degree of coupling coordination (D_CCIO) of input–output on arable land, this study pointed out different recessive transitions of arable land use among 31 provinces in mainland China. First, the total input of arable land and the input per unit area showed different spatial-temporal changes. Although the total input and input per unit area of nitrogen fertilizer, pesticide, and labor force all decreased, the total input and input per unit area of compound fertilizer, mulching film, potash fertilizer, and mechanical power increased. The total input of phosphorus fertilizer increased, while its input per unit area decreased. The spatial barycenter of the total inputs moved towards the northwest, while the input per unit area moved in another direction (southwest, except potash and compound fertilizers).

Second, the total output and the output per unit area also showed spatial-temporal disparities in arable land. The total output and yield per unit area of grain, vegetables, fruit, and oil crops increased, yet the total yield and yield per unit area of tobacco and cotton decreased. The total yield of sugar crops decreased, but the yield per unit area increased. The spatial barycenter of total grain output moved towards the northeast, but the grain yield per unit area moved to the northwest. The spatial barycenter of the total output of oil crops moved towards the southwest, per unit area of yield to the northwest. The total yield of tobacco moved towards the southwest, while the yield per unit area moved towards the southeast. The total cotton output moved towards the northwest, while the output per unit area moved towards the southwest.

Third, the recessive transition mechanism of arable land use was hidden in the internal change of the input–output structure. Although the D_CCIO increased overall, the output of most provinces lagged behind the input. Specifically, Henan and Shandong showed the highest D_CCIO, with a transition mechanism of "increasing compound fertilizer and machinery power input, and decreasing the overall input, accompanied by incomplete mechanization." Northeast China (Liaoning, Jilin, Heilongjiang) showed lower D_CCIO, with a recessive transition mechanism of arable land use of "increasing overall inputs and sparing labor force on the premise of expanding sown area, with lagging productivity on per unit area." Xinjiang in northwest China showed a transition mechanism of arable land use in this area characterized by "specializing plantation of predominant crop, with priority of resource input and production output."

Finally, this paper proposes a policy proposal for optimized arable land use patterns. On the one hand, the lagging of per unit area yield in Northeast China, the incomplete mechanization level on the Huang-Huai-Hai Plain, and the support of dominant characteristic crop planting in Xinjiang should be improved. On the other hand, considering the spatial decoupling total yield and yield per unit area, the planting structure should be adjusted according to local advantages. The grain of Xinjiang, the vegetables of Inner Mongolia, Xinjiang, Liaoning and Hebei, the cotton of Jilin, the oil crops of Hebei, Guangdong and Jilin, the fruits of Inner Mongolia, Heilongjiang, Liaoning and Hebei, and the tobacco of Inner Mongolia and Gansu with "high yield per unit and low total yield" should be given priority. In addition, in view of the human-land relationship, specialized large-scale arable land management should improve in regions with rich arable land resources (northeast China and Xinjiang agricultural reclamation corps). In the fragmented arable land on the North China Plain, stable contract rights should be improved. Smallholders should be encouraged to freely transfer their arable land to professional farmers with large-scale land patches to improve mechanization in this area. In the hilly and mountainous areas of Southwest China, it is necessary to avoid compulsory large-scale land consolidation to avoid the loss of flat arable land and land abandoned by farmers. Based on the above research conclusions, this paper provides a detailed scientific reference for the observation of arable land transition and land use optimization in China.

Limited as it is by provincial data, this study cannot explore the spatial-temporal pattern of recessive transition of arable land use in finer scale. In the future, deeper studies based on city-level or county level are highly recommended to reflect the possible different recessive transitions mechanisms of arable land use in China or other countries.

**Author Contributions:** Conceptualization, Y.L. and G.Y.; methodology, Y.L. and G.Y.; software, Y.L., G.Y. and Y.X.; validation, Y.L., G.Y. and Y.X.; formal Analysis, G.Y.; investigation, Y.L., S.L. and G.L.; resources, G.Y. and X.W.; data Curation, Y.L.; writing—original draft preparation, Y.L.; Y.X. and S.X.; writing—review and editing, Y.L. and G.Y.; visualization, Y.L. and Y.X.; supervision, G.Y.; project administration, G.Y. and Y.L.; funding acquisition, G.Y. All authors have read and agreed to the published version of the manuscript.

**Funding:** This paper was funded by the National Natural Science Foundation of China (Project No. 41701590); Humanities and Social Sciences Foundation of the Ministry of education, China (Project No. 17YJCZH228); the Natural Science Foundation of Shandong Province, China (Project No. ZR2017-BD004); the China Postdoctoral Science Foundation (Project No. 2017M612340); the Research project on the cultivation and reform of teaching achievements of Shandong Normal University (2019XM42) and the Shandong Social Science Planning Fund Program (Project No. 19BJCJ23).

**Data Availability Statement:** Data available in a publicly accessible repository that does not issue DOIs Publicly available datasets were analyzed in this study. This data can be found here: http://www.stats.gov.cn/tjsj/tjcbw/202008/t20200826_1785896.html.

**Acknowledgments:** The authors extend great gratitude to the anonymous reviewers and editors for their helpful review and critical comments.



## References

1.  Li, G.; Zhao, Y.; Cui, S. Effects of urbanization on arable land requirements in China, based on food consumption patterns. *FOOD Sec.* **2013**, *5*, 439–449. [CrossRef]
2.  Zhang, H.; Zhang, S.; Liu, Z. Evolution and influencing factors of China's rural population distribution patterns since 1990. *PLoS ONE* **2020**, *15*, e0233637. [CrossRef]
3.  Liang, Y.; Lu, W.; Wu, W. Are social security policies for Chinese landless farmers really effective on health in the process of Chinese rapid urbanization? A study on the effect of social security policies for Chinese landless farmers on their health-related quality of life. *Int. J. Equ. Health* **2014**, *13*. [CrossRef]
4.  Yang, X.J. China's Rapid Urbanization. *Science* **2013**, *342*, 310. [CrossRef]
5.  Chien, S.-S. Local farmland loss and preservation in China—A perspective of quota territorialization. *Land Use Pol.* **2015**, *49*, 65–74. [CrossRef]
6.  Jintao, L.; Li, Y. Influence measurement of rapid urbanization on agricultural production factors based on provincial panel data. *Soc.-Econ. Plan. Sci.* **2019**, *67*, 69–77. [CrossRef]
7.  Liu, X.; Zhao, C.; Song, W. Review of the evolution of cultivated land protection policies in the period following China's reform and liberalization. *Land Use Pol.* **2017**, 660–669. [CrossRef]
8.  Wu, Y.; Shan, L.; Guo, Z.; Peng, Y. Cultivated land protection policies in China facing 2030: Dynamic balance system versus basic farmland zoning. *HAB Int.* **2017**, *69*, 126–138. [CrossRef]
9.  Song, W.; Pijanowski, B.C. The effects of China's cultivated land balance program on potential land productivity at a national scale. *Appl. Geogr.* **2014**, *46*, 158–170. [CrossRef]
10. Zhang, C.; Wang, X.; Liu, Y. Changes in quantity, quality, and pattern of farmland in a rapidly developing region of China: A case study of the Ningbo region. *Landsc. Ecol. Eng.* **2019**, *15*. [CrossRef]

11. Jiao, C.; Chen, L.; Sun, C.; Jiang, Y.; Shen, Z. Evaluating national ecological risk of agricultural pesticides from 2004 to 2017 in China. *Environ. Pollut.* **2019**, *259*, 113778. [CrossRef]

12. Luo, Y.; Yue, H.F.; Li, S.H.; Yi, H.; Zhu, G.P. On the loss regularity of water and fertilizer in sandy soil with abundance of underground water but lack of cultivated layer. *IOP Conf. Ser. Earth Environ. Sci.* **2019**, *346*, 012044. [CrossRef]

13. Shi, C.; Li, J.; Cao, T. Evaluation of the Soil Nutrients of Added Cultivated Land Based on Entropy Weight-Fuzzy Comprehensive Method: A Case Study in Chengcheng County, Shaanxi Province. In *Advances in Engineering Research, Proceedings of the 2nd International Conference on Material Science, Energy and Environmental Engineering (MSEEE 2018), Chongqing, China, 16–17 August 2018*; Atlantis Press: Amsterdam, The Netherlands, 2018.

14. Xin, Y.; Yin, G.; Lou, L. Analysis of Cultivated Land Use Heterogeneity Based on the Three-Dimensional Differentiation of Rural Households in Shandong Province. *China Land Sci.* **2020**, *34*, 78–85. [CrossRef]

15. Zou, L.; Liu, Y.; Wang, Y.; Hu, X. Assessment and analysis of agricultural non-point source pollution loads in China: 1978–2017. *J. Environ. Manag.* **2020**, *263*, 110400. [CrossRef]

16. Liu, Y.; Song, W.; Mu, F. Changes in ecosystem services associated with planting structures of cropland: A case study in Minle County in China. *PCEA* **2017**, *102*, 10–20. [CrossRef]

17. Xie, H.; Liu, G. Spatiotemporal differences and influencing factors of multiple cropping index in China during 1998–2012. *J. Geog. Sci.* **2015**, *25*, 1283–1297. [CrossRef]

18. Long, H.; Qu, Y. Land use transitions and land management: A mutual feedback perspective. *Land Use Pol.* **2018**, *74*, 111–120. [CrossRef]

19. Chen, K.; Long, H.; Liao, L.; Tu, S.; Li, T. Land use transitions and urban-rural integrated development: Theoretical framework and China's evidence. *Land Use Pol.* **2020**, *92*, 104465. [CrossRef]

20. Lyu, X.; Niu, S.; Gu, G.; Peng, W. Conceptual cognition and research framework on sustainable intensification of cultivated land use in China from the perspective of the "New Agriculture, Countryside and Peasants". *J. Nat. Ressour.* **2020**, *35*, 2029–2043. [CrossRef]

21. Yin, G.; Jiang, X.; Sun, J.; Qiu, M. Discussing the regional-scale arable land use intensity and environmental risk triggered by the micro-scale rural households' differentiation based on step-by-step evaluation-a case study of Shandong Province, China. *Environ. Sci. Pollut. Res. Int.* **2020**, *27*, 8271–8284. [CrossRef]

22. He, G.; Zhao, Y.; Wang, L.; Jiang, S.; Zhu, Y. China's food security challenge: Effects of food habit changes on requirements for arable land and water. *JCP* **2019**, *229*, 739–750. [CrossRef]

23. Li, F.; Liu, W.; Lu, Z.; Mao, L.; Xiao, Y. A multi-criteria evaluation system for arable land resource assessment. *Environ. Monit. Assess.* **2020**, *192*. [CrossRef] [PubMed]

24. Xie, H.; Chen, Q.; Wang, W.; He, Y. Analyzing the green efficiency of arable land use in China. *Technol. Forecast. Soc. Chang.* **2018**, *133*, 15–28. [CrossRef]

25. Ouyang, W.; Cai, G.; Huang, W.; Hao, F. Temporal–spatial loss of diffuse pesticide and potential risks for water quality in China. *Sci. Total Environ.* **2015**, *541*, 551. [CrossRef] [PubMed]

26. Duncan, E.G.; O'Sullivan, C.A.; Roper, M.M.; Biggs, J.S.; Peoples, M.B. Influence of co-application of nitrogen with phosphorus, potassium and sulphur on the apparent efficiency of nitrogen fertiliser use, grain yield and protein content of wheat: Review. *Field Crops Res.* **2018**, *226*, 56–65. [CrossRef]

27. Zhao, J.; Ni, T.; Li, J.; Lu, Q.; Fang, Z.; Huang, Q.; Zhang, R.; Li, R.; Shen, B.; Shen, Q. Effects of organic–inorganic compound fertilizer with reduced chemical fertilizer application on crop yields, soil biological activity and bacterial community structure in a rice–wheat cropping system. *Appl. Soil Ecol.* **2016**, *99*, 1–12. [CrossRef]

28. He, H.; Wang, Z.; Guo, L.; Zheng, X.; Zhang, J.; Li, W.; Fan, B. Distribution characteristics of residual film over a cotton field under long-term film mulching and drip irrigation in an oasis agroecosystem. *Soil Tillage Res.* **2018**, *180*, 194–203. [CrossRef]

29. Xu, B.; Wang, F.; Wang, J.; Zhao, S. Establishment and verification of labor demand estimation model in planting industry. *Int. J. Agric. Biol. Eng.* **2017**. [CrossRef]

30. Luo, X.; Zhang, Z.; Lu, X.; Zhang, L. Topographic Heterogeneity, Rural Labor Transfer and Cultivated Land Use: An Empirical Study of Plain and Low Hill areas in China. *Pap. Reg. Sci.* **2019**. [CrossRef]

31. Li, X.; Liu, L.; Xie, J.; Wang, Z.; Li, Y. Optimizing the quantity and spatial patterns of farmland shelter forests increases cotton productivity in arid lands. *Agric. Ecosyst. Environ.* **2020**, *292*. [CrossRef]

32. Wu, L.; Wu, X.; Shaaban, M.; Zhou, M.; Zhao, J.; Hu, R. Decrease in the annual emissions of $CH_4$ and $N_2O$ following the initial land management change from rice to vegetable production. *Environ. Sci. Pollut. Res.* **2018**, *25*, 13014–13025. [CrossRef] [PubMed]

33. Tian, Y.; Zhao, L.; Meng, H.; Sun, L.; Yan, J. Estimation of un-used land potential for biofuels development in (the) People's Republic of China. *Appl. Energy* **2009**, *86*, S77–S85. [CrossRef]

34. Cheng, X.; Long, R.; Chen, H.; Li, Q. Coupling coordination degree and spatial dynamic evolution of a regional green competitiveness system—A case study from China. *Ecol. Indic.* **2019**, *104*, 489–500. [CrossRef]

35. Zhang, Z.; Li, Y. Coupling coordination and spatiotemporal dynamic evolution between urbanization and geological hazards—A case study from China. *Sci. Total Environ.* **2020**, *728*. [CrossRef] [PubMed]

36. Fu, S.; Zhuo, H.; Song, H.; Wang, J.; Ren, L. Examination of a coupling coordination relationship between urbanization and the eco-environment: A case study in Qingdao, China. *Environ. Sci. Pollut. Res.* **2020**, *27*, 23981–23993. [CrossRef]

37. Zhang, T.; Li, L. Research on temporal and spatial variations in the degree of coupling coordination of tourism–urbanization–ecological environment: A case study of Heilongjiang, China. *Environ. Dev. Sustain.* **2020**, 1–18. [CrossRef]

38. Lefever, D.W. Measuring Geographic Concentration by Means of the Standard Deviational Ellipse. *Am. J. Sociol.* **1926**, *32*, 88–94. [CrossRef]

39. Yin, S.; Guo, M.; Wang, X.; Yamamoto, H.; Ou, W. Spatiotemporal variation and distribution characteristics of crop residue burning in China from 2001 to 2018. *Environ. Pollut.* **2020**. [CrossRef]

40. Shi, Y.; Matsunaga, T.; Yamaguchi, Y.; Li, Z.; Gu, X.; Chen, X. Long-term trends and spatial patterns of satellite-retrieved PM2.5 concentrations in South and Southeast Asia from 1999 to 2014. *Sci. Total Environ.* **2018**, *615*, 177–186. [CrossRef]

41. He, Y.; Chen, Y.; Tang, H.; Yao, Y.; Yang, P.; Chen, Z. Exploring spatial change and gravity center movement for ecosystem services value using a spatially explicit ecosystem services value index and gravity model. *Environ. Monit. Assess.* **2010**, *175*, 563–571. [CrossRef]

42. Qu, Y.; Long, H. A framework of multi-disciplinary comprehensive research on recessive farmland transition in China. *Acta Geogr. Sin.* **2018**, *73*, 1226–1241. [CrossRef]

43. Tian, J.; Wang, B.; Zhang, C.; Li, W.; Wang, S. Mechanism of regional land use transition in underdeveloped areas of China: A case study of northeast China. *Land Use Pol.* **2020**, *94*, 104538. [CrossRef]

44. Long, H.; Qu, Y.; Tu, S.; Zhang, Y.; Jiang, Y. Development of land use transitions research in China. *J. Geog. Sci.* **2020**, *30*, 1195–1214. [CrossRef]

45. Zhu, F.; Zhang, F.; Ke, X. Rural industrial restructuring in China's metropolitan suburbs: Evidence from the land use transition of rural enterprises in suburban Beijing. *Land Use Pol.* **2018**, *74*, 121–129. [CrossRef]

46. Yin, G.; Lin, Z.; Jiang, X.; Yan, H.; Wang, X. Spatiotemporal differentiations of arable land use intensity—A comparative study of two typical grain producing regions in northern and southern China. *JCP* **2018**, *208*, 1159–1170. [CrossRef]

47. Li, S.; Lei, Y.; Zhang, Y.; Liu, J.; Shi, X.; Jia, H.; Wang, C.; Chen, F.; Chu, Q. Rational trade-offs between yield increase and fertilizer inputs are essential for sustainable intensification: A case study in wheat–maize cropping systems in China. *Sci. Total Environ.* **2019**, *679*, 328–336. [CrossRef]

48. Xu, Y.; Li, Z.; Wang, L. Temporal-spatial differences in and influencing factors of agricultural eco-efficiency in Shandong Province, China. *Ciência Rural.* **2020**, *50*. [CrossRef]

49. Kong, X.; Zhang, X.; Lal, R.; Zhang, F.; Chen, X.; Niu, Z. Chapter Two: Groundwater Depletion by Agricultural Intensification in China's HHH Plains, Since 1980s. *Adv. Agron.* **2016**, *135*, 59–106. [CrossRef]

50. Xin, F.; Xiao, X.; Dong, J.; Zhang, G.; Li, B. Large increases of paddy rice area, gross primary production, and grain production in Northeast China during 2000-2017. *Sci. Total Environ.* **2019**, *711*, 135183. [CrossRef]

51. Liu, C.; Zhou, L.; Jia, J.; Wang, L.; Si, J.; Li, X.; Pan, C.; Li, F. Maize yield and water balance is affected by nitrogen application in a film-mulching ridge–furrow system in a semiarid region of China. *Eur. J. Agron.* **2014**, *52*, 103–111. [CrossRef]

52. Zhang, P.; Xu, S.; Zhang, G.; Pu, X.; Wang, J.; Zhang, W. Carbon cycle in response to residue management and fertilizer application in a cotton field in arid Northwest China. *J. Integr. Agric.* **2019**, *18*, 1103–1119. [CrossRef]

53. Huang, C.; Yang, H.; Li, Y.; Zhang, M.; Lv, H.; Zhu, A.X.; Yu, Y.; Luo, Y.; Huang, T. Quantificational effect of reforestation to soil erosion in subtropical monsoon regions with acid red soil by sediment fingerprinting. *Environ. Earth Sci.* **2017**, *76*, 34. [CrossRef]

54. He, L.; Du, Y.; Wu, S.; Zhang, Z. Evaluation of the agricultural water resource carrying capacity and optimization of a planting-raising structure. *Agric. Water Manag.* **2021**, *243*, 106456. [CrossRef]

55. Wang, J.; Fu, B.; Wang, L.; Lu, N.; Li, J. Water use characteristics of the common tree species in different plantation types in the Loess Plateau of China. *Agric. For. Meteorol.* **2020**, 108020. [CrossRef]

56. Li, G.; Feng, Z.; You, L.; Fan, L. Re-examining the inverse relationship between farm size and efficiency: The empirical evidence in China. *China Agric. Econ. Rev.* **2013**. [CrossRef]

57. Jin, Y.; Qian, W.; Wu, B. Off-farm employment and grain production change: New evidence from China. *China Econ. Rev.* **2020**, *63*, 101519. [CrossRef]

58. Wang, Y.; Li, X.; Xin, L.; Tan, M. Farmland marginalization and its drivers in mountainous areas of China. *Sci. Total Environ.* **2019**, *719*, 135132. [CrossRef]

 *land*

*Article*

# The Impact of Farmland Transfer on Rural Households' Income Structure in the Context of Household Differentiation: A Case Study of Heilongjiang Province, China

Lili Chen [1,2], Hongsheng Chen [3,*], Chaohui Zou [4] and Ye Liu [1,2]

[1] School of Geography and Planning, Sun Yat-Sen University, Xingang Xi Road, Guangzhou 510275, China; chenlli8@mail.sysu.edu.cn (L.C.); liuye25@mail.sysu.edu.cn (Y.L.)
[2] Guangdong Key Laboratory for Urbanization and Geo-Simulation, Sun Yat-Sen University, Xingang Xi Road, Guangzhou 510275, China
[3] School of Architecture and Urban Planning, Shenzhen University, Shenzhen 518060, China
[4] Guangdong Guodi Land Planning Technology Co., Ltd., Guangzhou 510650, China; 1410035@stu.neu.edu.cn
* Correspondence: chenhongsh@szu.edu.cn

**Abstract:** Farmland transfer is an important factor affecting rural households' income and sustainable development of rural areas in developing countries. However, recent studies have reached controversial conclusions on how farmland transfer affects rural households' income because of ignoring the household differentiation and the difference in the impacts of farmland transfer-in and transfer-out on the income structure. Taking the Heilongjiang province, the major cereal production area in China, as the study area, the paper aims to estimate the impacts of farmland transfer-in or transfer-out of different rural households on income structure based on the Propensity Score Matching (PSM) model. Results showed that the total income of all rural households transferring-in farmland increased significantly while the income decreased after transferring-out farmland, and I part-time households have the largest increase, followed by pure-agricultural households and II part-time households, whereas I part-time households has the smallest reduction, followed by pure-agricultural households and II part-time households. Because the increase in the agricultural income and subsidies was greater than the decrease in the outworking income for I part-time households transferring-in farmland, while the outworking income not increasing but decreasing when II part-time households transferring-out farmland. We can conclude that (1) encouraging pure-agricultural and I part-time households to transfer farmland in and II part-time households to transfer out of farmland, and develop mutual assistance for the aged in rural areas should be strengthened. (2) Improving the farmland transfer market and promoting non agricultural employment of surplus-labor need to be synchronized. (3) Agricultural subsidies should be provided to cultivators.

**Keywords:** farmland transfer; household differentiation; income structure; rural households; land use transition



**Citation:** Chen, L.; Chen, H.; Zou, C.; Liu, Y. The Impact of Farmland Transfer on Rural Households' Income Structure in the Context of Household Differentiation: A Case Study of Heilongjiang Province, China. *Land* **2021**, *10*, 362. https://doi.org/10.3390/land10040362

Academic Editors: Hualou Long, Xiangbin Kong, Shougeng Hu and Yurui Li

Received: 7 March 2021
Accepted: 30 March 2021
Published: 1 April 2021

**Publisher's Note:** MDPI stays neutral with regard to jurisdictional claims in published maps and institutional affiliations.



## 1. Introduction

Farmland is one of the most important productive assets of rural households in many countries, and it can be sold, leased, or exchanged [1–3]. How farmland is owned, used, and exchanged has far-reaching implications for productivity, equity, and overall economic growth, and there is a large and growing body of literature on the impacts of land tenure security on farmland investments and agricultural productivity, and consider that the establishment of private land ownership and tenure security facilitates transactions in land rental and sales markets by reducing transaction costs, stimulates land investment by assuring investment returns [2,4,5]. However, how to realize the positive role of farmland transactions in the case of the inability to obtain land ownership remains to be explored in depth. Unlike in many countries, under the Household Contract Responsibility System of China (HCRS), village collectives own farmlands, and farming households contract farmland from collectives and receive the land contract and management rights, while rural households are not allowed to sell their farmland, and they only have farmland usage rights rather than ownership, and they can trade their farmland usage rights to other households or economic organizations for enlarging or reducing their farmland scale. The trade of farmland usage rights is often referred to as "farmland transfer" in China, and it includes transfer, lease, exchange, shareholding, etc. [6–8], which has been considered as a typical way of land use transitions [9], because it involves the changes in planting structure by solving the farmland fragmentation and changing the planting scale, helping to achieve agricultural modernization [8,10–12], particularly for China where agricultural production has dominated by the traditional small-scale household economies, and the significance of optimizing farmland achieving the optimal allocation of rural land resources and agricultural scale management through farmland transfer is particularly prominent. Thus, a series of policies have been issued to promote farmland transfer. According to the data of the Ministry of Agriculture of the People's Republic of China, the transferred area of farmland reached $2.69 \times 10^7$ hm$^2$ by the end of 2018, accounting for 30.4% of the total area of household-contracted farmland, approximately 20% higher than in 2012 [13], and a more important reason for farmland transfer flourishing in China is that the government expected to promote rural households' income through farmland transfer [14]. Because the income gap between urban and rural is the important cause of the large population migration in rural China, which further caused "hollowing village" and the countryside decline [15–17]. Hence, increasing the income of rural households through farmland transfer is a necessary prerequisite for retaining rural elites and realizing rural revitalization.

In this context, the No.1 Central Committee's Documenting China has addressed the farmland transfer with the aim of positively influence farmers' income, which has also received considerable attention from academia. Although farmland transfer undoubtedly has an impact on the income of rural households, the positivity or negativity of its effects has caused a lot of controversies. One view is that farmland fragmentation has a significant negative impact on farmland efficiency, while farmland transfer can promote fragmented land to be concentrated into the pure-agricultural with a large area of farmland and improve economies of scale, finally increasing farmers' income to a certain degree [18–20]. Empirical research shows that farmland transfer can increase the income of any households and farmer households who have participated in farmland transfer by 19% and 33%, respectively [10]. However, another point of view considers that farmland transfer has a significant negative on the increase of rural households' income [21,22]. Some scholars also found the average cost curve of farmers is "U" shaped; that is, the average cost first decreases then increases with the expansion of the farm-scale [23,24], and if considering the farmland cost, farmland transfer will have a negative impact on agricultural production efficiency in China [25].

Hence, there is no comprehensive knowledge concerning the circumstances under which farmland transfer derives significant negative or positive effects on the rural households' income. Some research has classified households' income, but these studies ignored the difference in the impact of farmland transfer-in and transfer-out on rural households' income [26,27]. A few studies further investigated the impact of farmland transfer on households' total income by dividing rural households transferring-in farmland and transferring-out farmland and by analyzing the changes of agricultural machinery value, non-agricultural income, and rental-land income [21], while it does not eliminate the impact of management without farmland transfer on households' income. It is worth noting that some research has adopted the propensity score matching (PSM) method to estimate the impacts [10,28], eliminating the deviation caused by natural changes in households' income. Nonetheless, most of these studies have ignored the heterogeneity of rural households, and they also have paid less attention to the causes of farmland transfer affecting the income structure of different rural households. Hence, putting farmland transfer-in and transfer-out, household differentiation, and household income structure in the same framework, and considering the differences that farmland transfer-in and transfer-out affecting the income structure in the context of rural household differentiation need to be further explored.

With the ongoing rural labor migration, the non-agricultural income of rural households has gradually become the main source of households' income [29]. However, agricultural income is still the main source of most rural households' income in Heilongjiang province as the grain production and food-commodity supply area. Especially under the current macroeconomic environment of economic decline in Northeast China [30], how to stabilize and strengthen agricultural production, improving agricultural productivity through farmland transfer, and ultimately increase rural households' income is extremely important. Therefore, this paper takes Heilongjiang province as the case, quantitatively estimates the impacts of farmland transfer-in or transfer-out on the total incomes in the context of household differentiation; secondly, this study further explores the underlying mechanisms and causes by analyzing the changes in the income structures of different rural households adopting the PSM model. Investigating the impacts of farmers regarding their production factors is of great significance for the study of the joint development of the labor force and land factors.

Based on the above, the contribution of this study mainly is that we have estimated the impacts of rural households' with farmland transfer-in and farmland transfer-out on households' total income in the case of the elimination of bias coming from the change of households' income when they do not transfer farmland, and further explored the difference in changes in income structure of different types of rural households after they transferring-in or transferring-out farmland based on detailed micro-data, which fills the gap that the current research fails to reasonably and accurately guide different types of rural households to carry out farmland transfer [10,24–26], and also provides a reference for guiding the farmland transfer of different types of rural households reasonably in such areas for main grain production and food-commodity supply. The remainder of this paper is structured as follows. Section 2 describes the theoretical framework and research hypotheses about the changes in income structure between households with transferring-in farmland and transferring-out farmland. Section 3 displays the source of data, the descriptive analysis of variables as well as the introduction of the empirical model (Propensity Score Matching model). Empirical results of the model are presented in Section 4, and Section 5 presents the discussion for the results, whereas Section 6 concludes with a summary of our main findings and a discussion of policy implications.

## 2. Theoretical Analysis and Research Hypotheses

Due to differences in the original state of agricultural operations and the stability of non-agricultural income, different rural households will make different decisions to transfer farmland [11], and also will obtain different agricultural productivity and agricultural income after transferring-out or transferring-in farmland [22,26]. Firstly, pure-agricultural households generally have large-scale farmland and rich agricultural production experience, and invested enough time and energy in agricultural production, and may be more inclined to transfer farmland in and expand production scale to optimize their endowment to maximize profits [31]. Empirical studies show that the application of organic fertilizer in some households has increased significantly compared to households with small-scale farmland after they expanding the size of their farmland by farmland transfer, which not only decreased the production cost, but also increased the agricultural production [32]. Meanwhile, small-scale households are reluctant to apply technology, and the expansion of farmland scale leads to more family resource inputs to agricultural production, increasing agricultural income [33]. While if they transfer their farmland out, more labor will be allocated to non-agricultural employment, leading to increase the non-agricultural income, but the surplus-labor might not be engaged in non-agricultural employment on time because of lacking non-agricultural employment experience, which is likely to lead to a decrease in total household income. Secondly, Ranis and Fei pointed out that the labor whose agricultural production efficiency is not zero but is lower than the non-agricultural wage will be absorbed by the modern industrial sector, and if the agricultural production efficiency is not improved accordingly, then agricultural production will be negatively affected [34]. Because I part-time labor conduct agricultural work seasonally and spend the rest of their time on non-agricultural work, which will further lower production efficiency, thus they would be likely to transfer their farmland out, engaging labor in the non-agricultural sector, which could subsequently increase outworking and rental-land incomes whilst decreasing the agricultural income. However, the less stability rural households have—for the non-employment—the less likely they will be to transfer their farmland out [35,36]. Thus, they may be worried that unstable non-agricultural income might enough to support them to completely separate from the farmland and live in the urban [16,37,38]. Meanwhile, also because of this, the non-agricultural income they may get is not very high even if they transferred farmland out. Because a rational household will naturally adjust farmland resources based on the principle of households' utility maximization [39,40]. As a result, they would not transfer farmland out or even transfer it in, and their non-agricultural income can be re-invested in agricultural production and used to expand their farmland scale. Theoretically, with a farmland size increase, they would increase agricultural inputs—such as fertilizers, pesticides, or agricultural machinery—which could significantly contribute to their agricultural and total incomes, thereby promoting farmland productivity [26,41–44]. Lastly, because the non-agricultural income is the main source of the total households' income, II part-time rural households are less dependent on farmland than I part-time households due to the fact of a more stable non-agricultural employment. They do not expect to increase income through agricultural production to a large extent [45,46]; thus, they are more likely to transfer their farmland out, which would allow them to spend more time and energy on their non-agricultural employment, thereby making more outworking income as well as some rental-land income eventually increasing their total income. Moreover, many studies also have proved that less efficient farm households that are more successful in non-agricultural employment can gradually opt-out of agriculture by renting out their land, thus increasing off-farm income [2,47]. Nevertheless, there might also be another situation with relatively small probability, where II part-time rural households might re-invest outworking income into agricultural production and further transfer some farmland in, expanding the scale of the farmland and increasing the agricultural income and subsidies received, but this cannot make up the decrease in outworking income due to the reduction of time and energy spent on non-agricultural employment, eventually decreasing the total income.

Based on the above, the farmland transfer-in and transfer-out have different impacts on three types of rural households. Therefore, the present study aimed to test the following hypotheses:

**Hypotheses (H1).** *The total income of all rural households who transferred-in their farmland will increase because the increase in agricultural income and subsidies will overcome the decrease in outworking income. Conversely, the total income of all rural households who transferred-out their farmland will decrease because the increase in the outworking and rental-land incomes will be lower than that in the agricultural income and subsidies.*

**Hypotheses (H2).** *Pure-agricultural households transferring-in their farmland will increase their total income by expanding the farmland scale and obtaining more agricultural income and subsidies, while non-agricultural income will decrease. However, the total income of pure-agricultural households transferring-out their farmland will decrease because of the reduction of the agricultural income and subsidies, while rental-land will increase, and non-agricultural income has hardly changed.*

**Hypotheses (H3).** *Both farmland transfer-in or transfer-out by I part-time rural households will increase the total income. If participating in farmland transfer-in, I part-time rural households will increase their agricultural income and subsidies whilst decreasing the non-agricultural income. Conversely, if participating in farmland transfer-out, their agricultural income and subsidies will decrease while the outworking and rental-land incomes will increase.*

**Hypotheses (H4).** *II part-time rural households participating in farmland transfer-out will increase the total income because of the increase of their outworking and rental-land incomes. Conversely, II part-time rural households transferring-in their farmland will increase their agricultural income and agricultural subsidies, yet not making up for the decrease in the outworking income and the eventual decrease in the total income.*

## 3. Materials and Methods

### 3.1. Study Area

The Heilongjiang province—located in the north-eastern part of China (Figure 1)—has a cultivated land area of 1.59 million km$^2$, that is, 11.7% of cultivated land in the whole country; it is regarded as a particularly important place in China for grain production and food-commodity supply, playing a vital role in safeguarding national food security. In this context, the Overall Program for the Comprehensive Reform of Modern Agriculture in the "two Great Plains" of the Heilongjiang province promulgated in 2013, aiming to improve agricultural production, guard food security, and increase farmers' income, addressed that farmland transfer is an important way for improving the income of rural households. Therefore, the exploration of the relationship between farmland transfer and households' income—and, subsequently, the rational guidance of the farmland transfer—is an essential step for national food security and social stability of China.

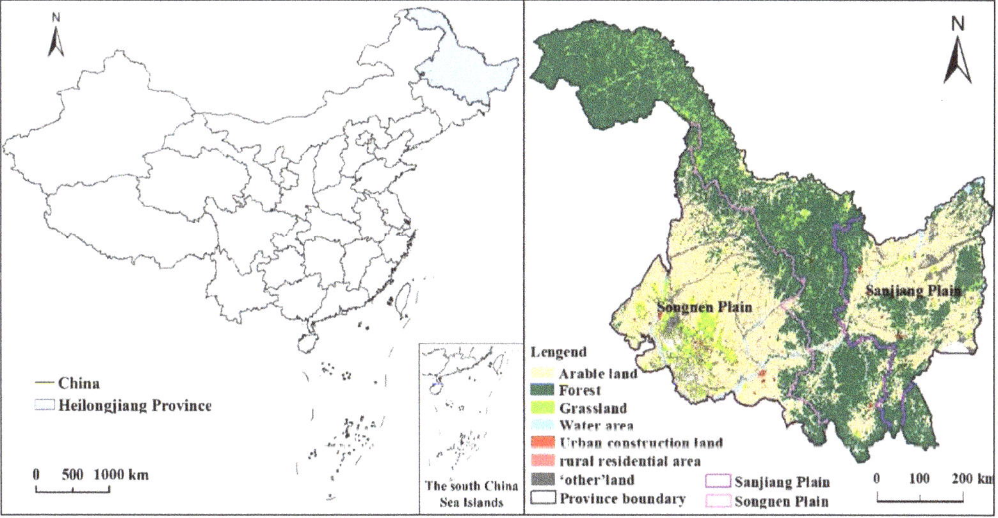


**Figure 1.** The geographical location of the Heilongjiang province and its land use in 2014.

### 3.2. Data Sources and Descriptive Statistics

The study data were derived from the Rural Fixed Observation Point Survey of the Heilongjiang province in 2014, which covered more than 1000 households distributed in 14 villages located in the study area. The Rural Fixed Observation Point Survey was established in 1984, guided by the policy research office of the central committee of the communist party of China and the ministry of agriculture and rural affairs of the People's Republic of China, which was established by the state to conduct a long-term monitoring survey on rural households and rural development issues. All survey samples include more than 360 villages and more than 24,000 rural households, covering 346 counties (cities, districts) in 31 provinces (regions, cities) across the whole country. The annual regular survey indicators include nearly 2000 items, involving many aspects of rural economy and society, and it is very representative to reflect the micro problems of households.

According to the criteria of rural-household differentiation in the Rural Fixed Observation Point Office of the Ministry of Agriculture, that is, rural households whose agricultural income accounts for more than 80% are classified as pure-agricultural households, and those with agricultural income accounting for 50% to 80% of total income are classified as I part-time rural households, and those with agricultural income accounting for less than 50% are classified II part-time rural households. Meanwhile, rural households participating in farmland transfer are divided into farmland transfer-in households (renting the farmland to other households or economic organizations) and transfer-out households (taking over this leased farmland from other households). Meanwhile, according to the income structure, the households' total income was divided into four types: agricultural income, outworking income (from non-agricultural employment), subsidy income (from government incentives and subsidies for agricultural production), and rental-land income (from farmland lease, farmland shareholding, and farmland exchange). We tried to compare changes in the total income between rural households who transferred their farmland either in or out, as well as to investigate the mechanisms underlying the impacts of farmland transfer on the income structure of different households based on the PSM model, and the detailed research framework is shown in Figure 2.

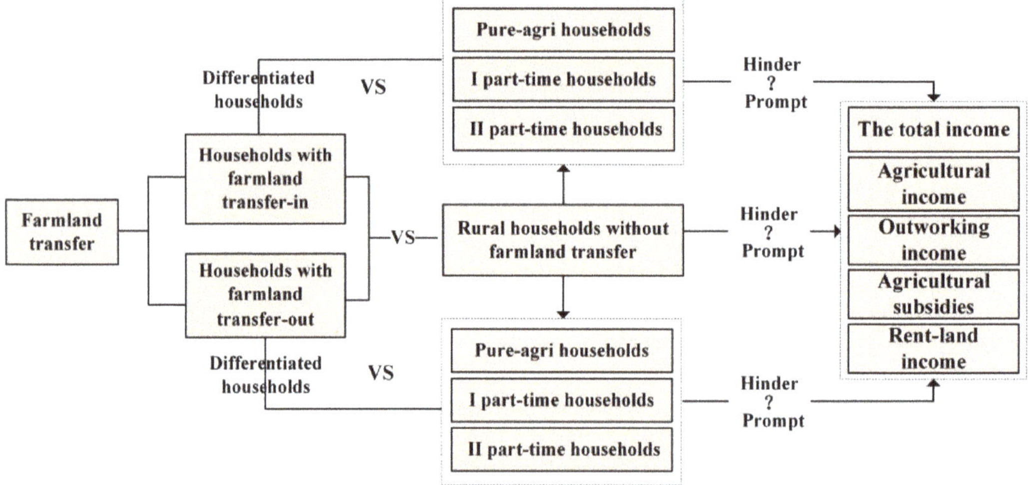


**Figure 2.** Structure of the analysis of the impacts of farmland transfer on rural households' income.

By eliminating the abnormal data of indicators needed in this study, we finally selected 484 households who did not transfer their farmland, and 304 who transferred their farmland; of these last, 178 and 126 were transferred-in and transferred-out, respectively. In terms of household type, our sample included 259 pure-agricultural, 119 I part-time, and 106 II part-time rural households.

Table 1 summarizes the average annual income structures. Rural households who transferred-in their farmland had the highest total income (111,394¥), followed by rural households who did not transfer their farmland (62,109¥) and rural households who transferred-out their farmland (39,522¥). Comparing the total income of different types of rural households not participating in farmland transfer, the pure-agricultural households had the highest income (69,040¥), followed by I part-time (57,149¥) and II part-time households (51,441¥).

**Table 1.** Descriptive statistical characteristics of variables.

| Variables | | Variable Definition | No-Transfer Households | | | | Transfer-in Households | Transfer-out Households |
|---|---|---|---|---|---|---|---|---|
| | | | Total Samples | Pure-Agricultural Households | I Part-Time Households | II Part-Time Households | | |
| Outcomes | TI | total incomes (¥) | 62,109 | 69,040 | 51,441 | 57,149 | 111,394 | 39,522 |
| | AI | agricultural incomes (¥) | 46,927 | 63,971 | 34,465 | 19,271 | 94,180 | 0 |
| | NFI | outworking incomes (¥) | 11,273 | 1164 | 13,564 | 33,403 | 8455 | 26,286 |
| | SI | subsidy incomes (¥) | 2212 | 2653 | 2101 | 1256 | 3991 | 2347 |
| | RI | rental-land incomes (¥) | 0 | 0 | 0 | 0 | 0 | 10,177 |
| Covariates | CL | contracted area (m$^2$) | 23,212 | 29,755 | 19,148 | 11,834 | 18,841 | 18,821 |
| | LA | laborers (n) | 2.4 | 2.3 | 2.3 | 2.6 | 2.6 | 1.7 |
| | APFA | agricultural productive fixed assets (¥) | 17,087 | 24,994 | 10,401 | 5274 | 21,241 | 2468 |
| | NAPFA | non-agricultural productive fixed assets (¥) | 6646 | 8903 | 4055 | 4037 | 26,081 | 1746 |
| | ELHH | education level of head of household (years) | 7 | 7.1 | 6.7 | 7 | 7.1 | 7.2 |

Note: the table refers to the average values of each type of rural households in the study area during 2014.

Comparing the income structure in different types of rural households, four noteworthy features have been identified:

(1) Rural households who transferred-in their farmland had the highest annual average agricultural income (94,180¥) followed by rural households who did not transfer their farmland (46,927¥) and rural households who transferred-out their farmland (0¥, since they did not conduct any agricultural production). Moreover, pure-agricultural households had the highest income (63,971¥), followed by I part-time (34,465¥) and II part-time households (19,271¥).

(2) Rural households who transferred-out their farmland had the highest annual average outworking income (26,286¥), followed by households who did not transfer their farmland (11,273¥) and households who transferred-in their farmland (8455¥). Among the different types of rural households who did not transfer their farmland, II part-time households had the highest outworking income (33,403¥), followed by I part-time (13,564¥) and pure-agricultural households (1164¥). Moreover, the outworking income of rural households who transferred-out their farmland is 7117¥ lower than II part-time households' income; this is the primary reason why the total income of rural households who transferred-out their farmland is lower than rural households who did not transfer it.

(3) Rural households who transferred-in their farmland had the highest subsidy income (3991¥). The difference in the annual average subsidy between rural households who transferred-out their farmland (2347¥) and those who did not transfer it (2212¥) was negligible (135¥). In addition, pure-agricultural households had the highest annual average subsidy (2653¥) among different rural households who did not transfer their farmland, followed by II part-time (2101¥) and I part-time households (1256¥).

(4) Only the rural households transferring-out farmland had a rental-land income (10,177¥).

Further, it is worth noting that farmland transfer is a "self-selection"; thus, the statistical differences of all the indicators shown in Table 1 might not be the results of farmland transfer only, but they might have been influenced by other factors. Therefore, we need to objectively consider the results in Table 1 and to perform a causal analysis to test the impacts of farmland transfer on households' income structures.

*3.3. Propensity Score Matching Model*

The households' decision to participate in farmland transfer is not completely random, yet closely related to the different characteristics of the households themselves (a virtual endogenous variable), which would imply selection bias [18,48,49]. Nevertheless, the PSM model based on a counterfactual analysis framework can deal with such bias [50–52].

The counterfactual analysis framework concerning the impacts of farmland transfer on households' income structures is a comparison between the factual and counterfactual results [53].

First, the factual result is based on the actually observed income of farmland-transferring households in order to extrapolate a first expected income, whereas the counterfactual result is based on the hypothetical income of farmland-transferring households if they would not participate in farmland transfer to calculate a second expected income; then, the impacts of farmland transfer on rural households' income can be obtained by comparing the two expected incomes. Second, the central "matching" idea of the PSM model is to use a control group to emulate a randomized experiment. The matching involves treatment units and comparison units, turning them into observable characteristics except for the selection of farmland transfer. Thus, the samples of rural households without farmland transfer behaviors are taken as the control groups and used to simulate the "counterfactual situation" of rural households participating in farmland transfer.

The specific steps of the PSM model based on the counterfactual analysis framework are the following:

(1) Covariates selection: Relevant variables likely affecting the households' decision-making process of farmland transfer were included to ensure the validity of the conditional independence assumption). Related factors were selected as covariates, namely rural households who contracted farmland areas (CL), household laborers (LA), agricultural productive fixed assets (APFA), non-agricultural productive fixed assets (NAPFA), and the education level of the rural householder (ELHH).

(2) Propensity scores estimation: The Logit model was used to estimate the possibility of transferring farmland of rural households (i.e., estimation of the propensity scores) [52,54].

(3) PSM model implementation: Households who transferred-in or transferred-out their farmland were matched to those who did not participate in farmland transfer, subsequently constructing matching groups. To ensure the robustness of the matching results, we chose two kinds of matching algorithms, namely, the radius matching method and kernel matching method.

(4) Matching quality assessment: First, because we did not condition on all covariates but the estimated propensity score, we checked whether the matching procedure could balance the distribution of the relevant variables in both the control and treatment groups [51]. Second, the common support condition, as conventionally measured, is a major source of evaluation bias [48,55]. There are two methods of estimating balancing property: the first method is to compare the situation before and after matching. If there is no systematic difference after conditioning on the propensity score, and a likelihood ratio test on the joint significance of all regressors can be performed in the Logit model and should be rejected before but after matching, meanwhile, the Pseudo-$R^2$ may lower, which indicates matching on the score is successful; the other method consists of requiring that the standardized deviation of samples after matching cannot be greater than 20; otherwise it would imply the failure of the matching process [56].

(5) Calculation of the average treatment effect on the treated (ATT) [57]: After having identified the matching samples using the radius and kernel matching methods, the *ATT* of rural households with transferred farmland can be calculated to determine the impacts of farmland transfer on their income; thus, according to the matched samples, the counterfactual result is obtained for each rural household with transferred farmland under the assumption of not transferring it. This result is then compared with the factual result calculated by the actual observable income.

## 4. Results

*4.1. Analysis of Covariates Affecting the Households' Decision-Making for Farmland Transfer*

In this study, the Logit model was used to construct the decision-making equation of farmland transfer to compare the differences in terms of income structure among different rural households and then to calculate the propensity score. The treatment groups of the two types of equations were households with farmland transfer-in and transfer-out, whereas the control groups were households without farmland transfer. Then, each type of equation was further divided into four equations with different making-decisions processes, according to the three types of different households without farmland transfer in the control group. The control group of equation 1 comprises households without farmland transfer, while the control groups of equations 2, 3, and 4 comprise pure-agricultural, I part-time, and II part-time households, respectively.

The estimated results of decision-making equations on households' farmland transfer using the Logit model are shown in Table 2. Firstly, the impact of covariates on the decision-making of farmland transfer showed that CL was negatively associated with the decision-making process of farmland transfer-in, yet positively associated with that of farmland transfer-out. Further, the relationship between CL and the decision-making process on farmland transfer was significantly different for the three types of different rural households. For I part-time and II part-time households, the more CL they had, the more likely they were to participate in farmland transfer; furthermore, with the same CL, II part-time households were more likely to participate in farmland transfer than I part-time households, which means to some extent that the higher the proportion of non-agricultural income, the more likely rural households were to participate in farmland transfer. Secondly, the number of LA was positively associated with the decision-making process of farmland transfer-in yet negatively associated with the decision-making process of farmland transfer-out. Comparing the three types of different rural households, the more LA they had, the more likely pure-agricultural households had, the more likely they will transfer farmland in, and less likely they will transfer farmland out; and for the other two types of rural households, the higher non-agricultural incomes and the more LA rural households had, the less likely they were to participate in farmland transfer-in or transfer-out. Thirdly, APFA had no significant relationship with the decision-making process of farmland transfer-in but it was negatively associated with the decision-making process on farmland transfer-out. While the more APFA II part-time households had, the more likely they will transfer farmland in, and less likely will transfer their farmland out. Fourthly, NAPFA either had no significant association with the decision-making process on farmland transfer, while among three types of rural households, those with the higher non-agricultural income and the more NAPFA were more likely to transfer their farmland in, meaning that II part-time households were more willing to participate in farmland transfer than I part-time households, and the latter in turn were more willing to participate in farmland transfer than pure-agricultural households. Fifthly, ELHH had no significant relationship with the decision-making on farmland transfer.

**Table 2.** Estimation results of decision-marking of equations of farmland transfer.

| Variables | Farmland Transfer-in Equation | | | | Farmland Transfer-out Equation | | | |
|---|---|---|---|---|---|---|---|---|
| | Equation 1 | Equation 2 | Equation 3 | Equation 4 | Equation 1 | Equation 2 | Equation 3 | Equation 4 |
| CL | −0.0104 ** | −0.0261 *** | 0.0029 | 0.0918 *** | 0.0217 *** | −0.0004 | 0.0427 *** | 0.138 *** |
| LA | 0.201 ** | 0.269 ** | 0.281 * | −0.305 * | −0.836 *** | −0.694 *** | −0.728 *** | −1.2 *** |
| APFA | 0 | 0 | 0 | 0.0001 *** | −0.0001 *** | −0.0002 *** | −0.0002 *** | −0.0001 ** |
| NAPFA | 0 | 0 | 0.0001 *** | 0.0001 *** | 0 | 0 | 0 | −0.0001 ** |
| ELHH | 0.0041 | 0.0283 | 0.0164 | −0.0297 | 0.0716 | 0.0972 | 0.104 | −0.0133 |
| Intercept | −1.665 | −0.617 | −1.339 ** | −1.749 ** | 0.177 | 1.365 ** | 0.399 | 0.456 |
| LR chi2(5) | 89.12 *** | 79.93 *** | 76 *** | 133.68 *** | 145.34 *** | 173.68 *** | 69.91 *** | 112.83 *** |
| Pseudo $R^2$ | 0.1156 | 0.1353 | 0.19 | 0.3562 | 0.2339 | 0.3568 | 0.206 | 0.3527 |

Note: ***, **, and * represent statistical significance at 1%, 5%, and 10%, respectively. CL, contracted farmland areas; LA, household laborers; APFA, agricultural productive fixed assets; NAPFA, non-agricultural productive fixed assets; ELHH, education level.

### 4.2. Matching Effect Estimation

Finding out whether the matched results could be used as counterfactual results requires matching effect estimation. This section mainly estimated and tested the common support of matched samples and balancing properties of matched results by adopting standardized bias, pseudo-$R^2$, and a likelihood ratio test on the joint significance.

As shown in Tables 3 and 4, the results of testing the region of common support of the matched samples indicate that most of the matched samples were in the region of the common support by adopting either radius or kernel matching, as well as that the proportion of lost individuals was small, meaning that the matching quality posed a few problems and could be better guaranteed.

**Table 3.** Results of common support and data balancing estimation for farmland transfer-in equations.

| Decision-Making Equations | Matching Approaches | Equation of Farmland Transfer-in | | | |
| --- | --- | --- | --- | --- | --- |
| | | Common Support | Pseudo $R^2$ | LR chi2 ($p$) | Standardized Bias |
| Equation 1 | Pre-matching | 662 | 0.11 | 84.63 (0) | 27.6 |
| | Radius matching | 656 | 0.007 | 3.16 (0.675) | 8.6 |
| | Kernel matching | | 0.007 | 3.57 (0.613) | 8.9 |
| Equation 2 | Pre-matching | 437 | 0.13 | 76.88 (0) | 34.1 |
| | Radius matching | 391 | 0.002 | 0.91 (0.969) | 3.8 |
| | Kernel matching | | 0.002 | 0.95 (0.966) | 3.5 |
| Equation 3 | Pre-matching | 297 | 0.185 | 74.07 (0) | 36.2 |
| | Radius matching | 272 | 0.018 | 7.84 (0.165) | 11.0 |
| | Kernel matching | | 0.021 | 8.80 (0.117) | 11.4 |
| Equation 4 | Pre-matching | 284 | 0.358 | 134.34 (0) | 43.7 |
| | Radius matching | 165 | 0.003 | 0.71 (0.982) | 3.4 |
| | Kernel matching | | 0.008 | 1.69 (0.890) | 6.1 |

**Table 4.** Results of common support and data balancing estimation for farmland transfer-out equations.

| Decision-Making Equations | Matching Approaches | Equation of Farmland Transfer-out | | | |
| --- | --- | --- | --- | --- | --- |
| | | Common Support | Pseudo $R^2$ | LR chi2 ($p$) | Standardized Bias |
| Equation 1 | Pre-matching | 610 | 0.230 | 84.63 (0) | 40.4 |
| | Radius matching | 516 | 0.002 | 3.16 (0.675) | 3.6 |
| | Kernel matching | | 0.003 | 3.57 (0.613) | 3.7 |
| Equation 2 | Pre-matching | 385 | 0.349 | 170.08 (0) | 52.4 |
| | Radius matching | 264 | 0.01 | 3.07 (0.689) | 7.4 |
| | Kernel matching | | 0.009 | 2.72 (0.743) | 6.0 |
| Equation 3 | Pre-matching | 245 | 0.204 | 69.18 (0) | 39.1 |
| | Radius matching | 222 | 0.004 | 1.32 (0.933) | 5.6 |
| | Kernel matching | | 0.005 | 1.55 (0.907) | 6.5 |
| Equation 4 | Pre-matching | 232 | 0.353 | 112.93 (0) | 49.4 |
| | Radius matching | 176 | 0.006 | 1.37 (0.928) | 5.4 |
| | Kernel matching | | 0.004 | 0.87 (0.973) | 4.2 |

Note: "Pre-matching" refers to the original samples without matching, and "Radius matching and Kernel matching" refers to the groups after matching.

Related research showed that the smaller the absolute value of standardized bias, the better the matching effect. The results of the testing balancing property of matched results showed that the maximum value of the average standardized bias is less than 11.4% and much less than the average standardized bias of the pre-matching group, which greatly reduces the total bias of the matched group. In addition, comparing the pseudo-$R^2$s before and after matching, the estimation results show that pseudo-$R^2$s after matching are fairly low, which indicates there is no obvious difference between treatment units and comparison units after matching, and the results of a likelihood ratio test on the joint significance (LR chi2 ($p$)) of matched groups show that it was not rejected before matching but be rejected after matching. All of these indicate the appropriateness of the matching effects, as well as that the matching results passed the balancing test.

*4.3. Analysis of the Impacts of Farmland Transfer on the Income Structure of Different Rural Households*

4.3.1. Results of Farmland Transfer by Differentiated Rural Households Based on the ATTs

We tested the *ATTs* for total, agricultural, and outworking incomes as well as for subsidies of rural households who participated in farmland transfer (Table 5). The obtained estimation results are the same after having matched with the radius and kernel matching methods, indicating that the estimation results are robust. Therefore, the average values in the following analysis had been to be adopted. Further, the standard errors of the *ATT* results were calculated using a 200-replication bootstrap method.

Based on Table 5, combining the income structure of rural households before farmland transfer, the changes in the income structure of different rural households after farmland transfer-in or transfer-out are estimated and some important indicators also are calculated, such as the changes in the ratios between either agricultural income or non-agricultural income and the total income (Table 6).

4.3.2. Analysis of Impacts of Farmland Transfer of Pure-Agricultural Households on the Income Structure

(1)     Impacts of farmland transfer on the income structure of all rural households

As shown in Tables 5 and 6, compared with all rural households without farmland transfer, the total income of all rural households after participating in farmland transfer-in increased, and the ratio increased as much as 61.39% with an average increase of 38,129¥. About the income structure, the average agricultural and outworking incomes significantly increased by 39,176¥ and decreased by 3736¥, respectively, indicating that the ratio of the agricultural in total incomes increased by 10.34% while the share of non-agricultural income decreased 10.63%. In addition, agricultural subsidies also significantly increased by 1709¥ on average. However, the coefficients of the total income of all rural households who transferred-out their farmland were negatively significant at $p < 0.01$, indicating that the ratio of total income significantly decreased, specifically by 16.25% with an average increase of 10,094¥. Meanwhile, the agricultural income significantly decreased by 33,366¥, highlighting that the ratio of the agricultural in total incomes decreased by 49.48%. By contrast, the average outworking income increased 14,104¥, and its share increased by 30.64%. In addition, the rent-land income also significantly increased, with an average increase of 9843¥. However, a change in agricultural subsidies was not observed. Therefore, H1 is supported by empirical evidence.

(2)     Impacts of farmland transfer on the income structure of pure-agricultural households

Compared with pure-agricultural households without farmland transfer, the ratio of the total income of pure-agricultural households after farmland transfer-in significantly increased by 31.32%, with an average increase of 21,626¥. Further, agricultural income was significant at $p < 0.05$ or $p < 0.1$. The agricultural income increased by 12,341¥, but the average ratio of the agricultural and total incomes decreased by 8.49%. Conversely, the outworking income decreased by 6843¥ and its share in the total income decreased by 7.95%. In addition, the agricultural subsidies also have significantly increased by 1423¥ on average. However, the ratio of total income and the ratio of agricultural income in total income after pure-agricultural households transferring-out their farmland has significantly decreased by 27.96% and 77.50%, with an average decrease of 19,305¥ and 56,433¥, respectively. By contrast, the average share of the outworking income in households' total income greatly increased by 53.61%, with an increase of 26,335¥. In addition, the rent-land income also significantly increased by 10,684¥. However, changes in agricultural subsidies were not observed. Therefore, empirical evidence partially confirmed the H2, because the outworking income increased significantly after their farmland transfer-out, hence the refusal of the assumption according to which the outworking income would have changed hardly. Since rural households in China often have small pieces of arable land and the actual production scale of these households is often lower than it could be [58], and some rural labor may be the surplus-labor, so they are optimally allocated when households transferring-in farmland.

(3)   Impacts of farmland transfer on the income structure of I part-time rural households

Compared with I part-time households without farmland transfer, the ratio of total income significantly increased by 56.45% after I part-time households transferring-in farmland, with an average increase of 29,038¥. Meanwhile, the coefficients of the agricultural and outworking incomes were significant at 1% level, and their shares separately increased by 22.81% and 21.59%, with an average increase of 37,811¥ and 9716¥. By contrast, the outworking income decreased by 9716¥, and its share in the total income decreased by 21.59%. In addition, agricultural subsidies did not show any significant changes. However, the coefficients of the total income after they transferring-out farmland was negative and significant at $p < 0.05$, showing its ratio decreased by 24.56%, with a decrease of 12,636¥. Meanwhile, the average agricultural, outworking and rent-land incomes were significant at $p < 0.01$. The average agricultural income decreased by 34,329¥ and its average ratio in the total income decreased greatly by 66.65%. Conversely, the average share of the outworking income and rent-land income increased by 12,823¥ and 9591¥, respectively, and the former's ratio in the total income increased by 41.63%. As a result, empirical evidence leads to refusing the H3. In fact, it is not that both farmland transfer-in and transfer-out could increase the total income. Not all I part-time labor's non-agricultural income can offset the loss caused by giving up agricultural production after farmland transfer, particularly for low-skilled and low-educated labor.

(4)   Impacts of farmland transfer on the income structures of II part-time households

Compared with II part-time households without farmland transfer, the ratio of the total income of II part-time households who transferring-in farmland increased by 27.80%, with an average increase of 15,889¥. Meanwhile, the average agricultural income increased by 39,654¥ and its average share in the total income increased by 46.96%, and agricultural subsidies also significantly increased by 920¥. By contrast, the outworking income decreased by 24,160¥, and its share in the total income decreased by 45.79%. While the total income and agricultural income after they transferring-out farmland separately decreased by 22,709¥ and 18,486¥, and their ratio decreased by 39.74% and 31.44%, but the rent-land income increased by 7082¥. However, inconsistently with a part of hypothesis H4, although the share of outworking income in the total income also increased by 18.68%, the average outworking income decreased by 6840¥. Therefore, empirical evidence leads to refusing the H4, because there is a clear division of labor between two generations in II part-time households, and most II part-time labor is generally undertaken by younger people who are not original participants in agricultural production, whereas the middle-aged and elderly are the mainstays of agricultural production with lower opportunity costs due to limited opportunities for outworking [59]. When they transferring-out their farmland, the middle-aged and elderly are most unlikely to be engaged in another non-agricultural employment and find a non-agricultural job. Instead, they are likely to live with young children, which will more or less affect the outworking (non-agricultural) income of the latter since they will have to take care of their parents. Consequently, the outworking income slightly decreased after II part-time households transferred their farmland out; meanwhile, due to the reduction in their agricultural activities, their agricultural incomes fell sharply.

**Table 5.** The comparison of average treatment effects on the treated (*ATTs*) for Propensity Score Matching (PSM) of rural households with transferring farmland.

| Equations of Decision-Marking | Matching Methods | Equations of Farmland Transfer-In | | | | Equations of Farmland Transfer-Out | | | | |
|---|---|---|---|---|---|---|---|---|---|---|
| | | Total Incomes | Agricultural Incomes | Outworking Incomes | Subsidy Incomes | Total Incomes | Agricultural Incomes | Outworking Incomes | Subsidy Incomes | Rental-Land Incomes |
| Equation 1 | Radius matching | 38,307 *** | 39,316 *** | −3652 ** | 1682 *** | −10,080 *** | −33,480 *** | 14,244 *** | 241 | 9843 *** |
| | Kernel matching | 37,951 *** | 39,035 *** | −3819 ** | 1735 *** | −10,108 *** | −33,251*** | 13,963 *** | 244 | 9843 *** |
| | The average value | 38,129 | 39,176 | −3736 | 1709 | −10,094 | −33,366 | 14,104 | 243 | 9843 |
| Equation 2 | Radius matching | 21,170 *** | 11,918 * | 6808 *** | 1417 *** | −18,146 ** | −55,268 *** | 26,344 *** | 161 | 10,684 *** |
| | Kernel matching | 22,082 *** | 12,764 ** | 6877 *** | 1428 *** | −20,464 *** | −57,598 *** | 26,325 *** | 255 | 10,684 *** |
| | The average value | 21,626 | 12,341 | 6843 | 1423 | −19,305 | −56,433 | 26,335 | 208 | 10,684 |
| Equation 3 | Radius matching | 29,381 *** | 38,098 *** | −9746 *** | 429 | −12,459 ** | −34,147 *** | 12,759 *** | 77 | 9591 *** |
| | Kernel matching | 28,695 *** | 37,524 *** | −9686 *** | 354 | −12,813 ** | −34,511 *** | 12,886 *** | 48 | 9591 *** |
| | The average value | 29,038 | 37,811 | −9716 | 392 | −12,636 | −34,329 | 12,823 | 63 | 9591 |
| Equation 4 | Radius matching | 16,251 ** | 39,771 *** | −23,899 *** | 916 *** | −22,637 *** | −18,495 *** | −6716 ** | 288 | 7082 *** |
| | Kernel matching | 15,526 ** | 39,537 *** | −24,420 *** | 924 *** | −22,780 *** | −18,477 *** | −6963 ** | 299 | 7082 *** |
| | The average value | 15,889 | 39,654 | −24,160 | 920 | −22,709 | −18,486 | −6840 | 294 | 7082 |

Note: ***, **, and * represent statistical significance at 1%, 5%, and 10%, respectively.

**Table 6.** Comparison of changes in the income structure of different rural households after transferring-in or transferring-out of farmland.

| The Types of Rural Households | The Status of Farmland Transfer | The Total Income (¥) | Changes in the Ratio of the Total Income (%) | Agricultural Income (¥) | Changes in the Ratio of Agri-incomes in the Total Income (%) | Outworking Income (¥) | Changes in the Ratio of Non-Agri Incomes in the Total Income (%) | Agricultural Subsidies (¥) | Rental-Land Income (¥) |
|---|---|---|---|---|---|---|---|---|---|
| All rural households | Without transfer | 62,109 | / | 46,927 | / | 11,273 | / | 2212 | 0 |
| | Transfer-in | 38,129 *** | 61.39 | 39,176 *** | 10.34 | −3736 ** | −10.63 | 1709 *** | 0 |
| | Transfer-out | −10,094 *** | −16.25 | −33,366 *** | −49.48 | 14,104 *** | 30.64 | / | 9843 *** |
| Pure-agri households | Without transfer | 69,040 | / | 63,971 | / | 1164 | / | 2653 | 0 |
| | Transfer-in | 21,626 *** | 31.32 | 12,341 ** | −8.49 | −6843 *** | −7.95 | 1423 *** | 0 |
| | Transfer-out | −19,305 *** | −27.96 | −56,433 *** | −77.50 | 26,335 *** | 53.61 | / | 10,684 *** |
| I part-time household | Without transfer | 51,441 | / | 34,465 | / | 13,564 | / | 2101 | 0 |
| | Transfer-in | 29,038 *** | 56.45 | 37,811 *** | 22.81 | −9716 *** | −21.59 | / | 0 |
| | Transfer-out | −12,636 ** | −24.56 | −34,329 *** | −66.65 | 12,823 *** | 41.63 | / | 9591 *** |
| II part-time household | Without transfer | 57,149 | / | 19,271 | / | 33,403 | / | 1256 | 0 |
| | Transfer-in | 15,889 ** | 27.80 | 39,654 *** | 46.96 | −24,160 *** | −45.79 | 920 *** | 0 |
| | Transfer-out | −22,709 *** | −39.74 | −18,486 *** | −31.44 | −6840 ** | 18.68 | / | 7082 *** |

Note: The rows corresponding to "Farmland transfer-in" and "Farmland transfer-out" are the average values of the significant changes after farmland transfer-in or transfer-out reported in Table 5. The row corresponding to "Without farmland transfer" refers to the number of different types of income. "/" represents no data or no significant change. *** and ** represent statistical significance at 1% and 5%, respectively.

## 5. Discussion

This paper answers the question of how farmland transfer of different rural households impacts the income structure, filling the gap that the current research does not analyze the changes in income structure of different types of rural households after they transferring-in or out of farmland [10,26], and also providing a scientific basis for guiding different types of rural households to carry out farmland transfer accurately for maximizing benefits of the whole society that not be documented by current research [24–26].

(1)    The total income significantly increased after rural households transferring-in farmland, consistently with the results of the most recent studies [10,26,28,60]. The expansion of farmland scale and the significant increase in agricultural incomes are the primary reasons. However, the increased ratio in the agricultural income of different rural households showed some differences. China is a mountainous country, with 70% of its land area being hilly. However, unlike many other districts, Heilongjiang province is a typical plain area, its farmland area is vast and the connectivity among arable land plots is high, and the level of modern mechanized agriculture is also relatively high, an appropriate management scale should be larger than that in other regions in China [61]. Moreover, a large amount of empirical experience proves that an appropriate management scale could effectively promote grain production [62,63]. Because of this, regardless of the type of households in the study area, they can increase their agricultural income after transferring-in that proved in our study. Therefore, regardless of the type of rural household in the study area, transferring-in farmland will contribute to achieving an appropriate-scale operation and forming a scale economy, saving production cost, and improving agricultural productivity, which could greatly increase the agricultural income, particularly II part-time. Because there are still stable farmers in I part-time households engaged in agricultural production, and they still can basically maintain their agricultural operations even if they transfer in arable land due to limited availability of arable land area and, where there are large family sizes, the actual scale of production may be smaller than the scale of management appropriate to the situation, especially in Heilongjiang province [61,64,65], while II part-time needs to transfer more labor working in the non-agricultural sector to be more engaged in agricultural production when they transferring-in farmland; thus, the outworking labor in II part-time households is the most affected by farmland transfer-in, as well as the one which fell the most, followed by I part-time households, and pure-agricultural households. Therefore, the increase in the total income is, from high to low: I part-time households, pure-agricultural households, and II part-time households, and it highlights that pure-agricultural and I part-time households transferring-in farmland are more suitable than II part-time households, which further could contribute to achieving the appropriate-scale management of Heilongjiang province.

(2)    Regardless of different rural households, the total income decreased after they transferring-out farmland, and the drastic decrease in the agricultural income was greater than the increase in the outworking income. The decreased ratio in the total income is, from high to low: II part-time households, pure-agricultural households, and I part-time households, the important reason is the decrease in outworking income of II part-time households while it increased in other rural households, although the agricultural income of II part-time households had the smallest reduction. As analyzed in Section 4.3.2, the key is to handle the issue of support for the elderly who have quit farming, so that II part-time households can better perform their non-agricultural work. Meanwhile, it further indicates that the urbanization peace and economic level of the study area may not be consistent with the speed of the migration of rural laborers because even the increase in non-agricultural income cannot offset the decrease in agricultural income. Related studies showed that the Heilongjiang province has the highest rate of agricultural surplus-labor among the three provinces of north-eastern China. There were 4.75 million laborer surpluses by the end of 2012,

which is 60.3% of the total agricultural laborers in Heilongjiang province [66]. Therefore, promoting the non-employment of surplus-labor may greatly contribute to the increase of rural households' total and outworking income when they transferring-out farmland. However, the promotion of farmland transfer should be a gradual process. In fact, "pure-agricultural households → I part-time households → II part-time households" reflects the process of the rural households gradually moving towards non-agriculturalization, and the strengthening of this process is inevitable in the future [67], which also indicates there are different stages of farmland transfer correspondingly, and farmland transfer needs to be further promoted. Studies show that there is a mutual feedback mechanism for land use transition and the formulation of land management policies and institutions [63], it is, therefore, important to form and adjust related policies is based on the situation of farmland transfer and non-agriculture of households in different periods, preventing farmers' life or ecological environment problems caused by excessively promoting farmland transfer.

Interestingly, agricultural subsidies of different rural households when transferring-out farmland showed no significant changes, while it increased when households transferring-in farmland. Firstly, because the samples of transferring-in farmland and transferring-out farmland in this paper are not in one-to-one correspondence, this may also a shortcoming of this research, and how to select the samples that matching the rural households who transferring-in farmland and transferring-out needs to be further explored. Secondly, this may be related to the differences in the subjects of agricultural subsidies in different regions. Surveys show that the real targets of 69.2% of agriculture subsidies are rural households with land-contract right, are not the actual cultivator [68], although most of the policy documents stipulate that the subsidy is based on the actual planting area of grain, in practice, even if rural households with land-contract right transferred their farmland out, they still can obtain the same agricultural subsidies as before the transfer [69,70]. This could have a negative impact on rural-urban migration of surplus-labor, and may also reduce farmers' enthusiasm for agricultural production, which is not conducive for farmers to transfer to land and form large-scale operations [26,69]. Thus, the direct subsidies for growing grain such as the Generalized System of Preference (GSP) [71] should be issued, which will help the increase of the agricultural subsidies-related income, as well as that of the total income.

## 6. Conclusions and Policy Implications

The main contribution of this study lies in the inclusion of different rural households and income structures into the same framework, allowing for a thorough sectional exploration of the impacts of farmland transfer. Further, this study has introduced the PSM method based on the counterfactual analysis framework, solving the "self-selection" issues related to rural households' farmland transfer behaviors and the subsequent potential endogenous problems and selectivity bias, and the study provides a scientific basis to reasonably plan farmland-transfer guidance of different rural households. More specifically, the paths to increase rural households' income are summed up (Figure 3), and the major three conclusions and policy implications that have been drawn are as follows:

(1) Promoting farmland transfer-in to pure-agricultural and I part-time households and transfer-out to II part-time households. Despite all types of rural households after farmland transfer-out will experience a decrease in households' total income, some rural households must be prompted to transfer out of their farmland to consolidate small plots for large-scale farming, to form scale-management, to improve the overall income of farmland transfer, and to eventually boost economic growth in the entire rural area. Because the increase in the total income of II part-time households with farmland transfer-in was the smallest, while the outworking not increasing but decreasing is the main reason for the decrease in total income after II part-time households transferring-out farmland. Therefore, it is vital and reasonable to prompt II part-time households to transfer their farmland out, as well as to prompt pure-

agricultural and I part-time households to transfer their farmland in, and at the meantime to develop mutual assistance for the aged in rural areas for solving the problem of taking care of the elderly who quit agricultural production and promoting II part-time labor to be better engaged in non-agricultural work.

(2) Improving the farmland transfer market and promoting non-agricultural employment of surplus-labor need to be synchronized. No matter China or other countries, the great farmland transfer/rental market is an important condition for promoting transferring farmland and improve rural households' income [2,3,10], and a platform providing the farmland transfer information, price assessment and negotiation guidance should be set up to reduce the cost of farmland transfer and ensure farmland transfer-out or transfer-in smoothly and fairly. When farmland can be transferred smoothly, the surplus-labor will inevitably increase. Studies show that nonagricultural employment effectively promotes the development of the farmland transfer market [35], and in turn, the development of the farmland transfer market could promote nonagricultural employment of rural labor [72]. Only combining the transformation from agriculture to non-agriculture of rural labor with promoting farmland transfer market can lower farmland fragmentation, improve agricultural productivity, and achieve agricultural modernization. The one key to promoting non-agricultural employment of rural labor is the improvement of the capacity of attracting labor in rural areas. Because rural elites are crucial actors in the transformational development of relatively successful villages [73]. Firstly, the government could provide financial supports and policy services to encourage the establishment of the agricultural products processing industry. Secondly, the village collectives could implement and assist in the establishment of agricultural production services or products processing industries. Especially after this COVID-19, it is well-known that if villages and towns can provide enough non-agricultural employment opportunities for rural surplus-labor, and a part of outflowing rural labor could be engaged in non-agricultural industries located in nearby town or villages, which not only could reduce the spread of the epidemic, but also could make the impacts of the work of outflow rural labor (most of the part-time labor) and the economy of villages and towns less affected by the epidemic. Meanwhile, different villages can selectively develop related enterprises, such as leisure tourism, health care, shared farms, and rural e-commerce (Taobao villages) according to the villages or towns' geographical location, resource conditions, villagers' willingness, etc. Another key to promoting non-agricultural employment of rural labor is the improvement of the welfare of migrant workers working in urban areas so that they can gradually settle down in cities or towns. The special household registration system in China is regarded as the main factor affecting the non-agricultural transformation of rural labor, where rural labor engaged in non-agricultural work in urban areas (nongmingong) cannot enjoy the same welfare and benefits as urban residents, such as education and medical resources, pension, etc. [74,75]. Similar to China, some studies in other countries show that high-wage firms, which tie pension benefits to the earnings of the worker, avoid hiring low-wage workers, as they have to offer all full-time workers the same health benefits. As a result, health insurance is mostly offered to full-time high-wage workers rather than part-time low-wage workers [39,76]. Thus, related policies and measures should be formulated to lower the conditions and improve the welfare for rural labor working in urban areas (nongmingong) entering cities and gradually settle down.

(3) Improve the agricultural-subsidy system. The agricultural-subsidy is not only directly related to rural households' income, but also directly affect the enthusiasm of farmers in agricultural production, further affecting food security [69,70]. Thus, build a reasonable agricultural subsidy system is of vital importance. Firstly, agricultural subsidies should be provided to farmers with actually growing grain, ensuring benefits to farmers engaged in agricultural production. Therefore, follow-up mechanisms on the distribution of agricultural subsidies could be implemented. Secondly, a flex-

ible policy of the number of agricultural subsidies would be recommended; more specifically, since the number of agricultural subsidies depends on the farmland area cultivated by farmers, the total crops, and the market price of crops the agricultural subsidies per unit weight of the crops could be increased when the market price of crops drops and decreased when the latter rises. This could eventually incentivize subsidized agricultural management.

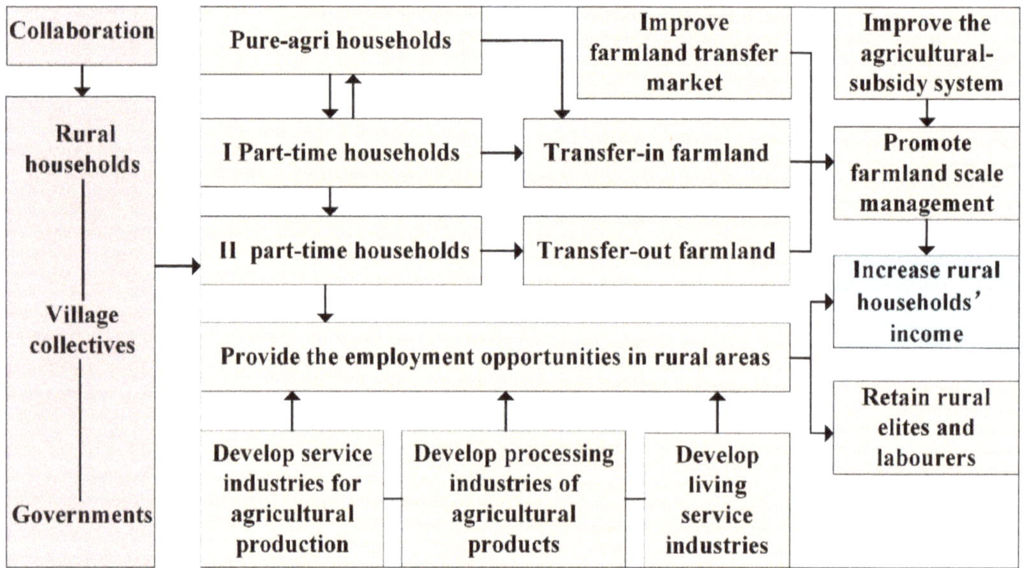


**Figure 3.** The paths to increase rural households' income.

Although our study showed that farmland transfer has different effects on different types of rural households, and drew the paths to increase households' income, the internal influence mechanism of farmland transfer on different types of rural households' income is not clear, for example, the increase in agricultural income of pure-agricultural households after they transferring-in farmland, is it because the expansion of the farmland scale improves the agricultural production efficiency or is it because the agricultural production cost is reduced? And have I part-time rural households after they transferring-in farmland promoted the application of agricultural mechanization, thereby increasing the agricultural production efficiency and improving the agricultural production efficiency? These question relate to how do different types of rural households participating in farmland land, affect agricultural production technology and production materials and further affect rural households' income need to be explored in future study.

**Author Contributions:** Conceptualization, L.C., C.Z., and H.C.; Data processing and Methodology, L.C. and C.Z.; Formal analysis, L.C., C.Z., and Y.L.; Funding acquisition, L.C. and H.C.; Project administration, H.C. and L.C.; Supervision, H.C.; Writing—original draft, L.C. and C.Z.; Writing—review and editing, L.C. and H.C. All authors have read and agreed to the published version of the manuscript.

**Funding:** This research was funded by the National Natural Science Foundation of China, grant number 51908114 and China Postdoctoral Science Foundation, grant number 2020M683149.

**Institutional Review Board Statement:** Not applicable.

**Informed Consent Statement:** Not applicable.

**Data Availability Statement:** Data used in this study were derived from the Rural Fixed Observation Point Survey in 2014 (http://www.moa.gov.cn/govpublic/zcggs/201911/t20191127_6332466.htm, accessed on 30 March 2021), which was guided by the policy research office of the central committee of the communist party of China and the ministry of agriculture and rural affairs of the People's Republic of China.

**Acknowledgments:** Thank you to Ge Song, at Northeast University for your advice throughout this project, and thank you to Hongsheng Zhao at the University of Cambridge for writing help throughout this project.

**Conflicts of Interest:** The authors declare no conflict of interest.

## References

1. Ranjan, P.; Wardropper, C.B.; Eanes, F.R.; Reddy, S.M.; Harden, S.C.; Masuda, Y.J.; Prokopy, L.S. Understanding barriers and opportunities for adoption of conservation practices on rented farmland in the US. *Land Use Policy* **2019**, *80*, 214–223. [CrossRef]
2. Jin, S.; Jayne, T.S. Land Rental Markets in Kenya: Implications for Efficiency, Equity, Household Income, and Poverty. *Land Econ.* **2013**, *89*, 246–271. [CrossRef]
3. Kijima, Y.; Tabetando, R. Efficiency and equity of rural land markets and the impact on income: Evidence in Kenya and Uganda from 2003 to 2015. *Land Use Policy* **2020**, *91*, 91. [CrossRef]
4. Do, Q.; Iyer, L. Land Titling and Rural Transition in Vietnam. *Econ. Dev. Cult. Chang.* **2008**, *56*, 531–579. [CrossRef]
5. Mwesigye, F.; Matsumoto, T.; Otsuka, K. Population pressure, rural-to-rural migration and evolution of land tenure institutions: The case of Uganda. *Land Use Policy* **2017**, *65*, 1–14. [CrossRef]
6. Gao, J.; Song, G.; Sun, X. Does labor migration affect rural land transfer? Evidence from China. *Land Use Policy* **2020**, *99*, 105096. [CrossRef]
7. Zhou, C.; Liang, Y.; Fuller, A. Tracing Agricultural Land Transfer in China: Some Legal and Policy Issues. *Land* **2021**, *10*, 58. [CrossRef]
8. Ye, J. Land Transfer and the Pursuit of Agricultural Modernization in China. *J. Agrar. Chang.* **2015**, *15*, 314–337. [CrossRef]
9. Ma, L.; Long, H.; Tu, S.; Zhang, Y.; Zheng, Y. Farmland transition in China and its policy implications. *Land Use Policy* **2020**, *92*, 104470. [CrossRef]
10. Mao, P.H.; Xu, J. Agricultural land system, the transfer of land management rights and the growth of farmers' income. *Manag. World* **2015**, 63–74+88. (In Chinese) [CrossRef]
11. Zhang, Y.; Halder, P.; Zhang, X.; Qu, M. Analyzing the deviation between farmers' Land transfer intention and behavior in China's impoverished mountainous Area: A Logistic-ISM model approach. *Land Use Policy* **2020**, *94*, 104534. [CrossRef]
12. Cao, Y.; Zou, J.; Fang, X.; Wang, J.; Li, G. Effect of land tenure fragmentation on the decision-making and scale of agricultural land transfer in China. *Land Use Policy* **2020**, *99*, 104996. [CrossRef]
13. Wang, Y.; Yang, Q.; Xin, L.; Zhang, J. Does the New Rural Pension System Promote Farmland Transfer in the Context of Aging in Rural China: Evidence from the CHARLS. *Int. J. Environ. Res. Public Health* **2019**, *16*, 3592. [CrossRef]
14. Yuan, X.; Du, W.; Wei, X.; Ying, Y.; Shao, Y.; Hou, R. Quantitative analysis of research on China's land transfer system. *Land Use Policy* **2018**, *74*, 301–308. [CrossRef]
15. Liu, Y.S.; Li, Y.H. Revitalize the world's countryside. *Nature* **2017**, *548*, 275–277. [CrossRef] [PubMed]
16. Xiao, W.; Zhao, G. Agricultural land and rural urban migration in China: A new pattern. *Land Use Policy* **2018**, *74*, 142–150. [CrossRef]
17. Long, H.; Tu, S.; Ge, D.; Li, T.; Liu, Y. The allocation and management of critical resources in rural China under restructuring: Problems and prospects. *J. Rural. Stud.* **2016**, *47*, 392–412. [CrossRef]
18. Zhu, J.J.; Hu, J.L. Analysis on the Impact of Farmland Transfer on Farmers' Income Distribution: Based on CHARL Data. *J. Nanjing Agric. Univ.* **2015**, *15*, 75–83.
19. Li, Q.H.; Li, R.; Wang, Z.H. The land rental market and its welfare effects. *China Econ. Q.* **2012**, *11*, 269–288.
20. Jin, S.; Deininger, K. Land rental markets in the process of rural structural transformation: Productivity and equity impacts from China. *J. Comp. Econ.* **2009**, *37*, 629–646. [CrossRef]
21. Guo, J.P.; Qu, S.; Xia, Y.; Lv, K.Y. Income distribution effect of rural land transfer. *China Popul. Resour. Environ.* **2018**, *28*, 160–169.
22. Xu, H.Z.; Shi, S.Q. Impact of Farmer Differentiation on Farm Households Willingness in Farmland Transference. *China Land Sci.* **2012**, *22*, 90–96.
23. Macdonald, J.M.; McBride, W.D.; O'Donoghue, E.; Nehring, R.F.F.; Sandretto, C.; Mosheim, R. Profits, Costs, and the Changing Structure of Dairy Farming. *SSRN Electron. J.* **2007**. [CrossRef]
24. Løyland, K.; Ringstad, V. Gains and structural effects of exploiting scale-economies in Norwegian dairy production. *Agric. Econ.* **2001**, *24*, 149–166. [CrossRef]
25. Liu, Y.; Yan, B.; Wang, Y.; Zhou, Y. Will land transfer always increase technical efficiency in China?—A land cost perspective. *Land Use Policy* **2019**, *82*, 414–421. [CrossRef]
26. Zhang, L.; Feng, S.; Heerink, N.; Qu, F.; Kuyvenhoven, A. How do land rental markets affect household income? Evidence from rural Jiangsu, P.R. China. *Land Use Policy* **2018**, *74*, 151–165. [CrossRef]

27. Xue, F.R.; Qiao, G.H.; Su, R.N. The effect evaluation of land transfer on farmers' incom—Based on DID model analysis. *China Rural Surv.* **2011**, 36–42. (In Chinese)

28. Yang, Z.; Ma, X.L.; Zhu, P.X.; Ma, D. Land transfer and income change of peasants. *China Popul. Resour. Environ.* **2017**, 27, 111–120.

29. Zhu, S.E.; Yang, R.D.; Wu, B. Rural household income flow in China: 1986~2017. *Manag. World* **2018**, 34, 63–72.

30. Yang, L.; Zhang, X.P. Age structure, population migration and economic growth in northeast China. *China Popul. Resour. Environ.* **2016**, 26, 28–35.

31. Xie, H.; Lu, H. Impact of land fragmentation and non-agricultural labor supply on circulation of agricultural land management rights. *Land Use Policy* **2017**, 68, 355–364. [CrossRef]

32. Xu, Z.G.; Zhang, J.Y.; Lv, K.Y. The scale of operation, term of land ownership and the adoption of inter-temporal agricultural technology: An example of "straw return to soil directly. *China Rural Econ.* **2018**, 61–74.

33. Arslan, A.; Belotti, F.; Lipper, L. Smallholder productivity and weather shocks: Adoption and impact of widely promoted agricultural practices in Tanzania. *Food Policy* **2017**, 69, 68–81. [CrossRef]

34. Ranis, G.; Fei, J.C. A theory of economic development. *Am. Econ. Rev.* **1961**, 51, 533–565.

35. Su, B.; Li, Y.; Li, L.; Wang, Y. How does nonfarm employment stability influence farmers' farmland transfer decisions? Implications for China's land use policy. *Land Use Policy* **2018**, 74, 66–72. [CrossRef]

36. Li, H.; Zhang, X.; Li, H. Has farmer welfare improved after rural residential land circulation? *J. Rural Stud.* **2019**. [CrossRef]

37. Xu, Q.; Liu, J.; Qian, Y.F. Labor mobility, farmland confirmation and farmland transfer. *J. Agrotech.* **2017**, 4–16. (In Chinese) [CrossRef]

38. Deininger, K.; Jin, S.; Xia, F.; Huang, J. Moving Off the Farm: Land Institutions to Facilitate Structural Transformation and Agricultural Productivity Growth in China. *World Dev.* **2014**, 59, 505–520. [CrossRef]

39. Bharadwaj, L.; Findeis, J.; Chintawar, S. US Farm households: Joint decision making and impact of health insurance on labor market outcomes. *Health Econ. Rev.* **2013**, 3, 16. [CrossRef]

40. Zang, L.; Araral, E.; Wang, Y. Effects of land fragmentation on the governance of the commons: Theory and evidence from 284 villages and 17 provinces in China. *Land Use Policy* **2019**, 82, 518–527. [CrossRef]

41. Gray, C.L.; Bilsborrow, R.E. Consequences of out-migration for land use in rural Ecuador. *Land Use Policy* **2014**, 36, 182–191. [CrossRef] [PubMed]

42. Liu, G.; Wang, H.; Cheng, Y.; Zheng, B.; Lu, Z. The impact of rural out-migration on arable land use intensity: Evidence from mountain areas in Guangdong, China. *Land Use Policy* **2016**, 59, 569–579. [CrossRef]

43. Caulfield, M.; Bouniol, J.; Fonte, S.J.; Kessler, A. How rural out-migrations drive changes to farm and land management: A case study from the rural Andes. *Land Use Policy* **2019**, 81, 594–603. [CrossRef]

44. Maharjan, A.; Kochhar, I.; Chitale, V.S.; Hussain, A.; Gioli, G. Understanding rural outmigration and agricultural land use change in the Gandaki Basin, Nepal. *Appl. Geogr.* **2020**, 124, 102278. [CrossRef]

45. Hennessy, T.; Brien, M.O. Is off-farm income driving on-farm investment? *Work. Pap.* **2007**, 13, 235–246.

46. Damon, A.L. Agricultural Land Use and Asset Accumulation in Migrant Households: The Case of El Salvador. *J. Dev. Stud.* **2010**, 46, 162–189. [CrossRef]

47. Deininger, K.; Jin, S. The potential of land rental markets in the process of economic development: Evidence from China. *J. Dev. Econ.* **2005**, 78, 241–270. [CrossRef]

48. Locke, C.M.; Butsic, V.; Rissman, A.R. Zoning effects on housing change vary with income, based on a four-decade panel model after propensity score matching. *Land Use Policy* **2017**, 64, 353–362. [CrossRef]

49. Fei, R.; Lin, Z.; Chunga, J. How land transfer affects agricultural land use efficiency: Evidence from China's agricultural sector. *Land Use Policy* **2021**, 103, 105300. [CrossRef]

50. Heckman, J.J.; Vytlacil, E.J. Chapter 71 Econometric Evaluation of Social Programs, Part II: Using the Marginal Treatment Effect to Organize Alternative Econometric Estimators to Evaluate Social Programs, and to Forecast their Effects in New Environments. *Handb. Econom.* **2007**, 6, 4875–5143. [CrossRef]

51. Rosenbaum, P.R.; Rubin, D.B. Constructing a Control Group Using Multivariate Matched Sampling Methods That Incorporate the Propensity Score. *Am. Stat.* **1985**, 39, 33–38. [CrossRef]

52. Sileshi, M.; Kadigi, R.; Mutabazi, K.; Sieber, S. Impact of soil and water conservation practices on household vulnerability to food insecurity in eastern Ethiopia: Endogenous switching regression and propensity score matching approach. *Food Secur.* **2019**, 11, 797–815. [CrossRef]

53. Martey, E.; Kuwornu, J.K.; Adjebeng-Danquah, J. Estimating the effect of mineral fertilizer use on Land productivity and income: Evidence from Ghana. *Land Use Policy* **2019**, 85, 463–475. [CrossRef]

54. Alix-Garcia, J.; Kuemmerle, T.; Radeloff, V.C. Prices, Land Tenure Institutions, and Geography: A Matching Analysis of Farmland Abandonment in Post-Socialist Eastern Europe. *Land Econ.* **2012**, 88, 425–443. [CrossRef]

55. Heckman, J. Instrumental Variables: A Study of Implicit Behavioral Assumptions Used in Making Program Evaluations. *J. Hum. Resour.* **1997**, 32, 441. [CrossRef]

56. Elizabeth, A.S. Matching methods for causal inference: A review and a look forward. *Stat. Sci.* **2010**, 25, 1–21. [CrossRef]

57. Teka, A.M.; Lee, S.-K. The impact of agricultural package programs on farm productivity in Tigray-Ethiopia: Panel data estimation. *Cogent Econ. Finance* **2019**, 7, 7. [CrossRef]

58. Chen, F.; Zhai, W.J. Land transfer incentive and welfare effect research from perspective of farmers' behavior. *Econ. Res. J.* **2015**, *50*, 163–177.
59. Huang, W.B.; Chen, F.B. Did non-farm employment inhibit land transfer?—Based on familiarity with farm work. *J. Agrotech.* **2020**, 44–58. (In Chinese) [CrossRef]
60. Han, X.; Zhang, A.L.; Zhu, Q.X.; Wang, K. Influence of land circulation on farmers' income growth and household's optimal management scale: An empirical study of Hubei and Jiangxi's mountainous and hilly regions. *Res. Agric. Mod.* **2015**, *36*, 368–373.
61. Chen, L.; Song, G.; Meadows, M.E.; Zou, C. Spatio-temporal evolution of the early-warning status of cultivated land and its driving factors: A case study of Heilongjiang Province, China. *Land Use Policy* **2018**, *72*, 280–292. [CrossRef]
62. Lu, H.; Xie, H.; He, Y.; Wu, Z.; Zhang, X. Assessing the impacts of land fragmentation and plot size on yields and costs: A translog production model and cost function approach. *Agric. Syst.* **2018**, *161*, 81–88. [CrossRef]
63. Long, H.; Qu, Y. Land use transitions and land management: A mutual feedback perspective. *Land Use Policy* **2018**, *74*, 111–120. [CrossRef]
64. Cao, J.H.; Wang, H.Y.; Huang, X.M. Study on supply and demand willingness of rural Land transaction and evaluation of transaction efficiency. *China Land Sci.* **2007**, *21*, 54–60. (In Chinese)
65. Chen, H.; Wang, X. Exploring the relationship between rural village characteristics and Chinese return migrants' participation in farming: Path dependence in rural employment. *Cities* **2019**, *88*, 136–143. [CrossRef]
66. Gao, S.; Gao, Y.; Chen, L.X. Research on the route of the rural surplus labor transferring in Northeastern area in the "limited surplus" stage. *Popul. J.* **2017**, *39*, 103–112.
67. Zeng, J.X.; Gao, L.L.; Wang, B.; Long, W.J. Who are the Chinese professional farmers? a comparative analysis of the characteristics of the Chinese and foreign agricultural labourers. *Issues Agric. Econ.* **2020**, 130–142. (In Chinese) [CrossRef]
68. Zhou, M.; Hu, B.X.; Zhang, Y. The Separation of Three Rights, Contest for Agricultural Subsidies and Incentive of Agricultural Management. *J. Huazhong Univ. Sci. Technol.* **2019**, *33*, 61–68.
69. Gale, F.H. *Growth and Evolution in China's Agricultural Support Policies*; Economic Research Report 153; United States Department of Agriculture: Washington, DC, USA, 2013.
70. Huang, J.; Wang, X.; Rozelle, S. The subsidization of farming households in China's agriculture. *Food Policy* **2013**, *41*, 124–132. [CrossRef]
71. Lin, G.B.; Zheng, C. The performance evaluation of the agricultural subsidy policy and the design of policy improvement in China. *Rev. Econ. Manag.* **2019**, *35*, 102–111. (In Chinese) [CrossRef]
72. Chernina, E.; Dower, P.C.; Markevich, A. Property rights, land liquidity, and internal migration. *J. Dev. Econ.* **2014**, *110*, 191–215. [CrossRef]
73. Li, Y.; Fan, P.; Liu, Y. What makes better village development in traditional agricultural areas of China? Evidence from long-term observation of typical villages. *Habitat Int.* **2019**, *83*, 111–124. [CrossRef]
74. Zhou, W.; Zhao, F.; Yang, F.; Li, L. Land transfer, reform of household registration system and urbanization in China: Theoretical and simulation test. *Econ. Res. J.* **2017**, *52*, 183–197. (In Chinese)
75. Wen, G.J.; Xiong, J. The Hukou and Land Tenure Systems as Two Middle Income Traps—The Case of Modern China. *Front. Econ. China* **2014**, *9*, 438–459.
76. Yuan, F.; Shi, Q.H. Re-examining Inequality: Capability and Income Inequality and the Welfare of Rural Migrant Workers'. *Manag. World* **2013**, 49–61. (In Chinese) [CrossRef]

*Article*

# Livelihood Capital and Land Transfer of Different Types of Farmers: Evidence from Panel Data in Sichuan Province, China

Huanxin Yang [1], Kai Huang [1], Xin Deng [2] and Dingde Xu [3,*]

[1]  College of Management, Sichuan Agricultural University, Chengdu 611130, China; yanghuanxin@stu.sicau.edu.cn (H.Y.); hk666@stu.sicau.edu.cn (K.H.)

[2]  College of Economics, Sichuan Agricultural University, Chengdu 611130, China; dengxin@sicau.edu.cn

[3]  Sichuan Center for Rural Development Research, College of Management, Sichuan Agricultural University, Chengdu 611130, China

*  Correspondence: dingdexu@sicau.edu.cn; Tel.: +86-13-4085-98819

**Abstract:** Farmers' livelihood and land have been the focus of academic and political attention for a long time. In the process of rapid urbanization in China, as farmers change their livelihood strategies and livelihood capital allocation driven by economic interests, farmland abandonment increases, which is not conducive to the guarantee of food security. This study aims to explore the characteristics of livelihood capital and land transfer of farmers under different livelihood strategies and the effect of livelihood capital on land transfer. Based on the data obtained from Sichuan Province in 2012, 2016 and 2019 by the China Rural Development Survey Group, this paper divides farmers into pure farmers, part-time farmers and non-farmers according to the proportion of non-agricultural income in total income, and constructed the panel binary Logit model and panel Tobit model. The analysis points to the following results: (1) pure farmers tend to shift other capitals toward natural capital, so their livelihood capital total index value decreased. The part-time farmers have different shift characteristics but their livelihood capital total index value both increased first and then decreased. Non-farmers tend to shift natural capital towards other livelihood capitals, so their livelihood capital total index value increased. (2) The higher the natural capital and human capital, the higher the probability of land transfers in. The higher the natural capital, the larger the area of land transfers in. The higher the financial capital, the higher the probability of land transfers out. The higher the financial capital and social capital, the larger the area of land transfers out. It is expected to provide suggestions for the policy of farmers' land transfer under different livelihood capital endowments.

**Keywords:** livelihood capital; livelihood strategy; land transfer; transfer scale; Sichuan Province



**Citation:** Yang, H.; Huang, K.; Deng, X.; Xu, D. Livelihood Capital and Land Transfer of Different Types of Farmers: Evidence from Panel Data in Sichuan Province, China. *Land* **2021**, *10*, 532. https://doi.org/10.3390/land10050532

Academic Editors: Hualou Long, Xiangbin Kong, Shougeng Hu and Yurui Li

Received: 14 April 2021
Accepted: 16 May 2021
Published: 17 May 2021

**Publisher's Note:** MDPI stays neutral with regard to jurisdictional claims in published maps and institutional affiliations.



## 1. Introduction

China, as a populous country [1,2], has faced a prominent dichotomy in its man–land relationship for a long time [3,4]. Rapid urbanization and agricultural modernization promote the rural labor force to go out for work in large numbers driven by economic interests [5,6], which changes the livelihood capital structure and livelihood strategy of farmers, and also changes the land structure [7–9]. According to existing studies, due to the large number of out-migrating labor forces and the low income of agricultural workers, many villages in hilly areas of China have idle and abandoned land [2,10]. Land use has been the focus of geography, economics and other disciplines for a long time [11,12]. However, land use is not immutable; it will be transformed with the economic and social development of a country or region [13,14]. Scholars have also carried out many studies on land use transition [15–22]. Among them, land use forms are the core content of land use transition research, including dominant and recessive forms. Dominant forms refer to the structure of the main land use types in a country or region during a specific period, including quantity and spatial structure. Recessive forms refer to the land use forms that are not easy to detect and can only be obtained after analysis and investigation [23]. As one of the key measures in the reform of land management, land transfer belongs to the category of hidden land use [24]. The reason why the phenomenon of land abandonment is increasing is closely related to the deep-rooted land complex of farmers. Land transfer can not only revitalize the abandoned farmland in rural areas, but also help to improve the scale and intensive management of land. However, farmers know little about the merits of land transfer policy and prefer to abandon their land rather than transfer their land to other individuals or organizations who are capable and willing to manage it. Therefore, paying attention to land transfer plays an important role in solving the problem of land abandonment [25,26]. Some scholars found that the reasonable transfer of land can reduce agricultural cost, achieving the appropriate scale of land management, and then solve the problem of land abandonment [27–29]. In China, ownership, contracting rights and management rights of rural land are separated. Land ownership belongs to public collectives, while contracting rights are granted to farmers, and the operating rights are controlled by capital (the contracted farmers transfer the land to other individuals or organizations for operation, and other individuals or organizations obtain the right to operate the land) [30]. Rural land transfer refers to the practice that rural families retain the contracting right and transfer the management right only to other farmers or economic organizations by subcontracting, transferring, becoming a shareholder, cooperating, leasing or exchanging the contracted land. However, in view of the current situation of China, due to the fragmentation of land, there is a low proportion of land transfer and the coexistence of land abandonment in many regions [31], which is not conducive to the guarantee of national food security [32].

As the most basic resource of famers, land and its utilization mode and structure will be directly affected by their livelihood strategies [28,33–35]. Therefore, the research on land transfer cannot be separated from research on farmers' livelihood capital and livelihood strategy. However, although there is some research on household livelihood and land use, there is relatively little research on the whole. For example, some scholars have explored the correlation between household livelihood and land use [36,37], sustainable livelihood and conversion of farmland to forest [38–40]. At the same time, a review of existing studies shows that most scholars explore the effect of livelihood capital on land transfer using static cross-section data [41,42]. However, in the existing studies, the analysis based on panel data mostly focuses on risk, new agricultural insurance and other aspects, and there is almost no study on the dynamic characteristic changes and the effect of livelihood capital on land transfer [43,44]. In the context of rapid urbanization and agricultural modernization, it is necessary to use dynamic panel data to systematically reveal the dynamic change characteristics and the effect of livelihood capital on land transfer under different livelihood strategies.

Based on this, using the survey data of Sichuan Province in 2012, 2016 and 2019, and using the sustainable livelihoods framework, the paper divides livelihood capital into five categories: human capital, natural capital, financial capital, physical capital and social capital. According to the proportion of non-agricultural income in the household income, farmers are divided into pure farmers, part-time farmers and non-farmers. Considering land transfer direction and area, the paper systematically analyzes the characteristics of household livelihood capital, livelihood strategies and land transfer, building panel econometric models to explore the effect of livelihood capitals on land transfer. Although this study only focuses on Sichuan Province of China, the index system, theoretical analysis framework design and research ideas of this study can provide references for other developing countries or developed countries. The two research questions to be answered in this study are as follows:

1.  What are the characteristics of household livelihood capital and land transfer under different livelihood strategies in different periods? What is the effect of livelihood capitals on land transfer?
2.  In answering these questions, the paper contributes to analyze the characteristics of farmers' capital and land transfer under different livelihood strategies, as well

as the effect of natural capital, human capital, financial capital, physical capital and social capital on land transfer. Livelihood capital, as the core of farmers' survival and development, has an effect on land transfer. It is expected to provide suggestions for the policy of farmers' land transfer under different livelihood capital endowments.

## 2. Materials and Methods

### 2.1. Data Sources

Based on the data collected by China Rural Development Survey Group in Sichuan Province in 2012, 2016 and 2019, the study mainly investigates the livelihood capital status and land use of farmers. In order to ensure the typicality and representativeness of the data, the principle of stratified equal probability random sampling was adopted to determine the sample households. Specifically, according to the research results of He et al. [45], Rozelle [46], Shui et al. [47] and Xue et al. [48], the index of per capita industrial gross output was used to cluster all districts and counties in Sichuan Province into 5 categories from high to low, and 1 district or county from each category was randomly selected as the sample. After the sample districts and counties were selected, the towns in the sample districts and counties were divided into high income group and low income group according to the order of per capita industrial output value, and then 1 township in each group was randomly selected as the sample town. After the sample towns were selected, the sample towns were divided into high income group and low income group according to the order of per capita industrial output value, and 2 sample villages were selected from each sample town. After the sample village was determined, 20 households were randomly selected according to the list of farmers. According to this principle, a sample of 400 households was obtained in 2012, 2016 and 2019. Simple data processing was carried out, and the famers' samples in all the three phases were retained. Finally, 299 famer household samples were obtained for subsequent analysis.

### 2.2. Theoretical Analysis Framework

The livelihood of farmers has been of wide concern in many countries and regions as well as academic circles [49–51]. Early livelihood studies mainly focused on poverty alone; that is, they focused on income level, consumption capacity and other factors related to basic living needs [52]. With the deepening of research, on the basis of a large number of poverty alleviation practices and theoretical development, income and consumption are no longer the only criteria to measure poverty [53]. In 1992, the United Nations Conference on Environment and Development proposed the concept of "sustainable livelihood" and advocated using the amount of livelihood capital of farmers to represent the strength of their capabilities [8]; thus the sustainable livelihood analysis method came into being [53]. Due to different understandings of livelihood, there are many different methods of livelihood sustainability analysis [52]. Among them, the sustainable livelihoods approach (SLA), established by the UK's Department for International Development (DFID), has been widely used by many organizations and scholars [6,29], which divides livelihood capital into five types: human capital, natural capital, physical capital, financial capital and social capital.

In this study, the sustainable livelihoods framework of DFID was slightly adjusted. We combined it with the direction and area of land transfer to construct a framework, as shown in Figure 1. This paper focuses on the effect of farmers' livelihood capital on land transfer, and adds a new solid line arrow of "livelihood capital→land transfer". In addition, livelihood strategy has indirect influence on land transfer through livelihood capital, and may also directly affect farmers' land transfer, which is not discussed in this study. So the path "livelihood strategy→land transfer" adopts the dashed arrow. Other paths of DFID are also shown with dashed lines.

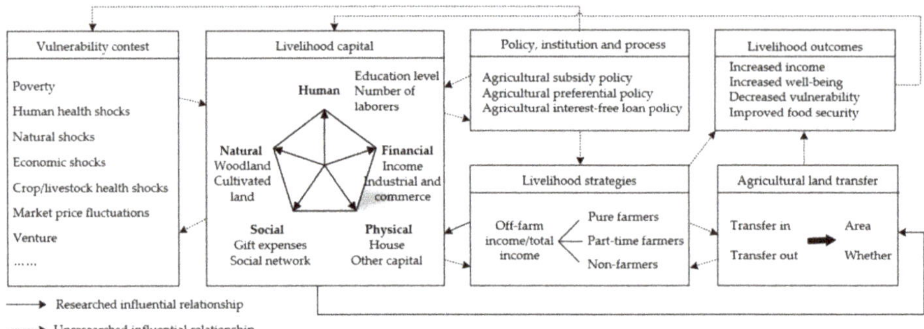



**Figure 1.** Analysis frame diagram of household livelihood capital and land transfer.

Farmers survive and develop against a background of fragility; they are affected and suppressed by internal and external dangers [54,55]. Under the effect of the external risk, farmers can mitigate the effect on their families through reasonable allocation of livelihood capital, selection of appropriate livelihood strategies, land transfer decisions, cooperation with agricultural policies, etc., so as to maximize utility and minimize risk [3,7,35], and achieve a positive livelihood output.

This study focuses on the characteristics of livelihood capitals and land transfer under different livelihood strategies and the effect of livelihood capitals on land transfer. Theoretically, different livelihood strategies adopted by farmers will lead to different allocations of livelihood capital, which will have an effect on land transfer decisions [56] (livelihood strategy→livelihood capital→land decision). Taking the natural capital and human capital owned by farmers as an example, when the natural capital of farmers is high and the human capital is low, the family labor force cannot meet the demand of labor required for the cultivation of the land, and when the cost of employing other labor forces is higher than the income from the cultivation of this part of land, the farmers are more likely to transfer part of their land out. When the natural capital is high and the human capital is also high, the family labor force meets or even exceeds the demand of labor required for the cultivation of the land; the farmers are more likely to transfer in the land of other farmers to realize the appropriate scale of land management, so as to maximize the utility. However, according to the existing research, there is no unified understanding of the effect of livelihood capitals on land transfer

In terms of natural capital, on the one hand, some scholars found that the increase of natural capital will promote land transfer in. For example, the empirical study by Ji et al. [57] and the theoretical study by Long et al. [58] showed that if the cultivated land is contiguous and the area is large, the soil quality is good, and the production efficiency is high, farmers will be inclined to transfer in land to realize the large-scale operation of land. On the other hand, some scholars reported that the increase of natural capital will promote land transfer out. For example, the empirical researches by the authors of [9,59,60] concluded that the land in many regions is fragmented and scattered, so it is difficult to realize the proper scale of centralized and continuous operation. At this time, the high input and low return of land will inhibit the enthusiasm of farmers in agricultural production, and may promote the transfer out of land. Although there is no unified understanding about the influence of natural capital on land transfer in the academic circle, the mainstream view is that the increase of natural capital will promote land transfer in. Based on this, this study proposes research Hypothesis H1:

**Hypothesis (H1).** *The higher the natural capital of farmers, the higher the probability of land transfer in and the larger the areas of land transfer in.*

In terms of human capital, on the one hand, some scholars found that the increase of human capital will promote land transfer in [61]. For example, the empirical studies by Zhang et al. [62], Zhu et al. [63] showed that in some regions, when the labor force of farmers meets the labor force of land demands, the farmers are more inclined to transfer land in due to their dependence on land; they are able and willing to cultivate the land. On the other hand, some scholars found that the increase of human capital will promote land transfer out. For example, the theoretical study by Ge et al. [64] and the empirical study by Yuan et al. [65] concluded that with the rapid development of secondary and tertiary industries, the income of non-agricultural industry is higher than that of agriculture, so the income loss of farmers who put labor into agricultural production instead of putting labor into non-agricultural production will increase; that is, the opportunity cost of agricultural production will increase. Farmers are likely to put more energy into non-agricultural industries to increase the overall income of families through the increase of wage income. Although there is no unified understanding of the influence of human capital on land transfer in the academic circle, but in most empirical studies, the increase of human capital tends to promote land transfer in. Based on this, this study proposes research Hypothesis H2:

**Hypothesis (H2).** *The higher the human capital of farmers, the higher the probability of land transfer in and the larger the areas of land transfer in.*

In terms of financial capital, on the one hand, some scholars found that the increase of financial capital will promote land transfer in [66]. For example, the empirical study by Liu et al. [67] and the theoretical study by Martin and Clapp [68] showed that high financial capital will increase farmers' investment in agriculture, such as adopting advanced agricultural technology and purchasing more agricultural machinery, and then increasing their income through proper scale operation of land. On the other hand, some scholars found that the increase of financial capital will promote land transfer out. For example, the empirical studies by Su et al. [69] and Xu et al. [8,9] reported that the income of non-agricultural industry is generally higher than that of agricultural industry, and the increase of financial capital will encourage farmers to transfer to the secondary and tertiary industries, and invest more capital and labor into the non-agricultural industry. Although there is no unified understanding about the influence of financial capital on land transfer in the academic circle, most scholars advocate that the increase of financial capital will promote land transfer out. Based on this, this study proposes research Hypothesis H3:

**Hypothesis (H3).** *The higher the financial capital of farmers, the higher the probability of land transfer out and the larger the areas of land transfer out.*

In terms of physical capital, on the one hand, some scholars found that the higher the farmers' physical capital is, the more inclined they are to transfer in land. For example, the empirical study by Wang et al. [70] concluded that in some areas, rural resources are well endowed and ecological environment is good. At the same time, if the type and structure of rural houses are good, many rural households will tend to live in the countryside and generate income through agriculture due to the influence of various factors such as enjoyment psychology. On the other hand, some scholars found that the increase of physical capital will encourage farmers to transfer out land. For example, the empirical study by Wen et al. [71] showed that for farmers with higher physical capital, the risk of transferring out land is low and they are more willing to participate in the outflow. Although there is no unified understanding about the influence of physical capital on land transfer in the academic circle, the view that the increase of physical capital will promote land transfer in occupies the mainstream position. Based on this, this study proposes research Hypothesis H1:

**Hypothesis (H4).** *The higher the physical capital of farmers, the greater the probability of land transfer in and the larger the areas of land transfer in.*

In terms of social capital, on the one hand, some scholars found that the increase of social capital will encourage farmers to transfer in land. For example, the empirical study by Deng et al. [10] reported that one of their relatives is a village cadre, which to some extent means that they have a great influence on the village. It is easier for such farmers to obtain technology, information and financial help, and they are more inclined to transfer in land to carry out large-scale agricultural production. On the other hand, some scholars found that the higher the social capital, the more land is transferred out. For example, the empirical study by Xu et al. [1] and the case study by Lu et al. [72] concluded that with the development of farmers' social network, they will have more contacts and can master more non-agricultural information, and then participate in other sideline businesses to realize the diversification of livelihood strategies and reduce livelihood risks. Although there is no unified understanding about the influence of social capital on land transfer in the academic circle, the results of most research show that the increase of social capital will promote land transfer out. Based on this, this study proposes research Hypothesis H5:

**Hypothesis (H5).** *The higher the social capital of farmers, the greater the probability of land transfer out and the larger the areas of land transfer out.*

### 2.3. Variable Measure

### 2.3.1. Measurement of Livelihood Capital

Referring to the framework to analyze sustainable livelihoods [73], the division of farmers' livelihood capital and the measurement of farmers' livelihood capital studied by Peng et al. [3], Guo et al. [27] and Kuang et al. [74], this study also divides livelihood capital into five categories: human capital, natural capital, financial capital, physical capital and social capital, and then sets up specific indexes to measure them (Table 1). Among them, natural capital refers to the natural resources and services that people rely on for survival and development [75]. In this paper, farmers' per capita cultivated land area and per capita forestland area are used to measure natural capital. Human capitals include the knowledge and skills mastered by farmers, as well as their physical health status and potential ability [76]. The number of their labor force and the education level of the head of the household are used to measure the human capital. Financial capital refers to the cash that farmers can independently manage and raise, and its sources mainly include their own income, loans and free assistance. Annual cash income and whether they run their own businesses are taken as two indexes to measure financial capital. Physical capital refers to the facilities and equipment used by farmers for production and living [7]. In order to reduce the interference of other factors, this paper converted all kinds of physical capital of farmers into present value for comparison, mainly including the converted present value of houses and other capital other than houses, such as farm tools, draft animals and furniture. Social capital refers to the social network owned by farmers [3]. This paper uses two indexes to measure social capital: annual gift expenditure and whether there are public officials among relatives and friends. All the indexes in the table are the original indexes for the subsequent calculation of the core independent variables.

**Table 1.** Household livelihood capital index system.

| Category | Index Attributes and Evaluation Methods | Obs | Mean | SD | Entropy | Weight [b] |
|---|---|---|---|---|---|---|
| 1. Natural capital | | | | | | |
| Cultivated land | Per capita operating cultivated land area (Mu [a]) | 299 | 1.356 | 2.599 | 0.897 | 0.062 |
| Woodland | Per capita operating forestland area (Mu [a]) | 299 | 0.260 | 0.722 | 0.602 | 0.236 |
| 2. Human capital | | | | | | |
| Labor | Number of labor force (people) | 299 | 2.928 | 1.428 | 0.971 | 0.017 |
| Education level | Years of education for head of household (years) | 299 | 5.894 | 3.238 | 0.966 | 0.020 |
| 3. Financial capital | | | | | | |
| income | Annual cash income (10,000 yuan) | 299 | 4.984 | 4.741 | 0.944 | 0.033 |
| Industrial and commerce | There are self-employed businesses (0 = No, 1 = Yes) | 299 | 0.235 | 0.424 | 0.743 | 0.153 |
| 4. Physical capital | | | | | | |
| house | Present value of the house (10,000 yuan) | 299 | 23.596 | 74.486 | 0.848 | 0.090 |
| Other physical capital | Present value of farm tools, draft animals and furniture other than houses (10,000 yuan) | 299 | 2.525 | 4.937 | 0.854 | 0.086 |
| 5. Social capital | | | | | | |
| Gift expenses | Annual gift expenditure (10,000 yuan) | 299 | 0.294 | 0.356 | 0.910 | 0.053 |
| Social network | Members or relatives serving as village cadres (0 = No, 1 = Yes) | 299 | 0.091 | 0.288 | 0.578 | 0.250 |

Note: [a] Mu is a unit of land area commonly used in rural China, 1 mu $\approx$ 0.067 ha; [b] the weights are calculated by the entropy method, the details of calculative process can be found in Appendix A.

### 2.3.2. Measurement of Land Transfer

Land transfer involves the direction and area. The direction of land transfer is measured by whether farmers have transferred land in or out, and the area of land transfer is measured by the area of farmers' land inflow and the area of land outflow [8,9].

### 2.4. Research Methods

#### 2.4.1. Entropy Value Method

One of the objectives of this study is to explore the relationship between farmers' livelihood capital and land transfer. In order to achieve this goal, it is necessary to obtain the information about the livelihood capital index value and land transfer of farmers. Referring to the studies of He et al. [77], Ning et al. [78] and Xu et al. [79], this study adopts the entropy value method of objective assignment method to determine the weight of each livelihood index and the comprehensive score of the five types of livelihood capital, so as to avoid the error of subjective judgment. The detailed calculation steps of the entropy method are in Appendix A.

#### 2.4.2. Regression Model

Considering that the data used in this study are three periods of balance panel data in 2012, 2016 and 2019, this study adopts the panel binary Logit and panel Tobit models according to the characteristics of the dependent variables, and uses Stata16.0 for processing.

Since the direction of land transfer (whether it transfers in/out) is a dichotomous dependent variable, this study adopts the panel binary Logit model to analyze the effect of livelihood capitals on land transfer. The study estimates the results of fixed effect and random effect, and finally, determined by the Hausman Test, that land transfer in is suitable for fixed effect estimation and land transfer out is suitable for random effect estimation. The main formula for panel binary Logit is as follows:

$$p(Y_{it} = 1 | X_{it}, \beta_i) = p(Z_{it} > 0) = p(u_{it} > -\beta_i X_{it}) = F(\beta_i X_{it}) \tag{1}$$

$$p(Y_{it} = 1 | X_{it}, \beta_i) = F(\beta_i X_{it} + u_{it}) = \frac{1}{1 + e^{-(\beta_i X_{it} + u_{it})}} \tag{2}$$

This study introduces qualitative variables as dependent variables (whether the farmer "$i$" had land transfer in/transfer out in the "$t$" period), and $Y_{it} = 1$ when the phenomenon under study occurs, $Y_{it} = 1$ when it does not occur. All explanatory variables are $X_{it}$ (the livelihood capital index values of the farmer "$i$" in the "$t$" period). Since the index value of the probability of occurrence of the event can only be 0~1, an unobserved reference variable $Z_{it}$ is introduced to replace the virtual dependent variable $Y_{it}$. When the estimated $Z_{it} > 0$, $Y_{it} = 1$; otherwise, $Y_{it} = 0$. $u_{it}$ is the random disturbance term.

Since the area of land transfer (transfer in/out area) is quite concentrated on the number 0, which belongs to the left merge data, this study adopts the panel Tobit model to re-estimate the trunking distribution of the restricted dependent variable, so as to make it conform to the actual distribution, and then analyze the effect of livelihood capitals on land transfer. Because the panel Tobit model cannot be used for fixed effect estimation in Stata 16.0, this study only estimates the results of random effect. The main formula of the panel Tobit model is as follows:

$$Y'_{it} = \alpha + \beta X'_{it} + \varepsilon \tag{3}$$

In this study, $Y'_{it}$ is introduced as the dependent variable (the area of land transfer in/out of the farmer "$i$" in the "$t$" period). All explanatory variables are $X'_{it}$ (the livelihood capital index values of the farmer "$i$" in the "$t$" period). $\alpha$ is the constant term, $\beta$ is the regression coefficient, $\varepsilon$ is the random disturbance term.

The ultimate purpose of this study is to explore the relationship between livelihood capitals and land transfer under different livelihood strategies. To achieve this goal, it is necessary to obtain the index value of livelihood capital and the index value of land transfer. The livelihood capital index values of farmers under different strategies are obtained by entropy method (Appendix A). The land circulation of different types of farmers can be obtained by simple summation.

## 3. Results

### 3.1. Livelihood Strategies

Referring to the classification of farmers by Zhang et al. [80], Zhou et al. [81], this study divides farmers into pure farmers, part-time farmers (including first part-time farmers and second part-time farmers) and non-farmers according to the proportion of non-agricultural income in total income. Among them, pure farmers refers to those with non-agricultural income below 20% of total household income, first part-time farmers refers to those with non-agricultural income accounting for 20–50% of total household income (including 20%), and second part-time farmers refers to those with non-agricultural income accounting for 50–80% of total household income (including 50%). Non-farmers refers to those with non-agricultural income accounting for 80% or more of total household income (Table 2).

**Table 2.** Farmer types and sample distribution in each year.

| Farmer Types | | Year | Sampling Number | Proportion |
|---|---|---|---|---|
| Pure farmers | | 2012 | 76 | 25.42% |
| | | 2016 | 76 | 25.42% |
| | | 2019 | 76 | 25.42% |
| Part-time farmers | First part-time farmers | 2012 | 21 | 7.02% |
| | | 2016 | 14 | 4.68% |
| | | 2019 | 7 | 2.34% |
| | Second part-time farmers | 2012 | 19 | 6.36% |
| | | 2016 | 28 | 9.36% |
| | | 2019 | 57 | 19.06% |
| Non-farmers | | 2012 | 183 | 61.20% |
| | | 2016 | 181 | 60.54% |
| | | 2016 | 159 | 53.18% |

As can be seen from Table 2, in the survey of three periods, non-farmers accounted for the largest proportion, followed by pure farmers, and part-time farmers were the least. Among them, the number of pure farmers in three periods of the survey did not change, while the proportion of part-time farmers (including first part-time farmers and second part-time farmers) and non-farmers changed to different degrees. Specifically, in the three period surveys in 2012, 2016 and 2019, the proportion of pure farmers was 25.42%, the proportion of part-time farmers was 13.38%, 14.04% and 21.40%, respectively, and the proportion of non-farmers was 61.20%, 60.54% and 53.18%, respectively. Among the part-time farmers, the proportion of first part-time farmers decreased by 7.02%, 4.68% and 2.34, respectively, while the proportion of second part-time farmers increased by 6.36%, 9.36% and 19.06%, respectively. It can be seen that in both 2016 and 2019, some of the first part-time farmers and non-farmers switched to become second part-time farmers.

### 3.2. Descriptive Statistical Analysis

### 3.2.1. Characteristics of Household Livelihood Capital in Different Periods

Table 3 shows the livelihood capital index value of farmers in three periods. It can be seen from Table 3 that, based on the changes in the three periods, the natural capital of pure farmers increased, while for part-time farmers and non-farmers it decreased. The human capital of non-farmers increased, while for pure farmers and part-time farmers it decreased. The social capital of part-time farmers and non-farmers increased, while for pure famers it decreased. The financial capital, physical capital and the total index value of livelihood capitals of non-farmers increased, while for pure farmers and part-time farmers it decreased.

**Table 3.** Household livelihood capital index value.

| Farmer Types | | Year | Natural Capital | Human Capital | Financial Capital | Physical Capital | Social Capital | Total |
|---|---|---|---|---|---|---|---|---|
| Pure farmers | | 2012 | 0.083 | 0.059 | 0.088 | 0.086 | 0.067 | 0.383 |
| | | 2016 | 0.115 | 0.052 | 0.065 | 0.058 | 0.055 | 0.344 |
| | | 2019 | 0.123 | 0.039 | 0.019 | 0.026 | 0.028 | 0.237 |
| Part-time farmers | First part-time farmers | 2012 | 0.076 | 0.072 | 0.162 | 0.120 | 0.108 | 0.538 |
| | | 2016 | 0.048 | 0.086 | 0.153 | 0.248 | 0.124 | 0.660 |
| | | 2019 | 0.058 | 0.064 | 0.126 | 0.117 | 0.162 | 0.527 |
| | Second part-time farmers | 2012 | 0.077 | 0.071 | 0.132 | 0.090 | 0.078 | 0.448 |
| | | 2016 | 0.060 | 0.070 | 0.114 | 0.072 | 0.133 | 0.451 |
| | | 2019 | 0.049 | 0.072 | 0.062 | 0.064 | 0.102 | 0.349 |
| Non farmers | | 2012 | 0.058 | 0.069 | 0.040 | 0.050 | 0.061 | 0.279 |
| | | 2016 | 0.049 | 0.071 | 0.054 | 0.056 | 0.057 | 0.287 |
| | | 2019 | 0.047 | 0.078 | 0.089 | 0.085 | 0.069 | 0.368 |

Note: the livelihood capital index values in this table are obtained from Table 1 by entropy method, and are dimensionless values between 0 and 1.

Natural capital includes the area of cultivated land and forestland. The average area of per capita cultivated land of pure farmers, first part-time farmers and second part-time farmers and non-farmers was 2.138 mu, 1.598 mu, 1.278 mu and 1.038 mu, respectively, indicating that pure farmers had higher natural capital. Human capital includes the age of the head of the household, the education level and the number of the labor force. In general, the longer the education and the larger the number of the labor force, the higher the human capital. Taking the education level of the head of the household as an example, the average education of the part-time farmers was 7 years, higher than that of non-farmers and pure farmers. Financial capital includes annual cash income and whether they have self-employed businesses. The average annual cash income of the first part-time farmers was 75,120, which was at least 18,220 higher than that of other farmers. The main factors to measure physical capital are the converted present value of houses, and the converted present value of other physical capital such as farm tools, draft animals and furniture. The average converted present value of other physical capital of the first part-time farmers was 52,080, at least 22,680 higher than that of other farmers. Annual gift expenditure of households is one of the important factors affecting social capital. Horizontal comparison shows that the annual gift expenditure of part-time farmers was 1.2–1.5 times that of non-farmers and pure farmers. Longitudinal comparison showed that based on 2012, the expenditure of non-farmers increased by 25.37%, while that of pure farmers and part-time farmers decreased by 17.88% and 15.56%, respectively.

The changes of total livelihood capital index value of farmers under different livelihood strategies show different characteristics. In 2012 and 2016, the total capital index value of the first part-time farmers was the highest, followed by that of the second part-time farmers, pure farmers and finally non-farmers. In 2019, the total capital index value was the highest for first part-time farmers followed by non-farmers, second part-time farmers and finally pure farmers. In general, the total livelihood capital index value of pure farmers decreased; the total livelihood capital index value of non-farmers increased; the total livelihood capital index value of first part-time farmers and second part-time farmers increased first and then decreased.

In order to more intuitively compare the changes of livelihood capital of different types of farmers in the three-year survey in 2012, 2016 and 2019, this paper analyzes the changes of the five livelihood capitals of pure farmers, part-time farmers and non-farmers through a radar chart. As can be seen from Figure 2, the pure farmers showed a shift of other livelihood capital towards natural capital (Figure 2a), the first part-time farmers' capital shift characteristics were not obvious (Figure 2b), the second part-time farmers showed a shift of natural capital, physical capital and financial capital towards social capital (Figure 2c), while the non-farmers showed a shift of natural capital towards other livelihood capital (Figure 2d).

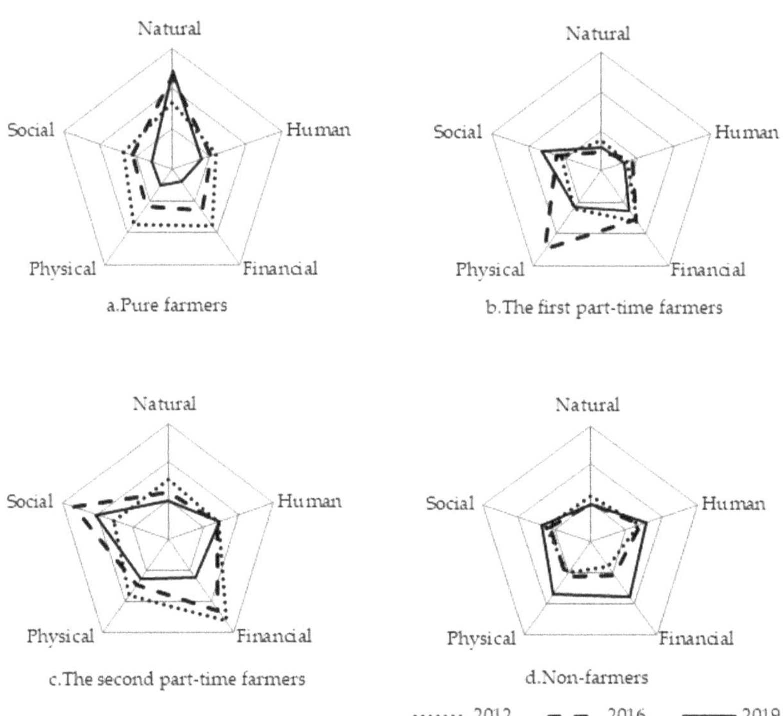


**Figure 2.** Changes in livelihood capital allocation of different types of farmer households. Note: (**a**) Table 3. (**b**) The maximum value of the coordinate axes in (**a–d**) is 0.15, the scale unit is 0.05; the maximum value of the coordinate axis in (**b**) is 0.3, the scale unit is 0.1; and the minimum value of the axes of all graphs is 0.

### 3.2.2. Characteristics of Farmers' Land Transfer in Different Periods

In order to grasp the situation of land transfer, land transfer is divided into land transfer in and land transfer out according to the direction, and the areas of land transfer in and out of different types of farmers in 2012, 2016 and 2019 are calculated respectively (Table 4).

The direction of land transfer shows that the total area of land transfer out exceeded the total area of land transfer in this phenomenon is based on the total area of land transfer of all investigated farmers. Specifically, the survey data in 2012 showed that the area of land transfer in of pure farmers, first part-time farmers, second part-time farmers and non-farmers was larger than the land transfer out area, and the areas of land transfer in were 33.60 mu, 9.70 mu, 17.17 mu and 117.50 mu, respectively. The areas of land transfer out were 30.30 mu, 8.10 mu, 6.40 mu and 32.13 mu, respectively. In 2016, the first part-time farmers had a phenomenon that the area of land transfer out was larger than the area of land transfer in: the area of land transfer in was 1.00 mu, and the area of land transfer out was 8.90 mu. In 2019, the area of land transfer out of all types of farmers was larger than the area of land transfer in. The areas of land transfer in of pure farmers, first part-time farmers, second part-time farmers and non-farmers were 22.89 mu, 1.30 mu, 11.16 mu and 11.57 mu, respectively. The areas of land transfer out were 29.99 mu, 2.57 mu, 16.76 mu and 35.20 mu, respectively.

From the area of land transfer, on the whole, the total area of land transfer shows a trend of increasing first and then decreasing. Specifically, in 2016, the areas of land transfer in of pure famers, second part-time farmers and non-farmers increased by 22.10 mu, 1.83 mu and 33.60 mu, respectively, compared with that in 2012; the areas of land transfer out of first part-time farmers, second part-time farmers and non-farmers increased by 0.80 mu, 6.30 mu and 22.87 mu, respectively, compared with that in 2012. In 2019, the areas of land transfer in of pure famers, second part-time farmers and non-farmers decreased by 32.81 mu, 7.84 mu and 139.53 mu, respectively, compared with that in 2016; while the areas of land transfer out of first part-time farmers and non-farmers decreased by 6.33 mu and 19.8 mu, respectively, compared with that in 2016.

**Table 4.** Farmers' total land transfer area [a][c] (*n* = 299).

| Farmer Types | | Year | Transfer in Area (Mu [b]) | Transfer out Area (Mu [b]) |
| --- | --- | --- | --- | --- |
| Pure farmers | | 2012 | 33.60 (18.88%) | 30.30 (39.39%) |
| | | 2016 | 55.70 (24.56%) | 29.46 (27.78%) |
| | | 2019 | 22.89 (35.48%) | 29.99 (35.48%) |
| Part-time farmers | First part-time farmers | 2012 | 9.70 (5.45%) | 8.10 (10.53%) |
| | | 2016 | 1.00 (0.44%) | 8.90 (8.39%) |
| | | 2019 | 1.30 (2.77%) | 2.57 (3.04%) |
| | Second part-time farmers | 2012 | 17.17 (9.64%) | 6.40 (8.32%) |
| | | 2016 | 19.00 (8.38%) | 12.70 (11.97%) |
| | | 2019 | 11.16 (23.79%) | 16.76 (19.83%) |
| Non-farmers | | 2012 | 117. 55 (66.03%) | 32.13 (41.76%) |
| | | 2016 | 151.10 (66.62%) | 55.00 (51.86%) |
| | | 2019 | 11.57 (24.66%) | 35.20 (41.65%) |

Note: [a] data in brackets are the proportion of such famers in the total area of land transfer in/transfer out in the current survey period. [b] Mu is a unit of land area commonly used in rural China, 1 mu ≈ 0.067 ha. [c] The land transfer in/transfer out area in this table is based on Table 2, simply summing up the land transfer in/transfer out areas of all farmers in the same type.

### 3.3. Analysis of Regression Results

Table 5 shows the regression results of livelihood capital and land transfer. Before building the model, we conducted correlation analysis on the independent variables of the model, and the coefficients of the correlation analysis were all less than 0.5, indicating that there was no problem of multicollinearity. In addition, although we cannot estimate the robust standard error of the panel binary Logit model and the panel Tobit model in Stata16.0, when we use the entropy method to determine the comprehensive index, the variables are standardized, and the deviation trend will not be very strong. Therefore, theoretically, there will be no heteroscedasticity problem. Due to the panel data including only 299 famers and the small number of some types of farmers, it is not suitable for further regression with the livelihood strategy, so this study only carried out overall regression. Model 1 and Model 2 show the regression results of household livelihood capital and whether there is land transfer or not, while Model 3 and Model 4 show the regression results of household livelihood capital and its land transfer area. According to the overall significance test results of the models, all the four models were statistically significant at the 0.05 level, indicating that at least one of the independent variables and the dependent variables have a statistically significant influence, which can be used for subsequent analysis.

**Table 5.** Regression results of livelihood capital and land transfer.

| Variable | Panel Logit Model | | Panel Tobit Model | |
| --- | --- | --- | --- | --- |
| | Model 1 | Model 2 | Model 3 | Model 4 |
| | Whether Transfer In | Whether Transfer Out | Area of Transfer In | Area of Transfer Out |
| Natural capital | 290.2499 *** (50.3092) | 26.1096 (17.2572) | 9.3034 *** (2.6800) | 1.2900 (1.4974) |
| Human capital | 165.6151 * (88.4463) | 41.9614 (54.3115) | 3.4148 (27.0636) | 14.4909 (14.6734) |
| Financial capital | 25.2693 (24.7348) | 55.3376 *** (17.8329) | 1.8055 (9.4044) | 14.3344 *** (5.1902) |
| Physical capital | 10.4350 (22.4749) | 10.4649 (14.3151) | 0.1215 (5.3519) | 3.1917 (2.9826) |
| Social capital | 15.1650 (16.2948) | 6.2124 (9.7411) | 7.0181 (5.3447) | 5.6876 * (2.9286) |
| LR/Wald chi2(5) | 75.56 *** | 13.55 ** | 13.74** | 13.74 ** |

Note: data in brackets are standard errors. ***, ** and * represent statistical significance at 1%, 5% and 10%, respectively.

It can be seen from Model 1 that natural capital and human capital had a statistically significant positive effect on whether farmers transferred land in, while financial capital, physical capital and social capital had no statistically significant obvious effect on whether farmers transferred land in. This indicates that the higher the farmer's natural capital and human capital score, the higher the probability of land transfer. According to Model 2, financial capital had a statistically significant positive effect on whether farmers transferred land out, while natural capital, human capital, physical capital and social capital had no statistically significant obvious effect on whether farmers transferred land out. This shows that the higher the financial capital scores of farmers, the greater the probability of their land transfer out. According to Model 3, natural capital had a statistically significant positive effect on the area of farmers' land transfer in, while human capital, financial capital, physical capital and social capital had no statistically significant obvious effect on the area of farmers' land transfer in. This indicates that the higher the farmers' natural capital score, the larger the areas of land transfer in. According to Model 4, financial capital and social capital had a statistically significant positive effect on the area of farmers' land transfer out, while the effect of natural capital, human capital and physical capital on the area of farmers' land transfer out was not statistically significant. This indicates that the higher the score of financial capital and social capital, the larger the areas of land transfer out.

## 4. Discussion

Farmer livelihood sustainability and land use have been the focus of research for a long time [82,83]. In fact, the United States, Europe, Australia, Spain and other developed countries as well as Latin America, Southeast Asia, China and other developing countries have reported varying degrees of farmland wastage [8,84–86], which is gradually evolving into a global social and economic phenomenon [87–89]. In China, a populous country, the dichotomy between human and land use has been prominent for a long time, which is representative to a certain extent. Since the mid-1980s, China has formulated a series of land policies to promote rational land transfer [79]. For example, the steady promotion of farmland ownership confirmation and certification can provide good conditions for land transfer [90,91]. With the promotion and advocacy of land transfer policy, the problem of land abandonment has been alleviated. Some scholars have also proved that land transfer is conducive to solving the problem of land abandonment. For example, Shao et al. [92] found that land transfer is conducive to the more effective use of cultivated land resources and the reduction of land abandonment. Zhang and Li [93] found that there was a strong negative correlation between the transfer ratio and the abandonment ratio of farmland. Zhu and Xu [94] advise strengthening of land contract management and accelerating the pace of land transfer, and put forward that land transfer is one of the important measures to solve the problem of land abandonment. Therefore, by studying the effect of livelihood capital on land transfer, this study can help better implement land transfer policy and provide a reference for solving the land abandonment problem. This study is based on the analysis of the survey data of five districts and counties in Sichuan Province in 2012, 2016 and 2019, using the sustainable livelihoods framework to establish the index system and analysis logic, and the panel binary Logit model and panel Tobit model according to the characteristics of dependent variables. This study explores the characteristics of livelihood capital and land transfer under different livelihood strategies and the effect of livelihood capitals on land transfer.

In terms of the characteristics of livelihood capital and land transfer, this study found that the characteristics of change were different for livelihood capitals of farmers under different livelihood strategies, and land transfer also changed. In terms of livelihood capital, the natural capital of pure farmers increased, while the human capital, financial capital, physical capital, social capital and total capital all decreased, which showed the characteristics of other livelihood capital shifted to natural capital. Pure farmers have been dependent on land for survival for a long time, so they were more inclined to transfer land in. In 2012, their average area of land was 1.227 mu, which increased by nearly 108.15% in 2019. First part-time farmers had no obvious shift characteristics. Different from pure farmers, first part-time farmers had started to engage in non-agricultural industries and did not completely depend on the land for survival. The farmers needed to consider whether it was beneficial for them to engage in non-agricultural industries. Therefore, these farmers were more inclined to maintain their original status, and the shift characteristics were not obvious. The second part-time farmers' natural capital, financial capital, physical capital and total capital decreased, social capital increased, and human capital did not change obviously. Second part-time farmers tended to shift natural capital, physical capital and financial capital towards social capital. Compared with the first part-time farmers, the second part-time farmers had a higher non-agricultural labor force. They had more vocational skills and a wider network of social connections. This was mainly reflected in expenditure on gifts, which increased from 2426 yuan in 2012 to around 3989 yuan in 2016 and 2019. The natural capital of non-farmers decreased while their human capital, financial capital, physical capital, social capital and total capital increased; they tended to shift natural capital towards other livelihood capital. Non-farmers got 80% or more of their total income from non-agricultural industries; their dependence on the countryside and land was very low. In addition, in the survey, the number of years of education (5.7→6.3), per capita annual cash income (3.8→7.5), the converted present value of houses (10.2→38.0) and annual gifts of family (2417.5→3644.0) of non-farmer households all increased. This promotes the increase of human capital, financial capital, physical capital and social capital. In terms of land transfer, it shows that the total area of land transfer out exceeded the total area of land transfer in. The reason is that among the 299 farmers in the sample, some farmers transferred land in and some farmers transferred land out. The land transferred out was not necessarily transferred to the 299 sample farmers in this study, but to other farmers in the same village. So the total area of land transferred out exceeded the total area of land transferred in. In addition, the obvious reason why the total area of land transfer could be logically different was that this study only analyzed 299 sample farmers and did not ask all farmers in a region. In the total population of farmers, the total area of land transfer in must logically equal the area of land transfer out.

In the results of this study, some hypotheses are supported and others are rejected. At the same time, there are similarities and differences with the results of existing studies. In terms of natural capital and financial capital, the research hypothesis H1 and H3 are supported. This study found that the higher the natural capital of farmers, the higher the probability of land transfers in and the larger the area of land transfers in. The higher the financial capital of farmers, the higher the probability of land transfers out and the larger the area of land transfers out. The results are also supported by Xu et al. [8,9], Ji et al. [57], Long et al. [58], Su et al. [69], Bian et al. [95] and Peng et al. [96]. In terms of human capital, the research hypothesis H2 is partly supported; human capital had a statistically significant positive effect on the probability of land transfer in; however, it had no statistically significant effect on the area of land transfer in. The results of this study are different from those of Zhang et al. [60], Muchomba [61], Liu et al. [97], who concluded that human capital has a statistically significant positive effect on land transfer in, and different from those of Ge et al. [64], Yuan et al. [65] and Peng et al. [96], who concluded that human capital has a statistically significant positive effect on land transfer out. This may be due to the differences of samples in the study area. The samples investigated in this study were all samples from the hilly area. Generally speaking, most of them were small-scale farmers with a small land holding in areas where young rural laborers were forced to go out for work in large numbers driven by economic interests. In addition to the farmers who stayed in the countryside to conduct moderate-scale land management, most of the farmers who stayed in the countryside were the elderly. Due to the limitation of the elderly's labor ability, they are difficult to engage in non-agricultural industries. So they are more inclined to rely on the land they are familiar with for survival; they prefer to transfer land in. Meanwhile, in order to facilitate farming, most of the lands they transferred in are close to home and of high quality. In addition, in order to match their labor capacity, they will not transfer much land in. Therefore, for farmers with higher human capital, the government can take different training measures according to different groups. For farmers who want to engage in proper scale operation of land, the government can encourage farmers to transfer land in more, and further strengthen farmers' training in agricultural planting, harvesting, disease and insect control, so as to improve farmers' production capacity in agriculture. For the elderly who just want to maintain their basic living needs, the government can encourage farmers to transfer land in within the scope of their labor force, giving agricultural subsidies to elderly farmers. In terms of physical capital, the research hypothesis H4 is rejected. This study finds that the physical capital of farmers has no statistically significant effect on the probability and area of land transfer. The results of this study are different from those of Ji et al. [57], Zhu et al. [63] and Wang et al. [70], who concluded that physical capital has a statistically significant positive effect on land transfer in, and different from those of Wen et al. [71], who concluded that physical capital has a statistically significant positive effect on land transfer out. Possibly because China's agricultural technology has improved in recent years, farmers have been able to further liberate labor by using farm machinery, which means less labor is needed for land than before. Sichuan is a big agricultural province, agriculture has a long history, and the development of agricultural technology training is relatively good. Therefore, whether farmers change their livelihood strategy to transfer land out or transfer land in to realize the appropriate scale management of land, both can help families maintain their livelihood and output. Therefore, for the farmers with high physical capital, different encouragement measures can be taken. If the urban population is under great pressure and the employment rate is low, the government can encourage farmers to transfer land in and provide farmers with training on agricultural planting. If there are many jobs in the secondary and tertiary industries in local towns and cities, and many jobs in non-agricultural industries, the government can encourage farmers to transfer their land out and provide vocational skills and training in the non-agricultural industries, so as to enhance their adaptability to the cities. In terms of social capital, the research hypothesis H5 is partly rejected. This study finds that the social capital of farmers has no statistically significant effect on the probability of land transfer out; however, it has a statistically significant positive effect on the area of land transfer out. The results of this study are different from those of Deng et al. [10] and Zhu et al. [63], who concluded that social capital has a statistically significant positive effect on land transfer in, and different from those of Xu et al. [1], Lu et al. [72] and Peng et al. [96], who concluded that social capital has a statistically significant positive effect on land transfer out. This may be due to differences in variable measures and research methods. The indexes used in this study to measure social capital are annual gift expenditure and members or relatives serving as village cadres. Additionally, the panel binary Logit model and panel Tobit model were constructed for regression, which was different from other studies. For example, Zhu et al. [63] measured the social capital by the number of households of the village relatives, the degree of trust in people around them and the favor expenditure, then constructed the binary Logit model for regression. Xu et al. [1] measured the social capital by the social network of relatives and friends available for assistance when seeking non-farm work, the social network of relatives and friends available for assistance when in urgent need of a lot of money and whether farmers participated in a farming association, then an ordered Logistic model was contrasted for regression. The wider the social network of farmers, the more people will accept their land, and the more the area of land is transferred out. Therefore, for farmers with high social capital, the government can encourage farmers to transfer their land out, and encourage farmers to use their strong social network to actively expand income channels. Through the above discussion, this study finds that different types of livelihood capitals have different statistically significant effects on the direction and area of land transfer. In the future land transfer market, the difference of farmers' capital structure should be considered to make land decisions.

Compared with the existing research, this research has a core innovation point: in the research design, this study combines the sustainable livelihoods framework and land transfer, and uses dynamic panel data from the perspective of dynamic analysis to explore the characteristics of farmers' livelihood capital and land transfer under different livelihood strategies, and the effect of livelihood capital on land transfer. In addition, there are some limitations in this study, which need to be solved in future studies. Firstly, this study is only based on the panel data of Sichuan Province of China for analysis and discussion. Although it can provide a reference for other relevant studies and regions, it remains to be verified to what extent it can be generalized to other regions and whether it is applicable to other developing or developed countries. Secondly, this study only focuses on the status of the three survey time periods in 2012, 2016 and 2019, without considering the change of the intermediate years, which can be further explored in future studies. Finally, in terms of causality, this study only analyzes one direction, that is, the effect of farmers' livelihood capital on land transfer. It does not discuss the effect of land transfer on livelihood capital, or the effect of the same third factor on livelihood capital and land transfer, which can be further discussed in future studies.

## 5. Conclusions

Through analysis, this study mainly draws the following two conclusions:

1. The livelihood capital and land transfer of farmers under different livelihood strategies show different characteristics. In terms of livelihood capital, pure farmers tend to shift other livelihood capital towards natural capital, so their total index value of livelihood capital decreased. First part-time farmers had no obvious shift characteristics and strong dependence on the original path, so their total index value of livelihood capital increased first and then generally decreased. Second part-time farmers tended to shift natural capital, physical capital and financial capital towards social capita, so their total index value of livelihood capital increased first and then generally decreased. Non-farmers tended to shift natural capital towards other livelihood capital, so their total index value of livelihood capital increased. In terms of land transfer, from the direction of land transfer, the total area of land transfer out

exceeded the total area of land transfer in; from the area of land transfer, all types of farmers showed a trend of increasing first and then decreasing.

2. Livelihood capital affects the direction and area of land transfer. The higher the natural capital and human capital, the higher the probability of land transfer in. The higher the natural capital, the larger the area of land transfers in. The higher the financial capital, the higher the probability of land transfers out. The higher the financial capital and social capital, the larger the area of land transfers out.

This study is expected to provide suggestions for the policy of farmers' land transfer under different livelihood capital endowments. Based on the above analysis, the study has two policy implications: (1) Suggesting the government strengthen the support of pure farmers and part-time farmers in agricultural production and promote the return of a rural labor force. The study found that in the three periods of the survey, only the livelihood capital total index value of non-farmers increased, while the livelihood capital total index value of both pure farmers and part-time farmers decreased, indicating that only the living of non-farmers was improving, which is one of the reasons for the large rural labor force emigration. Based on this, we suggest that the government take appropriate measures to strengthen the support and encouragement of pure farmers and part-time farmers, so as to increase their livelihood capital total index value. (2) Suggesting that the government take into account the difference of livelihood capital endowment to encourage farmers to transfer land. Considering that natural capital and human capital have a statistically significant positive effect on land transfer, we suggest that the government encourage farmers with higher natural capital and human capital to increase the operating area of land according to their own needs and ability, promoting land transfer in. Considering that financial capital and social capital have a statistically significant negative impact on land transfer, we suggest that the government encourage farmers with higher financial capital and social capital to use their financial capital advantages and social network to broaden income channels and transfer land out. Considering that physical capital has no statistically significant effect on land transfer, we suggest that the government should take different measures for farmers with higher physical capital according to the employment pressure of local non-agricultural industries.

**Author Contributions:** Conceptualization, H.Y. and D.X.; methodology, K.H. and X.D.; software, X.D.; formal analysis, H.Y. and K.H.; investigation, D.X.; writing—original draft preparation, H.Y. and D.X.; writing—review and editing, H.Y. and D.X.; supervision, D.X.; funding acquisition, D.X. All authors have read and agreed to the published version of the manuscript.

**Funding:** We gratefully acknowledge financial support from National Natural Science Foundation of China (41801221), Special Program for Cultivating Excellent Young Talents under the Dual Support Plan of Sichuan Agricultural University and Undergraduate Research Interest Cultivation Program in 2021 of Sichuan Agricultural University (2021489; 2021488).

**Institutional Review Board Statement:** Not applicable.

**Data Availability Statement:** Not applicable.

**Acknowledgments:** The authors also extend great gratitude to the anonymous reviewers and editors for their helpful review and critical comments.

**Conflicts of Interest:** The authors declare no conflict of interest.

## Appendix A

The process of measuring the weight of each livelihood index and the comprehensive score of the five types of livelihood capital by the entropy value method is as follows:

1.  Dimensionless treatment

In order to avoid the adverse impact of different measurement units on the measurement of livelihood capital, the indexes are treated as dimensionless. Since the 10 indexes in this paper are all positive indexes, the calculation formula is as follows (A1):

$$X_{ij} = \frac{x_{ij} - m_j}{M_j - m_j}, i = 1, 2, 3, \ldots, p; \; j = 1, 2, 3, \ldots, q \tag{A1}$$

Among them, $x_{ij}$ is the original value of the farmer "$i$" in the "$j$" index, $M_j$ is the maximum value of $x_{ij}$, $m_j$ is the minimum value of $x_{ij}$, and $X_{ij}$ is the standardized value of the farmer "$i$" in the "$j$" index. Meanwhile, in order to eliminate zero and make the data operation processing meaningful, the standardized value is moved overall; that is, $X'_{ij} = X_{ij} + \alpha$, in this paper, $\alpha = 0.0001$.

2.  Calculation of numerical proportion

To calculate the proportion of the index"$j$" of the farmer "$i$" in the total number of the index"$j$", the formula is as follows (A2):

$$P_{ij} = \frac{X'_{ij}}{\sum_{i=1}^{p} X'_{ij}}, i = 1, 2, 3, \ldots, p \tag{A2}$$

3.  Calculation of entropy

To calculate the entropy value of the index "$j$", the formula is as follows (A3):

$$E_j = -\frac{1}{lnp} \sum_{i=1}^{p} P_{ij} \ln(P_{ij}), i = 1, 2, 3, \ldots, p \tag{A3}$$

4.  Calculation of index difference coefficient

To calculate the index difference coefficient of the index"$j$", the formula is as follows (A4):

$$D_j = 1 - E_j \tag{A4}$$

5.  Calculation of weight

To calculate the weight of the index"$j$", "$q$" is the number of livelihood capital indexes. The formula is as follows (A5):

$$w_j = \frac{D_j}{\sum_{j=1}^{q} D_j}, \; j = 1, 2, 3, \ldots, q \tag{A5}$$

6.  Calculation of farmer household single index evaluation score

To calculate the single index evaluation score of the farmer "$i$" in the "$j$" index, the formula is as follows (A6):

$$S_{ij} = w_j * X_{ij} \tag{A6}$$

7.  Calculation of farmers' livelihood capital index value

After determining the weight of each index and the evaluation score, the scores of natural capital, human capital, financial capital, physical capital and social capital of farmers can be obtained by adding the comprehensive scores of each index in the same dimension.

## References

1. Xu, D.; Zhang, J.; Rasul, G.; Liu, S.; Xie, F.; Cao, M.; Liu, E. Household Livelihood Strategies and Dependence on Agriculture in the Mountainous Settlements in the Three Gorges Reservoir Area, China. *Sustainability* **2015**, *7*, 4850–4869. [CrossRef]
2. Xu, D.; Guo, S.; Xie, F.; Liu, S.; Cao, S. The impact of rural laborer migration and household structure on household land use arrangements in mountainous areas of Sichuan Province, China. *Habitat Int.* **2017**, *70*, 72–80. [CrossRef]
3. Peng, L.; Xu, D.; Wang, X. Vulnerability of rural household livelihood to climate variability and adaptive strategies in landslide-threatened western mountainous regions of the Three Gorges Reservoir Area, China. *Clim. Dev.* **2018**, *11*, 469–484. [CrossRef]
4. Xie, Y.; Jiang, Q. Land arrangements for rural–urban migrant workers in China: Findings from Jiangsu Province. *Land Use Policy* **2016**, *50*, 262–267. [CrossRef]
5. Deng, X.; Xu, D.; Zeng, M.; Qi, Y. Does Internet use help reduce rural cropland abandonment? Evidence from China. *Land Use Policy* **2019**, *89*, 104243. [CrossRef]
6. Deng, X.; Xu, D.; Zeng, M.; Qi, Y. Does early-life famine experience impact rural land transfer? Evidence from China. *Land Use Policy* **2019**, *81*, 58–67. [CrossRef]
7. Xu, D.; Deng, X.; Guo, S.; Liu, S. Sensitivity of Livelihood Strategy to Livelihood Capital: An Empirical Investigation Using Nationally Representative Survey Data from Rural China. *Soc. Indic. Res.* **2018**, *144*, 113–131. [CrossRef]
8. Xu, D.; Deng, X.; Guo, S.; Liu, S. Labor migration and farmland abandonment in rural China: Empirical results and policy implications. *J. Environ. Manag.* **2019**, *232*, 738–750. [CrossRef] [PubMed]
9. Xu, D.; Deng, X.; Huang, K.; Liu, Y.; Yong, Z.; Liu, S. Relationships between labor migration and cropland abandonment in rural China from the perspective of village types. *Land Use Policy* **2019**, *88*, 104164. [CrossRef]
10. Deng, X.; Zeng, M.; Xu, D.; Qi, Y. Does Social Capital Help to Reduce Farmland Abandonment? Evidence from Big Survey Data in Rural China. *Land* **2020**, *9*, 360. [CrossRef]
11. Long, H.; Wu, X.; Wang, W.; Dong, G. Analysis of Urban-Rural Land-Use Change during 1995-2006 and Its Policy Dimensional Driving Forces in Chongqing, China. *Sensors* **2008**, *8*, 681–699. [CrossRef]
12. Deng, X.; Huang, J.; Rozelle, S.; Zhang, J.; Li, Z. Impact of urbanization on cultivated land changes in China. *Land Use Policy* **2015**, *45*, 1–7. [CrossRef]
13. Grainger, A. National land use morphology: Patterns and possibilities. *Geography* **1995**, *80*, 235–245. Available online: https://www.jstor.org/stable/40572668 (accessed on 14 April 2021).
14. Long, H. Land Use Transition: A New Integrated Approach of Land Use/Cover Change Study. *Geogr. Territorial Res.* **2003**, *19*, 87–90. (In Chinese)
15. Long, H.; Li, X. Cultivated-land Transition and Land Consolidation and Reclamation in China: Research Progress and Frame. *Prog. Geogr.* **2006**, *25*, 67–76. (In Chinese)
16. Long, H.; Tu, S. Land Use Transition and Rural Vitalization. *Chin. Land Sci.* **2018**, *32*, 1–6. (In Chinese)
17. Long, H.; Qu, Y. Land use transitions and land management: A mutual feedback perspective. *Land Use Policy* **2018**, *74*, 111–120. [CrossRef]
18. Long, H.; Ge, D.; Wang, J. Progress and prospects of the coupling research on land use transitions and rural trans-formation development. *Acta Geogr. Sinica.* **2019**, *74*, 2547–2559. (In Chinese)
19. Long, H.; Chen, K. Urban-rural integrated development and land use transitions: A perspective of land system science. *Acta Geogr. Sinica.* **2021**, *76*, 295–309. (In Chinese)
20. Grainger, A. The Future Role of The Tropical Rain Forests in The World Forest Economy. Ph.D. Thesis, University of Oxford, Oxford, UK, 1986.
21. Guo, S.; Zhang, P. The need for land use transition near Shenzhen special economic zone from Guanlan. *Plan. Perspecitves.* **2008**, *24*, 72–77. (In Chinese)
22. Yang, Y.; Yang, X. Research on urban spatial expansion and land use inner structure transformation of the large valley-basin cities in China from 1949 to 2005 & mdash | A case study of Lanzhou. *J. Nat. Resour.* **2009**, *24*, 37–49. (In Chinese)
23. Long, H. Land Use Transition and Rural Transformation Development. *Prog. Geogr.* **2012**, *31*, 131–138. (In Chinese)
24. Long, H. Land use transition and land management. *Geogr. Res.* **2015**, *34*, 1607–1618. (In Chinese)
25. Pi, X.; Lou, B. Reasons analysis and countermeasures of rural land abandonment. *Seeker* **2002**, *1*, 25–27. (In Chinese)
26. Milenov, P.; Vassilev, V.; Vassileva, A.; Radkov, R.; Samoungi, V.; Dimitrov, Z.; Vichev, N. Monitoring of the risk of farmland abandonment as an efficient tool to assess the environmental and socio-economic impact of the Common Agriculture Policy. *Int. J. Appl. Earth Obs. Geoinf.* **2014**, *32*, 218–227. [CrossRef]
27. Guo, S.; Lin, L.; Liu, S.; Wei, Y.; Xu, D.; Li, Q.; Su, S. Interactions between sustainable livelihood of rural household and agricultural land transfer in the mountainous and hilly regions of Sichuan, China. *Sustain. Dev.* **2019**, *27*, 725–742. [CrossRef]
28. Long, H.; Tu, S.; Ge, D.; Li, T.; Liu, Y. The allocation and management of critical resources in rural China under restructuring: Problems and prospects. *J. Rural. Stud.* **2016**, *47*, 392–412. [CrossRef]
29. Xu, D.; Ma, Z.; Deng, X.; Liu, Y.; Huang, K.; Zhou, W.; Yong, Z. Relationships between Land Management Scale and Livelihood Strategy Selection of Rural Households in China from the Perspective of Family Life Cycle. *Land* **2020**, *9*, 11. [CrossRef]
30. Hong, Z.; Sun, Y. Power, capital, and the poverty of farmers' land rights in China. *Land Use Policy* **2020**, *92*, 104471. [CrossRef]
31. Huang, K.; Deng, X.; Liu, Y.; Yong, Z.; Xu, D. Does off-Farm Migration of Female Laborers Inhibit Land Transfer? Evidence from Sichuan Province, China. *Land* **2020**, *9*, 14. [CrossRef]

32. Jiang, S.; Su, Q. Off-farm Employment of Rural Labor Force and Land Transfer—A Study Based on a Dynamic Perspective. *Econ. Surv.* **2012**, *2*, 110–114. (In Chinese)

33. Gao, J.; Liu, Y.; Chen, J. China's initiatives towards rural land system reform. *Land Use Policy* **2020**, *94*, 104567. [CrossRef]

34. Liu, Y.; Zou, L.; Wang, Y. Spatial-temporal characteristics and influencing factors of agricultural eco-efficiency in China in recent 40 years. *Land Use Policy* **2020**, *97*, 104794. [CrossRef]

35. Zhou, Y.; Li, X.; Liu, Y. Rural land system reforms in China: History, issues, measures and prospects. *Land Use Policy* **2020**, *91*, 104330. [CrossRef]

36. Davis, J.; Lopez-Carr, D. Migration, remittances and smallholder decision-making: Implications for land use and livelihood change in Central America. *Land Use Policy* **2014**, *36*, 319–329. [CrossRef] [PubMed]

37. Rignall, K.; Kusunose, Y. Governing livelihood and land use transitions: The role of customary tenure in southeastern Morocco. *Land Use Policy* **2018**, *78*, 91–103. [CrossRef]

38. Soini, E. Land use change patterns and livelihood dynamics on the slopes of Mt. Kilimanjaro, Tanzania. *Agric. Syst.* **2005**, *85*, 306–323. [CrossRef]

39. Xie, F.; Xu, D.; Zhang, J.; Yang, Y. Change and Influencing Factors of Farmers' Livelihood Based on the Policy of Returning Farmland to Forest. *Guizhou Agric. Sci.* **2015**, *43*, 220–226. (In Chinese)

40. Yi, F.; Chen, Z. Analysis of the influence of returning farmland to forest on non-agricultural employment. *Soft Sci.* **2006**, *8*, 31–40. (In Chinese)

41. Xu, D.; Yong, Z.; Deng, X.; Zhuang, L.; Qing, C. Rural-Urban Migration and its Effect on Land Transfer in Rural China. *Land* **2020**, *9*, 81. [CrossRef]

42. Zhao, L.; Kang, X.; Shi, J. The Empirical Analysis on the Effects of the Farmland Transfer on Household Trans-formation. *Chin. J. Agric. Resour. Reg. Plann.* **2017**, *38*, 158–162. (In Chinese)

43. Wang, Q.; Guan, R.; Yu, J. Impacts of Risk Attitude and Risk Perception on Land Transfer—Based on Panel Data of 1429 Households in Henan, Shandong, Hebei, Anhui, and Jiangsu. *J. Huazhong. Agric. Univ.* **2019**, *2019*, 149–158. (In Chinese)

44. Zhang, Y.; Bai, Y.; Zhen, L.; Xin, L. Could the new rural social pension insurance promote farmers' land transfer: Based on three waves panel data of CHARLS. *J. Nat. Resour.* **2019**, *34*, 1016–1026. (In Chinese)

45. He, C.; Peng, L.; Liu, S.; Xu, D.; Xue, P. Factors influencing the efficiency of rural public goods investments in mountainous areas of China—Based on micro panel data from three periods. *J. Rural. Stud.* **2016**, *47*, 612–621. [CrossRef]

46. Rozelle, S. Stagnation Without Equity: Patterns of Growth and Inequality in China's Rural Economy. *China J.* **1996**, *35*, 63–92. [CrossRef]

47. Shui, Y.; Xu, D.; Liu, Y.; Liu, S. The Influence of Human Capital and Social Capital on the Gendered Division of Labor in Peasant Family in Sichuan, China. *Soc. Indic. Res.* **2021**, *18*, 1–18. [CrossRef]

48. Xue, K.; Xu, D.; Liu, S. Social Network Influences on Non-Agricultural Employment Quality for Part-Time Peasants: A Case Study of Sichuan Province, China. *Sustainability* **2019**, *11*, 4134. [CrossRef]

49. Barrett, C.; Reardon, T.; Webb, P. Nonfarm income diversification and household livelihood strategies in rural Africa: Concepts, dynamics, and policy implications. *Food Policy* **2001**, *26*, 315–331. [CrossRef]

50. Donohue, C.; Biggs, E. Monitoring socio-environmental change for sustainable development: Developing a Multidimensional Livelihoods Index (MLI). *Appl. Geogr.* **2015**, *62*, 391–403. [CrossRef]

51. Liu, Z.; Liu, L. Characteristics and driving factors of rural livelihood transition in the east coastal region of China: A case study of suburban Shanghai. *J. Rural. Stud.* **2016**, *43*, 145–158. [CrossRef]

52. Su, F.; Xu, Z.; Shang, H. An Overview of Sustainable Livelihoods Approach. *Adv. Earth Sci.* **2009**, *24*, 61–69. (In Chinese)

53. He, R.; Liu, S.; Chen, G.; Xie, F.; Yang, X.; Liang, L. Research progress and tendency of sustainable livelihoods for peasant household in China. *Adv. Earth Sci.* **2003**, *32*, 657–670. (In Chinese)

54. Mensah, E.J. The sustainable livelihood framework: A reconstruction. In *Global Youth Alliance*; l'Oasis Development Group: Accra, Ghana, 2011.

55. Scoones, I. Sustainable rural livelihoods: A framework for analysis. In *IDS Working Paper 52*; IDS: Brighton, UK, 1998.

56. Ma, L.; Long, H.; Tu, S.; Zhang, Y.; Zheng, Y. Farmland transition in China and its policy implications. *Land Use Policy* **2020**, *92*, 104470. [CrossRef]

57. Ji, H.; Cai, Y. Influence of heterogeneous livelihood capital on farm households' land transfer behavior—A case study of 516 respondents in Wuhan suburb. *Changjiang Liuyu Ziyuan Yu Huanjing* **2017**, *26*, 220–226. (In Chinese)

58. Long, H.; Qu, Y.; Tu, S.; Zhang, Y.; Jiang, Y. Development of land use transitions research in China. *J. Geogr. Sci.* **2020**, *30*, 1195–1214. [CrossRef]

59. Zhang, G. Analysis on Current Situation of Livelihood Capital of Farmers on Land Transfer in Jilin Province under the Framework of Sustainable Livelihood Analysis. *Rural Econ. Sci. Technol.* **2018**, *29*, 142–143. (In Chinese)

60. Zhang, Y.; Long, H.; Li, Y.; Ge, D.; Tu, S. How does off-farm work affect chemical fertilizer application? Evidence from China's mountainous and plain areas. *Land Use Policy* **2020**, *99*, 104848. [CrossRef]

61. Muchomba, F.M. Women's Land Tenure Security and Household Human Capital: Evidence from Ethiopia's Land Certification. *World Dev.* **2017**, *98*, 310–324. [CrossRef]

62. Zhang, Y.; Halder, P.; Zhang, X.; Qu, M. Analyzing the deviation between farmers' Land transfer intention and behavior in China's impoverished mountainous Area: A Logistic-ISM model approach. *Land Use Policy* **2020**, *94*, 104534. [CrossRef]

63. Zhu, L.; Cai, Y. The Impacts of Farmer Households' Livelihood Endowment on Farmland Transfer: Cases in Different Types of Functional Areas of Hubei Province. *J. Nat. Resour.* **2016**, *31*, 1526–1539. (In Chinese)
64. Ge, D.; Long, H.; Qiao, W.; Wang, Z.; Sun, D.; Yang, R. Effects of rural–urban migration on agricultural transformation: A case of Yucheng City, China. *J. Rural. Stud.* **2020**, *76*, 85–95. [CrossRef]
65. Yuan, D.; Chen, M.; Liao, C.; Xie, X.; Liao, X.; Yao, D. Differentiation of Livelihood Capital of Farmers and the Change of Livelihood Strategy. *Res. Soil Water Conserv.* **2019**, *26*, 349–354. (In Chinese)
66. Liu, M.; Yang, L.; Bai, Y.; Min, Q. The impacts of farmers' livelihood endowments on their participation in eco-compensation policies: Globally important agricultural heritage systems case studies from China. *Land Use Policy* **2018**, *77*, 231–239. [CrossRef]
67. Liu, Z.; Chen, Q.; Xie, H. Comprehensive Evaluation of Livelihood Assets of Farmers in Western Mountains—Taking Zunyi City as an Example. *J. Resour. Ecol.* **2018**, *9*, 154–163. (In Chinese)
68. Martin, S.J.; Clapp, J. Finance for Agriculture or Agriculture for Finance? *J. Agrar. Chang.* **2015**, *15*, 549–559. [CrossRef]
69. Su, C.; Bi, R.; Liu, H.; Guo, Y. Land Transfer and Abandonment Behavior of Rural Households in Poor Mountainous Areas: A Study of 275 Farmers in Heshun County, Shanxi Province. *J. Agric.* **2019**, *9*, 89–96. (In Chinese)
70. Wang, Q.; Gen, S.; Gui, M. Empirical Analysis of Land Leasing Behavior. *Neimenggu Nongye Daxue Xuebao Ziran Kexueban* **2007**, *4*, 45–47. (In Chinese)
71. Wen, B.; Li, Z. A Study on the Departure between Farmers' Willing and Behavior of Land Transfer under Livelihood Capital. *South. Agric. Mach.* **2020**, *51*, 40–62. (In Chinese)
72. Lu, J.; He, Q. Analysis of Farmers' Willingness to Land Transfer and Its Influencing Factors from the Perspective of Livelihood—Based on Field Investigation of LongCun, a County in Henan Province. *Rural Econ.* **2016**, *2*, 39–43. (In Chinese)
73. The UK's Department for International Development. *Sustainable Livelihoods Guidance Sheets*; DFID: London, UK, 1999; p. 445.
74. Kuang, F.; Jin, J.; He, R.; Ning, J.; Wan, X. Farmers' livelihood risks, livelihood assets and adaptation strategies in Rugao City, China. *J. Environ. Manag.* **2020**, *264*, 110463. [CrossRef]
75. Ekins, P.; Simon, S.; Deutsch, L.; Folke, C.; De Groot, R. A framework for the practical application of the concepts of critical natural capital and strong sustainability. *Ecol. Econ.* **2003**, *44*, 165–185. [CrossRef]
76. Zeng, X.; Guo, S.; Deng, X.; Zhou, W.; Xu, D. Livelihood risk and adaptation strategies of farmers in earthquake hazard threatened areas: Evidence from sichuan province, China. *Int. J. Disaster Risk Reduct.* **2021**, *53*, 101971. [CrossRef]
77. He, A.; Yang, X.; Chen, J.; Wang, Z. Impact of Rural Tourism Development on Farmer'S Livelihoods—A Case Study of Rural Tourism Destinations in Northern Slop of Qinling Mountains. *Econ. Geogr.* **2014**, *34*, 174–181. (In Chinese)
78. Ning, Z. Rural Household's Sustainable Livelihood Capitals and Targeting Poverty. *Huanan Nongye Daxue Xuebao* **2017**, *16*, 86–94. (In Chinese)
79. Xu, D.; Liu, E.; Wang, X.; Tang, H.; Liu, S. Rural Households' Livelihood Capital, Risk Perception, and Willingness to Purchase Earthquake Disaster Insurance: Evidence from Southwestern China. *Int. J. Environ. Res. Public Health* **2018**, *15*, 1319. [CrossRef]
80. Zhang, C.; Peng, C.; Kong, X. Evolutionary Logic, Historical Evolution and Future Prospect of Rural-household Dif-ferentiation. *Reform* **2019**, *300*, 5–16. (In Chinese)
81. Zhou, W.; Guo, S.; Deng, X.; Xu, D. Livelihood resilience and strategies of rural residents of earthquake-threatened areas in Sichuan Province, China. *Nat. Hazards* **2021**, *106*, 255–275. [CrossRef]
82. Li, Y.; Li, Y.; Westlund, H.; Liu, Y. Urban–rural transformation in relation to cultivated land conversion in China: Implications for optimizing land use and balanced regional development. *Land Use Policy* **2015**, *47*, 218–224. [CrossRef]
83. Yan, X.; Bauer, S.; Huo, X. Farm Size, Land Reallocation, and Labour Migration in Rural China. *Popul. Space Place* **2013**, *20*, 303–315. [CrossRef]
84. Alonso-Sarría, F.; Martínez-Hernández, C.; Romero-Diaz, A.; Cánovas-García, F.; Gomariz-Castillo, F. Main Environmental Features Leading to Recent Land Abandonment in Murcia Region (Southeast Spain). *Land Degrad. Dev.* **2015**, *27*, 654–670. [CrossRef]
85. Baumann, M.; Kuemmerle, T.; Elbakidze, M.; Ozdogan, M.; Radeloff, V.C.; Keuler, N.S.; Prishchepov, A.V.; Kruhlov, I.; Hostert, P. Patterns and drivers of post-socialist farmland abandonment in Western Ukraine. *Land Use Policy* **2011**, *28*, 552–562. [CrossRef]
86. Queiroz, C.; Beilin, R.; Folke, C.; Lindborg, R. Farmland abandonment: Threat or opportunity for biodiversity conservation? A global review. *Front. Ecol. Environ.* **2014**, *12*, 288–296. [CrossRef]
87. Cao, Z.; Li, Y.; Liu, Y.; Chen, Y.; Wang, Y. When and where did the Loess Plateau turn "green"? Analysis of the tendency and breakpoints of the normalized difference vegetation index. *Land Degrad. Dev.* **2017**, *29*, 162–175. [CrossRef]
88. Levers, C.; Schneider, M.; Prishchepov, A.V.; Estel, S.; Kuemmerle, T. Spatial variation in determinants of agricultural land abandonment in Europe. *Sci. Total Environ.* **2018**, *644*, 95–111. [CrossRef] [PubMed]
89. Li, S.; Li, X.; Sun, L.; Cao, G.; Fischer, G.; Tramberend, S. An estimation of the extent of cropland abandonment in mountainous regions of China. *Land Degrad. Dev.* **2018**, *29*, 1327–1342. [CrossRef]
90. Zheng, J. A preliminary study on the interaction mechanism between farmland right confirmation and farmland transfer. *Rural Econ.* **2009**, *8*, 23–26. (In Chinese)
91. Liu, Y.; Xu, H. Influence of farmland right to rural land circulation—Based on the perspective of farmers differen-tiation. *J. Arid Land Resour. Environ.* **2016**, *30*, 25–29. (In Chinese)
92. Shao, J.; Wei, C.; Xie, D. Farmers' explanations of land transfer under the household responsibility system: The results from seven villages' analysis in Chongqing. *Geogr. Res.* **2007**, *26*, 275–286. (In Chinese)

93. Shao, J.; Zhang, S.; Li, X. Effectiveness of farmland transfer in alleviating farmland abandonment in mountain regions. *J. Geogr. Sci.* **2015**, *26*, 203–218. [CrossRef]
94. Zhu, D.; Xu, D. Investigation and analysis of the problem of rural land abandonment. *Issues Agric. Econ.* **2000**, *12*, 10–13. (In Chinese)
95. Bian, Q.; Zhou, S.; Ge, J. Analysis on the Influencing Factors of Land Transfer in Developed Region—Case of Zhejiang, China. *J. Agrotech. Econ.* **2010**, *6*, 28–36. (In Chinese)
96. Peng, Q.; Wang, C.; Deng, C. Potential land transfer-out contracted households, the spatial organization of agricul-tural management and livelihood capital. *Resour. Sci.* **2017**, *39*, 1477–1487. (In Chinese)
97. Liu, K.; Lin, J. Farmland Rental Markets Circulation and Land Redistribution: Theoretic and Empirical Analysis. *J. Quant. Tech. Econ.* **2005**, *11*, 99–111. (In Chinese)

*Article*

# Analyzing Characteristics and Implications of the Mortgage Default of Agricultural Land Management Rights in Recent China Based on 724 Court Decisions

Hongguang Zheng [1,2,*] and Zhanbin Zhang [1,3]

1  School of Business Administration, Northeastern University, Shenyang 110819, China; zhangzhanbin@263.net
2  School of Economic Crime Investigation, Criminal Investigation Police University of China, Shenyang 110854, China
3  School of Marxism, Party School of the Central Committee of C.P.C. (National Academy of Governance), Beijing 100089, China
*  Correspondence: zhenghongguang@cipuc.edu.cn

**Abstract:** The transfer of rural land contractual management rights belongs to the recessive transition of land use. The mortgage of rural land management rights is a way of rural land circulation, and has an important impact on the transformation of land use. Rural land management rights mortgage loans can enable farmers to obtain more credit funds, which is conducive to agricultural development and Rural Revitalization. However, with the development of rural land mortgage financing, the associated risk has become increasingly prominent. The most typical risk is the default risk of farmers' mortgage loans. Based on court decisions regarding rural land mortgage default during 2014–2020, this paper analyzes the characteristics of farmers' default in different periods and locations. The empirical results reveal that the time and space of rural land mortgage default cases are widely distributed in China, especially in Heilongjiang Province. In the default judgement, the loan amount of CNY 50,000 to CNY 100,000 and the loan periods of 1 year accounted for the highest proportion. When making mortgage loan policies for rural land management rights, financial institutions should give farmers the most preferential treatment regarding the amount, term and interest rate of loans. Farmers' social security should be improved, and agricultural insurance should be strengthened. Meanwhile, the credit review of small and short-term loan farmers should be heightened.

**Keywords:** land use transition; rural land management right; mortgage default; default characteristics; China

**Citation:** Zheng, H.; Zhang, Z. Analyzing Characteristics and Implications of the Mortgage Default of Agricultural Land Management Rights in Recent China Based on 724 Court Decisions. *Land* **2021**, *10*, 729. https://doi.org/10.3390/land10070729

Academic Editors: Hualou Long, Xiangbin Kong, Shougeng Hu and Yurui Li

Received: 9 June 2021
Accepted: 9 July 2021
Published: 12 July 2021

## 1. Introduction

China is a large agricultural country where agricultural land is the foundation of the rural social and economic system [1,2]. The system of agricultural land is not only closely related to agriculture, rural areas and farmers, but also directly affects the overall development level of the national economy. China's agricultural land system has its own particularities. Land belongs to the state or collective, and private ownership is not allowed. In the past, the law prohibited the mortgage of agricultural land. In recent years, the separation of the ownership, contracting rights and management rights of agricultural land was proposed, and mortgage financing of land management rights was allowed in China [3–6]. The rural land mortgage system has Chinese characteristics. The Central Committee of the Communist Party of China (CPC) has long promoted comprehensive rural reform and supply-side structural reform, and the reform of the agricultural land system has been the core content [7–9]. After a long period of practice, the system of agricultural land mortgaging in developed countries is relatively mature [10,11]. With the introduction of the concept of land use transition into China [12–15], woodland and cultivated land have been the hot spots of land use transition research [16,17]. The transfer of rural land contractual management rights belongs to the recessive transformation of land use. The mortgage of rural land management rights is a way of rural land circulation, which will have an important impact on the transformation of land use. Rural land management rights mortgage loans can enable farmers to obtain more credit funds, which is conducive to agricultural development and rural revitalization.

The mortgage financing of farmland management rights is an important means for the government to support agriculture through the financial market and plays a positive and effective role in the development of rural finance [18–20]. However, as a new financial product, there are still many obstacles and restrictive factors in the financing of farmland mortgages in China. Among them, the risk problem is the greatest obstacle, which restricts the development of farmland mortgage financing and affects the implementation of mortgage financing through farmland management rights. Therefore, based on promoting land circulation, preventing and controlling the default risk of the mortgage of farmland management rights, and minimizing the cost of financial institutions supporting agriculture, rural areas and farmers have become a topic of wide concern to the state and all sectors of society.

In this regard, many studies have addressed risk types, empirical cases and the risk control of farmland management right mortgages. Agricultural land mortgage financing entails many types of risks, such as credit, nature, market (operation) and policy (system) [21–25]. The risks of agricultural land mortgage financing are reflected in the risk of farmers' livelihood and the repayment source risk of banks at the micro level and in the rural social risk and rural financial risk at the macro level [26]. The regression analysis method and AHP method were used to demonstrate the factors that affect credit risk and predict the probability of default [27–29]. Studies have shown that the bank credit system, relevant systems, mortgage and disposal conditions, and risk compensation and sharing mechanisms were the key points of risk management [26,30].

In addition, the unclear property rights of farmland as collateral and high market transaction costs were the main causes of the risks perceived by financial institutions, such that the institutions did not actively lend to farmers who applied for loans with such collateral [31,32]. A prerequisite for effective agricultural land mortgage development is the development of effective instruments for the risk management of creditors in the pledging of agricultural land [33]. Yin [34] conducted empirical research on the risk measurement of mortgage loans on rural land contracts and management rights in Heilongjiang Province.

Previous studies on rural land mortgages have mainly focused on the willingness of actors on the supply and demand sides, financial innovation mechanisms and performance, and loan risk evaluation systems, and these studies mainly used the questionnaire method or model prediction within a certain area. The content of such surveys reflects the ideas of the respondents, not the objective situation, and conclusions based on such information lack scientific support. The above studies are important, but there is no precedent for statistics of rural land mortgage default cases nationwide.

In recent years, with the help of big data, legal judgment documents are increasingly applied to many fields [35,36]. This article uses the empirical research method to study the cases of rural land mortgage default judged by the first instance of the national court during 2014–2020. The 724 default cases in this article are all confirmed cases by the court and the data are true and reliable. According to the phenomenon of rural mortgage default, the formation mechanism of rural land mortgage loans is analyzed. Along with the court cases, the competent department of agricultural land mortgage finance of Heilongjiang Province is investigated. The research method of this article is highly objective and rigorous. It is of great significance to understand the characteristics of rural land mortgage default from all over the country and to reduce the risk of default.

## 2. Empirical Approach and Data

### 2.1. Sources of Data

Based on the key words of "rural land mortgage", 868 civil judgements of the first instance of the national court were retrieved from the "China judgement documents website", of which 724 were effective without repeated judgements. The "China judgment documents website" is a national platform for publishing court judgement documents established by the Supreme People's Court of the People's Republic of China. According to the requirements of the Supreme People's court, the Supreme People's court, all higher people's courts and intermediate people's courts across the country must publish judgement documents on the "China judgement documents website" from 1 January 2014. In addition, the basic people's courts of 10 eastern provinces, including Beijing, Tianjin and Liaoning, and three central and western provinces, Henan, Guangxi and Shaanxi, should publish their judgement documents online. Since the end of June 2015, the courts at all three levels in 31 provinces (autonomous regions and municipalities) and Xinjiang Production and Construction Corps have all published effective judgement documents online. Therefore, the data source of this paper is authoritative, and the cases retrieved in this paper are comprehensive. In this paper, the court judgement time of farmland mortgage default cases is continuous from 2014 to 2020.

*2.2. Empirical Approach*

We adopted the empirical analysis method. Our empirical analysis focused on three issues. First, we combed each case and set specific indicators such as judgment court, judgment time, natural situation of borrower, name of lender, loan amount, loan term, loan interest rate, etc. Second, we summarized the indicators and find the common characteristics. Finally, we analyzed the causes of default and propose solutions.

## 3. Results

The comprehensive quality of borrowers plays a decisive role in the operation and management ability of farmers' families, and can directly affect the use behavior of farmers' credit funds. Among them, age is an important factor to reflect the repayment ability of the lender. Therefore, this paper analyzed the age characteristics of the defaulter. Similarly, the characteristics of the rural land mortgage also play an important role in the analysis of farmers' default characteristics, so this article also analyzes the loan characteristics of default farmers.

*3.1. Trend of the Default Cases*

Figure 1 shows the proportion of default cases of farmland mortgages in each year from 2014 to 2020 in China. China's law once prohibited the mortgage of farmland management rights, and there were few judgements on mortgage default cases before 2016. On 27 December 2015, authorized by the committee of the National People's Congress (NPC), the State Council implemented "The Property Law" and "The Guarantee Law", outlining the provisions that the right to use collectively owned cultivated land shall not be mortgaged in the administrative areas of 232 pilot counties (cities and districts). Since 2016, the "two-right" mortgage loan pilot project has been authorized by law, which has significantly stimulated the rural financial market. In 2018, the rural land contract law was amended to allow the mortgage of rural land management rights nationwide. With the guidance and publicity of local governments and financial departments, the number of rural land management right mortgage loans has gradually increased since 2016. In the early stage of loans, some borrowers had the impulse to borrow. The impulse of borrowing is the borrower's cognitive deviation. The borrower does not consider his own actual situation, has the herd mentality when borrowing and has no proper use after borrowing, which leads to the failure to repay the loan on time; thus, there was a large number of default cases that peaked in 2018. The court decision shows that the financial institutions in the pilot areas sued the court, and the farmers who had borrowed money protested on the grounds that the mortgage violated legal provisions. As the loan review of financial institutions became stricter and borrowers began to make wiser decisions, loan default cases started a downward trend in 2019. Moreover, financial institutions have explored other ways to address risk, such as requiring the government to provide guarantees, setting up risk funds, and adopting multiple guarantees [37]. These measures have effectively addressed the default risk of borrowers to a certain extent.

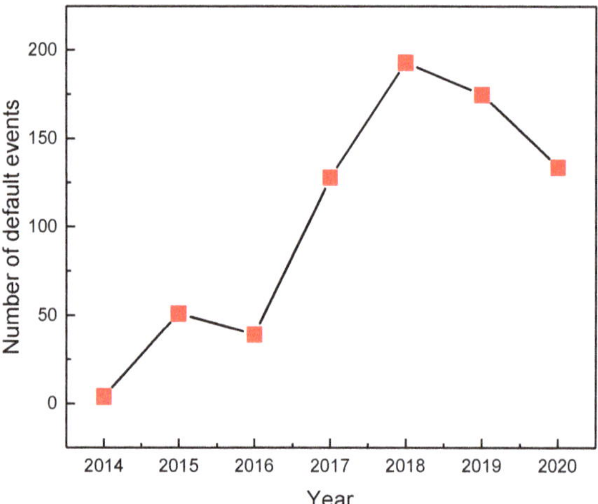


**Figure 1.** Annual distribution of 724 farmland mortgage default events in China from 2014 to 2020.

### 3.2. Spacial Characteristics of the Default Cases

Figure 2 shows the regions where farmers defaulted on agricultural land mortgage loans in China from 2014 to 2019, with 19 provinces and autonomous regions affected; thus, the coverage area was relatively wide. From Figure 2, the provinces with a high number of default cases are Heilongjiang, Jilin and Inner Mongolia. There were 556 defaulting households in Heilongjiang Province, accounting for 77% of the total defaulting households. Heilongjiang Province presents the largest number of farmers defaulting on farmland mortgage loans. Heilongjiang Province is a large agricultural province in China, with 239 million mu of arable land, accounting for one ninth of the arable land in the country. The per capita arable land of the agricultural population is more than 10 mu, ranking first in the country [38]. Thus, Heilongjiang has the material basis for farmland mortgage loans. The scale of agricultural land mortgage loans in Heilongjiang Province is far greater than that in other provinces, as is the number of default cases. On the other hand, as early as 2010, Heilongjiang Province formulated the "Heilongjiang Province rural land management right mortgage loan method (Trial)", selecting four cities and six counties to carry out the pilot work. In 2015, Heilongjiang was identified as the pilot area of land management right mortgage by the State Council, with 15 pilot districts and counties, ranking first in the country. By the end of 2017, the balance of rural land loans in Heilongjiang Province accounted for more than 30% of the total scale of the national pilot areas [39].

### 3.3. Loan Amount of the Default Cases

Table 1 reveals the relationship between the number of defaulting farmers and the amount of default. As shown in Table 1, the defaulting households with a loan amount of 50,000 to 100,000 account for the largest proportion, up to 33.7%, followed by the defaulting households with a loan amount of less than 50,000, accounting for 21%, and the defaulting households with a loan amount of 500,000 to 1 million represent the smallest proportion, accounting for only 6.5% in total.

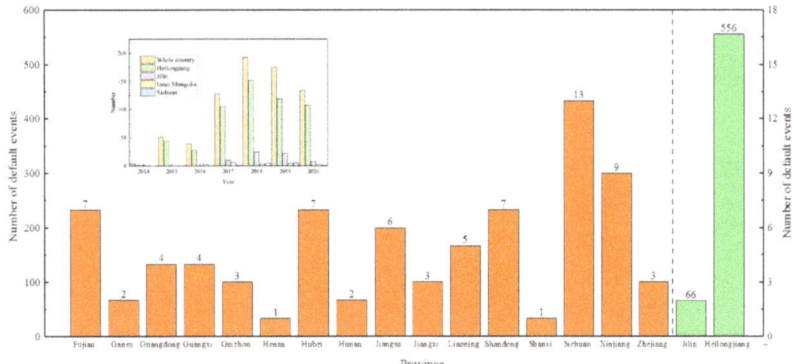

**Figure 2.** Number of default events of farmland mortgages during 2014–2020.

**Table 1.** The default amount and proportion of farmland mortgages.

| Loan Amount, CNY 10,000 | Numbers | Ratio, % |
| --- | --- | --- |
| $0 < x \leq 5$ | 152 | 21 |
| $5 < x \leq 10$ | 244 | 33.7 |
| $10 < x \leq 20$ | 139 | 19.2 |
| $20 < x \leq 50$ | 84 | 11.6 |
| $50 < x \leq 100$ | 47 | 6.5 |
| $100 < x$ | 58 | 8 |
| Total | 724 | 100 |

In all 724 default cases, 694 borrowers were natural persons, and 30 borrowers were companies or agricultural operation organizations. This is related to the area of mortgageable land owned by peasant households. The mortgage loan amount of rural land contracting and management rights is generally between 50 and 80% of the recognized value of the land assessed (including the attached objects on the ground), with different regulations in different regions.

Table 2 indicates that the default cases in which the borrowers are companies or cooperatives account for 4% of all default cases. Companies or cooperatives are the borrowers in 70% of the default cases with a loan amount of more than CNY 1 million.

**Table 2.** Default amount of farmland mortgage loans with the borrower being a company or agricultural operation organization.

| Loan Amount, CNY 10,000 | Numbers | Ratio, % |
| --- | --- | --- |
| $\leq 10$ | 1 | 3.33 |
| $20 \leq x \leq 50$ | 4 | 13.33 |
| $50 < x \leq 100$ | 4 | 13.33 |
| $>100$ | 21 | 70 |
| Total | 30 | 100 |

### 3.4. Loan Term of the Default Cases

Figure 3 shows the distribution characteristics of the defaulting farmers' age and the loan years in Heilongjiang Province from 2014 to 2020. As shown in Figure 3, the number of households with a loan term equal to 1 year is the largest, accounting for more than 40% of the total number of households with loans. Due to the high risk and volatility of agricultural operations, the loan term is relatively short. Generally, the mortgage loan term of rural land management rights is 1 year and, in principle, no more than 3 years. Terms of 5–10 years or more are also available in some areas, but they are few. From the characteristics of loan age, the largest number of farmers defaulting on farmland mortgage loans are between 40 and 50 years old, accounting for approximately 40% of each age group, followed by farmers between 50 and 60 years old, accounting for approximately 30% of the total. This is consistent with the results in previous studies [40]. The survey data [41] show that the age of agricultural labor force is middle-aged, with an average age of 48.5 years old. Men nearly 50 years old have become the main force in farming, and more than 60% of them are full-time agricultural producers. Most households with borrowing needs and borrowing behaviors are households whose heads are older than 40. Therefore, these households also account for the highest proportion of default events.

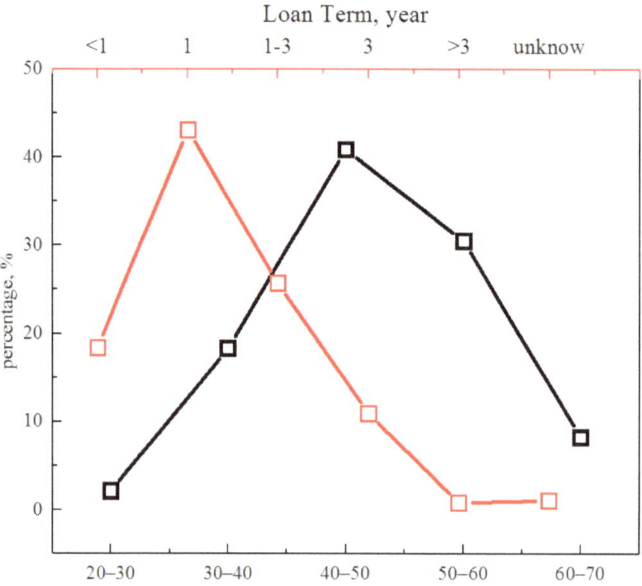


**Figure 3.** Age of the household head and the loan term of default farmland mortgage loans in Heilongjiang Province, starting time of loan, 2014–2020.

### 3.5. Loan Institutions and Interest Rates of the Default Cases

Figure 4 shows the financial institutions and the number of farmland mortgage default cases during 2014–2020. Figure 4 indicates that rural credit cooperatives and village commercial banks are involved in the most default cases, followed by the Agricultural Bank of China. The first consideration of financial institutions issuing loans is the security, profitability, and liquidity of loans. Farmland mortgage loans entail high risk, have a long cycle and offer poor profit-making, so financial institutions are often not willing to carry out such businesses. Rural credit cooperatives, village commercial banks and the Agricultural Bank of China have been engaged in the rural market for a long time, with their main business being related to agricultural funds, but other financial institutions rarely participate.

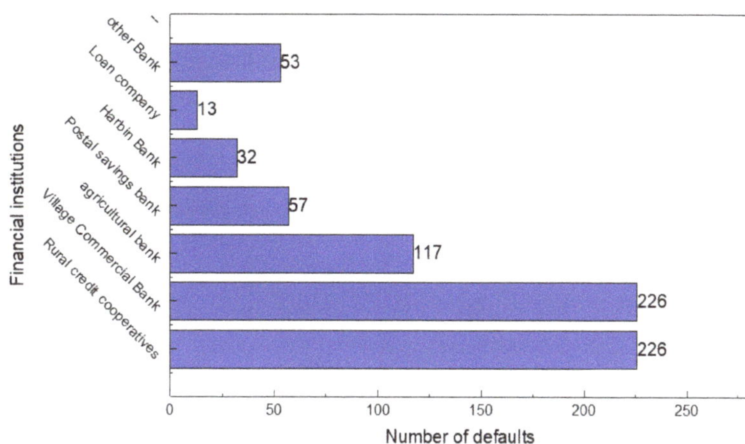


**Figure 4.** Financial institutions and the number of farmland mortgage default cases during 2014–2020.

In the statistical cases of farmland mortgage default, the monthly interest rate is concentrated between 7‰ and 10‰. Figure 5 shows the proportion of default cases of farmland mortgage with monthly interest rate between 7‰ and 10‰. The default cases with monthly interest rate of 7‰ and 9.6‰ accounted for 20.14% and 25.12%, respectively. In addition, the highest monthly interest rate is 14.895‰, and the lowest is 3.9887‰. Most loan contracts stipulate that after overdue, the monthly interest rate will be charged 50% as penalty interest based on the original interest rate. Additionally, there are eight contracts provide for a 30% rise in lending rates at the same level of benchmark lending rates at the People's Bank of China over the same period.

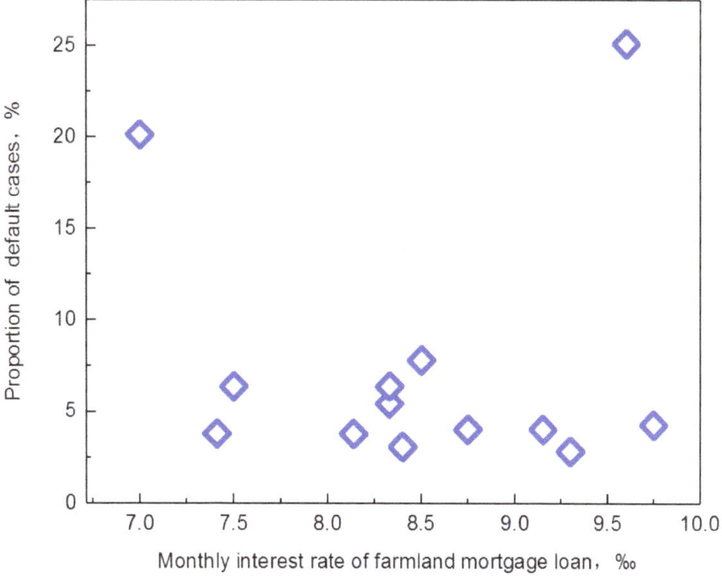


**Figure 5.** Farmland mortgage monthly interest rate.

## 4. Discussion

### 4.1. Reasons for Farmland Mortgage Default

#### 4.1.1. Frequent Agriculture Natural Disasters

China is a traditional agricultural country. As a developing country, China's agricultural infrastructure construction has received increasing attention, but it is still relatively weak. Agricultural natural disasters are the main cause of farmers' losses. Changes in natural conditions bring many uncertain factors to agricultural production and management [42]. Only from the data since the beginning of the new century, the annual loss of grain due to drought in China is as high as more than 30 billion kg, about 6% of the total grain output in the same period [43]. Failures in agricultural land management make farmers unable to repay loans. Turvey and Norton [44] proposed that the core assumption is that there exists a covariate relationship between the underlying weather event and crop loss. Its general form is given as follows:

$$H(Y(x_T), x_T) = \int_l^u h(Y(x_T), x_T) f(x_T) \mathrm{d}x_T \tag{1}$$

where

$Y(\vec{x}|x_T) =$ production function defined by the vector of inputs x $\sim$, and the random weather component $x_T$; and

$f(x_T) =$ probability distribution function capturing the universe of weather related risks.

$x_T =$ the covariate risk.

Natural disasters can significantly reduce agricultural output.

Table 3 [45] indicates that the losses caused by agricultural natural disasters are great. Agricultural insurance is not common in China, and agricultural risk mainly depends on farmers' self-relief. At the current stage, agricultural income is still the main economic source of agricultural operators and the first source of funds to repay mortgage loans. Once a large-scale natural disaster occurs, agricultural operators suffer great economic losses, are unable to repay loans and therefore violate the land management right mortgage contract.

**Table 3.** Agricultural natural disasters in China during 2010–2019.

| Index, $10^3$ HA | 2010 | 2011 | 2012 | 2013 | 2014 | 2015 | 2016 | 2017 | 2018 | 2019 |
|---|---|---|---|---|---|---|---|---|---|---|
| Covered Area | 37,426 | 32,471 | 24,962 | 31,350 | 24,891 | 21,770 | 26,221 | 18,478 | 20,814 | 19,257 |
| Covered by Flood | 17,525 | 6863 | 7730 | 8757 | 4718 | 5620 | 8531 | 5415 | 3950 | 6680 |
| Covered by Drought | 13,259 | 16,304 | 9340 | 14,100 | 12,272 | 10,610 | 9873 | 9875 | 7712 | 7838 |
| Covered by wind and hail | 2180 | 3309 | 2781 | 3387 | 3225 | 2918 | 2908 | 2268 | 2407 | |
| Covered by freezing | 4121 | 4447 | 1618 | 2320 | 2133 | 900 | 2885 | 525 | 3413 | |
| Affected Area | 18,538 | 12,441 | 11,475 | 14,303 | 12,678 | 12,380 | 13,670 | 9201 | 10,569 | 7913 |
| Affected by Flood | 7024 | 2840 | 4145 | 4859 | 2704 | 3327 | 4338 | 3022 | 2551 | 2612 |
| Affected by Drought | 8987 | 6599 | 3509 | 5852 | 5677 | 5863 | 6131 | 4444 | 2621 | 3332 |
| Affected by Hail | 916 | 1348 | 1368 | 1682 | 2193 | 1825 | 1424 | 1238 | 1548 | |
| Affected by Freezing | 1444 | 1291 | 795 | 885 | 933 | 474 | 1179 | 312 | 1870 | |

#### 4.1.2. Huge Agricultural Market Risk

Agricultural system is not only highly dependent on the natural environment, but also highly dependent on the market [46,47]. Once the market environment changes adversely, it will bring serious uncertainty and uncontrollable to agricultural production.

The Chinese agricultural market is underdeveloped, and the distribution is different. Eastern China has a large number and large scale of production market, which has a significant role in promoting industry and agricultural products circulation, while the central and western regions have a small number of production market and low construction standards. The trading and settlement of the origin market is relatively backward, inefficient and risky, and it is difficult to form an open and fair transaction price.

Under the background of economic globalization, agriculture is faced with not only domestic market risks, but also international market risks. The uncertainty of market risks increases (Figure 6), many factors are often superimposed, and the price fluctuates greatly. International grain price fluctuations and impact on Chinese grain prices (Figure 7).

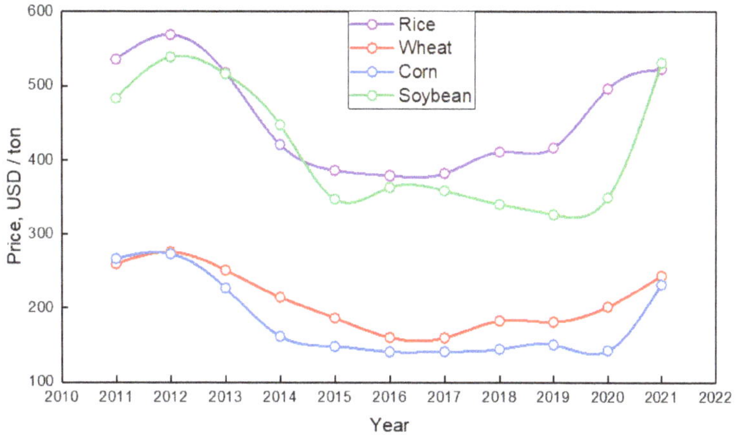


**Figure 6.** International grain prices, 2011–2021 (Ministry of Agriculture and Rural Affairs of the People's Republic of China).

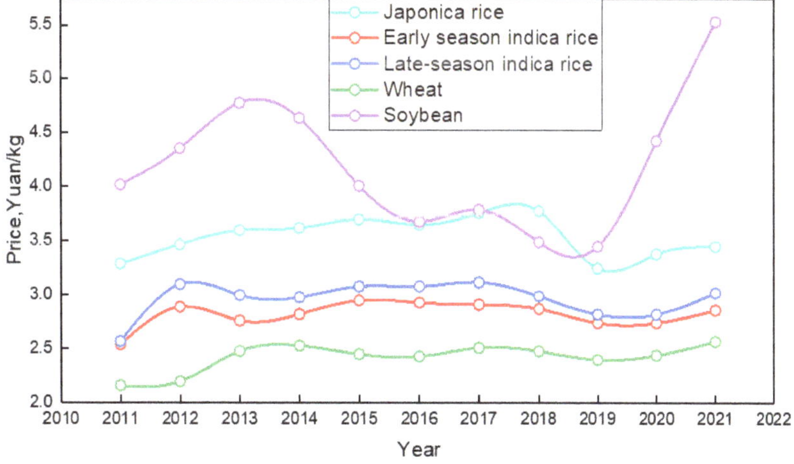


**Figure 7.** Annual changes of major grain purchase prices in China, 2011–2021 (Ministry of Agriculture and Rural Affairs of the People's Republic of China).

### 4.1.3. High Cost of Agricultural Production

Chinese agriculture has entered the era of high production cost. In the increment of agricultural production input, the increase of direct production cost is the main factor to promote the increase of total agricultural production cost. Costs of seeds, fertilizers, pesticides, agricultural films, machinery operations, irrigation and drainage, land rent, labor, etc., accounting for over 80% of total costs [48].

High production costs have pushed up food prices, reduced agricultural operating income and damaged agricultural competitiveness. As shown in Table 4, the domestic grain price is close to or even higher than the international grain import to the shore tax price, the grain trade deficit situation, and this situation is expanding year by year.

**Table 4.** Changes in Foreign Trade of Agricultural Products in China from 2011 to 2019 (billion USD) [1].

| Year | Import and Export Volume | Exports | Imports | Exports − Imports |
|---|---|---|---|---|
| 2011 | 1556.23 | 607.51 | 948.72 | −341.21 |
| 2012 | 1757.68 | 632.89 | 1124.79 | −491.90 |
| 2013 | 1866.92 | 678.25 | 1188.67 | −510.42 |
| 2014 | 199.29 | 719.60 | 1225.38 | −505.78 |
| 2015 | 1875.62 | 706.82 | 1168.81 | −461.99 |
| 2016 | 1845.55 | 729.86 | 115.69 | −385.83 |
| 2017 | 2013.88 | 755.32 | 1258.56 | −503.24 |
| 2018 | 2177.08 | 804.48 | 1372.60 | −568.12 |
| 2019 | 2300.68 | 790.98 | 1509.70 | −718.72 |

[1] Ministry of Agriculture and Rural Affairs of the People's Republic of China.

### 4.1.4. Low Efficiency of Agricultural Production

In the process of promoting the large-scale operation of agricultural services in China, a series of problems, such as the small scale of agricultural land, the high degree of fragmentation, the small farmers as the main body of agricultural management and the insufficient supply of services, limit the full stimulation of the large-scale benefits of agricultural services.

As of July 2020, the cultivated land area under household contracts was 1,545,766,706 mu and the number of farmers under household contracts was 220,040,147. The average land contracted by each household was 2.48 mu. The details are as follows in Table 5.

**Table 5.** Farmland scale of farmers in China, 2020 [1].

| Farmland Scale (mu) | Farm Household (Ten Thousand) |
|---|---|
| <10 | 23,661.7 (2.561 million households not operating cultivated land) |
| 10–30 | 2966.7 |
| 30–59 | 706.5 |
| 50–100 | 283.6 |
| 100–200 | 104.9 |
| >200 | 47.2 |

[1] China Rural Policy and Reform Statistics Annual Report 2019.

There are more than 200 million agricultural operators in China, and the average cultivated land area is only 7 mu, which is only 1/40 of the European Union and 1/400 of the United States [49] Agricultural labor productivity is approximately 47% of the world average, 2% of high-income countries and 1% of the United States. China's current land per labor and household arable land is no more than 10 mu, which is not only significantly below the world average, but also significantly below the Asian average [50]. Scholars have studied the functional equation of grain yield and its influencing factors in China for a long time [51,52]. The law of diminishing returns is in operation as more physical inputs are applied to shrinking land. Small-scale agriculture results in low agricultural productivity high unit production costs, low agricultural income and weaken the ability of farmers to resist natural disasters. When farmers' input is greater than output, they may be unable to repay the loan and default.

### 4.1.5. Poor Credit Environment in Rural Areas

Of the 724 judgements, 570 were judged by default because the defendant (the borrower) did not appear in court. As the whereabouts of the defendant (the borrower) were unknown, three cases of prosecution were rejected by the court. Some agricultural operators do not actively communicate with the lenders or appear in court when they breach the contract. Instead, they take a negative attitude and let the court dispose of the mortgaged land management right. This reflects a lack of contract spirit and legal thinking among agricultural operators and suggests that the agricultural operators do not value the land.

Since the reform and opening-up in 1978, with the development of China's social economy, people's demands regarding the legal environment have become increasingly urgent. The legal environment of China has been greatly improved, as has citizens' legal consciousness. However, due to the imbalance of China's social and economic development, the legal environment in the vast rural areas is relatively poor, which is manifested in the poor legal awareness of farmers, the weak legislation in rural and agricultural areas, the insufficient legal popularization in rural areas, the greater use of power than law in the management of rural affairs, etc. [53]. In the current situation, farmers' awareness of contracts is relatively poor, the cost of enforcing farmers' performance is very high and there are high social risks. At present, there are no effective measures to solve this problem.

### 4.1.6. Gradually Weakening the Land Restrictions for Farmers

In recent years, a large number of rural laborers have transferred to cities. Table 6 shows that the rural population of migrant workers reached 287 million in 2017, as massive farmers left their homes and went to cities for employment. The phenomenon of rural land transfer and the separation of people and land has become very common. The new generation of farmers accounts for 49.7% of the total number of migrant workers; they hardly participate in agricultural production, have long been accustomed to urban life, and are unwilling to engage in farm work. In addition, due to the low income from farming and the fragmentation of cultivated land, some agricultural areas have been abandoned. In the mountainous areas of Southern Henan and Western Hunan, the proportion of abandoned farmland is close to one quarter [54]. The restriction of land to farmers is gradually weakening. Land is no longer important to agricultural operators. Some agricultural operators choose to give up the right of land management and do not repay the loan. Farmers become part-time farmers and even urban workers. The cost of default to agricultural operators is low, which leads to high moral hazard.

**Table 6.** Number of migrant workers in China during 2015–2019.

| Year | 2015 | 2016 | 2017 | 2018 | 2019 |
|---|---|---|---|---|---|
| Number of migrant workers, 10,000 | 27,747 | 28,171 | 28,652 | 28,836 | 29,077 |
| Number of migrant workers in the province, 10,000 | 9139 | 9268 | 9672 | 9510 | 9917 |

Data source: from 2015 to 2019 "migrant workers monitoring survey report".

The sharp decrease in the agricultural labor force will have a great impact on Chinese farms, making it a very serious problem. The Chinese government has paid attention to this problem. In 2014, the State Council proposed cultivating new agricultural operators, focusing on those whose land management scale is equivalent to 10 to 15 times the average contracted land area of local households.

### 4.1.7. High Interest Rates of Loans

The interest rate of rural land mortgage loans in China is higher than that in developed countries [55,56]. With the penalty interest after loans become overdue, the interest and penalty interest of some loan cases exceed the loan principal. When applying for loans, some rural land operators do not carefully read the terms of the contract or do not seriously consider the consequences of the interest rate and penalty interest. Once default occurs, farmers have a sense of being deprived of value and then turn to negative non-cooperation, allowing the court to decide.

*4.2. Methods for Reducing Farmland Mortgage Default*

Farmers are the main demanders of the rural financial market, and their credit default constitutes the main source of the credit risk of rural cooperative financial institutions. Therefore, improving the loan repayment rate of farmers and reducing default are fundamental to realize the virtuous cycle of rural land mortgages.

### 4.2.1. Strengthen Agricultural Insurance

Agricultural insurance is an effective means to disperse and resolve agricultural risks and has become an important part of many countries' agricultural policy systems [57–60]. In addition, agricultural insurance is a "green box" policy in line with the provisions of the agricultural agreement and is an important non-price agricultural protection tool. However, in recent years, the efficiency of China's agricultural insurance premium subsidies has been weakened, and there are unsustainable risks in policy agricultural insurance. There are still other problems in China's agricultural insurance, such as a low level of security, narrow insurance liability, and claim conditions. Therefore, we need to improve agricultural insurance policy, expand the scope of insurance, and increase the number of claims. When agricultural land operators encounter natural or market risks in agricultural production, they should reduce losses through insurance and increase the source of funds for the repayment of mortgages based on land management rights.

Through various publicity methods, farmers can be encouraged to correctly understand the role of agricultural insurance and the related policies of agricultural insurance to improve their recognition of agricultural insurance and effectively protect their own interests.

### 4.2.2. Cultivation of Farmers' Contract Spirit

The market economy is both a contract economy and a legal economy, and rural land operators should strengthen their contract consciousness and legal spirit. When disasters affect agricultural production and farmers cannot repay loans in time, we should actively negotiate with the lender to formulate a practical and feasible repayment plan. Based on the unique geographically based relationships in the countryside, village committees should play a role in improving the rural credit environment, collect comprehensive credit information about farmer households, establish a complete credit information database, and employ professionals to systematically manage the credit information database to enhance the binding force of credit on farmer households. The evaluation of farmers' credit should be carried out under a unified standard to ensure fairness and transparency.

### 4.2.3. Reduce the Loan Interest Rate

Rural land finance needs the government to provide subsidies through credit. At present, China's rural land mortgage interest rate is generally high, which is not conducive to the development of this business and increases the burden of rural land operators. By comparison, it can be seen that in the United States, the monthly interest rate of land mortgage loans is usually 4%~6%, while in Germany, the monthly interest rate of land mortgage loans is less than 5% [56,57]. It is urgent to reduce the mortgage interest rate of land management rights and reduce the burden of agricultural land operators. This can also effectively reduce the risk of adverse selection.

### 4.2.4. "Project Pool" Mode

The practice of Wucheng County in Shandong Province is worth learning from. Wucheng County initiated the "project pool" mechanism for undertaking mortgage loans for land management rights and establishing risk prevention [61]. In the project pool, high-quality subjects with good operating conditions and high reputation are selected from the new agricultural operating subjects in the county and given preferential policies, such as agricultural project support and financial support. When a borrower is unable to repay a loan through normal operation, the mortgaged land management right is undertaken by other subjects in the project pool, who continue to pay the farmers' land rent and repay the bank loan with part of the aboveground facilities. This reduces the loan risk of the bank and enables the bank to dispose of the land management rights of a borrower who violates the rules in the later stage. In addition, the county took the lead in the development and construction of a rural comprehensive property rights information sharing system, which helps banks understand the operating situation of collateral and borrowers, thus solving the problem of information asymmetry and reducing the risk to the bank.

## 5. Conclusions

The reform of rural land ownership, contract rights and management rights not only represent an innovation of the rural land system with Chinese characteristics but is also the only way to develop modern agriculture. At present, rural land mortgage has been carried out all over the country, but the empirical research on rural land mortgage default is few. According to the court's judgment, this paper comprehensively analyzed the characteristics of defaults in different regions of China during 2014–2020 and explained them. This can provide a reference for the governance of default risk of rural land mortgage. In the field of recessive transformation of agricultural land use, the topic is also worth studying.

The number of farmland mortgage defaults reached a peak in 2018, and since then, the value declined year by year, which confirmed that after the separation of the management rights of contracted rural land from the management rights of contracted land, farmers' farmland mortgage loans could be protected by law, and the default risk of farmland mortgage still exists, but it has been reduced.

The mortgage loan defaults for rural land management rights amounting to less than CNY 100,000 accounted for the largest proportion, 54.7%. A small loan amount can promote a balanced distribution between the loan amount and the borrower's income and effectively disperse the liquidity risk. These borrowers may have a weak ability in avoiding risk and could be prone to moral hazard. For these borrowers, more preferential loans or financial assistance should be considered. The default events of mortgage loans for rural land management rights concerned mainly 1-year short-term loans, which was consistent with the actual situation. To reduce the risk of farmers' default, the term of bank loans was generally limited to one year.

China's financial institutions mainly issue short-term agricultural loans (within 3 years), with a typical loan term of 12 months. Default cases are concentrated in the loan term of 6–12 months, of which defaults in the loan term of 12 months accounts for more than 40%.

The average age of agricultural labor force is 48.5 years old. Nearly 50-year-old men have become the main force of agriculture, of which more than 60% are full-time agricultural producers. Therefore, most of the households with loan demand and behavior are those whose head of household is over 40 years old, which also leads to the farmers in this age group may become the main body of default.

Natural disasters are the main cause of farmers' default. The annual loss of grain caused by drought alone in China is as high as more than 30 billion kg, about 6% of the total grain output in the same period. The failure of agricultural land management caused by natural disasters makes farmers unable to repay loans and result in default.

There are few financial institutions involved in farmland mortgage, which are not easy to share risks. It is recommended to expand financial institutions involved in farmland mortgage. These findings are not only a summary of the current situation of rural land mortgage default in China, but also the first-hand information on empirical research on rural land mortgage default, which can provide reference for the governance of rural land mortgage default risk.

The deficiency of this article lies in that this article makes only descriptive statistics on the default judgments of rural land management right mortgage loans from 2014 to 2020 in China, the data obtained from the court judgments can truly reflect the farmers' default. There are also limitations regarding cases of defaults that were addressed by the court. For example, some cases of "de facto default" have not been granted a trial, so this part of the data cannot be obtained from the court. We discussed only the default cases judged by the court in this article. This is a work that needs to be further promoted. In follow-up research, it is necessary to conduct in-depth interviews to explore the institutional, individual, and natural causes of rural land management right mortgage default in China. It will be more helpful to reveal the formation mechanism of the default risk of rural land management rights and mortgage loans in China, clarify the current situation and characteristics of the default of rural land mortgage loans, and put forward suggestions for preventing the default risk of rural land mortgage loans.

China's land system reform needs to pay attention to some problems. The first is clearing property relations. At present, the question as to whether rural land management right is a property right or creditor's right is controversial, which is directly related to the protection of property right or creditor 's right. In addition, the content of land contract management right and land management right is not clear, which affects the practical effect of land contract management right and land management right. The second problem is a sound assessment system. At present, due to the lack of professional evaluation institutions and scientific evaluation standards, the real value of collateral cannot be accurately reflected in agricultural land mortgage. Therefore, it is necessary to improve the evaluation institutions, cultivate professional talents and improve the evaluation methods to effectively protect the legitimate rights and interests of all parties in the process of agricultural land mortgage. The third is supervision of land use. The Food and Agriculture Organization of the United Nations has set the warning line for arable land at 0.8 mu per capita, and no mortgage is allowed for arable land below 0.8 mu per capita. In the process of farmland mortgage, the tendency of farmland's "non-agriculturalization" and "non-grain growing" should be eliminated to ensure that "the land use is not changed and the comprehensive agricultural production capacity is not destroyed" and that the red line of 1.8 billion mu of farmland will not be broken.

**Author Contributions:** This is the independent work of the authors; investigation, original draft, writing and editing, H.Z.; theoretical direction, Z.Z. Both authors have read and agreed to the published version of the manuscript.

**Funding:** This research was supported by Ministry of education Youth Fund for Humanities and Social Sciences Research (Grant No. 15YJCZH239).

**Institutional Review Board Statement:** Not applicable.

**Informed Consent Statement:** Not applicable.

**Conflicts of Interest:** The authors declare no conflict of interest.

## References

1. Li, J. Analysis of Agricultural Product Brand Marketing Strategy. *Financ. Eng. Risk Manag.* **2020**, *3*, 61–64.
2. Zhuo, Y.F.; Xu, Z.G.; Li, G.; Liao, R.; Christiaan, L.; Wu, C.F.; Wu, Y. LADM-based profile for farmland Tripartite Entitlement System in China. *Land Use Policy* **2020**, *92*, 104459. [CrossRef]
3. Wang, Q.; Zhang, X. Three rights separation: China's proposed rural land rights reform and four types of local trials. *Land Use Policy* **2017**, *63*, 111–121. [CrossRef]
4. Xu, Y.; Huang, X.; Bao, H.X.; Ju, X.; Zhong, T.; Chen, Z.; Zhou, Y. Rural land rights reform and agro-environmental sustainability: Empirical evidence from China. *Land Use Policy* **2018**, *74*, 73–87. [CrossRef]
5. Zhou, Y.; Guo, L.; Liu, Y. Land consolidation boosting poverty alleviation in China: Theory and practice. *Land Use Policy* **2019**, *82*, 339–348. [CrossRef]
6. Zhou, Y.; Guo, Y.; Liu, Y.; Wu, W.; Li, Y. Targeted poverty alleviation and land policy innovation: Some practice and policy implications from China. *Land Use Policy* **2018**, *74*, 53–65. [CrossRef]
7. Yang, Z.; Xunhuan, L.; Yansui, L. Rural land system reforms in China: History, issues, measures and prospects. *Land Use Policy* **2020**, *91*, 104330.
8. Qing-ling, H. On Agricultural Land System Reform—From the Perspective of Land Order and Land Cognition. *Soc. Sci. Beijing* **2016**, *5*, 22–30.
9. Zou, Q.R. *Property Right and Institutional Change: An Empirical Study of China's Reform*; Peking University Press: Beijing, China, 2004.
10. Aditya, R.K.; Omobolaji, O. Rural Finance, Capital Constrained Small Farms, and Financial Performance: Findings from a Primary Survey. *J. Agric. Appl. Econ.* **2020**, *52*, 288–307.

11. Chizoba, P.A.; Nneamaka, O.T.N.; Cynthia, O. Relationship between Rural Finance Institution Services and Standard of Living of Rural Farming Households in Anambra State, Nigeria. *Asian J. Agric. Ext. Econ. Sociol.* **2020**, 102–109. [CrossRef]
12. Long, H.L.; Li, X.B. Analysis on regional land use transition: A case study in transect of the Yangze River. *J. Nat. Resour.* **2002**, *17*, 144–149.
13. Long, H.L.; Qu, Y.; Tu, S.; Zhang, Y.; Jiang, Y.F. Development of land use transitions research in China. *J. Geogr. Sci.* **2020**, *30*, 1195–1214. [CrossRef]
14. Long, H.L.; Zhang, Y. Rural planning in China: Evolving theories, approaches, and trends. *Plan. Theory Pract.* **2020**, *21*, 782–786. [CrossRef]
15. Long, H.L. *Land Use Transitions and Rural Restructuring in China*; Springer Nature: Singapore, 2020.
16. Li, M.; Long, H.L.; Tu, S.S.; Zhang, Y.N.; Zheng, Y.H.; Yang, R. Farmland transition in China and its policy implications. *Land Use Policy* **2020**, *92*, 104470.
17. Long, H.L.; Zou, J.; Pykett, J.; Li, Y.R. Analysis of rural transformation development in China since the turn of the new millennium. *Appl. Geogr.* **2011**, *31*, 1094–1105.
18. Mishra, A.K.; Moss, C.B.; Erickson, K.W. The Role of Credit Constraints and Government Subsidies in Farmland Valuations in the US: An Options Pricing Model Approach. *Empir. Econ.* **2008**, *34*, 285–297. [CrossRef]
19. Johnson, J. Rural Economic Development in the United States: An Evaluation of the US Department of Agriculture's Business and Industry Guaranteed Loan Program. *Econ. Dev. Q.* **2009**, *23*, 229–241. [CrossRef]
20. Fang, Q.M.; Luo, J.C.; Cai, Q.H. The Optimal Land Scale under the Maximum Willingness of Rural Land Mortgage Financing. *J. South China Agric. Univ. Soc. Sci. Ed.* **2016**, *6*, 49–57.
21. Zhan, Z.M.; Luo, J.C. Logistic-DEA-Model-Based Assessment of Risk Control Effect of Rural Land Contract and Management Rights Mortgage Pilot. *Wuhan Univ. J.* **2016**, *69*, 47–54.
22. Lin, J.W. Risks and precaution of mortgage of farmland management right. *J. Fujian Agric. For. Univ.* **2016**, *19*, 14–19.
23. Zhao, Y.Z.; Wang, Q. The Risk Management of the Rural Land Contracted Management Right Mortgage Loan: From the Perspective of Agriculture-related Financing Institutions. *J. Anhui Agric. Univ.* **2015**, *24*, 12–16.
24. Hui, X. Exploration and path selection of rural land mortgage financing practice mode—Based on the empirical observation of rural land finance pilot. *Southwest Financ.* **2014**, *3*, 66–71.
25. Peng, Y.; Liu, W.B.; Tan, C. Study on Risks in Mortgage Financing of Rural Land Management Right. In Proceedings of the 2017 6th International Conference on Social Science, Education and Humanities Research (SSEHR 2017), Jinan, China, 18 October 2017.
26. Pan, W. On the Risks of Rural Land Management Right Mortgage Loan. *J. Nanjing Agric. Univ.* **2015**, *5*, 104–113.
27. Lihong, Y. Credit risk evaluation of rural land management right mortgage—Based on AHP analysis. *Rural Econ.* **2014**, *11*, 79–82.
28. Chao, W.; Liao, Y.; Wang, M. Research on Farmland Mortgage Loan Risk Evaluation Based on AHP-Fuzzy Comprehensive Evaluation Method. *J. Hebei Norm. Univ. Sci. Technol.* **2014**, *13*, 38–43.
29. Dehong, L.; Wuke, Z. Influential Factors and Measuring of Credit Risk on Rural Land Contract Right Mortgage Loan—Estimation Based on CreditRisk+ Model. *J. Huazhong Agric. Univ.* **2018**, *4*, 137–147.
30. Qicai, Y.; Xie, L.; Wenlong, H. The Realization and Risk of Mortgage Loan Based on Agricultural Land Using Rights: A Comment on the Practices and Case. *Probl. Agric. Econ.* **2015**, *10*, 4–11.
31. Liu, Y.X.; Min, J.; Liu, Y. Identification and Evaluation of Mortgage Financing Risk of Agricultural Land Management Right under the 'Threepower Split': An Empirical Study Based on Structural Equation Model. *Macroeconomics* **2019**, *1*, 158–175.
32. Christen, R.P.; Pearce, D. *Managing Risks and Designing Products for Agricultural Microfinance: Features of an Emerging Model*; Consultative Group to Assist the Poor (CGAP). Washington, DC, USA, 2005; pp. 1–44.
33. Ruslana, S.; Oksana, A.; Olha, H.; Kateryna, M. Mortgage Lending in the Agricultural Economy: Opportunities and Risks. *Financ. Credit Act. Probl. Theory Pract.* **2019**, *1*, 225–233.
34. Yin, K.X.; Lan, Q.G.; Kan, S.S.; Su, Y.D. Empirical Research on the Risk Mea surement of Mortgage Loans on Rural Land Contract and Management Rights: A Case Study of Heilongjiang Province. *J. Coast. Res.* **2020**, *103*, 226–230. [CrossRef]
35. Ge, Y.; Yu, L.; Shu, Z.; Chen, Z. Data Augmentation for Deep Learning of Judgment Documents. In Proceedings of the IScIDE 2019 Intelligence Science and Big Data Engineering, Big Data and Machine Learning, Nanjing, China, 17–20 October 2019; pp. 232–242.
36. Yuan, J.; Wei, Z.; Gao, Y.; Chen, W.; Song, Y.; Zhao, D.; Ma, J.; Hu, Z.; Zou, S.; Li, D.; et al. Overview of SMP-CAIL2020-Argmine: The Interactive Argument-Pair Extraction in Judgement Document Challenge. *Data Intell.* **2021**, *3*, 287–307.
37. The State Council's Summary Report on the Pilot Project of Operating Right of Contracted Land and Mortgage Loan of Farmers' Housing Property Right in Rural Areas. Available online: http://www.npc.gov.cn/npc/c12435/201812/2067ecc784a8437cbe878 0a32bcf48ac.shtml (accessed on 23 December 2018).
38. Overview of Agricultural and Rural Areas in Heilongjiang Province, Agricultural and Rural Department of Heilongjiang Province. Available online: http://nynct.hlj.gov.cn/jggk/nygk/2020/index.htm (accessed on 1 January 2021).
39. The Balance of Rural Contracted Land Management Right Mortgage Loan Reached 7.955 Billion Yuan. The "Two Rights" Mortgage Loan Revitalized the "Sleeping" Resources. Available online: https://www.hlj.gov.cn/zwfb/system/2018/01/08/010 859158.shtml (accessed on 8 January 2018).
40. Zhang, Y.Y. Research on the Influencing Factors and Control of Credit Risk of Rural Cooperative Financial Institutions in Shanxi Province. Ph.D. Thesis, Northwest Agriculture & Forestry University, Xianyang, China, 2015.

41. The Voice of Chinese Countryside. Available online: http://country.cnr.cn/gundong/20181231/t20181231_524467209.shtml (accessed on 29 December 2018).
42. State Statistical Yearbook. Available online: https://data.stats.gov.cn/easyquery.htm?cn=C01 (accessed on 11 July 2021).
43. Ding, Y.G.; Sun, Q.X. Does agricultural insurance mitigate the negative impac of natural disaster on the agricultural economy? *Theory Pract. Financ. Econ.* **2021**, *42*, 43–49.
44. Turvey, C.G.; Norton, M. An internet-based tool for weather risk management. *Agric. Resour. Econ. Rev.* **2008**, *37*, 63–78. [CrossRef]
45. Rosenzweig, M.R.; Binswanger, H.P. Wealth, Weather Risk and the Composition and Profit Ability of Agricultural Investments. *Econ. J.* **1993**, *103*, 56–78. [CrossRef]
46. Yoji, K.; Gen, S.; Toshichika, I. Systemic Risk in Global Agricultural Markets and Trade Liberalization under Climate Change: Synchronized Crop-Yield Change and Agricultural Price Volatility. *Sustainability* **2020**, *12*, 10680.
47. Moschini, G.; Hennessy, D.A. Chapter 2 Uncertainty, risk aversion, and risk management for agricultural producers. *Handb. Agric. Econ.* **2001**, *1*, 87–153.
48. Wan, B.R. Chinese agriculture has entered the era of high production cost. *People's Daily*, 13 March 2014.
49. Han, C.F. Rural land reform in China. In Proceedings of the 40-Year Symposium on Rural Reform of the Ministry of Agriculture and Rural Affairs, Beijing, China, 4 December 2018.
50. Lin, W.L. Farmland operation scale: International experience and Chinese realistic choice. *Probl. Agric. Econ.* **2017**, *7*, 33–42.
51. Battese, G.E.; Coelli, T.J. Prediction of Firm-Level of Technical efficiencies with a Generalised Frontier Production Function and Panel Data. *J. Econom.* **1988**, *38*, 387–399. [CrossRef]
52. Yao, S.; Liu, Z. Determinants of Grain Production and Technical Efficiency in China. *J. Agric. Econ.* **1998**, *49*, 171–184. [CrossRef]
53. The Survey Report of Migrant Workers in 2019. Available online: http://www.stats.gov.cn/tjsj/zxfb/202004/t20200430_1742724.html (accessed on 30 April 2019).
54. Wei, H.K.; Yan, K. *Report on China's Rural Development—To Stimulate New Energy of Rural Development by Comprehensively Deepening Reform*; China Social Science Press: Beijing, China, 2017; p. 17.
55. Cheng, Y.; Wang, B. Foreign farmland mortgage loan. *Land Resour.* **2016**, *1*, 50–53.
56. Babić, L. Comparison of Agricultural Insurance Development of Croatia, EU and USA. *Poljoprivreda* **2014**, *20*, 49–52.
57. Ministry of Finance of the PRC. Available online: http://jrs.mof.gov.cn/zhengcefabu/201910/t20191012_3400537.htm (accessed on 19 September 2019).
58. Yarmolenko, V.V. The Factors' Influence on the Functioning of the Agricultural Insurance Market. *Bìznes Ìnform* **2019**, *9*, 144–151.
59. Rokicki, T. The agricultural insurance market in Poland. *Rocz. Nauk. Stowarzyszenia Ekon. Rol. Agrobiz.* **2018**, *1*, 117–122. [CrossRef]
60. Ranjan, K.G.; Shweta, G.; Vartika, S. Demand for Crop Insurance in Developing Countries: New Evidence from India. *J. Agric. Econ.* **2021**, *72*, 293–320.
61. China Rural Network. Available online: http://journal.crnews.net/ncgztxcs/2020/dsbq/dc/139687_20200929111724.html (accessed on 29 September 2020).

*Article*

# Scale Transition and Structure–Function Synergy Differentiation of Rural Residential Land: A Dimensionality Reduction Transmission Process from Macro to Micro Scale

Yanbo Qu [1], Xiaozhen Dong [1], Lingyun Zhan [1], Hongyun Si [1,*], Zongli Ping [2] and Weiya Zhu [2]

[1] School of Public Administration and Policy, Shandong University of Finance and Economics, Jinan 250014, China; 20126008@sdufe.edu.cn (Y.Q.); 202109032@mail.sdufe.edu.cn (X.D.); 192109026@mail.sdufe.edu.cn (L.Z.)

[2] Shandong Institute of Territorial and Spatial Planning, Jinan 250014, China; pingzongli123@163.com (Z.P.); weiyazhu111@163.com (W.Z.)

* Correspondence: sihongyun@tongji.edu.cn

**Abstract:** In order to enhance the scientific understanding of the transition law of rural residential areas and enrich the theory and method system of land use transition research, this article takes Shandong Province as an example and constructs a comprehensive research framework of rural residential land scale, structure, and function from the perspective of the combination of the macro and micro scales based on differences between the rural residential areas in the region and the village scale forms. Using model quantitative analysis and horizontal comparative analysis methods, this paper explores the process characteristics of rural residential land use scale transition and the corresponding stage differentiation law of spatial structure and system function. Research has shown that the stage characteristics of the scale transition of rural residential land use in Shandong Province in the past 10 years are significant. The five transition stages—from the primary stage, low stage, intermediate stage, advanced stage, to the stable stage—show obvious spatial agglomeration and spatial autocorrelation, which are mainly driven by the positive and negative interactions of economic development, the policy environment, natural conditions, and population. With the gradual upgrading of the land use scale transition in rural residential areas, the spatial pattern of rural residential areas has been continuously optimized, the land use structure has tended to be balanced and complicated, and the living-production-ecological function as a whole has been strengthened. The essence of this type of differentiation is the differential performance of rural residential areas adapting their own conditions to the external environment. The transition of the rural residential area from the macro to the micro scale is also the process of realizing rural reconstruction and rural revitalization. In the future, under the framework of the "element–structure–function" system of rural residential areas, the rural transition and development should be continuously promoted through the support, organization, guidance, and promotion mechanisms of internal and external factors.

**Keywords:** land use transition; rural residential areas; land use scale; structure; function; rural reconstruction; Shandong Province

---

## 1. Introduction

Land use transition (LUT) is a new topic in the international frontier research of Land Change Science (LCS) [1]. Since the proposal of Mather and Grainger et al. [2–4], rapid development has been achieved in the land use transition of woodland [5,6], farmland [7,8], urban and rural construction land [9,10], and regional land [11,12]. In recent years, the depth of research on land use transition, especially arable land use transition from the point of view of land-scale research, has expanded to green transition [13], function transition [14], intensive transition [15], and arable land use transition [16–19] from the new perspectives of labor factor changes, the relationship between input and output, the organizational structure, and rural revitalization. At the same time, the scope of land use

**Citation:** Qu, Y.; Dong, X.; Zhan, L.; Si, H.; Ping, Z.; Zhu, W. Scale Transition and Structure–Function Synergy Differentiation of Rural Residential Land: A Dimensionality Reduction Transmission Process from Macro to Micro Scale. *Land* **2021**, *10*, 647. https://doi.org/10.3390/land10060647

Academic Editors: Hualou Long, Xiangbin Kong, Shougeng Hu and Yurui Li

Received: 8 May 2021
Accepted: 11 June 2021
Published: 17 June 2021

**Publisher's Note:** MDPI stays neutral with regard to jurisdictional claims in published maps and institutional affiliations.

transition research has also been widened, and a large number of studies have emerged on the relationship between land use transition and rural reconstruction [20], farmers' livelihoods [21], and the impact of land use transition on ecological environment change [22], continuously enriching the theory and method system of land use transition to form a research theory and method system including the comprehensive transition of explicit and implicit forms of land use [11].

Studies have shown that the governance of rural areas is mainly related to land issues [23]. As a highly important land use type that exists widely in rural areas, rural residential areas have been in a state of steady evolution since they emerged from the stable natural geographical environment and location conditions, with the influence of the natural increases and decreases in rural population and the slow growth of the rural economy. However, with the acceleration of urbanization, the phenomenon of idle rural residential areas and the hollowing out of villages has stood out, coupled with the insufficient supply of newly added construction land under the rigid control of construction land indicators. In this context, policies and activities such as linking urban-rural construction land increase and decrease in the hook, the reform of the rural land system, and comprehensive land improvement across the whole region have been implemented. Meanwhile, under the joint guidance of the rural revitalization strategy and rural construction actions, a drastic reduction in the sizes of rural residential areas has taken place or will take place over the whole area, which reflects the basic characteristics and general laws of land use transition and has aroused widespread concern from all of society.

At present, the research on rural residential areas mainly focuses on its quantity change [24,25], spatial distribution [26,27], utilization state [28,29], regulation mode [30,31], etc., which all belong to the research category of land use and land-cover change (LUCC) in the early stage and rarely involve research on the transition of rural residential areas. In 2006, Long [32] proposed the basic concept and measurement method of rural homestead transition and applied it to specific areas. Then, based on this theory, some scholars conducted a first empirical exploration of the transition of rural residential areas in Shandong Province and the Beijing-Tianjin-Hebei region; revealed the spatial differentiation characteristics of the transition of regional rural residential areas [33,34]; and further explored the process and mechanism of the transition of the human–land relationship in rural residential areas, using per capita rural residential land area (PCRA) as the characterization index [35]. On this basis, related studies were carried out on the relationship between rural residential areas and many types of land transition [36,37], the influencing factors of the transition of rural residential areas [38], and the effect of rural residential area transition [39], preliminarily forming a general theoretical system of rural residential area transition [40–42]. However, these studies mainly focused on the transition analysis of the scale and quantity of rural residential land, and there have been few studies carried out on the morphological evolution trend or transition characteristics of the land structure and system function under the spatial carrier attribute of rural residential land. The basic theories, research perspectives, and methods of rural residential area transition need to be further explored.

To enhance the understanding of the theory and law of rural residential transition, and enrich the methodological system and practical application of land use transition, this study takes Shandong Province, which can be considered as a microcosm of China, as the research area. Based on the manifestations and characteristics of rural residential areas at village and regional levels, and combining both macro and micro levels, the method of amalgamating model measurement analysis and horizontal comparison is used to explore the process mechanisms of rural residential land scale transition and the corresponding gradual differentiation trend of the spatial structure and system function. Then, the framework and strategy of rural reconstruction are created from the comprehensive perspective of "element-structure-function". The structure of this paper is as follows: Section 2 proposes the theoretical framework and research ideas. Section 3 introduces the relevant research methods and describes the data source and processing. Section 4 analyzes the scale transition, structure and function differentiation of rural residential land on the macro and micro scales. Sections 5 and 6 discuss the theoretical contributions and limitations of this paper, put forward a rural reconstruction strategy, and draw relevant conclusions.

## 2. Theoretical Framework and Research Ideas

### 2.1. Theoretical Framework

Rural residential areas are a type of a complex system that is widely distributed in rural areas and corresponds to the locations of cities. At the national and regional levels, rural residential areas are usually regarded as the "containers" that accommodate the rural population. The scale of rural residential areas, especially their increase or decrease, which occurs with the movement of rural populations, is the key to their role in economic and social development [35], and also the main means by which administrative departments at all levels carry out statistical analyses. However, in certain areas affected by external factors such as geographical environment, traffic location, and industrial basic conditions, rural residential areas appear as patches of different shapes embedded in the agricultural landscape. These patches differ in density and arrangement, with obvious spatial structure characteristics of complexity of patch boundaries and agglomeration or dispersion of spatial patterns, which has become the main issue in territory spatial planning and village planning. In specific villages, a large number of houses, horizontal and longitudinal roads, scattered factories, concentrated public service facilities, and natural or artificial vegetation green spaces are distributed, which reflects that the patches of rural residential areas also contain various land types, buildings, facilities, industries, and other elements. These elements provide the most basic living guarantee, necessary production conditions, and special ecological environments for rural residents in different combination forms, reflecting the element composition, structure organization, and functional value form of rural residential areas at the micro level. Therefore, the understanding of rural residential areas needs to be examined from a multi-scale perspective [33]. At the macro level, with the advancement of urbanization, the rural population outflow increases and the corresponding scale of rural residential land should decrease accordingly. The transition of rural residential land mainly shows that the incremental proportion of land use scale decreases gradually and tends to be stable; at the micro level, there will always be some people living in rural areas. Rural housing security, equal public services, and improvement in living environments are also important factors in the transition of rural residential areas. Therefore, from the perspective of combining macro and micro scales, the transition of rural residential areas is not a simple scale reduction and spatial transition, but also a process of the adjustment or redistribution of elements to drive the optimization of the system structure and functions.

In general, the structure and function of rural residential areas are also transited in the process of the land scale transition of rural residential areas [40] (Figure 1). In the initial stage of the rural residential land scale transition, the scale increment is relatively large. On the one hand, affected by external factors such as transportation, nearby cities, and service facilities, the newly added land for rural residential areas is mostly distributed outside villages, and some of the land will be far away from the center of the village, causing the overall layout of rural residential areas to spread out, resulting in the shape of rural residential areas becoming more complex and the patch density becoming more scattered. On the other hand, due to the different uses of newly added land (generally residence, industry, and public service facilities), the structural combination and carrier functions of rural residential areas have shown a diversified trend. When the newly added land is mainly the residential land, the land structure will tend to be singular, and the dominance of the residential security function will accordingly be stronger. When the newly added land is mainly based on the construction of the village and township industrial parks, the land structure of production and residence in rural residential areas will tend to be balanced and the rural non-agricultural production functions will also be improved. When the amount of land allotted to public service facilities increases, the land use structure of rural residential areas will likely be diversified, the life service functions will be enhanced, and the system functions of rural residential areas will become more complex.

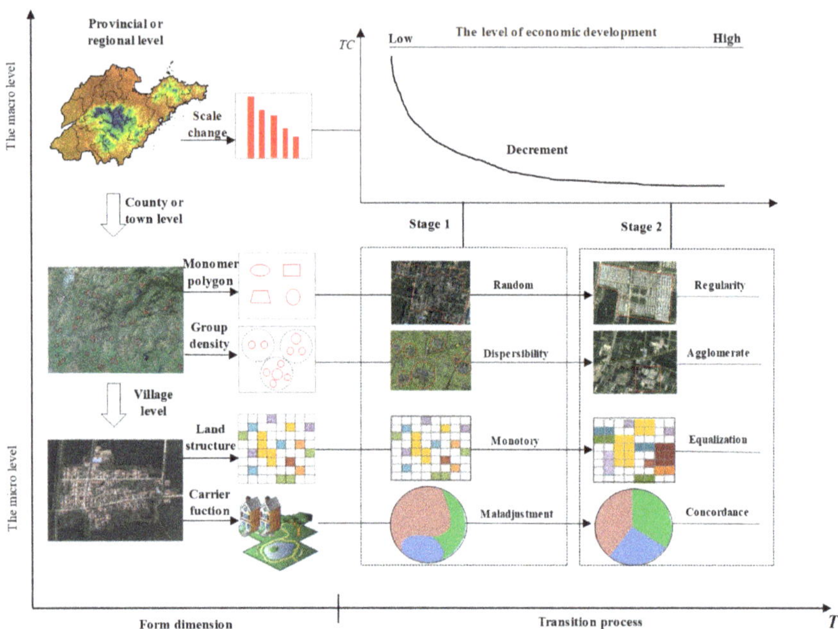



**Figure 1.** Theoretical framework of rural residential transition based on dimension reduction.

With the strict promotion and implementation of village planning and control policies, the scale of rural residential land tends to be stable or even shrink, and the amount of newly added land is relatively reduced. In the process of the unified planning and construction of new rural communities in scattered villages, the shape of rural residential areas tends to be regular, and the spatial layout of rural residential areas within a certain range is more concentrated. At the same time, in order to improve the equalization of urban and rural public services and the quality of the human settlement environment, rural areas also require increased education, medical care, transportation, leisure, ecology, and other basic service facilities, encouraging the structure of rural residential land to be balanced, while the corresponding living, production, and ecological functions are continuously optimized and improved. In essence, the transition of rural residential areas is a synchronous process of the decreasing of the land use scale, with a regular and compact spatial layout, balanced land use structure, and coordination of living-production-ecological function.

### 2.2. Research Ideas

Land use transition research is generally based on the longitudinal comparative study of long-term series data, but the land use statistics in China concern a relatively short time-scale, especially with regard to micro data such as data on land use, buildings, facilities, population, and industry in a large range of rural residential areas. However, affected by the regional differences in the natural environment and social and economic development, rural residential areas show obvious regional characteristics [43]. These characteristics show that the rural residential areas in different regions are in different transition stages due to the differences in regional socioeconomic development in the same period of time between areas. This provides weight to the feasibility of our idea to use the horizontal comparison method to carry out structural and functional changes in rural residential areas.

Taking the above factors into consideration, the research idea of this study is shown in Figure 2. First, based on the change data of rural residential land scale, this paper uses the homestead transition measurement method proposed by Long [32] to identify the transition stage of the rural residential land scale in different counties and urban areas at the macro provincial level and further explores the influencing factors and mechanisms of the transition. Then, this study considers the characteristics of the influencing factors of the transition of rural residential land scale and selects the typical sample areas (counties/cities) from different transition stages. By adopting the idea of "point" mapping "surface" [44], this study further selects typical sample points (villages) in typical sample areas. Through sample site surveys and in-depth interviews, micro-data on land, population, industry, buildings, and facilities inside rural residential areas were obtained. Using the diversified econometric model, the differentiation characteristics of the structure and function of rural residential areas in different transition stages were analyzed at the micro-level. Finally, based on the mechanism of rural residential land scale transition and the characteristics of structural and functional differentiation, the paper puts forward relevant strategies for promoting rural reconstruction through rural residential land transition.

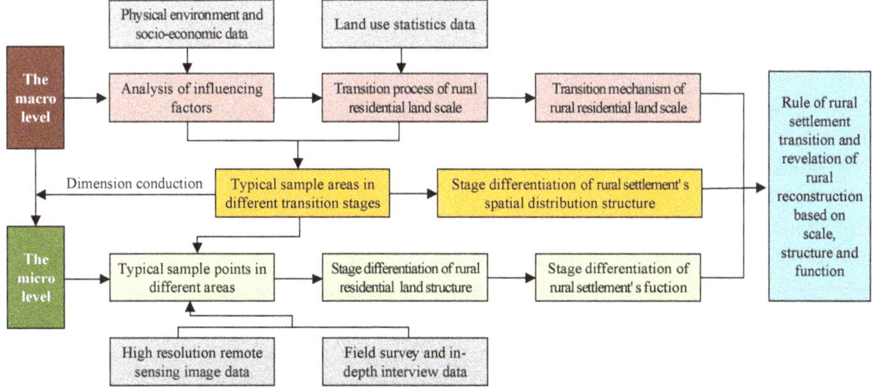


**Figure 2.** Research framework of rural residential land transition.

## 3. Research Method

### *3.1. Measurement of Rural Residential Land Scale Transition*

### 3.1.1. Transition Model

Based on the characteristics of rural homestead scale change, Long [32] proposed the theoretical hypothesis of rural homestead transition—that is, rural homestead transition refers to the fact that the proportion of rural homestead out of the total amount of increased construction land will gradually decrease with the change in social and economic development stage, until it tends towards a fixed value. The rural homestead he refers to is the land use type with the largest share of the rural residential area, and its essence is the rural residential area in a narrow sense. Therefore, this article uses this idea to calculate the transition index of the rural residential land use scale, and the calculation method is as shown in Equation (1).

$$TC_r = \frac{IA_{rl}}{IA_{cl}} \times 100\%. \tag{1}$$

In the formula, $TC_r$ is the transition index of the rural residential land use scale with power index distribution characteristics. $IA_{rl}$ is the area increment in rural residential land, $hm^2$; $IA_{cl}$ is the increment of the total scale of construction land, $hm^2$.

### 3.1.2. Analysis of Influencing Factors

Traditional regression models (OLS models) can be used to estimate samples and parameters globally. However, due to the influence of the spatial pattern relationship, if there is a spatial autocorrelation of independent variables, the independence assumption of residuals in the OLS model cannot be satisfied, and the parameter estimation of the ordinary least square method is no longer applicable. The geographically weighted regression (GWR) model can estimate the influence of different regions and reflect the non-stationarity of parameter estimation in different spaces, making the results more in line with reality [35]. Therefore, in order to comprehensively analyze the influencing factors and the spatial differentiation of rural residential land transition, the GWR model was introduced on the basis of OLS model analysis so as to reflect the impact of influencing factors in different spaces. The model structure is shown in Formula (2):

$$y_i = \beta_0(\mu_i, v_i) + \sum_{k=1}^{n} \beta_k(\mu_i, v_i) x_{ik} + \varepsilon_i. \tag{2}$$

In the formula, $y_i$ is the PCRA score of the $i$-th spatial unit; $\beta_0$ $(\mu_i, v_i)$ is the regression coefficient of the $i$-th spatial unit, indicating the influence degree of the independent variable on the dependent variable; $(\mu_i, v_i)$ are the geographic center coordinates of the $i$-th spatial unit; $\beta_k$ $(\mu_i, v_i)$ is the score of the continuous function $\beta_k$ $(\mu, v)$ in the $i$-th spatial unit; $x_{ik}$ is the score of the $k$-th explanatory variable in the $i$-th spatial unit; $\varepsilon_i$ is random error. The accuracy of the GWR model is greatly affected by the weight and bandwidth. Considering the differences between the freedom degrees of different models, this paper uses the Gaussian weighting method and "Optimization to Minimize AICc" bandwidth method for local estimation.

Changes to the scale of rural residential land use are usually affected by many factors. We referred to related literature [32,35,44] and initially selected altitude (A1), distance from urban resident (A2), transportation convenience (A3), GDP change rate (B1), local fiscal revenue change rate (B2), fixed asset investment change rate (B3), rural population change rate (C1), change rate of rural residents' per capita income (C2), change rate of per capita arable land area (C3), urban and rural construction land planning control scale (D1), urban and rural construction land increase or decrease linkage scale (D2), land supply rate of construction land planning (D3), and 12 other indicators as influencing factors for the transition of rural residential land use scale from the perspective of reflecting the regional natural background conditions, economic development level, social living conditions, and policy and institutional environment.

### 3.2. Measurement of Rural Residential Areas' Structure

From a macro point of view, in an overall pattern in the form of a plaque, rural residential areas are manifested in various shapes, agglomerations, and dispersions in a certain area, with structural differences in their spatial distribution. On the micro scale, rural residential areas include residential land, production land, public service land, roads, and greening land. These land use types form the whole rural residential space with different combinations and have different characteristics of land use structure. Therefore, the structural characteristics of rural residential areas were measured from the two aspects of spatial distribution structure and internal land use structure.

#### 3.2.1. Spatial Distribution Structure

Previous studies have shown that the landscape ecology index can effectively measure land use layout [45]. Generally, land use layout is described by selecting relevant indexes from two aspects: individual characteristics and community characteristics of landscape patches. The landscape pattern index measurement software Fragstats contains nearly 30 optional indicators, but most of these indicators are collinear. To reflect the synthesis of indicators and eliminate nonlinearity, this study refers to the screening results of principal component analyses carried out in existing studies [46,47]. Meanwhile, considering the characteristics of the patch size, irregular shape, and scattered distribution of rural residential land, five indicators (i.e., area-weighted mean patch shape index (AWMSI), area-weighted mean patch fractal index (AWMPFD), patch density(PD), patch association index (COHESION) and interspersion juxtaposition index(IJI)) were selected to measure the differentiation law of the spatial distribution pattern. The analyses are conducted from two aspects, i.e., shape complexity and spatial agglomeration degree. The meanings and calculation methods of each index are shown in Table 1.

**Table 1.** The indicator system of the spatial distribution structure.

| Indicator | Metric | Equation | Description |
|---|---|---|---|
| Degree of Shape Complexity | Area-Weighted Mean Patch Shape Index (AWMSI) | $AWMSI = \sum_{j=1}^{n} [(\frac{0.25 P_{ij}}{\sqrt{a_{ij}}})(\frac{a_{ij}}{\sum_{j=1}^{n} a_{ij}})]$ <br> $P_{ij}$ = perimeter (m) of patch$_{ij}$. <br> $a_{ij}$ = area (m$^2$) of patch$_{ij}$. | The value reflects the complexity of different plaque shapes, the value range is $\geq 1$, and it increases as the irregularity of the plaque shape increases with no upper limit. When AWMSI = 1, the patch is square. |
| | Area-Weighted Mean Patch Fractal Dimension (AWMPFD) | $AWMPFD = \sum_{j=1}^{n} [(\frac{2ln(0.25 P_{ij})}{ln(a_{ij})})(\frac{a_{ij}}{\sum_{j=1}^{n} a_{ij}})]$ <br> $P_{ij}$ = perimeter (m) of patch$_{ij}$. <br> $a_{ij}$ = area (m$^2$) of patch$_{ij}$. | The value reflects the impact of human activities on the landscape pattern, and the value range is between 1 and 2. The larger the value, the more complex the shape, indicating that the plaque is affected more by nature and less by humans. |
| Degree of Spatial Agglomeration | Patch Density (PD) | $PD = n_i / A$ (10,000) (100) <br> $n_i$ = number of patches in the landscape of patch type (class) $i$. <br> $A$ = total landscape area (m$^2$) | The value reflects the spatial distribution of plaques. This is $\geq 0$ with no upper limit. The higher the value, the more fragmented the plaque. |
| | Patch Cohesion Index (COHESION) | $COHESION =$ <br> $\left[1 - \frac{\sum_{i=1}^{m} \sum_{j=1}^{n} P_{ij}}{\sum_{i=1}^{m} \sum_{j=1}^{n} P_{ij} \cdot \sqrt{a_{ij}}}\right] \left[1 - \frac{1}{\sqrt{A}}\right]^{-1} (100)$ <br> $P_{ij}$ = perimeter (m) of patch$_{ij}$. <br> $a_{ij}$ = area (m$^2$) of patch$_{ij}$. <br> $A$ = total landscape area (m$^2$) | Reflects the physical connectivity of similar patches. The value ranges from 0 to 100. The larger the value, the higher the connectivity between patches. |
| | Interspersion Juxtaposition Index (IJI) | $IJI = \frac{-\sum_{i=1}^{m} \sum_{j=1}^{n} \left[(\frac{e_{ij}}{E}) \cdot ln(\frac{e_{ij}}{E})\right]}{ln(0.5[m(m-1)])} (100)$ <br> $e_{ij}$ = Boundary type length <br> $E$ = Sum of the length of the boundary type <br> $m$ = Number of plaques | Reflects the corresponding type of adjacent focus and dispersion under a specific random distribution. The value ranges from 0 to 100. The higher the value, the more scattered the plaques. |

### 3.2.2. Internal Land Use Structure

According to the classification standard of land use status (GB-T21010-2007), the internal land use structure of rural residential areas is divided into six categories: homestead, public service land, industrial land, commercial service land, road, and idle land. The Gibbs–Martin diversity index and concentration index [48], which are often used to express the element structure in landscape ecology, were used to conduct the quantitative analysis of the internal land use structure of rural residential areas in this research, and the measurement indexes and calculation methods are shown in Formulas (3) and (4).

$$GM = 1 - \frac{\sum_{i=1}^{6} X_i^2}{\left(\sum_{i=1}^{6} X_i\right)^2}. \tag{3}$$

where $GM$ is the Gibbs–Martin diversification index and $X_i$ is the area or proportion of different land use types. Generally speaking, the higher the $GM$ score, the more diverse the land use type, the greater the balance of land use, and the more complex the structure.

When the area of each land use structure type is equal, the *GM* score reaches its theoretical maximum value.

$$I = \frac{A - R}{M - R}.$$ (4)

In the formula, *I* is the concentration index of the regional land use structure; *A* is the sum of the cumulative percentages of various land use types in the region; *M* is the sum of the maximum cumulative percentages assuming that the land use in the region is concentrated in a certain type, which is 600; *R* is the sum of the cumulative percentages of various land use types in the upper level area of the sample area, which is 532.

### 3.3. Measurement of Rural Residential Areas' Function

As an important carrier of rural residents' life and production, rural residential areas comprise a complex system formed by the interaction and organization of population, land, industry, buildings, facilities, and other elements, playing an obvious multi-functional part in meeting the needs of residents [44]. According to the territorial system of human–environment interaction of the logical train of thought of expression function elements, rural development takes moderately livable life space, intensive and efficient production space and ecological space of picturesque scenery as the goals. This study evaluates the functions of rural residential areas from the three aspects: living, production, and ecological functions. In accordance with the coupling, compatibility, and symbiosis processes of various elements, this study further subdivides the functions into six sub-functions of residence guarantee, basic services, agricultural production, non-agricultural production, ecological conservation, and environmental maintenance to comprehensively measure the strength and coordination of various functions of rural residential areas (Table 2). First, according to the principle of the leading, comprehensibility, conciseness, and substitution of the evaluation index, the multi-functional evaluation index system of rural residential land was constructed. Residential guarantee and basic services functions are provided by residential buildings, public service facilities, and roads. Agricultural production and non-agricultural production functions are provided by factors such as productive buildings, employment methods, industrial types, and land resource allocation. Ecological conservation and environmental maintenance functions are provided by elements such as ecological infrastructure and environmental governance facilities. Then, the range standardization method was adopted to quantify the values of each index between 0 and 1, and the intensity score of each function of rural residential areas was measured by referring to the idea and method of quantifying farmers' livelihood assets [44]. Finally, the comprehensive coordination degree model (Formulas (5) and (6)) was used to measure the comprehensive strength of rural residential functions and the coordination between them.

$$F = \alpha F_l + \beta F_p + \lambda F_e$$ (5)

$$C = 3\left[\frac{(F_l \times F_p \times F_e)}{(F_l + F_p + F_e)^3}\right]^{1/3}$$ (6)

Here, *F* represents the comprehensive intensity index of living, production, and ecological functions and *C* represents the degree of coordination among living, production, and ecological functions. $\alpha$, $\beta$, and $\lambda$ represent the functional coefficients, and their sum is 1. Considering that the significance and value of life, production, and ecological functions to rural residents are different, in this paper, $\alpha$, $\beta$, and $\lambda$ are set as 0.4, 0.35, and 0.25, respectively. For example, living function is the basic life guarantee function of rural residents, and living conditions and social security directly determine the improvement of farmers' happiness; as the employment guarantee function of rural residents, production function reflects the important impacts that the extension of the rural industrial chain and the cultivation of new business forms shows on the survival guarantee of rural households in the context of rural revitalization; the dual differentiation between rural and urban areas makes rural areas better than urban areas in terms of ecological aspects, such as vegetation coverage and air purification, which further weakens rural households' demand for ecological functions relative to living and production functions.

**Table 2.** Multi-functional index system and calculation method of rural residential land.

| Type of Functions | | Metrics | Calculation Method | Comprehensive Calculation Formula |
|---|---|---|---|---|
| Living function (Fl) | Residence guarantee (Flr) | Per capita housing area ($r_1$) | $r_1$ = residential floor area/rural population | $Flr = (r_1 + r_2 + r_3)/3$ |
| | | Building quality ($r_2$) | $r_2$ = the number of buildings with brick and concrete structure/total number of village houses | |
| | | The proportion of buildings ($r_3$) | $r_3$ = the number of residential buildings/total number of rural households | |
| | Basic services (Flf) | Completeness of public service facilities ($f_1$) | $f_1$ = the proportion of the number of 7 public service facilities in the village (including water supply system, drainage system, garbage disposal equipment, health room, school, cultural station, fitness place) | $Flf = (f_1 + f_2 + f_3)/3$ |
| | | Road area per capita ($f_2$) | $f_2$ = rural road area/rural population | |
| | | Rural road quality ($f_3$) | $f_3$ = hardened rural road area/total area of rural road | |
| Production function (Fp) | Agricultural Production (Fpa) | Cultivated land area per capita ($a_1$) | $a_1$ = arable land area/rural population | $Fpa = (a_1 + a_2 + a_3)/3$ |
| | | Agricultural employment ratio ($a_2$) | $a_2$ = number of people engaged in agricultural production/total population | |
| | | Agricultural income ratio ($a_3$) | $a_3$ = agriculture income/total income | |
| | Non-agricultural Production (Fpna) | Per capita area of commercial building land ($na_1$) | $na_1$ = rural industrial land area/rural population | $Fpna = (na_1 + na_2 + na_3)/3$ |
| | | Non-agricultural employment ratio ($na_2$) | $na_2$ = number of people engaged in non-agricultural production/total rural population | |
| | | Non-agricultural income ratio ($na_3$) | $na_3$ = non-agriculture income/total income | |
| Ecological function (Fe) | Ecological conservation (Fec) | Green area ratio ($c_1$) | $c_1$ = green land area in village/total area of village | $Fec = (c_1 + c_2)/2$ |
| | | Ecological landscape land area ratio ($c_2$) | $c_2$ = ecological land area such as forest, grass, and water in the village/total area of village | |
| | Environmental maintenance (Fem) | Sewage treatment rate ($m_1$) | $m_1$ = number of households with centralized sewage treatment/total number of rural households | $Fem = (m_1 + m_2)/2$ |
| | | Waste treatment rate ($m_2$) | $m_2$ = number of households with centralized garbage disposal/total number of rural households | |

Comprehensive Calculation Formula (overall): $Fl = (Flr + Flf)/2$; $Fp = (Fpa + Fpna)/2$; $Fe = (Fec + Fem)/2$

*3.4. Data Sources and Processing*

The research data involve two aspects: statistical data and survey data. The statistical data include land use change data, basic geographic information data, and economic and social statistical data, mainly from the National Basic Geographic Information System database and the "Shandong Province 2010~2020 Statistical Yearbook". Survey data refer to the micro-basic information of typical samples. Aiming to assess the differences in the economic development level and topographical conditions of the study area, this study selected villages with the assistance and recommendation of local natural resources departments according to the principle of "economically developed counties choose villages with relatively strong internal economic strength, underdeveloped counties choose their internally developed villages, plain counties choose internally flat villages, mountainous counties choose villages with more complex internal terrain". From July 2019 to December 2020, three survey teams investigated 123 villages five times. Assisted by field observations of sample villages and questionnaire interviews with village committees or village cadres, we obtained data on the scale, land use structure, population structure, building structure, economic income, facility construction of each village, among other types of data. Based on the above basic data, the ArcGIS10.2 operating platform was used to establish the basic database of rural residential land transition research in Shandong Province, including county-level administrative units and rural residential land plaques.

## 4. Results and Analysis

*4.1. The Transition Process of Rural Residential Land Scale at the Macro Level*

### 4.1.1. Division and Distribution Characteristics of Transition Stages

As can be seen from Figure 3 and Table 3, the transition index of rural residential land at each county level in Shandong Province gradually decreases with the fitting curve characteristic of power function ($R^2$ = 0.9117), and the sequence mutations between each unit are obvious. Taking the mutation point as the critical value, the transition stage of rural residential area was divided into five stages from the primary stage to stable stage by the breakpoint method. This indicates that from the lower stage to the higher stage, the balance between rural residential land and other construction land gradually tends to a new state, which is basically consistent with the theory of rural homestead transition proposed by Long [32] and its application results along the Yangtze River [49].

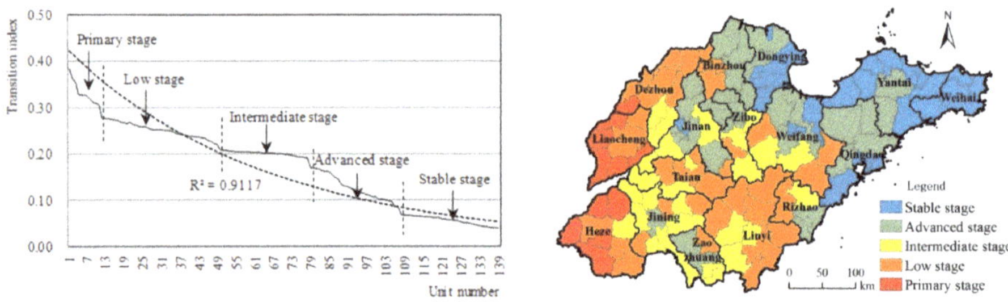


**Figure 3.** The division of rural residential land transition stages in Shandong Province.

At the same time, the coefficient of variation of the transition index of rural residential areas corresponding to the five transition stages of Shandong Province is relatively low, indicating that the deviation of the transition index of rural residential areas within each stage is small and the concentration is strong. It can also be seen from Figure 3 that the spatial distribution of rural residential areas in the transition stage of Shandong Province is relatively concentrated, with the primary stage containing 11 administrative units, mainly distributed in Heze City and Liaocheng City on Luxi Plain. The low stage contains 38 county-level units, which are the most numerous and widely distributed, including Linyi City, Laiwu City, Tai'an City, and other places adjacent to the primary stage in the West

Shandong Plain and the middle of the Shandong mountainous area. The intermediate stage contains 29 county-level units, which are mainly distributed in the piedmont plains from the central Shandong mountainous area to the southwest and north of Shandong, mostly in Jinan City, Jining City, and Zibo City. The advanced stage also contains 29 county-level units, which are mainly located in Binzhou City, Dongying City, Weifang City, and the hinterland of Jiaodong Peninsula in the coastal area of northern Shandong Province, while a small number of units are distributed in Zaozhuang City and Rizhao City in the hilly area of southern Shandong Province. In the stable stage, there are 30 county-level units, which is a relatively large number, which are concentrated in the coastal cities of Jiaodong Peninsula and the core area of the Yellow River Delta.

**Table 3.** Mathematical statistics of the rural residential transition index in Shandong Province.

| Transition Stage | Index Threshold | Mean Value | Mutation Point | Coefficient of Variation | Numbers of Counties |
|---|---|---|---|---|---|
| Primary stage | 0.3068~0.3838 | 0.3324 | 0.3068 | 0.0246 | 11 |
| Low stage | 0.2201~0.2775 | 0.2519 | 0.2201 | 0.0160 | 38 |
| Intermediate stage | 0.1920~0.2089 | 0.2015 | 0.1920 | 0.0048 | 29 |
| Advanced stage | 0.0845~0.1701 | 0.1275 | 0.1701 | 0.0276 | 29 |
| Stable stage | 0.0396~0.0678 | 0.0552 | 0.0678 | 0.0094 | 30 |

### 4.1.2. Identification and Action Pattern of Key Influencing Factors

(1)　Identification of key influencing factors

In order to avoid the influence of index multicollinearity on the local estimation, the principal component analysis method was first used to reduce the amount of index data and transform the variables to eliminate the overlapping parts of much of the data, and a few new variables were used to represent the data structure of the original variables. The analysis found that the KMO test value of the original variable factor was 0.822, while the concomitant probability of the Bartlett sphericity test was 0.000, which was less than the significance level of 0.05, so it was suitable for factor analysis. Due to the fact that the eigenvalue was greater than 1, four principal component variables (Tables 4 and 5) were extracted. The first principal component basically reflected the regional natural conditions of the transition of rural residential areas, the second principal component reflected the regional economic development level of the transition of rural residential areas, the third principal component reflected the social conditions of the transition of rural residential areas, and the fourth principal component reflected the policy environment of the transition of rural residential areas. Therefore, in accordance with the principle of independence and simplification, the maximum correlation coefficient of each principal component was selected as the final influencing factor, which were, respectively, regional altitude (A1), GDP change rate (B1), rural population change rate (C1), and land supply rate of construction land planning (D3).

(2)　Analysis of the action pattern of key influencing factors.

Generally speaking, $R^2$ and $R_{adj}^2$ are effective parameters for analyzing the performance of the quantitative evaluation model, and their values vary between 0 and 1. The larger the score, the better the fitting effect. Compared with the OLS model (Table 6), the $R^2$ and $R_{adj}^2$ of the GWR model reached 0.726 and 0.741, respectively, indicating that the selected variables of the model can explain about 75% of the transition differentiation of regional rural residential areas, which was 15% higher than that of the OLS model. In addition, the AICc score of the GWR model was significantly smaller than that of the OLS model, indicating that the former had a greater advantage in estimating local differences and could better explain the spatial heterogeneity between rural residential area transition and influencing factors.

**Table 4.** Total variance explained.

| Component | Initial Eigenvalues | | | Extraction Sums of Squared Loadings | | | Rotation Sums of Squared Loadings | | |
|---|---|---|---|---|---|---|---|---|---|
| | Total | % of Variance | Cumulative % | Total | % of Variance | Cumulative % | Total | % of Variance | Cumulative % |
| 1 | 5.525 | 42.352 | 42.352 | 5.525 | 42.352 | 42.352 | 3.894 | 25.42 | 25.42 |
| 2 | 3.82 | 26.044 | 68.396 | 3.82 | 26.044 | 68.396 | 3.738 | 23.153 | 48.573 |
| 3 | 2.041 | 13.503 | 81.899 | 2.041 | 13.503 | 81.899 | 3.57 | 22.846 | 71.419 |
| 4 | 1.738 | 10.274 | 92.173 | 1.738 | 10.274 | 92.173 | 3.216 | 20.754 | 92.173 |
| 5 | 0.765 | 5.322 | 97.495 | | | | | | |
| 6 | 0.414 | 1.003 | 98.498 | | | | | | |
| 7 | 0.125 | 0.872 | 99.37 | | | | | | |
| 8 | 0.063 | 0.365 | 99.735 | | | | | | |
| 9 | 0.032 | 0.136 | 99.871 | | | | | | |
| 10 | 0.017 | 0.083 | 99.954 | | | | | | |
| 11 | 0.006 | 0.042 | 99.996 | | | | | | |
| 12 | 0.001 | 0.004 | 100 | | | | | | |

**Table 5.** Rotated component matrix.

| Indexes | Component | | | |
|---|---|---|---|---|
| | 1 | 2 | 3 | 4 |
| A1 | 0.921 | −0.038 | −0.242 | −0.176 |
| A3 | −0.882 | 0.319 | 0.182 | 0.211 |
| A2 | −0.847 | 0.330 | 0.137 | 0.067 |
| B1 | 0.167 | −0.962 | 0.207 | 0.326 |
| B2 | −0.328 | −0.944 | 0.086 | 0.153 |
| B3 | −0.040 | −0.897 | 0.063 | 0.180 |
| C1 | −0.089 | −0.159 | 0.914 | −0.065 |
| C3 | −0.305 | 0.129 | −0.887 | −0.231 |
| C2 | −0.155 | −0.236 | 0.845 | −0.084 |
| D3 | −0.093 | −0.165 | −0.214 | −0.908 |
| D1 | 0.088 | 0.163 | 0.208 | −0.854 |
| D2 | −0.235 | 0.202 | 0.091 | −0.836 |

**Table 6.** GWR and OLS model estimation result for rural residential land transition.

| Variables | Regression Coefficient of OLS Model | Regression Coefficient of GWR Model | | | | |
|---|---|---|---|---|---|---|
| | | Minimum | 1/4 Median | Median | 3/4 Median | Maximum |
| Intercept | – | 0.152 | 0.244 | 0.351 | 0.473 | 0.675 |
| Altitude | 0.511 *** | 0.208 | 0.310 | 0.516 | 0.627 | 0.818 |
| GDP Change rate | −0.634 *** | −0.481 | −0.564 | −0.762 | −0.811 | −0.938 |
| Rural population change rate | −0.484 *** | −0.264 | −0.389 | −0.554 | −0.676 | −0.818 |
| Land supply rate of construction land planning | −0.612 *** | −0.311 | −0.478 | −0.594 | −0.717 | −0.886 |
| Local $R^2$ | – | 0.311~0.887 | | | | |
| $R^2$ | 0.557 | 0.726 | | | | |
| $R_{adj}^2$ | 0.572 | 0.741 | | | | |
| AICc | −101.43 | −178.54 | | | | |

Note: *** indicate that the regression coefficients are significant at the level of 1%.

The GWR model is a local estimation model, and each sample data point has a set of local parameter estimates. From the perspective of multivariate synthesis effects (Figure 4), it can be seen that the parameter estimation results and regression coefficients of the control variables of each county-level unit in Shandong Province are not the same, indicating that instability of the geographic space exists. The model determination coefficient $R^2$ is between 0.311 and 0.887, with an average value of 0.615, which shows a differentiation pattern of "high in the west and low in the east, and abrupt changes in the southeast" as a whole. Among them, Heze City, Liaocheng City, Dezhou City, Jining City, Weifang City, Rizhao City, and other regions have higher $R^2$ values, indicating that these regions have been better simulated. The $R^2$ values of Yantai City, Weihai City, Dongying City, and Qingdao City are relatively low, indicating that the fitting optimization of these areas is slightly poor, and the differentiation of rural residential areas' transition stages is also affected by other factors outside the model. In terms of univariate effects (Figure 4), the order of influence on the transition index of rural residential areas is: GDP change rate > land supply rate of construction land planning > altitude > rural population change rate. In addition to altitude, other factors generally have a negative effect.

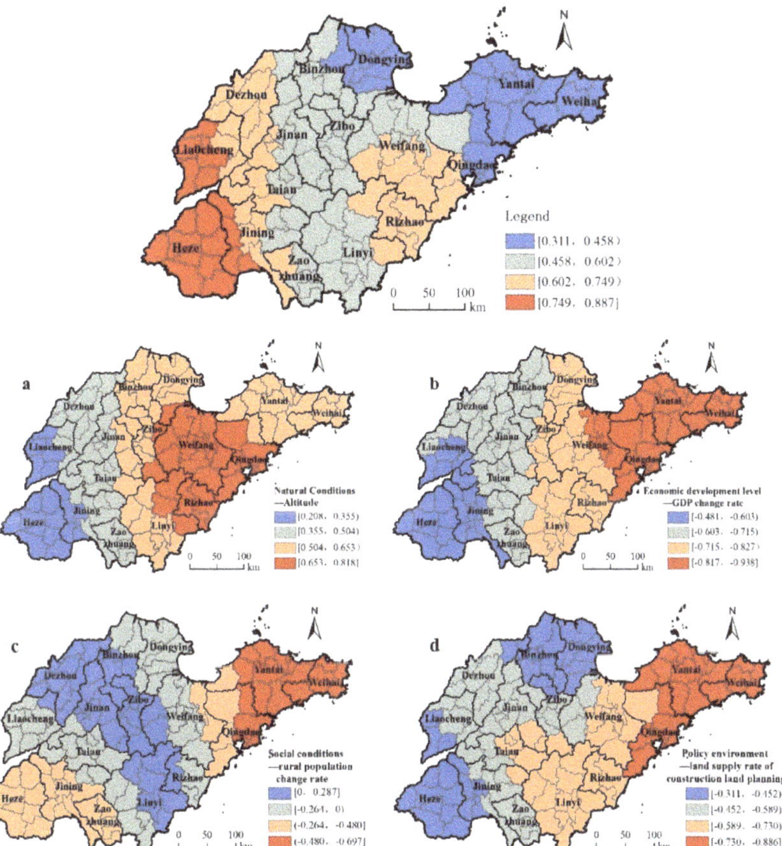


**Figure 4.** Spatial distribution map of $R^2$ in the GWR model and distribution maps for the regression coefficients of independent variables in the GWR model. Among them, (**a**) represents the spatial distribution of the impact of altitude on rural residential area transition; (**b**) represents the spatial distribution of the impact of GDP change rate on rural residential area transition; (**c**) represents the spatial distribution of the impact of rural population change rate on rural residential area transition; (**d**) represents the spatial distribution of the impact of land supply rate of construction land planning on rural residential area transition.

Altitude has the least impact on the transition of rural residential areas in southwestern Shandong. The terrain in this area is simple, with a minimal topography, and the entire area is composed of plains. The effects of terrain conditions on the transition index of rural residential areas are very similar. In contrast, altitude has a greater impact on central and eastern Shandong. The terrain in this region is relatively complex, and is a transition zone from mountainous areas to piedmont plains. The new construction land is preferentially arranged in the flat terrain area with a low construction cost, and mostly in the area where the municipal or county-level government is stationed. From plains to hills and mountains, the newly increased scale of construction land is gradually decreasing, which will help to increase the proportion of rural residential areas and promote the transition of rural residential areas in the region.

The GDP change rate has less of an impact on the transition of rural residential areas in eastern Shandong and has a greater impact on western Shandong. The differences in economic levels between the county-level units in eastern Shandong are relatively small, the scale of rural residential areas is relatively stable, and the structure and function of the rural residential areas are relatively stable. The economic development in western Shandong is relatively backward and the differences between the regions are large, especially from provincial capital cities to provincial cities, and from urban areas to counties. From the high administrative level to the low administrative level, the level of economic development gradually decreases. As a result, the scale of newly added construction land has increased. Administrative levels and economic conditions have a significant reverse effect on the transition index of rural residential areas.

The rural population change rate has positive effects on the transition index of rural residential areas in eastern Shandong. The rural population flow in this area is mainly characterized by the characteristics of "leave the soil and never leave the hometown, and enter the factory and not enter the city". The rapid development of the local economy has driven the employment transformation of local farmers from agriculture to industry. The increase in income has driven people to return to their hometown and build new homes. This phenomenon has led to the continuous expansion of the rural housing area and promoted an increase in the transition index of rural residential areas. The impact of the rural population's mobility in central Shandong has a negative effect. Faced with the problem of hollowing in rural residential areas, the region vigorously promotes rural land improvement projects and linking urban–rural construction land increase and decrease in the hook projects to transfer rural populations to cities for the purpose of living, finding employment and retiring to vacant and over-standard rural residential land. The structure of construction land was adjusted by tapping the potential of the stock—that is, the rural population flow of "leaving the soil and leaving the hometown" is conducive to promoting the transition of rural residential areas.

The land supply rate of construction land planning also has the largest negative effect on the transition of rural residential areas in eastern Shandong, especially in Qingdao City. The scale of newly added construction land in Qingdao City is directly approved by the state, and the land supply rate is significantly higher than that in other regions. In addition, the newly added construction land is mainly used for urban development and the construction of infrastructure. The scale of rural residential areas is basically stable, which is conducive to the continuous progress of its transition. However, the level of urbanization in southwestern Shandong is low, and the allocation of planning indicators for newly added construction land is relatively small. Many rural houses in the region are in violation of regulations. Coupled with the incomplete management system and high management costs, the scale of rural construction land continues to expand, which is extremely detrimental to the transition of rural residential areas.

*4.2. Staged Differentiation of Structure and Function of Rural Residential Areas at the Micro Level*

4.2.1. Selection of Typical Sample Areas and Sample Points in Different Transition Stages

Comprehensively considering the social, economic, and natural environment and other leading factors that affect the transition of rural residential areas, we selected for analysis five typical county-level sample areas in different transition stages in Longkou

County, Huantai County, Gaotang County, Mengyin County, and Cao County and their internal 123 sample villages (Table 7 and Figure 5).

**Table 7.** Basic profile of rural residential land in typical areas.

| Name | Transition Index | Transition Stage | Economic Development Stage | Geographical Conditions | Per Capita Land Area of Rural Residential Areas (m$^2$) | Sample Numbers |
|---|---|---|---|---|---|---|
| Longkou County | 0.0581 | Stable stage | Economically developed stage | Ludong Hills | 220.50 | 31 |
| Hengtai County | 0.1187 | Advanced stage | Advanced stage of industrialization | Lubei Plain | 232.03 | 21 |
| Gaotang County | 0.2020 | Intermediate stage | Intermediate stage of industrialization | Luxi Plain | 278.76 | 22 |
| Mengyin County | 0.2616 | Low stage | Initial stage of industrialization | Luzhong Mountains | 304.05 | 25 |
| Cao County | 0.3605 | Primary stage | Primary production stage | Lunan Plain | 366.23 | 24 |

Note: The economic development stage is divided according to the standard of per capita GDP of 1200, 2400, 4800, 9000 yuan/person in the literature [11].

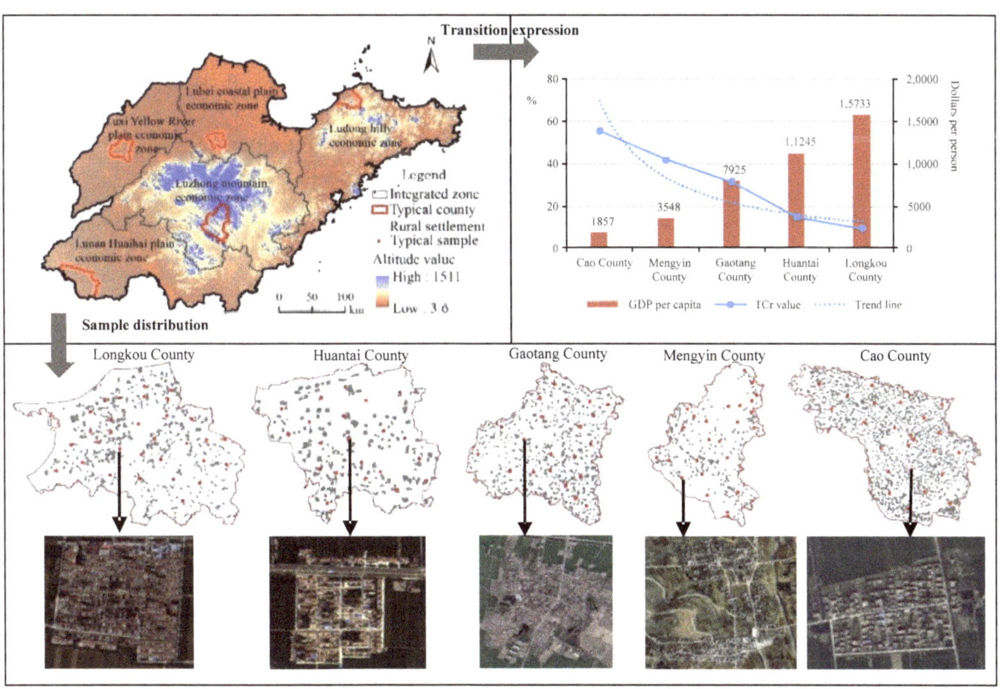


**Figure 5.** Distribution of typical sample areas and points in different transition stages.

From the perspective of the sampling areas, on the one hand, the five typical sample areas and counties are located in the Ludong hilly economic zone, the Lubei coastal plain economic zone, the Luxi yellow river plain economic zone, the Luzhong mountain economic zone, and the Lunan Huai-hai plain economic zone, reflecting the differences in comprehensive geographical conditions and social and economic development. On the other hand, the transition index of rural residential land scale in the five sample counties decreased gradually from Cao County, through Mengyin County, Gaotang County, and Huantai County, to Longkou County, showing the characteristic of power exponent consistent with the theoretical hypothesis and representing the different stages of rural residential land transition. From the perspective of samples, on the one hand, the number of typical samples within each sample area is equivalent, distributed in each township within each sample area and corresponding to the economic level and topographic conditions of the sample area, ensuring the representativeness of the sample. On the other hand, the land use structure, spatial distribution structure, and living-production-ecological function of rural residential areas in different typical samples of different areas have certain similarities and differences, providing the conditions for us to explore the differentiation law of the structure and function of rural residential areas in different transition stages.

### 4.2.2. Stage Differentiation Characteristics of Rural Residential Structure

(1)    Differentiation of spatial distribution structure in rural residential areas

From the perspective of the plaque shape complexity of rural residential areas (Table 8), the changes in the AWMSI and AWMPFD indices of the five samples are consistent. Overall, with the continuous upgrade from lower stage to higher stage in the transition, the shape of the rural residential areas changes from complex to regular, which also indicates that the fractal characteristics of rural residential areas tend to be simplified and that the influence degree of human activity is gradually increasing. This is mainly due to the influence of the mountainous terrain, the scattered rural residential areas, and the irregular shape of the village's periphery.

**Table 8.** Calculation results of the landscape metrics of rural residential land in a typical area.

| Sample Area | Shape Complexity | | Spatial Agglomeration | | |
| --- | --- | --- | --- | --- | --- |
| | AWMSI | AWMPFD | PD | COHESION | IJI |
| Longkou County | 14.66 | 1.07 | 11.56 | 68.85 | 26.21 |
| Hengtai County | 19.08 | 1.17 | 9.02 | 60.40 | 26.32 |
| Gaotang County | 23.09 | 1.29 | 9.21 | 53.93 | 40.32 |
| Mengyin County | 30.91 | 1.33 | 14.43 | 45.90 | 46.23 |
| Cao County | 28.09 | 1.31 | 9.46 | 43.52 | 47.34 |

From the perspective of the spatial agglomeration degree of plaques in rural residential areas (Table 8), with the continuous upgrading of the transition stage, the fragmentation degree of rural residential plaques is gradually reduced, but the characteristics of this change are greatly affected by the terrain. On the other hand, the plaque connectivity and dispersion of the five samples show an opposite trend of increases and decreases. Among them, the COHESION indexes of Longkou County and Huantai County are significantly higher than those of the other three regions, and their IJI indexes are significantly lower than those of the other three regions. This shows that with the continuous upgrading of the transition stage, the degree of spatial connection and agglomeration of rural residential areas has gradually increased, and the spatial pattern of rural residential areas has gradually evolved from extensive to intensive.

(2)    Differentiation of land use structure in rural residential areas

According to the average statistics of the area proportion of land use types in the rural residential areas of the typical sample villages in various areas (Figure 6), the homestead is the main part of the rural residential areas, and shows a trend of gradually increasing from high-level stage areas to low-level stage areas. Road occupies the second place in terms of size in rural residential areas, presenting the opposite characteristics to those found with regard to homestead change. Among them, the proportion of street land area in Mengyin County is slightly higher. The reason for this is that rural houses in mountainous areas are scattered and the area of roads connecting with each other increases; public service land, industrial land and commercial land have a low overall scale configuration and show the opposite characteristics to those found with regard to the changes in homesteads. These characteristics indicate that the level of infrastructure allocation in rural residential areas at each stage remains to be improved. Idle land refers to vacant and unused land in a village, which makes up a certain proportion in the five typical sample areas. Affected by terrain, the proportion of idle land in the plain area is significantly higher than that in the hilly area and the mountainous area, which shows that the internal land use structure conversion in the transition process of rural residential areas is not yet sufficient and the level of intensive land use needs to be improved, especially in plain areas.

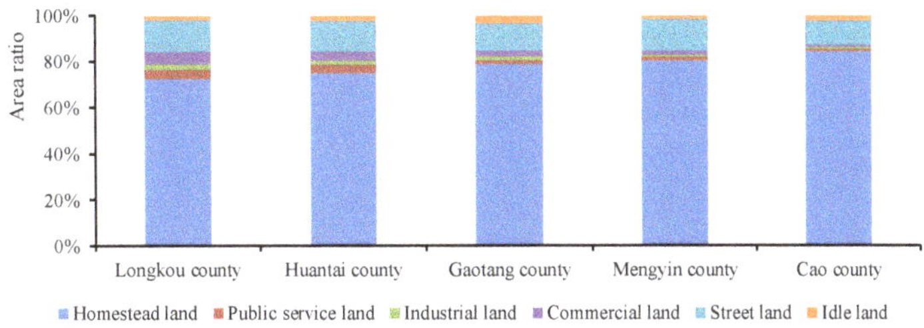


**Figure 6.** The mean statistics of the proportion of rural residential land types in different areas.

In terms of the diversity and concentration of the internal land use structure of rural residential areas (Figure 7), both the *GM* score and the *I* score show opposite changes at the sample point and area level. First of all, there are certain differences in the internal land use structure of rural residential areas among the various sample points in the five sample areas. Among them, the maximum and minimum values of *GM* continue to increase with the upgrading of the transition stage, while the threshold range of *GM* gradually decreases from mountains and hills to the plain areas—that is, as the complexity of the terrain increases, the diversity of land use types within rural residential areas becomes more significant. The change in the *I* value is basically the opposite of the *GM* score. The proportions of samples with an *I* value of less than 0 continues to increase with the upgrading of the transition stage, showing that the internal land use types of rural residential areas in the high-level transition stage tend to be more diversified, while the internal land use types of rural residential areas in the low-level transition stage tend to be a single type dominated by homesteads. The differences in the internal land use structure of rural residential areas as a whole in the five sample areas are also significant. With the gradual escalation of the transition stage of rural residential areas, the average *GM* value continues to increase. The mean value of *I* decreases, and the law of stage differentiation is obvious. This phenomenon has a strong relationship with the regional socio-economic conditions. The higher the level of socio-economic development, the higher the demand for improved living conditions, while the higher the intensity of industrial activities, the more diverse functions are carried by rural residential areas. Therefore, it is necessary to provide more land use options for people's living and production activities so as to make the internal land use structure of rural residential areas more diversified. The more land types there are and the more balanced the structure ratio is, the lower the possibility of there being a single or small amount of land use types is and the weaker the concentration is.

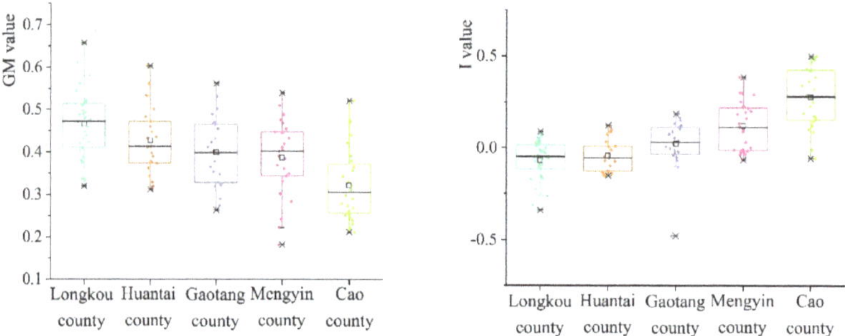

**Figure 7.** Statistics on the internal structure characteristics of rural residential land in different areas.

### 4.2.3. Stage Differentiation Characteristics of Rural Residential Functions

(1)  Intensity differentiation of single function

The living, production, and ecological functions of rural residential areas in the five sample areas all showed obvious gradient differences (Figure 8). Among them, the living function presents a "three-stage" gradient change, which is due to the high-level rural economy, the relatively large number of buildings, and the complete configuration of public services and infrastructure in the advanced transition stage. In terms of the production function of rural residential areas, the "four-stage" gradient change characteristics of the five sample areas are relatively significant, and the differences between different transition stages are more obvious. Among them, the non-agricultural production function in the primary stage is marginally better than that of the low stage. The reason for this is that the amount of per capita arable land area is relatively small. More rural laborers go out to work, and long-term or short-term non-agricultural production also drive the increase in non-agricultural income, which affects the changes in production and income structure. The ecological function of rural residential areas shows obvious characteristics of "three-stage" gradient changes. More specifically, the change trajectories of the ecological conservation function and environmental maintenance function are quite different. Among them, the overall difference in the scores for the ecological conservation functions is not large, showing a phased increase, which is closely related to the per capita green area in the village. Villages in the advanced stage have relatively complete rural home renovation and infrastructure construction, especially as the road hardening rate increases and the number of original trees in front of doors and at the roadside gradually decreases. In the low stage villages, a large number of trees and other types of vegetation are still preserved on both sides of the farmers' courtyards and at the side of dirt roads. In this case, the overall score of the environmental maintenance function is quite different, and it maintains the characteristic of gradual decline. The ecological facilities of the advanced stage villages are more complete and sound than those of the low stage villages, showing that they are important influencing factors.

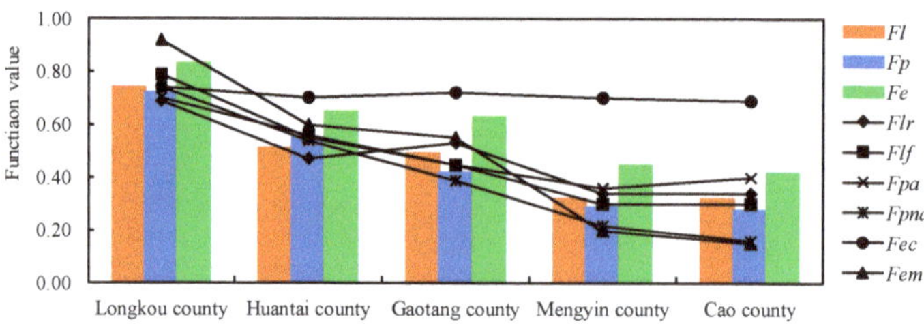

**Figure 8.** Single function value of rural residential areas in different areas.

(2)　Multi-functional integration and coordinated differentiation

From the perspective of the comprehensive degree of the living-production-ecological function of rural residential areas (Figure 9), there are certain differences between the five sample areas and their typical sample points. Among them, the highest value of the multi-functional comprehensiveness of rural residential areas in the advanced transition stage is close to 1.0, the threshold interval is about 0.45, and the average value is above 0.6. However, the intermediate stage and low stage have the highest value of functional comprehensiveness at around 0.8, the threshold interval is around 0.55, and the average value is between 0.3 and 0.5. This shows that with the upgrading of the transition stage, the comprehensive functions of rural residential areas have been continuously enhanced, and the differences within the samples have gradually slowed down. From the perspective of the multi-functional coupling relationship of rural residential areas, the five sample areas and their typical sample sites all have certain process fluctuations—that is, with the continuous upgrading of the transition stage, the mean coupling degrees of the living-production-ecological function of rural residential areas decrease first and then increase. This volatility shows a certain terrain difference—that is, from the plain area to the hilly area to the mountain area, the multi-functional coupling degree of rural residential areas gradually decreases. In the plain area, with the upgrading of the transition stage, the multi-functional coupling degree of the rural residential areas gradually increases.

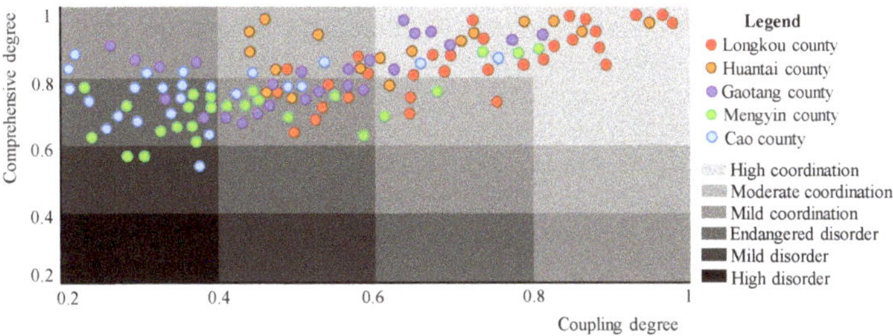


**Figure 9.** Comprehensive and coupling degree of rural residential area functions in different areas.

From the perspective of the coordination of the multi-functional comprehensive degree and coupling degree of rural residential areas and the downgrading of the transition stage, the multi-functional coordination of rural residential areas gradually decreases and the types of coordination tends to become more complicated. The areas in the advanced transition stage have a high-level of multi-functional coordination. Among them, more than 50% of the samples in Longkou County are in the high coordinated state. Gaotang County, which is in the intermediate transition stage, has a relatively general multi-functional coordination, with mild coordination as the mainstay, accounting for about 60% of the total number of typical samples. However, Mengyin County and Cao County, which are in the low transition stage, have low multi-functional coordination. In these two counties, less than 10% of the samples are coordinated moderately and highly, while more than 80% of the samples are in mild coordination and endangered disorder.

## 5. Discussion

### 5.1. The General Law and Formation Mechanism of Rural Residential Transition

The scale transition of rural residential land is the primary manifestation of the transition of rural residential areas. The transition of rural residential areas is actually a process of synergy and differentiation of structure and function on the basis of scale transition. This process is the result of the combined effect of multiple driving forces. Among them, the natural background condition is the basic driving force for the transition of rural residential areas, which has a long-term influence on the evolution of rural residential areas and plays a decisive role in the transition of early rural residential areas. With the development of economy and urbanization, the influence of natural conditions has been surpassed by other non-natural factors to some extent. Affected by economic development, population mobility, and government regulation [11], the transition of rural residential land scale has become more intense [29].

The transition of rural residential areas has stage characteristics. From the low-level transition stage to the high-level transition stage, the scale transition of rural residential areas is manifested as a slowdown or even as negative growth according to the land use scale. Meanwhile, the structures and functions of rural residential areas are gradually optimized and tend to be coordinated. In the low-level transition stage, it is less affected by urbanization and policy-related factors and retains significant rurality. The transition of rural residential land largely comprises the renewal and reconstruction of villages. With the gradual development of the rural economy, farmers have a strong desire to expand and build houses to improve their living conditions. The scale of rural residential areas has increased significantly. The spatial distribution is affected by natural conditions and lacks planning guidance, often presenting a scattered and disorderly distribution. The functions of rural residential areas are single, with traditional living function as the mainstay. In the mid-level transition stage, driven by urbanization and the market economy, the orientation of rural industrialization is obvious. Rural construction has shifted from housing to infrastructure and factories, and the rurality is gradually weakened. The spatial form of rural residential areas is greatly affected by population migration. At this stage, the spatial structure presents a trend of differentiation, while regional agglomeration and internal empty and disused areas coexist. The internal land structure of rural residential areas tends to be balanced and decentralized, which is reflected in the gradual decrease in the proportion of homestead areas, and the gradual increase in public service facility land, industrial land, and roads. Rural residential areas mainly have living function and production function. In the high-level transition stage, the urban–rural integration becomes closer due to the effects of urban development and the government's policy regulation, while the rurality gradually weakens or even disappears locally. Strictly controlled by policies such as land use planning and management systems, the scale of rural residential areas has remained stable or even been reduced, the spatial form has begun to pursue fairness and justice, the spatial distribution has become more intensive and reasonable, the amount of empty and disused areas has decreased, and artificial buildings have become dense and regular. The internal structure of rural residential areas tends to be more complicated and their functions tend to be diversified. It can be seen that the transition stage differentiation of rural residential land scale consists in the difference in its own conditions to adapting to the external environment. For rural residential areas with superior conditions and gradual improvement, their ability to adapt to changes in the external environment is relatively strong and they are gradually upgraded from the lower stage to the higher stage. For rural residential areas with poor conditions and continuous degradation, they do not adapt to the development of the external environment and remain at a low stage until they die out [40]. This finding further enriches the existing single research content on the scale, structure, and functional transition of rural residential land use [29,32,50].

*5.2. Rural Reconstruction Strategy Based on the Comprehensive Framework of "Elements–Structure–Function"*

As a basic land use type and an important carrier of rural life and production, rural residential areas have comprehensively embodied the systematic characteristics of "element–structure–function" in the process of continuous evolution and transition. Rural reconstruction is a process of adapting to changes in rural internal factors and the external environment. By optimizing the allocation of elements and strengthening management methods, the reconstruction of the rural social form and regional spatial pattern is aimed at achieving the optimization of the internal structure and function of the rural regional system [51]. It can be seen that the spatial distribution pattern, internal land structure, and stage differentiation and upgrading of the system functions of rural residential areas driven by the transition of land element attributes are essentially the process of rural reconstruction [11,52]. Therefore, combined with the transmission characteristics of the rural residential area transition and dimensionality reduction proposed in this paper, this research constructs a comprehensive rural restructuring framework of "element–structure–function" (Figure 10) and proposes corresponding implementation strategies.

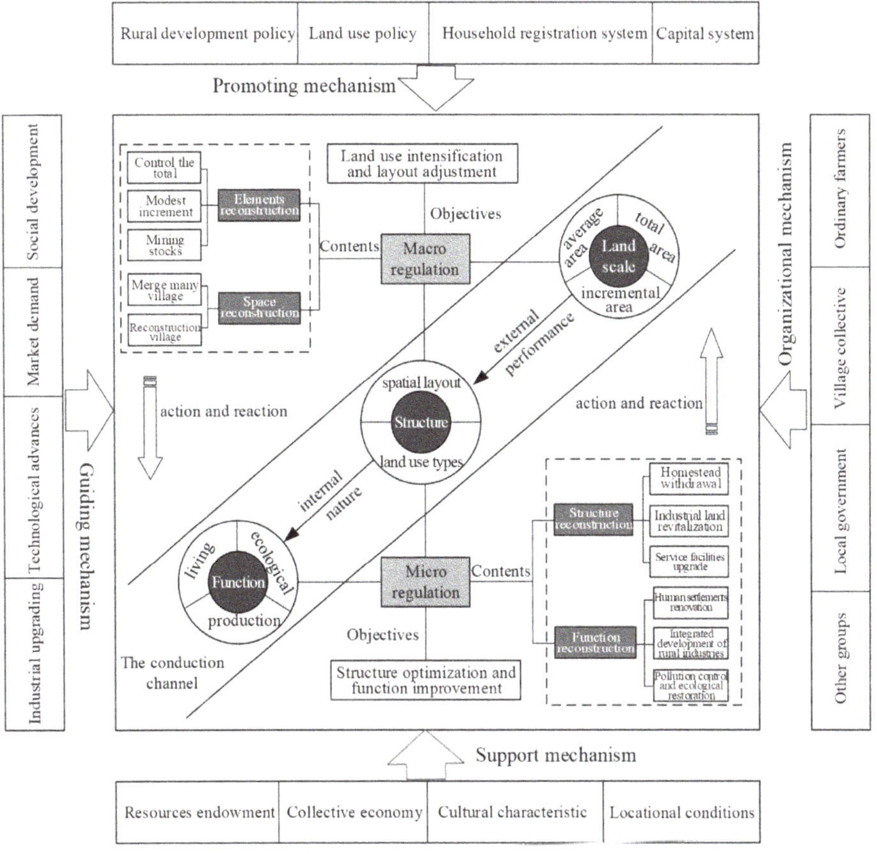


**Figure 10.** The framework of rural reconstruction based on the integration of "factor, structure, and function".

At the macro level, the land element attributes and spatial distribution pattern of rural residential areas are the main manifestations of this process. Under the influence of the urbanization process and the lack of a control system, the total scale of rural residential land is large, the phenomenon of hollowing and illegal construction is serious, and the spatial distribution is irregular and scattered. Changing these undesirable forms is the basic premise for promoting the transition of rural residential land. Therefore, under the joint action of a series of guiding mechanisms, such as external social development, market demand, technological progress, industrial upgrading, and policy innovation, the reconstruction of rural elements and space should be emphasized [53]. On the one hand, the village planning should be formulated scientifically. Based on the concept of equal emphasis on intensification and development, the village development boundary should be reasonably delimited and the overall land use scale of the village should be strictly controlled. At the same time, it is necessary to give full play to the blanking mechanism of planning and reserve a certain amount of land for uncertain projects in the process of future rural revitalization. On the other hand, it is also necessary to promote the land comprehensive consolidation and vigorously carry out hollow village governance. Taking advantage of the opportunity for rural land system reform, it is necessary to strengthen the orderly removal of the remaining rural construction land, such as idle and abandoned rural homesteads and industrial plants, so as to fully tap the potential of rural residential land and make use of land elements to promote the integrated development of rural industries and optimize the comprehensive effect of rural governance capacity. Moreover, with the effective planning and regulation of favorable policies concerning urban–rural integrated development and rural revitalization, the rural population should be appropriately concentrated in rural areas and non-agricultural industries should be appropriately structured to develop according to local conditions. Through different means, such as the relocation of villages and towns and the intensive internal organization of large-scale villages, the reconstruction of the rural spatial layout could be realized.

At the micro level, the internal land use structure of rural residential areas and the diverse functions that meet the needs of villagers are the main manifestations. The improvement and promotion of awareness of the problems of simple land use structures within villages, unbalanced land use for supporting service facilities and industrial development, poor living conditions, low production income, and pollution of the ecological environment are the fundamental factors for promoting the transition of rural residential areas. In this regard, it is necessary to combine the resource endowment conditions, location conditions, economic foundation, and subject behavior characteristics; encourage the internal support and organizational mechanism of the rural system; and focus on promoting the optimization and reorganization of the rural land structure and the restructuring of system functions. Among them, the optimization of land use structure should focus on rural homesteads, industrial land, and public service land. On the one hand, according to the principle of "one household has one house" and "every house has people living in it", some excess homestead land could be traded to users who need houses through ownership adjustment by means of policy incentives so as to solve the problem of the insufficient supply of homesteads. The other part of the surplus homesteads could be used for the development of rural industry through conversion to solve the problem of the lack of non-agricultural industrial land. On the other hand, according to the revitalization needs of rural industries, the planning of rural industrial parks should be promoted and the integrated development of rural industries should be promoted through the market transactions of stock collective construction land and appropriate incremental replenishment so as to provide opportunities for local employment and urban capital to the countryside. In addition, according to the requirements of the equalization of urban and rural public services, the allocation of rural public service resources should be improved. In particular, for villages with priority development, land investment and a centralized layout of basic service facilities such as education, health, parks, and squares should be strengthened to continuously improve the rural service capacity.

It is necessary to combine different aspects of life, production, and ecology with a definite target to improve the function of the rural system. On the one hand, under the guidance of the rural revitalization strategy, appropriate human intervention measures should be taken to improve the traditional rural development mode. With the help of policies and institutions, it is possible to improve the organizational capacity of rural communities and provide diversified sources of livelihoods. Using engineering technology to strengthen the renovation of rural dilapidated houses and the treatment of environmental pollution, we can gradually change rural employment methods, lifestyles, and family consumption concepts and ideologies and comprehensively improve the life quality of rural residents. On the other hand, based on planning guidance and market mechanisms and according to the principle of agglomeration development and intensive management, the development layout of rural industries should be coordinated, non-agricultural industries should be integrated into parks, and point land supply should be combined so as to promote the mutual flow of urban and rural populations and capital and accelerate the reconstruction of rural industries. In addition, based on the characteristics of the rural ecosystem, an ecological interception system should be established to absorb and purify non-point source pollution; strengthen the comprehensive treatment project of pollutants; and form a system of source reduction, flow interception, and treatment. At the same time, it is also important to follow the principles of landscape ecology to strengthen the background matrix of rural woodland and farmland ecosystems, improve the corridor of the rural ecosystem, protect the habitat environment of species, and maintain the diversity of biological resources.

In this way, rural reconstruction can be placed in the integrated frame of the "element–structure–function" of the rural residential system. Under the guidance and promotion mechanism of the external environment of the system and through the macro reconstruction of land elements and spatial patterns, the structure and function of rural residential land will be continuously upgraded from the aspects of resource optimal allocation and planning control. With the support of the internal elements of the system and the action of the organizational mechanism, the micro land use structure optimization and function improvement can meet the needs of the improvement of residents' living quality and rural transition, and then feed back to the macro element allocation and spatial reconstruction. Finally, an evolutionary process of interaction between the guidance of macro-control and feedback of micro-control is formed to promote rural reconstruction [54].

### 5.3. Contributions, Limitations, and Future Work

In theory, we have established a multi-level framework with comprehensive scale, structure, and function to discuss the transition mechanism and stage differentiation characteristics of rural residential areas, and—to some extent—solve the limitations of rural residential land use transition research from a single perspective. The scale transition of rural residential land and the synergistic differentiation of structure and function revealed by this research will help enrich and improve the existing rural land use transition theories. The proposed theoretical framework is designed for research into the transition of different regions and different types of rural residential areas. It has a wide range of promotional value and universality. At the same time, we adopted Long's classic model of rural homestead transition measurement [32], a landscape index model, a spatial statistical analysis method, and a multi-factor comprehensive analysis method. This study gradually reduces the dimensionality from the macro-provincial scale to the micro-village scale, and conducts a systematic empirical analysis of the transition process of rural residential land use. We also put forward a feasible strategy for achieving rural restructuring, realizing an effective combination of theory and application. Therefore, the systematic research logic and the multiple analysis methods of the dimensionality reduction process adopted by this research provide a new perspective for multidisciplinary cross-integration research, which will help to enrich the theoretical research and practical applications of land use transition and rural development worldwide.

However, although this study puts forward new ideas and a more effective method for understanding the transition problems of rural residential areas, there are certain potential uncertainties in the results. On the one hand, land use transition is a regional and even global issue, and different regions have different transition characteristics due to various factors [43]. This research mainly focuses on the transition characteristics of rural residential areas with different natural conditions and social and economic development levels. The consideration of human factors, such as policy system, culture, and subject behavior, is still lacking. On the other hand, the various and complicated methods of land use transition are also a shortcoming of the current research, which is related to the land use system, including various types of land, such as productive land, public welfare land, and ecological land. This research mainly focuses on the scale characteristics of rural residential areas and proposes a land use transition analysis method system for use in the process of dimensionality reduction. The analysis of long-term historical sequence evolution must be strengthened. Meanwhile, various detailed indicators are more suitable for the characteristics of the study area, and are not yet fully popularized. For use in other regions, they should be adjusted and supplemented. Therefore, the application of the research framework constructed in this article to China as a whole, and indeed the world in the long-term rural residential area transition research still requires further research and discussion. In the future, comprehensive analysis of the driving factors behind all elements of nature, society, the economy, culture, and institutions, the research theory of the interaction between subject and object, and the horizontal, vertical, and universal method systems will be an important research direction with regard to the transition of rural residential areas and even land use transition, which have important scientific value and practical significance in terms of developing a deep understanding of the evolutionary process of rural residential areas in different regions of the world.

## 6. Conclusions

Based on the process of dimensionality reduction transmission from the macro pattern to the micro sample points, this paper used the index of scale transition, the landscape pattern, the diversity and concentration index, and the living-production-ecological function index to divide the transition stages of the rural residential land use scale and comparatively analyze the influential factors of typical areas and sample points. On this basis, the procedural characteristics of the scale transition of rural residential land in Shandong Province over the past 10 years and the stage differentiation law with the corresponding structure and function are discussed in depth.

(1) The transition index of rural residential area in each county (district) of Shandong Province ranges from 0.0396 to 0.3838, with significant stage characteristics. The transition stage of rural residential areas in Shandong Province can be divided into five stages: a primary stage, low stage, intermediate stage, advanced stage, and stable stage. From the lower stage to the higher stage, the rural residential land gradually tends toward a new balance, and the spatial distribution shows an obvious agglomeration and autocorrelation. The driving factors and intensity of the effects are expressed as GDP change rate > land supply rate of construction land planning > altitude > rural population change rate. Altitude has a positive effect on the transition of rural residential areas, while other factors have a negative effect on the whole.

(2) With the gradual upgrading of the transition stage, the spatial distribution pattern of rural residential areas has been continuously integrated and optimized and the spatial distribution has become more intensive. The internal land use structure of rural residential areas tends to be balanced and decentralized, and the system functions of rural residential areas are gradually changing. The comprehensive functions of living, production, and ecology are gradually increasing, while the ecological conservation function is weakening. The stage differentiation of the structure and function of rural residential areas shows the difference in the conditions of rural residential areas adapting to the external environment. For rural residential areas with superior conditions and gradual improvement, their ability to adapt to changes in the external environment is relatively strong and they are gradually upgraded from the lower stage to the higher stage. Meanwhile, rural residential areas with poor conditions and continuous degradation cannot adapt to the development of the external environment and will remain in the lower stage until they die out.

**Author Contributions:** Conceptualization, Y.Q.; methodology, Y.Q., X.D. and L.Z.; formal analysis, Y.Q., X.D. and L.Z.; data curation, Y.Q., H.S., Z.P. and W.Z.; writing—original draft preparation, Y.Q., X.D. and L.Z.; writing—review and editing, Y.Q., X.D. and H.S.; visualization, Y.Q., X.D. and L.Z.; supervision, Y.Q. and H.S.; funding acquisition, Y.Q. All authors have read and agreed to the published version of the manuscript.

**Funding:** This research was funded by the National Natural Science Foundation of China, grant numbers 42077434, 41771560, and the Shandong Provincial Institutions of Higher Learning "Youth Innovation Team Development Plan" Project, grant number 2019RWG016.

**Institutional Review Board Statement:** Not applicable for studies not involving humans or animals.

**Informed Consent Statement:** Not applicable for studies not involving humans.

**Conflicts of Interest:** The authors declare no conflict of interest.

## References

1. Turner, B.L.; Lambin, E.F.; Reenberg, A. The emergence of land change science for global environmental change and sustainability. *Proc. Natl. Acad. Sci. USA* **2007**, *104*, 20666–20671. [CrossRef]
2. Mather, A.S. *Global Forest Resources*; Belhaven: London, UK, 1990.
3. Mather, A.S. The forest transition. *Area* **1992**, *24*, 367–379.
4. Grainger, A. National land use morphology: Patterns and possibilities. *Geography* **1995**, *80*, 235–245.
5. Bae, J.S.; Joo, R.W.; Kim, Y.-S. Forest transition in South Korea: Reality, path and drivers. *Land Use Policy* **2012**, *29*, 198–207. [CrossRef]
6. Yeo, I.Y.; Huang, C. Revisiting the forest transition theory with historical records and geospatial data: A case study from Mississippi (USA). *Land Use Policy* **2013**, *32*, 1–13. [CrossRef]
7. Bertoni, D.; Aletti, G.; Ferrandi, G.; Micheletti, A.; Cavicchioli, D.; Pretolani, R. Farmland Use Transitions After the CAP Greening: A Preliminary Analysis Using Markov Chains Approach. *Land Use Policy* **2018**, *79*, 780–800. [CrossRef]
8. Ma, L.; Long, H.; Tu, S.; Zhang, Y.; Zheng, Y. Farmland transition in China and its policy implications. *Land Use Policy* **2020**, *92*, 104470. [CrossRef]
9. Ernstson, H.; Van der Leeuw, S.E.; Redman, C.L.; Meffert, D.J.; Davis, G.; Alfsen, C.; Elmqvist, T. Urban transitions: On urban resilience and human-dominated ecosystems. *Ambio* **2010**, *39*, 531–545. [CrossRef] [PubMed]
10. Kong, X.; Liu, Y.; Jiang, P.; Tian, Y.; Zou, Y. A novel framework for rural homestead land transfer under collective ownership in China. *Land Use Policy* **2018**, *78*, 138–146. [CrossRef]
11. Long, H.; Qu, Y.; Tu, S.; Zhang, Y. Development of land use transitions research in China. *J. Geogr. Sci.* **2020**, *30*, 1195–1214. [CrossRef]
12. Drummond, M.A.; Auch, R.F.; Karstensen, K.A.; Sayler, K.L.; Taylor, J.L.; Loveland, T.R. Land change variability and human–environment dynamics in the United States Great Plains. *Land Use Policy* **2011**, *29*, 710–723. [CrossRef]
13. Lu, X.; Qu, Y.; Sun, P.; Yu, W.; Peng, W. Green Transition of Cultivated Land Use in the Yellow River Basin: A Perspective of Green Utilization Efficiency Evaluation. *Land* **2020**, *9*, 475. [CrossRef]
14. Song, X.; Li, X. Theoretical explanation and case study of regional cultivated land use function transition. *Acta Geogr. Sin.* **2019**, *74*, 992–1010.
15. Ge, D.; Wang, Z.; Tu, S.; Long, H.; Yan, H.; Sun, D.; Qiao, W. Coupling analysis of greenhouse- led farmland transition and rural transformation development in China's traditional farming area: A case of Qingzhou City. *Land Use Policy* **2019**, *86*, 113–125. [CrossRef]
16. Liao, L.; Long, H.; Ma, E. Rural Labor Change and Farmland Use Transition. *Econ. Geogr.* **2021**, *41*, 148–155.
17. Lou, Y.; Yin, G.; Xin, Y.; Xie, S.; Li, G.; Liu, S.; Wang, X. Recessive Transition Mechanism of Arable Land Use Based on the Perspective of Coupling Coordination of Input–Output: A Case Study of 31 Provinces in China. *Land* **2021**, *10*, 41. [CrossRef]
18. Zhou, X.; Li, X.; Song, W.; Kong, X.; Lu, X. Farmland Transitions in China: An Advocacy Coalition Approach. *Land* **2021**, *10*, 122. [CrossRef]
19. Niu, S.; Fang, B.; Cui, C.; Huang, S. The spatial-temporal pattern and path of cultivated land use transition from the perspective of rural revitalization: Taking Huaihai Economic Zone as an example. *J. Nat. Resour.* **2020**, *35*, 1908–1925.
20. Long, H.; Tu, S.; Ge, D.; Li, T.; Liu, Y. The Allocation and Management of Critical Resources in Rural China under Restructuring: Problems and Prospects. *J. Rural Stud.* **2016**, *47*, 392–412. [CrossRef]
21. Barbieri, A.F.; Guedes, G.R.; dos Santos, R.O. Land use systems and livelihoods in demographically heterogeneous frontier stages in the amazon. *Environ. Dev.* **2020**, *38*, 100587. [CrossRef]
22. Long, H.; Ge, D.; Zhang, Y.; Tu, S.; Qu, Y.; Ma, L. Changing man-land interrelations in China's farming area under urbanization and its implications for food security. *J. Environ. Manag.* **2018**, *209*, 440–451. [CrossRef]
23. Cowell, M.; Eckerd, A.; Smart, H. The rural identity and the encroaching city: Governance, policy and development in Northern Virginia's Wine Country. *Growth Chang.* **2020**, *51*, 79–101. [CrossRef]
24. Liu, Y.; Yang, R.; Long, H.; Gao, J.; Wang, J. Implications of land use change in rural China: A case study of Yucheng, Shandong Province. *Land Use Policy* **2014**, *40*, 111–118. [CrossRef]
25. Wang, Y.; Bi, G.; Yang, Q.; Wang, Z. Analyzing land use characteristics of rural settlements on the urban fringe of Liangjiang New Area, Chongqing, China. *J. Mt. Sci.* **2016**, *13*, 1855–1866. [CrossRef]
26. Sharma, A. Urban Proximity and Spatial Pattern of Land Use and Development in Rural India. *J. Dev. Stud.* **2016**, *52*, 1593–1611. [CrossRef]
27. Yang, R.; Xu, Q.; Long, H. Spatial distribution characteristics and optimized reconstruction analysis of China's rural settlements during the process of rapid urbanization. *J. Rural Stud.* **2016**, *47*, 413–424. [CrossRef]
28. Mazloum, B.; Pourmanafi, S.; Soffianian, A.; Salmanmahiny, A.; Prishchepov, A.V. The fate of rangelands: Revealing past and predicting future land-cover transitions from 1985 to 2036 in the drylands of Central Iran. *Land Degrad. Dev.* **2021**, *14*. [CrossRef]
29. Ma, W.; Jiang, G.; Zhang, R.; Li, Y.; Jiang, X. Achieving rural spatial restructuring in China: A suitable framework to understand how structural transitions in rural residential land differ across peri-urban interface? *Land Use Policy* **2018**, *75*, 583–593. [CrossRef]

30. Muyombano, E.; Espling, M. Land use consolidation in Rwanda: The experiences of small-scale farmers in Musanze District, Northern Province. *Land Use Policy* **2020**, *99*, 105060. [CrossRef]
31. Zhang, F.R.; Zhou, J.; Zhang, B.L. Construction land consolidation potential analysis for rural settlements from aspects of land use structure and their functions. *J. China Agric. Univ.* **2016**, *21*, 155–160.
32. Long, H. Rural housing land transition in China: Theory and verification. *Acta Geogr. Sin.* **2006**, *61*, 1100.
33. Qu, Y.; Jiang, G.; Zhang, B. Spatial characteristics of rural residential land transition and its economic gradient differentiation. *Acta Geogr. Sin.* **2017**, *72*, 1845–1858.
34. Wen, Y.; Zhang, Z.; Liang, D.; Xu, Z. Rural residential land transition in the Beijing-Tianjin- Hebei region: Spatialtemporal patterns and policy implications. *Land Use Policy* **2020**, *96*, 104700. [CrossRef]
35. Shang, R.; Qu, Y.; Jiang, H. Spatiotemporal characteristics and formation mechanism of rural residential land transition from the perspective of human-land relationship. *Resour. Sci.* **2020**, *42*, 672–684.
36. Qu, Y.; Jiang, G.; Li, Z.; Tian, Y.; Wei, S. Understanding rural land use transition and regional consolidation implications in China. *Land Use Policy* **2019**, *82*, 742–753. [CrossRef]
37. Qu, Y.; Jiang, G.; Tian, Y.; Shang, R.; Wei, S.; Li, Y. Urban-Rural construction land Transition (URCLT) in Shandong Province of China: Features measurement and mechanism exploration. *Habitat Int.* **2019**, *86*, 101–115.
38. Silva, R.J.; Nardoto, G.B.; Schor, T.; da Silva, M.R.F.; Martinelli, L.A. Impacts of market economy access and livelihood conditions on agro-food transition in rural communities in three macro-regions of Brazil. *Environ. Dev. Sustain.* **2021**, 1–21. Available online: https://link.springer.com/article/10.1007/s10668-021-01480-3 (accessed on 5 February 2021).
39. Qu, Y.; Jiang, G.; Ma, W.; Li, Z. How does the rural settlement transition contribute to shaping sustainable rural development? Evidence from Shandong, China. *J. Rural Stud.* **2021**, *82*, 279–293.
40. Qu, Y. Transition of rural settlements: Concept, feature, mechanism and path. *Sci. Geogr. Sin.* **2020**, *40*, 572–580.
41. Chen, K.; Long, H.; Liao, L.; Tu, S.; Li, T. Land Use Transitions and Urban-Rural Integrated Development: Theoretical Framework and China's Evidence. *Land Use Policy* **2020**, *92*, 104465. [CrossRef]
42. Chen, R.; Ye, C.; Cai, Y.; Xing, X.; Chen, Q. The impact of rural out-migration on land use transition in China: Past, present and trend. *Land Use Policy* **2014**, *40*, 101–110. [CrossRef]
43. Foley, J.A.; Defries, R.; Asner, G.P.; Barford, C.; Bonan, G.; Carpenter, S.R.; Chapin, F.S.; Coe, M.T.; Daily, G.C.; Gibbs, H.K.; et al. Global Consequences of Land Use. *Science* **2005**, *309*, 570–574. [CrossRef]
44. Qu, Y.; Jiang, G.; Zhao, Q.; Ma, W.; Zhang, R.; Yang, Y. Geographic identification, spatial differentiation, and formation mechanism of multifunction of rural settlements: A case study of 804 typical villages in Shandong Province, China. *J. Clean. Prod.* **2017**, *166*, 1202–1215. [CrossRef]
45. Zhang, Q.; Zhou, B.; Mo, Y.; Wu, Q. Ecological Landscape Evaluation for Regional Sustainability of Land-use. *Chin. J. Eco-Agric.* **2008**, *3*, 741–746.
46. McGarigal, K.; Cushman, S.A.; Ene, E. FRAGSTATS, v4: Spatial Pattern Analysis Program for Categorical and Continuous Maps. 2012. Available online: https://www.umass.edu/landeco/research/fragstats/fragstats.html (accessed on 5 February 2021).
47. Lemoine-Rodríguez, R.; Inostroza, L.; Zepp, H. The global homogenization of urban form. An assessment of 194 cities across time. *Landsc. Urban Plan.* **2020**, *204*, 103949. [CrossRef]
48. Jiang, G.; Zhang, F.; Zhou, D. Analyzing the land use structure characteristics of rural residential area in Beijing city. *Resour. Sci.* **2007**, *29*, 109–116.
49. Long, H.; Heilig, G.K.; Li, X.; Zhang, M. Socio-economic development and land-use change: Analysis of rural housing land transition in the Transect of the Yangtse River, China. *Land Use Policy* **2005**, *24*, 141–153. [CrossRef]
50. Zhu, F.; Zhang, F.; Li, C.; Zhu, T. Functional transition of the rural settlement: Analysis of land-use differentiation in a transect of Beijing, China. *Habitat Int.* **2014**, *41*, 262–271. [CrossRef]
51. Long, H. *Land Use and Rural Transformation Development in China*; Science Press: Beijing, China, 2012.
52. Long, H.; Liu, Y. Rural restructuring in China. *J. Rural Stud.* **2016**, *47*, 387–391. [CrossRef]
53. Long, H. *Land Use Transitions and Rural Restructuring in China*; Springer: Singapore, 2020.
54. Long, H.; Qu, Y. Land use transitions and land management: A mutual feedback perspective. *Land Use Policy* **2018**, *74*, 111–120. [CrossRef]

*Article*

# Rural-Spatial Restructuring Promoted by Land-Use Transitions: A Case Study of Zhulin Town in Central China

Dong Han [1,2], Jiajun Qiao [1,2,*] and Qiankun Zhu [2]

[1] Key Laboratory of Geospatial Technology for the Middle and Lower Yellow River Regions (Henan University), Ministry of Education, Kaifeng 475004, China; dhan89@henu.edu.cn
[2] College of Environment and Planning, Henan University, Kaifeng 475004, China; Zhuqk@henu.edu.cn
* Correspondence: jjqiao@henu.edu.cn

**Abstract:** Rural-spatial restructuring involves the spatial mapping of the current rural development process. The transformation of land-use morphologies, directly or indirectly, affects the practice of rural restructuring. Analyzing this process in terms of the dominant morphology and recessive morphology is helpful for better grasping the overall picture of rural-spatial restructuring. Accordingly, this paper took Zhulin Town in Central China as a case study area. We propose a method for studying rural-spatial restructuring based on changes in the dominant and recessive morphologies of land use. This process was realized by analyzing the distribution and functional suitability of ecological-production-living (EPL) spaces based on land-use types, data on land-use changes obtained over a 30-year observation period, and in-depth research. We found that examining rural-spatial restructuring by matching the distribution of EPL spaces with their functional suitability can help to avoid the misjudgment of the restructuring mode caused by the consideration of the distribution and structural changes in quantity, facilitating greater understanding of the process of rural-spatial restructuring. Although the distribution and quantitative structure of Zhulin's EPL spaces have changed to differing degrees, ecological- and agricultural-production spaces still predominate, and their functional suitability has gradually increased. The spatial distribution and functional suitability of Zhulin are generally well matched, with 62.5% of the matched types being high-quality growth, and the positive effect of Zhulin's spatial restructuring over the past 30 years has been significant. We found that combining changes in EPL spatial area and quantity as well as changes in functional suitability is helpful in better understanding the impact of the national macro-policy shift regarding rural development. Sustaining the positive spatial restructuring of rural space requires the timely adjustment of local actors in accordance with the needs of macroeconomic and social development, and a good rural-governance model is essential.

**Keywords:** ecological-production-living spaces; spatial distribution; functional suitability evaluation; land-use transition; rural-spatial restructuring

**Citation:** Han, D.; Qiao, J.; Zhu, Q. Rural-Spatial Restructuring Promoted by Land-Use Transitions: A Case Study of Zhulin Town in Central China. *Land* **2021**, *10*, 234. https://doi.org/10.3390/land10030234

Academic Editor: Hualou Long

Received: 3 February 2021
Accepted: 22 February 2021
Published: 25 February 2021



## 1. Introduction

Over the past several decades, under the influence of urbanization, industrialization, informatization and globalization, the social, economic and spatial structures of rural areas have undergone significant changes [1,2]. Space is a carrier of various elements of the rural territorial system, and the reshaping of the socioeconomic structure of rural areas will inevitably lead to changes in this carrier, which in turn will lead to the restructuring of rural space [3–5]. Global industrialization and urbanization have triggered and will continue to trigger dramatic changes in rural space. In the Western world, the transition from a preindustrial economy to a knowledge-based economy took centuries [6–10], while in newly industrialized countries such as China, this is a more compressed process [11–14]. China's rural villages have undergone a transformational rural development process in just a few decades. Rapid urbanization and industrialization have led to significant changes in the industrial, employment and social structures of rural territories [15,16]. While contributing to urban and rural economic development, this has resulted in the conversion of large amounts of arable land into urban built-up land, leading to a drastic reduction in rural land and a shift in land use and ownership. The disorderly expansion of urban and rural built-up land has caused environmental pollution, ecological damage, and inefficient land use [17]. At the same time, unevenness in urban–rural development has led to serious rural population loss, rural aging, rural poverty and village depopulation [18]. The continued opening of the political, economic and cultural spheres to the outside world has allowed China to enjoy the dividends brought by globalization [19], while at the same time exacerbating the complexity of problems related to geographic change in rural China.

Theoretical research on rural-spatial restructuring is the foundation for solving rural problems. Moreover, the optimal reorganization of rural space is an important means for implementing a strategy for promoting rural revitalization and realizing the integrated development of both urban and rural areas. In this process, land-use transitions play an important role in promoting rural-spatial restructuring [17].

Due to the problems and challenges of China's urban and rural development process, in 2012, China began to implement the strategy of ecological civilization construction. As a result, China proposed changing rural-spatial development from a production-space-led model to an ecological-production-living (EPL) space model. The EPL space model is more comprehensive than the previous model and is an effective means for optimizing the spatial development pattern of China [20]. Promoting rural-spatial restructuring to optimize land use in this way will transform traditional rural areas through land concentration and large-scale operation, as well as optimizing village and industrial layouts. This will promote industrialization and the modernization of agriculture, optimize urban-rural-spatial patterns, build new urban-rural relationships and achieve integrated urban-rural development [21], thus alleviating urban-rural development conflicts [22]. Therefore, a profound analysis of the process of rural-spatial restructuring represented by the changes in EPL space morphologies, and the mechanism behind them, could enable correct human interventions to guide positive rural development. The restructuring of rural space is closely related to three important rural issues in China. Rural space provides a resource base and physical space for rural development. The optimization and reorganization of rural space is an important means for implementing a strategy of rural revitalization and realizing the integrated development of urban and rural areas.

Land-use morphology includes the dominant morphology and recessive morphology [17]. The dominant morphology refers to the structure of land use in a particular area over a fixed period, including characteristics such as the number (area and proportion) and spatial patterns of land-use types. The recessive morphology refers to a special form based on the explicit form and can be observed only by means of analysis, testing, monitoring and investigation, including the quality and function of land use [23]. Current research on rural-spatial restructuring is based mainly on the area, structure, distribution and other dominant morphological characteristics of regional land-use types [24], but not enough research focuses on quality, function, management methods and other recessive morphologies [17]. Many scholars focus on case studies, with an emphasis on the spatial needs of rural populations and socioeconomic development, as well as concentrating on the optimization and reorganization of rural space. For example, many researchers identify the time points when changes in the spatial morphology of farmland occur to analyze the transformation and restructuring process for agricultural-production space [25–27]. In terms of spatial patterns, agricultural land tends to be scattered and at a lower level of socioeconomic development. Socioeconomic development concentrates this phenomenon; thus, business patterns and landscape patterns are two main indicators for studying the spatial transformation and restructuring of agriculture [28]. The change in rural housing area per capita is a direct way to reflect the restructuring of rural life, which is influenced by both changes in total residential area and population migration [29]. There is substantial empirical evidence for the spatial distribution of rural living space in China, involving multiple scales and national/regional, municipal, county and village levels [30,31]. Spatial restructuring strategies for different types of village have been proposed to promote different models of rural-spatial restructuring [32,33], and empirical studies on the spatial restructuring of rural settlements in different regions have been conducted [34–36]. Long proposed that rural restructuring should be considered an integrated approach to optimizing urban-rural-spatial organization and promoting coordinated urban–rural development, recommending a land remediation approach to realizing rural-spatial restructuring [3]. It is thought that this will help to solve the problem of hollowed-out villages [37] and provide a comprehensive platform for rural revitalization [38]. These research results propose differing models of rural-spatial restructuring, specify future research ideas for rural-spatial restructuring, enrich the literature on rural-spatial restructuring, provide technical support and methodological guidance for different types of rural development planning, and have a beneficial impact on the practice of rural development in China.

However, limiting rural-spatial restructuring to the perspective of change in the explicit morphological characteristics of EPL spaces will lead to a deviation in our understanding of the patterns of rural-spatial restructuring, which will then lead to deviations in policy formulation and implementation. Undeniably, changes in quantity are an important characteristic of change in EPL spaces, and the disorder of the quantity structure of EPL spaces often leads to unsustainable spatial development and conflicting landscape functions [39]. For example, Yang et al. [22] analyzed the impact of change in the quantity structure of EPL spaces on the quality of the ecological environment, concluding that the transformation of agricultural-production land to urban and rural living land is the main factor behind the deterioration of regional ecological environmental quality. These authors propose that ecological space be further expanded through land reclamation and greening to provide a solid ecological barrier for urban development. From a macro-policy perspective, the results of that study undoubtedly provide guidance. However, the change in the quantitative structure of EPL spaces is only one of their characteristics, and analyzing the restructuring process only from this perspective will lead to a biased understanding of regional development issues, which in turn will lead to decisions that are inappropriate for regional development based on the requirements of macro-policies. For example, agricultural land in East Asian countries, which have large populations and limited arable land, is characterized by fragmentation and small-scale family operations. In the process of industrialization and urbanization, agriculture is sidelined, and agricultural workers are aging, which has led to a decline in the efficiency of arable land and an increase in the proportion of abandoned arable land. In this context, Japan, Korea and China have adjusted their policies and laws to promote rural land management to ensure food security and promote large-scale rural land operations. However, the overuse of land area as a measurement standard has led to problems in policy implementation. For example, in Japan, under their policy stipulating the distribution system for rice-cultivation areas, the government has taken compulsory measures to make different regions have the same proportions of rice-production areas, which has harmed the interests of some large-scale farmers with higher production efficiency. China's local governments, forced by pressure for arable land protection, strongly dominate rural-spatial restructuring. However, they focus only on increasing arable land area, leading to the opening of sloping fields, fencing of lakes and land reclamation, resulting in new conflicts in the relationship between people and land. Given social and economic changes and innovations, it is difficult to adequately study rural-spatial restructuring in the context of the relationship between people and land by focusing only on the quantitative and spatial structural characteristics of land-use patterns [40,41]. Therefore, we examined rural-spatial restructuring based on land-use transformation via two aspects: quantitative changes and qualitative changes. The quantitative aspect manifests mainly in explicit morphological changes in land use, i.e., changes in the data and spatial patterns of land use. The qualitative aspect manifests as changes in implicit land-use patterns, especially in the evaluation of systemic functions [42].

Accordingly, this paper proposes a hypothesis of rural-spatial restructuring based on the evaluation of EPL spaces in terms of the changes in the dominant and recessive morphologies of land use, using Zhulin Town of Henan Province as an example. The objectives of this paper were as follows: (1) to analyze the changes in the dominant morphology of land use by identifying the distribution characteristics of the elements of EPL spaces and analyzing the structural changes in EPL spaces over the past 30 years; (2) to construct a framework for objectively evaluating the functional suitability of EPL spaces in different periods, starting from the natural background constraints and socioeconomic development incentives, to characterize the changes in the recessive morphology of land use; and (3) to combine the distribution and structural changes in EPL spaces with their functional suitability in different periods to determine any relations between them in order to judge the process of rural-spatial restructuring and lay a solid foundation for the next step in rural revitalization.

## 2. Materials and Methods

### 2.1. Study Area

Zhulin is located in the central region of China, Henan Province, in the transition zone between mountains and hills, with National Highway 310 passing through the territory (Figure 1). Originally an administrative village, Zhulin was established as a town in 1994 and became the first all-resident town in Henan Province in 2010, when all villages under its jurisdiction were converted into neighborhood committees (In China's rural areas, towns and administrative villages are two different administrative levels, and villages are subordinate to towns. The neighborhood committees are the smallest administrative units of the city, and their administrative level is the same as that of the village. The change from village to neighborhood community means that the people in Zhulin have changed from villagers to urban residents). After several large-scale zoning adjustments in 1994, 2006 and 2012, the town now has a total area of about 20 km$^2$. In 2019, the town had a resident population of 21,000, total social output value of CNY 10 billion, tax revenue of CNY 300 million and per capita income of CNY 40,300.

Following the past 40 years of reform and opening up, Zhulin is one of the few inland mountain villages to have evolved from a small village with no industry and far from any city to a modern town with a focus on industrial and tourism development. It is also a pilot town for sustainable development in China established by the UNDP and has won the Dubai International Award for the Best Practice in Improving the Living Environment, established by UN-Habitat. Thus, the rural restructuring of Zhulin can be regarded as a condensed version of China's rural development, and the restructuring trajectory of Zhulin from an inland agricultural village to a modern town makes it a perfect model for study. There is an element of chance in Zhulin's development process, as the direction of development of any geographic system cannot be purely inevitable, and an element of chance is understandable [43]; rural territorial systems are no exception. A study of the spatial restructuring history of Zhulin to determine the objective factors regarding the chance and necessity in its development process is of strong significance for guiding the development of other villages.

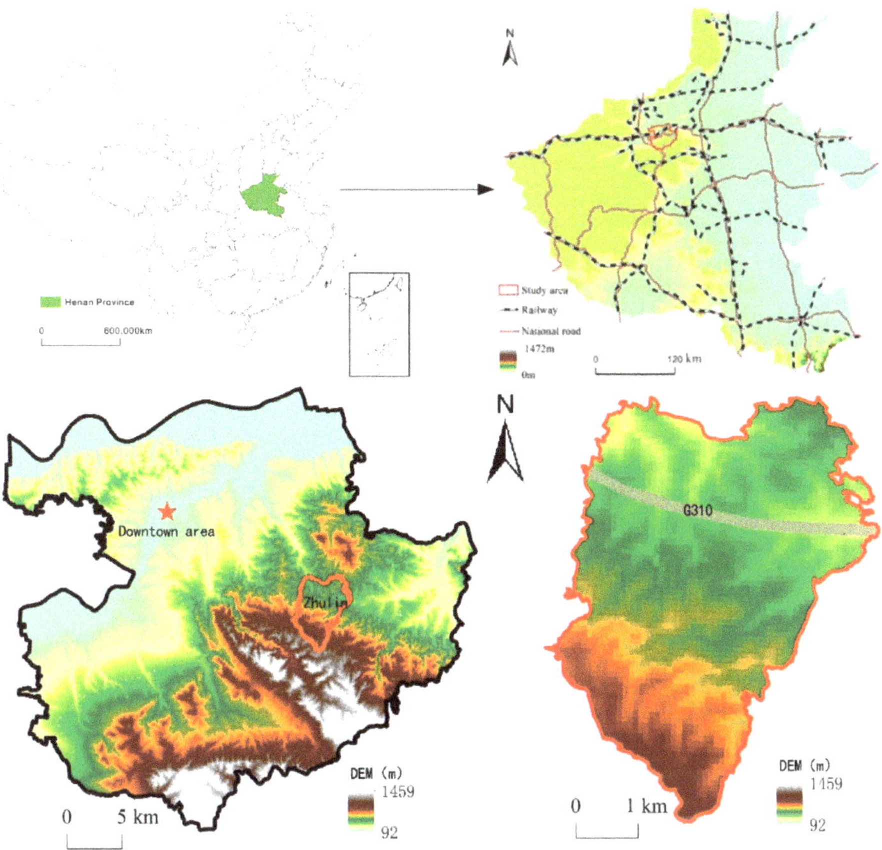



**Figure 1.** Zhulin location and topographic map (2018).

### 2.2. Data Source

It is difficult to obtain land-use data for long-term series at the village and township levels, and our land-use data come from a survey of Zhulin initiated in 2016. Additional land-use data from 1990 and 1995 were obtained by digitizing hand-drawn maps in village files and spatially matching important markers. A map from the end of 2005 was obtained from the current land-use map provided by the land department, and the maps for the ends of 2010 and 2018 were obtained from the current land-use map of the general plan of Zhulin prepared in those years. On this basis, through in-depth research, old village cadres, retired government or enterprise managers of the village, and current town leaders who had experienced Zhulin's complete development cycle were asked to assist in conducting participatory assessment surveys and record the time points of construction and changes in important surface structures, roads and facility sites to calibrate and match the above base maps by inverting the land-use situation over the historical period. Furthermore, the elevation, slope and other related data used in the functional suitability analysis were extracted from 30 m-precision TM remote-sensing data provided by the National Science and Technology Infrastructure Platform, National Earth System Scientific Data Sharing Platform, Lower Yellow River Scientific Data Center. Data on the multiyear average vegetation net primary productivity, multiyear average precipitation, multiyear average temperatures, soil infiltration factors, rainfall erosion forces, and conditions for accumulated temperatures, and the NDVI vegetation index dataset involved in the ecological suitability analysis were obtained using the Resource and Environment Data Cloud Platform of the Chinese Academy of Sciences and calculated using relevant conversion formulas. The soil physicochemistry-related data were mainly obtained from the Chinese Soil Dataset V1.1 based on the Harmonized World Soil Database (HWSD) of the Cold and Arid Regions Science Data Center. Data on forest park and ecological reserve delineation were obtained from information provided by the Natural Resources and Planning Bureau of Gongyi City. Data related to agricultural water supply facilities and major agricultural infrastructure were provided by the Agriculture and Rural Bureau of Gongyi City. Data on geological hazard susceptibility assessment came from the assessment in the Thirteenth Five-Year Plan for Geological Hazards of Gongyi City.

### 2.3. Research Methods

### 2.3.1. Identification and Classification of EPL Spaces

In addition to the natural properties of land, the classification of EPL spaces should be based on the subjective land-use intentions of the actors [22]. Among these, production space refers to a land-use system that provides a material space carrier for human production and business activities [44]. Living space is a land-use system that carries and protects human residential life and social activities. Ecological space is a land-use system that regulates, maintains and protects ecological security functions [45]. There are two common models for identifying and classifying EPL spaces. One is the index system measurement method, which classifies EPL spaces mainly by establishing a comprehensive evaluation index system [46]. Due to data availability, this method is studied mainly at the medium and macro scales, such as cities and counties (districts). Another method is spatial merging based on land-use type, i.e., merging and reclassifying land-use types according to the dominant functions of the land [47]. For the relatively micro village and township scales, the latter method can quickly identify the spatial distribution of land-use spaces and reflect the functions of land from the perspective of land-use structure. Accordingly, this paper took the land-use classification standards of the People's Republic of China Current Land Use Classification (GB/T21010-2017) and Standard for Planning of Town (GB50188-2007) as a basis and, together with actual land-use patterns in Zhulin, identified the quantity and distribution of EPL spaces in Zhulin. In particular, note that production space carries human production activities and contains primary industry land, secondary industry land and tertiary industry land. Primary industry land can also be agricultural-production space, the main form of rural production space in the early stage of development, while secondary and tertiary industry land can also be nonagricultural-production space, which gradually emerges in the countryside only after industrialization and urbanization [20], which have clear distinctions in terms of land type and functional suitability [47]. Therefore, production space is further classified into agricultural-production space and nonagricultural-production space (Table 1).

**Table 1.** Classification of ecological–production–living spaces.

| Ecological–Production–Living Space Classification | Level 1 Land-Use Type | Level 2 Land-Use Type |
|---|---|---|
| Ecological space | Green space | Public green space<br>Protected green space |
| | Water and other land | Water<br>Forest land in agricultural and forestry land<br>Unused land<br>Pasture for grazing |
| Agricultural-production space | Production facility land | Land for agricultural-production services |
| | Land for agriculture and forestry | Arable land, vegetable land, garden land, nursery |
| Nonagricultural-production space | Production facility land | Class 1 industrial land<br>Class 2 industrial land<br>Class 3 industrial land |
| | Land for storage facilities | Land for general storage<br>Land for storing hazardous materials |
| | Land for public facilities | Commercial and financial land<br>Market land |
| Living space | Land for residential facilities | Class 1 residential land<br>Class 2 residential land |
| | Land for public facilities | Land for administration<br>Land for educational institutions<br>Land for culture, sports and technology<br>Land for healthcare |
| | External transportation land | Land for highways<br>Land for other transportation |
| | Engineering facilities land | Land for public engineering<br>Land for sanitation facilities<br>Land for disaster prevention facilities |

### 2.3.2. Evaluation Index System for Functional Suitability of EPL Spaces

The functional suitability of land refers to its degree of suitability for a particular use within a range under certain conditions [48], and the suitability evaluation of land function is essentially a concept formed by the interaction between the natural environment and the human social system; the exploration of its theoretical connotations returns to the theory of the territorial system of human-land interaction itself to find its basis [49]. Therefore, based on the concept of land function suitability evaluation, the suitability evaluation of EPL spaces can be defined as the suitability of land for production, living and ecological functions in a specific range under specific conditions [50]. Specifically, the evaluation of the functional suitability of EPL spaces should ultimately return to the service of human development, which includes three aspects [51]. (1) It must ensure the safety of the ecological environment base, which is the basis of human survival, requiring the development activities and scale to be coordinated with the carrying capacity of regional resources and the environment, maintaining surface ecosystem service functions, and emphasizing the protection of important ecosystems. (2) It must ensure the safety of human society. Urbanization and industrial development should occur in areas with stable and good natural conditions to ensure human security and avoid using built-up land in disaster risk areas. (3) It must provide certain economic benefits. From the perspective of human society, land function needs to meet the laws of spatial economy, requiring the siting of construction to consider the impact of land conditions on engineering costs, while facilitating the external linkage of people and socioeconomic activities carried out in the construction space and making socioeconomic services easily accessible. Factors (1) and (2) can be regarded as constraining conditions based on the natural background, while factor (3) can be regarded as an incentive condition in line with the laws of socioeconomic development. Accordingly, we took the natural background as the constraining condition and the factors that are conducive to economic and social development as the incentive condition to establish a corresponding index system based on the actual situation of the case study area and data availability. We then comprehensively evaluated the functional suitability of the EPL spaces based on these two aspects. The spatial suitability evaluation of individual factors under the three functions of ecology, production and living was performed based on five levels: suitable, moderately suitable, average, moderately unsuitable, and unsuitable; the weights were finally determined through expert consultation and an AHP (analytical hierarchy process) (Eight experts in related fields scored the evaluation system. Three experts in the first round of scoring could not pass the logical consistency test, so three more experts were invited for the second scoring evaluation. The evaluation matrix was established based on the scores, in which the ecological space judgment matrix λmax was 5.4451, with a CR value of 0.0994 < 0.1; the agricultural-production space judgment matrix λmax was 6.2079, with a CR value of 0.0330 < 0.1; the nonagricultural-production space judgment matrix λmax was 6.2593, with a CR value of 0.0412 < 0.1; and the living space judgment matrix λmax was 6.1689, with a CR value of 0.0268 < 0.1, all of which passed the consistency test). After rasterizing the evaluation indexes, $100 \times 100$ m raster cells were used as the basic evaluation units, and the ArcGIS spatial analysis function was used for weighted superposition. The weighted summation method of factor scores was adopted to obtain comprehensive evaluation scores of the spatial suitability of the EPL spaces. The calculation formula was:

$$S_i = \sum_j^n w_{ij} V_{ij} = \{S_1, S_2, S_3, S_4\} \tag{1}$$

$\{S_1, S_2, S_3, S_4, S_5\}$ = {suitable, moderately suitable, average, moderately unsuitable, unsuitable, respectively} = {(5,4), [4,3), [3,2), [2,1), [1,0)}, where $S_i$ is the functional suitability evaluation value of class *i* space in the EPL spaces (the larger the value, the higher the suitability of the corresponding space); $w_{ij}$ and $V_{ij}$ are the weight and role score of the *j*th factor in class *i* space, respectively; and *n* is the total number of influencing factors.

1.  Evaluation of the functional suitability of ecological space

The constraining factors include the biodiversity-maintenance function, water-conservation function, soil-and-water-conservation function and vegetation cover. The biodiversity-maintenance function is the role played by ecosystems in maintaining genetic, species and ecosystem variability, and is one of the most important functions provided by ecosystems. The water-conservation function is an important part of the ecological space function, which is a direct manifestation of the interaction between vegetation and water in the field of ecological services and an important function in meeting human water demand. The soil-and-water-conservation function is also an important part of the ecological space function, and is conducive to the full use of the economic and social benefits of soil and water resources, reducing flood and drought disasters, and establishing a good ecological environment. We adopted these three indicators through comprehensive evaluation of the variables, including the multiyear mean vegetation net primary productivity, multiyear average precipitation, multiyear average temperature, soil infiltration factor, rainfall erosion force and soil erodibility, as collected from the Resource and Environment Data Cloud Platform of the Chinese Academy of Sciences. Other variables, including the elevation, slope and slope direction ere obtained from digital elevation model (DEM) data analysis. The indicators were evaluated in a hierarchical manner according to the number of accumulated service functions in the evaluation results. The calculation basis for these three indicators was as follows:

Biodiversity-maintenance function:

$$V_{bio} = NPP_{mean} \times F_{pre} \times F_{tem} \times (1 - F_{ele})\tag{2}$$

where $V_{bio}$ is the value of the biodiversity-maintenance function, $NPP_{mean}$ is the multiyear mean vegetation net primary productivity, $F_{pre}$ is the multiyear average precipitation factor, $F_{tem}$ is the multiyear average temperature factor, and $F_{ele}$ is the elevation factor.

Water-conservation function:

$$V_{wr} = NPP_{mean} \times F_{sic} \times F_{pre} \times (1 - F_{slo})\tag{3}$$

where $V_{wr}$ is the value of the water-conservation function, $NPP_{mean}$ is the multiyear mean vegetation net primary productivity, $F_{sic}$ is the soil infiltration factor, $F_{pre}$ is the multiyear average precipitation factor, and $F_{slo}$ is the slope factor.

Soil-and-water-conservation function:

$$V_{sw} = NPP_{mean} \times (1 - K) \times (1 - F_{slo})\tag{4}$$

where $V_{sw}$ is the value of the soil-and-water-conservation function, $NPP_{mean}$ is the multiyear mean vegetation net primary productivity, $K$ is the soil erodibility factor, and $F_{slo}$ is the slope factor.

Vegetation is the primary component and functional body of the ecosystem, and the study of the spatiotemporal characteristics of the fraction of vegetation cover (FCV) is the basis for evaluating ecological spatial quality. The annual NDVI vegetation index dataset for the period of 1990–2018 was obtained using the Resource and Environment Data Cloud Platform of the Chinese Academy of Sciences. There is a significant linear correlation between vegetation cover and the NDVI, and vegetation cover information is usually extracted directly by establishing the conversion relationship between the two [52]. The multiyear average vegetation cover of the study area was obtained using the pixel dichotomy method [53,54], and the evaluation was performed according to vegetation cover.

The incentive factors are mainly ecological control factors, namely, the three protection facility systems established in situ nationwide, including nature reserves, forest parks and scenic spots. Nature reserves are mainly for absolute protection, and forest parks and scenic spots are for both protection and development. The suitability of the ecological spatial incentive factors was evaluated hierarchically according to whether they belonged to these three systems (Table 2).

**Table 2.** Evaluation of functional suitability of ecological space.

| Target Layer | Criterial Layer | Index Layer | Suitable | Moderately Suitable | Average | Moderately Unsuitable | Unsuitable | Weight |
|---|---|---|---|---|---|---|---|---|
| Evaluation of functional suitability of live space | Constraining factors | Biodiversity-maintenance function | Top 30% of cumulative service function volume | Top 30–50% | Top 50–70% | Top 70–85% | Below top 85% | 0.3398 |
| | | Water-conservation function | Top 30% of cumulative service function volume | Top 30–50% | Top 50–70% | Top 70–85% | Below top 8% | 0.3319 |
| | | Soil-and-water-conservation function | Top 30% of cumulative service function volume | Top 30–50% | Top 50–70% | Top 70–85% | Below 85% | 0.1567 |
| | | Fraction of vegetation cover | $0.7 < \mathrm{FVC} \leq 1$ | $0.5 < \mathrm{FVC} \leq 0.7$ | $0.3 < \mathrm{FVC} \leq 0.5$ | $0.1 < \mathrm{FVC} \leq 0.3$ | $\mathrm{FVC} \leq 0.1$ | 0.1010 |
| | Incentive factors | Ecological control factors | National nature reserve | Provincial nature reserve | National forest park/scenic spot | Provincial forest park/scenic spot | Local protection facilities | 0.0706 |

2. Evaluation of the functional suitability of agricultural-production space

The constraining factors include the slope, soil texture, agricultural water supply conditions and light and heat conditions. The slope affects the water and fertilizer uptake by crops, along with the light conditions for crops. In addition, the greater the slope, the more likely it is that agricultural activities will cause soil erosion. Soil texture is one of the physical properties of soil and is closely related to the conditions of soil aeration, fertilizer and water retention, and ease of cultivation. Agricultural water is the most basic condition for agricultural production. Light and heat conditions affect the distribution of crop species, the replanting system, and yield. Accordingly, this distribution is based on the slope, soil texture, average rainfall and surface water supply over time; the elevation-corrected active accumulated temperature of the multiyear average daily temperature $\geq 0\,^\circ$C serves as the evaluation index. According to the grade evaluation from the Technical Regulations of Land Use Status Survey issued by the China Agricultural Zoning Commission and China's National Standard Cultivated Land Quality Grade (GB/T 33469-2016) for the slope, soil type, water supply conditions and temperature accumulation conditions, the classification was performed according to the actual situation of the case study area, and we evaluated the suitability of the spatial constraining factors of the agricultural-production space.

The incentive factors include, mainly, the cultivation radius and distance to major agricultural facilities. The quality of agricultural-production space is related not only to natural endowments, but also to human-made factor inputs. Under equal conditions, the closer the arable land is to a settlement, the greater the ease of cultivation and the higher the suitability in comparison. Major agricultural facilities can greatly increase the efficiency of land-based production, improve the utilization of resources and labor productivity, and thus increase the efficiency, quality and competitiveness of agriculture. Therefore, the suitability of the agricultural-production space was evaluated by the nearest distance method according to the distance from settlements. It was also evaluated hierarchically using the Jenks natural-breaks classification method. The suitability of the incentive factors for the agricultural-production space were obtained by evaluation using buffer analysis according to the distance from major agricultural facilities (Table 3).

**Table 3.** Evaluation of functional suitability of agricultural-production space.

| Target Layer | Criteria Layer | Index Layer | Suitable | Moderately Suitable | Average | Moderately Unsuitable | Unsuitable | Weight |
|---|---|---|---|---|---|---|---|---|
| Evaluation of functional suitability of agricultural space | Constraining factors | Gradient | 0°–2° | 2°–6° | 6°–15° | 15°–25° | Above 25° | 0.2293 |
| | | Soil texture | Loam soil | Clay loam soil, powdered clay loam, sandy clay loam | Powdered clay, sandy clay, sandy loam, powdered loam | Loamy sandy soil, clay | Sandy soil, chalky soil | 0.2278 |
| | | Agricultural water supply conditions | Fully satisfied | Satisfied | Basically satisfied | Inadequately satisfied | Unsatisfied | 0.3032 |
| | | Light and heat conditions | Above 4500 | 3400–4500 | 1600–3400 | N/A | N/A | 0.0943 |
| | Incentive factors | Farming radius | Within 40.63 | 40.63–126.76 | 126.76–255.26 | 255.26–427.79 | Above 427.79 | 0.0665 |
| | | Distance to major agricultural facilities | 500 m | 500–1000 m | 1000–1500 m | 1500–2000 m | Above 2000 m | 0.0790 |

3.     Evaluation of the functional suitability of nonagricultural-production space

The constraining factors include the slope, elevation, topographical relief and geological conditions. Nonagricultural-production space is mainly urban built-up land, and the slope has a strong impact on this space. Engineering construction costs increase with increased terrain slopes, and steep terrain is prone to geological phenomena such as landslides and mudslides. Elevation is an important factor affecting land for urban and rural construction; low-elevation areas are generally more suitable for human habitation than high-elevation areas, and the suitability for human habitation decreases with increasing elevation [55]. Topographic relief refers to the difference between the elevation of the highest point and the elevation of the lowest point in a specific area, and to a certain extent, it reflects the difficulty of engineering construction. The stability of engineering geological conditions is the basis of site selection for nonagricultural-production space. Thus, according to the slope classification standard for industrial land selection of the National Standard

Code for Vertical Planning on Urban and Rural Development Land (CJJ 83-2016) of the People's Republic of China, and the altitude of elevation, we calculated the topographical relief and regional geological hazard susceptibility using grids ($30 \times 30$ m grid cells) and evaluated the suitability of the constraining factors for nonagricultural-production space in the context of the local situation.

The incentive factors include, mainly, industrial agglomeration and convenient transportation. Rural nonagricultural industries include, mainly, the secondary industry, the production-support service industry, the living service industry to meet residents' consumption demand and the tourism industry. Due to external economies and diseconomies, industrial agglomeration develops from the relative concentration of industrial and commercial enterprises in geographic and spatial locations, bringing corresponding costs and benefits to enterprises, and thus further influencing the spatial layout of industries. Therefore, we used the hot-spot analysis method in ArcGIS to determine the hot and cold spots through the distribution of natural break points and the influence of the industrial agglomeration effect on the industrial layout. Based on the distance from traffic arteries, we used the buffer analysis in ArcGIS to evaluate the transportation convenience. Different levels of buffer were established according to the levels of national roads, provincial roads, county roads and township roads to determine traffic convenience; this was used to evaluate the suitability of the incentive factors for nonagricultural-production space (Table 4).

**Table 4.** Evaluation of functional suitability of nonagricultural-production space.

| Target Layer | Criteria Layer | Index Layer | Suitable | Moderately Suitable | Average | Moderately Unsuitable | Unsuitable | Weight |
|---|---|---|---|---|---|---|---|---|
| Evaluation of functional suitability of nonagricultural space | Constraining factors | Gradient | 0°–5° | 5°–8° | 8°–15° | 15°–25° | Above 25° | 0.1174 |
| | | Elevation | Under 200 m | 200–300 m | 300–500 m | 500–1000 m | Above 1000 m | 0.0528 |
| | | Topographical relief | 0–50 | 50–100 | 100–150 | 150–200 | Above 200 | 0.0621 |
| | | Geological conditions | No susceptibility | Low susceptibility | Medium susceptibility | High susceptibility | Extreme susceptibility | 0.0526 |
| | Incentive factors | Industrial agglomeration | Hot spots | Sub-hot spots | Mild spots | Sub-cold spots | Cold spots | 0.3999 |
| | | Transportation convenience | Convenient | Moderately convenient | Average | Moderately inconvenient | Inconvenient | 0.3152 |

4. Evaluation of the functional suitability of living space

Living space also belongs to the category of urban built-up land, and its constraining conditions are similar to those of nonagricultural-production space, including four factors: the slope, elevation, topographical relief and geological conditions. The slope conditions were evaluated for their functional suitability according to the slope grading criteria for urban and rural residential land selection in the National Standard of the People's Republic of China Code for Vertical Planning on Urban and Rural Development Land (CJJ 83-2016). Since the elevation, topographic relief and engineering geology were judged on the same basis as nonagricultural-production space, they will not be described separately.

The incentive factors include, mainly, living convenience and transportation. In addition to the function of living, corresponding social activities and social functions are essential for meeting the daily living needs of human beings. Therefore, using the buffer analysis method in ArcGIS, and based on the delineation of the community's 15 min living circle, buffer zones with radii of 350, 700 and 1000 m were established, with the village and town administrative service center, hospitals, schools and cultural and sports facilities serving as the center for comprehensively evaluating the spatial living convenience. This resulted in five classification levels: convenient, moderately convenient, average, moderately inconvenient and inconvenient. Since the evaluation criteria for transportation convenience were the same as those for nonagricultural-production space, they are not described separately (Table 5).

**Table 5.** Evaluation of functional suitability of living space.

| Target Layer | Criteria Layer | Index Layer | Suitable | Moderately Suitable | Average | Moderately Unsuitable | Unsuitable | Weight |
|---|---|---|---|---|---|---|---|---|
| Evaluation of functional suitability of living space | Constraining factors | Gradient | 0°–5° | 5°–8° | 8°–20° | 20°–25° | Above 25° | 0.1074 |
| | | Elevation | Under 200 | 200–300 | 300–500 | 500–1000 | Above 1000 | 0.0573 |
| | | Topographical relief | 0–50 | 50–100 | 100–150 | 150–200 | Above 200 | 0.0662 |
| | | Geological conditions | No susceptibility | Low susceptibility | Medium susceptibility | High susceptibility | Extreme susceptibility | 0.0449 |
| | Incentive factors | Living convenience | Convenient | Moderately convenient | Average | Moderately inconvenient | Inconvenient | 0.4184 |
| | | Transportation convenience | Convenient | Moderately convenient | Average | Moderately inconvenient | Inconvenient | 0.3058 |

### 2.3.3. Evaluation of Match between Distribution and Function of EPL Spaces

Based on the distribution and changes in the functional suitability of EPL spaces, we established a coordinate system to comprehensively judge the match between structural changes in EPL spaces and their functional suitability to reflect Zhulin's spatial restructuring process. In this coordinate system (Figure 2), the X-axis is the change in the area structure of EPL spaces and the Y-axis is the change in the functional suitability of EPL spaces. The values of $X$ and $Y$ were calculated by Formulas (5) and (6).

$$X = \Delta X_i(t)/X_i(t-\Delta t) = [X_i(t) - X_i(t-\Delta t)]/X_i(t-\Delta t). \tag{5}$$

$$Y = \Delta Y_i(t)/Y_i(t-\Delta t) = [Y_i(t) - Y_i(t-\Delta t)]/Y_i(t-\Delta t) \tag{6}$$

where $\Delta X_i(t)$ is change in the area structure of EPL spaces; $X_i(t)$ is the spatial area of type $i$ in period $t$; $X_i(t-\Delta t)$ is the spatial area of type $i$ in the last cycle of change in period $t$; $\Delta Y_i(t)$ is the change in the functional suitability of EPL spaces; $Y_i(t)$ is the average functional suitability of space of type $i$ in period $t$; $Y_i(t-\Delta t)$ is the average functional suitability of space of type $i$ in period $t$ during the last cycle of change in period $t$; and $i$ = 1, 2, 3, 4 denotes ecological space, agricultural-production space, nonagricultural-production space and living space, respectively.

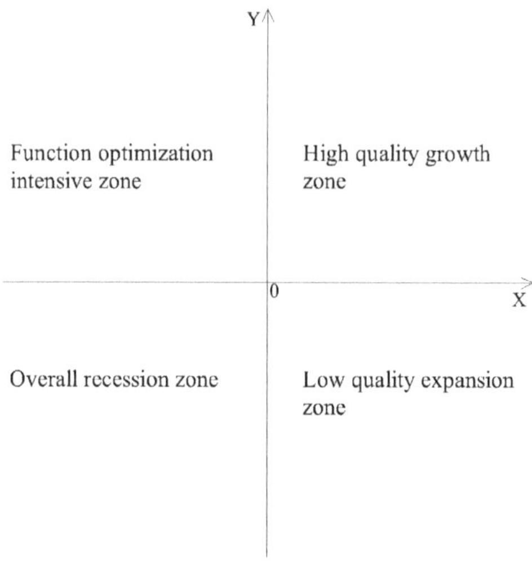


**Figure 2.** Schematic diagram of spatial distribution and function matching of ecological–production–living (EPL) spaces.

According to the values of X and Y, four quadrants are described as follows:

1. Quadrant I: $\Delta X_i(t) > 0$ indicates that space of type *i* has grown, and $\Delta Y_i(t) > 0$ indicates that the functional suitability of space of type *i* has improved. This indicates that, along with the growth of space of that type, the functional suitability of space of that type has also improved. Therefore, the first quadrant is a high-quality growth area.

2. Quadrant II: $\Delta X_i(t) < 0$ indicates that space of type *i* has decreased, and $\Delta Y_i(t) > 0$ indicates that the functional suitability of space of type *i* has improved. This indicates that even with the decrease in the area of space of that type, its functional suitability has improved. Therefore, the second quadrant is the functional optimization intensification area.

3. Quadrant III: $\Delta X_i(t) < 0$ indicates that space of type *i* has decreased, and $\Delta Y_i(t) < 0$ indicates that the functional suitability of space of type *i* has started to decrease. This indicates that, along with the decrease in the area of space of this type, its functional suitability has decreased. Therefore, the third quadrant is a full-scale decline area.

4. Quadrant IV: $\Delta X_i(t) > 0$ indicates that space of type *i* has increased, and $\Delta Y_i(t) < 0$ indicates that the functional suitability of space of type *i* has begun to decrease. This indicates that the functional suitability of space of type *i* decreases with increases in its area. Therefore, the fourth quadrant is a low-quality expansion area.

## 3. Results

### 3.1. Spatial Distribution and Structural Change of EPL Spaces

#### 3.1.1. EPL Spatial Distribution

Originally a village-level administrative unit, Zhulin was abolished in 1994 and established as a town. It went through several adjustments of administrative divisions and expanded its administrative area from an initial 4.2 km$^2$ to 20.5 km$^2$. The main changes were as follows: (1) In 1994, the village became a town, and the administrative area of Zhulin was expanded from 4.2 to 6.4 km$^2$; (2) in 2006, according to the development requirements of the United Nations Sustainable Development Pilot Town and the opinions of the China Small Town Reform and Development Center of the State Council, the administrative area was further expanded to 16.45 km$^2$; (3) in 2012, to cooperate with the construction of Zhulin Industrial Park and further promote local economic development, the total area of Zhulin was further expanded to 20.5 km$^2$, which is the current administrative division of Zhulin (Figure 3). With the rapid economic and social development of Zhulin and the continuous changes in administrative areas over the decades, the structure of Zhulin's EPL spaces has also changed dramatically. To reflect the distribution structure of its EPL spaces within different administrative boundaries and in different periods of economic and social development, based on data availability, we used 1990 (representing the development period before the establishment of Zhulin Town), 1995 (representing the development period at the beginning of Zhulin Town in 1994), 2005 (representing the development period before the administrative division of Zhulin in 2006), 2010 (representing the development period before Zhulin's administrative division in 2012) and 2018 (representing the development period after Zhulin's administrative division in 2012) as time points to analyze the process of structural changes in Zhulin's EPL spaces.

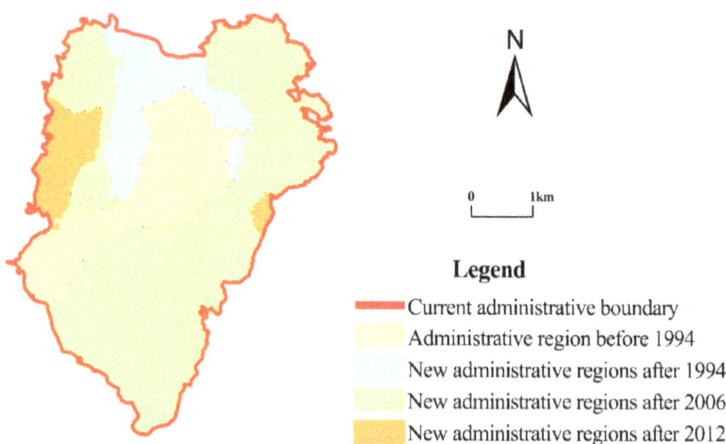


**Figure 3.** Zhulin administrative area change map.

With the continuous expansion of administrative divisions, the absolute quantity of Zhulin's EPL spaces has grown significantly. From 1990 to 2018, its ecological space grew from 148.26 to 803.46 hm$^2$, its agricultural-production space grew from 217.73 to 954.66 hm$^2$, its nonagricultural-production space grew from 21.23 to 111.43 hm$^2$, and its living space grew from 36.11 to 176.27 hm$^2$. The area of ecological space shows some fluctuation, while all the other spaces show a linear growth trend (Table 6).

**Table 6.** Classification of areas of Zhulin's EPL spaces.

|  | Ecological Space (hm$^2$) | Agricultural-Production Space (hm$^2$) | Nonagricultural-Production Space (hm$^2$) | Living Space (hm$^2$) | Total (hm$^2$) |
|---|---|---|---|---|---|
| 1990 | 148.26 | 217.73 | 21.23 | 36.11 | 423.33 |
| 1995 | 271.92 | 292.74 | 24.24 | 52.51 | 641.42 |
| 2005 | 137.79 | 351.47 | 49.22 | 103.41 | 641.89 |
| 2010 | 824.08 | 814.63 | 69.53 | 165.04 | 1873.28 |
| 2018 | 803.46 | 954.66 | 111.43 | 176.27 | 2045.82 |

The regional characteristics of the spatial distribution of Zhulin's ecological space and agricultural-production space have become increasingly clear. In 1990 and 1995, Zhulin's ecological space was concentrated mainly in the southern mountainous area and the northern region, but its ecological space and agricultural-production space generally showed a cross-distribution (Figure 4a,b). By 2005, the ecological space in the north began to shrink, and the south became the main distribution area of Zhulin's ecological space (Figure 4c). By 2010 and 2018, with the expansion of Zhulin's administrative boundaries, the extent of mountainous areas in the south further increased, while the extent of ecological space in the north further shrank. Eventually, a distribution pattern formed with ecological space dominating in the south and agricultural-production space dominating in the north (Figure 4d,e). As for living space and nonagricultural-production space, the early living space was concentrated near the main traffic routes, especially on both sides of National Highway 310, but there were also many scattered distributions in the mountainous areas in the south and north. Since 2005, the clustering of living space, in particular, and nonagricultural-production space has strengthened. Many scattered living spaces began to disappear, while the living and nonagricultural-production spaces along the main traffic routes continued to expand.

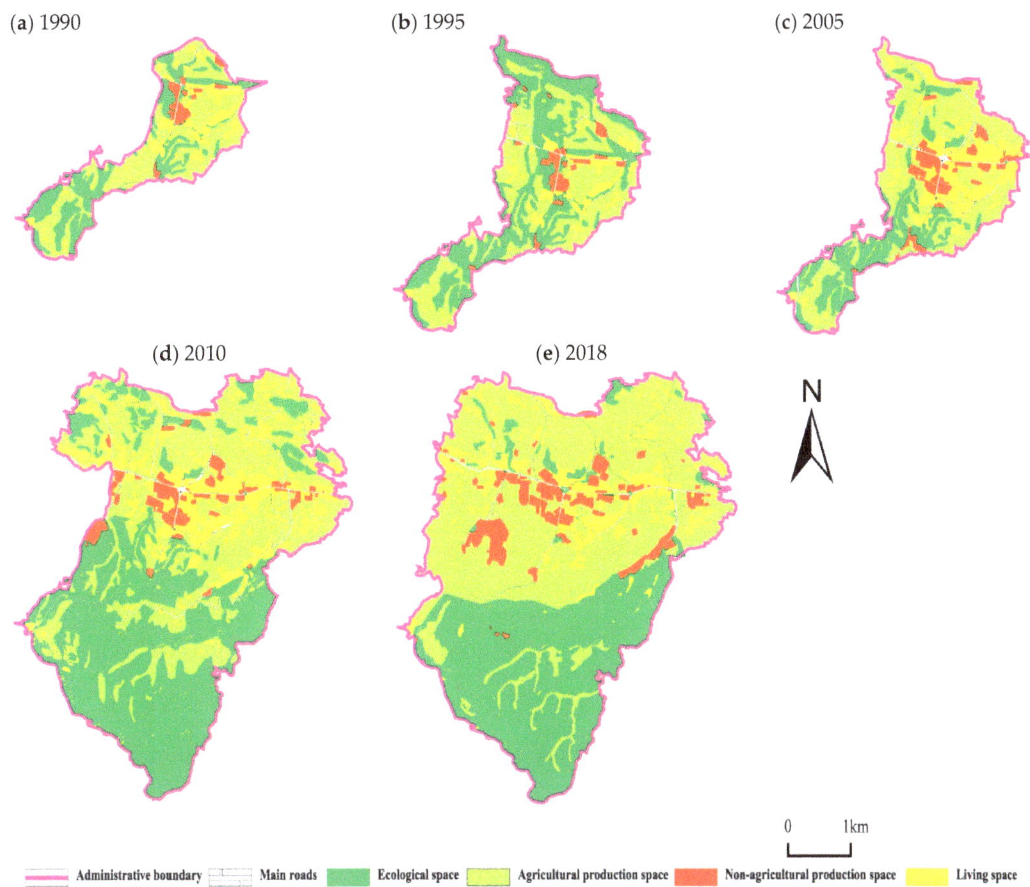



**Figure 4.** Distribution of Zhulin's EPL spaces, 1990–2018: (**a**) 1990, (**b**) 1995, (**c**) 2005, (**d**) 2010, (**e**) 2018.

### 3.1.2. Structural Change in EPL Spaces

In terms of the proportion of EPL spaces, from 1990 to 2018, agricultural-production space and ecological space constituted the largest proportions in Zhulin, with average proportions of 48.40% and 36.43%, respectively, followed by living space, with an average proportion of 10.05%. Nonagricultural-production space was smallest, with an average proportion of 5.12%. Overall, Zhulin is still dominated by agricultural-production space and ecological space.

From the structural changes in Zhulin's EPL spaces by year (Figure 5), we observe that the proportion of production space (including agricultural-production space and nonagricultural-production space) and living space decreased and then increased, and then decreased and increased again, showing a W-shaped fluctuation curve in general, while the change in the proportion of ecological space shows an opposite pattern to the change in production and living space, showing an M-shaped fluctuation curve. The proportion of production and living space peaked in 2005, when the proportion of ecological space was the lowest; the proportion of ecological space peaked in 2010.

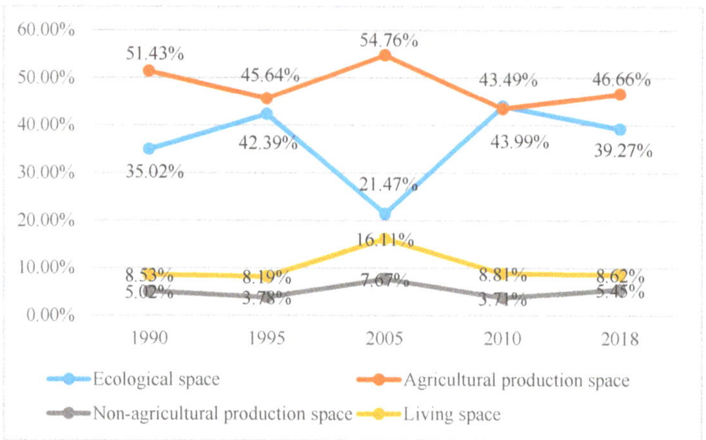



**Figure 5.** Structural change of Zhulin's EPL spaces.

One reason for this is the change in the area of EPL spaces brought about by the expansion of administrative space. The other reason is that the original EPL spaces have also undergone mutual transformation. In most periods, the increase in the area of EPL spaces was brought about mainly by the increase in administrative boundaries (Table 7). By contrast, the decrease in EPL spaces was mainly due to interconversion among EPL spaces that occurred in different periods (Table 7). From 1990 to 1995, EPL spaces shifted mainly from agricultural-production space to ecological space, nonagricultural-production space and living space. This period was accompanied by decreased agricultural-production space and growth in ecological space, nonagricultural-production space and living space. From 1995 to 2005, EPL spaces were transferred mainly from ecological space to agricultural-production space, nonagricultural-production space and living space. From 2005 to 2010, EPL spaces were transferred mainly from agricultural space, nonagricultural space and living space to ecological space. From 2010 to 2018, EPL spaces were transferred mainly from ecological space and living space to nonagricultural-production space and agricultural-production space.

**Table 7.** Changes in areas of Zhulin's EPL spaces.

| Unit: hm² | 1990–1995 | | | 1995–2005 [2] | 2005–2010 | | | 2010–2018 | | |
|---|---|---|---|---|---|---|---|---|---|---|
| | Total | External [1] Increase | Internal Changes | Internal Changes | Total | External Increase | Internal Changes | Total | External Increase | Internal Changes |
| Ecological space | 123.66 | 108.24 | 15.42 | −134.13 | 686.29 | 654.08 | 32.21 | −20.62 | 0.00 | −20.62 |
| Agricultural-production space | 75.01 | 96.60 | −21.59 | 58.73 | 463.16 | 495.89 | −32.73 | 140.03 | 124.78 | 15.25 |
| Nonagricultural-production space | 3.01 | 2.73 | 0.28 | 24.98 | 20.31 | 22.31 | −2.00 | 41.90 | 3.27 | 38.63 |
| Living space | 16.40 | 11.04 | 5.36 | 50.90 | 61.63 | 62.33 | −0.70 | 11.23 | 65.33 | −54.10 |
| Total | 218.09 | 218.61 | −0.52 | 0.47 | 1231.39 | 1234.61 | −3.22 | 172.54 | 193.38 | −20.84 |

[1] External increase refers to new areas of various types of land due to expansion of the administrative division, and internal change refers to change in the areas of EPL spaces in the same area as the previous administrative area. [2] The size of the administrative area in 2005 was the same as that in 1995, so there was no external or internal change.

### 3.2. Evaluation of Functional Suitability of EPL Spaces

### 3.2.1. Changes in Functional Suitability of Ecological Space

An improvement in the functional suitability of the ecological space is clear. In general, our evaluation shows that the functional suitability of Zhulin's ecological spaces deteriorated first and then improved continuously. The mean value of the ecological space suitability evaluation decreased from 1.81 to 1.59, and then continued to rise to 3.52. That means the overall suitability increased from moderately unsuitable, represented by values of (1–2] to moderately suitable, represented by values of (3–4] (Table 8). In terms of intragroup variation, the maximum value of the ecological space suitability evaluation showed a large change, while the minimum value remained relatively stable and unchanged. The standard deviations of the indicators ranged from 0.51 to 0.98, showing a general trend of gradual increase, indicating that the differences between samples gradually increased. The main reason for this is that the proportion of areas evaluated as moderately unsuitable for the functional suitability of ecological space gradually decreased, while the proportion of areas evaluated as average, moderately suitable or suitable gradually increased (Figure 6).

**Table 8.** Overall evaluation of functional suitability of ecological space.

| Year | Max. | Min. | Mean | Standard Deviation |
|------|------|------|------|--------------------|
| 1990 | 2.67 | 1.13 | 1.81 | 0.51 |
| 1995 | 2.67 | 1.13 | 1.59 | 0.46 |
| 2005 | 2.67 | 1.13 | 1.76 | 0.52 |
| 2010 | 3.50 | 1.03 | 2.32 | 0.79 |
| 2018 | 4.50 | 1.13 | 3.52 | 0.98 |

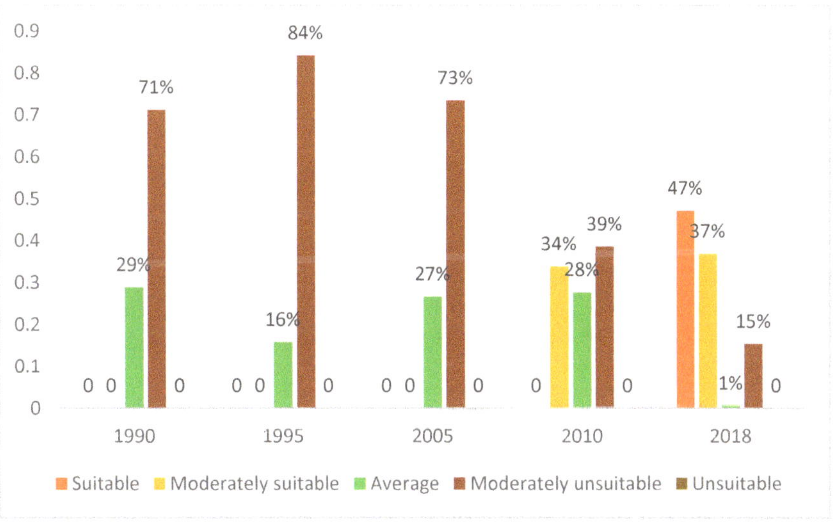


**Figure 6.** Proportional structure of number of grid cells for evaluation of functional suitability of ecological space.

### 3.2.2. Changes in Functional Suitability of Agricultural-Production Space

The lowest variability was found in the functional suitability of the agricultural-production space. In general, the suitability index of Zhulin's agricultural-production has not changed much, showing a gradual increase from an initial 3.29 in 1990 to 3.90 in 2018, and its overall suitability has remained in the moderately suitable interval. Based on the intragroup variation, the maximum value of the suitability index of Zhulin's agricultural-production space continued to increase, while the minimum value fluctuated only slightly. The standard deviations of the indicators ranged from 0.21 to 0.26, and the overall variation was small, indicating no significant difference (Table 9). The main reason for this small variation is that the proportion of areas evaluated as moderately suitable for agricultural-production space in Zhulin remained high, and the gradual increase in the level of functional suitability was due to the upgrading of some moderately suitable areas to suitable areas (Figure 7).

**Table 9.** Overall evaluation of functional suitability of agricultural-production space.

| Year | Max. | Min. | Mean | Standard Deviation |
|------|------|------|------|--------------------|
| 1990 | 3.77 | 2.73 | 3.29 | 0.21 |
| 1995 | 3.77 | 2.63 | 3.32 | 0.21 |
| 2005 | 4.08 | 3.07 | 3.63 | 0.21 |
| 2010 | 4.38 | 3.28 | 3.93 | 0.23 |
| 2018 | 4.38 | 2.78 | 3.90 | 0.26 |

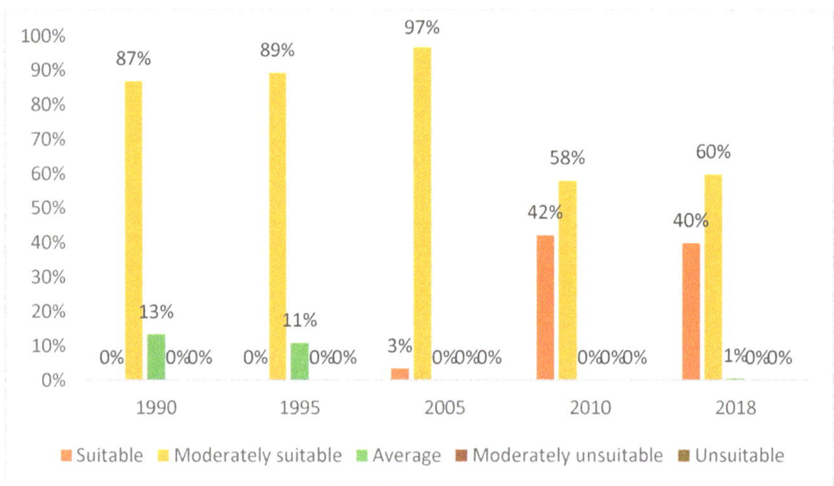


**Figure 7.** Proportional structure of number of grid cells for evaluation of functional suitability of agricultural-production space.

### 3.2.3. Changes in Functional Suitability of Nonagricultural-production Space

The mean value of the functional suitability for the nonagricultural-production space was the highest and most stable. In general, the average suitability index of Zhulin's nonagricultural-production space had a smaller variation, showing a small fluctuation and increasing trend, and the evaluation of its overall suitability remained in the interval of moderately suitable. Based on the intragroup variation, the maximum value of the suitability index for the nonagricultural-production space remained almost unchanged, while the minimum value showed a fluctuating decreasing trend. The standard deviations of the indicators ranged from 0.40 to 0.60, and the overall sample dispersion was not high and showed a small fluctuating increase, indicating that the difference between the samples showed a small fluctuation (Table 10). The reason for this phenomenon is that the suitability evaluation of nonagricultural-production space was mainly moderately suitable and suitable in general, with no grid cells of moderately unsuitable and unsuitable. The proportion of grid cells evaluated as suitable declined between 2005 and 2010, but then improved and exceeded the proportion of moderately suitable grid cells, becoming dominant (Figure 8).

**Table 10.** Overall evaluation of functional suitability of nonagricultural-production space.

| Year | Max. | Min. | Mean | Standard Deviation |
|------|------|------|------|--------------------|
| 1990 | 4.73 | 2.97 | 3.88 | 0.47 |
| 1995 | 4.78 | 2.14 | 3.80 | 0.60 |
| 2005 | 4.73 | 2.09 | 3.82 | 0.46 |
| 2010 | 4.73 | 2.85 | 3.86 | 0.40 |
| 2018 | 4.78 | 2.31 | 4.00 | 0.51 |

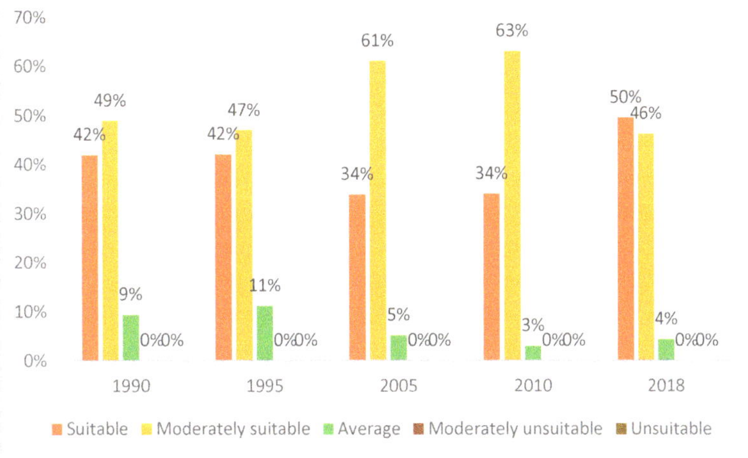


**Figure 8.** Proportional structure of number of grid cells for evaluation of functional suitability of nonagricultural-production space.

### 3.2.4. Changes in Functional Suitability of Living Space

The greatest difference was found in the functional suitability of the living space. In general, the average suitability indexes of Zhulin's living space first increased, then decreased, and finally increased gradually. Overall, the changes were minor, showing a slight increase with fluctuation. The evaluation of the overall suitability was always in the interval of moderately suitable (Table 11). Based on the intragroup variation, the maximum value of the living space suitability index continued to increase slightly, while the minimum value showed a small decrease. The standard deviations of the indicators ranged from 0.94 to 1.13, the highest dispersion in the EPL spaces, and showed a small fluctuating increase, indicating that the difference between samples was relatively large in the EPL spaces. According to the structure of the grid distribution, although moderately suitable and suitable were also dominant in general, the average and moderately unsuitable grid cells in different years accounted for a notable proportion. Despite the overall suitability remaining at a relatively high level and the proportion of grid cells evaluated as suitable and moderately suitable increasing, internal differentiation has not improved much (Figure 9).

**Table 11.** Overall evaluation of functional suitability of living space.

| Year. | Max. | Min. | Mean | Standard Deviation |
|-------|------|------|------|--------------------|
| 1990 | 4.31 | 0.90 | 3.18 | 1.02 |
| 1995 | 4.73 | 0.90 | 3.29 | 1.04 |
| 2005 | 4.73 | 0.90 | 3.83 | 0.94 |
| 2010 | 4.79 | 0.60 | 3.46 | 1.13 |
| 2018 | 4.79 | 0.60 | 3.58 | 1.04 |

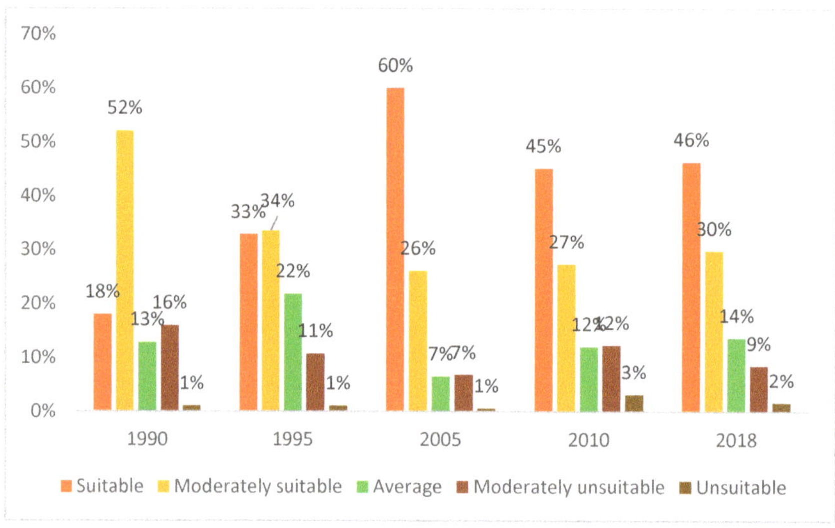



**Figure 9.** Proportional structure of number of grid cells for evaluation of functional suitability of living space.

### 3.3. EPL Analysis of Matching between Distribution of EPL Spaces and Their Functional Suitability

Since the administrative area of Zhulin changed several times during the study period, we analyzed the match between the distribution of the EPL spaces and their functional suitability in different time periods, in a comprehensive manner, by using the overall estimated administrative area of each period (Figure 10) and considering only the match between the distribution of the internal EPL spaces and their functional suitability, excluding external new areas (Figure 11).

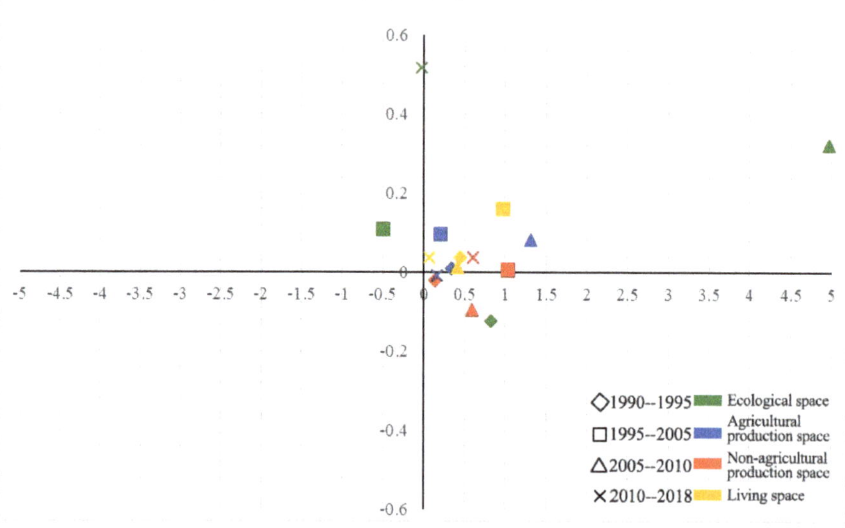



**Figure 10.** Match between overall distribution and functional suitability of Zhulin's EPL spaces.

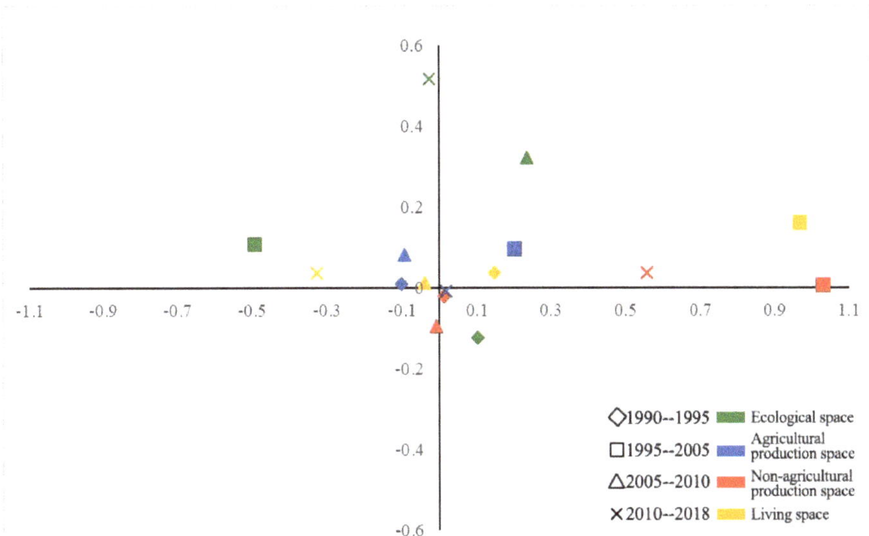


**Figure 11.** Match between internal distribution and functional suitability of Zhulin's EPL spaces.

In general, due to changes in administrative divisions, economic development and infrastructure construction, the following features emerged in the matching patterns between the distribution and functional suitability of Zhulin's EPL spaces from 1990 to 2018:

1.  The restructuring of Zhulin's EPL spaces has generally achieved a positive effect. Owing to the continuous growth of the administrative area of Zhulin, the overall expansion trend for various types of spaces is significant, except for ecological space in some individual periods, and 62.5% of the matching cases fall into the first quadrant, a high-quality growth type. The restructuring of Zhulin's EPL spaces has achieved a positive effect. However, the quality of expansion differs in different periods.

2.  Most of the spatial high-quality growth occurred in the period 1995–2005. The administrative area of Zhulin did not change during this period. The distribution and functional suitability of the agricultural-production space, nonagricultural-production space and living space in this period all fell into the first quadrant, which is in the high-quality growth area. The ecological space, however, fell into the second quadrant, in the area of functional optimization and intensification.

3.  Spatial low-quality expansion occurred mainly in the 1990–1995 period. In this period, along with the expansion of Zhulin's administrative area, both ecological space and nonagricultural-production space grew, but their functional suitability declined both in general and in terms of internal changes. Furthermore, the match between the distribution and functional suitability of its ecological space and nonagricultural-production space fell in the fourth quadrant, in the low-quality expansion area.

4.  The overall match for the agricultural-production space and living space was good in different years. In terms of both overall changes and internal changes, the living space was in the first and second quadrants of continuous functional optimization, and entirely in the high-quality growth area. The agricultural-production space was also in the first and second quadrants of continuous functional optimization. Except for 2018, when there was low-quality expansion, the changes in all the other periods were in the high-quality growth area. This is mainly due to the fact that economic development has made it possible for Zhulin to vigorously improve water supply conditions, road facilities and other basic infrastructure conditions, thus increasing the suitability for living and agricultural production.

5.  Low-quality expansion of the nonagricultural-production space occurred most frequently. Low-quality expansion of nonagricultural industrial space appeared in the periods 1990–1995 and 2005–2010, corresponding to the period of early development and the period of administrative area expansion, respectively. Led by economic interests, governments' management of land policy began to loosen, and enterprises developed in a disorderly manner in the pursuit of profits, resulting in the low-quality expansion of nonagricultural industrial space. However, in the 2010–2018 period, the nonagricultural industrial space (after the administrative area expansion) witnessed high-quality growth, mainly because Zhulin began to orient its development toward tourism so that the location decisions of enterprises no longer pursued size alone, leading to relatively higher-quality development.

## 4. Discussion

The main purpose of analyzing the history of rural-spatial restructuring is to grasp the patterns of its changes, which can ultimately be applied to the improvement of rural land space, with the goal of optimizing the spatial pattern of rural land, improving the efficiency of land resource utilization and enhancing its spatial quality. However, most of the literature takes the perspective of EPL's spatially explicit attributes, i.e., changes in area, structure or spatial distribution, while few studies address whether structural changes match functional suitability. This makes it impossible to grasp the overall picture of the functional evolution of spatial structure, which affects our scientific judgment and the implementation and formulation of policies. For example, in the process of the spatial restructuring of Zhulin, from 1995 to 2005, ecological spaces suffered a large reduction simply in terms of quantitative and structural changes. According to the traditional analysis paradigm [25–27], this point, a turning point in ecological spatial area change, often heralds a sudden change in ecological space, which may lead to its dysfunction. However, the results of the functional suitability evaluation show that the functional suitability of Zhulin's ecological, living, agricultural-production and nonagricultural-production spaces continued to increase during this period. This indicates that the path of development in this period was substantially in line with the actual local situation. The integration of land-use distribution has led to the continuous optimization of local land-use functions, resulting in an overall increase in the efficiency of land and space use. This is why Zhulin was awarded the title of China's Sustainable Small Town Pilot by UNDP during this period, and why it was also awarded the Dubai International Award for Best Practice in Improving the Living Environment by UN-Habitat in 2002.

Clearly, the process of land-use transformation influenced by land-resource allocation and management is complex, because the value of a land-use type relative to its competing uses changes over time [42]. Therefore, analysis of EPL spatial restructuring with a one-sided emphasis on area change may not fully capture the functional suitability of rural spaces, which may lead to difficulty in adequately matching the restructuring of EPL spaces to the territorial spaces in which they are located. If so, the actual value of allocated land may be far from optimal, distorting economic and political incentives and undermining the sustainability of the countryside [56]. This will result in the previous practice of simply pursuing increases in the area of certain types of land use while ignoring whether functional suitability is satisfied, risking the repetition of tragedies in human–land relations such as the enclosure of lakes to create fields, the reclamation of land, steep slope clearing and deforestation (grass clearing). Therefore, in analyzing rural-spatial restructuring, it is necessary to consider not only changes in explicit morphological characteristics but also changes in implicit characteristics, represented by quality and function [23].

Taking Zhulin Town as a case study, this paper sets forth a spatial-restructuring analysis method based on the evaluation of the land distribution and functional suitability of EPL spaces and analyzed the spatial-restructuring process over the past 30 years of reform and opening up.

We propose a method for studying rural-spatial restructuring based on changes in the dominant and recessive morphologies of land use. That will enable us to understand the process of rural-spatial restructuring more comprehensively from the changes in both the dominant and recessive morphologies of land use. The development of Zhulin from a remote mountain village to a modern town is essentially a condensed version of the development of China's rural urbanization, and its rural-spatial restructuring process essentially reflects the impact of the shift in national macro development policies on rural development. Before 2000, China was still in a period of economic construction, and the goal of economic development overruled all other considerations. At that time, no clearly delineated land-use guidelines were issued at the national level, and land-use and development guidelines were set by local governments [57]. Under the policy orientation of overriding economic development, local governments tended to blindly expand production space to bring about more economic benefits without considering other factors. This is why Zhulin's low-quality expansion occurred mainly between 1990 and 1995 and the functional suitability of EPL spaces declined in this period. The early 21st century witnessed the conversion of a lot of arable land by urbanization and industrialization, resulting in a significant decrease in the rural labor force engaged in agricultural production; the output efficiency of arable land decreased, affecting national food security. Therefore, the Ministry of Land and Resources of China launched the first ten-year (2001–2010) national plan focusing on maintaining a dynamic balance of arable land to rearrange and reclaim, and to develop idle, abandoned and damaged land to increase the quantity and improve the quality of arable land [58]. Top-down policy constraints enabled the recovery of arable land area in various places during this period. It was also during this period that Zhulin's proportion of agricultural-production space peaked. At the same time, the ecological and environmental problems associated with China's rapid urbanization became increasingly apparent. The expansion of land for construction led to landscape fragmentation, environmental degradation and the destruction of ecosystem services [59]. Faced with these problems and challenges, since 2012, China has proposed a shift in the spatial development model for its land from a production-space orientation to ecological–production–living coordination to optimize the spatial layout of land and alleviate the contradiction between urban and rural development [22]. This is why Zhulin's ecological spatial area decreased from 2010 to 2018, while its functional suitability was significantly enhanced in this period. The same is true for the functional suitability of other space types.

Macro development policy is a powerful mechanism by which to influence regional and rural development [60]. The spatial restructuring of Zhulin over the past 30 years reflects a good fit with the national policy shift. This is not easy for local governments to achieve and is the most important reason why Zhulin's economic and social development is sustainable. The timely adjustment of Zhulin's development pattern within its administrative division in response to the national policy shift is due to its good rural-governance model. Within the framework of China's grassroots self-governance system, Zhulin has developed a governance model system that includes two aspects, which are named "San Ping" and "Shi Ping" in Chinese. The "San Ping" refers to a process of evaluating the governance of village leaders and cadres through a mechanism of public supervision. The "Shi Ping" refers to a practice of evaluating top performers among the masses to motivate them to make contributions in the process of Zhulin's development. This mode of governance can actively mobilize the grassroots to participate in politics, and help to unify the ideologies of all the villagers, so that local actors can grasp development opportunities in a timely manner and correct problems that arise in the development process, thus promoting Zhulin's high-quality development.

With the transformation of the rural economy, the economic importance of agricultural land will continue to decline, while the function of land as a space carrier, an asset and capital will increase. This trend will eventually lead to a return to the stable state of the agricultural-production function and ecological landscape function. This trend will be mapped to land-use pattern changes. During periods when land-use patterns cannot adapt to rural development needs, there will be various land-use problems, which will require the reasonable intervention of local actors to adjust the land-use patterns to, again, adapt them to development needs. The most important reason for this is that, under the influence of the current stage of rural grassroots autonomy policy, the impact of external national or regional macro policies and changes in the market environment on the development of villages needs to be determined more through the awareness of elected leaders and competent people in the village concerning the relevant policies and changes in the market environment [61,62]. Local actors are often the centerpiece of the rural development process [63]. It is therefore crucial for local actors to intervene reasonably and to respond proactively in the interactive process of land-use and village development [64,65].

It should be pointed out that there are still some problems worthy of further discussion, such as the in-depth analysis of the evolution mechanism of spatial restructuring promoted by land-use transitions. We need to consider more complex and diverse issues when we take a relatively micro area, such as Zhulin Town, as a study area to perform a long-term study. We not only need to consider the influence of its own development conditions and external development conditions, but also need to consider the influence of some accidental factors. Therefore, it is difficult to study its evolution mechanism. Research on its mechanism is of great significance for promoting the coordinated development of urban and rural areas. It needs to be further strengthened. In addition, the suitability of function is only one of the characteristics of the recessive morphology of land-use morphologies. The recessive morphologies also include land-use management systems, property rights and so on. As the impact of recessive morphologies on land-use transformation will gradually increase [17,23], more in-depth research should be carried out in the future.

## 5. Conclusions

Based on land-use type, this paper has established a framework for the identification of EPL spaces and constructed an evaluation system for the functional suitability of EPL spaces. Based on this framework, we propose criteria for judging the match between the distribution of EPL spaces and their functional suitability in terms of changes in their quantitative distribution and functional suitability characteristics. We used Zhulin Town in Central China as a case study area to analyze its rural-spatial restructuring process since 1990. The regional characteristics of the spatial distribution of the ecological and agricultural-production space were made clear. The clustering trend of the living space and nonagricultural-production space has strengthened, especially for the living space. Many scattered living spaces have begun to die out, while living and nonagricultural-production spaces along main transportation routes have continued to expand. Overall, Zhulin is dominated by agricultural-production and ecological space. The structure of its EPL spaces has fluctuated over the past 30 years, with the proportion of production space (including agricultural-production and nonagricultural-production space) and living space first decreasing and then increasing, and then decreasing again, yielding a W-shaped fluctuation curve. The change in the proportion of Zhulin's ecological space is opposite to that in its production and living spaces, showing an M-shaped fluctuation curve. The fluctuation in EPL spaces is due to the expansion of Zhulin's administrative space as well as its own internal transformation. Although the distribution and structure of EPL spaces have been changing over the past 30 years, their overall functional suitability has steadily increased. Analysis of the distribution and functional suitability of Zhulin's EPL spaces revealed a significant trend of expansion for various types of spaces in general, and 62.5% of the matching types were of high-quality growth, which shows that the restructuring of Zhulin's EPL spaces has achieved a more positive effect. High-quality spatial growth occurred mainly in the 1995–2005 period, and low-quality spatial expansion occurred mainly in the 1990–1995 period. The overall match for Zhulin's agricultural-production space and living space was good in some years, and nonagricultural-production space showed the most low-quality expansion.

This case study of Zhulin shows that a spatial restructuring study combining changes in spatial area, the quantity of EPL spaces, and changes in functional suitability generates a more comprehensive understanding of the process of restructuring. It also helps to improve understanding of the impact of the shift in national macro development policies on rural development. Thus, future studies on rural-spatial restructuring should consider changes in the explicit morphological characteristics of the land as well as changes in the implicit characteristics represented by quality and function. Rural-spatial restructuring requires local actors to make timely adjustments to meet the needs of macroeconomic and social development, while a good rural-governance model serves as an important guarantee to ensure that rural-spatial functions can be optimized. In the future, we should strengthen the research on the mechanism of land-use transitions promoting rural restructuring and the influence of recessive morphologies on land-use transformation.

**Author Contributions:** Conceptualization, D.H. and J.Q.; methodology, D.H. and J.Q.; formal analysis, D.H.; data curation, D.H. and Q.Z.; writing—original draft preparation, D.H.; writing—review and editing, J.Q. and Q.Z.; visualization, Q.Z.; supervision, J.Q.; funding acquisition, J.Q. All authors have read and agreed to the published version of the manuscript.

**Funding:** This research was funded by National Natural Science Foundation of China (42071220 and 41671172) and Philosophy and Social Science Innovative Talent Project of Higher Education in Henan Province (2021-CX-016).

**Acknowledgments:** We wish to thank the kind local villagers who were willing to be interviewed and the local government departments of Zhulin town and Gongyi City for providing relevant information.

**Conflicts of Interest:** The authors declare no conflict of interest.

## References

1. Elshof, H.; Haartsen, T.; Van Wissen, L.J.G.; Mulder, C.H. The influence of village attractiveness on flows of movers in a declining rural region. *J. Rural Stud.* **2017**, *56*, 39–52. [CrossRef]
2. Woods, M. *Rural Geography: Processes, Responses and Experiences in Rural Restructuring*; Sage: London, UK, 2005; pp. 29–41. [CrossRef]
3. Long, H. Land consolidation: An indispensable way of spatial restructuring in rural China. *J. Geogr. Sci.* **2014**, *24*, 211–225. [CrossRef]
4. Woods, M. Engaging the global countryside: Globalization, hybridity and the reconstitution of rural place. *Prog. Hum. Geogr.* **2007**, *31*, 485–507. [CrossRef]
5. Galani-Moutafi, V. Rural space (re)produced-practices, performances and visions: A case study from an Aegean island. *J. Rural Stud.* **2013**, *32*, 103–113. [CrossRef]
6. Hedlund, M.; Lundholm, E. Restructuring of rural Sweden: Employment transition and out-migration of three cohorts born 1945–1980. *J. Rural Stud.* **2015**, *42*, 123–132. [CrossRef]
7. Ellis, F.; Biggs, S. Evolving themes in rural development 1950s–2000s. *Dev. Policy Rev.* **2001**, *19*, 437–448. [CrossRef]
8. Lobley, M.; Potter, C. Agricultural change and restructuring: Recent evidence from a survey of agricultural households in England. *J. Rural Stud.* **2004**, *20*, 499–510. [CrossRef]
9. Nelson, P.B. Rural restructuring in the American West: Land use, family and class discourses. *J. Rural Stud.* **2001**, *17*, 395–407. [CrossRef]
10. Wilson, O.J. Rural restructuring and agriculture-rural economy linkages: A New Zealand study. *J. Rural Stud.* **1995**, *11*, 417–431. [CrossRef]
11. Li, Y.; Westlund, H.; Liu, Y. Why some rural areas decline while some others not— An overview of rural evolution in the world. *J. Rural Stud.* **2019**, *68*, 135–143. [CrossRef]
12. Echánove, F. Globalization and restructuring in rural Mexico: The case of fruit growers. *Tijdschr. Econ. Soc. Geogr.* **2005**, *96*, 15–30. [CrossRef]
13. Tu, S.; Long, H.; Zhang, Y.; Ge, D.; Qu, Y. Rural restructuring at village level under rapid urbanization in metropolitan suburbs of China and its implications for innovations in land use policy. *Habitat Int.* **2018**, *77*, 143–152. [CrossRef]
14. Liu, Y.; Long, H.; Chen, Y.; Wang, J.; Li, Y.; Li, Y.; Yang, Y.; Zhou, Y. Progress of research on urban-rural transformation and rural development in China in the past decade and future prospects. *J. Geogr. Sci.* **2016**, *26*, 1117–1132. [CrossRef]
15. Liu, Y.; Yan, B.; Wang, Y. Urban-rural development problems and transformation countermeasures in the new period in China. *Econ. Geogr.* **2016**, *36*, 1–8. (in Chinese). [CrossRef]
16. Tu, S.; Long, H. Rural restructuring in China: Theory, approaches and research prospect. *J. Geogr. Sci.* **2017**, *27*, 1169–1184. [CrossRef]

17. Long, H. *Land Use Transitions and Rural Restructuring in China*; Springer Nature: Singapore, 2020; pp. 31–160. [CrossRef]
18. Long, H.; Li, Y.; Liu, Y.; Woods, M.; Zou, J. Accelerated restructuring in rural China fueled by 'increasing vs. decreasing balance' land-use policy for dealing with hollowed villages. *Land Use Policy* **2012**, *29*, 11–22. [CrossRef]
19. Long, H.; Woods, M. Rural Restructuring under Globalization in Eastern Coastal China— What can we learn from Wales? *J. Rural Community Dev.* **2011**, *6*, 70–94.
20. Huang, A.; Xu, Y.; Lu, L.; Liu, C.; Zhang, Y.; Hao, J.; Wang, H. Research progress of the identification and optimization of production-living-ecological spaces. *Prog. Geogr.* **2020**, *39*, 503–518. (in Chinese). [CrossRef]
21. Long, H. Land consolidation and rural spatial restructuring. *Acta Geogr. Sin.* **2013**, *68*, 1019–1028. (In Chinese)
22. Yang, Y.; Bao, W.; Li, Y.; Wang, Y.; Chen, Z. Land use transition and its eco-environmental effects in the Beijing–Tianjin–Hebei urban agglomeration—A production–living–ecological perspective. *Land* **2020**, *9*, 285. [CrossRef]
23. Long, H.; Qu, Y.; Tu, S.; Zhang, Y.; Jiang, Y. Development of land use transitions research in China. *J. Geogr. Sci.* **2020**, *30*, 1195–1214. [CrossRef]
24. Zhou, X.; Li, X.; Song, W.; Kong, X.; Lu, X. Farmland transitions in China: An advocacy coalition approach. *Land* **2021**, *10*, 122. [CrossRef]
25. Bertoni, D.; Aletti, G.; Ferrandi, G.; Micheletti, A.; Cavicchioli, D.; Pretolani, R. Farmland use transitions after the CAP greening—A preliminary analysis using Markova chains Approach. *Land Use Policy* **2018**, *79*, 789–800. [CrossRef]
26. Ge, D.; Long, H.; Zhang, Y.; Ma, L.; Li, T. Farmland transition and its influences on grain production in China. *Land Use Policy* **2018**, *70*, 94–105. [CrossRef]
27. You, H.; Hu, X.; Wu, Y. Farmland use intensity changes in response to rural transition in Zhejiang province, China. *Land Use Policy* **2018**, *79*, 350–361. [CrossRef]
28. Long, H.; Heilig, G.K.; Li, X.; Zhang, M. Socio-economic development and land-use change—Analysis of rural housing land transition in the Transect of the Yangtse River, China. *Land Use Policy* **2007**, *24*, 141–153. [CrossRef]
29. Qu, Y.; Jiang, G.; Li, Z.; Tian, Y.; Wei, S. Understanding rural land use transition and regional consolidation implications in China. *Land Use Policy* **2019**, *82*, 742–753. [CrossRef]
30. Ma, W.; Jiang, G.; Zhang, R.; Li, Y.; Jiang, X. Achieving rural spatial restructuring in China—A suitable framework to understand how structural transitions in rural residential land differ across peri-urban interface? *Land Use Policy* **2018**, *75*, 583–593. [CrossRef]
31. Yu, Z.; Xiao, L.; Chen, X.; He, Z.; Guo, Q.; Vejre, H. Spatial restructuring and land consolidation of urban-rural settlement in mountainous areas based on ecological niche perspective. *J. Geogr. Sci.* **2018**, *28*, 131–151. [CrossRef]
32. Sun, J.; Lu, B.; Chen, R.; He, C. The approach of village system spatial reconstruction on the background of urban-rural integration—A case study of Jiuduhe town in Huairou district of Beijing. *Urban Stud.* **2009**, *16*, 75–81, 107. (In Chinese) [CrossRef]
33. Li, T.; Long, H.; Liu, Y.; Tu, S. Multi-scale analysis of rural housing land transition under China's rapid urbanization: The case of Bohai Rim. *Habitat Int.* **2015**, *48*, 227–238. [CrossRef]
34. Xiao, J.; Ou, W. Study on the spatial restructure in the coordinated urban and rural development—A case study of Suqian city. *China Land Sci.* **2013**, *27*, 54–60, 97. (In Chinese) [CrossRef]
35. Du, X.; Wang, C.; Jiang, W.; Wan, Q.; Liu, Y. Rural settlements restructuring based on analysis of the gravity model at village level—The case of Dazhu village, Chongqing, China. *Econ. Geogr.* **2015**, *35*, 154–160. (In Chinese) [CrossRef]
36. Li, C.; Liang, S.; Che, Q. Spatial reconstruction of the rural settlements in Wuhu city based on the function oriented zoning. *Resour. Environ. Yangtze Basin* **2015**, *24*, 1736–1743. (In Chinese) [CrossRef]
37. Sun, H.; Liu, Y.; Xu, K. Hollow villages and rural restructuring in major rural regions of China—A case study of Yucheng City, Shandong Province. *Chin. Geogr. Sci.* **2011**, *21*, 354–363. [CrossRef]
38. Li, Y.; Liu, Y.; Long, H.; Cui, W. Community-based rural residential land consolidation and allocation can help to revitalize hollowed villages in traditional agricultural areas of China—Evidence from Dancheng County, Henan Province. *Land Use Policy* **2014**, *39*, 188–198. [CrossRef]
39. Gökyer, E. *Understanding Landscape Structure Using Landscape Metrics*; InTech: London, UK, 2013; pp. 663–676. [CrossRef]
40. Lambin, E.F.; Meyfroidt, P. Land use transitions: Socio-ecological feedback versus socio-economic change. *Land Use Policy* **2010**, *27*, 108–118. [CrossRef]
41. Long, H.; Li, T. The coupling characteristics and mechanism of farmland and rural housing land transition in China. *J. Geogr. Sci.* **2012**, *22*, 548–562. [CrossRef]
42. Long, H.; Qu, Y. Land use transitions and land management—A mutual feedback perspective. *Land Use Policy* **2018**, *74*, 111–120. [CrossRef]
43. Fang, C. Dissipative structure theory and geography system. *Arid Land Geogr.* **1989**, *12*, 53–58. (In Chinese)
44. De, Z.; Xu, J.; Lin, Z. Conflict or coordination?—Assessing land use multi-functionalization using production-living-ecology analysis. *Sci. Total Environ.* **2017**, *577*, 136–147. [CrossRef]
45. Yu, Z.; Xu, E.; Zhang, H.; Shang, E. Spatio-temporal coordination and conflict of production-living-ecology land functions in the Beijing-Tianjin-Hebei Region, China. *Land* **2020**, *9*, 170. [CrossRef]
46. Yang, Y.; Bao, W.; Liu, Y. Coupling coordination analysis of rural production-living-ecological space in the Beijing-Tianjin-Hebei region. *Ecol. Indic.* **2020**, *117*, 106512. [CrossRef]
47. Liu, J.; Liu, Y.; Li, Y. Classification evaluation and spatial-temporal analysis of "production-living-ecological" spaces in China. *Acta Geogr. Sin.* **2017**, *72*, 1290–1304. (In Chinese) [CrossRef]

48. Steiner, F. Resource suitability: Methods for analyses. *Environ. Manag.* **1983**, *7*, 401–420. [CrossRef]
49. Fan, J.; Wang, Y.; Ouyang, Z.; Li, L.; Xu, Y.; Zhang, W.; Wang, C.; Xu, W.; Li, J.; Yu, J. Risk forewarning of regional development sustainability based on a natural resources and environmental carrying index in China. *Earths Future* **2017**, *5*, 196–213. [CrossRef]
50. Huang, J.; Lin, H.; Qi, X. A literature review on optimization of spatial development pattern based on ecological-production-living space. *Prog. Geogr.* **2017**, *36*, 378–391. (In Chinese) [CrossRef]
51. Yu, Z.; Zhang, W.; Liang, J.; Zhuang, L. Progress in evaluating suitability of spatial development and construction land. *Prog. Geogr.* **2015**, *34*, 1107–1122. (In Chinese) [CrossRef]
52. Mu, S.; Li, J.; Chen, Y.; Gang, C.; Zhou, W.; Ju, W. Spatial differences of variations of vegetation coverage in inner mongolia during 2001–2010. *Acta Geogr. Sin.* **2012**, *67*, 1255–1268. (In Chinese) [CrossRef]
53. He, B.; Ding, J.; Zhang, Z.; Abduwasit, G. Experimental analysis of spatial and temporal dynamics of fractional vegetation cover in Xinjiang. *Acta Geogr. Sin.* **2016**, *71*, 1948–1966. (In Chinese) [CrossRef]
54. Hansen, M.C.; DeFries, R.S.; Townshend, J.R.G.; Sohlberg, R.; Dimiceli, C.; Carroll, M. Towards an operational MODIS continuous field of percent tree cover algorithm—Examples using AVHRR and MODIS data. *Remote Sens. Environ.* **2002**, *83*, 303–319. [CrossRef]
55. Xu, Y.; Tang, Q.; Fan, J.; Bennett, S.J.; Li, Y. Assessing construction land potential and its spatial pattern in China. *Landsc. Urban Plan.* **2011**, *103*, 207–216. [CrossRef]
56. Barbier, E.B.; Burgess, J.C.; Grainger, A. The forest transition: Towards a more comprehensive theoretical framework. *Land Use Policy* **2010**, *27*, 98–107. [CrossRef]
57. Huang, Q.; Li, M.; Chen, Z.; Li, F. Land consolidation—An approach for sustainable development in rural China. *AMBIO* **2011**, *40*, 93–95. [CrossRef]
58. Tian, Y.; Liu, Y.; Liu, X.; Kong, X.; Liu, G. Restructuring rural settlements based on subjective well-being (SWB)—A case study in Hubei province, central China. *Land Use Policy* **2017**, *63*, 255–265. [CrossRef]
59. Qu, Y.; Long, H. The economic and environmental effects of land use transitions under rapid urbanization and the implications for land use management. *Habitat Int.* **2018**, *82*, 113–121. [CrossRef]
60. Brown, C.G.; Waldron, S.A.; Longworth, J.W. Rural development in China: Industry policy, regionalism, integration and scale. *Int. J. Soc. Econ.* **2005**, *32*, 17–33. [CrossRef]
61. Ren, G.; Zhu, X.; Heerink, N.; Feng, S.; van Ierland, E.C. Persistence of land reallocations in Chinese villages—The role of village democracy and households' knowledge of policy. *J. Rural Stud.* **2019**. [CrossRef]
62. Li, Y.; Westlund, H.; Zheng, X.; Liu, Y. Bottom-up initiatives and revival in the face of rural decline: Case studies from China and Sweden. *J. Rural Stud.* **2016**, *47*, 506–513. [CrossRef]
63. Skinner, M.W.; Joseph, A.E.; Kuhn, R.G. Social and environmental regulation in rural China: Bringing the changing role of local government into focus. *Geoforum* **2003**, *34*, 267–281. [CrossRef]
64. Douglas, D.J.A. The restructuring of local government in rural regions: A rural development perspective. *J. Rural Stud.* **2005**, *21*, 231–246. [CrossRef]
65. Liu, Z.; Müller, M.; Rommel, J.; Feng, S. Community-based agricultural land consolidation and local elites: Survey evidence from China. *J. Rural Stud.* **2016**, *47*, 449–458. [CrossRef]

*Article*

# The Transition Mechanism and Revitalization Path of Rural Industrial Land from a Spatial Governance Perspective: The Case of Shunde District, China

Lianma Zhang [1,2], Dazhuan Ge [2,*], Pan Sun [2] and Dongqi Sun [3]

[1]  Bureau of Natural Resources and Planning in Siyang County, Suqian 223700, China; 201301011@njnu.edu.cn
[2]  School of Geography, Nanjing Normal University, Nanjing 210023, China; 09429@njnu.edu.cn
[3]  Institute of Geographic Sciences and Natural Resources Research, Chinese Academy of Sciences, Beijing 100101, China; sundq@igsnrr.ac.cn
*   Correspondence: gedz@njnu.edu.cn; Tel.: +86-25-8589-1347

**Abstract:** The transition of rural industrial land has a critical role to play in rural revitalization. The study of rural spatial governance is an important starting point for analyzing the processes and exploring the paths through which the transition of rural industrial land takes place. This study takes the case of Shunde District, China, a typical semi-urbanized area, as its research object and constructs an analytical framework for rural industrial land transition based on spatial governance; it uses this case to conduct an analysis of the spatiotemporal processes and dilemmas involved in rural industrial land transition. Hengding Industrial Park is taken as a specific example to study how the processes and mechanisms involved in the transition of rural industrial land work in practice from a spatial governance perspective, and the path of rural revitalization based on rural spatial governance is discussed. The conclusions are as follows: (1) the fragmentation of rural space, the difficulty of renewing rural industrial land, the chaos of ownership, and the incomplete mechanism of the differentiation and game of multiple subjects, are the main obstacles in the process of rural industrial land transition in Shunde District; (2) since the 1990s, the rural industrial land dominant morphology—including quantity, structure, and so on—and the recessive morphology, including property rights, organizational systems, and input–output efficiency, have all undergone significant changes; (3) the comprehensive governance of rural space under the analytical framework of "matter-ownership-organization," is an important starting point for analyzing the process of transition of rural industrial land. The "top-down" and "bottom-up" approaches, combining rural spatial governance strategy and the effective participation of multiple subjects, are important means of promoting the transition of rural industrial land; (4) rural spatial governance is conducive to promoting the transition of rural land use and the healthy development of rural space. The experience of semi-urbanized regions with rural revitalization is of vital significance for other regions.

**Keywords:** rural industrial land; rural spatial governance; land-use transition; rural revitalization

**Citation:** Zhang, L.; Ge, D.; Sun, P.; Sun, D. The Transition Mechanism and Revitalization Path of Rural Industrial Land from a Spatial Governance Perspective: The Case of Shunde District, China. *Land* **2021**, *10*, 746. https://doi.org/10.3390/land10070746

Academic Editor: Frank Vanclay

Received: 29 June 2021
Accepted: 15 July 2021
Published: 16 July 2021

**Publisher's Note:** MDPI stays neutral with regard to jurisdictional claims in published maps and institutional affiliations.

## 1. Introduction

With the rapid development of urbanization and industrialization, land use has undergone a dramatic transition in both space and function. The concept of land-use transition, first proposed by Graninger [1], is based on the forest transition hypothesis model [2–5] and originally referred to the change in the morphology of regional land use during regional socio-economic development [6,7]. It is now understood to refer to the process whereby regional land use changes from one morphology (including the dominant and recessive morphologies) to another driven by economic and social changes and innovations during a period of time corresponding to a stage of transition in economic and social development [8]. In the existing research, studies of the driving forces of industrial land transition account for a large proportion of papers published, and this topic is growing in popularity. There is, however, a lack of analysis of typical cases on the micro-scale.

The object of land-use transition studied in this paper is rural industrial land. In the context of urbanization, there are currently rapid changes in aspects of both urban and rural development. Rural industrial land, as the carrier with which human economic activity and rural industrial development are most closely related, has changed significantly in its spatial distribution characteristics, mode of use, and scale. Rural industrial land refers to construction land occupied by industrial enterprises in a village [9], which is mainly used collectively for various construction purposes, including industrial production, material transfer, professional acquisition, and storage. According to the spatial distribution pattern and spatiotemporal evolution characteristics of rural industrial land at different scales, in-depth exploration of the unique characteristics and laws of the transformation of rural industry can provide a basis for the transition and upgrading of rural industry. The existing research on rural industrial land mainly focuses on the qualitative analysis of property rights [10], business models [11], and land ownership [12].

The transition process of rural industrial land is inseparable from the promotion of multiple subjects. Since China's reform and opening up (an economic reform initiative implemented in 1978, featuring a socialist market economy and opening to the outside world), with the rapid development of urbanization and industrialization in the Pearl River Delta region, innovation in the "bottom-up" land shareholding system has played a positive role in the development of rural industry [13], which usually participates in the process of rural industrialization through land and factory leasing [14]. However, rural industrialization led by towns and villages is based on collective land. Against the background of the transition in the mode of economic development, the disadvantages of the original land-use methods have begun to emerge, such as inefficient use methods, lack of planned extensive development of rural industrial land, scattered collective construction land, and weak government macro-control capabilities [15,16]. So far, scholars have analyzed the "bottom-up" policy system for rural industrial land and have explored its operating mechanisms and implementation effects [17–21]. Related studies have also quantitatively described the evolutionary mechanism of the distribution pattern of rural industrial land, explored the dynamic changes in land use, and examined the reasons for scattered and inefficient land use [22–25]. The existing research has mainly focused on the role of the government [26–28] but less on the role of the multiple groups involved in rural industrial land transition. There is a lack of in-depth analysis of the power distribution and game mechanism among multiple subjects.

Rural space governance is an important means to promote the transition of land use, and its related theories are an important tool to analyze the process of land use transition. Rural spatial governance attempts to manage the social relations embodied in material space [29], thereby optimizing the organization and ownership of rural space and forming a joint force to promote rural development. This research field has gradually expanded to the study of the governance of rural ownership relationships and spatial organization modes; it attempts to optimize spatial relationships to create conditions for the integrated development of urban and rural areas [30]. Rural spatial governance research provides an important starting point for analyzing the internal mechanism of land-use transition, exploring the internal relationship between spatial governance and land-use transition, and providing references for the promotion of sustainable rural development. Based on this, this article starts with the theory of land-use transition oriented by rural spatial governance and shows how it can be applied to a case study of the transition process of rural industrial land in Shunde District; it deeply analyzes the transition mechanism of rural industrial land oriented by spatial governance and discusses how rural industrial transition and development can be perfected in semi-urbanized regions to further rural revitalization in other regions.

## 2. An Analytical Framework for Understanding Rural Industrial Land Transition from a Spatial Governance Perspective

### 2.1. Challenges in the Transition of Rural Industrial Land in the Pearl River Delta

Rural industrial land has played an important role in the process of rural transformation and development in the Pearl River Delta and has become an important feature in shaping the regional model. Since the 1980s, in the process of rural industrialization in the Pearl River Delta (an economic/geographical area located in the middle of Guangdong Province, China), the land transfer represented by the village collective establishment of joint-stock cooperatives and the rural industrial development model characterized by the leasing to village-level industrial enterprises have rapidly promoted rural industrial development. Over this period, village-level industrial parks have played a special and important role. The rural industrialization model of "every household in every village participates in industrialization" has greatly promoted the development of rural society and the rural economy. Basic village collective organizations have changed the use of collective land in rural areas (a large amount of agricultural land has been converted into industrial land), making the value of rural space substantially higher; the farmers generally earn dividends on their shares and effectively participate in the waves of industrialization. Against the background of the gradual withdrawal of township and village enterprises from the stage of history, the Pearl River Delta is unlike other regions in that rural industrial land still occupies an important position in its industrial development today. Rural industrial land plays an important role in the "bottom-up" urbanization of the Pearl River Delta. However, with the national demand for high-quality development and with the continuous improvement of the rule of law concerning rural space use, the rural industrial land transition in the Pearl River Delta is facing many challenges.

The fragmentation of rural space and renewal of rural industrial land have become new problems restricting rural development. The types of industry in rural areas are mainly labor-intensive, and there are problems of inefficient use of land and difficulties in industrial upgrading. The rural industrial land pattern presents a development form dominated by rural collective industrial land, and the spatial layout is characterized by a high proportion and scattered distribution, and the "fragmented" spatial pattern is prominent. The village collective members are the core land stock cooperatives. After more than 20 years of operation, their interests are intertwined, and industrial land has become the norm across villages and regions. Under the "Transformation of the Three Olds" (renovation measures for old towns, old factories, and old villages) due to the fragmentation of distribution, mixing of space, and intersection of interests, rural industrial land has become difficult to update, and there are few successful examples of attempts to do so. This has resulted in a conundrum for the revitalization of stock construction land.

Ownership disorder is a distinctive feature of the rural industrial land in the Pearl River Delta and is mainly manifested by unclear ownership of collective land and multi-layered transfer leases. Rural industrial land is based on collective land, and the illegal conversion of agricultural land to construction land has become a fait accompli in the era of weak land management laws and rural space-use control. In the process of gradual improvement of urban and rural planning laws, only a small part of this type of construction land has completed the required legal procedures, and there is still a considerable amount of rural land whose ownership relationships are difficult to determine legally (Figure 1). The notable feature is that the ownership relationships of rural industrial land are largely undetermined, and only a small amount has been confirmed as being owned by the state or the collective. In addition, the village land-share cooperative manages the collective land (use rights) of the village in a unified manner, and the village committee is responsible for attracting investment. The land of the joint-stock company was leased to early local enterprises (houses) for industrial activities in the form of "lease instead of sale" and "lease instead of levy," and the rent distribution plan was established. With the pursuit of short-term rent as the core goal, land-share cooperatives entered into agreements in which enterprises do not have any real rights when they expire. Moreover, rent is adjusted every three to five years, leading to the phenomenon of rural industrial land subleasing and multilayered circulation, which further aggravates the chaos of rural industrial land ownership.

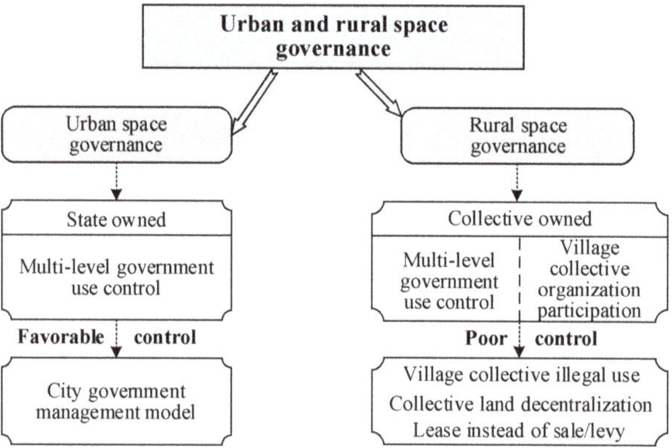


**Figure 1.** Different ownership of urban and rural spaces in China.

The differentiation of multiple subjects and the unsound mechanism of interest game have become the organizational background factors hindering the renewal of rural industrial land. The vigorous development of rural industrial land in the rural areas of the Pearl River Delta continues to this day for reasons closely related to the village community formed by the bond of village clan relationships. There is a strong patriarchal concept that, superimposed on consistent economic interests, gives rise to a solid grassroots community of shared interests; grassroots village collective organizations with a stock of rural industrial land, a strong sense of rights and interests, and strong grassroots organizational capabilities; economic organizations such as land stock cooperatives, which further bolster the strength of the farmers' organizations in the region [16,22]. Since China's reform and opening up, in the process of pursuing economic development, the government's ability to control grassroots villages has weakened, and the space-use control policy has changed significantly. The notable feature is that the development of rural industrial land has changed from early encouragement to subsequent restrictions and then to the current guidance and renewal. A grassroots governance pattern of "weak government and strong society" has gradually formed. Local entrepreneurs and new foreign entrepreneur groups closely related to the development and leasing of rural industrial land are intertwined between the grassroots village collective organizations and multilevel governments, seeking to maximize benefits, thus making it normal for multiple entities to participate in the transformation and renewal of rural industrial land. In the face of the government's implementation of the "Transformation of the Three Olds", the "Double Compliance" (remediation measures for rural industrial parks to make environmental protection and safe production up to standards), and other rural industrial land renewal policies, the degree of differentiation of the interests of multiple subjects has further deepened (Figure 2). How to effectively promote the "top-down" linkage, fairness, and justice of the interests of multiple subjects has become a core issue.

**Figure 2.** Rural industrial land landscape in Shunde District (Source: Urban Renewal Bureau of Shunde District).

### 2.2. An Analytic Framework for Understanding Rural Industrial Land Transition from a Spatial Governance Perspective

Rural spatial governance starts with the restructuring of rural material space, the reshaping of spatial ownership relationships, and the reorganization of spatial relations, thus re-allocating the key resources of rural space, realizing the core goals of government space-use control and grassroots orderly governance, and mobilizing multiple subjects to participate actively in the process of spatial development. On the basis of promoting the fair and just distribution of space rights and interests, the implementation of a multi-subject effective game is realized. Rural industrial land has reached a special epoch in its development that is a product of the times, when solving the problems of rural employment and industrial development is of the utmost importance. Rural industrial land is embedded in the rural regional system and has a significant particularity in that its spatial economic value is significantly different from that of other rural lands. Rural industrial land has also become an important springboard for rural development in the region. Through the development of rural industry, the economic value of rural space has been significantly enlarged, and the means and capabilities of farmers to participate in economic development have been enhanced. In the new era, facing realistic demands for high-quality industrial development and the transition of ecological civilization, how to promote the transformation of a combination of "top-down" and "bottom-up" approaches and how to enable multiple subjects' participation in rural industrial land have become the core goals of rural spatial governance.

The combination of "top-down" and "bottom-up" approaches to rural spatial governance will be a desirable solution to the dilemma of the transition of rural industrial land. The combination of "top-down" and bottom-up" approaches to spatial governance means realizing the national "top-down" territorial space-use control and space-governance goals in the transition of rural industrial land, strengthening the government's spatial governance capabilities, and improving governance capabilities in areas where rural industrial land is widely distributed. In addition, through "bottom-up" grassroots governance, the rural spatial governance system will be improved, the self-organizing capabilities of governance entities will be coordinated, and a combination of "rigid restraint" and "flexible regulation" in rural spatial governance will be promoted (Figure 3). By strengthening the spatial governance path combining "top-down" and "bottom-up" approaches, the governance drawbacks of the "weak government and strong society" phenomenon in the transition of rural industrial land will be changed, and a benefits plan will be provided for improving the mechanism of rural industrial transition. Through the combination of "top-down" and "bottom-up" approaches to spatial governance logic, the entire development and regional revitalization of rural industrial land will be promoted, breaking the existing "fragmented" spatial distribution pattern, and also providing broad space for the implementation of the government's macro-industry layout plan [6,30]. Therefore, the "top-down" spatial governance path promotes the improvement of the rural industrial land structure in the governance of physical space, emphasizes "planning and negotiation" in spatial organization and governance, and clarifies the spatial attributes of spatial ownership governance, which is conducive to promoting the transition and upgrading of rural industrial land as a whole.

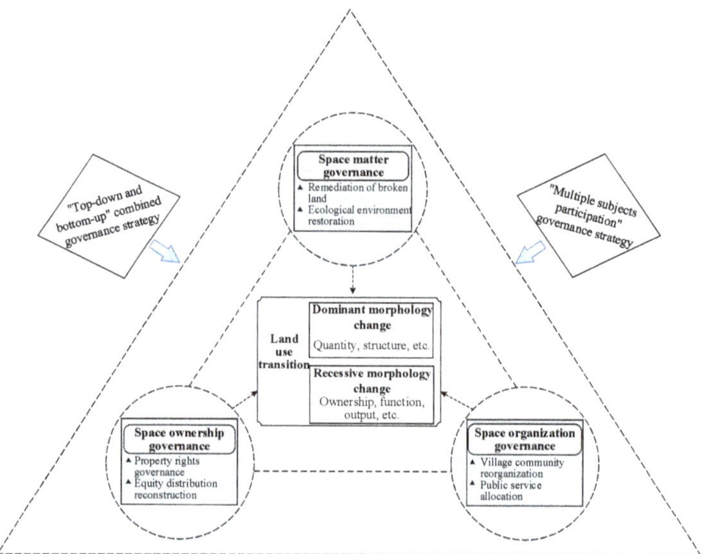


**Figure 3.** Analysis framework of rural industrial land transition oriented by space governance.

The effective participation of multiple subjects is an important guarantee for promoting the transition of rural industrial land. The disorder of ownership in rural spatial governance, the fragmentation of organizational systems, and the changes in and diversity of stakeholders involved determine whether the transition of rural industrial land can be successful [15]. Therefore, it is necessary to carry out targeted governance to address the above problems. Among the multiple subjects, multilevel governments, property rights issues, and market issues constitute the core issues in the organizational governance and ownership governance of rural industrial land. Any contradictions between the distribution of rights and interests and the governance plan for the transition of industrial land will result in a failure to achieve renewal, and the transition of rural industrial land will be put into a difficult situation. The effective participation of multiple entities, based on the government's establishment, will clarify the understanding of the property rights and interests of the various entities on rural industrial land, promote space ownership governance, and clarify the ownership of rural industrial land. In addition, the participation of market players is also an important factor in ensuring the transition of industrial land. Without the extensive participation of market players, it will be difficult to begin the transition of rural industrial land. The effective participation of multiple entities should ensure "legalization" and "standardization", reduce rent-seeking behavior in the process of spatial governance, strengthen the supervisory role of non-governmental organizations, establish smooth communication and discussion mechanisms to coordinate the differentiated interests of grassroots governance entities, and promote the establishment of a multi-round game mechanism.

The comprehensive management of physical space, space ownership, and spatial organization need to be strengthened with respect to rural industrial land. The management of the physical space of rural industrial land is mainly needed to change the quantitative and spatial structural characteristics of rural industrial land, promote its agglomeration, tap its potential, and reverse its fragmentation, which will help further improve the efficiency of rural industrial land use. Rural industrial land ownership governance and organizational governance are the keys to ensuring the success of rural industrial land transition. The sustainable transition of the recessive morphology of property rights, organizational models, utilization efficiency, and functional characteristics of rural industrial land is the key to ensuring the success of the renewal of rural industrial land. In response to the disorder in the ownership of rural industrial land, it is necessary to amend the stock-cooperation charter and supervision mechanism, strengthen the implementation of space-use control, and control the chaos of space ownership. In the context of the continuous differentiation of market entities, we should start with the organization of property rights entities, standardize the circulation of rural industrial land, strengthen the coordinating role of multilevel governments in the organization of rural industrial land, and establish a game-based supervision mechanism for market entities and property rights entities in the renewal of rural industrial land to prevent damage to public interests.

## 3. An Analysis of the Spatiotemporal Process of the Transition of Rural Industrial Land in Shunde District

Shunde District, one of five districts of Foshan, is located in the core area of the Guangdong–Hong Kong–Macao Greater Bay Area. The district currently governs 4 streets, 6 towns, and 205 villages, with a total area of 806 km$^2$ (Figure 4). After more than 40 years of development since the beginning of China's reform and opening up, it has been ranked first among the top 100 regions in China for eight consecutive years and has created two 300 billion industrial clusters of home appliances and machinery. In 2018, the GDP of the district was RMB 316.393 billion, and the third industrial structure was 1.4%:56.1%:42.5%. At the end of 2018, the permanent population of the district was 2,704,700, and the registered population was 1,452,600. In 1993, Shunde was one of the earliest districts in China to implement the system of rural joint-stock cooperative economic cooperatives; it embarked on a rural industrialization path based on the development of village-level industrial parks and merged more than 2000 production teams into 197 village-resident joint-stock cooperatives. Taking the lead in development means taking the lead in facing new development issues. By the end of 2017, 382 village-level industrial parks were scattered across 205 villages, covering a total area of 78.36 km$^2$ and accounting for 55.65% of the current industrial land in the district. There are more than 19,000 enterprises in rural industrial land, but they only contribute 27% of the output value and 4.3% of the tax revenue in the whole area. The average floor area ratio is only 0.78, and the fragmentation of the spatial pattern of industrial land is extremely prominent (Figure 5). Moreover, dilapidated factory buildings, low-end industry, and safety hazards have become prominent problems in the process of high-quality development in Shunde District.

The social forces represented by the grassroots village community and active market entities in the Pearl River Delta have jointly shaped a grassroots social and economic governance system with regional characteristics. The scope of government power is limited to a relatively small domain, and the government's ability to manage village-level affairs is relatively weak [16]. The bottom of the society is linked by village collectives and share cooperatives, and clan-like interest groups with small village communities are gradually formed. The village community has a strong sense of identity and belonging. The core members of the community have a close relationship with the two village committees or assume important positions in them, which further strengthens the grassroots rights system with "geographical relationships, blood relationships, industry relationships" as links, thus shaping a grassroots social organization and governance logic with local characteristics. The fact that the village community has mastered the stock construction land indicator shows that the village community has strong negotiation capital and game ability in the renewal of rural industrial land.

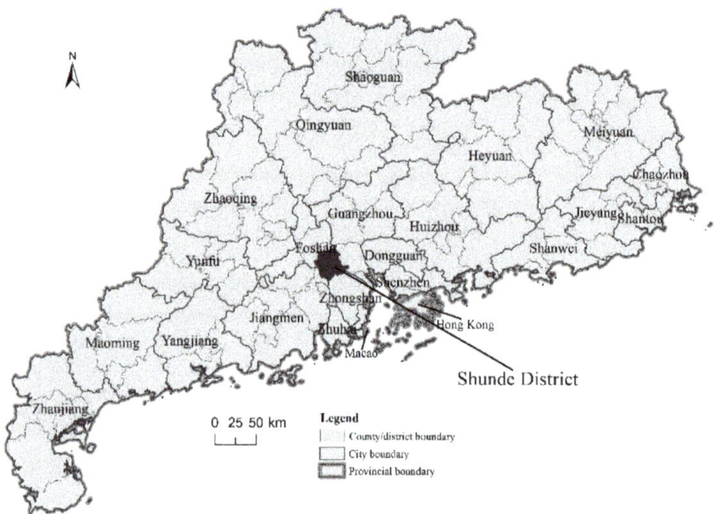



**Figure 4.** Location of Shunde District in Guangdong Province.

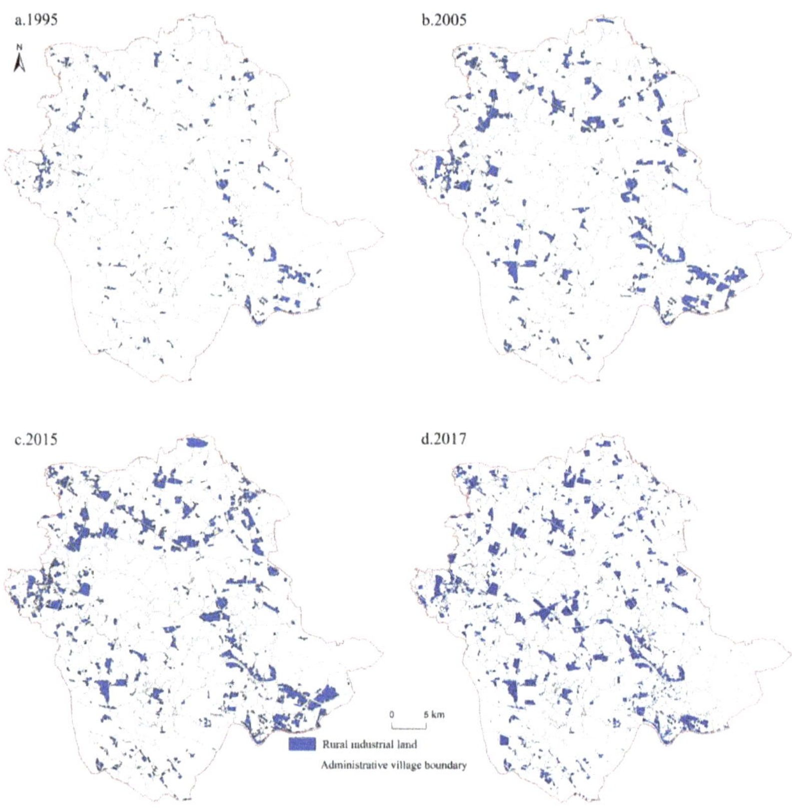



**Figure 5.** Rural industrial land of Shunde District in 1995 (**a**), 2005 (**b**), 2015 (**c**), 2017 (**d**).

### 3.1. The Process of Evolution of the Dominant Morphology of Rural Industrial Land in Shunde District

Since the 1990s, the dominant morphological changes to rural industrial land in Shunde District have mainly been manifested in three aspects: quantitative changes, spatial expansion, and structural changes. This process has, on the whole, resulted in the quantity of rural industrial land in Shunde District and its spatial scope increasing and the structural characteristics becoming more complex (Figure 5). In the 1990s, during the implementation of the land shareholding system, collectively contracted land continued to be transformed into industrial land. A large number of village-level industrial parks appeared in Shunde, but these were generally small in scale and scattered in distribution, and most were based on village groups. The rural land-share cooperation system established a rural land property rights system in which collective land rights are shared by collectives and farmers, creating conditions for village collectives to carry out rural industrialization through land leases and factory leases. However, there is a strong sense of community and clan relationships within the villages in the district. Rural industrialization often occurs within the villages, resulting in an excessive number of joint-stock cooperatives that are too small and the fragmentation of industrial land.

After 2000, with the further development of rural industrialization, the quantity of rural industrial land continued to increase, and the structural characteristics of industrial land became more complex. Contiguous industrial land has emerged, but its ownership still belongs to the original village collective or developer. From 1995 to 2005, the total area of rural industrial land in Shunde District increased from about 22.71 km$^2$ to about 62.00 km$^2$. The number of industrial sites rose sharply, while the scale of existing industrial sites continued to expand. In structure, in the process of expansion of scale, rural industrial land is restricted by the scale of village land use, resulting in the interconnection of land with different ownerships, and a large number of cross-village industrial parks have appeared. During this period, the number of smaller industrial parks started to decrease, and the number of larger industrial parks started to increase.

At the beginning of the implementation of the "Transformation of the Three Olds" policy, the amount of rural industrial land in Shunde District was still on the rise, but the restrictions on new development land resulted in fewer new industrial sites, and the scale of new industrial sites was generally small. Land-use restrictions have kept the scale of existing rural industrial sites unchanged, and only a small number of industrial parks have expanded in scale. By 2015, the total area of rural industrial land in Shunde District was 80.93 km$^2$. With the deepening of the implementation of the "Transformation of the Three Olds" and other policies, the amount of rural industrial land in Shunde District began to decline, and its scale began to decrease. The reason is that part of the industrial land has undergone land use conversion (such as conversion to residential land and commercial land) and ownership conversation (such as conversion to provincial industrial land). By 2018, the total area of rural industrial land in Shunde District had dropped to 78.36 km$^2$. Land consolidation reduced the number and scale of existing industrial sites, significantly changing the structure of rural industrial land, and at the same time promoting greater land-use efficiency.

### 3.2. The Process of Evolution of the Recessive Morphology of Rural Industrial Land in Shunde District

Changes in the recessive morphology of rural industrial land in Shunde District mainly involve property rights, organizational systems, and input–output efficiency. The implementation of the land shareholding system in the villages of Shunde District and the "Transformation of the Three Olds" carried out in the later period have significantly changed the property rights to rural industrial land, and prompted changes in the quantity, scale, and type of rural industrial land. In the process of rural industrialization, a diverse set of actors based on the three main groups of social entities (the original landowners, village collectives, etc.), market entities (developers, investors, consumers, etc.), and government entities (cities, districts, town governments, village committees, etc.) continue to compete with each other and divide, and their respective powers ebb and flow, which jointly promote the changes in the rural industrial land organization system in Shunde District. Extensive development and scale restrictions have caused problems, such as inefficient input and output of rural industrial land.

The land shareholding transition in Shunde District has promoted the transition of the recessive morphology of rural industrial land. The land shareholding system further separates collective land's contract right and management rights (the ownership of land belongs to the state or collective, and citizens have the contract right to own collective land through a contract and the management right to use the land and enjoy the benefits), allowing collective land and farmers' contracted land in the village to participate in the process of rural industrialization through a form of "land shareholding," changing the way of realizing property rights to village land and strengthening the trend toward commercialization of land property rights. In addition, in the process of rural industrialization, external entities have continuously participated in the development of rural industrial land, and the land-use organization system has changed from one of single to multiple differentiation. Market players participate in the development and construction of rural industrial land. In the process, foreign entrepreneurs have further strengthened the gaming role of diversified market players in the development of rural industrial land. The multilevel government entities are leaders of changes in the organizational system of rural industrial land, the social entities are both providers of industrial land and participants in the process of industrialization, and the market entities are one of the core forces promoting this process. From the perspective of input–output efficiency, with the conversion of local land-use properties from agricultural land and residential land to industrial land, the input–output efficiency of land has been greatly improved, but disordered expansion has also brought with it waste of land resources and inefficient land use.

The "Transformation of the Three Olds" has profoundly changed the morphology of rural industrial land in Shunde District. The "Transformation of the Three Olds" allows for the improvement of the historical land-use procedures according to the status quo, the use of agreements for land transfer [21], and the value-added benefits of land redevelopment to be shared between the government and the rights-holders. From the perspective of property rights, the "Transformation of the Three Olds" clarified the property rights to existing land, rectified the land that did not meet the development plan, integrated part of the fragmented land, and re-developed low-efficiency land, which promoted clarity of property rights (Figure 6). From the perspective of the organizational system, the "Transformation of the Three Olds" uses a combination of "top-down" and "bottom-up" governance methods to build a multiple-game mechanism of multiple subjects. In addition, the "Transformation of the Three Olds" has effectively improved the input and output efficiency of rural industrial land, and the efficiency of land use and degree of intensification have been continuously improved.

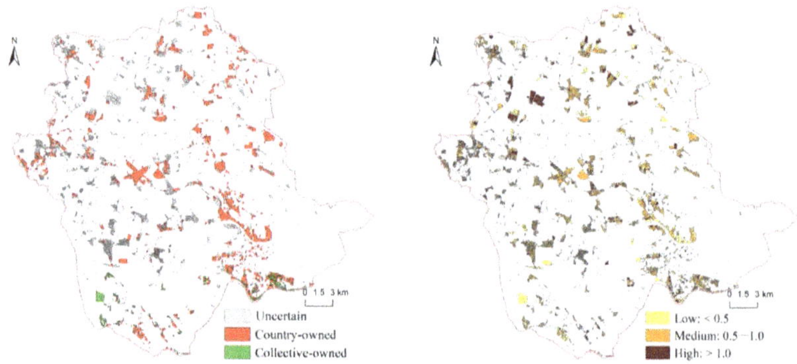


**Figure 6.** Ownership and plot ratio of rural industrial land in Shunde District (2018).

### 3.3. An Analysis of the Dilemma Involved in the Transition of Rural Industrial Land in Shunde District from the Perspective of Spatial Governance

In the process of rural industrialization, the transition of rural industrial land in Shunde District is faced with problems such as fragmentation of the land pattern, vague property rights, and entanglement of interests. The fragmented pattern caused by the decentralized development of village-level units has caused problems such as inefficient input and output of land, waste of land resources, limited development scale, and pollution of the natural environment of human settlements. Various types of ownership of industrial land are intertwined with each other, compete with each other in a disordered manner, and are often restricted in different ways. The property rights to some lands are not clear, which has created huge difficulties for the sound development of industrial land in the region. At the same time, the interests of government entities, social entities, and market entities are entangled with each other, and the collective consciousness of local clans is strong, further highlighting the urgency of developing comprehensive spatial governance. The large amount and complex structure of rural industrial land in Shunde District and the intertwining of the interests of multiple subjects have become important obstacles restricting the transition and upgrading of rural industrial land. In the face of these problems, Shunde District has adopted diversified governance ideas and achieved certain results, but there are still many problems. The problems faced in the transition of rural industrial land can be summarized as the "incomplete comprehensive governance system," "inadequate integration of upper and lower levels," and "inadequate participation of multiple subjects".

Faced with the problems of the large number of village-level industrial parks, their diversity of types, scattered layout, inefficient use of land, and weak industrial upgrading, the existing policy of the "Transformation of the Three Olds" alone cannot achieve comprehensive management of large amounts of rural industrial land. The disordered expansion of rural industrial land and the infringement and occupation of land have been effectively curbed in the gradual strengthening of land-use control. However, the scale of industrial parks with village-level units is often greatly restricted, and large-scale development cannot be achieved. Measures such as land-use improvement and restoration of the ecological environment based on material effective space governance cannot play a positive role. Complicated property rights and a scattered organizational system for rural industrial land have become important obstacles acting as a check on the large-scale governance of rural industrial land. The "Transformation of the Three Olds" is based on a plan to identify village-level current land use. It is a recognition of the unreasonable land use in the past, which further strengthens the village community's awareness of its rights and increases the difficulty of managing property rights relationships with respect to industrial land. The lack of coordination between the governance of industrial land property rights and the organization and governance of industrial land stakeholders further highlights the problem of rural industrial land governance in Shunde District. In addition, the inability to break through the scattered distribution of village-level industrial land means that there is still a long way to go in the governance of rural industrial land.

In the process of implementing rural industrial land governance, the divergence between "top-down" rigid control and "bottom-up" flexible governance has become an important factor restricting the transition of rural industrial land. It is difficult for the government to implement the relevant national standards for the "double compliance" of industrial land. Due to the interleaved layout of "production and living spaces" between residential buildings and industrial plants, it is difficult to meet the standard that the distance between residential buildings and industrial plants should exceed 50 m. As rural industrial areas cannot connect to the municipal pipe network, it is more difficult to achieve environmentally friendly discharge of industrial wastewater. The strongly conflicting relationship between the government and the grassroots entities in the implementation of space-control policies makes it difficult to promote industrial land governance. The asymmetry of information has increased the villagers' distrust of government policies. In today's game the village community has the negotiation advantage about land governance, which makes it difficult to advance industrial land governance. The core of the above problems is insufficient "linkage of upper and lower levels" of industrial land governance. It would be beneficial to promote the realization of industrial land governance by paying attention to realistic demands for flexible governance at the grassroots level and avoiding the governance logic of "one-size-fits-all".

Insufficient participation of multiple subjects and unsmooth game mechanisms are important factors influencing the dilemma of rural industrial land transition in Shunde District. Even in response to the call for the "Transformation of the Three Olds," the original problems of rural industrial land still exist. Moreover, the "Transformation of the Three Olds" itself faces many problems. For example, 80% of the renovation projects are of a real estate development type, and most of the projects implemented are the renovation of old factories. Among the approved "Transformation of the Three Olds" projects, 78% of the projects are demolitions and reconstructions, resulting in a substantial increase in development intensity, and there are difficulties in "organic renewal". An in-depth study of the internal mechanisms behind this would need to pay close attention to the difficulty of effectively balancing the interests of the multiple subjects involved in spatial governance. Against the background that the land finance is unsustainable, the main body of the government hopes to revitalize the existing land resources but lacks sufficient funds for renewal and transformation. Therefore, it is necessary to introduce market capital for governance actions, but then the governance model of retreating from industrial land to real estate land with a higher rate of return emerges. With the characteristics of pursuing interests, market entities, for enterprises (households), terms due to issues such as land, contracts, leases, etc., it is impossible to invest capital for upgrading and transformation under the current land lease mechanism. However, social capital is rational with respect to economic benefits, and social capital in the form of high subsidies under the drive of interests is used to promote the development of real estate projects. Different village collectives and different interest groups within the village collectives are entangled with each other due to their internal clan ties, community consciousness, and interest relations, which have created huge obstacles to the renovation of local industrial land and the large-scale and collaborative development of rural industry. The dividends due to stockholders on account of industrial land governance may be stranded or damaged in a short period of time, making the villagers more resistant. It is difficult for multiple subjects to achieve a balance among the demands for their interests in industrial land, and it is difficult to resolve contradictions; this makes it difficult to achieve a breakthrough in the governance dilemma of rural industrial land in a short period of time.

## 4. The Transition and Upgrading of Rural Industrial Land Based on Spatial Governance: Taking Hengding Industrial Park as an Example

### 4.1. The Process of Transition and Upgrading of Rural Industrial Land in Hengding Industrial Park

Hengding Industrial Park is located within the first phase of Ronggui Huakou Environmental Protection Industrial Park and is located in the neighborhood committee of Huakou Community, Ronggui Street, with a project area of 43,147.23 m$^2$. Its predecessor was Ronggui Huakou Electroplating City, which was built around 2000 and was one of the designated electroplating production bases in Shunde. After years of development, the 26 metal-surface processing enterprises in the electroplating city had become scattered. The electroplating city's sewage and waste gas treatment system was unable to meet modern environmental protection requirements, causing risks to the environmental protection of the surrounding environment. The average plot ratio was 0.43, and the land-use efficiency was relatively low. Due to the relatively dispersed distribution and small scale of enterprises, the lack of rigorous management of wastewater discharge within the enterprise, and the unclassified or imperfect classification of electroplating wastewater in some enterprises, the cost and difficulty of wastewater treatment increased (Figure 7). Excessive discharge occurred from time to time and caused dissatisfaction among the surrounding people. The land used in the park belongs to the collective land of the village. During the development of the park, there were phenomena such as land subleases, and the situation of the "second landlord" became more common. In the face of the assessment pressure of "double compliance" and the awakening of public awareness of environmental protection, how to effectively promote the transition and upgrading of the park became the common goal of all parties.

**Figure 7.** Before and after comparison of the renewal of Hengding Industrial Park (photo source: Urban Renewal Bureau of Shunde District).

In 2012, Hengding Industrial Park realized the reconstruction of industrial land property rights through the "Transformation of the Three Olds" policy and improved the procedures for historical land use. In August 2013, it was transferred to Hengding Investment Co., Ltd. through a public transaction for development by the Hengding Company. The high-standard factory buildings were built only to be rented but not sold. On the one hand, the successful transformation of the project solved the remaining problems and integrated scattered land into high-quality industrial land. High-standard factory buildings were built, and industry was moved into buildings, greatly improving the efficiency of land use and providing sufficient space for industrial transformation and upgrading. On the other hand, through the construction of high-standard wastewater and waste gas treatment facilities, the wastewater and waste gas in the park are treated in a unified and centralized manner, and the environmental protection requirements are strictly implemented, which fundamentally solves the environmental protection problems of electroplating enterprises. After completion, the Hengding Industrial Park has attracted a large number of electroplating companies to settle in. Currently, 16 electroplating companies have set up operations, using 58 floors of factory buildings. Hengding Industrial Park has become a successful example of rural industrial land in Shunde District retreating from use as industrial land to serve as an industrial development space. An analysis of the transition of the park's industrial land may serve as a representative example for other districts to emulate.

### 4.2. An Analysis of the Mechanism of Transition of the Hengding Industrial Park from a Spatial Governance Perspective

The spatial governance process combining "top-down" and "bottom-up" approaches is an important means for the transformation and upgrading of industrial land. Hengding Industrial Park did not have legal land-use procedures before the implementation of the "Transformation of the Three Olds" policy. It belongs to the self-use state-owned construction land of Ronggui Street Huakou Cooperative Economic Cooperative. The ownership relationship is mixed. The land-use procedures of superimposing multiple layers of subleasing have resulted in complicated industrial land property rights. In 2012, the land-use procedures of Hengding Industrial Park were improved through the enhancement of land acquisition compensation procedures, and the historical land-use procedures were supplemented and perfected, and transformed into legal state-owned industrial land (the government applies the current construction land for approval for the land included in the scope of "Transformation of the Three Olds," exempting the formalities for agricultural conversion). A successful transition of land use is the result of the comprehensive effect of the "top-down" policy transmission and the "bottom-up" multi-subject game. "Top-down" use-control transmission includes both the "rigid constraints" of policies (such as environmental protection assessment, use control, etc.) and "flexible guidance" (such as supplementary land-use procedures, plot ratio compensation, etc.). The "bottom-up" space governance is mainly manifested in the protection of basic-level utilization demands and rights protection under reasonable circumstances. By guiding the development of the whole block of regional land, the government has carried out the upgrading plans of industrial parks to realize the transition and upgrading of industrial land. The "top-down" and "bottom-up" negotiation and communication mechanisms in the transition and upgrading of Hengding's industrial land are the guarantee for the smooth development of spatial governance. In this process, the property rights of rural industrial land are more clearly defined, and the fragmentation and inefficient characteristics are ameliorated.

The effective participation of "multiple subjects" is one of the internal reasons for the successful transition of Hengding Industrial Park. For a long time, village-level industrial land has had problems such as difficulty in transformation, few successful cases, and differences from the expectations of the "Transformation of the Three Olds" policy. The success of Hengding Industrial Park is closely related to the introduction of multiple subjects in the transformation process to participate in spatial governance. In the process of optimizing the industrial structure for rural industrial land, it is first necessary to coordinate the relationship among stakeholders such as multiple levels of government, new land developers, members of joint-stock cooperatives, village collective economic organizations, plant contractors (or second or third landlords). In the transition of industrial land into a park, the property-dividend mechanism is given to the president of the stock company. The Hengding Company, the park developer, has strengthened land output and environmental protection control by obtaining land-use rights, effectively attracting electroplating companies to settle in. Electroplating enterprises have effectively reduced their production costs due to the optimal environmental protection and industrial facilities in the park. Moreover, the stable land-use rights have also increased the investment confidence of the settled enterprises. The upgrading and transition of the industrial park have concentrated on solving the environmental protection problems of the enterprises so that there need be no concerns regarding their development. Multiple levels of government play important roles in negotiation and communication in the transition of industrial parks. They are not only participants in the game of multiple subjects but also active guides and promoters. The successful transition of the industrial park has also become a business card for the local government to publicize, inspiring other regions to learn from and imitate it. The effective participation of "multiple subjects" changed the rural industrial land organization system and mode of operation and promoted the effective implementation of spatial ownership governance and organizational governance.

Hengding Industrial Park's land transition process through the comprehensive management of space showed different characteristics in its various stages. Before the comprehensive management of the space was carried out, the land structure of the plot was broken, the ownership relationship was mixed, and the organizational system was chaotic, and other morphological characteristics had become important obstacles to rural development. Through the implementation of spatial governance measures combining "top-down" and "bottom-up" approaches, the use of the land has been significantly changed, especially the changes in land property rights, structural characteristics, and organizational models, which have brought an improvement in the efficiency of land-use inputs and outputs (Figure 8). Correspondingly, the face of rural development has also been significantly improved. The improvement of village ecological environment quality and the reconstruction of the social network are advancing in parallel, and the rural governance system and the level of rural industrial development have reached new levels. The benign interaction between land-use transition and rural development can be realized through spatial governance.

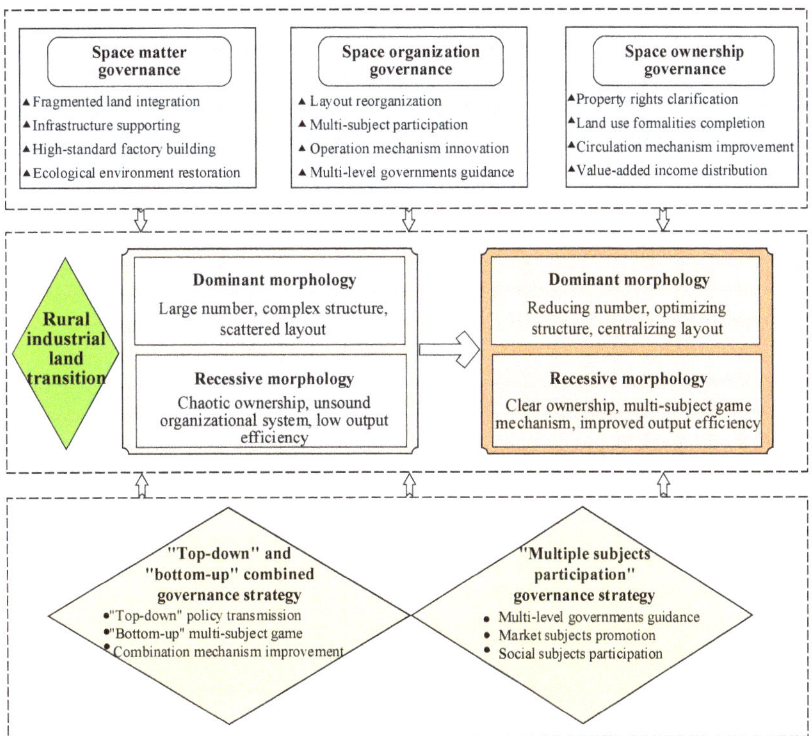


**Figure 8.** The transition mechanism of Hengding Industrial Park under spatial governance.

## 5. Discussion

*5.1. The Experience and Enlightenment of Rural Revitalization in the Transition of Industrial Land in Semi-Urbanized Areas*

This article first offers a theoretical analysis of the transition of rural industrial land from a spatial governance perspective and then attempts to analyze the process of transition and the existing problems concerning rural industrial land in Shunde District. It analyzes the internal mechanism of the comprehensive treatment of space represented by the "Transformation of the Three Olds" policy in combination with case studies of typical areas to promote the transition of rural industrial land. The study found that, as the frontier area of China's urbanization, rural development in semi-urbanized areas—represented by the development of rural industrial land in Shunde District—is constantly demanding breakthroughs in the face of challenges. Through spatial governance, we can change the morphology of rural land use, optimize the structure and functional system of rural space use, and fully explore the value characteristics and realization methods of rural space, which has brought sufficient funds for rural development [6]. The semi-urbanization of rural areas is a notable feature of the process of development of rural space in the Pearl River Delta. It is closely related to the urbanization and industrialization process of the area and is also closely linked to the rural governance system represented by clan governance [1]. Summarizing the revitalization experience represented by rural industrialization in semi-urbanized areas is of practical significance for inspiring other regions to carry out rural revitalization [31].

Formal management by the government and non-confrontational policy breakthroughs at the bottom appear historically to be necessary for unconventional development and breakthroughs in rural areas [32]. The historical process of the development of rural industrial land in the Pearl River Delta is accompanied by the emergence of a large amount of illegal land. In the context of a specific historical development, the government's ineffective control—or even negligence in control or acquiescence in the social and economic environment—no longer exists [33]. The policy improvements and remedial measures represented by the "Transformation of the Three Olds" have alleviated the current "strong conflict" or contradiction between government control and grassroots development to a certain extent. At present, China is striving to build a spatial governance system represented by a "single picture" of all aspects of land and space management while strengthening the transmission of "top-down" use control and strengthening the control and restraint of the underlying space [34]. Therefore, in trying to tackle "non-compliant" development, reducing the development space gradually, absorbing the reasonable development demands of the bottom in the process of high-quality urban and rural development, honoring the spatial rights of rural development, and connecting "bottom-up" innovation and "top-down" management have important guiding significance for the development of rural revitalization in other areas [8].

The development path of rural areas in semi-urbanized areas fully demonstrates that enhancing and manifesting the value of rural space is an important means of achieving rural revitalization [35]. The industrial land that was developed from a large amount of agricultural land in semi-urbanized areas served as the basic space for rural leapfrog development in the Delta area. Unlike the Yangtze River Delta, in which rural industrial land is gradually becoming concentrated in towns and parks, there is still a large amount of rural industrial land distributed within the villages in this area. The core differences are related to the management and control methods and implementation efforts of different regions, as well as the ability and operational methods of rural industrial land. The strong gaming and bargaining capabilities of the village community represented by the clans in the Pearl River Delta region are important driving factors in ensuring that rural industry continues to bring local benefits. Rural industrial land has become an important source of income and game capital for farmers, and it also enhances the value of rural space to a certain extent [36]. The spatial governance process represented by the "Transformation of the Three Olds" brings about the process of transition of industrial land and is also an important manifestation of the evolution of the rural spatial structure system, value characteristics, and functional effects. In the process of rural revitalization, learning from semi-urbanized areas, promoting the manifestation of rural space value, and innovating the realization of rural space value provide an important material basis for ensuring the rural revitalization strategy.

### 5.2. A Discussion of the Path of Spatial Governance Leading to Rural Revitalization

The key to the governance of rural space is coordinating multiple strategies through multiple channels to implement and encourage multiple entities to participate in governance to ensure the achievement of tangible results. The governance of rural material space is the "fulcrum", and the spatial organization and ownership governance are the "levers". Rural spatial reconstruction, organizational reconstruction, and ownership remodeling can be realized through spatial governance. Coordinated, comprehensive governance of the three is the guarantee for realizing the improvement of rural space in terms of its ability to add value, organizational perfection, and efficiency. In the current national governance system, there is still a need for further improvement of the governance system for rural space. In the process of improving the system and its mechanisms, opening up a path for multiple subjects to participate in the governance of rural space and building a spatial organization model that serves rural revitalization will help consolidate the foundation for rural development [4].

Rural spatial governance should tap the potential for spatial development, consolidate the material foundation for rural revitalization, and cultivate the rural endogenous development momentum, internal organizational strength, and resilience of the system [37]. Rural spatial governance aims to reduce the irrational use of space such as homesteads, agricultural land, collective operating land, rural industrial land, public service land, and ecological land through optimizing the trend of land-use transition so that the adjustment of land-use structures and optimization of functions are realized. The governance of rural spatial organization mobilizes farmers' enthusiasm for participating in rural revitalization by rebuilding the rural spatial relationship network and reorganizing the spatial organizational and operational system [38]. Rural spatial organization strengthens the leadership of talent, the linkage of organizations, and the interaction between urban and rural areas, which is conducive to promoting industrial development, cultivation of personnel, organizational revitalization, cultural inheritance, and the implementation of rural revitalization goals [39]. By clarifying the relationship of space property rights, the governance of rural space ownership clarifies the economic interests of multiple subjects to establish the distribution mechanism of rural development rights, defines the boundary of public and private space, and builds a rural space ownership system with clear rights and responsibilities [40]. Through ownership governance, rural spatial governance improves the rural space value system, expands the ways in which space value can be realized, and enhances the efficiency of distribution of the value of space.

The benign process of interaction between land-use transition and rural development based on spatial governance will help promote the realization of rural revitalization goals. This study found that the benign coupling state of land-use patterns and rural development status promotes rural development, whereas the reverse inhibits rural progress. In the process of urban–rural transition and development, coordinating the relationship between urbanization and rural revitalization will be an important part of ensuring the sustainable development of rural areas and the social stability of the transition. Rural spatial governance aims to tackle the problems that arise in the development and utilization of rural space, starting with a variety of governance methods, strengthening the comprehensive governance of rural space, promoting the optimization of the coupling state of rural land-use transition and rural development, and ensuring the implementation of rural revitalization goals.

At present, there is insufficient analysis of typical cases of industrial land at the microscale. The existing research on rural industrial land lacks an analysis of spatio-temporal and characteristics and transition laws. Furthermore, it pays less attention to the role of multiple subjects in the transition of rural industrial land. This article starts with the theory of land use transition oriented by rural space governance, combined with the transition process of rural industrial land in Shunde District, deeply analyzes the internal mechanism of the transition of rural industrial land oriented, and discusses the enlightenment of the transition and development of rural industry in semi-urban areas. The paper analyzes the governance mechanism of the combination of "top-down" and "bottom-up", while enriching the theoretical connotation of rural spatial governance.

## 6. Conclusions

Starting with the construction of an analytical framework for understanding rural land-use transition based on rural spatial governance, in combination with case studies of the process of rural industrial land transition in Shunde District and Hengding Industrial Park, this article deeply analyzed the internal mechanism of rural industrial land transition from the perspective of rural spatial governance and discussed the promise that the transition and development of rural industries in semi-urbanized regions can bring to rural revitalization in other regions.

The conclusions are as follows:

(1)  In the process of "bottom-up" urbanization and industrialization, problems such as the fragmentation of rural space, difficulties in the renewal of rural industrial land,

(2) Since the 1990s, measures such as the "land shareholding system" and the "Transformation of the Three Olds" have significantly changed the dominant and recessive morphology of rural industrial land in Shunde District. The changes in the dominant morphology are reflected in the quantity and structure, and the changes in the recessive morphology are reflected in the property relations, organizational system, and input–output efficiency.

(3) The analytical framework based on the "matter-ownership-organization" comprehensive management of rural space is an important starting point for analyzing the process of transition of rural industrial land and exploring its transition path. The combination of "top-down" and "bottom-up" rural spatial governance strategies and the effective participation of multiple subjects are important means of promoting the transition of rural industrial land.

(4) Rural spatial governance is ultimately conducive to promoting the transition of rural land use and the healthy development of rural areas by promoting rural spatial reconstruction, organizational reconstruction, and ownership remodeling. The revitalization experience represented by rural industrialization in semi-urbanized areas can serve as an important example for the transition and development of other rural areas.

> The text begins at top: disorder of ownership, and an incomplete mechanism for the differentiation and game of multiple subjects, are the problems of rural areas in Shunde District. These typical characteristics mean that there are many challenges in the transition of rural industrial land in this area.

**Author Contributions:** Conceptualization, D.G. and L.Z.; methodology, L.Z. and D.G.; formal analysis, L.Z., D.G., P.S. and D.S.; investigation, L.Z. and D.S.; writing—original draft preparation, L.Z. and D.G.; writing—review and editing, D.G. and P.S.; project administration, D.G.; funding acquisition, D.G. All authors have read and agreed to the published version of the manuscript.

**Funding:** This research was funded by "the National Natural Science Foundation of China" (grant number 41901204), "The Foundation of Humanity and Social Sciences of the Ministry of Education of China" (grant number 19YJCZH036), "China Postdoctoral Science Foundation" (grant number 2019M660109, 2021T140303), "Jiangsu Provincial Science Foundation" (grant number BK20190717), and "Jiangsu Provincial Social Science Foundation" (grant number 19GLC002).

**Institutional Review Board Statement:** Not applicable.

**Informed Consent Statement:** Not applicable.

**Data Availability Statement:** Not applicable.

**Conflicts of Interest:** The authors declare no conflict of interest.

## References

1. Grainger, A. National land use morphology: Patterns and possibilities. *Geography* **1995**, *80*, 235–245.
2. Long, H.; Qu, Y. Land use transitions and land management: A mutual feedback perspective. *Land Use Policy* **2017**, *74*, 111–120. [CrossRef]
3. Mather, A. Forest theory and the reforesting of Scoltland. *Scolttish Geogr. J.* **2004**, 83–98. [CrossRef]
4. Lambin, E.F.; Meyfroidt, P. Land use transitions: Socio-ecological feedback versus socio-economic change. *Land Use Policy* **2010**, *270*, 108–118. [CrossRef]
5. Zhou, L.; Dang, X.; Haowei, M.U.; Wang, B.; Wang, S. Cities are going uphill: Slope gradient analysis of urban expansion and its driving factors in China. *Sci. Total Environ.* **2020**, *709*, 1–16.
6. Ge, D.; Zhou, G.; Qiao, W.; Yang, M. Land use transition and rural spatial governance: Mechanism, framework and perspectives. *J. Geogr. Sci.* **2020**, *30*, 1325–1340. [CrossRef]
7. Long, H.; Zhang, Y.; Tu, S. Rural vitalization in China: A perspective of land consolidation. *J. Geogr. Sci.* **2019**, *29*, 517–530. [CrossRef]
8. Ge, D.; Wang, Z.; Tu, S.; Long, H.; Yang, H.; Sun, D.; Qiao, W. Coupling analysis of greenhouse-led farmland transition and rural transformation development in China's traditional farming area: A case of Qingzhou City. *Land Use Policy* **2019**, *86*, 113–125. [CrossRef]
9. Cong, Y.; Wei, L. Review of land issues since 2003. *City Plan. Rev.* **2008**, *9*, 57–62.

10. Long, H.; Qu, Y.; Tu, S.; Zhang, Y.; Jiang, Y. Development of land use transitions research in China. *J. Geogr. Sci.* **2020**, *30*, 1195–1214. [CrossRef]
11. Long, H.; Tu, S.; Ge, D.; Li, T.; Liu, Y. The allocation and management of critical resources in rural China under restructuring: Problems and prospects. *J. Rural Stud.* **2016**, *47*, 392–412. [CrossRef]
12. Kong, X.; Liu, Y.; Jiang, P.; Tian, Y.; Zou, Y. A novel framework for rural homestead land transfer under collective ownership in China. *Land Use Policy* **2018**, *78*, 138–146. [CrossRef]
13. Zou, B.; Mishra, A.K.; Luo, B. Grain subsidy, off-farm labor supply and farmland leasing: Evidence from China. *China Econ. Rev.* **2020**, *62*. [CrossRef]
14. Yin, J.; Zhao, X.; Zhang, W.J.; Wang, P. Rural land use change driven by informal industrialization: Evidence from Fengzhuang Village in China. *Land* **2020**, *9*, 190. [CrossRef]
15. Jiang, G.; Ma, W. Agglomeration or dispersion? Industrial land-use pattern and its impacts in rural areas from China's township and village enterprises perspective. *J. Clean. Prod.* **2017**, 207–219. [CrossRef]
16. Yang, R.; Chen, Y.; Xu, Q. Evolution of rural industrial land use in semi-urbanized areas and its multi-dynamic mechanism: A case study of Shunde district in Foshan city. *Sci. Geogr. Sin.* **2018**, *38*, 511–521.
17. Hansen, A.J.; Brown, D.G. Land-use change in rural America: Rates, drivers, and consequences. *Ecol. Appl.* **2005**, *15*, 1849–1850. [CrossRef]
18. Zhang, C.; Kuang, W.; Wu, J.; Liu, J.; Tian, H. Industrial land expansion in rural China threatens environmental securities. *Front. Environ. Sci. Eng.* **2021**, *15*, 29. [CrossRef]
19. Liu, Z.; Huang, H.; Werners, S.E.; Yan, D. Construction area expansion in relation to economic-demographic development and land resource in the Pearl River Delta of China. *J. Geogr. Sci.* **2016**, *37*, 1394–1402. [CrossRef]
20. Guo, Y.; Yuan, Q.; Qiu, J. Hold-up problems in urban regeneration and implication for urban planning: A case study from Pearl River Delta region. *Urban Plan. Int.* **2016**, *31*, 95–101. [CrossRef]
21. Tian, L.; Luo, C. Land share-holding system and land-use in industrialization of countryside: A case study on Shunde. *City Plan. Rev.* **2012**, *36*, 25–31.
22. Tian, L.; Liang, Y.; Zhang, B. Measuring residential and industrial land use mix in the peri-urban areas of China. *Land Use Policy* **2017**, *69*, 427–438. [CrossRef]
23. Huang, X.; Li, H.; Zhang, X.; Zhang, X. Land use policy as an instrument of rural resilience—The case of land withdrawal mechanism for rural homesteads in China. *Ecol. Indic.* **2018**, *87*, 47–55. [CrossRef]
24. Yang, R.; Liu, Y.; Long, H.; Qiao, L. Spatio-temporal characteristics of rural settlements and land use in the Bohai Rim of China. *J. Geogr. Sci.* **2015**, *25*, 559–572. [CrossRef]
25. Yang, R. Spatial differentiation and mechanisms of typical rural areas in the suburbs of a metropolis: A case study of Beicun Village, Baiyun District, Guangzhou. *Acta Geogr. Sin.* **2019**, *74*, 1622–1636.
26. Tian, L.; Ge, B. Land use characteristics and driving forces of Peri-urban area in the transitional economy. *Urban Plan. Forum* **2011**, *35*, 66–73.
27. Shen, D.; Lv, X.; Wang, B. The role of government in rural construction land marketization: Based on the content analysis of contracts. *China Land Sci.* **2019**, *33*, 34–41.
28. Gedefaw, A.A.; Atzberger, C.; Seher, W.; Agegnehu, S.K.; Mansberger, R. Effects of land certification for rural farm households in Ethiopia: Evidence from Gozamin District, Ethiopia. *Land* **2020**, *9*, 421. [CrossRef]
29. Sun, P.; Zhou, L.; Ge, D.; Lu, X.; Sun, D.; Lu, M.; Qiao, W. How does spatial governance drive rural development China's farming areas? *Habitat Int.* **2021**, *109*, 102320. [CrossRef]
30. Ge, D.; Long, H. Rural spatial governance and urban- rural integration development. *Acta Geogr. Sin.* **2020**, *75*, 1272–1286.
31. Tu, S.; Long, H. Rural restructuring in China: Theory, approaches and research prospect. *J. Geogr. Sci.* **2017**, *27*, 1169–1184. [CrossRef]
32. van Vliet, J.; Eitelberg, D.A.; Verburg, P.H. A global analysis of land take in cropland areas and production displacement from urbanization. *Glob. Environ. Chang.* **2017**, *43*, 107–115. [CrossRef]
33. Nguyen, H.L.; Duan, J.; Zhang, G.Q. Land politics under market socialism: The state, land policies, and rural-urban land conversion in China and Vietnam. *Land* **2018**, *7*, 51. [CrossRef]
34. Ge, D.; Long, H.; Zhang, Y.; Li, M.; Li, T. Farmland transition and its influences on grain production in China. *Land Use Policy* **2018**, *70*, 94–105. [CrossRef]
35. Long, H.; Zhang, Y. Rural planning in China: Evolving theories, approaches, and trends. *Plan. Theory Pract.* **2020**, *21*, 782–786.
36. Navarro, F.A.; Woods, M.; Cejudo, E. The LEADER initiative has been a victim of its own success. The decline of the bottom-up approach in rural development programmes. The cases of Wales and Andalusia. *Sociol. Rural.* **2016**, *56*, 270–288. [CrossRef]
37. Chen, Y.; Liu, Y.; Li, Y. Agricultural development status and industrial prosperity path under the relationship between urban and rural areas. *Geogr. Res.* **2019**, *38*, 632–642.
38. Liu, Y.; Li, Y. Revitalize the world's countryside. *Nature* **2017**, *548*, 275–277. [CrossRef]
39. Castro-Arce, K.; Vanclay, F. Transformative social innovation for sustainable rural development: An analytical framework to assist community-based initiatives. *J. Rural Stud.* **2020**, *74*, 45–54. [CrossRef]
40. Xiao, P. A research on right structure of tripartite rural land entitlement system. *China Land Sci.* **2018**, *32*, 24–29.

*Article*

# Land-Use Transition of Tourist Villages in the Metropolitan Suburbs and Its Driving Forces: A Case Study of She Village in Nanjing City, China

Yibei Chai [1,2], Weifeng Qiao [1,2,3,*], Yi Hu [1,2], Tianqi He [1,2], Kaiyang Jia [1,2], Ting Feng [1,2] and Yahua Wang [1,2,3]

[1] School of Geography, Nanjing Normal University, Nanjing 210023, China; 191302139@njnu.edu.cn (Y.C.); 191302141@njnu.edu.cn (Y.H.); 191302143@njnu.edu.cn (T.H.); 181302148@njnu.edu.cn (K.J.); 181302144@stu.njnu.edu.cn (T.F.); 09123@njnu.edu.cn (Y.W.)

[2] College of Rural Vitalization, Nanjing Normal University, Nanjing 210023, China

[3] Jiangsu Center for Collaborative Innovation in Geographical Information Resource Development and Application, Nanjing 210023, China

[*] Correspondence: qiaoweifeng@njnu.edu.cn

**Abstract:** In the context of the transition from "Native-rural China" to "Urban-rural China", suburban villages have undergone rapid reconstruction of format, industry, and function. Aiming to reveal the evolution characteristics and driving forces, this study selected She Village, located in suburban areas of Nanjing, to analyze the changes of both dominant and recessive morphology of land use by employing participatory rural appraisal, remote sensing, and geographic information systems. The results showed that She Village witnessed three stages, including industrial development, ecological restoration, and service industry development, from 1980 to 2018, with more diversified management modes, multifunctional land use, and intensified land fragmentation. The drivers included natural resources, population growth, policy of Grain for Green, urban market demand, etc., the intensity of which showed trends of "increase–increase–increase", "increase–decrease–decrease", "periodically intermittent", and "increase–decrease–increase" in turn. The tourist villages undergo three stages of industrial development, agricultural development, and service industry development, with periodical characteristics driven by top-down policies, the endogenous force of the village, and the radiation and diffusion of the city. This research deepens the understanding of the development process of suburban villages and provides a reference for land policy making and planning in other similar villages.

**Keywords:** rural transformation; land-use morphology; rural multifunction; spatial differentiation; impact intensity

**Citation:** Chai, Y.; Qiao, W.; Hu, Y.; He, T.; Jia, K.; Feng, T.; Wang, Y. Land-Use Transition of Tourist Villages in the Metropolitan Suburbs and Its Driving Forces: A Case Study of She Village in Nanjing City, China. *Land* **2021**, *10*, 168. https://doi.org/10.3390/land10020168

Academic Editor: Xiangbin Kong

Received: 19 January 2021
Accepted: 3 February 2021
Published: 6 February 2021

**Publisher's Note:** MDPI stays neutral with regard to jurisdictional claims in published maps and institutional affiliations.

## 1. Introduction

With industrialization, urbanization, informatization, and agricultural modernization, the once-solidified urban–rural dualistic structure is being deconstructed [1]. The production factors between rural and urban areas have turned into a two-way flow from a one-way flow, and China has been transferred from "Native-rural China" to "Urban-rural China" [2], in which the rural system has been characterized by rapid reconstruction of economic, social, and comprehensive dimensions [3]; heterogeneity [4]; consumerization and capitalization [5] of rural space; and multifunctionalization of agriculture and village [6], etc. After the introduction of the rural revitalization strategy, many topics have been put forward on rural transformation [7–10]. In research of rural transformation, land-use change, as a projection of social and economic development spatially, can directly show the stages and issues of rural transformation and development [11,12]. Today, land-use transition has become an important aspect to study rural transformation, thanks to deepened exploration in geography, land-use morphology, and other relevant disciplines. On one hand, the current theoretical research on land-use transition at a village scale largely focuses on the coupling relationship between land-use transition and rural transformation [13] and rural economy [14], on the relationship between land-use transition and rural revitalization [15] and rural spatial governance [16], and also on analysis framework of land-use transition in certain regions [17], etc. On the other hand, empirical research mainly includes the evolution and mechanism of spatial differentiation and reconstruction [3,18,19], the land-use change of typical villages [20–22], and transformation of land-use function [23–25]. However, despite great emphasis put on the dominant morphology of land use in the study of rural land-use transition, the current pool of research on recessive morphology is limited to the function of land use and ignores the roles played by other recessive factors that also contribute to rural transformation, such as the management modes and land ownership. In addition, the current research has only been committed to single land-use morphology, but produced hardly any research results in a deepened exploration of land-use transition when it comes to a dual perspective of dominant and recessive morphology. Therefore, an analysis with a dual perspective at the village scale can further contribute to the research on rural land-use transition.

Suburban villages, located in marginal areas of a city, where the urban and rural production factors are actively allocated, are affected by both the rural and urban systems [26]. Due to their characteristics of mixed population composition, diverse industrial structures, and large gradients of landscape space [27], these villages have emerged as the most active area of rural transformation and development. In the context of rapid agricultural modernization and industrialization, the types of land use are complex, and some suburban villages are facing challenges such as disorderly expansion of land use, ambiguous land ownership [12], and lagging improvement of function [19]. All these factors, to a certain extent, have hindered the transition and development of suburban villages. Nanjing is a typical megacity in eastern China, where the land-use morphology of some suburban villages has seen dramatic changes after the implementation of the Beautiful Countryside Construction policy. This paper focused on the spatial-temporal evolution of the dominant and recessive morphology of land-use transition in rural areas, and analyzed the driving forces of rural land-use transition. This paper, taking She Village in the suburbs of Jiangning, Nanjing City as an example, analyzed the changing course and driving forces of dominant and recessive morphology of land use of the village from 1980 to 2018, and explored the transformation mechanism of land use in a typical village, which revealed the law of land use in the process of suburban rural development. It is also conducive to deeper exploration of the land-use transition under the rural revitalization strategy, and at the same time, provides suggestions and references for the development of format and land policy making and planning in suburban villages.

## 2. Analysis of Land-Use Transition in Tourist Villages in the Suburbs

### 2.1. Division of Transformation Stages and Characteristics of Land-Use Morphology of Tourist Villages in the Suburbs

Land-use transition is a long-term event [28] that can be materialized through various channels. It has no clear boundary between the beginning and the end of the land-use transition. However, suburban tourist villages have seen notable variation characteristics at different rural development stages in land-use morphology, which are caused by rapid industrialization and urbanization. The division of stages involved in relevant studies is mainly based on leading industries, key events, moving T-test techniques, and the Mann–Kendall test [29]. Based on different leading industries in different stages, the development of suburban tourist villages can be divided into three stages: industrial development, agricultural development, and service industry development.

The stage of agricultural development mainly features the expansion of cultivated land, the slowed growth of residential land, and a relatively high proportion of unused land within the village. Land is collectively owned and collectively managed after transferring from petty-farmer management. At this stage, the suburban villages play the role of a production base of agricultural and sideline products for the entire city, as its land mainly undertakes the functions of residence and agricultural production. At the early stage of industrial development, land-use change is mainly manifested as rapid expansion of industrial land and the decrease of agricultural land. The change of agricultural land specifically features reduction of food crops, increase of orchards and economic forests, and the increasing commercialization of food products. With the slowdown in the growth rate of industrial land, the area of newly added industrial land has reached its limit. At the end of this stage, the area of industrial land no longer grows or even shows negative growth, usually accompanied with ecological restoration and environmental remediation. The investment of capital and labor in land decreases obviously, the enthusiasm for farming decreases gradually, and the land appears to have extensive utilization to a certain extent. Meanwhile, urban land appears after massive land expropriation, occupying agricultural land and other types of land. Land-use rights are transferred on a small-scale based on contracted management rights, with the coexistence of household operations and scale operations. At this stage, the village, as the city's industrial development zone, commodity distribution center, and warehouse logistics center, undertakes the functions transferred from urban areas, including logistics and warehousing, science, education, culture, and health. The stage of the service industry development is mainly manifested in the rapid increase of commercial land, public service land, and ecological land such as forest land. The distribution of urban and rural land are interlaced, with more fragmented and intensive land and more appropriate internal layout of villages. The property rights of rural land are further differentiated, and household operation, collective operation, enterprise operation, and cooperative operation coexist. Ecological protection and commercial service functions become prominent functions of suburban villages at this stage.

### 2.2. The Driving Forces of Land-Use Transition in Tourism Villages

Land-use transition always happens within the three-fold framework of natural systems, economic and social systems, and institutional systems [15]. Suburban villages are subject to the joint impact by both rural and urban–rural systems due to their special geographical location—on the margin of a city [30]. There are three major factors driving rural land-use transition: First, national and regional policies. By implementing various strategies at different levels including land policies, village planning, household registration policies, etc., governments can impose control on the direction and mode of land-use transition. However, it is hardly feasible to make any adjustment due to its comprehensiveness, thoroughness, high efficiency, and coerciveness [31]. Second, the rural endogenous power. With various factors, including natural resources, cultural characteristics, population growth, and income levels, such factors can indirectly affect the process and results of land-use transition through the behavioral decisions made by subjects of land use, which have a certain degree of fixity and uncontrollability. As a leading factor in rural land-use transition at a certain stage, this second driving force features notable characteristics in each stage in terms of intensity, scope, and duration. Third, the radiation and diffusion of cities. Consisting of the land market, economic development level, locational conditions, market demand, technological progress, and other factors, these driving forces have, directly or indirectly, changed the use of rural land through investment in rural space or the transmission of market demand. With the advancement of the rural development stage, the impact intensity of such driving forces has been further strengthened and has become a key factor for the development of tourism-oriented villages. In addition, land-use subjects, such as ordinary farmers, rural organizations, business enterprises, and urban residents, play a dominant role in rural land-use transition by using and combining various factors that can deliver impact on land-use transition. Simply put, all these factors interact to determine the degree of rural land-use change, as well as the direction and speed of transition (Figure 1).

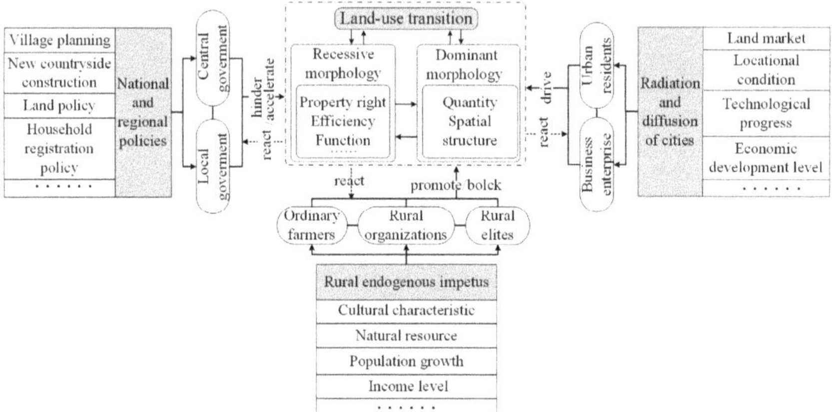

**Figure 1.** The driving forces of land-use transition in tourist villages.

## 3. Materials and Methods

### 3.1. Study Area

Located at the western foot of Qinglong Mountains in Jiangning District, Nanjing, She Village is about 13 km away from the downtown area of the city, the topography of which is high in the northeast and low in the southwest. She Village enjoys convenience in transportation, as it is connected to 104 National Highway in the south and Nanjing Ring Expressway in the west (Figure 2). Also, the village abounds in tourism resources, such as the tombs of Sun Wu in Shangfang and the ancient Dou Village, located within 3 km around She Village. Covering a total area of 17.09 hm$^2$, She Village has 583 farmer households with a total population of 2091 in 2018, among which the population of permanent residents is about 1680. The rate of nonagricultural employment is 100% with a per capita income of 37,000 yuan per year. Huanglong Mountains on the west side of She Village are rich in limestone, with good quality and abundant reserves. At present, the development of She Village is largely driven by ecological construction and rural tourism. In the future, it is expected to become a multifunctional and idyllic village that can offer tourists rich experience in farming and leisure sports.

### 3.2. Date Sources

The relevant land-use data used in this paper consisted of remote sensing images of She Village in 2000 and 2016 with a spatial resolution of 30 m, results of the third national land survey of Jiangning District, Nanjing in 2018. Rural socioeconomic data were mainly derived from the Protection and Development Planning of Traditional Villages, The Overall Plan for Rural Tourism Development in Nanjing, and first-hand information from field research.

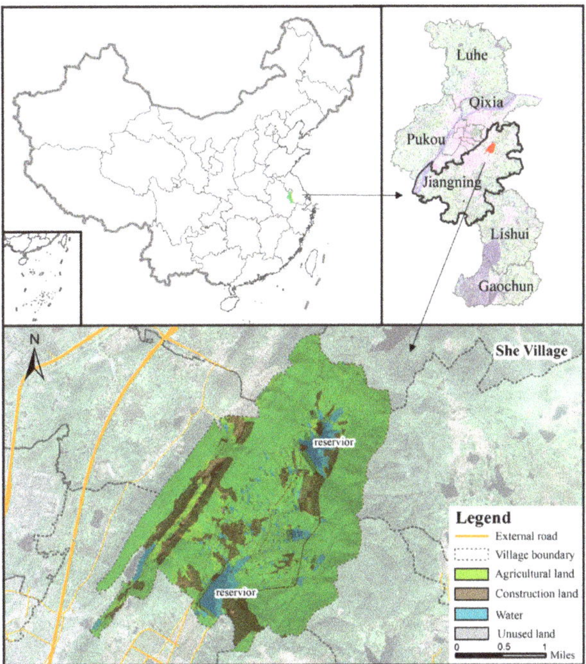



**Figure 2.** The location of She Village.

### 3.3. Methods

Land-use classification. Based on the classification proposed by the third national land survey, this paper divided the land-use types into 8 categories in light of the resolution of remote sensing images, including cultivated land, forest land, garden land, grassland, residential land, commercial and service facility land, industrial and mining land, public management service land, transportation land, water and water conservancy facilities land, and other land.

Stage division. This paper mainly adopted the qualitative division method based on key events. Industrial development of She Village began in 1980. The Grain for Green Project, an ecological restoration project in the whole region, and land consolidation were implemented successively since 2000, which led to the shutdown of industrial enterprises. Beautiful Countryside Construction and tourism development were carried out in 2016. The three clear-cut time periods of 1980–2000, 2000–2016, and 2016–2018, which witnessed industrial development, ecological restoration, and service industry development, respectively, were established, shedding light on the characteristics of She Village's development stage.

Analysis of land-use change. The participatory rural appraisal (PRA) was introduced to collect historical land-use data to invert the process of land spatial expansion from 1980 to 2018 and draw land-use maps of 1980, 2000, 2016, and 2018. ArcGIS spatial analysis tool and the land-use transfer matrix [32] were used to visualize spatial changes in land-use types and quantify the conversion among different land-use categories, respectively. Net change ($N_j$) was introduced to measure the absolute difference between the increased area and decreased area in the land-use matrix. The calculation formula of net change ($N_j$) is as follows [33]:

$$N_j = \mathrm{MAX}\left(S_{j+} - S_{jj},\ S_{+j} - S_{jj}\right) - MIN\left(S_{j+} - S_{jj},\ S_{+j} - S_{jj}\right) = \left|S_{j+} - S_{j+}\right| \quad (1)$$

where $N_j$ stands for net change of the $j$-th land-use type, $S_{j+}$ refers to the total area of the $j$-th land-use type converted to other land-use types, and $S_{j+}$ refers to the total area of the $j$-th land-use type converted from other land-use types. In order to reveal the case in which the net change was 0 due to the mutual transformation of different land-use types of the equal area at different locations, swap change ($D_j$) was used to describe the relative area change of a certain land-use type transferred to other land-use types in situ and other land-use types transferred to the $j$-th land-use type at other locations [20]. The formula is as follows:

$$D_j = 2 \times MIN\left(S_{j+} - S_{jj} \, , \, S_{+j} - S_{jj}\right) \tag{2}$$

where $D_j$ represents the swap change of the $j$-th land-use type. The net change and swap change of each land-use type together constitute the total land-use change ($W_j$), and its formula is as follows:

$$W_j = N_j + D_j = MAX\left(S_{j+} - S_{jj}, \, S_{+j} - S_{jj}\right) + MIN\left(S_{j+} - S_{jj}, \, S_{+j} - S_{jj}\right) \tag{3}$$

in which the total land-use change is equal to the sum of the increased area and decreased area. It is worth noting that the total net change and swap change of the region are 1/2 of the sum of net change and exchange change of each land type, since the total area of the region is certain. The expansion intensity index (M) was introduced to quantify the expansion speed of construction land in different periods [34] and its formulas is as follows:

$$M = U \times \frac{100}{A \times \Delta t} \tag{4}$$

where $U$ refers to the area converted from nonconstruction to construction land, $A$ is the total area of the study area, and $\Delta t$ is the duration of a certain period.

## 4. Results

### 4.1. Land-Use Transition of She Village

#### 4.1.1. Dominant Morphology of Land-Use Transition in She Village

Before 1980, given the restrictions imposed by the planned economy and production factors, such as capital, the land-use structure in the village was mainly based on agricultural land, such as forest land and cultivated land, supplemented by construction land, such as industrial and mining land and residential land. Over the last 40 years or so, She Village witnessed a rapid transformation of land-use quantity and spatial structure (Table 1, Figure 3 ). ① During 1980 to 2000, the land use change of She Village mainly featured the conversion between forest land and other land-use types. The proportion of forest land decreased by 13.81%, while that of cultivated land, grassland, and industrial and mining land increased by 3.73%, 2.33%, and 7.13%, respectively. Residential land and grassland increased by 25.45 hm² and 25.84 hm², with an amplification rate of 231.79% and 147.50%. The area of the remaining land-use types all increased slightly. Forest land had the greatest total land-use change of 317.75 hm², and was the only land-use type with a reduction in net change, mainly transferring to industrial and mining land, cultivated land, and grassland by 120.34 hm², 84.14 hm², and 36.80 hm², respectively, with an account for more than 75% of the total reduced area of forest land. Moreover, the swap change of construction land consisted of residential land, commercial and service facility land, industrial and mining land, public management service land, and transportation land was 0, while net change increased, which indicated the accelerated expansion of village in the study area. ② During 2000 to 2016, land-use change was contrary to the previous stage, but the internal conversions between land-use types were more complex. Forest land increased by 5.48%, while cultivated land, grassland, and industrial and mining land decreased by 5.27%, 0.32%, 4.49%, and 1.26%, respectively. Residential land continued to grow, mainly occupying forest land, garden land, and other land, with an increase of 25.26 hm². Due to the allotment of the collectively owned land, 29.71 hm² of forest land were transferred into public management service land which had an explosive growth. Forest land was still the land type with the greatest total change—the area transferred to 206.11 hm$^2$ from 112.47 hm$^2$. It is worth noting that the swap change of other land was 20.15 hm$^2$, accounting for 91.05% of the total land-use change. The swap change was similar to the total land-use change, which means that a large number of spatial location shifts had taken place in the case of a small change in the total amount of other land. ③ During 2016 to 2018, the land-use structure of She Village tended to be stable, and slight changes happened in different land-use types, mainly concentrated in the north of the study area. The main changes were the decrease of 7.06 hm$^2$ and 9.92 hm$^2$ of industrial and mining land and other land, and the increase of 7.8 hm$^2$ of forest land mainly transferred from other land, accounting for 85.28% of the area converted to forest land. On the one hand, the net change of garden land and grassland increased, and there was a certain amount of swap change on the other hand, and the swap change was large, which indicated that the garden land and grassland mainly changed in location. The commercial and service facility land, by continuously replacing or recombining industrial land and residential land, expanded to14.19 hm$^2$, with a growth rate of 55.93%, thus leading to commercial–residential mixed and industrial–commercial mixed land-use patterns, and the diversified land-use characteristics (Table 2).

**Table 1.** Quantity change of various types of land in She Village from 1980 to 2018.

| Land-Use Type | 1980 | | 2000 | | 2016 | | 2018 | |
|---|---|---|---|---|---|---|---|---|
| | Area/hm$^2$ | Percent | Area/hm$^2$ | Percent | Area/hm$^2$ | Percent | Area/hm$^2$ | Percent |
| CL | 118.40 | 6.62% | 176.85 | 10.34% | 86.69 | 5.07% | 85.27 | 4.94% |
| FL | 1359.81 | 75.99% | 1062.98 | 62.18% | 1156.63 | 67.66% | 1164.43 | 67.40% |
| GD | 17.52 | 0.98% | 43.36 | 2.54% | 37.92 | 2.22% | 40.53 | 2.35% |
| GL | 55.98 | 3.13% | 93.31 | 5.46% | 16.61 | 0.97% | 14.18 | 0.82% |
| RL | 10.98 | 0.61% | 36.43 | 2.13% | 61.69 | 3.61% | 64.84 | 3.75% |
| IM | 32.11 | 1.79% | 152.53 | 8.92% | 131.04 | 7.67% | 123.99 | 7.18% |
| CSF | 1.06 | 0.06% | 3.29 | 0.19% | 13.58 | 0.79% | 14.19 | 0.82% |
| PMS | 0.54 | 0.03% | 1.51 | 0.09% | 31.37 | 1.83% | 32.30 | 1.87% |
| TL | 9.68 | 0.54% | 20.30 | 1.19% | 34.60 | 2.02% | 36.70 | 2.12% |
| WL | 96.03 | 5.37% | 108.45 | 6.34% | 126.91 | 7.42% | 130.52 | 7.56% |
| OL | 7.41 | 0.41% | 10.52 | 0.62% | 12.49 | 0.73% | 2.57 | 0.15% |

CL: cultivated land; FL: forest land; GD: garden land; GL: grassland; RL: residential land; CSF: commercial and service facility land; IM: industrial and mining land; PMS: public management service land; TL: transportation land; WL: water and water conservancy facilities land; OL: other land.

**Table 2.** Land-use change in study area, 1980–2018.

| Land-Use Type | | CL | FL | GD | GL | RL | IM | CSF | PMS | TL | WL | OL |
|---|---|---|---|---|---|---|---|---|---|---|---|---|
| | ATI | 85.44 | 10.46 | 26.30 | 37.32 | 25.45 | 120.41 | 2.22 | 0.96 | 10.62 | 13.41 | 9.05 |
| | ATO | 26.99 | 307.29 | 0.46 | 0 | 0 | 0 | 0 | 0 | 0 | 0.99 | 5.94 |
| 1980–2000 | TC | 112.42 | 317.75 | 26.77 | 37.32 | 25.45 | 120.41 | 2.22 | 0.96 | 10.62 | 14.40 | 14.99 |
| | NC | 58.45 | 296.82 | 25.84 | 37.32 | 25.45 | 120.41 | 2.22 | 0.96 | 10.62 | 12.42 | 3.11 |
| | SC | 53.97 | 20.93 | 0.92 | 0 | 0 | 0 | 0 | 0 | 0 | 1.98 | 11.88 |
| | ATI | 2.83 | 206.11 | 28.07 | 16.06 | 25.91 | 22.06 | 10.60 | 29.86 | 14.94 | 21.08 | 12.05 |
| | ATO | 92.99 | 112.47 | 33.51 | 92.76 | 0.65 | 149.55 | 0.31 | 0 | 0.64 | 2.63 | 10.07 |
| 2000–2016 | TC | 95.82 | 318.59 | 61.58 | 108.81 | 26.56 | 171.62 | 10.91 | 29.86 | 15.58 | 23.71 | 22.13 |
| | NC | 90.16 | 93.64 | 5.44 | 76.70 | 25.26 | 127.49 | 10.30 | 29.86 | 14.29 | 18.45 | 1.98 |
| | SC | 5.66 | 224.95 | 56.14 | 32.11 | 1.29 | 44.13 | 0.61 | 0 | 1.29 | 5.26 | 20.15 |
| | ATI | 1.84 | 9.24 | 4.08 | 2.38 | 3.64 | 0.03 | 0.61 | 0.94 | 2.09 | 3.64 | 0 |
| | ATO | 3.26 | 1.43 | 1.47 | 4.81 | 0.49 | 7.0 | 0 | 0 | 0 | 0 | 9.93 |
| 2016–2018 | TC | 5.10 | 10.67 | 5.55 | 7.19 | 4.13 | 7.10 | 0.61 | 0.94 | 2.09 | 3.64 | 9.93 |
| | NC | 1.42 | 7.81 | 2.61 | 2.43 | 3.16 | 7.04 | 0.61 | 0.94 | 2.09 | 3.64 | 9.93 |
| | SC | 3.68 | 2.87 | 2.94 | 4.76 | 0.98 | 0.06 | 0 | 0 | 0 | 0 | 0 |

ATI: area transferred from other land-use types; ATO: area transferred to other land-use types; TC: the total land-use change; NC: net change; SC: swap change.

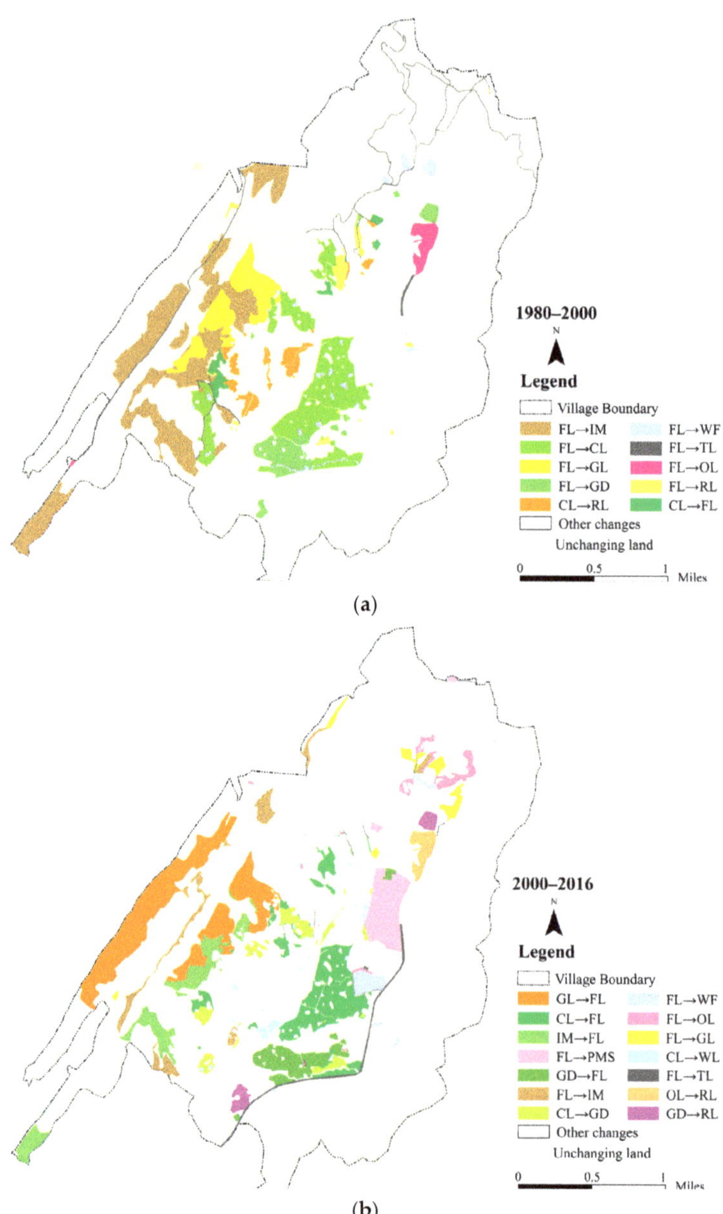


**Figure 3.** *Cont.*

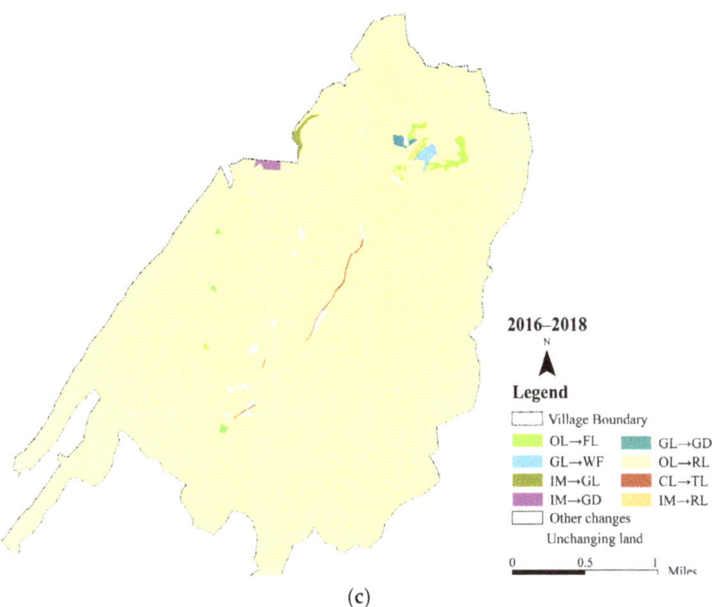


(c)

**Figure 3.** (a) Major internal conversions between land-use types from 1980 to 2000; (b) major internal conversions between land-use types from 2000 to 2016; (c) major internal conversions between land-use types from 2016 to 2018.

The diversification of land-use types and quantity structure led to the change of spatial layout. The village is located between two mountains and surrounded by vast forests and farmland, forming a ladder layout composed of houses, streets and alleys, reservoirs, farmlands, and mountains (Figure 4). ① In 1980, the spatial layout of land-use in She Village consisted of public management service land, residential land, cultivated land, and forest land from center to edge, forming a concentric circle structure of "public service area—traditional residential area—agricultural production area—forestry area". Pan's ancestral hall, Pan's ancient building complex, and Jiulong ridge, together, constituted the center of the rural settlement in She Village, among which the ancestral hall had the function of public service by providing office space for the village committee. The traditional residence maintained an ancient style of cyan bricks and black tiles, extending to the south and north from the center of the settlement, and connected to the reservoir. The periphery of the traditional residence was contiguous cultivated land, garden land, and forest land. She Village presented a centralized layout pattern with a compact internal structure. Buildings were distributed along contour lines, high in the north and low in the south, and formed a diverse spatial layout together with the latticed streets and lanes. ② In 2000, rapid industrialization led to an explosive expansion of the industrial and mining land. Eight industrial and mining enterprises were put into operation in She Village over the past two decades, which formed an industrial belt along Huanglong and Qinglong Mountain on the west side of She Village and occupied a large expanse of agricultural space. The development of mining industries caused degradation of forests and the increasing area of bare mountains, which further squeezed the space for agricultural production eastward. Significant changes also took place in the internal structure of agricultural land, where the cultivated land expanded eastward, occupying the original forest land, and creating a new fan-shaped area eastward based on the original ring structure. In the late 1990s, with the population growth and households division, a vast expanse of residential land increased. Modern residences expanded along the periphery of traditional ones, gathering together at each node of the road network. Also, they were distributed in a discrete way, breaking the tight spatial layout of the village. During this period, the spatial change of land use in She Village featured the establishment of industrial areas, expansion of mining areas, scattered growth of residential areas, and the emergence of a small portion of commercial land, contributing to the more mixed spatial layout of She Village. ③ In 2018, after ecological restoration and the development of the tertiary industry, the spatial layout of land use was further diversified. Through the listing and auction of land parcels, urban residential land emerged and squeezed part of the agricultural production area and forestry area. Forest area was restored in the east of the village after the policy of Grain for Green came into effect, and the mining land, once occupied forests and farmland, was shrunk with the implementation of the comprehensive rehabilitation of mining area. The space for agricultural production tended to be contracted, fragmented, and decentralized. Currently, the layout of She Village is a concentric circle structure, with public service facilities as the core, traditional and modern residences as the inner circle, agricultural production areas as the intermediate circle, and forestry and industrial areas as the outer circle. Commercial service areas, part of modern residential areas and public service areas, have replaced part of agricultural production area and are distributed between the agricultural circle and the forest circle along the rural road, in the form of clumps (Figure 4). During the construction of demonstration villages, She Village repaired and protected historic buildings and demolished the sheds, toilets, and temporary houses etc., which were done by villagers privately. The internal structure of the village was upgraded by optimizing the rural road network system and strengthening the bonds between primary and secondary roads.

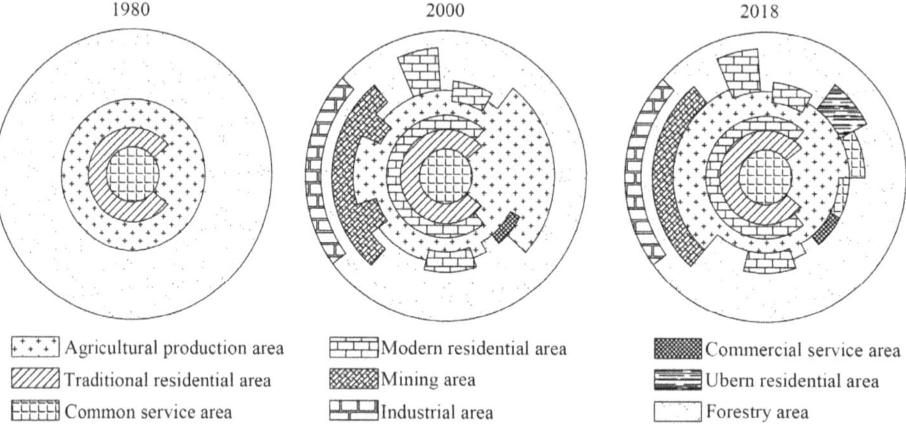


**Figure 4.** Land-use spatial structure in 1980, 2000 and 2018.

### 4.1.2. Recessive Morphology of Land-Use Transition in She Village

With the rapid changes in the dominant morphology of land use, recessive morphology of land use has transferred in She Village, especially cultivated land, forest land, industrial and mining land, and residential land, in terms of ownership, management modes and entities, input and output, and functions. From 1980 to 2018, land-use intensity of She Village increased from 3.18% to 15.90%, and the expansion intensity of construction land in the three stages was 46.70%, 37.79%, and 21.38% respectively (Table 3). In view of the limited new construction land quotas, the expansion of construction land will gradually stabilize. The road network in the village was further improved and the road network density increased from 0.56% to 2.14%.

**Table 3.** The expansion intensity of construction land in She Village from 1980 to 2018.

| Stages | Expansion Area (m$^2$) | Expansion Intensity (%) |
|---|---|---|
| 1980–2000 | 159.68 | 46.70% |
| 2000–2016 | 103.38 | 37.79% |
| 2016–2018 | 7.31 | 21.38% |

In the early 1980s, the implementation of the household contract responsibility system led to the separation of contracted management rights of cultivated land from farmers' collective land ownership. At the same time, the management of cultivated land transferred from cooperative operation to household operation, and the cultivated land, mainly undertook agricultural production functions. With the outflow of local labor force who should have been engaged in agricultural production, the cultivated lands were gradually marginalized, some of which were abandoned and others subcontracted between farmers. Grains were replaced by cash crops, such as coleseeds and tea trees, and the proportion of planting areas of grain crops decreased from 86.70% in 1980 to 45.34% in 2018. At the beginning of 2016, more than one-third of the cultivated land in this village was left deserted. Except for the land reserved by the farmers, the village transferred 90% of the cultivated land to large farms for unified management at a price of 700 yuan/mu, and the rest was leased. With the development of landscape agriculture, cultivated land had both the multifunction of agricultural production and ecology conservation. In addition to the expansion and contraction in quantity and area, the function of industrial and mining land changed remarkably. Some abandoned mining areas, suffering from rain erosion all year round, formed a special landform similar to the "Yadan landform", which has become a new business card for external publicity of She Village. She Village, taking advantage of the vertical drop of some mine pits, created a rafting tourism product, resulting in the transformation of industrial production function to tourism function. Due to the rapid and disorderly expansion of industrial land, large-scale exploitation activities led to the deterioration of the quality of surrounding soil, serious degradation of 47.15 hm$^2$ forest land to grassland, and damage to ecological functions. In 2003, 2000 mu of cultivated land was returned to forest land, which was allocated to the forest farm at a price of 700 yuan/mu for ecological restoration and management. The ecological function of the forest land was restored year by year. In addition, part of the collectively owned forest land was allocated for state-owned land to build a football training base without any compensation, leading to the land-use function transferring to public service function. Forest land developed the compound function of ecology and tourism by combining health and wellness tourism. The residential land in the village initially assumed the residential function, with a small proportion for self-employment to provide commodities to the villagers. But later, production functions were added to it because of the diversification of targeted groups and management entities. Under such circumstances, rural households made great efforts to develop agritainments, homestay inns, and other activities through independent operation and commissioned operation. Some merchants directly rented idle rural houses for commercial production. All these led to the bifunction of residence and production of rural housing land.

### 4.2. Driving Forces of Land-Use Transition in She Village

#### 4.2.1. Main Driving Forces of Each Stage

Since the beginning of the 1980s, the land-use morphology of She Village has experienced rapid transformation and development, which has been led by local organizations and rural elites, and driven by the dual influence of both urban and rural systems. She Village has undergone a shift from passive development to active development due to its favorable geographical location, effective policies, and abundant resources.

During the stage of industrial development from 1980 to 2000, mineral resources, particularly limestone, and population growth were the dominant factors that shaped the land-use morphology of She Village. Relying on such abundant limestone, many township enterprises kept cropping up, including several quarries and lime factories. Meanwhile, large amounts of raw materials were continuously transported to Nanjing, Ma'an shan, and other places to support national construction. Industrial enterprises obtained high profits by mining and processing industrial raw materials. The flourishing of mining industry also profoundly changed the landform morphology of Huanglong Mountain in She Village. With the constant growth of population in She Village, villagers cultivated new land and built new houses in order to meet their agricultural production and living needs, leading to the expansion of cultivated and residential areas in the village. The production and operation activities by farmers and industrial enterprises became the main driver for the shrinkage of forest land.

During the stage of ecological restoration from 2000 to 2016, the national macro policy and the transfer of urban functions were the main driving forces for the changes in land-use morphology of She Village. In the early 2000s, the policy of Grain for Green was put into practice across the board, which led to the closure of industrial and mining enterprises in She Village. In line with the land consolidation campaign, the village carried out an all-round ecological restoration project, including mine rehabilitation, which reshaped the land-use morphology of She Village. The implementation of these policies resulted in the loss of industrial jobs and restrictions on agricultural production. Since the income from agricultural production alone could not meet their living needs, a majority of farmers turned to nonagricultural sectors to make a livelihood transformation and looked for jobs in nearby towns and cities, which caused a buildup of desolated farm land. At the same time, some public services and residential functions of cities were transferred to the suburbs due to reduced costs in transportation and land price advantage, leading to the mixture of rural and urban land in the suburbs.

During the stage of service industry development from 2016 to 2018, various behavior subjects, including local government and organizations, urban residents, and commercial business operators, together with other factors, such as market demand, location advantages and regional policies, promoted the reallocation of production factors of She Village, creating a new direction for the development of the village and the evolution of land-use morphology. Thanks to rapid industrialization and urbanization, the tourism industry in She Village, only 13 km away from the downtown area of Nanjing, enjoyed inherent advantages to prosper, as urban residents had a stronger demand for pastoral landscape and rural life experience. In 2017, the tourism-oriented development orientation of She Village was completely settled, when the Beautiful Countryside Construction and other relevant policies and blueprints were put into place to support tourism development of this village in many aspects, including construction of sewage pipelines, cleaning of river ponds, improvement of infrastructure, etc. Commercial tenants and other service business operators changed the property rights structure, management modes, and land functions of rural land by renting houses and lands. The local organizations acted as key intermediaries to negotiate with the government, business entities, and the nonindigenous farmers on behalf of the villagers in regards to the price of land acquisition, land transfer, and land leasing. All the cultivated land in the village was transferred and managed by the large farmer households in a unified manner, which transformed the decentralized management mode of cultivated land.

### 4.2.2. Changes in the Impact Intensity of Driving Forces

The driving forces that dominated the land-use transition of different stages in She Village were different, and each driving force showed different intensity in each stage of land-use transition of She Village. The intensities of these driving forces of land-use transition in She Village were mainly manifested in four changing trends: "periodically intermittent", "increase–increase–increase", "increase–decrease–increase", and "increase–decrease–decrease" (Figure 5).

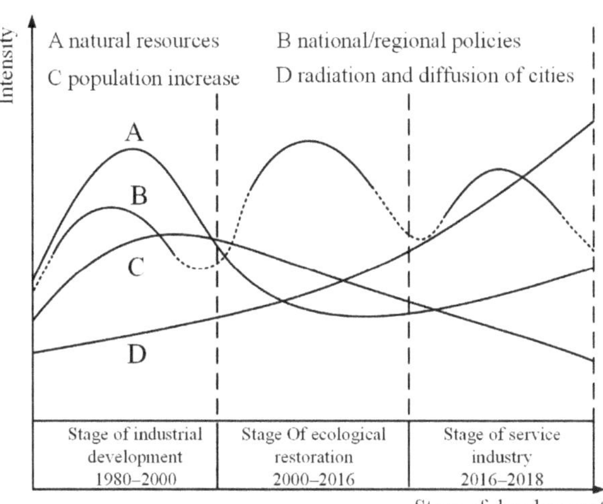



**Figure 5.** Changes in the impact intensity of driving forces.

The periodically intermittent trend has mainly been presented by the impact intensity of the implementation of regional and national policies, including the household contract responsibility system, Grain for Green, and Beautiful Countryside Construction, which injected impetus to the evolution of land-use morphology into the three time stages of She Village. These policies can deliver rapid and strong impacts on relevant land parcels in a certain period of time, with an intensity increasing first and then decreasing. The changes in relevant lands reach saturation and gradually stabilize. The influence process of the policies and systems have stopped, but the influence result is continuous.

The increasing–increasing–increasing trend has mainly manifested as the impact intensity of urban radiation and diffusion. With the improvement of She Village's external transportation, the flow of production factors between the downtown area of Nanjing and She Village has been strengthened. The stage of industrial development features the outflow of industrial raw materials. The stage of ecological restoration features the outflow of labor forces and inflow of real-estate capital. The two-way flow of population between urban and rural areas, together with an influx of industrial and commercial capital, is the characteristic of the rural–urban flow of production factors during stage of service industry development. As a result, She Village's production mode has been changed, with functional transformation facilitated and land-use transition strengthened under the radiation of the downtown area.

The trends of increase–decrease–increase and increase–decrease–decrease have mainly manifested as the intensity of rural endogenous impetus, such as natural resources and population, in the land-use transition. Mineral resource is the primary factor of industrial development and the most representative natural resource in She Village. At the stage of ecological restoration, the exploitation of mineral resources was restricted and the expansion of mining land slowed down due to social–ecological negative feedback, leading to a reduced impact on land-use change. At the stage of service industry development, landscapes in rural areas, such as mountains, water, forests, and fields, have become new consumer goods, a magnet for urban residents. This trend, therefore, gave birth to the development of tourism in this village, which triggered further transformation in its land use. The population in She Village showed a trend of increasing first and then decreasing during three stages. The increase in population in the industrial development period brought about the expansion of cultivated land and homesteads. During the stage of ecological restoration and service industry development, the population drain and the decreasing dependence on land lead to the decreasing direct reconstruction of the land by villagers.

## 5. Discussion

### 5.1. Problems in the Process of Land-Use Transformation in She Village

Rural transformation and development is not only an opportunity for the adjustment of rural industries and improvement of living environments, but also a typical period during which man–land contradictions and conflicts between people keep cropping up [35]. Some problems stood out during the transformation and development of She Village.

The land-use transition in She Village, dominated by national and regional policies, has caused instability of the development in She village due to its rapid and little-adjustable implementation process. At the stage of ecological restoration, the policy of Grain for Green directly led to the elimination of the mining industry in She Village, which greatly slashed the financial revenue. In the following decade, the economic development of the village was stagnant, delivering a negative impact on the income of ordinary farmers and the development of the village. In addition, rapid transformation and development will inevitably cause unsustainable land use [36]. She Village preliminarily completed the Beautiful Countryside Construction program according to the village planning within two years. Field research found that some of the lands in the village were in a deserted state due to the lack of effective follow-up management, such as the untended flower beds that had overgrown with weeds, leading to the failure of the realization of the expected land functions (Figure 6). Moreover, after the development of tourism, villagers' consciousness of service was weak, which hadn't shifted from meeting their own needs to those of tourists, reflected by the fact that some villagers cleaned up the flowers and plants used for landscaping around their houses for vegetable planting and stacked the square with junk. The public space of the village has transferred from the simple production and living space of the villagers to the consumption and operating space of tourists and operators [20]. Villagers' market consciousness had not yet been established in the rapid rural transformation.

**Figure 6.** Unrealized leisure function.

Unable to share rural development dividends with villagers. At the stage of industrial development, the collective economic organization could not afford the cost of mining due to the weak economic foundation of She Village. As a result, other village collectives obtained mining rights at a lower price and almost occupied all profits, while only some employment opportunities with low incomes were left for villagers in She Village. At the stage of ecological restoration, the real estate developers acquired land-use rights for commercial buildings. With the improved environment of She Village and growing demand of urban residents for a second house, the land price in She Village rose, but the value-added part was grabbed by real-estate developers instead of villagers. At the stage of service industry development, a small number of rural households who left the village became rentiers by collecting rents. However, more villagers who stayed in the village suffered from the lack of knowledge in operating modes and commodity packaging, as well as the high cost of house renovation [37], which made it difficult for them to compete with commercial tenants in the development of tertiary industries, such as accommodation and catering. The development dividend of She Village is mainly occupied by industrial and commercial capitalists, while the villagers are excluded from the rentiers.

Lagging reconstruction of local organizations. At present, the tourism-oriented land-use transition has been basically completed, and the prosperity of tertiary industry has also taken shape. However, the organizational restructuring in the structure of "people–land–industry" is greatly lagging behind in She Village. The increasing mobility of the rural population and the growing participation of urban residents and industrial and commercial operators are undermining the traditional rural social network and the community system, which are based on blood and geographical relationship [38]. The traditional Chinese village is a manners-controlled society, which is governed by the village covenant, and is on the verge of collapse under the impact of modernism. The mode of "rural governance", with the government as its core, under which village committees, social organizations, and rural residents are marginalized and subordinated, is no longer suitable for rural development.

### 5.2. Implications of Suburban Land-Use Transformation for Rural Transformation

Suburban villages, as the front belt of urbanization, the pilot area of agricultural modernization, the coordination area of urban and rural relations, and the ecological barrier of the city [22], have a general trend of weakening rurality and strengthening urbanism [39]. The village not only provides production and living space for the villagers, but also undertakes the function of ecological conservation, part of which is transferred from urban areas [40]. Due to their miscellaneous population, industry. and land use, the study on the trend and influence mechanism of land-use change in suburban villages is enlightening to the transformation and development of other villages.

Multifunctional land use is an effective way of rural transformation. The essence of rural recession is the alienation or degradation of rural regional functions. However, the rural revitalization is the optimization or enhancement of rural regional functions [41]. Two types of transformations of She Village have been caused by the degradation of industrial production function and the enhancement of tourism service function. Rural land-use transition is not a process of linear replacement. Specifically, the transition should promote the compound use of agricultural land through the application of new technology, large-scale management, agricultural tourism development, and other ways, giving full play to its ecological and production functions. The construction land mainly undergoes functional expansion on the basis of the original residential and production functions combined with the development of commercial retail, catering and accommodation, and new entertainment projects.

The improvement of the property rights system and land revenue distribution institution is the key to rural transformation. The contradiction between the fixity of land and the mobility of population leads to the waste of land resources [42], reflected by the abandonment of homesteads and cultivated land, which is a major obstacle to rural transformation. At the same time, unclear property rights of some land prevent companies and large planters from promoting the transfer of farmland [43]. Under the current rural land contract management system, the cash crop production model with family management as the mainstay faces many challenges, such as difficulty in expanding the scale of land, improving the industrial level, and extending the industrial chain. In addition, in the income distribution link of land circulation, the subjects participating in income distribution are not clear and the proportion of income distribution is diverse with no clear stipulation, which leads to little enthusiasm of farmers participating in the marketization of collective commercial construction land. Therefore, the improvement of the property rights system and land revenue distribution institution is an important driving force for the current rural transformation and development.

*5.3. Research Shortage*

Compared with previous research, this paper studied land-use transition from a dual perspective of dominant and recessive morphology and emphasizes the influence of urban radiation and diffusion on suburban villages, which can fully show the evolution of land use in suburban villages and contribute to a better understanding of rural urbanization. Given the complexity of the rural regional system, and the diversity of rural types and land-use transition channels, this study has the following deficiencies: due to the long time span, it was difficult to obtain land-use data on village scale, and part of the research content can only be supported by information obtained through a participatory evaluation method for qualitative analysis rather than quantitative analysis. The tourist villages in suburbs have different development paths [44] under different dominant factor, including government, commercial capital, or villagers [45]. This paper only provided analysis for the tourism-oriented rural land-use transition led by the government. Therefore, it is necessary to carry out the comparative study on tourist villages under different dominant factors.

Moreover, in the context of rural transformation, tourism development has become an important way for rural transformation in suburbs, while problems such as industrial homogenization, excessive commercialization of agriculture, and gentrification of social space [46] come one after another. A number of issues need further study, including how to effectively combine the driving forces of land-use transition and promote the transformation and development of suburban villages, how to properly develop a multifunctional countryside that can benefit villagers, and how to avoid rural space developing into a pure consumption space while protecting the spatial development rights of villagers in rural tourism transformation to ensure the sustainability of land-use transition.

## 6. Conclusions

During rapid urbanization and industrialization, this tourism village has witnessed a clear-cut transition in their land use, which can be divided into three stages, namely, the stages of agricultural development, industrial development, and service industry development. The change of land-use morphology shows distinct characteristics at each stage, which is led by top-down government policies and driven by the endogenous force of a village and the radiation and diffusion of a city. The overall trend is concluded as follows. The residential land first expanded, then the growth rate slowed down to zero. The area of land for agricultural production and industrial production first increased and then decreased. The area of public services and commercial service land increased continuously. She Village tended to have diversified land-use types, intensive and reasonable land-use layout, activated land ownership, complicated functions, and diverse management modes and entities.

From 1980 to 2018, She Village underwent two transformations "industrial development—ecological restoration—rural tourism development". The quantity and spatial structure underwent a drastic restructuring, and the degree of land fragmentation was intensified. The land-use characteristics of "commercial–residential mixed" and "tourism redevelopment of industrial land" emerged. Now, a concentric circle structure has been formed, which is "public service area—traditional residential area—modern residential area—agricultural production area—forest area and industrial area" from inside to outside. The diversification of land property rights and business entities, from local farmers to nonlocal rich farmers and commercial enterprises, have led to multifunctional and complex land issues. The multifunctional land for production and ecology, or for residence and production, has become the mainstream. Rural space has experienced the transformation from living and production space to consumption space.

Over the past four decades, the dominant factors driving rapid land-use transition in She Village were different at each stage. The land-use transition during the stage of industrial development was driven by natural resources and population growth; during ecological restoration, driven by the national macro policies and the transfer of urban functions; and during service industry development, driven by market demand, location advantages, regional policies, and the interaction between various subjects of land use. The national and regional policies played a decisive role in the two periods of transformation and development of She Village. Different driving forces had different impact intensities at each stage of the land-use transition of She Village. Natural resource, urban radiation and diffusion, population growth, and national and regional policies showed four changing trends of periodically intermittent, increase–increase–increase, increase–decrease–increase, and increase–decrease–decrease, respectively.

**Author Contributions:** Conceptualization, K.J.; data curation, Y.C. and Y.H.; funding acquisition, W.Q.; investigation, Y.W.; methodology, Y.C.; resources, W.Q.; software, T.H.; supervision, T.F. writing—review and editing, Y.C. All authors have read and agreed to the published version of the manuscript.

**Funding:** This research was funded by National Natural Science Foundation of China (41871178, 41901204) and Natural Resources and Technology Program of Jiangsu Province (KJXM2019012, KJXM2019005, KJXM2019009).

**Data Availability Statement:** No new data were created or analyzed in this study. Data sharing is not applicable to this article.

**Acknowledgments:** We especially thank the National Natural Science Foundation of China for funding this study.

**Conflicts of Interest:** The authors declare no conflict of interest.

## References


1. Hoggart, K.; Paniagua, A. What rural restructuring? *J. Rural Stud.* **2001**, *17*, 41–62. [CrossRef]
2. Liu, S.; Wang, Y. From Native Rural China to Urban-rural China: The Rural Transition Perspective of China Transformation. *Manag. World* **2018**, *34*, 128–146. [CrossRef]
3. Tu, S.; Long, H.; Zhang, Y.; Ge, D.; Qu, Y. Rural restructuring at village level under rapid urbanization in metropolitan suburbs of China and its implications for innovations in land use policy. *Habitat Int.* **2018**, *77*, 143–152. [CrossRef]
4. Lange, A.; Piorr, A.; Siebert, R.; Zasada, I. Spatial differentiation of farm diversification: How rural attractiveness and vicinity to cities determine farm households' response to the CAP. *Land Use Policy* **2013**, *31*, 136–144. [CrossRef]
5. Holmes, J. Impulses towards a multifunctional transition in rural Australia: Gaps in the research agenda. *J. Rural Stud.* **2006**, *22*, 142–160. [CrossRef]
6. Jia, K.; Qiao, W.; Chai, Y.; Feng, T.; Wang, Y.; Ge, D. Spatial distribution characteristics of rural settlements under diversified rural production functions: A case of Taizhou, China. *Habitat Int.* **2020**, *102*, 102201. [CrossRef]
7. Yansui, L.; Long, H.; Chen, Y.; Wang, J.; Li, Y.; Li, Y.; Yang, Y.; Zhou, Y. Progress of research on urban-rural transformation and rural development in China in the past decade and future prospects. *J. Geogr. Sci.* **2016**, *26*, 1117–1132. [CrossRef]
8. Meyfroidt, P.; Chowdhury, R.R.; de Bremond, A.; Ellis, E.C.; Erb, K.-H.; Filatova, T.; Garrett, R.D.; Grove, J.; Heinimann, A.; Kuemmerle, T.; et al. Middle-range theories of land system change. *Glob. Environ. Chang.* **2018**, *53*, 52–67. [CrossRef]
9. Long, H.; Zou, J.; Pykett, J.; Li, Y. Analysis of rural transformation development in China since the turn of the new millennium. *Appl. Geogr.* **2011**, *31*, 1094–1105. [CrossRef]
10. Ma, W.; Jiang, G.; Wang, D.; Li, W.; Guo, H.; Zheng, Q. Rural settlements transition (RST) in a suburban area of metropolis: Internal structure perspectives. *Sci. Total Environ.* **2018**, *615*, 672–680. [CrossRef]
11. Long, H.; Qu, Y. Land use transitions and land management: A mutual feedback perspective. *Land Use Policy* **2018**, *74*, 111–120. [CrossRef]
12. Ge, D.; Wang, Z.; Tu, S.; Long, H.; Yan, H.; Sun, D.; Qiao, W. Coupling analysis of greenhouse-led farmland transition and rural transformation development in China's traditional farming area: A case of Qingzhou City. *Land Use Policy* **2019**, *86*, 113–125. [CrossRef]
13. Long, H.; Ge, D.; Wang, J. Progress and Prospects of the Coupling Research on Land Use Transitions and Rural Transformation Development. *Acta Geogr. Sin.* **2019**, *74*, 2547–2559. [CrossRef]
14. Su, K.; Yang, Q.; Zhang, B.; Zhang, Z. The Coupling Mechanism between Rural Land Use Transition and Small-scale Peasant Economy Change in Mountainous Areas. *Geogr. Res.* **2019**, *38*, 399–413. [CrossRef]
15. Long, H.; Tu, S. Land Use Transition and Rural Vitalization. *China Land Sci.* **2018**, *32*, 1–6. [CrossRef]
16. Ge, D.; Zhou, G.; Qiao, W.; Yang, M. Land use transition and rural spatial governance: Mechanism, framework and perspectives. *J. Geogr. Sci.* **2020**, *30*, 1325–1340. [CrossRef]

17. Zhang, B.; Sun, P.; Jiang, G.; Zhang, R.; Gao, J. Rural land use transition of mountainous areas and policy implications for land consolidation in China. *J. Geogr. Sci.* **2019**, *29*, 1713–1730. [CrossRef]

18. Yang, R. Spatial Differentiation and Mechanisms of Typical Rural Areas in the Suburbs of a Metropolis: A Case Study of Beicun Village, Baiyun District, Guangzhou. *Acta Geogr. Sin.* **2019**, *74*, 1622–1636. [CrossRef]

19. Tu, S.; Long, H.; Zhang, Y.; Zhou, X. Process and Driving Factors of Rural Restructuring in Typical Villages. *Acta Geogr. Sin.* **2019**, *74*, 125–141. [CrossRef]

20. Jiang, K.; Liu, Z.; Li, Y.; Wang, Y.; Wang, Y. Land use change of typical villages in the loess hilly and gully region and implications for regional rural transformation and development. *Prog. Geogr.* **2019**, *38*, 1305–1315. [CrossRef]

21. Liu, Y.; Long, H. Land use transitions and their dynamic mechanism: The case of the Huang-Huai-Hai Plain. *J. Geogr. Sci.* **2016**, *26*, 515–530. [CrossRef]

22. Xi, J.; Zhao, M.; Ge, Q.; Kong, Q. Changes in Land Use of a Village Driven by over 25 Years of Tourism: The Case of Gougezhuang village, China. *Land Use Policy* **2014**, *10*, 119–130. [CrossRef]

23. Zhu, L.; Li, L.; Liu, S.; Li, Y. The evolution of village land use function in the metropolitan suburbs and its inspiration to rural revitalization: A Case Study of Jiangjiayan Village in Chengdu City. *Geogr. Res.* **2019**, *38*, 535–549. [CrossRef]

24. Li, B.; Zhou, X.; Liu, P.; Chen, C.; Liu, Y. Function Transformation and Spatial Reconstructuring of Zhangguying Village in Urbanization. *Sci. Geogr. Sin.* **2018**, *38*, 1310–1318. [CrossRef]

25. Zou, L.; Liu, Y.; Yang, J.; Yang, S.; Wang, Y.; Zhi, C.; Hu, X. Quantitative identification and spatial analysis of land use ecological-production-living functions in rural areas on China's southeast coast. *Habitat Int.* **2020**, *100*, 102182. [CrossRef]

26. Chen, Y.; Zhou, J. The Evolution Process and Spatial Layout Patterns of Land Use in Urban Fringe Areas. *Urban Plan. Overseas* **1998**, *2*, 10–16.

27. Chen, Y.; Wu, W. An Analysis on the Man-land System and its Dynamics in Urban-rural Interlocking Belt. *Sci. Geogr. Sin.* **1998**, *18*, 418–424.

28. Lapola, D.M.; Martinelli, L.A.; Peres, C.A.; Ometto, J.P.H.B.; Ferreira, M.E.; Nobre, C.A.; Aguiar, A.P.D.; Bustamante, M.M.C.; Cardoso, M.F.; Costa, M.H.; et al. Pervasive transition of the Brazilian land-use system. *Nat. Clim. Chang.* **2014**, *4*, 27–35. [CrossRef]

29. Liang, X.; Jin, X.; Ren, J.; Gu, Z.; Zhou, Y. A research framework of land use transition in Suzhou City coupled with land use structure and landscape multifunctionality. *Sci. Total Environ.* **2020**, *737*, 139932. [CrossRef] [PubMed]

30. Luo, Y.; Zhou, C.S. The Review and Outlook on Urban-rural Fringe Studies in China. *Urban Stud.* **2005**, *12*, 25–30.

31. Hu, S.; Tong, L.; Long, H. Land Use Transition Potential and its Assessment Framework. *Geogr. Res.* **2019**, *38*, 1367–1377. [CrossRef]

32. Zhu, H.; Li, X. Discussion on the Index Method of Regional Land Use Change. *Acta Geogr. Sin.* **2003**, *58*, 643–650.

33. Zhu, D. Methods for Detecting Land Use Changes Based on the Land Use Transition Matrix. *Resour. Sci.* **2010**, *32*, 1544–1550.

34. Liu, S.; Wu, C.; Shen, H. A GIS based model of urban land use growth in Beijing. *Acta Geogr. Sin.* **2002**, *55*, 407–416.

35. Yang, N.; Long, H.; Liu, Y.; Zhang, Y. Research Progress and Prospect of Rural Transformation and Reconstruction in China: Paradigms and Main Content. *Prog. Geogr.* **2015**, *34*, 1019–1030. [CrossRef]

36. Ristic, D.; Vukoicic, D.; Milincic, M. Tourism and Sustainable Development of Rural Settlements in Protected Areas Example NP Kopaonik (Serbia). *Land Use Policy* **2019**, *89*, 104231. [CrossRef]

37. Tao, H.; Huang, Z.; Ran, F. Rural Tourism Spatial Reconstruction Model from the Perspective of ATV: A Case Study of Mufu Township, Hubei Province, China. *Sustainability* **2018**, *10*, 2675. [CrossRef]

38. Yao, G.; Xie, H. Rural spatial restructuring in ecologically fragile mountainous areas of southern China: A case study of Changgang Town, Jiangxi Province. *J. Rural. Stud.* **2016**, *47*, 435–448. [CrossRef]

39. Zhang, J.; Jiang, K.F. Analyzing Capital Logic Behind Rural Construction Boom in Contemporary China. *Mod. City Stud.* **2016**, *31*, 2–8.

40. Tian, J.; Wang, B.; Cheng, L.; Wang, S. The Process and Mechanism of Regional Land Use Transition Guided by Policy: A Case Study of Northeast China. *Geogr. Res.* **2020**, *39*, 805–821. [CrossRef]

41. Zhou, Y.; Li, X.; Liu, Y. Rural land system reforms in China: History, issues, measures and prospects. *Land Use Policy* **2020**, *91*, 104330. [CrossRef]

42. Ge, D.; Long, H.; Zhang, Y.; Ma, L.; Li, T. Farmland transition and its influences on grain production in China. *Land Use Policy* **2018**, *70*, 94–105. [CrossRef]

43. Ge, D.; Long, H.; Qiao, W.; Wang, Z.; Sun, D.; Yang, R. Effects of rural–urban migration on agricultural transformation: A case of Yucheng City, China. *J. Rural Stud.* **2020**, *76*, 85–95. [CrossRef]

44. He, T.; Qiao, W.; Jia, K.; Chai, Y.; Hu, Y.; Sun, P.; Wang, Y.; Feng, T. Selecting Rural Development Paths Based on Village Multifunction: A Case of Jingjiang City, China. *Complexity* **2020**, *2020*, 1–15. [CrossRef]

45. Huang, J. Rural Revitalization: Rural Transformation, Structural Transformation and Government's Functions. *Issues Agric. Econ.* **2020**, *481*, 4–16. [CrossRef]

46. Zasada, I. Multifunctional peri-urban agriculture—A review of societal demands and the provision of goods and services by farming. *Land Use Policy* **2011**, *28*, 639–648. [CrossRef]

*Article*

# What Role(s) Do Village Committees Play in the Withdrawal from Rural Homesteads? Evidence from Sichuan Province in Western China


Peng Tang [1,2,3], Jing Chen [1], Jinlong Gao [4,5,*], Min Li [1] and Jinshuo Wang [3]

[1] Department of Land Resource and Real Estate Management, Sichuan University, Chengdu 610065, China; tp@scu.edu.cn (P.T.); ckj1217@foxmail.com (J.C.); limingg@scu.edu.cn (M.L.)

[2] Center for Social Development and Social Risk Control of Sichuan University, Chengdu 610065, China

[3] Department of Geography, Planning and Environment, Radboud University, 6525 Nijmegen, The Netherlands; j.wang@fm.ru.nl

[4] Key Laboratory of Watershed Geographic Sciences, Nanjing Institute of Geography and Limnology, Chinese Academy of Sciences, Nanjing 210008, China

[5] Department of Planning and Landscape Architecture, University of Wisconsin-Madison, Madison, WI 53706, USA

[*] Correspondence: jlgao@niglas.ac.cn



Received: 25 October 2020; Accepted: 25 November 2020; Published: 27 November 2020



**Abstract:** Village committees, as grassroots spontaneously formed by rural collective members in China's hierarchy system, play an irreplaceable role in the management of rural public affairs. Based on the filed survey dataset taken from three pilot counties/districts in Sichuan province, we explored the significant role that village committees played in farmers' withdrawal from rural homesteads (WRH). Our empirical results, according to binary logistic regression (BLR) modelling, indicated that the WRH was significantly affected by the triple roles of village committees, among which information intermediary was the most effective followed by the trust builder and then the coordinated manager. Firstly, village committees' involvement facilitated the WRH by improving policy transparency and decreasing information cost. Secondly, the depth of village committees' participation (i.e., being involved in multiple phases) positively affected the WRH given its signification of the participation of farmers. Whereas the breadth of participation (i.e., considering various demands of different participants) negatively affected the process of WRH by reducing the decision-making efficiency. Thirdly, farmers' trust in institutions played a positive role in the WRH, but their confidence in village cadres had limited impact. We therefore argue that promising village committees should act as "all-round stewards" in the decision-making of rural households, which not only includes the transmit of information between those above and those below, but also needs to actively strive for farmers' trust by letting their voice heard. Based on our empirical findings, this paper finally proposed some policy suggestions, such as strengthening mutual communication, empowerment of rural grassroots, encouraging farmers' participation and improving formal institutions.

**Keywords:** rural homestead reform; grassroots village; farmers' willingness; land use transition; rural revitalization


## 1. Introduction

The depopulation of rural areas (excluding exurbs along the urban–rural continuum) is observed across the globe [1–3]. Since the reform and opening-up in the late 1970s, China has urbanized at an unprecedented speed and has perhaps experienced the world's greatest rural-to-urban migration ever [4–6]. Nevertheless, construction land in rural China has increased steadily since the 1990s with the registered population declining, which is recognized as the 'paradox of development' or 'dilemma of governance' [7–10]. This phenomenon has inevitably led to a far-reaching restructuring of physical and human landscapes in rural China and the inefficiency of land use and 'village hollowing' as well [11–14]. As Zhou et al. [8] report, there had been 2.7 million rural settlements in China, covering an area of 19.13 million hectares, of which about one-ninth is underutilized in 2015.

To address the issues of 'village hollowing' and achieve sustainable rural development, the Chinese government has launched a set of rural land-use reforms, which is called rural revitalization [7,8,15]. Withdrawal from rural homesteads (WRH) is expected to be helpful in accelerating the restructuring of rural areas and balancing urban–rural development [11,15,16]. Given the dualistic nature of urban–rural land use system, the WRH in China is widely considered as a hybrid and contested process involving numerous stakeholders [14,15,17,18].

Generally, two strands of literature are related to the local and external studies of WRH in China. In the first strand, researchers focus on analyzing and modeling farmers' willingness to withdraw from rural homesteads and spatial heterogeneity features with the help of pilot survey datasets, geographical information, and socio-economic variables from statistical yearbooks [17–20]. Relying on the technical advancements in data mining technology, GIS spatial modeling, remote sensing, and spatial econometrics, these geographical studies have identified individual attributes (e.g., age, gender, education, occupation), household characteristics (e.g., family size, homestead area, total family income, and urban housing status), and regional configurations (e.g., village location, average arable land area, urban–rural income gap) as key factors influencing WRH in China [16,19–22]. In the second strand, processes of withdrawal from rural homesteads are examined from the perspective of institutions and governance mostly through qualitative methods (e.g., text analysis, social network approach) [11,14,15,23,24]. This line of political science literature has demonstrated that WRH in China is inextricably intertwined with the separation of the "three rights of rural land" (i.e., collective ownership, land contract right, and land use right), dual-track household register (hukou) and social security systems, as well as the community cohesion and rural identity [25–30].

These parallel but distinct research traditions limit a comprehensive understanding of the WRH in China, because of their neglecting of the role played by village committees—agents of collective economic organizations. Given that homesteads in rural China are collectively owned and serve as the last safety net for survival by the majority of farmers with the Chinese bifurcated social security system [11,27,31], it is reasonable to fully understand the reason why farmer's willingness represents a key factor in the withdrawal process [16,17,19]. Wang et al. [32] find that farmers can hardly understand the legal regulations regarding property rights of rural homesteads and any other institutional arrangements, while the village committee, as a grassroots organization having deep daily contact with rural dwellers, plays an irreplaceable role in local governance [33,34]. In this sense, it is important to analyze village committees' role(s) in the WRH.

Stemming from both theories on behavioral economics and the rural system reform reality, this research broadened the debate about different roles played by village committees in the WRH by developing a comprehensive framework in a testbed of rural China. While most of the research foci have been attached to the WRH in China, the other side of the nexus—village committees—has often been reduced to the observed intermediary and intervention effects [35]. There seems to be a lack of knowledge about the roles and impacts of village committees concerning the WRH. Furthermore, the WRH—a key part of land use transition—plays an important role in facilitating land use transition [4]. Specifically, the WRH involves changes in both dominant and recessive morphologies of land use. This work can to some extent enrich our understanding of the trajectory and impetus of land use transition, especially in the aspect of recessive morphologies of land use.

Bearing the aforementioned background in mind, this research investigates the role(s) of village committees influencing farmers' willingness to withdraw from rural homesteads, and explores how village committees affect households' decisions regarding the WRH. Then we conduct an empirical test by drawing on the case of Sichuan Province in western China. In the following sections, we first propose a conceptual framework to analyze the influencing mechanisms of village committees in WRH.

Section 3 provides a brief review of the study area and explains the data and methodology. In Section 4, we examine the influencing factors of WRH in 24 villages in Sichuan Province. Section 5 discusses the major findings and policy implications. Finally, Section 6 concludes.

## 2. Background and Conceptual Framework

WRH refers to the transfer of homestead use right to collective economic organizations under farmers' voluntary application with compensation from the government. After transferring, farmers can either live in concentrated settlements nearby or directly migrate to cities. However, house-losing farmers tend to resettle in concentrated residential zones rather than migrating to cities, given that most farmers can hardly afford urban housing and have to earn their lives by engaging in agricultural activities [22]. In China, WRH is mostly dominated by local governments at the county/township-levels. These local authorities tend to give grassroots (village committees) greater autonomy regarding the withdrawal procedure, compensation standards and resettlement planning, and introduce enterprises to implement the project of land remediation. Within the context of administrative decentralization and community empowerment, the leading role of village committees in rural governance is increasingly recognized by both scholars and policymakers [36–38]. The participation of village committees is essential for WRH in at least the following three aspects (Figure 1, drawn by authors).

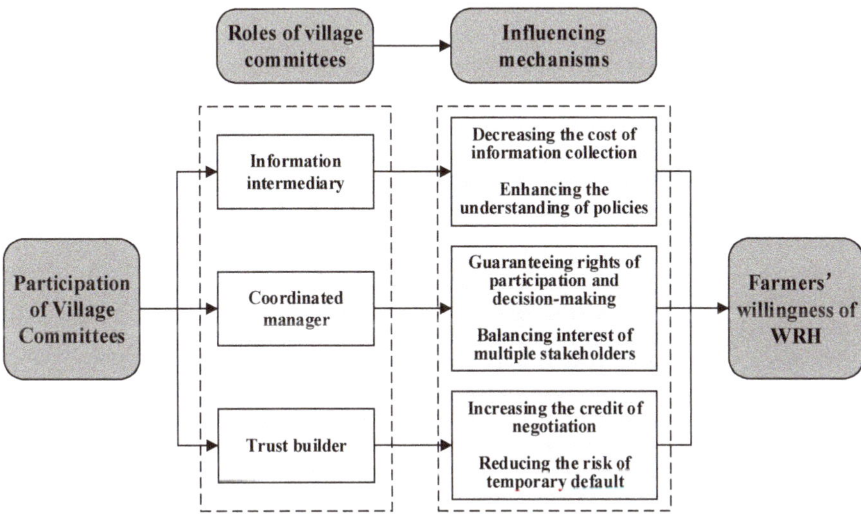


**Figure 1.** Influencing mechanisms of village committees on the withdrawal from rural homesteads (WRH) in China.

### 2.1. Information Intermediary

In recent years, development research has increasingly focused on the importance of access to information and found it a critical role in farmers' better understanding of governmental policies and in reducing their uncertainty about eventual benefits [39]. Despite recent advances in communication techniques, access to information in some rural communities is still restricted by geographical distance and administrative areas [40]. Besides, rural communities in developing countries are naturally oral societies, and thus prefer information delivered through face-to-face communication [41]. Specifically, farmers in China rely heavily on village cadres as sources of authoritative information about rural land use due to relevancy in context and content [42,43]. Therefore, village committees, as grassroots in China's hierarchy administrative system, as an intermediary, play a significant role in delivering information to villagers.

In a nutshell, the village committees' role, as the information intermediary, is mainly shown through propagating approaches and information transparency. First, policy propagating approaches used by village committees affect farmers' understanding of relevant information. Diversified approaches of propaganda do not only advance the access to information, it may also enhance farmers' understanding and acceptance through exposure to new ideas and an increased amount of information [44]. Secondly, information transparency is the key to policy implementation. In theory, information transparency can be understood from two perspectives, namely procedure transparency and legitimacy transparency. The former is a transparent flow of information, through which participants need to be informed about the course of the process and expected inputs. The latter means transparency on the legitimacy of decision-making [45].

### 2.2. Coordinated Management

As noted at the outset, the withdrawal of homesteads in rural China is a complex process involving a multitude of activities conducted by numerous participants/stakeholders [14,18]. Village committees also play an irreplaceable role in the management of rural daily affairs. Given that successful self-organization by farmers can greatly reduce holdout problems and costs of negotiation than the public organized structure of governance, involving village committees in WRH is usually more efficient and effective [46]. In a neoliberal governance regime, village committees can increase the legitimacy of farmers' participation in WRH and guarantee their interests by giving villagers a voice and making it heard by superior authorities [47].

In practice, village committees take the lead in formulating implementation plans of WRH and convening villagers to select their preferred one(s), and take part in setting standards of housing and compensation for house-losing farmers as well. Village committees can transmit farmers' demands for community remediation to up-level authorities in a timely manner and maximize their compensation. Meanwhile, village committees, as the agent for all villagers, are responsible for selecting developers for resettlement (valid candidates are usually selected in means of public bidding) and supervising the progress and quality of related projects. With local knowledge and farmers' lifestyle in mind, village committees can also coordinate and balance the interests of different stakeholders. In this regard, village committees' involvement in WRH can not only guarantee farmers' rights of participation in decision-making but also help to protect the interests of those house-losing farmers. Inferentially, the more deeply village committees participate and/or the more democratic their decision-making is, the higher degree of farmers' willingness is to withdraw from their homesteads.

### 2.3. Trust Building

As a precondition for a well-managed contract, trust can help to reduce transaction costs and maintain stable relations among participants [48]. Specifically, mutual trust can save bargaining time before the transaction, and decrease resources in the implementation of a contract [49]. Farmers' trust in village committees can form closer social relations in communities and significantly improve the efficiency of response to problems [50]. Scholars have pointed out that low trust and experiences with low reciprocity could lead to a decrease in farmers' willingness to participate in community management schemes [51].

In theory, farmers' trust in village committees can be divided into interpersonal trust (i.e., trust in village cadres and their daily works) and institutional trust (i.e., trust in political institutions and village committees' management schemes) [52]. From the perspective of interpersonal trust, negotiation cost can be decreased on WRH, if the interpersonal relationship between farmers and village cadres is harmonious. Otherwise, more procedural works would be needed because of the asymmetrical distribution of power [53] and farmers' willingness to sign the contract of WRH would decrease as a corresponding increase in the transaction cost [54]. In other words, the higher level of trust can increase the sense of credit in negotiation processes and good expectations of the results, while reducing negotiation resistance and stimulating more farmers to withdraw from the rural homesteads. Besides, institutional trust depends on the "reputation mechanism" of village committees which is developed from the long-term cooperation between the village committee and farmers [55]. Farmers' trust in village committees leads to a higher degree of acceptance of relevant policies and a lower risk of temporary default. In summary, farmers' trust in grassroots policies (i.e., institutional trust) and village cadres (i.e., interpersonal trust) may have a positive influence on the WRH.

## 3. Research Setting

### 3.1. Study Area and Data Collection

As illustrated in Figure 2, a total of 24 administrative villages in three counties/districts (i.e., Luxian, Pidu, Qionglai) in Sichuan Province were selected as samples. The primary reason to take these regions as examples is that these areas have carried out pilot projects of rural homesteads reform in China on one hand. On the other hand, these pilot regions are at different levels of economic development and topographical characteristics, which might be helpful in increasing the representativeness of samples To be specific, Luxian County is one of the 15 national level pilots on rural homestead acquisition and approval, paid use of homesteads, and mortgage loans [7]. Situated in Chengdu, the provincial capital of Sichuan, Pidu District and Qionglai City are exploring rural homesteads exit mechanisms within the context of urban–rural integration, comprehensive land remediation, high-standard farmland conservation, new countryside building, and rural vitalization [56]. Topographically, the sample villages are located in the southern mountain area (Luxian and Qionglai) and Chengdu plain (Pidu), respectively. Additionally, the level of economic development in the three regions varies. Pidu is an economically developed region with its per capita GDP over RMB 73,000 in 2019; Luxian is relatively underdeveloped with a per capita GDP of RMB 43,400 and the lowest urbanization rate; Qionglai is in the middle of its two counterparts with a per capita GDP of RMB 53,413.

Based on the population size and geographic location (e.g., distance to urban areas and terrain conditions), we chose a total of 24 sample villages, most of which were pilots of homestead reform. Considering the differentiated progress of reform in villages, we also took local governments' suggestions regarding the collection of sample villages into account. Concerning the specific method of household sampling, we decided to randomly select 20–30 households in each village and finally interviewed a total of 310 households from August to October 2018. After the test of the validity of questionnaires, we received 308 valid responses (Table 1). Firstly, the gender ratio was balanced, and male farmers had a higher degree of willingness to withdraw from homesteads than their female counterparts. In terms of age, respondents between 40 and 70 years old were the majority (70.8%), and farmers aged 30 to 39 had the lowest willingness of WRH. Regarding the educational level, most respondents received the education of primary school or below, and only 37.0% received a secondary or above education. The data showed that farmers with higher education levels tend to have a higher degree of willingness. For the household income, the majority ranged from RMB 15,000 to RMB 30,000, and 89.3% of the total households earned more than RMB 10,000 annually. Coincidently, we hereby assume that farmers with higher incomes are more willing to withdraw from homesteads. Moving to the family size, the majority was made up of five or six members, followed by three/four-person families. Only 8.1% were single or double families. Households with more family members were more inclined to withdraw from their homesteads.

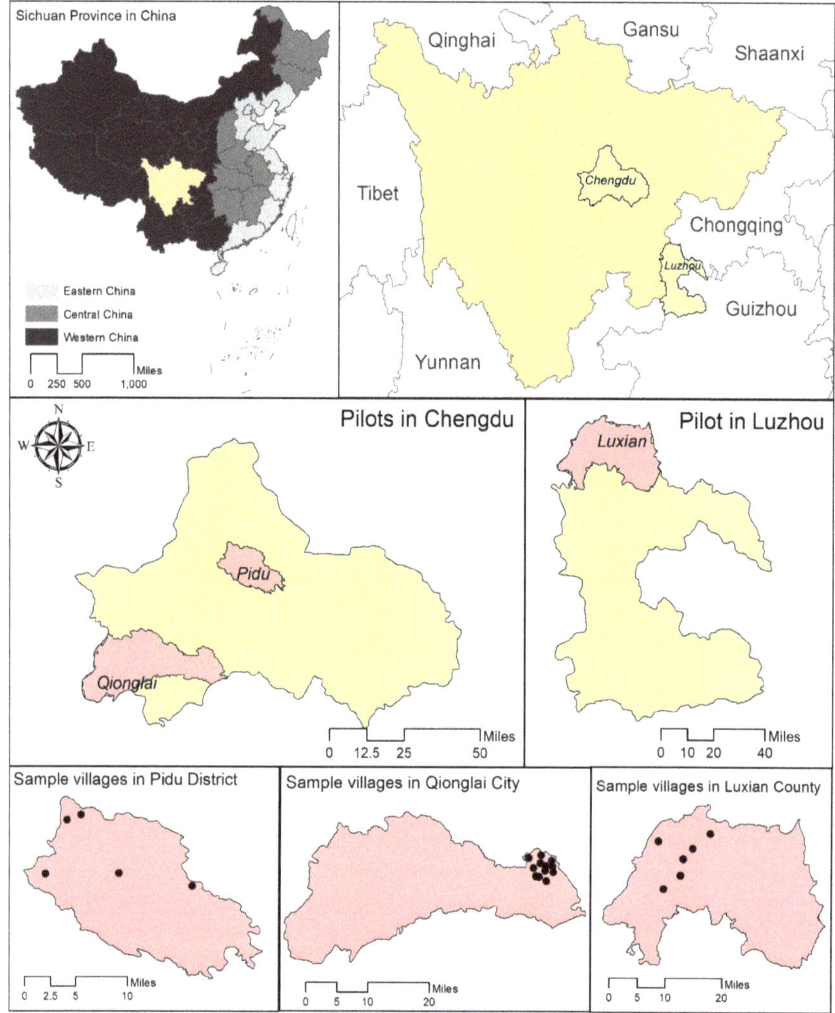


**Figure 2.** Study area and location of sample villages.

**Table 1.** Profile of survey respondents in sample villages (*n* = 308).

| Feature | Options | Percentage (%) | Willing of WRH (%) | Feature | Options | Percentage (%) | Willing of WRH (%) |
|---|---|---|---|---|---|---|---|
| Gender | Male | 42.9 | 83.3 | | <0.5 | 10.7 | 65.4 |
| | Female | 57.1 | 79.5 | | [0.5, 1) | 13 | 75.0 |
| Age | <30 | 5.7 | 83.3 | Household annual net income (RMB 10k) | [1, 1.5) | 11.7 | 86.4 |
| | 30–39 | 6.4 | 56.3 | | [1.5, 3) | 24.7 | 81.6 |
| | 40–49 | 24.2 | 80.8 | | [3, 5) | 22.1 | 84.2 |
| | 50–59 | 20.8 | 79.8 | | [5, 7) | 9.7 | 100.0 |
| | 60–69 | 25.8 | 81.3 | | [7, 9) | 3.6 | 100.0 |
| | 70–79 | 14.9 | 91.3 | | ≥9 | 4.5 | 100.0 |
| | ≥80 | 2.2 | 85.7 | | | | |
| Education | Primary school and below | 63 | 82.5 | | 1–2 | 8.1 | 80.0 |
| | Junior high school | 25.3 | 82.1 | | 3–4 | 35.1 | 81.5 |
| | Senior high school | 6.5 | 75.0 | Family size | 5–6 | 48.7 | 80.0 |
| | Vocational/technical college | 1.6 | 80.0 | | 7–8 | 6.8 | 85.7 |
| | Undergraduate and above | 3.6 | 63.6 | | 9–10 | 1.3 | 100.0 |

### 3.2. Empirical Models and Variable Specifications

We first employed binary logistic regression (BLR) to estimate the impact of village committees on farmers' willingness of WRH. BLR is a nonlinear statistical method of regression analysis for binary dependent variables [57,58]. In this study, farmers' willingness of WRH (Y) is a binary dependent variable with values of 0 (unwilling) and 1 (willing). Given the multiple impacts of China's transition on WRH at multiple levels, the overall probability of WRH is $P(Y = 1)$. Thus, the BLR model can be framed as follows:

$$P(Y = 1|x_1, x_2 \cdots, x_m) = \exp\left(\beta_0 + \sum \beta_i x_i\right) / \left(1 + \exp\left(\beta_0 + \sum \beta_i x_i\right)\right) \tag{1}$$

$$\mathrm{logit}P(Y = 1|x_1, x_2 \cdots, x_m) = \ln\left(\frac{P}{1 - P}\right) = \beta_0 + \sum_{i=1}^{m} \beta_i x_i \tag{2}$$

Here, $x_i$ is the explanatory variables, and logit $P(Y = 1)$ is a linear combination function of the explanatory variables. Parameters $\beta_i$ denote the partial regression coefficient to be estimated. After variable standardization, the obtained $\beta_i$ may reflect the relative influence of each independent variable on dependent variables. Drawing upon the aforementioned conceptual framework, four groups of explanatory variables were selected to model the influence of village committees on farmers' willingness of WRH (Table 2).

**Table 2.** Definition and descriptive statistics of each variable.

| Variables | Definitions | Mean | S.D. | Expected Sign |
|---|---|---|---|---|
| **Information intermediary** | | | | |
| Diversity of publicity | The number of specific ways for village committees to promote policies about WRH. The value ranges from 1 to 6. | 1.675 | 0.908 | + |
| Information transparency | Level of publicity and transparent of WRH policies and other related information: very opaque = 1; opaque = 2; general = 3; transparent = 4; very transparent = 5 | 3.299 | 0.976 | + |
| **Coordinated management** | | | | |
| Workload of village committees | The number of specific works that village committees participate in during the process of WRH. The value ranges from 0 to 6. | 3.240 | 2.003 | + |
| Degree of democratic decision-making | The frequency of the consultation organized by village committees during the process of WRH: never = 1; hardly = 2; often = 3; usually = 4 | 3.010 | 1.087 | + |
| **Trust building** | | | | |
| Interpersonal trust | Whether village committees are trusted by farmers in daily work: distrust = 1; less trust = 2; general = 3; trust = 4; very trust = 5 | 2.750 | 1.0734 | + |
| Institutional trust | Whether the rules and regulations established regarding the WRH can be trusted by farmers: distrust = 1; less trust = 2; general = 3; trust = 4; very trust = 5 | 3.273 | 1.057 | + |
| **Control variables** | | | | |
| Age | Age of all respondents | 55.536 | 13.614 | − |
| Gender | Male = 1; female = 2 | 1.571 | 0.4957 | − |
| Educational level | Primary school and below=1; junior high school = 2; senior high school = 3; vocational or technical secondary college = 4; Undergraduate and above = 5 | 1.575 | 0.950 | + |
| Proportion of migrants | The ratio of migrant workers in households | 0.402 | 0.238 | + |
| Household annual net income | [0, 5k) = 1; [5k, 10k) = 2; [10k, 15k) = 3; [15k, 30k) = 4; [30k, 50k) = 5; [50k, 70k) = 6; [70k, 90k) = 7; [90k, +∞) = 8 | 4.000 | 1.808 | + |

**Table 2.** *Cont.*

| Variables | Definitions | Mean | S.D. | Expected Sign |
|---|---|---|---|---|
| **Control variables** | | | | |
| Area of the contracted land | The actual total area of contracted land owned by households | 3.021 | 1.898 | +/− |
| Area of the homestead | The actual area of the original homestead owned by households or having been withdrawn. | 289.698 | 299.368 | +/− |
| Land used period of the homestead | The land used period of the original homestead owned by households or having been withdrawn | 24.666 | 11.103 | − |

## 4. Results

### 4.1. General Impacts of Village Committees on Farmers' Willingness of WRH

In association with theoretical and contextual issues, the influencing mechanisms of farmers' willingness of WRH are examined in a BLR model with consideration of the heterogeneities of governance attributes and individual characteristics. We first analyzed the overall preferences of different respondents with the participation of village committees. The result is reported in Table 3.

**Table 3.** Farmers' willingness of WRH with the participation of village committees.

| Variables | | WRH (%) | | Variables | | WRH (%) | |
|---|---|---|---|---|---|---|---|
| | | Willing | Unwilling | | | Willing | Unwilling |
| Diversity of publicity | 1 | 72.3 | 27.7 | Degree of democratic decision-making | 1 | 72.3 | 27.7 |
| | 2 | 91.4 | 8.6 | | 2 | 89.7 | 10.3 |
| | 3 | 97.2 | 2.8 | | 3 | 74.4 | 25.6 |
| | 4 | 88.2 | 11.8 | | 4 | 86.0 | 14.0 |
| | 5 | 100.0 | 0.0 | | Average | 3.1 | 2.8 |
| | Average | 1.8 | 1.3 | | | | |
| Information transparency | 1 | 15.4 | 84.6 | Interpersonal trust | 1 | 31.8 | 68.2 |
| | 2 | 55.8 | 44.2 | | 2 | 63.0 | 37.0 |
| | 3 | 86.5 | 13.5 | | 3 | 87.1 | 12.9 |
| | 4 | 91.1 | 8.9 | | 4 | 94.1 | 5.9 |
| | 5 | 100.0 | 0.0 | | 5 | 82.1 | 17.9 |
| | Average | 3.5 | 2.3 | | Average | 3.4 | 2.4 |
| Workload of the village committee | 0 | 64.7 | 35.3 | Institutional trust | | | |
| | 1 | 63.6 | 36.4 | | 1 | 21.1 | 78.9 |
| | 2 | 87.5 | 12.5 | | 2 | 63.0 | 37.0 |
| | 3 | 94.7 | 5.3 | | 3 | 87.6 | 12.4 |
| | 4 | 80.0 | 20.0 | | 4 | 94.8 | 5.2 |
| | 5 | 78.1 | 21.9 | | 5 | 80.0 | 20.0 |
| | 6 | 93.3 | 6.7 | | Average | 3.5 | 2.4 |
| | Average | 3.4 | 2.4 | | | | |

First, farmers' willingness to withdraw from rural homesteads generally increases with the way publicity is diversifying. This implies the important role of village committees in farmers' better understanding of policies on WRH, as well as of their future development [29,59,60]. Similarly, higher policy transparency significantly improves the degree of farmers' willingness of WRH, whereas lower transparent policies tend to have negative effects. This suggests the importance of information availability in the process of WRH.

Compared with farmers under the management of village committees with less workload, their counterparts coordinated by those 'hardworking' committees were not necessarily more inclined to withdraw from their homesteads, suggesting that diligent leaders neither necessarily improve work efficiency in general nor improve farmers' willingness of WRH in particular. Instead, the workload of village committees only makes sense on average. Coincidently, farmers' willingness of WRH fluctuated with an increasing degree of democracy in village committees' decision-making process.

This shows that the frequency of consultations with villagers may not directly have an impact on their final decisions regarding WRH. Rather, the details of consultations may matter.

From the perspective of trust building, farmers who are willing to withdraw from their homesteads have an average interpersonal trust level of 3.4, which is 42% higher than their unwilling counterparts, reflecting the importance of village committees' daily image in farmers' minds. Consistent with the impact of policy transparency, the changing trend of institutional trust echoes that of farmers' willingness of WRH excluding those at the highest level of trust. Yet, the average level of institutional trust indicates a similar pattern as interpersonal trust.

*4.2. Roles Played by Village Committees in WRH: A Multi-Mechanism Perspective*

To shed further light on the role of village committees, we applied the BLR model to investigate the determinants of WRH. The results are presented in Table 4. The significance value of the H-L test is 0.896 and most coefficients are statistically significant as expected, indicating that the BLR model is a satisfactory fit.

**Table 4.** Results of the binary logistic regression (BLR) model regression.

| Variables | Coefficient | Standard Error | Significance |
|---|---|---|---|
| **Information intermediary** | | | |
| Diversity of publicity | 0.638 | 0.289 | 0.027 ** |
| Information transparency | 0.866 | 0.245 | 0.000 *** |
| **Coordinated management** | | | |
| Workload of the village committee | 0.192 | 0.099 | 0.053 * |
| Degree of democratic decision-making | −0.377 | 0.188 | 0.045 * |
| **Trust building** | | | |
| Interpersonal trust | −0.141 | 0.302 | 0.639 |
| Institutional trust | 0.578 | 0.314 | 0.066 * |
| **Control variables** | | | |
| Age | −0.067 | 0.318 | 0.833 |
| Gender | 0.018 | 0.015 | 0.238 |
| Educational level | −0.164 | 0.194 | 0.398 |
| Proportion of migrant workers | 1.287 | 0.671 | 0.055 * |
| Household annual net income | 0.217 | 0.092 | 0.019 ** |
| Total area of the contracted land | 0.222 | 0.103 | 0.031 ** |
| Total areas of the original homestead | 0.000 | 0.000 | 0.578 |
| Land used period of the original homestead | 0.003 | 0.014 | 0.839 |
| Constant | −4.67 | 1.685 | 0.006 *** |
| −2 log likelihood | | 207.666 | |
| Cox & Snell R Square | | 0.254 | |
| Nagelkerke R Square | | 0.410 | |
| H-L Test | Chi-Square = 3.546, Df = 8, Significance = 0.896 | | |

Note: ***, **, and * represent significance at 1%, 5%, and 10%, respectively.

Variables representing information intermediary are the most significant variables related to farmers' willingness of WRH. Specifically, the degree of farmers' willingness increases by 0.638 and 0.866 when one unit increases concerning publicity diversity and information transparency, respectively. These indicate that the more diverse the publicity approaches of village committees are, the stronger the farmers' intention to withdraw from their homesteads is. The reason may be that village committees, as intermediaries for higher-level governments to transmit documents on WRH to rural households, which are much more skilled in publicizing those policies can significantly reduce the cost of villagers' accessibility to crucial information and enhance farmers' intention to withdraw from their homesteads.

Regarding the impact of coordinated management, the workload of village committees shows a significant positive relationship with the farmers' willingness to WRH at the statistical level of 10%, which is consistent with the hypothesis that village committees play an important role as leaders, organizers, and coordinators in the process of rural restructuring [32,60]. In contrast, the democratic decision-making of village committees has a negative influence on farmers' willingness of WRH, which deviates from the theoretical hypothesis. The possible reason is that, villagers with strong and weak democratic preferences tend to coexist in the same community, which might be a puzzle for collective decision-making and to some extent reduce the efficiency of decision execution [61,62].

We also use two variables to describe the village committees' role in trust building, and the results show that farmers' willingness of WRH is affected by the institutional trust, namely farmers' trust in institutions is an impetus underlying WRH. The coefficient of "institutional trust" is significant and positive. However, the coefficient of "interpersonal trust" is negative and not statistically significant, suggesting a limited impact of farmers' trust in local cadres on their final decisions. The results show that, compared with interpersonal trust, institutional trust could provide a more standardized and reliable participation atmosphere for farmers, which increases the farmers' willingness on WRH.

*4.3. Impacts of Sample-Specific Characteristics: Explanation of Control Variables*

Given that rural homestead is nowadays not only a space for living but also a symbol of rural identity in China [27], social affiliation affects farmers' willingness of WRH more significantly than homestead characteristics. Thus, neither total area nor using period of land as homestead features is statistically significant. Additionally, the significant coefficients in the range from 0.217 to 1.287 for household attributes (i.e., the proportion of migrant workers, household annual net income, and the total area of contracted land) indicate that family background rather than individual characteristics of the respondents (mostly the head of rural household) played a vital role in WRH at the case level. This could be explained that WRH can greatly change the way of whole families' production and lifestyle and is a collective decision by all family members. Specifically, villagers are more inclined to withdraw from their homesteads with an increase in the proportion of migrant workers and a rise in the total net household income. Contrasting with the agriculture-based families, those with more members engaging in nonagricultural activities tend to have higher income and depend less on the rural economy [63,64]. Moreover, an increase in the area of contracted land also improves the degree of farmers' willingness to leave the homestead. This finding has been confirmed by Hao and Tang [65].

## 5. Discussion

*5.1. Main Findings: Triple Roles of Village Committees in the WRH*

The results showed that there is a close and significant relationship between the triple roles of village committees and the farmers' willingness on WRH, based on the descriptive statistical methods and BLR models. First, the role of village committees as information intermediaries has the most significant and positive impact, which echoes the study on village committees' role as information intermediary in farmland transfer by Shi et al. [35]. The participation of village committees can significantly reduce farmers' information costs, facilitate their understanding of policies.

Second, the role of village committees—as coordinated managers—matters in the WRH. As Zhang and Han [16] noted, both the breadth and depth of farmers' involvement in the WRH were limited given the complexity of rural land consolidation. The involvement of village committees is, therefore, significant in protecting farmers' interests. The deeper village committees participate in those processes, the more likely farmers can receive reasonable compensation and make positive decisions on the WRH. On the other hand, the efficiency depends heavily on the degree of villagers' participation. In rural China, the participation of villagers is restricted to the levels of "tokenism", which locates at the middle rungs of the "ladder of citizen participation" [66]. In other words, villagers, as powerless have-nots, lack the power to ensure their views are being heeded, though they are proffered to hear according to procedures. In this view, if farmers believe that the democratic decision-making of village committees is relatively "troublesome" and has limited impact, their enthusiasm to participate will be reduced.

Third, village committees as policy transmitters in the WRH can gain villagers' trust in related policies. In other words, farmers tend to trust in policies issued by higher-level governments rather than local village cadres. On one hand, Chinese peasants are far more rational than being expected in making decisions related to their interests; on the other hand, most villagers remain skeptical about what local cadres say [52], due to inherent constraints on resources and power in contemporary China. Ironically, during our field visit to sample villages, we found that villagers were still willing to withdraw from their homesteads, even if they did not trust local cadres. The reason is that farmers appear to appreciate and trust the policies formulated by higher-level governments. This confirms the finding by Yang and Tang [67]—that institutional trust is more than a product of traditional values in China. Instead, it is resultant from an individual's rational choice based heavily on the evaluation of institutional performance and government-controlled politicization.

### 5.2. Adaptabilities and Limitations of Village Committee Roles Initiatives

In addition to the rural homesteads system reform, village committees tend to play an essential role in almost all village public affairs decisions. The explanation framework of village committees' roles thus could be widely used in the interpretation of related issues, such as rural land transfer, construction of village public roads, publicity, and implementation of major policies. Firstly, improving the authenticity and transparency of information will solve the issue of the information asymmetry between village committees and farmers. Secondly, we need to pay more attention to the effectiveness of democratic decision-making and local participation. Democratic consultation and participation in form will not bring the expected results. Thirdly, creating a more harmonious and trust relationship between the cadres and the peasantry by changing communication ways in daily life will not only affects the progress of the homestead system reform, but also smooth decision-making processes of village affairs. Fourthly, improving the institutional trust will help remove practical difficulties faced by village cadres.

The research limitations are as follows: (1) it is difficult to establish a theoretical framework to cover all roles of village committees, because the natural, cultural, social, and economic conditions of villages are quite different. We attempt to build this framework based on the basic functions of village committees; (2) Variables need to be adjusted in accordance with local conditions, to reflect the different roles of village committees in other public affairs and regions. (3) We mainly use the Likert scale method to develop indicators. This method cannot measure the efficiency of democratic decision-making.

### 5.3. Policy Implications

From a policy perspective, we argue that more efforts are still needed to facilitate the reform of rural homesteads and to implement the national strategies of rural revitalization. Firstly, various approaches and channels of communication between the village committee and farmers should be developed to enhance the intermediary role of village committees. Particularly, information disclosure systems need to be enhanced regarding relevant policies and information of public affairs. It should be noted that there is currently a lack of market information platforms for WRH, and the development of agricultural land transfer and transaction information platforms is still in the early stage. Thus, building a public market platform should be prioritized in the future. Secondly, considering the differences among villages, even within a county-level jurisdiction, in terms of both economic and social conditions, village committees should be empowered in decision-making regarding WRH. Thirdly, the involvement of farmers is essential for WRH and the development of the countryside [68]. Fourthly, we suggest trust can be built in the countryside by various collective activities and constructing formal policies on administrative credit.

## 6. Conclusions

Based on the field survey data in three pilot counties/districts of Sichuan province, China, we in this study analyzed the triple roles of village committees and investigated their impacts on the WRH. The results indicate that village committees' participation significantly affected farmers' willingness to withdraw from rural homesteads. Theoretically, the work reported in this paper initiated a conceptual framework on the roles of village committees, which might be beneficial to our better understanding of the village cadre-farmers nexus and more conducive to rural land system reform.

Taking the role of information intermediary, village committees can significantly improve villagers' willingness to withdraw from rural homesteads through effectively propagating information and explaining policies clearly. With respect to the role of coordinated managers, village committees are helpful in protecting the interest and benefit of those house-losing farmers, which is conducive to the orderly implementation of WRH. Yet, the democratic coordination of village committees is not always effective given the occasional occurrence of symbolically soliciting villagers' opinions. Regarding the role of trust builders, village committees are of great help in reducing the risk and cost of policy negotiation. Compared with interpersonal trust, institutional trust has a significantly positive role in WRH, indicating that the personal charm of village cadres has a limited impact and more formal institutional arrangements are needed in contemporary China.

Furthermore, we conclude that this study can be improved in the future by more case comparisons and analyses, particularly the trajectories of WRH in the eastern coast. Specifically, the following two directions are worthy of scholar attentions: (1) it is important to analyze the different characteristics of village committees in homestead exit, considering the differences in natural conditions and institutional backgrounds; (2) it is meaningful to identify factors such as the personal characteristics of village leaders and the acquaintances of rural communities, given that the personal characteristics and leadership styles of village cadres can affect the relationship between farmers and village committees.

**Author Contributions:** Conceptualization, P.T. and J.G.; methodology, P.T. and J.C.; validation, P.T., J.G., and M.L.; formal analysis, P.T., J.C., and J.G.; investigation, P.T. and J.C.; resources, P.T. and M.L.; data curation, P.T. and J.C.; writing—original draft preparation, P.T., J.C., and J.G.; writing—review and editing, P.T., J.G., and J.W.; project administration, P.T.; funding acquisition, P.T. All authors have read and agreed to the published version of the manuscript.

**Funding:** This research was financially supported by Sichuan Science and Technology Program (2020JDR0106), Sichuan Center for Rural Development Research Program (CR1917), the China Scholarship Council (201906245025), and National Natural Science Foundation of China (41971215).

**Conflicts of Interest:** The authors declare no conflict of interest.

## References

1. Liu, Y.; Li, Y. Revitalize the world's countryside. *Nature* **2017**, *548*, 275–277. [CrossRef] [PubMed]
2. Lambin, E.F.; Meyfroidt, P. Land use transitions: Socio-ecological feedback versus socio-economic change. *Land Use Policy* **2010**, *27*, 108–118. [CrossRef]
3. Kates, R.W.; Parris, T.M. Long-term trends and a sustainability transition. *Proc. Natl. Acad. Sci. USA* **2003**, *100*, 8062. [CrossRef] [PubMed]
4. Long, H. *Land Use Transitions and Rural Restructuring in China*; Springer: Singapore, 2020.
5. Bai, X.; Shi, P.; Liu, Y. Realizing China's urban dream. *Nature* **2014**, *509*, 158–160. [CrossRef]
6. Li, L.; Liu, Y. Understanding the gap between de facto and de jure urbanization in China: A perspective from rural migrants' settlement intention. *Popul. Res. Policy Rev.* **2020**, *39*, 311–338. [CrossRef]
7. Gao, J.; Liu, Y.; Chen, J. China's initiatives towards rural land system reform. *Land Use Policy* **2020**, *94*, 104567. [CrossRef]
8. Zhou, Y.; Li, X.; Liu, Y. Rural land system reforms in China: History, issues, measures and prospects. *Land Use Policy* **2020**, *91*, 104330. [CrossRef]
9. Yu, Z.; Wu, C.; Tan, Y.; Zhang, X. The dilemma of land expansion and governance in rural China: A comparative study based on three townships in Zhejiang Province. *Land Use Policy* **2018**, *71*, 602–611. [CrossRef]
10. Gao, J.; Jiang, W.; Chen, J.; Liu, Y. Housing-industry symbiosis in rural China: A multi-scalar analysis through the lens of land use. *Appl. Geogr.* **2020**, *124*, 102281. [CrossRef]
11. Long, H.; Li, Y.; Liu, Y.; Woods, M.; Zou, J. Accelerated restructuring in rural China fueled by 'increasing vs. decreasing balance' land-use policy for dealing with hollowed villages. *Land Use Policy* **2012**, *29*, 11–22. [CrossRef]

12. Liu, Y.; Liu, Y.; Chen, Y.; Long, H. The process and driving forces of rural hollowing in China under rapid urbanization. *J. Geogr. Sci.* **2010**, *20*, 876–888. [CrossRef]

13. Liu, Y.; Fang, F.; Li, Y. Key issues of land use in China and implications for policy making. *Land Use Policy* **2014**, *40*, 6–12. [CrossRef]

14. Zhu, X. Impact of the household registration system on farmers' rural housing land use decisions in China. *Land* **2017**, *6*, 75. [CrossRef]

15. Huang, X.; Li, H.; Zhang, X.; Zhang, X. Land use policy as an instrument of rural resilience—The case of land withdrawal mechanism for rural homesteads in China. *Ecol. Indic.* **2018**, *87*, 47–55. [CrossRef]

16. Zhang, X.; Han, L. Which factors affect farmers' willingness for rural community remediation? A tale of three rural villages in China. *Land Use Policy* **2018**, *74*, 195–203. [CrossRef]

17. Cao, Q.; Sarker, M.N.I.; Sun, J. Model of the influencing factors of the withdrawal from rural homesteads in China: Application of grounded theory method. *Land Use Policy* **2019**, *85*, 285–289. [CrossRef]

18. Wang, Q.; Zhang, M.; Cheong, K.-C. Stakeholder perspectives of China's land consolidation program: A case study of Dongnan Village, Shandong Province. *Habitat Int.* **2014**, *43*, 172–180. [CrossRef]

19. Fan, W.; Zhang, L. Does cognition matter? Applying the push-pull-mooring model to Chinese farmers' willingness to withdraw from rural homesteads. *Pap. Reg. Sci.* **2019**, *98*, 2355–2369. [CrossRef]

20. Gao, X.; Xu, A.; Liu, L.; Deng, O.; Zeng, M.; Ling, J.; Wei, Y. Understanding rural housing abandonment in China's rapid urbanization. *Habitat Int.* **2017**, *67*, 13–21. [CrossRef]

21. Chen, H.; Zhao, L.; Zhao, Z. Influencing factors of farmers' willingness to withdraw from rural homesteads: A survey in Zhejiang, China. *Land Use Policy* **2017**, *68*, 524–530. [CrossRef]

22. Zhang, L.; Fan, W. Rural homesteads withdrawal and urban housing market: A pilot study in China. *Emerg. Mark. Financ. Trade* **2020**, *56*, 228–242. [CrossRef]

23. Liu, Y.; Li, J.; Yang, Y. Strategic adjustment of land use policy under the economic transformation. *Land Use Policy* **2018**, *74*, 5–14. [CrossRef]

24. Zhao, Y. When guesthouse meets home: The time-space of rural gentrification in southwest China. *Geoforum* **2019**, *100*, 60–67. [CrossRef]

25. Kong, X.; Liu, Y.; Jiang, P.; Tian, Y.; Zou, Y. A novel framework for rural homestead land transfer under collective ownership in China. *Land Use Policy* **2018**, *78*, 138–146. [CrossRef]

26. Li, Y.; Liu, Y.; Long, H.; Cui, W. Community-based rural residential land consolidation and allocation can help to revitalize hollowed villages in traditional agricultural areas of China: Evidence from Dancheng County, Henan Province. *Land Use Policy* **2014**, *39*, 188–198. [CrossRef]

27. Gao, J.; Wu, Z.; Chen, J.; Chen, W. Beyond the bid-rent: Two tales of land use transition in contemporary China. *Growth Chang.* **2020**, *51*, 1336–1356. [CrossRef]

28. Chen, C.; Woods, M.; Chen, J.; Liu, Y.; Gao, J. Globalization, state intervention, local action and rural locality reconstitution—A case study from rural China. *Habitat Int.* **2019**, *93*, 102052. [CrossRef]

29. Long, H.; Tu, S.; Ge, D.; Li, T.; Liu, Y. The allocation and management of critical resources in rural China under restructuring: Problems and prospects. *J. Rural Stud.* **2016**, *47*, 392–412. [CrossRef]

30. Sargeson, S. Subduing "the rural house-building craze": Attitudes towards housing construction and land-use controls in four Zhejiang villages. *China Q.* **2002**, *172*, 927–955. [CrossRef]

31. Wang, M. Impact of the Global Economic Crisis on China's Migrant Workers: A Survey of 2700 in 2009. *Eurasian Geogr. Econ.* **2010**, *51*, 218–235. [CrossRef]

32. Wang, Y.; Chen, L.; Long, K. Farmers' identity, property rights cognition and perception of rural residential land distributive justice in China: Findings from Nanjing, Jiangsu Province. *Habitat Int.* **2018**, *79*, 99–108. [CrossRef]

33. Ma, S.; Mu, R. Forced off the farm? Farmers' labor allocation response to land requisition in China. *World Dev.* **2020**, *132*, 104980. [CrossRef]

34. Chen, C.; Gao, J.; Chen, J. Behavioral logics of local actors enrolled in the restructuring of rural China: A case study of Haoqiao Village in northern Jiangsu. *J. Rural Stud.* **2019**, in press. [CrossRef]

35. Shi, X.; Chen, S.; Ma, X.; Lan, J. Heterogeneity in interventions in village committee and farmland circulation: Intermediary versus regulatory effects. *Land Use Policy* **2018**, *74*, 291–300. [CrossRef]

36. Herbert-Cheshire, L. Contemporary strategies for rural community development in Australia: A governmentality perspective. *J. Rural Stud.* **2000**, *16*, 203–215. [CrossRef]

37. Landini, F.; Vargas, G.; Bianqui, V.; y Rebolé, M.I.M.; Martínez, M. Contributions to group work and to the management of collective processes in extension and rural development. *J. Rural Stud.* **2017**, *56*, 143–155. [CrossRef]

38. Iles, K.; Ma, Z.; Erwin, A. Identifying the common ground: Small-scale farmer identity and community. *J. Rural Stud.* **2020**, *78*, 25–35. [CrossRef]

39. Bukchin, S.; Kerret, D. The role of self-control, hope and information in technology adoption by smallholder farmers—A moderation model. *J. Rural Stud.* **2020**, *74*, 160–168. [CrossRef]

40. Onitsuka, K.; Hoshino, S. Inter-community networks of rural leaders and key people: Case study on a rural revitalization program in Kyoto Prefecture, Japan. *J. Rural Stud.* **2018**, *61*, 123–136. [CrossRef]

41. Msoffe, G.E.P.; Ngulube, P. Information sources preference of poultry farmers in selected rural areas of Tanzania. *J. Librariansh. Inf. Sci.* **2016**, *49*, 82–90. [CrossRef]

42. Hillman, B. The rise of the community in rural China: Village politics, cultural identity and religious revival in a Hui hamlet. *China J.* **2004**, *51*, 53–73. [CrossRef]

43. Gong, H. Study on policy "suspension" in rural area. *J. Northwest AF Univ. (Soc. Sci. Ed.)* **2017**, *17*, 51–57. Available online: http://www.xnxbs.net/xbnlkjdxskb/ch/reader/view_abstract.aspx?file_no=20170207&flag=1 (accessed on 26 November 2020).

44. Bodin, Ö.; Crona, B.I. The role of social networks in natural resource governance: What relational patterns make a difference? *Glob. Environ. Chang.* **2009**, *19*, 366–374. [CrossRef]

45. Rogge, E.; Dessein, J.; Verhoeve, A. The organisation of complexity: A set of five components to organise the social interface of rural policy making. *Land Use Policy* **2013**, *35*, 329–340. [CrossRef]

46. Tan, R.; Heerink, N. Public and self-organized land readjustment in rural China—A comparison. *J. Rural Stud.* **2017**, *53*, 45–57. [CrossRef]

47. Wang, J.; Hochman, Z.; Taylor, B.; Darbas, T.; van Rees, H.; Carberry, P.; Ren, D. Governing through representatives of the community: A case study on farmer organizations in rural Australia. *J. Rural Stud.* **2017**, *53*, 68–77. [CrossRef]

48. Joop, K.; Klijn, E.-H. *Managing Uncertainties in Networks: A Network Approach to Problem Solving and Decision Making*; Routledge: London, UK, 2004.

49. Guo, L. Influence of transaction costs, trusting relationship on farmer's compliance behavior—Based on survey data from 286 apple growers in Shandong Province. *J. Huazhong Agric. Univ. (Soc. Sci. Ed.)* **2015**, 56–61. Available online: http://hnxbw.cnjournals.net/hznydxsk/article/abstract/20150409 (accessed on 26 November 2020).

50. McCrea, R.; Walton, A.; Leonard, R. Rural communities and unconventional gas development: What's important for maintaining subjective community wellbeing and resilience over time? *J. Rural Stud.* **2019**, *68*, 87–99. [CrossRef]

51. Prazan, J.; Theesfeld, I. The role of agri-environmental contracts in saving biodiversity in the post-socialist Czech Republic. *Int. J. Commons* **2014**, *8*, 1–25. [CrossRef]

52. Lo, A.Y.; Cheung, L.T.O.; Lee, A.K.-Y.; Xu, B. Confidence and trust in public institution natural hazards management: Case studies in urban and rural China. *Prof. Geogr.* **2016**, *68*, 475–484. [CrossRef]

53. Prazan, J.; Dumbrovsky, M. Soil conservation policies: Conditions for their effectiveness in the Czech Republic. *Land Degrad. Dev.* **2011**, *22*, 124–133. [CrossRef]

54. Ieong, M.U. 'Know Who' may be better than 'Know How': Political connections and reactions in administrative disputes in China. *J. Chin. Gov.* **2019**, *4*, 233–251. [CrossRef]

55. Williamson, O.E. *The Economic Institutions of Capitalism: Firms, Markets, Relational Contracting*; Free Press: New York, NY, USA, 1985.

56. Liu, R.; Jiang, J.; Yu, C.; Rodenbiker, J.; Jiang, Y. The endowment effect accompanying villagers' withdrawal from rural homesteads: Field evidence from Chengdu, China. *Land Use Policy* **2020**, 105107, in press. [CrossRef]

57. Shu, B.; Zhang, H.; Li, Y.; Qu, Y.; Chen, L. Spatiotemporal variation analysis of driving forces of urban land spatial expansion using logistic regression: A case study of port towns in Taicang City, China. *Habitat Int.* **2014**, *43*, 181–190. [CrossRef]

58. Gao, J.; Yuan, F. Economic transition, firm dynamics, and restructuring of manufacturing spaces in urban China: Empirical evidence from Nanjing. *Prof. Geogr.* **2017**, *69*, 505–519. [CrossRef]

59. Tian, C.; Fang, L. The Impossible in China's Homestead Management: Free Access, Marketization and Settlement Containment. *Sustainability* **2018**, *10*, 798. [CrossRef]

60. Tu, S.; Long, H.; Zhang, Y.; Ge, D.; Qu, Y. Rural restructuring at village level under rapid urbanization in metropolitan suburbs of China and its implications for innovations in land use policy. *Habitat Int.* **2018**, *77*, 143–152. [CrossRef]

61. Pils, E. Assessing evictions and expropriations in China: Efficiency, credibility and rights. *Land Use Policy* **2016**, *58*, 437–444. [CrossRef]

62. Wang, Q. Public consultation: Settlement mechanism of preference intensity problems. *J. Cent. China Norm. Univ. (Humanit. Soc. Sci.)* **2010**, *49*, 158. Available online: http://journal.ccnu.edu.cn/sk/CN/abstract/abstract4720.shtml (accessed on 26 November 2020).

63. Liu, L.; Gao, X.; Zhuang, J.; Wu, W.; Yang, B.; Cheng, W.; Xiao, P.; Yao, X.; Deng, O. Evaluating the lifestyle impact of China's rural housing land consolidation with locational big data: A study of Chengdu. *Land Use Policy* **2020**, *96*, 104623. [CrossRef]

64. Gao, J.; Liu, Y.; Chen, J.; Cai, Y. Demystifying the geography of income inequality in rural China: A transitional framework. *J. Rural Stud.* **2019**, in press. [CrossRef]

65. Hao, P.; Tang, S. Floating or settling down: The effect of rural landholdings on the settlement intention of rural migrants in urban China. *Environ. Plan. A* **2015**, *47*, 1979–1999. [CrossRef]

66. Arnstein, S.R. A Ladder of Citizen Participation. *J. Am. Inst. Plan.* **1969**, *35*, 216–224. [CrossRef]

67. Yang, Q.; Tang, W. Exploring the Sources of Institutional Trust in China: Culture, Mobilization, or Performance? *Asian Politics Policy* **2010**, *2*, 415–436. [CrossRef]

68. Wang, X. Mutual empowerment of state and peasantry: Grassroots democracy in rural China. *World Dev.* **1997**, *25*, 1431–1442. [CrossRef]

**Publisher's Note:** MDPI stays neutral with regard to jurisdictional claims in published maps and institutional affiliations.

*Article*

# Measuring the Ecological Safety Effects of Land Use Transitions Promoted by Land Consolidation Projects: The Case of Yan'an City on the Loess Plateau of China

Weilun Feng and Yurui Li *

Institute of Geographic Science and Natural Resources Research, Chinese Academy of Sciences, Beijing 100101, China; fengwl@igsnrr.ac.cn
* Correspondence: liyr@igsnrr.ac.cn

**Abstract:** Land consolidation projects play an important role in promoting agricultural land use transitions, ensuring national food security, and accelerating the construction of ecological civilization. The Loess Plateau in China is a typical ecologically fragile area, where the Gully Land Consolidation Project (GLCP) has been implemented recently and had a major impact on local ecological safety. In this study, we established a quantitative evaluation model for ecological safety effects from the four aspects of dam safety, slope stability, efficient farmland, and effective management, and then scientifically measured the ecological safety effects of land use transitions promoted by land consolidation projects. Three small watersheds (Gutun, Yangjuangou and Luoping) within the GLCP area were employed to verify the evaluation model for ecological safety effects. The results showed that the GLCP can effectively improve the ecological environment and promote the development of modern agriculture, but the ecological safety of gullies and slopes in some areas may also facing a series of threats due to improper project management measures. Among them, Gutun had the highest ecological safety evaluation value, followed by Yangjuangou, while Luoping had the lowest value. The indicator system and evaluation method established in this research could be helpful to systematically diagnose the problems and scientifically guide the implementation of the GLCP from the perspective of ecological safety.

**Keywords:** land consolidation; land use transition; ecological safety; fuzzy comprehensive evaluation; Loess Plateau

**Citation:** Feng, W.; Li, Y. Measuring the Ecological Safety Effects of Land Use Transitions Promoted by Land Consolidation Projects: The Case of Yan'an City on the Loess Plateau of China. *Land* **2021**, *10*, 783. https://doi.org/10.3390/land10080783

Academic Editor: Diane Pearson

Received: 20 June 2021
Accepted: 19 July 2021
Published: 26 July 2021

**Publisher's Note:** MDPI stays neutral with regard to jurisdictional claims in published maps and institutional affiliations.

## 1. Introduction

Since the beginning of the new century, the rapid advancement of urbanization and industrialization has promoted the rapid transformation of land use, which has brought about considerable land-use problems worldwide, such as the large-scale occupation of cultivated land, soil degradation and land pollution, and severely restricted the sustainable development of the social economy [1–4]. According to the data released by the secretariats of the Intergovernmental Science-Policy Platform on Biodiversity and Ecosystem Services (IPBES), about 80% of the agricultural land, 10%~20% of the pastures, and 87% of the wetlands around the world have seen ecological function degradation by 2020 [5]. At the same time, with the widespread application of science and technology in the fields of agriculture and land use, large-scale agricultural land consolidation projects have been carried out around the world on degraded, inefficient, and unused land [6,7]. Especially in China, a wide range of land engineering constructions have been carried out across the country, focusing on land improvement, water and soil allocation, ecological conservation, and high-standard farmland construction [8–11]. Representative projects include the Gully Land Consolidation Project (GLCP) on the Loess Plateau, the foreign soil reconstruction project in Three Gorges Reservoir Region, the barren hillside consolidation project in the Taihang Mountains, the comprehensive consolidation project in the Mu Us sandy land,

and the rocky desertification control project in the karst area [12–16]. These projects have greatly changed the characteristics of local land use, and promoted the rapid transformation of agricultural land use, which has had a significant impact on the local ecological environment [17,18].

The Loess Plateau is located in the middle and upper reaches of the Yellow River, which is susceptible to serious soil erosion and one of the most ecological vulnerable areas in the world [19]. Since the ecological protection and high-quality development of the Yellow River basin was promoted to the top of China's development strategy in 2019, it has become the focus of research recently [20,21]. Since 1998, the Grain-for-Green Project (GFGP) has significantly improved the eco-environment of the Loess Plateau [22,23]. However, this project led to a shortage of farmland and grain in the region. To stabilize the return of farmland to forests, the GLCP in Yan'an has been implemented since 2014, which greatly changed the topographic conditions of the region [16,20,24]. Taking "Increase arable land, protect ecological environment and ensure the livelihood of farmers" as the theme, it provides an opportunity to optimize the structure of agricultural production, and improve land use efficiency and the ecological environment, which has an important impact on local ecological safety [25]. Therefore, accurate and quantitative determination of the ecological safety effects of the GLCP on the region is of great significance for systematically diagnosing the problems in the project construction process and scientifically guiding the effective implementation of subsequent projects.

Ecological safety refers to the health and integrity of the ecosystem, which is the guaranteed degree to which human beings are not affected by ecological damage and environmental pollution in terms of production, living standards, and health [26,27]. Land use transition refers to the change of regional land use forms in time sequence, which is closely related to regional ecological security. It is characterized by stages, regionality, subjectivity and comprehensiveness, including both dominant morphological characteristics such as quantity and space attributes, and recessive morphological characteristics such as quality, property rights and mode of operation [28]. Scientific measurement and accurate evaluation of the ecological safety effects of land use transitions promoted by land consolidation projects are helpful to standardize and guide the practical activities of project construction and are of great significance to the continued safety of the regional ecological environment [29]. At present, studies on the ecological safety effects of land consolidation projects mainly focus on the regional ecological environment, cultivated land changes, landscape patterns, soil erosion, climate change, farmers' income, and economic development [30–32]. In general, the research on the impact of land consolidation projects is mainly based on independent project data demonstration, and there is a lack of evaluation research under the same category and unified evaluation system [33]. In terms of research methods, the current research on the effects of land consolidation projects mainly focuses on the fuzzy comprehensive evaluation (FCE) method, ecological footprint model, extension matter element model evaluation method, multi-objective comprehensive decision-making method, system dynamics method and remote sensing model method [34–41]. Among them, the evaluation system based on the FCP can fully cover all aspects of the ecological safety effects of major land consolidation projects, which is more scientific and reasonable [42].

Taking the GLCP in Yan'an City as an example, this paper aims to: (1) establish a quantitative evaluation model to measure the ecological safety effects of land use transitions promoted by GLCP; (2) compare and analyze the ecological safety effects of GLCP in Gutun, Yangjuangou, and Luoping project areas; (3) systematically diagnose the problems of GLCP in different project areas based on field research and household interviews, and deeply explore the reasons for the differentiation of ecological safety effects; and (4) provide suggestions and countermeasures for the scientific implementation of subsequent GLCP.

## 2. Materials and Methods

### 2.1. Study Area

Yan'an City is located in the middle reaches of the Yellow River, which belongs to the hilly and gully region of the Loess Plateau (Figure 1). The terrain of Yan'an is high in the northwest and low in the southeast, with an average altitude of 1200 m. The climate of this region is semi-humid and semi-arid with an average annual precipitation of 390–700 mm and an average annual temperature of 7.8–10.6 °C. Since 2013, Yan'an has implemented the GLCP with a construction scale of 33,300 hectares, involving 13 districts and counties including Baota District, Yanchang County, and Yanchuan County. By the end of 2020, a series of engineering measures in Yan'an City had greatly improved the farmland production capacity, farmland quality, and land use efficiency in this region. Considering the terrain and landform of the GLCP area and the progress of engineering renovation, we chose the three typical project areas of Yangjuangou, Gutun, and Luoping as the research subjects, as they have strong representativeness and feasibility.

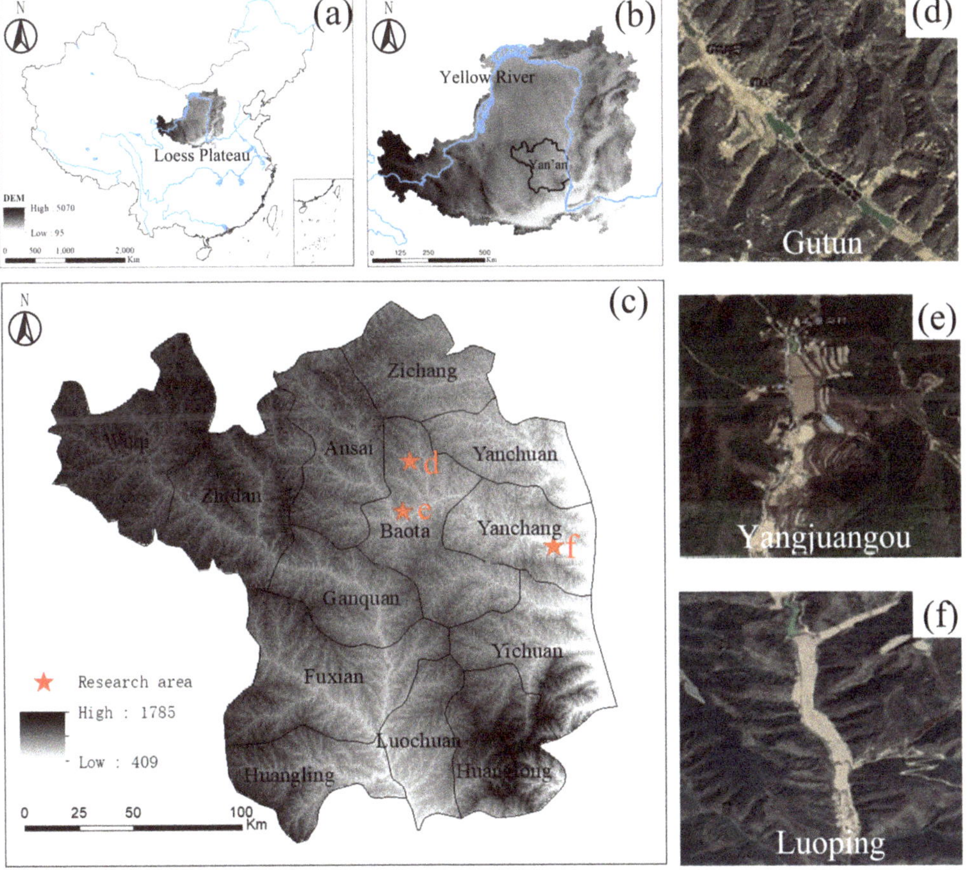


**Figure 1.** Location of Yan'an City (**a–c**) and the general situation of Gutun, Yangjuangou, and Luoping (**d–f**).

### 2.2. Data Sources

The basic data used in the research, such as the population, rural economy, and input-output of the agricultural section, were mainly from the Yan'an Statistical Yearbook and Yearbook of Yan'an. The total area of the project area, arable land area, total road length, and other data of each case study watershed were from the land use map and completion acceptance map of the project area. The standard sub-frame map and remote sensing image of the second land survey in the project area were taken as the base map. The ecological safety effects of the project were comprehensively evaluated based on the topographic information, the current factors of farmland water conservancy, the willingness of farmers to construct projects, as well as the requirements of relevant regulations and technical standards. From 2017 to 2019, a survey team was organized to carry out field surveys on the ecological safety status of the project areas several times to obtain relevant research data.

*2.3. Methodology*

The GLCP in Yan'an city is a systematic, comprehensive, and regional work, focusing on ecological safety effects including those of dam bodies, slopes, farmland and other aspects, and each aspect affects and restricts others. Thus, the FCE method was adopted to construct an ecological safety evaluation model to evaluate the ecological safety effects of GLCP, which included the following steps: the construction of a comprehensive index system, determination of the index weight, and establishment of an evaluation model.

2.3.1. Construction of the Index System

This study constructed an evaluation index system of ecological safety effects from the four aspects of dam safety, slope stability, efficient farmland, and effective management, and 23 relevant indicators were selected (Figure 2). The dam safety index mainly measured the construction stability of large, medium, and small dams, including project quality, layout rationality, anti-risk capability, damaged condition, channel safety, slope stability, and supporting facilities. The slope stability index mainly measured the stability of the newly added slopes and the vegetation protection condition, including design rationality, slope stability, vegetation protection, slope erosion, multistage slopes, and drainage performance. The efficient farmland index mainly measured the situation and stability of the newly added cultivated land, including increase of cultivated land, land levelness, quality of cultivated land, traffic accessibility, irrigation and drainage, crop yield, and convenience of production. The effective management index mainly refers to the management and protection situation after the GLCP, including management mechanism, management situation, and management benefit.

2.3.2. Determination of the Index Weight

The Analytic Hierarchy Process (AHP) was used to calculate the weight between each index [43]. Firstly, the Delphi method was used to make pairwise judgments between the indicators at the same level to form a comparison matrix of the importance of the indicators, where the importance comparison is obtained by the comparison scale method of 1–9. Secondly, the importance comparison matrix was used to calculate its maximum eigenvalue and the corresponding eigenvector to obtain the standard pair matrix. Then the arithmetic average of each row in the matrix was calculated to obtain the weight of each level factor. Finally, the consistency test was used to determine whether the obtained index weight value meets the requirements. The weight system of evaluation indexes in this paper is shown in Table 1.

2.3.3. Establishment of the Evaluation Model

This article mainly adopted the FCE method to construct the quantitative evaluation model. Fuzzy mathematics theory was used to assign evaluation objects according to different types and obtain different evaluation scores to achieve the purpose of artificial comprehensive evaluation. According to the engineering practice of the GLCP, the scoring standards of the secondary indicators were divided into five levels, which were excellent, good, fair, poor, and extremely poor. Accordingly, the evaluation scores of these five levels were assigned 0–2, 2–4, 4–6, 6–8 and 8–10 points. The higher the evaluation score, the better the ecological safety effect of this indicator, and the more beneficial it is to engineering safety and ecological protection. The fuzzy comprehensive evaluation model of ecological safety effect established in this paper is:

$$B = W \times R = (w_1, w_2, w_3, \cdots, w_m) \times \begin{vmatrix} r_{11} & r_{12} & \cdots & r_{1n} \\ r_{21} & r_{22} & \cdots & r_{2n} \\ \vdots & \vdots & \vdots & \vdots \\ r_{m1} & r_{m2} & \cdots & r_{mn} \end{vmatrix} \tag{1}$$

where m denotes the evaluation index; n denotes the engineering project area; B refers to the evaluation result vector of the ecological safety effect; W refers to the weight vector of the evaluation index; R refers to the fuzzy relation vector, that is, the average value of different indexes corresponding to the expert score.

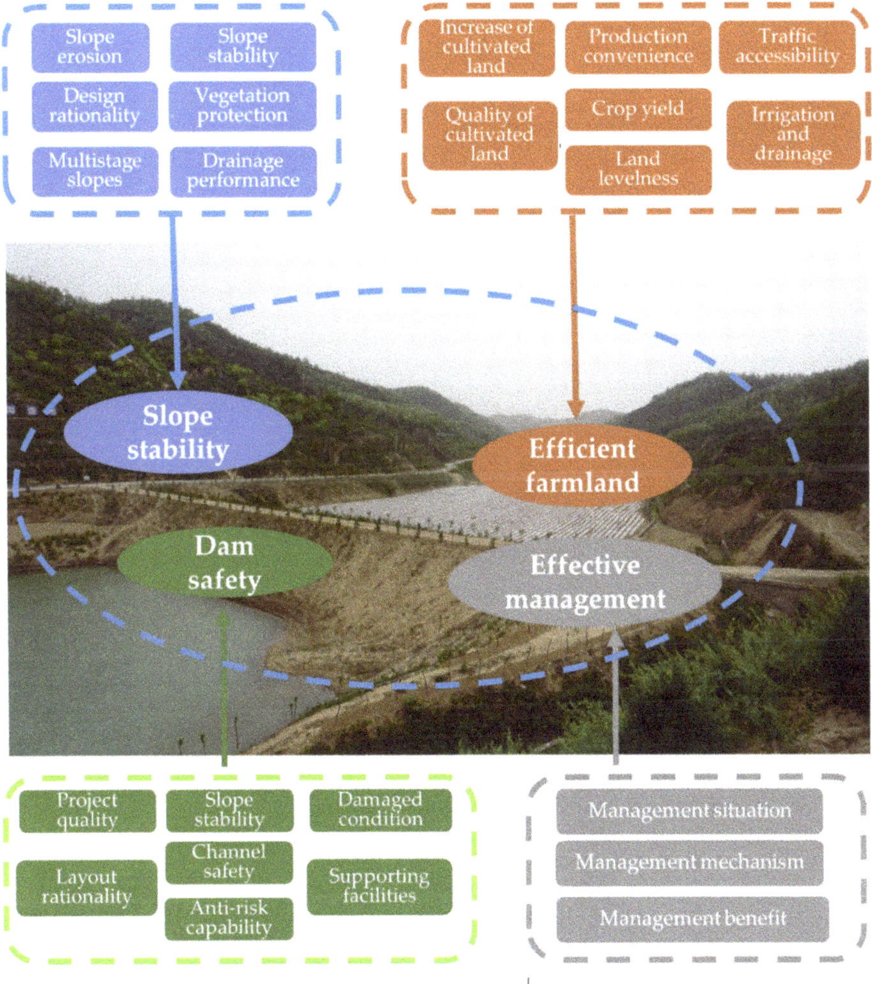


**Figure 2.** Index selection of ecological security effects and its main connotation.

**Table 1.** Index weight of the evaluation index system.

| Criteria | First-Level Weight | Index | Second-Level Weight | Combined Weight |
|---|---|---|---|---|
| Dam safety (B1) | 0.35 | Project quality (C11) | 0.22 | 0.077 |
| | | Layout rationality (C12) | 0.15 | 0.053 |
| | | Anti-risk capability (C13) | 0.15 | 0.053 |
| | | Damaged condition (C14) | 0.13 | 0.046 |
| | | Channel safety (C15) | 0.12 | 0.042 |
| | | Slope stability (C16) | 0.12 | 0.042 |
| | | Supporting facilities (C17) | 0.11 | 0.039 |
| Slope stability (B2) | 0.28 | Design rationality (C21) | 0.2 | 0.056 |
| | | Slope stability (C22) | 0.2 | 0.056 |
| | | Vegetation protection (C23) | 0.2 | 0.056 |
| | | Slope erosion (C24) | 0.1 | 0.028 |
| | | Multistage slopes (C25) | 0.2 | 0.056 |
| | | Drainage performance (C26) | 0.1 | 0.028 |
| Efficient farmland (B3) | 0.22 | Increase of cultivated land (C31) | 0.15 | 0.033 |
| | | Land levelness (C32) | 0.1 | 0.022 |
| | | Quality of cultivated land (C33) | 0.1 | 0.022 |
| | | Traffic accessibility (C34) | 0.2 | 0.044 |
| | | Irrigation and drainage (C35) | 0.2 | 0.044 |
| | | Crop yield (C36) | 0.1 | 0.022 |
| | | Convenience of production (C37) | 0.15 | 0.033 |
| Effective management (B4) | 0.15 | Management mechanism (C41) | 0.3 | 0.045 |
| | | Management situation (C42) | 0.3 | 0.045 |
| | | Management benefit (C43) | 0.4 | 0.060 |

## 3. Results

### 3.1. Evaluation Results of Ecological Safety Effects

Fuzzy comprehensive evaluation is a compound operation between the weight of each evaluation index and the value of the evaluation index. Operator M ($\cdot$, +) was used for calculation. The weight matrix W1 was composed of the second-level index weights in Table 1, and the benefit value of the second-level index was composed of the single-factor benefit evaluation matrix R1. Thus, the evaluation values of dam safety of Yangjuangou, Gutun, and Luoping project areas were calculated:

B1G = W1·R1G = [0.545, 0.413, 0.399, 0.382, 0.342, 0.352, 0.331]

B1Y = W1·R1Y = [0.461, 0.340, 0.307, 0.297, 0.307, 0.289, 0.291]

B1L = W1·R1L = [0.333, 0.262, 0.254, 0.215, 0.230, 0.236, 0.196]

Similarly, the evaluation values of first-level evaluation indexes of slope stability, efficient farmland and effective management in different project areas were calculated:

B2G = W2·R2G = [0.396, 0.370, 0.390, 0.208, 0.404, 0.208]

B2Y = W2·R2Y = [0.308, 0.334, 0.348, 0.190, 0.345, 0.193]

B2L = W2·R2L = [0.237, 0.243, 0.263, 0.123, 0.243, 0.144]

B3G = W3·R3G = [0.233, 0.159, 0.137, 0.312, 0.257, 0.158, 0.238]

B3Y = W3·R3Y = [0.222, 0.158, 0.122, 0.279, 0.220, 0.143, 0.208]

B3L = W3·R3L = [0.205, 0.118, 0.099, 0.183, 0.185, 0.122, 0.155]

B4G = W4·R4G = [0.347, 0.317, 0.517]

B4Y = W4·R4Y = [0.302, 0.311, 0.432]

B4L = W4·R4L = [0.248, 0.204, 0.330]

In conclusion, the evaluation values of ecological safety effects in different project areas were calculated as follows:

BG = [2.764, 1.977, 1.495, 1.181]

BY = [2.292, 1.716, 1.353, 1.045]

BL = [1.726, 1.253, 1.066, 0.783]

*3.2. Analysis of the First-Level Index of Ecological Safety Effects*

In general, Gutun had the highest evaluation value of ecological safety effects at 7.42, followed by Yangjuangou with a score of 6.41, and Luoping had the lowest score of only 4.83 (Figure 3). The evaluation scores of Gutun and Yangjuangou were significantly higher than that of Luoping, which shows that the ecological safety of Gutun and Yangjuangou was obviously better than that of Luoping. In terms of sub-indices, among the dam safety indicators, Gutun had the highest evaluation score (2.76), followed by Yangjuangou (2.29), and Luoping had the lowest score, with only 1.73. Among the slope stability indicators, Gutun had the highest evaluation score of 1.98, followed by Yangjuangou with 1.72, and Luoping had the lowest score of only 1.25. Among the efficient farmland indicators, Gutun has the highest evaluation score of 1.49, followed by Yangjuangou with 1.35, and Luoping had the lowest score of 1.07. Among the effective management indicators, Gutun's evaluation score was the highest at 1.18, the next was Yangjuangou, at 1.05, and Luoping had the lowest score at 0.78.

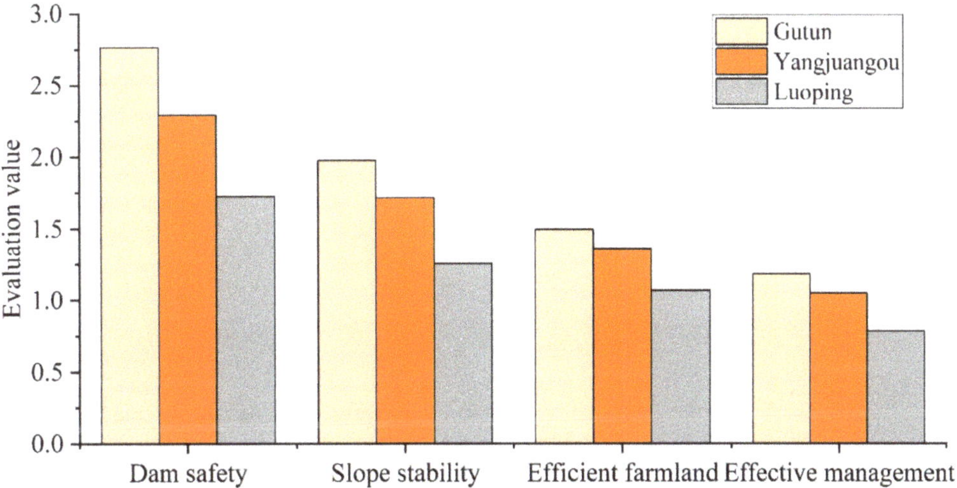


**Figure 3.** Evaluation value of the first-level indexs of ecological safety effects in different gully land consolidation project (GLCP) area.

*3.3. Analysis of Secondary Index of Ecological Safety Effects*

There are obvious differences among different project areas in the secondary index of ecological safety effects (Figure 4). The scores of Gutun in most secondary indexes were significantly higher, with an average value of 0.322. The second was Yangjuangou, with an average of 0.279. The score of Luoping was significantly lower than that of the other two project areas, with an average of 0.210. In terms of average value, Gutun was 16% higher than Yangjuangou, while Yangjuangou was 33% higher than Luoping, which indicated that the average values of both Gutun and Yangjuangou were significantly higher than that of Luoping. In addition, in terms of individual indicators, the average scores of irrigation and drainage, soil quality, design rationality, multistage slopes, drainage performance, and slope stability were relatively low, while the average scores of cultivated land area, convenience of production, management benefit, land levelness, and supporting facilities were relatively high.

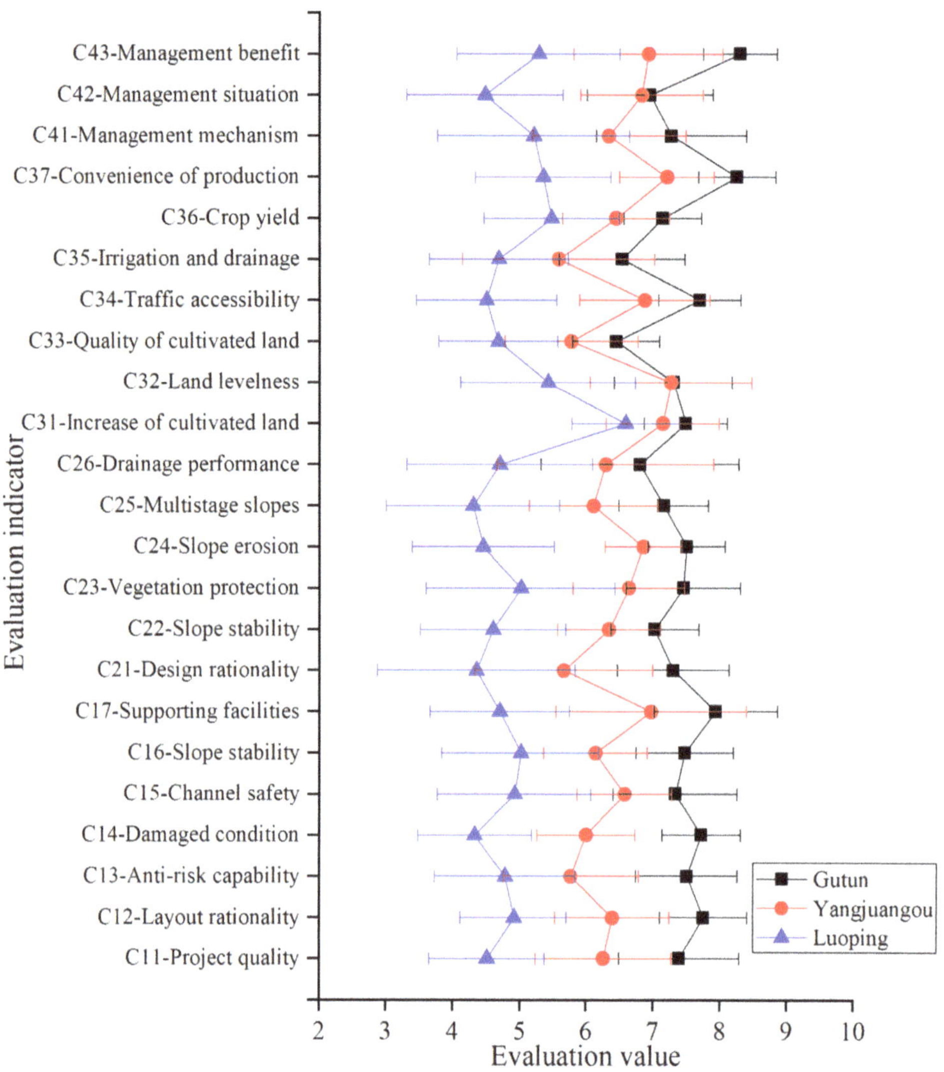



**Figure 4.** Evaluation value of the secondary indexs of ecological safety effects in different GLCP area.

In terms of the difference between Gutun and Yangjuangou, among the 13 indicators of dam safety and slope stability, Gutun scored significantly higher than the other two project areas, especially on the indexes of anti-risk capability, damaged condition, and layout rationality. In terms of efficient farmland and effective management, the difference between Gutun and Yangjuangou was relatively small on indicators such as cultivated land area, land levelness, management situation, etc. In terms of the difference between Yangjuangou and Luoping, it was obviously greater than that between Yangjuangou and Gutun. Especially on the indexes of dam supporting facilities, slope erosion, multistage slopes, and management situation, Yangjuangou scored significantly higher than Luoping. On the index of cultivated land area, the difference between Yangjuangou and Luoping was small—only 8%. The difference between Gutun and Luoping is concerned, the difference between them is the biggest. Especially in the dam damaged condition, traffic accessibility and other indicators, Gutun's evaluation scores were even 70 percent higher than Luoping. However, in the index of cultivated land area, the difference between Gutun and Luoping is small—only 14%.

## 4. Discussion

### 4.1. Differentiation Analysis of Ecological Safety Effects

In general, the ecological safety effects of the GLCP in Gutun, Yangjuangou, and Luoping project areas showed obvious differences. Regardless of the overall indicators of comprehensive ecological safety effects, or the sub-indices such as dam safety, slope stability, efficient farmland, and effective management, Gutun's evaluation scores were all significantly higher, which indicates that the local ecological safety effect of Gutun's project was obviously better than that of the other two project areas (Figure 5). Through several field investigations in Gutun from 2017 to 2019, it was found that the cultivated land quality was improved by 2–3 levels and the crop yield was significantly increased after the GLCP, which effectively promoted the intensification of agricultural resources and provided favorable conditions for the development of modern agriculture including cropping pattern transformation, large-scale operation, and agricultural mechanization. After the implementation of the project, the average maize yield was about 11,250 kg/ha, nearly double that before the project implementation. At the same time, the agricultural landscape, supporting facilities, and other aspects of Gutun have been greatly improved. In addition, the Yangjuangou project area is close to urban area of Yan'an, and the overall situation of the project implementation was relatively good. The flatness of the cultivated land, the quality of the cultivated land and the convenience of agricultural production had been significantly improved, and it also had positive effect on the local ecological safety.

**Figure 5.** Differentiation analysis of the ecological safety effects in different GLCP area.

However, the ecological safety of Luoping was at a relatively poor level, and the ecological safety of gullies and slopes was facing serious threats. Based on the field survey carried out in 2018, it was found that some areas of the project had problems such as roads washed out, slope sliding, and farmland subsidence. Specifically, some agricultural roads had been severely damaged by rainfall, resulting in traffic difficulties. Some slopes near cultivated land had collapsed and slid, which affected the normal drainage function of the canal [44]. In addition, there were many rocks in the cultivated land after partial regulation, which seriously affected the growth and yield of field crops. What is more serious is that some dams washed out and collapsed after repeated rainfall, and faced great danger of further deterioration, which seriously affected the ecological safety of the project in Luoping project area.

The differences of ecological safety effects in different watersheds reflect the common problems of land consolidation projects. In general, a land consolidation project involves survey, planning, design, engineering construction, engineering supervision, follow-up management and other aspects. Different participants are often responsible for different stages, and the completion quality of the previous stage may affect the implementation and quality of the follow-up stages. This requires the effective connection of different stages and the cooperation of key participants. In the three watersheds, field research showed that the main reason for the difference of ecological safety effects lay in the difference of the completion quality of engineering construction, engineering supervision, and follow-up management.

### 4.2. Comprehensive Benefits and Improvement Suggestions

The GLCP can effectively solve the long-standing bottleneck of rural agricultural development, which is conducive to ensure a stable production environment in the gully (Figure 6). Firstly, the project can significantly promote appropriate-scale land management and modern agricultural development, by simultaneously improving the quantity, quality, and flatness of farmland, as well as the construction and improvement of agricultural production facilities. Secondly, the project can consolidate the achievements of the Grain-for-Green Project (GFGP) and provide better conditions for multi-functional agricultural development models such as "planting + breeding + sightseeing" mode, which can help increase the economic benefits of cultivated land and farmers' income [45]. Finally, the project can significantly enhance the resistance of dam system to floods, and greatly enhance the ability to deal with the threats of climate change, such as extreme precipitation.

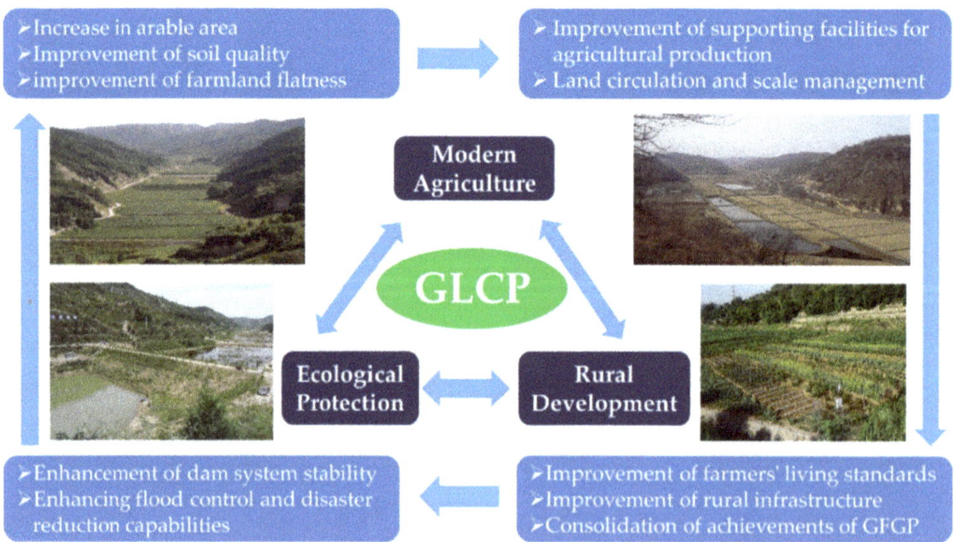


**Figure 6.** Comprehensive benefits of the GLCP to regional development.

However, there may also be some problems in the implementation of the project. For example, some areas in Luoping have experienced poor soil quality of cultivated land, slope collapse, and erosion of silt dams, due to imperfect supervision mechanism, unclear main responsibility, and unqualified engineering technology. These problems have seriously affected local agricultural production and threatened the ecological safety of the project area. Therefore, there is an urgent need to take corresponding measures to effectively manage and restrict the implementation of engineering projects to ensure the quality of engineering construction and the safety of the regional ecological environment. Specific measures are as follows:

(1) The legal and regulatory system for project quality supervision should be further improved, and a comprehensive project quality supervision mechanism should be established. In addition, the responsibilities of supervisory entities should be reasonably defined and clarified, and a responsive accountability system should be established to impose severe penalties on non-conforming projects.

(2) It is necessary to improve the project quality supervision methods to increase the sense of participation of various entities in society in supervision and feedback [46]. Specifically, various entities should be organically integrated to improve the macro supervision of local government, the direct supervision of construction entities, and the timely feedback of social entities.

(3) Follow-up investigations should be strengthened after the land consolidation project, and problems existing in the project construction should be fed back in time, so as to solve the problems in a targeted way and avoid the further deterioration of the project construction problems.

*4.3. Implications for Land Policy Innovation*

Since the beginning of the new century, with the rapid advancement of urbanization and the steady implementation of the GFGP, the land use pattern and the relationship between people and land in Yan'an have undergone tremendous changes. To make full use of the abundant channel resources and promote the development of modern agriculture, the local government implemented the GLCP to protect the ecology and benefit people's livelihoods which, as well, brought about some problems. The proper settlement of regional agricultural problems is a comprehensive and systematic project, which needs to make overall plans for the diagnosis of land use problems, scientific experiments and remediation, sustainable land use, and land policy innovation [33] (Figure 7). For land use nature, management and technical issues, the government should attach importance to policy innovation on land technology innovation, land consolidation project, land circulation management, and so on [47]. In addition, the government should further strengthen the guiding and supporting role of land use policy innovation in ecological protection, economic development, and people's livelihood security based on highlighting the key position of land use policy in national ecological civilization, targeted poverty alleviation, and rural revitalization [48,49].

Agricultural land consolidation engineering is the application of engineering and techniques to land use and development, which aims at increasing the amount of arable land, improving land use efficiency, and actively achieving a harmonious human-land relationship [50,51]. By combining theoretical and engineering technology research, it can effectively solve problems in regional agricultural development and support rural revitalization and modernization [52]. Facing the new era of agricultural supply-side reforms and agricultural and rural modernization strategies, relevant land policies should focus on technology research and development and coordination mechanism innovation for projects such as global land consolidation, agricultural resource utilization, and farmland system conservation [53].

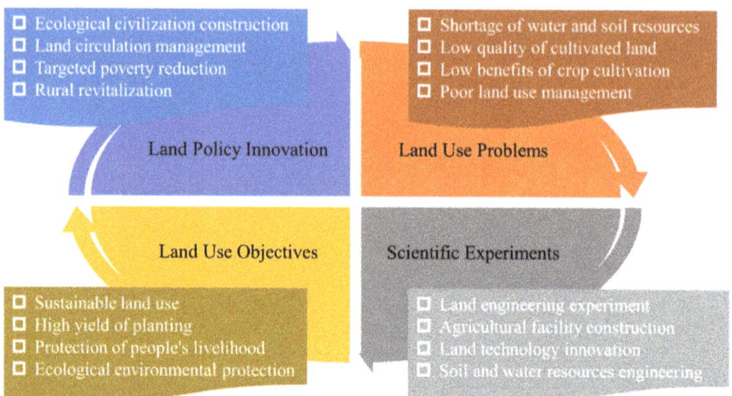


**Figure 7.** Logical diagram of land use management, engineering, and policy.

## 5. Conclusions

Taking the GLCP in Yan'an City as an example, this study established a quantitative evaluation model for ecological safety effects from the four aspects of dam safety, slope stability, efficient farmland, and effective management, and scientifically determined the ecological safety effects of the renovation project.

(1) The ecological safety effects of the Gutun, Yangjuangou, and Luoping project areas show obvious differences. Regardless of the sub-indices such as dam safety, slope stability, efficient farmland and effective management, or the overall indicators of comprehensive ecological safety effects, Gutun's evaluation scores were significantly higher than those of Yangjuangou and Luoping. After the implementation of GLCP, the agricultural production conditions have been greatly improved, and the comprehensive agricultural production capacity has been steadily improved, which is conducive to consolidating the results of returning farmland to forests, improving the basic level of agricultural development, and promoting harmony between man and land and the construction of ecological civilization.

(2) In general, agricultural land engineering is the key factor of modern agricultural development, ecological construction, and sustainable land use. It can solve land management spatial issues and is regarded as an indispensable method for land use transition and rural spatial restructuring [54,55]. However, our evaluation and investigation also showed that unreasonable engineering design and imperfect management mechanisms may have negative impacts on the ecological safety of the project area. This shows the need to strengthen the innovation of engineering technology, supervision of construction quality, and follow-up management and protection of the project.

(3) In recent years, lots of work on ecological protection and restoration has been highly valued all over the world under the background of climate change, and large-scale ecological construction projects have been widely implemented, gradually becoming a research hotspot. However, due to the fragile ecological background, poor site conditions, variable climatic conditions, imperfect management mechanisms and weak long-term observational research, these projects may face certain ecological safety issues after implementation. The effective implementation of engineering renovation requires further strengthening of engineering technology-related research and improvement of engineering supervision mechanisms. Therefore, there is an urgent need to carry out comprehensive and in-depth research on the ecological safety mechanism, process and effects of land use transitions promoted by land consolidation projects, so as to guide the implementation of the land consolidation projects from the perspective of ecological safety.

**Author Contributions:** Conceptualization, W.F. and Y.L.; methodology, W.F.; data analysis, W.F. and Y.L.; writing—original draft, W.F.; writing—review and editing, Y.L.; supervision, Y.L.; funding acquisition, W.F. and Y.L. Both authors have read and agreed to the published version of the manuscript.

**Funding:** This research was supported by National Key Research and Development Program of China, grant number 2017YFC0504701; National Natural Science Foundation of China, grant number 41931293 and 41801175; China Postdoctoral Science Foundation, grant number 2020M680658.

**Institutional Review Board Statement:** Not applicable.

**Informed Consent Statement:** Not applicable.

**Data Availability Statement:** Not applicable.

**Acknowledgments:** The insightful and constructive comments and suggestions from the three anonymous reviewers are greatly appreciated.

**Conflicts of Interest:** The authors declare no conflict of interest.

## References


1. Liu, Y.S.; Li, Y.H. Revitalize the world's countryside. *Nature* **2017**, *548*, 275–277. [CrossRef] [PubMed]
2. Chen, M.X.; Gong, Y.H.; Lu, D.D.; Ye, C. Build a people-oriented urbanization: China's new-type urbanization dream and Anhui model. *Land Use Policy* **2019**, *80*, 1–9. [CrossRef]
3. Deng, X.Z.; Huang, J.K.; Rozelle, S.; Zhang, J.P.; Li, Z.H. Impact of urbanization on cultivated land changes in China. *Land Use Policy* **2015**, *45*, 1–7. [CrossRef]
4. Liu, Y.S.; Fang, F.; Li, Y.H. Key issues of land use in China and implications for policy making. *Land Use Policy* **2014**, *40*, 6–12. [CrossRef]
5. Bai, X.M.; McPhearson, T.; Cleugh, H.; Nagendra, H.; Tong, X.; Zhu, T.; Zhu, Y.G. Linking urbanization and the environment: Conceptual and empirical advances. *Annu. Rev. Environ. Resour.* **2017**, *42*, 215–240. [CrossRef]
6. Allahyari, M.S.; Damalas, C.A.; Masouleh, Z.D.; Ghorbani, M. Land consolidation success in paddy fields of northern Iran: An assessment based on farmers' satisfaction. *Land Use Policy* **2018**, *73*, 95–101. [CrossRef]
7. Janus, J.; Markuszewska, I. Forty years later: Assessment of the long-lasting effectiveness of land consolidation projects. *Land Use Policy* **2019**, *83*, 22–31. [CrossRef]
8. Li, P.Y.; Qian, H.; Wu, J.H. Accelerate research on land creation. *Nature* **2014**, *510*, 29–31. [CrossRef]
9. Feng, W.L.; Liu, Y.S.; Qu, L.L. Effect of land-centered urbanization on rural development: A regional analysis in China. *Land Use Policy* **2019**, *87*, 104072. [CrossRef]
10. Liu, Y.S.; Feng, W.L.; Li, Y.R. Modern agricultural geographical engineering and agricultural high-quality development: Case study of loess hilly and gully region. *Acta Geogr. Sin.* **2020**, *75*, 2029–2046.
11. Liu, Y.S.; Zhang, Z.W.; Wang, J.Y. Regional differentiation and comprehensive regionalization scheme of modern agriculture in China. *Acta Geogr. Sin.* **2018**, *73*, 203–218.
12. Liu, Q.; Wang, Y.Q.; Zhang, J.; Chen, Y.P. Filling gullies to create farmland on the loess plateau. *Environ. Sci. Technol.* **2013**, *47*, 7589–7590. [CrossRef] [PubMed]
13. Fu, B.J.; Wu, B.F.; Lu, Y.H.; Xu, Z.H.; Cao, J.H.; Niu, D.; Yang, G.S.; Zhou, Y.M. Three Gorges project: Efforts and challenges for the environment. *Prog. Phys. Geogr.* **2010**, *34*, 741–754. [CrossRef]
14. Han, J.C.; Liu, Y.S.; Zhang, Y. Sand stabilization effect of feldspathic sandstone during the fallow period in Mu Us Sandy Land. *J. Geogr. Sci.* **2015**, *25*, 428–436. [CrossRef]
15. Wang, Y.S.; Li, Y.H.; Liu, Y.S. China's sandy land consolidation engineering and regional agricultural sustainable development practice under water resource constraint: Case study of Yulin city in Shaanxi province, China. *Bull. Chin. Acad. Sci.* **2020**, *35*, 1408–1416.
16. Li, Y.R.; Li, Y.; Fan, P.C.; Long, H.L. Impacts of land consolidation on rural human-environment system in typical watershed of the Loess Plateau and implications for rural development policy. *Land Use Policy* **2019**, *86*, 339–350.
17. Jin, Z. The creation of farmland by gully filling on the Loess Plateau: A double-edged sword. *Environ. Sci. Technol.* **2014**, *48*, 883–884. [CrossRef]
18. Wang, Y.S.; Li, Y.H.; Liu, Y.S. Principle and method of modern agricultural two-dimension optimization engineering experiment. *Eng. Sci.* **2019**, *21*, 48–54.
19. Wang, S.A.; Fu, B.J.; Piao, S.L.; Lu, Y.H.; Ciais, P.; Feng, X.M.; Wang, Y.F. Reduced sediment transport in the Yellow River due to anthropogenic changes. *Nat. Geosci.* **2016**, *9*, 38–41. [CrossRef]
20. Li, Y.R.; Zhang, X.C.; Cao, Z.; Liu, Z.J.; Lu, Z.; Liu, Y.S. Towards the progress of ecological restoration and economic development in China's Loess Plateau and strategy for more sustainable development. *Sci. Total Environ.* **2021**, *756*, 143676.
21. Fu, B.J.; Wang, S.; Liu, Y.; Liu, J.B.; Liang, W.; Miao, C.Y. Hydrogeomorphic ecosystem responses to natural and anthropogenic changes in the Loess Plateau of China. *Annu. Rev. Earth Planet. Sci.* **2017**, *45*, 223–243. [CrossRef]

22. Cao, Z.; Li, Y.R.; Liu, Y.S.; Chen, Y.F.; Wang, Y.S. When and where did the Loess Plateau turn "green"? Analysis of the tendency and breakpoints of the normalized difference vegetation index. *Land. Degrad. Dev.* **2018**, *29*, 162–175. [CrossRef]
23. Cao, S.X.; Sun, G.; Zhang, Z.Q.; Chen, L.D.; Feng, Q.; Fu, B.J.; McNulty, S.; Shankman, D.; Tang, J.W.; Wang, Y.H.; et al. Greening China Naturally. *Ambio* **2011**, *40*, 828–831. [CrossRef]
24. Liu, Y.S.; Li, Y.H. China's land creation project stands firm. *Nature* **2014**, *511*, 410. [CrossRef] [PubMed]
25. Liu, Y.S.; Li, Y.R. Engineering philosophy and design scheme of gully land consolidation in Loess Plateau. *Trans. Chin. Soc. Agric. Eng.* **2017**, *33*, 1–9.
26. Han, Y.; Yu, C.Y.; Feng, Z.; Du, H.C.; Huang, C.S.; Wu, K.N. Construction and optimization of ecological security pattern based on spatial syntax classification—Taking Ningbo, China, as an Example. *Land* **2021**, *10*, 380. [CrossRef]
27. Li, Z.T.; Yuan, M.J.; Hu, M.M.; Wang, Y.F.; Xia, B.C. Evaluation of ecological security and influencing factors analysis based on robustness analysis and the BP-DEMALTE model: A case study of the Pearl River Delta urban agglomeration. *Ecol. Indic.* **2019**, *101*, 595–602. [CrossRef]
28. Long, H.L.; Qu, Y. Land use transitions and land management: A mutual feedback perspective. *Land Use Policy* **2018**, *74*, 111–120. [CrossRef]
29. Gu, X.K.; Dai, B.; Chen, B.M. Landscape effects of land consolidation projects in Central China—A case study of Tianmen City, Hubei Province. *Chin. Geogr. Sci.* **2008**, *18*, 41–46. [CrossRef]
30. Han, X.L.; Lv, P.Y.; Zhao, S.; Sun, Y.; Yan, S.Y.; Wang, M.H.; Han, X.N.; Wang, X.R. The effect of the Gully land consolidation project on soil erosion and crop production on a typical watershed in the Loess Plateau. *Land* **2018**, *7*, 113. [CrossRef]
31. Hong, C.Q.; Jin, X.B.; Fan, Y.T.; Xiang, X.M.; Cao, S.; Chen, C.C.; Zheng, G.; Zhou, Y.K. Determining the effect of land consolidation on agricultural production using a novel assessment framework. *Land Degrad. Dev.* **2020**, *31*, 356–371. [CrossRef]
32. Kolis, K.; Hiironen, J.; Riekkinen, K.; Vitikainen, A. Forest land consolidation and its effect on climate. *Land Use Policy* **2017**, *61*, 536–542. [CrossRef]
33. Liu, Y.S. Integrated land research and land resources engineering. *Resour. Sci.* **2015**, *37*, 1–8.
34. Cay, T.; Iscan, F. Fuzzy expert system for land reallocation in land consolidation. *Expert Syst. Appl.* **2011**, *38*, 11055–11071. [CrossRef]
35. Cay, T.; Ayten, T.; Iscan, F. Effects of different land reallocation models on the success of land consolidation projects: Social and economic approaches. *Land Use Policy* **2010**, *27*, 262–269. [CrossRef]
36. Guo, B.B.; Fang, Y.L.; Jin, X.B.; Zhou, Y.K. Monitoring the effects of land consolidation on the ecological environmental quality based on remote sensing: A case study of Chaohu Lake Basin, China. *Land Use Policy* **2020**, *95*, 104569. [CrossRef]
37. Guo, B.B.; Jin, X.B.; Yang, X.H.; Guan, X.; Lin, Y.A.; Zhou, Y.K. Determining the effects of land consolidation on the multi-functionality of the cropland production system in China using a SPA-fuzzy assessment model. *Eur. J. Agron.* **2015**, *63*, 12–26. [CrossRef]
38. Liu, Y.S.; Guo, Y.J.; Li, Y.R.; Li, Y.H. GIS-based effect assessment of soil erosion before and after gully land consolidation: A case study of Wangjiagou project region, Loess Plateau. *Chin. Geogr. Sci.* **2015**, *25*, 137–146. [CrossRef]
39. Zhu, G.M.; Li, D.L. A method for land consolidation progress assessment based on GPS and PDA. *Math. Comput. Sci. Eng.* **2009**, *8*, 429–438.
40. Lu, S.S.; Li, J.P.; Guan, X.L.; Gao, X.J.; Gu, Y.H.; Zhang, D.H.; Mi, F.; Li, D.D. The evaluation of forestry ecological security in China: Developing a decision support system. *Ecol. Indic.* **2018**, *91*, 664–678. [CrossRef]
41. Yang, Y.; Cai, Z.X. Ecological security assessment of the Guanzhong Plain urban agglomeration based on an adapted eco-logical footprint model. *J. Clean. Prod.* **2020**, *260*, 120973. [CrossRef]
42. Gong, L.; Jin, C.L. Fuzzy comprehensive evaluation for carrying capacity of regional water resources. *Water Resour. Manag.* **2009**, *23*, 2505–2513. [CrossRef]
43. Pang, Y.J.; Liu, L.M.; Li, W.G. Fuzzy comprehensive evaluation model of ecological security based on distinguishable weight of indexes. *Prog. Environ. Sci. Technol.* **2009**, *51*(Pts A and B), 99–103.
44. He, C.X. The present situation, characteristics and functions of the project of treating ditches and constructing land in Yan'an. *J. Earth Environ.* **2015**, *15*, 255–260.
45. Liu, Y.S.; Chen, Z.F.; Li, Y.R. The planting technology and industrial development prospects of forage rape in the loess hilly area: A case study of newly-increased cultivated land through gully land consolidation in Yan'an, Shaanxi Province. *J. Nat. Resour.* **2017**, *32*, 2065–2074.
46. Gedefaw, A.A.; Atzberger, C.; Seher, W.; Mansberger, R. Farmers willingness to participate in voluntary land consolidation in Gozamin District, Ethiopia. *Land* **2019**, *8*, 148. [CrossRef]
47. Long, H.L.; Li, Y.R.; Liu, Y.S.; Woods, M.; Zou, J. Accelerated restructuring in rural China fueled by 'increasing vs. decreasing balance' land-use policy for dealing with hollowed villages. *Land Use Policy* **2012**, *29*, 11–22. [CrossRef]
48. Li, Y.R.; Liu, Y.S.; Long, H.L.; Cui, W.G. Community-based rural residential land consolidation and allocation can help to revitalize hollowed villages in traditional agricultural areas of China: Evidence from Dancheng County, Henan Province. *Land Use Policy* **2014**, *39*, 188–198. [CrossRef]
49. Zhu, M. New development of agricultural engineering in China. *Trans. Chin. Soc. Agric. Eng.* **2005**, *21*, 1–11.

50. Feng, W.L.; Liu, Y.S.; Chen, Z.F.; Li, Y.R.; Huang, Y.X. Theoretical and practical research into excavation slope protection for agricultural geographical engineering in the Loess Plateau—A case study of China's Yangjuangou catchment. *J. Rural Stud.* **2019**, *8*, 18–23. [CrossRef]
51. Jin, X.B.; Shao, Y.; Zhang, Z.H.; Resler, L.M.; Campbell, J.B.; Chen, G.; Zhou, Y.K. The evaluation of land consolidation policy in improving agricultural productivity in China. *Sci. Rep.* **2017**, *7*, 2792. [CrossRef] [PubMed]
52. Han, J.C.; Zhang, Y. Land policy and land engineering. *Land Use Policy* **2014**, *40*, 64–68. [CrossRef]
53. Lu, D.D.; Liu, Y.S.; Fang, C.L.; Chen, M.X.; Wang, J.E.; Xi, J.C. Development and prospect of human-economic geography. *Acta Geogr. Sin.* **2020**, *75*, 2570–2592.
54. Colombo, S.; Manuel, P.V. A practical method for the ex-ante evaluation of land consolidation initiatives: Fully connected parcels with the same value. *Land Use Policy* **2019**, *81*, 463–471. [CrossRef]
55. Long, H.L.; Tu, S.S.; Ge, D.Z.; Li, T.T.; Liu, Y.S. The allocation and management of critical resources in rural China under restructuring: Problems and prospects. *J. Rural. Stud.* **2016**, *47*, 392–412. [CrossRef]

*land*

MDPI

*Article*

# Land Use Change and Farmers' Sense of Place in Typical Catchment of the Loess Hilly and Gully Region of China

Yi Li [1], Yurui Li [2], Bin Fang [1,*], Lulu Qu [3], Chongjing Wang [1] and Wubo Li [1]

[1] School of Geography, Nanjing Normal University, Nanjing 210023, China; 201301009@njnu.edu.cn (Y.L.); wangchongjing@nnu.edu.cn (C.W.); 201302163@njnu.edu.cn (W.L.)

[2] Institute of Geographic Sciences and Natural Resources Research, Chinese Academy of Sciences, Beijing 100101, China; liyr@igsnrr.ac.cn

[3] School of Public Policy and Administration, Chongqing University, Chongqing 400044, China; qululu91@cqu.edu.cn

* Correspondence: 09199@njnu.edu.cn

**Abstract:** Clarifying the relationship between land use and farmers' sense of place on a micro scale is significant for enriching the perspective of research on human–environment relationships. Therefore, this paper analyzed land use change and the sense of place of farmers and further explored the interaction between them in the Yangjuangou catchment of Liqu Town in Baota District, Shaanxi Province from 1984 to 2020. The results indicated that: (1) the change in croplands was the most significant, i.e., its share in the total area decreased by 40%, and the decrease in sloping fields was the highest. The average relative altitude of croplands has decreased. The change in ecological land was also more significant, showing an increasing trend. Overall, the exploitation of land resources has declined; (2) the intensity of the sense of place of local farmers fluctuated downwards. The intensity of place attachment and place dependence decreased, and the intensity of the place identity increased; and (3) the decline in the intensity of the place attachment and place dependence promoted the reduction of sloping fields, the growth of ecological land and abandoned fields. By comparison, the increase in ecological land and check dam land promoted an increase in the intensity of place identity for local farmers. This paper suggested that rural areas in the Loess Hilly and Gully Region should strengthen innovation in land use patterns and focus on sustainability of farmers' livelihoods, in order to promote the harmonious development of human-environmental relations.

**Keywords:** land use; sense of place; catchment; Loess Hilly and Gully Region



**Citation:** Li, Y.; Li, Y.; Fang, B.; Qu, L.; Wang, C.; Li, W. Land Use Change and Farmers' Sense of Place in Typical Catchment of the Loess Hilly and Gully Region of China. *Land* **2021**, *10*, 810. https://doi.org/10.3390/land10080810

Academic Editor: Nicholas Magliocca

Received: 29 June 2021
Accepted: 29 July 2021
Published: 1 August 2021

**Publisher's Note:** MDPI stays neutral with regard to jurisdictional claims in published maps and institutional affiliations.



## 1. Introduction

The Loess Hilly and Gully Region (LHGR) represents the core area of the Loess Plateau in China, accounting for 33% of the total area of the Loess Plateau. It is an important area for the transition of agriculture from farming to pastoral areas, as well as a densely populated area in Northwest China [1]. Due to the loose loess texture, sparse vegetation, concentrated and frequent extreme rainfall, LHGR has become one of the most serious regions for soil erosion in China and globally [2–6]. It is full of dense gullies that are usually U-shaped or V-shaped. The broken terrain and limited available area led to increased land reclamation by local residents and the land became increasingly barren, forming a vicious circle. Over-exploitation of natural resources has greatly damaged the ecological environment [7–9]. To prevent environmental deterioration and improve production and living conditions in the region, a number of ecological projects have been implemented since the founding of the People's Republic of China, including the check dam land Construction, the Comprehensive Management of Small Watersheds (CMSC), the Grain for Green Project (GGP) and the Gully Land Consolidation Project (GLCP). These projects had a significant impact on ecological restoration, agricultural production and diversification of farmers' livelihood

Small catchment areas are part of the River Basin System of the Loess Plateau. This is the main region of human activities in the LHGR and occupies the most extensive area [10]. Small catchments in the LHGR are often rural areas, so they are also typical rural systems [11]. Small catchments include almost all geomorphic types of LHGR, which are a source of soil and water loss on the Loess Plateau and the main region of return of farmland to forest and ecological management [12]. The LHGR small catchments investigation is also a response to China's Rural Revitalization Strategy.

Land is not only a complex of natural resource, but also the main space carrier of human activities [13,14]. Land use is the long-term or periodic management of land by people, which has certain economic and social purposes [15,16]. Land use change is a direct reflection of the interaction between human activities and the natural environment on the land surface [17,18], which refers to the change in the quantity (area, proportion) and spatial pattern of land use types, i.e., dominant morphology changes of land use transition [19]. Related research combining land use change has been extensively conducted, which included the process, pattern, dynamic driving mechanism, and environmental effects of land use change [20–29]. Among them were more abundant research on the driving mechanism and environmental effects. The former mainly focused on the physical, social, and economic driving forces of land use change [20–22], while the latter mainly included the effects of land use change on hydrologic processes, climate change, and soil carbon stocks [27–29]. Land use change has become an important concern in research into global environmental evolution and sustainable regional development [30–32]. In LHGR, the implementation of a series of development strategies has led to an intense change in land use, and contradictions have been highlighted in the process of change. For example, extensive land reclamation and severe soil erosion, return of cultivated land to the forest and food security, leaving cultivated land and hollowing out of rural areas [33–36]. Therefore, the study of land use change is the focus of sustainable development in LHGR.

For centuries, human beings have had rich feelings for land. In rural regions, the land is not only the carrier of rural residents' emotion, but also the ultimate material guarantees of farmers' livelihood. The emotional connection between humans and land is called the sense of place in geography [37–39]. The sense of place is people's value and the cognition of land that comes from the interaction between people and the land and that is constantly evolving and changing [40,41]. The sense of place is in line with the people-oriented concept, which is an essential idea of human geography and can help people to further explore the social and cultural process of changing rural land use [42]. Referring to relevant research by domestic and foreign scholars [43,44], this paper will explore the change in sense of place from three dimensions, including place identity (PI), place dependence (PD) and place attachment (PA). PI represents people's recognition of the affections of local natural and cultural features on individual development. PD represents the recognition of local production and living functions by people, while PA represents the emotional attachment of human beings to the land [45,46].

Research on the sense of place is more abundant in Western countries and has become one of the important theories of geography. Research includes environmental science, land use change, rural development, urban research, leisure tourism, economic geography and other fields [42,44,47–49]. For example, Soini et al. used the concept of sense of place to investigate the relationship between people and landscape on the rural-urban fringe of Nurmijärvi in southern Finland [42]. Tapsuwan et al. predicted resident's intention to accept or reject land use planning decisions and groundwater policies by studying the sense of place (including notions of identity, attachment and dependence) of social groups living in Perth for many years [44]. Cross et al. believed that the place identity, the conservation ethic and economic dependence are distinct dimensions of the sense of place among the agricultural landowners, which can be used to predict the adoption of conservation easements. Chinese scholars have begun to focus on the sense of place of rural residents since the 21st century [49]. Xue et al. assessed the similarities and differences between land-lost and normal farmers' sense of place in land dependence and local attachment under the background of land circulation in Weibei Upland of Shaanxi Province [50]. Zhu et al. took Guangzhou's "Artist Village", Xiaozhou Village, as an example and assessed the evolution of the identities of local villagers and newcomer artists in the context of Guangzhou's urban spatial transformation [51]. In comparison, our knowledge of rural development from a comprehensive perspective of sense of place and land use remains weak, especially in LHGR.

In view of this, this study aims to analyze land use change and the sense of place of farmers and investigated the interaction between them in the Yangjuangou catchment of Liqu Town in Baota District, Shaanxi Province, where the relationship between people and environment has changed significantly. The research years selected in the study were 1984, 1998, 2012 and 2020, combined with remote sensing images and the different engineering impacts on land use. The study will provide a scientific reference for ecological construction, rural revitalization and harmonious development between people and environment in LHGR.

## 2. Materials and Methods

### 2.1. Research Design

This study was conducted as follows: (1) analysis of land use changes in the catchment based on remote sensing images during different periods. Cropland has the closest relationship with local emotions of farmers, so the evolutionary characteristics of the type structure, relative altitude and abandoned degree of cropland were analyzed. Furthermore, the degree of land use of the entire catchment was quantitatively analyzed in order to comprehensively evaluate the land use; (2) quantitative analysis of the sense of place from the perspective of PI, PD and PA; and (3) interaction between land use and sense of place, conducting a comprehensive study on the interaction between land use and sense of place based on quantitative data on land use and sense of place.

### 2.2. Study Area

The Yangjuangou catchment is located 14 km east of Yan'an city, with an area of 2.64 km$^2$. The study region is the secondary branch gully of the Yanhe River and the primary tributary of the Nianzhuanggou catchment. The loess soil is the main soil developed from the loess parent material. It has loose and uniform texture and high erodibility. The area is crossed with gullies, and the gully density is 2.74 km/km$^2$, the altitude of the area is from 1050 m to 1295 m [52]. The construction of the check dam land caused to some extent a change in the shape of the valley, including a decrease in the vertical gradient of the channel, an increase in the base level of erosion and a decrease in the relative altitude difference. The catchment has a semiarid continental monsoon climate. The average multi-annual precipitation is 535 mm and it is mainly concentrated from July to September, accounting for more than 60% of total annual precipitation [53]. The total annual solar radiation is 5800 KJ cm$^{-2}$, the annual sunshine is 2563 h, and the mean annual temperature is 9.4 °C. Artificial Robinia pseudoacacia is the main vegetation in the area (Figure 1).

In 2020, there were 186 registered inhabitants, with about 40 permanent residents with average age of over 45. The average income of each household was about 30,000 CNY. Check dam land and terraces were the main types of croplands with the main crops of foxtail millet, corn and potatoes. The main types of land use were forests, shrubs, grassland, check dam land and terraces. Considering the following reasons, the Yangjuangou catchment was selected as the case study area: (1) the Yangjuangou catchment, a traditional agricultural village, is characterized by a large number of labor migrations, rural hollowing and aging, which can be considered a microcosm of many rural areas in LHGR; (2) since economic reforms and open door policy, the catchment has experienced a number of ecological and livelihood projects, including the check dam land Construction, the Comprehensive Management of Small Watersheds, the Grain for Green Project, the Gully Land Consolidation Project, and the Village Combination Project[1]. As a result, there have been major changes in the natural and cultural landscape of the case study area; and (3) the research team has been conducting continuous field observations and household surveys in the catchment since 2012, so the context of the catchment development is clearly understood.

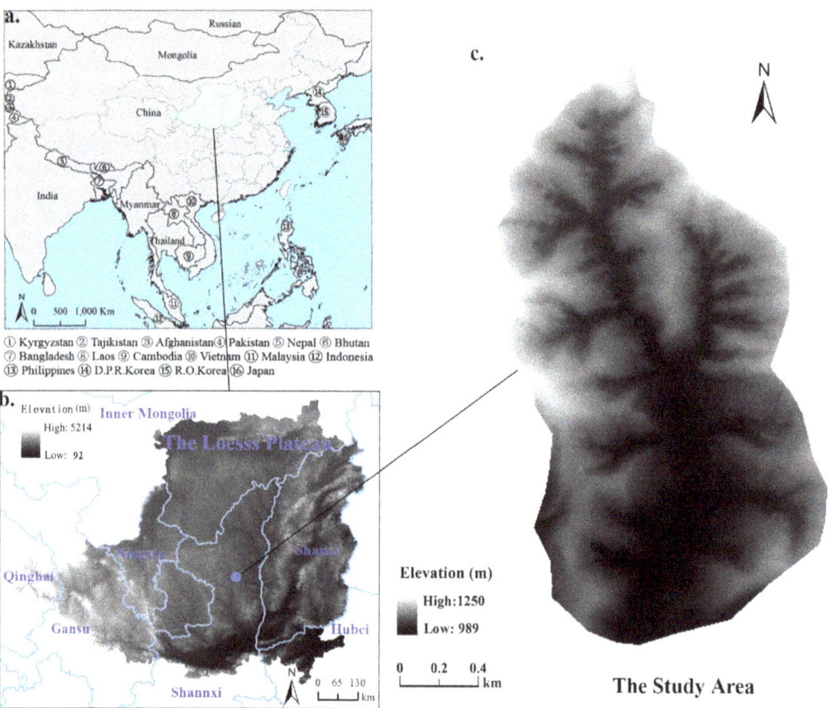


**Figure 1.** The geographical location and DEM of the study area. (**a**) Location of the Loess Plateau; (**b**) location of the study area; (**c**) boundary and altitude of the study area.

*2.3. Data Sources and Preprocessing*

2.3.1. Sources of Land Use Data

Land use data were obtained from remote sensing images, and data sources selected in the study include: (i) aerial photograph with a resolution of 100 m from 1984, (ii) satellite images of Google Earth with a resolution of 15 m from 1998, and (iii) high-resolution satellite images of Google Earth with a resolution of 2 m from 2012 and 2020. ArcMap 10.2 software was used to perform geometric rectification, coordinate registration, visual interpretation, image vectorization, while verification and correction of obtained data can be performed by field investigation. Further, ArcScene 10.2 was used to create a three-dimensional land use map through overlapping vector land use data and Digital Elevation Model data, which can clearly show the land use evolution in hilly areas. Image interpretation from 1984 and 1998 was combined with the participation of the villagers due to the poor quality of images. Based on the criteria for classifying land use in China and the actual land use characteristics in the study area, land use types were divided into thirteen second-class types, including check dam land, terraces, sloping fields, abandoned fields, forests, shrub land, grassland, industrial and mining land, hydraulic land, rural residential land, rural roads, water and bare land, which were merged in four one-class types, including ecological land, croplands, construction land, and bare land (Table 1).

**Table 1.** Classification of land use types of the study area.

| One-Class Types | Second-Class Types | Specification |
|---|---|---|
| Ecological land | Forests | Refers to big arbor, middle arbors and small arbors |
| | Shrub land | Refers to natural shrub, artificial shrub |
| | Grassland | Refers to natural grassland, artificial grassland |
| | Water | Refers to rivers, reservoirs, pond |
| Croplands | Sloping fields | Refers to croplands on slope over six degrees |
| | Terraces | Terraces are a series of flat areas built like steps on the side of a hill |
| | check dam land | Refers to farmland from silting up soil in a gully by the silt dam and farmland from filling the gully with soil removed from the gentle slopes of gully or steep hills with slip risk by the GLCP |
| | Abandoned fields | Refers to farmland that was abandoned over 12 months |
| Construction land | Rural residential land | Land for rural residents to build houses and life ancillary facilities |
| | Rural roads | Roads serving rural residents for production and life |
| | Industrial and mining land | Land for industry, mining, warehousing and poultry farming |
| | Hydraulic land | Land for flood protective bank, flood discharge trench and other water conservancy facilities |
| Bare land | Bare land | Refers to barren lands and exposed rock land |

### 2.3.2. Estimation of the Relative Altitude of Croplands

Digital Elevation Model (DEM) data were obtained from the website of Geospatial Data Cloud in China (http://www.gscloud.cn/ (accessed on 31 July 2021)), with a spatial resolution of 30 m. The Raster to Point tool of ArcMap 10.2 was used to extract the elevation of the center point of each grid in DEM data. Then, the ArcMap 10.2 spatial analysis tool was used to assign land class attributes to each point with elevation value, by connecting the point date with vector land use data. Then, the average altitude of each type of cropland will be calculated. The difference between the average altitude of cropland and the average altitude of gully was considered as the relative altitude of each type of cropland. The check dam land is located in the gully, so the relative altitude of the check dam land is considered as zero.

### 2.3.3. Land-Use Degree Index

Land-use degree was used to analyze the extent of land use. It can not only quantitatively reveal the natural properties of land, but also reflect the interaction between human activities and the natural environment [54]. Based on a comprehensive method of land-use degree analysis developed by Prof. Liu Jiyuan et al. [55], the land-use degree is divided into four degrees depending on the stable condition of the land affected by nature and human society, and given a grading index (Table 2). Comprehensive quantified indices of land-use degree are derived from mathematical synthesis of these four degrees and the formula is as follows:

$$La = 100 \times \sum_{i=1}^{n} Pi \times Qi \tag{1}$$

where, $La$ is a comprehensive quantified index of land-use degree (LUDI), $Pi$ is the grading index of $i$ land-use degree and $Qi$ is the percentage of area of $i$ land-use degree.

**Table 2.** The standard of dividing land use degree.

| Items | Unused Land | Natural Regeneration Land | Artificial Regeneration Land | Rural–Urban Construction Land |
|---|---|---|---|---|
| Degree index | 1 | 2 | 3 | 4 |
| Land-use pattern | bare land | forests, shrub land, grassland, water, abandoned fields | check dam land, terraces, sloping fields | hydraulic land, industrial and mining land, rural roads, rural residential land |

### 2.3.4. Data Sources and Quantitative Analysis of the Sense of Place

Data on the sense of place were collected by a questionnaire survey. The questionnaire included basic information about the respondents, PI, PD and PA, and each dimension of the sense of place included different variables. A 5-point Likert scale was applied to quantify the sense of place. For each variable, 5 meanings were used as follows: 5 means "strongly approve", 4 means "approve", 3 means "neutral" attitude, 2 means "disapprove" and 1 means "strongly disapprove". Each variable was scored by the respondents. We conducted a questionnaire survey lasting three days from 30 April to 2 May 2018, while a change in the intensity of the sense of place before 2018 was taken. However, due to the impact of COVID-19, data in 2020 were obtained through online questionnaire and telephone interviews. The average time required to complete each questionnaire was about 1 h. A total of 20 households from Yangjuangou village or the new rural community were selected for the questionnaire survey. Due to the similarity of sense of place among family members, one respondent was involved in each household. Eventually, 20 valid questionnaires were collected. The respondents selected in this study were older than 40 years, so they can provide a clear picture of the development of the catchment in the past 30 years. Moreover, in the process of analysis, we confirmed the information with village cadres and villagers over the telephone many times to ensure the accuracy and reliability of the data.

Data from the questionnaire survey were assessed for reliability. The Cronbach's $\alpha$ value was applied to evaluate the reliability of the data by SPSS 22.0 statistic package, with the higher value representing the better reliability. If the value is greater than 0.7, the data reliability is good. The test showed that four variables reduced the overall reliability of the scale, namely the identity degree of local culture of PI, neighbors' dependence of PD, growth memory attachment and place rootedness of PA. The standardized Cronbach's value of the three dimensions ranged from 0.708 to 0.950 after the removal of these four variables, suggesting that the reliability of each dimension became good (Table 3).

**Table 3.** The reliability of each dimension of the sense of place.

| Items | Cronbach's $\alpha$ | | | | Standardized Cronbach's $\alpha$ | | | |
|---|---|---|---|---|---|---|---|---|
| | 2020 | 2012 | 1998 | 1984 | 2020 | 2012 | 1998 | 1984 |
| place identity | 0.850 | 0.842 | 0.789 | 0.706 | 0.847 | 0.846 | 0.789 | 0.708 |
| place dependence | 0.939 | 0.837 | 0.715 | 0.720 | 0.942 | 0.851 | 0.726 | 0.730 |
| place attachment | 0.941 | 0.903 | 0.830 | 0.817 | 0.950 | 0.912 | 0.833 | 0.852 |

Finally, each dimension of the sense of place consisted of four variables. PD included dependence on natural resources, land production dependence, living dependence and lifestyle dependence. PA included social attachment, a sense of belonging to the material environment, a sense of familiarity with the material environment, and attachment to the local culture. PI contained the residents' recognition of beautiful natural landscape, agricultural productive conditions, rural infrastructure and standardized rural management system. The arithmetic means of the scores of all variables were calculated as an index of different items of the sense of place. According to the rule of Likert scale, the average values can be divided into three types: (i) the intensity of the sense of place for the respondents was negative (0~2.5), (ii) the intensity of the sense of place for the respondents was intermediate (2.5~3.5) and (iii) the intensity of the sense of place for the respondents was positive (3.5~5).

## 3. Results

### 3.1. Land Use Changes

During the study period, the application of ecological engineering significantly promoted the optimization of land use structures and the improvement of ecological environment in the study area (Figure 2). The change in croplands was the most significant, with a decrease of 83.39 ha, and its share in the total area decreased from 46.92% in 1984 to 15.33% in 2020. These changes were mainly due to the reduction of sloping fields from 30.92% in 1984 to less than 0.1% in 2020 due to the GGP. The growth of ecological land was the largest, with an increase of 81.19 ha, and its share in total area increased from 51.27% in 1984 to 82.02% in 2020. Among them, the percentage of forests in the total area increased the most from 11.06% in 1984 to 66.01% in 2020. In addition, the area of construction land increased by 5.45 ha, and its share in the total area increased from 0.49% in 1984 to 2.56% in 2020. The area of bare land decreased by 3.25 ha. Land use changes were analyzed for different periods (Table 4).

In the CMSC period (1984–1998), the reduction of croplands was the highest with a decrease of 27.60 ha, and its share in the total area decreased by 10.45%. Among them, the sloping fields decreased the most by 36.68 ha due to farmers who voluntarily returned the sloping fields to the forests. The area of terraces and check dam land increased by 7.31 ha and 1.77 ha, respectively. The growth of ecological land was the highest, with an increase of 27.02 ha, and its share in the total area increased by 10.23%. This was mainly due to the growth of forests, which had an increase of 19.89 ha. The change of construction land and bare land was not particularly significant.

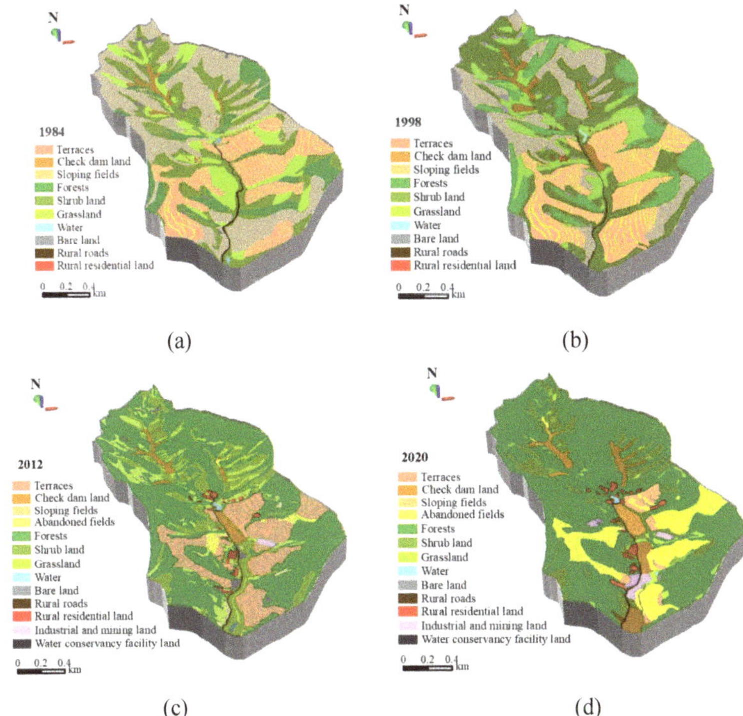


(a)  (b)

(c)  (d)

**Figure 2.** Land use types of the study area in 1984 (**a**), 1998 (**b**), 2012 (**c**) and 2020 (**d**).

**Table 4.** Area of land use types in the study area from 1984 to 2020 (ha).

| Types | 1984 | 1998 | 2012 | 2020 |
|---|---|---|---|---|
| Check dam land | 5.12 | 6.89 | 6.87 | 16.44 |
| Sloping fields | 81.65 | 44.97 | 0.97 | 0.09 |
| Terraces | 37.10 | 44.41 | 29.21 | 5.22 |
| Abandoned fields | — | — | 2.97 | 18.73 |
| Grassland | 36.89 | 42.52 | 25.67 | 8.39 |
| Shrub land | 68.92 | 70.47 | 42.90 | 33.50 |
| Forests | 29.19 | 49.08 | 150.40 | 174.29 |
| Water | 0.37 | 0.32 | 0.42 | 0.38 |
| Rural roads | 1.06 | 0.74 | 0.75 | 0.98 |
| Industrial and mining land | — | — | 0.51 | 3.12 |
| Hydraulic land | — | — | 0.25 | 0.09 |
| Rural residential land | 0.26 | 0.31 | 0.91 | 2.58 |
| Bare land | 3.47 | 4.32 | 2.20 | 0.22 |

In the GGP period (1998–2012), the reduction of croplands was still the highest with a decrease of 56.25 ha, and its share in the total area decreased by 21.30%, mainly due to a significant reduction of sloping fields. The area of sloping fields decreased by 44 ha. The increase in ecological land was still the largest, with an increase of 57 ha, and its share in the total area increased by 21.59%. Therefore, the ecological benefits of the GGP were significant. In addition, the area of bare land reduced by 2.12 ha. The area of construction land increased by 1.37 ha due to the expansion of rural residential and industrial and mining lands, and a chicken farm was constructed on terraces in 2010.

In the GLCP period (2012–2020), the area of ecological land decreased by 2.83 ha, due to the conversion from shrub land and grassland into check dam land, rural residential land and industrial and mining land. However, the forest area increased significantly, with an increase of 23.89 ha. In the same period, croplands increased by 0.46 ha. Among them, the increase in the check dam land area was the largest with an increase of 9.77 ha. For comparison, the reduction of terraces was significant, with a reduction of 8.44 ha, due to the conversion from terraces to shrub land and forests. However, the abandonment of croplands became more frequent and abandoned fields reached 18.73 ha in 2020. In addition, the area of construction land increased by 4.35 ha, while the area of rural residential land and industrial and mining land increased by 1.67 ha and 2.61 ha, respectively. At the same time, bare land decreased by 1.98 ha.

### 3.2. Characteristics of the Evaluation of Croplands

In terms of different types of land use, the changes in croplands were the most typical and were reflected in changes in area, structure and spatial distribution. It was noteworthy that the abandonment of croplands in the study area became widespread, and agricultural production became more depressed. The share of crops area in the total area decreased from 46.17% in 1984 to 8.23% in 2020. Moreover, the evolution of croplands is a direct reflection of the sense of place of farmers on the land and is closely linked to sustainable livelihoods of rural residents. Therefore, the evolutionary characteristics of croplands will be analyzed in detail from the aspect of the structure of cropland types, relative altitude and abandonment of croplands (Figure 3).

At the beginning of the study period, the Yangjuangou catchment was characterized by a traditional way of extensive cultivation. The total area of croplands was 123.87 ha, and sloping fields were the main type of croplands, which accounted for 65.92% of the total area of croplands. The sloping fields were mainly distributed on the top of a hill, slope of a hill, steep slope of a gully and gentle slope of a gully, with an average relative altitude of 66.95 m, which was the highest in investigated period. Secondly, the terraces, with an average relative altitude of 41.13 m, accounted for 29.95% of the total area of croplands. They were relatively fragmented and mainly located on gentle slope of the gully. The check dam land accounted for only 4.03% of the total croplands area and was mainly located in the north of the gully.

Grain yield per unit area of the sloping fields was less than one-third of that of the check dam land, and poor soil and water conservation performance of the sloping fields would affect crop yields. During the CMSC period, farmers spontaneously returned the sloping fields to the forests in 1987 and 1990, which were relatively far from the villages and located on steep slopes of hills and gullies. In addition, the check dam land in the middle of gully was renovated, and the sloping fields at low altitude were turned into terraces. In 1998, the total area of croplands decreased to 96.27 ha, and the sloping fields accounted for 46.71% of the total croplands. However, sloping fields were still the main type of croplands, and they were mainly distributed on the top of a hill, where the slope is gentle and the soil erosion is not intensive, so the average relative altitude of sloping fields increased by 12.80 m. The percentage of terraces in the total croplands increased to 46.13%, and the average relative altitude of terraces decreased by 7.4 m, because terraces at high altitude were returned to ecological land. The percentage of the check dam land in the total croplands was increased to 7.16%.

With the implementation of the GGP, the total area of croplands was reduced to 40.02 ha in 2012. Terraces have become the main type of croplands, and the percentage of terraces in the total croplands has increased to 72.99%, while the average relative altitude of terraces has decreased by 10.94 m due to the abandonment of terraces at higher altitude. The proportion of the check dam land in the total croplands has been increased to 17.17%. In contrast, the percentage of sloping fields in the total croplands decreased to 2.42%, the average relative altitude of the sloping fields decreased by 56.99 m, which was lower than the altitude of the terraces and was mostly located on the toe of the slope with low and gentle terrain. Abandonment of croplands occurred in 2012, which accounted for 7.42% of the total croplands. In particular, the abandoned terraces and the abandoned check dam land accounted for 87.21% and 12.79% of the total abandoned fields, respectively.

The GLCP was implemented in 2013 in the study area, so check dam land in the middle of the gully with flat terrain was significantly increased. In addition, the check dam land with poor farming conditions in the northern gully was renovated. In 2020, the total area of croplands increased to 40.48 ha, check dam land became the main type of croplands, and the percentage of the check dam land area in the total croplands area increased to 40.61%. However, the percentage of terraces in the total area of croplands decreased to 12.90% and its average relative altitude decreased by 12.87 m, due to the fact that some terraces at higher altitude were abandoned. In addition, the percentage of sloping fields in the total croplands decreased to 0.22%, and the average relative altitude decreased by 14.29 m. In the later period of the study, 46.27% of croplands were abandoned. In particular, the percentage of abandoned terraces in all abandoned fields increased to 96.88%. The sown terraces are wide terraces with good farming conditions and close to the residence. The percentage of abandoned check dam land in the total abandoned fields decreased to 3.11%, which was mainly located in the northern gully far from the residential area.

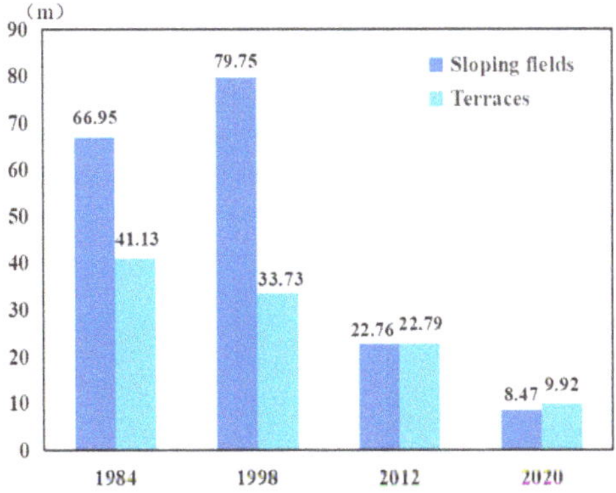


**Figure 3.** Average relative altitude of croplands from 1984 to 2020.

In general, guided by ecological restoration strategies, the study area has changed from an extensive cultivation phase to a terrace-dominated cultivation mode, and finally to a cultivation mode dominated by check dam land. In addition, the relative altitude of croplands has decreased. These changes reflected an effective improvement in ecological management and agricultural production conditions in LHGR. However, the abandonment of croplands reflected the change in farmers' livelihood from living on the land to working outside, i.e., the land was no longer the main source of farmers' livelihood.

### 3.3. Change of Land Use Degree

LUDI of Yangjuangou catchment decreased by 30 during the study period, which indicated that the availability of natural land attributes has improved, i.e., the exploitation of land resources has declined (Figure 4). From 1998 to 2012, the rate of decline of LUDI was the fastest, with an average annual decline of 1.43, mainly due to the accelerated reduction of croplands with a higher land use degree under the influence of the GGP. In addition, from 1984 to 1998, LUDI declined rapidly, with an average annual decline of 0.78, mainly due to the conversion of sloping fields into ecological land under the influence of the CMSC. From 2012 to 2020, LUDI declined slightly, with an average annual decline of 0.25, indicating that increasing attention was gradually being paid to the rational use of land resources. Although the abandonment of croplands was most severe from 2012 to 2020, the GLCP significantly increased the check dam land area with better production conditions, while rural residential land and industrial and mining land with the highest land use degree has increased due to improved rural infrastructure. Therefore, the rate of decline in land use degree has slowed in this period. Overall, the decline in land use degree was mainly the result of the reduction of croplands due to ecological projects and the abandonment of croplands because of the increased migration of farmers.

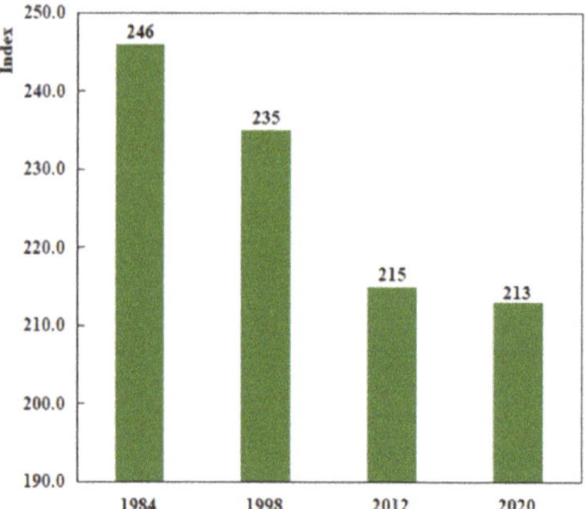


**Figure 4.** Comprehensive quantified index of land-use degree in the study area from 1984 to 2020.

### 3.4. Interpretation of the Change of Sense of Place

The intensity of the sense of place of local farmers showed a fluctuating declining trend during the study period (Figure 5). From 1984 to 1998, the sense of place index decreased by 0.19, mainly due to a decrease in PD and PA, which accounted for 77.65% and 22.35% of the decrease, respectively. However, the average value of PI increased by 0.28. From 1998 to 2012, the total value of the sense of place increased by 0.04, mainly due to an increase in the average value of PI, which increased by 1.27. In comparison, the average value of PD and PA decreased by 0.72 and 0.42, respectively. From 2012 to 2020, the decline in the total value of sense of place was the largest with 0.25. The decrease in PD and PA accounted for 80.15% and 19.85% of the decrease in the sense of place, yet the average value of PI increased by 0.61. We provide a detailed analysis of the changes in each dimension of the sense of place in the following subsections.

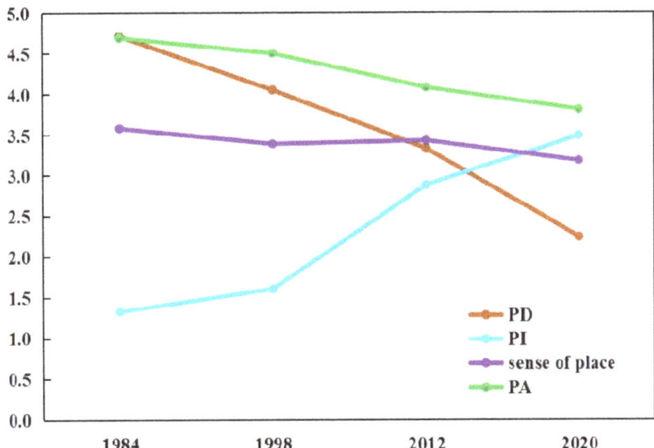



**Figure 5.** Intensity index of the sense of place of local farmers. PI = Place identity; PD = Place dependence; PA = Place attachment.

### 3.4.1. Changes in the Place Dependence Index

The PD index decreased the most, from 4.71 in 1984 to 2.24 in 2020, which indicated that the intensity of PD of local farmers has changed from positive dependence to negative dependence. Four variables of PD showed a declining trend (Figure 6). Among them was a significant decrease in the indices of natural resources dependence, land production dependence and living dependence, which decreased by 2.86, 2.64 and 2.54, respectively, and changed from positive dependence to negative dependence. However, the index of lifestyle dependence declined slightly and changed from positive dependence to intermediate dependence.

The decline in PD of local farmers was mainly influenced by natural and human factors, such as policy implementation, market economy, and natural disasters. In particular, the GGP banned farmers from exploiting ecological resources, which was one of the key factors in the decline in dependence on natural resources. The decline in land production dependence was influenced not only by ecological projects, but also by market economy factors. On the one hand, ecological projects have promoted the transformation from a large number of sloping fields into ecological land, which has changed the source of livelihood for farmers. On the other hand, rising costs of agricultural production have encouraged an increasing number of rural workers to start working in the non-agricultural production sector, leading to marginalization and abandonment of croplands. In addition, farmers' confidence in the development of planting industry has been affected by extreme weather conditions and other natural disasters. Through household interviews, we learned that almost all crops on terraces were destroyed in 2013 due to extreme rainstorms, so many farmers decided to migrate. Moreover, the number of wild animals has increased due to ecological restoration, which has a negative impact on the crops on the terraces, leading to more rural residents deciding to abandon the terraces. The widespread abandonment of croplands was the visual response of the decline in land production dependence. Decline in living dependence was affected by rural labor migration and the Village Combination Project in 2013. The index of lifestyle dependence decreased, as farmers gradually adopted a new concept of life, with rural inhabitants going out to work and moving to cities and towns.

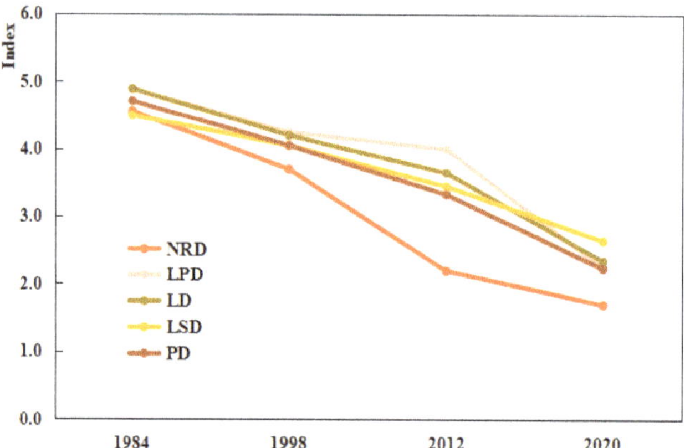

**Figure 6.** Intensity index of the place dependence of local farmers. NRD = Natural resources dependence; LPD = Land production dependence; LD = Living dependence; LSD = Lifestyle dependence; PD = Place dependence.

### 3.4.2. Changes in the Place Attachment Index

The PA index decreased from 4.69 in 1984 to 3.81 in 2020, and its intensity remained positive. Four variables of PA decreased and among them, the value of the index of social attachment decreased the most (Figure 7). Then, the sense of belonging to the material environment and the sense of familiarity with the material environment decreased by 0.97 and 0.94, respectively. The index of attachment to the local culture decreased by only 0.44. The intensity of the other three variables was positive, except that the intensity of the sense of belonging to the material environment has changed from positive to intermediate.

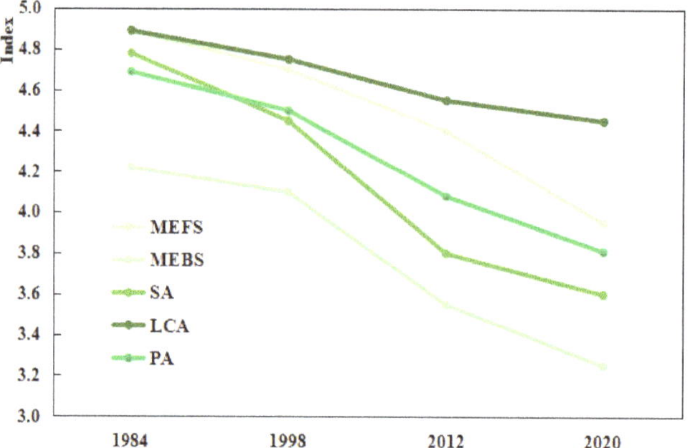

**Figure 7.** Intensity index of the place attachment of local farmers. MEFS = The sense of familiarity with the material environment; MEBS = The sense of belonging to the material environment; SA = Social attachment; LCA = Attachment to the local culture; PA = Place attachment.

The change in PA reflected the contradiction of psychological changes of farmers. On the one hand, if farmers have certain abilities or life pressure, they decide to engage in the non-agricultural industries with higher incomes. On the other hand, the land is still very important for the rural residents, as it is the ultimate guarantee for farmers' livelihood. According to the field investigation, local farmers believed that when they are too old and could not to go out to work, they still wanted to rely on the land to survive. This is in line with the Chinese cultural tradition of attachment to the homeland, i.e., although farmers are busy working far from home, they still have a strong topophilia. Therefore, although the intensity of PA of local farmers decreased, it was still positive.

### 3.4.3. Changes in the Place Identity Index

The PI index increased from 1.33 in 1984 to 3.49 in 2020, indicating that the PI intensity of farmers changed from negative to intermediate. Four variables of PI increased (Figure 8). Among them, the residents' recognition of beautiful natural landscape and agricultural productive conditions increased by 2.76 and 2.48, respectively. The residents' recognition of rural infrastructure and standardized rural management system increased by 1.83 and 1.54, respectively. The intensity of all variables changed from negative to intermediate.

The improvement in PI intensity was mainly the result of the promotion of ecological projects. The increase in the residents' recognition of beautiful natural landscape has been the most significant since 1998. It can be seen that the GGP has an obvious positive effect on ecological restoration and farmers have unanimously recognized it. In addition, the agricultural production conditions of the catchment have been significantly improved. In particular, the GLCP has significantly promoted the growth of the check dam land with high-yield, so the rate of increase in the residents' recognition of agricultural production conditions was the highest from 2012 to 2020. Moreover, with social progress and increased government investment in rural construction, the residents' recognition of rural infrastructure and the standardized rural management system has improved, but had intermediate value at the end of the study period and still needs to be improved.

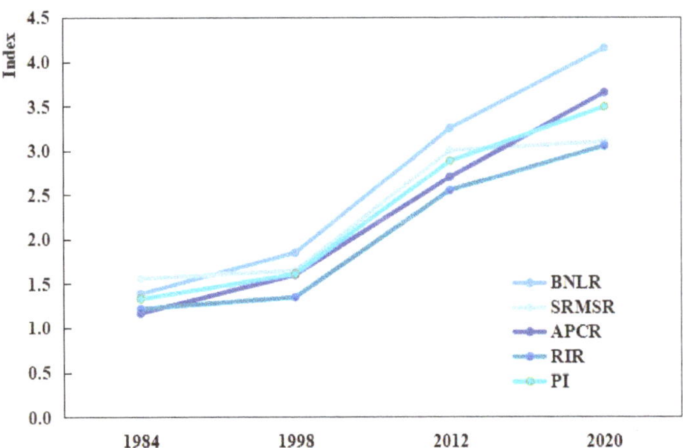


**Figure 8.** Intensity index of the place identity of local farmers. BNLR = The recognition of residents on beautiful natural landscape; SRMSR = The recognition of residents on standardized rural management system; APCR = The recognition of residents on agricultural productive conditions; RIR = The recognition of residents on rural infrastructure; PI = Place identity.

## 4. Interaction between Land Use and the Sense of Place of Farmers

The sense of place of farmers is an inherent expression of the relationship between humans and environment in traditional society [56–58]. There is an interaction between farmers' sense of place and land use (Figure 9), which we have qualitatively evaluated based on a quantitative change in the sense of place and land use.

As a living space, the change of land in the relationship between humans and environment is dominated by a change in the sense of place of humanity [50]. Prior to the 1980s, livelihoods of the rural residents were tied to a land and characterized by subsistence agriculture. Therefore, the main land use patterns in the study area had long been sloping fields influenced by the strong intensity of PD and PA of local farmers. Then, the intensity of PD and PA of local farmers gradually decreased due to the deterioration of the natural environment and low comparative benefit of agricultural production, which promoted the conversion of sloping fields to ecological land in the period of the CMSC. During the GGP period, PD and PA of local farmers further declined due to the GGP financial subsidies aimed at converting more sloping fields into forests. In addition, natural disasters have accelerated the reduction of PD and PA of local farmers. The confidence of local farmers in the development of agriculture has been undermined by natural disasters, and an increasing number of rural workers decided to migrate, leading to the abandonment of check dam land and terraces with poor production conditions. Moreover, the Village Combination Project also led to the decline of PD and PA of local farmers, and more croplands were abandoned. In the end, the terraces with better cultivation conditions were abandoned.

Under the influence of land use changes, farmers reconsider the land, which affects the sense of place of farmers. Ecological and livelihood projects have increased the land use proportion of check dam land, ecological land, roads and other infrastructure and optimized the land use structure. Conclusively, the PI intensity of local farmers has obviously improved. At the same time, the premise of increasing the ecological land was to give up more opportunities of agricultural production, which led to a decrease in the intensity of PA and PD of local farmers. It is notable that, although the area of the check dam land with high yield increased, which accounted for only 6.23% of the study area, farmers were not enthusiastic about traditional agriculture due to limited area for farming, which has no significant impact on rural development, and the residents' recognition of agricultural production conditions has yet to be further improved.

In LHGR, the decline in land use degree has revealed a decline in farmers' dependence on the production and living functions of land and a weakening of farmers' attachment to rural areas to a certain extent. In general, with the decline in land use degree, the sense of place of farmers also showed a declining trend. It showed that farmers are not traditional farmers and land is no longer the only material dependence of farmers, as farmers' livelihood patterns have become diverse, and rural areas are in the process of transformation from traditional agricultural stage to non-agricultural stage.

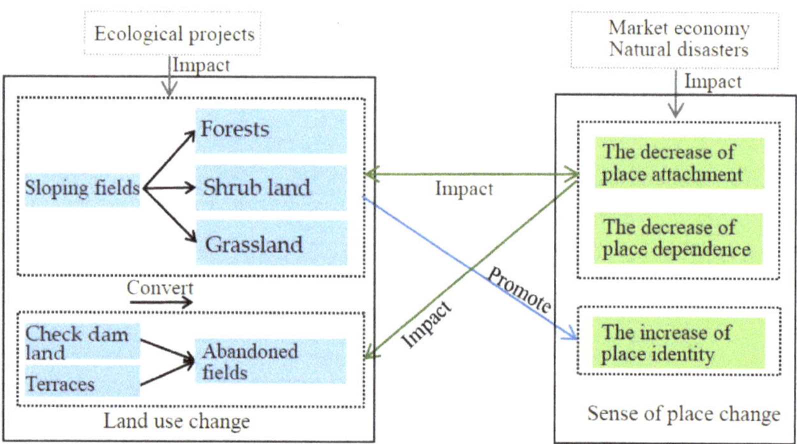


**Figure 9.** The interaction between land use and sense of place in the study area.

## 5. Discussion

This paper comprehensively explored land use changes and farmers' sense of place and the interaction between them, with the implementation of ecological and livelihood projects in a typical small catchment in LHGR from 1984 to 2020. The study of land use changes mainly involved the change in ecological land, construction land, croplands and bare land. The change in croplands was the most significant and is closely related to the change in the sense of place of farmers. Therefore, the evolution of croplands was analyzed in detail from the aspect of structure, relative altitude and abandonment of croplands. Then, the land use degree in the study area was quantitatively analyzed. In addition, the farmers' sense of place was quantitatively analyzed from the perspective of PI, PD and PA. That is, it was studied from following aspects: farmers' recognition of production and living functions of land, farmers' recognition of natural and human environment, and the emotional connection between farmers and land. Furthermore, the interactive mechanism between land use and the sense of place of farmers was explored. It is noticeable that there has been a lot of research on land use changes in LHGR, and it was mostly concentrated on the process of land use change and its ecological benefits at the county and regional level, and was rarely related to the sense of place of farmers at the village level [58–60]. Therefore, the influence of human subjectivity on land use change has been ignored. The novelty of this paper was that not only the land use changes from the micro level of the small catchment were explored, but also the interaction between the land use changes and the sense of place of farmers was explored in detail, which was helpful for people to deeply understand the evolution of rural space from a perspective of residents' cognition and provided a scientific reference for the construction of a highly satisfactory rural settlement. We will further discuss the implications derived from the above study.

*5.1. Innovating Land Use Patterns and Improving the Efficiency of Resource Utilization*

From the case study of this paper, although PI of farmers increased through the implementation of ecological and livelihood projects, PD and PA of farmers have decreased significantly. These changes were mainly manifested in a significant decline in the LUDI. For example, the abandonment of croplands in the study area was widespread, and almost 30% of croplands were abandoned. It can be seen that, although the implementation of different ecological projects has significantly improved the ecological environment and rural infrastructure, most of the land resources were still idle. That is, the socio-economic system has not yet adapted to the evolution of the geographic environment system in the catchment and there are still possibilities for optimizing the land use structure. According to research by Li et al., the function of ecosystem services has increased significantly on the Loess Plateau, but farmers could only benefit from the construction and management of ecological resources through ecological compensation. Moreover, due to the virtual property rights, unreasonable industrial structure, and imperfect land market mechanisms, the development of rural industries related to the exploitation of natural resources has lagged behind. As a result, a large number of rural workers turned to migrant workers, which led to extensive management and abandonment of croplands [36]. Lyu Changhe et al. believed that crop sown area decreased by 5% on the whole plateau and by 10% in the hilly region during 1998–2014 [61]. Our research was in line with the findings of previous research, i.e., low efficiency of the utilization of natural resources is a common problem in rural areas of the Loess Plateau. It will therefore be of great significance for the rural vitalization of LHGR to deeply explore innovative land use patterns to consider optimal land use as a key activity, to consider land engineering and rural land circulation as means, and to consider high agricultural efficiency as a breakthrough, to revitalize the idle land.

*5.2. Diversification of Farmers' Livelihoods to Ensure Their Sustainability*

With the implementation of the Cropland Reconversion Program, the Development of West China, and the Urbanization Development Strategy of China, the number of migrant workers from rural areas in LHGR has increased. However, due to the low level of education of farmers, most migrant workers are engaged in low-tech low-wage jobs, and the potential for increasing their incomes is limited [62]. There are great challenges in the sustainability of farmers' livelihoods in LHGR. The difference between farmers in the Yangjuangou catchment was obvious in terms of agricultural production mode, livelihood capital allocation and risk coping ability. Specifically, 21.87% of farmers were engaged in agricultural production, 57.66% of farmers were migrant workers and 14.60%

of farmers were engaged in non-farm self-employment. Regarding the future livelihood plan of farmers, 70% of them want to adjust the cropping structure and expand the scope of planting crops; 60% of farmers who rely on non-agricultural income believe that they still need land to make a living when they are older and not recognized by the society; and 40% of farmers living in new rural communities want to rebuild their households, mainly because they cannot afford to buy a new house in the cities. This showed that the current livelihood mode of most farmers has not reached the expected standard of living or they have a better and more stable livelihood mode to choose from, and the land remains a strong guarantee of the security of farmers' livelihoods. Therefore, the overall value of the sense of place of local farmers was still higher. The instability of farmers' livelihoods has become one of the key problems limiting the sustainable development of LHGR. In the LHGR, on the one hand, governments should help farmers build an efficient social capital network by expanding employment channels to diversify farmers' sources of income, as well as improve the cultural level and professional skills of farmers in order to improve their adaptability and anti-risk capacity for the market economy. On the other hand, the agricultural structure of planting should be optimized and adjusted in order to stimulate farmers' enthusiasm for agricultural production and promote rural development. For example, the authorities need to pay more attention to specialty and branded agricultural products, moreover, the added value of agricultural products needs to be improved.

## 6. Conclusions

This paper investigated the change and interaction of land use and the sense of place of farmers in a typical watershed of LHGR from 1984 to 2020, based on high-resolution remote sensing image data, GIS technique and household surveys. The main results were as follows: (1) the changes in ecological land and croplands were the most significant. The area of ecological land increased significantly, and the growth of forests was the highest. On the contrary, the area of croplands has decreased significantly, especially the sloping fields, and the abandonment of terraces has become more frequent. However, the area of the check dam land increased with the promotion of the GLCP. Generally speaking, the land use degree has been gradually declining; (2) the total value of the sense of place of local farmers showed a fluctuating downward trend. In particular, PD and PA showed a downward trend, the former declining the most, while PI increased; and (3) the land use change was dominated by the sense of place of farmers, while the land use change influenced the adjustment of the sense of place of farmers as it led the farmers to reconsider the land. This paper suggested that the rural areas in LHGR should innovate land use patterns and promote high efficiency and intensification of agriculture through land engineering and rural land circulation. Moreover, while ensuring the effects of ecological restoration, the government should also pay attention to extending livelihood strategies and improving the adaptability and anti-risk capability of farmers to a market economy, in order to promote the harmonious development of human-environmental relations in LHGR.

This paper was a preliminary study on the links between land use change and the farmers' sense of place in rural areas of the LHGR, and further research is needed. Firstly, due to the limitations of remote sensing images and data acquisition, only a typical catchment was selected in this study, and a comparative study of multiple catchments should be further conducted to fully understand the human-environment relationship in LHGR. Secondly, the theoretical framework of the interaction between land use and the sense of place of farmers, i.e., the dynamic concept of interaction between humans and environment needs to be improved with more examples. In addition, the research objects selected in this study were middle-aged and older population, so attention should be paid to the sense of place of farmers of different ages.

**Author Contributions:** Conceptualization, Y.L. (Yi Li), Y.L. (Yurui Li) and B.F.; methodology, Y.L. (Yi Li); software, Y.L. (Yi Li); validation, Y.L. (Yi Li), Y.L. (Yurui Li) and B.F.; formal analysis, Y.L. (Yi Li); investigation, Y.L. (Yi Li), C.W. and W.L.; resources, Y.L. (Yi Li); data curation, Y.L. (Yi Li) and Y.L. (Yurui Li); writing—original draft preparation, Y.L. (Yi Li); writing—review and editing, Y.L. (Yurui

Li), B.F. and L.Q.; visualization, Y.L. (Yi Li), Y.L. (Yurui Li) and B.F.; supervision, Y.L. (Yurui Li) and B.F.; project administration, Y.L. (Yi Li); funding acquisition, Y.L. (Yurui Li). All authors have read and agreed to the published version of the manuscript.

**Funding:** This research was funded by the National Key Research and Development Program of China, grant number 2017YFC0504701 and the National Natural Science Foundation of China, grant number 41271189, 41671174).

**Data Availability Statement:** The Sources and preprocessing of data are in Section 2.3. Other relevant data to support this study are available from the authors upon request.

**Acknowledgments:** The authors would like to thank the anonymous reviewers for their comments and suggestions which contributed to the further improvement of this paper.

**Conflicts of Interest:** The authors declare no conflict of interest.

## References


1. Cao, Z.; Li, Y.; Liu, Z.; Yang, L. Quantifying the vertical distribution pattern of land-use conversion in the loess hilly region of northern Shaanxi Province 1995–2015. *J. Geogr. Sci.* **2019**, *29*, 730–748. [CrossRef]
2. Liu, Y.; Guo, Y.; Li, Y.; Li, Y. GIS-based effect assessment of soil erosion before and after gully land consolidation: A case study of Wangjiagou project region, Loess Plateau. *Chin. Geogr. Sci.* **2015**, *25*, 137–146. [CrossRef]
3. McVicar, T.R.; Li, L.; Van Niel, T.; Zhang, L.; Li, R.; Yang, Q.; Zhang, X.; Mu, X.; Wen, Z.; Liu, W.; et al. Developing a decision support tool for China's re-vegetation program: Simulating regional impacts of afforestation on average annual streamflow in the Loess Plateau. *For. Ecol. Manag.* **2007**, *251*, 65–81. [CrossRef]
4. Liu, Y.S.; Li, Y.R. Engineering philosophy and design scheme of gully land consolidation in Loess Plateau. *Trans. CSAE* **2017**, *33*, 1–9. [CrossRef]
5. Cao, Z.; Li, Y.R.; Liu, Y.S.; Chen, Y.; Wang, Y. When and where did the Loess Plateau turn "green" Analysis of the tendency and breakpoints of normalized difference vegetation index. *Land Degrad. Dev.* **2017**, *29*, 1085–3278. [CrossRef]
6. Fu, B.; Liu, Y.; Lü, Y.; He, C.; Zeng, Y.; Wu, B. Assessing the soil erosion control service of ecosystems change in the Loess Plateau of China. *Ecol. Complex.* **2011**, *8*, 284–293. [CrossRef]
7. Li, Y.; Li, Y.; Fan, P.; Sun, J.; Liu, Y. Land use and landscape change driven by gully land consolidation project: A case study of a typical watershed in the Loess Plateau. *J. Geogr. Sci.* **2019**, *29*, 719–729. [CrossRef]
8. Jiang, N.; Shao, M.A. Characteristics of Soil and Water Loss of Different Slope Land Uses in Small Watershed on the Loess Plateau. *Trans. CSAE* **2011**, *27*, 36–41. [CrossRef]
9. Lu, X.-L.; Wu, C.-Y.; Xiao, G.-R. Fuzzy synthetic evaluation on resident's perceptions of tourism impacts. *Chin. Geogr. Sci.* **2006**, *16*, 87–94. [CrossRef]
10. Zhou, Q.H. *Settlements in the Valley of the Loess Plateau*; China Architecture & Building Press: Beijing, China, 2009.
11. Li, Y.R.; Li, Y.; Fan, P.C.; Liu, Y.S. Impacts of Land Consolidation on Rural Human–environment System in Typical Watershed of Loess Hilly and Gully Region. *Trans. CSAE* **2019**, *35*, 241–250. [CrossRef]
12. Liu, G.B.; Yang, Q.K.; Zheng, F.L. Small Watershed Management and Eco-rehabilitation on the Loess Plateau of China. *Sci. Soil Water Conserv.* **2004**, *2*, 11–15.
13. Perring, M.; De Frenne, P.; Baeten, L.; Maes, S.L.; Depauw, L.; Blondeel, H.; Carón, M.M.; Verheyen, K. Global environmental change effects on ecosystems: The importance of land-use legacies. *Glob. Chang. Biol.* **2016**, *22*, 1361–1371. [CrossRef]
14. Güneralp, B.; Seto, K.C.; Ramachandran, M. Evidence of urban land teleconnections and impacts on hinterlands. *Curr. Opin. Environ. Sustain.* **2013**, *5*, 445–451. [CrossRef]
15. Shi, P.J.; Gong, P.; Li, X.B. *Research Method and Practice of Land Use/Land Cover*; Science Press: Beijing, China, 2000.
16. Liu, Y.S. Introduction to land use and rural sustainability in China. *Land Use Policy* **2018**, *74*, 1–4. [CrossRef]
17. Skokanová, H.; Havlíček, M.; Borovec, R.; Demek, J.; Eremiášová, R.; Chrudina, Z.; Mackovčin, P.; Rysková, R.; Slavík, P.; Stránská, T.; et al. Development of land use and main land use change processes in the period 1836–2006: Case study in the Czech Republic. *J. Maps* **2012**, *8*, 88–96. [CrossRef]
18. Betts, R. Implications of land ecosystem-atmosphere interactions for strategies for climate change adaptation and mitigation. *Tellus B Chem. Phys. Meteorol.* **2007**, *59*, 602–615. [CrossRef]
19. Long, H.; Qu, Y. Land use transitions and land management: A mutual feedback perspective. *Land Use Policy* **2018**, *74*, 111–120. [CrossRef]
20. Bazame, R.; Tanrıvermis, H.; Kapusuz, Y.E. Land management and sustainable use of land resources in the case of Burkina Faso. *Land Degrad. Dev.* **2019**, *30*, 608–621. [CrossRef]
21. Mottet, A.; Ladet, S.; Coqué, N.; Gibon, A. Agricultural land-use change and its drivers in mountain landscapes: A case study in the Pyrenees. *Agric. Ecosyst. Environ.* **2006**, *114*, 296–310. [CrossRef]
22. Wang, J.; Chen, Y.; Shao, X.; Zhang, Y.; Cao, Y. Land-use changes and policy dimension driving forces in China: Present, trend and future. *Land Use Policy* **2012**, *29*, 737–749. [CrossRef]

23. Li, Y.; Cao, Z.; Long, H.; Liu, Y.; Li, W. Dynamic analysis of ecological environment combined with land cover and NDVI changes and implications for sustainable urban–rural development: The case of Mu Us Sandy Land, China. *J. Clean. Prod.* **2017**, *142*, 697–715. [CrossRef]

24. Halmy, M.W.A.; Gessler, P.E.; Hicke, J.A.; Salem, B. Land use/land cover change detection and prediction in the north-western coastal desert of Egypt using Markov-CA. *Appl. Geogr.* **2015**, *63*, 101–112. [CrossRef]

25. Song, X.-P.; Hansen, M.C.; Stehman, S.V.; Potapov, P.V.; Tyukavina, A.; Vermote, E.F.; Townshend, J.R. Global land change from 1982 to 2016. *Nature* **2018**, *560*, 639–643. [CrossRef] [PubMed]

26. Dadashpoor, H.; Azizi, P.; Moghadasi, M. Land use change, urbanization, and change in landscape pattern in a metropolitan area. *Sci. Total Environ.* **2019**, *655*, 707–719. [CrossRef]

27. Chu, H.; Lin, Y.-P.; Huang, C.-W.; Hsu, C.-Y.; Chen, H.-Y. Modelling the hydrologic effects of dynamic land-use change using a distributed hydrologic model and a spatial land-use allocation model. *Hydrol. Process.* **2010**, *24*, 2538–2554. [CrossRef]

28. Dirmeyer, P.; Niyogi, D.; De Noblet-Ducoudré, N.; Dickinson, R.E.; Snyder, P.K. Impacts of land use change on climate. *Int. J. Clim.* **2010**, *30*, 1905–1907. [CrossRef]

29. Sharma, G.; Sharma, L.; Sharma, K.C. Assessment of land use change and its effect on soil carbon stock using multitemporal satellite data in semiarid region of Rajasthan, India. *Ecol. Process.* **2019**, *8*, 42. [CrossRef]

30. Foley, J.A.; DeFries, R.; Asner, G.P.; Barford, C.; Bonan, G.; Carpenter, S.R.; Chapin, F.S.; Coe, M.T.; Daily, G.C.; Gibbs, H.K.; et al. Global Consequences of Land Use. *Science* **2005**, *309*, 570–574. [CrossRef]

31. Guerra, C.A.; Rosa, I.M.D.; Pereira, H.M. Change versus stability: Are protected areas particularly pressured by global land cover change? *Landsc. Ecol.* **2019**, *34*, 2779–2790. [CrossRef]

32. Levine, N.S.; Kaufman, C.C. Land Use, Erosion, and Habitat Mapping on an Atlantic Barrier Island, Sullivan's Island, South Carolina. *J. Maps* **2008**, *4*, 161–174. [CrossRef]

33. Shi, H.; Shao, M. Soil and water loss from the Loess Plateau in China. *J. Arid. Environ.* **2000**, *45*, 9–20. [CrossRef]

34. Fu, B.; Chen, L.; Ma, K.; Zhou, H.; Wang, J. The relationships between land use and soil conditions in the hilly area of the loess plateau in northern Shaanxi, China. *Catena* **2000**, *39*, 69–78. [CrossRef]

35. Cao, Z.; Liu, Y.; Li, Y. Rural transition in the loess hilly and gully region: From the perspective of "flowing" cropland. *J. Rural. Stud.* **2019**. [CrossRef]

36. Li, Y.; Zhang, X.; Cao, Z.; Liu, Z.; Lu, Z.; Liu, Y. Towards the progress of ecological restoration and economic development in China's Loess Plateau and strategy for more sustainable development. *Sci. Total Environ.* **2021**, *756*, 143676. [CrossRef]

37. Davenport, M.A.; Anderson, D.H. Getting From Sense of Place to Place-Based Management: An Interpretive Investigation of Place Meanings and Perceptions of Landscape Change. *Soc. Nat. Resour.* **2005**, *18*, 625–641. [CrossRef]

38. Johnston, R.J. *The Dictionary of Human Geography*; The Commercial Press: Beijing, China, 2004.

39. Jorgensen, B.S.; Stedman, R.C. A comparative analysis of predictors of sense of place dimensions: Attachment to, dependence on, and identification with lakeshore properties. *J. Environ. Manag.* **2006**, *79*, 316–327. [CrossRef] [PubMed]

40. Zhu, H.; Liu, B. Concepts analysis and research implications: Sense of place, place attachment and place identity. *J. S. China Norm. Univ.* **2011**, *1*, 1–8.

41. Moore, T. Planning for place: Place attachment and the founding of rural community land trusts. *J. Rural. Stud.* **2021**, *83*, 21–29. [CrossRef]

42. Soini, K.; Vaarala, H.; Pouta, E. Residents' sense of place and landscape perceptions at the rural–urban interface. *Landsc. Urban Plan.* **2012**, *104*, 124–134. [CrossRef]

43. Qian, J.; Zhu, H.; Liu, Y. Investigating urban migrants' sense of place through a multi-scalar perspective. *J. Environ. Psychol.* **2011**, *31*, 170–183. [CrossRef]

44. Tapsuwan, S.; Leviston, Z.; Tucker, D. Community values and attitudes towards land use on the Gnangara Groundwater System: A Sense of Place study in Perth, Western Australia. *Landsc. Urban Plan.* **2011**, *100*, 24–34. [CrossRef]

45. Hidalgo, M.; Hernández, B. Place attachment: Conceptual and empirical questions. *J. Environ. Psychol.* **2001**, *21*, 273–281. [CrossRef]

46. Kyle, G.; Mowen, A.J.; Tarrant, M. Linking place preferences with place meaning: An examination of the relationship between place motivation and place attachment. *J. Environ. Psychol.* **2004**, *24*, 439–454. [CrossRef]

47. Deutsch, K.; Goulias, K. Investigating the Impact of Sense of Place on Travel Behavior Using an Intercept Survey Methodology. *St. Barbar. Univ. Calif.* **2009**, 887–908.

48. Stedman, R.C. Is It Really Just a Social Construction?: The Contribution of the Physical Environment to Sense of Place. *Soc. Nat. Resour.* **2003**, *16*, 671–685. [CrossRef]

49. Cross, J.E.; Keske, C.M.; Lacy, M.G.; Hoag, D.L.; Bastian, C.T. Adoption of conservation easements among agricultural landowners in Colorado and Wyoming: The role of economic dependence and sense of place. *Landsc. Urban Plan.* **2011**, *101*, 75–83. [CrossRef]

50. Xue, D.Q.; Chen, Q.; Lv, Y.Q. Land dependence and local attachment of rural residents in Weibei Upland under the background of land circulation: A comparative study of land-lost and normal farmers in Huangling County. *J. Shaanxi Norm. Univ.* **2019**, *4*, 31–39. [CrossRef]

51. Nan, M.-X.; Sen, Z.-Y.; Lei, Z.; Ying, X. Population structure and understory species diversity of different aged Pinus sylvestris var. mongolica plantations in Nenjiang Sandy Land of Northeast China. *Chin. J. Appl. Ecol.* **2012**, *23*, 2332. [CrossRef]

52. Fu, B.J.; Chen, L.D.; Ma, K.M. The Effect of Land Use Change on the Regional Environment in the Yangjuangou Catchment in the Loess Plateau of China. *Acta Geogr. Sin.* **1999**, *3*, 51–56. [CrossRef]
53. Liu, Y.S.; Chen, Z.F.; Li, Y.R.; Feng, W.L.; Cao, Z. The Planting Technology and Industrial Development Prospects of Forage Rape in the Loess Hilly Area—A Case Study of Newly-increased Cultivated Land Through Gully Land Consolidation in Yan'an, Shaanxi Province. *J. Nat. Resour.* **2017**, *32*, 2065–2074. [CrossRef]
54. Wang, X.L.; Bao, Y.H. Study on the methods of land use dynamic change research. *Prog. Geogr.* **1999**, *1*, 3–5. [CrossRef]
55. Zhuang, D.; Liu, J. Modeling of regional differentiation of land-use degree in China. *Chin. Geogr. Sci.* **1997**, *7*, 302–309. [CrossRef]
56. Devine-Wright, P. Rethinking NIMBYism: The role of place attachment and place identity in explaining place-protective action. *J. Community Appl. Soc. Psychol.* **2009**, *19*, 426–441. [CrossRef]
57. Gieling, J.; Vermeij, L.; Haartsen, T. Beyond the local-newcomer divide: Village attachment in the era of mobilities. *J. Rural. Stud.* **2017**, *55*, 237–247. [CrossRef]
58. Zhu, H.; Wu, J.; Guo, S.; Huang, D.; Zhu, Q.; Ge, T.; Lei, T. Land use and topographic position control soil organic C and N accumulation in eroded hilly watershed of the Loess Plateau. *Catena* **2014**, *120*, 64–72. [CrossRef]
59. Zhang, X.; Deng, Y.; Hou, M.; Yao, S. Response of Land Use Change to the Grain for Green Program and Its Driving Forces in the Loess Hilly-Gully Region. *Land* **2021**, *10*, 194. [CrossRef]
60. Zhang, S.; Yang, D.; Yang, Y.; Piao, S.; Yang, H.; Lei, H.; Fu, B. Excessive Afforestation and Soil Drying on China's Loess Plateau. *J. Geophys. Res. Biogeosci.* **2018**, *123*, 923–935. [CrossRef]
61. Lyu, C.; Xu, Z. Crop production changes and the impact of Grain for Green program in the Loess Plateau of China. *J. Arid. Land* **2020**, *12*, 18–28. [CrossRef]
62. Tang, Q.; Xu, Y.; Li, Y. Assessment of farmers' sustainable livelihoods and future strategies on the Loess Plateau: Based on a survey of 1076 farmers in Yan'an City in Shaanxi Province and Guyuan City in Ningxia Hui Autonomous Region. *Prog. Geogr.* **2013**, *32*, 161–169.

*Article*

# Spatio-Temporal Evolution of Land Use Transition and Its Eco-Environmental Effects: A Case Study of the Yellow River Basin, China

**Dengyu Yin** [1,2,3], **Xiaoshun Li** [1,2,3,*], **Guie Li** [1,2], **Jian Zhang** [1,2,3] and **Haochen Yu** [1,3]

[1] School of Public Policy and Management, China University of Mining and Technology, Xuzhou 221116, China; dengyu.yin@cumt.edu.cn (D.Y.); Guieli@whu.edu.cn (G.L.); zhangjian89@cumt.edu.cn (J.Z.); haochen.yu@cumt.edu.cn (H.Y.)
[2] Research Center for Transition Development and Rural Revitalization of Resource-Based Cities in China, China University of Mining and Technology, Xuzhou 221116, China
[3] Observation and Research Station of Jiangsu Jiawang Resource Exhausted Mining Area Land Restoration and Ecological Succession, Ministry of Education, Xuzhou 221116, China
* Correspondence: lxshun@cumt.edu.cn; Tel.: +86-0516-8359-1322

Received: 21 November 2020; Accepted: 10 December 2020; Published: 11 December 2020

**Abstract:** Human activities and environmental deterioration have resulted in land use transition (LUT), which seriously affects the ecosystem service value (ESV) of its region. Therefore, relevant policy measures are urgently needed. Nevertheless, research on the relationships between LUTs and ESVs from the overall watershed scale is lacking. Thus, the geo-information Tupu method was applied to analyze the dynamic patterns of LUT based on land use data from 1990, 2000, 2010, and 2018 of the Yellow River Basin (YRB). Then, a newly revised ecosystem services calculation method was utilized to the responses of ESV to LUTs. The results indicated that the Tupu units of the LUT were mainly based on the mutual transformation of grassland and unused land, and cultivated land and forestland, which were widely distributed in the upper and middle reaches of the basin. The spatial distribution was concentrated, and the expansion's trend was also obvious. Moreover, the conversion of cultivated land into construction land was mainly distributed in the lower reaches of the basin. During 1990–2018, the total ESV fluctuated and increased ($+10.47 \times 10^8$ USD) in the YRB. Thereinto, the ESV of grassland (45%) and forestland (30%) made the greatest contribution to the total ESV. As for different reaches, the ESV increased in the upstream, but decreased in the midstream and the downstream. In terms of contribution rate, the conversion of unused land into grassland (12.477%) and grassland into forestland (9.856%) were the main types to enhance the ESV in the YRB, while the conversion of forestland into grassland (−8.047%) and grassland to unused land (−7.358%) were the main types to reduce the ESV. Furthermore, the range of ecological appreciation zones was widely distributed and scattered, while the range of ecological impairment zones was gradually expanded. These findings could have theoretical support and policy implications for land use planning and environmental services in the YRB.

**Keywords:** land use transition (LUT); ecosystem services value (ESV); geo-informatic Tupu; equivalent factor; the Yellow River Basin (YRB); China

---

## 1. Introduction

A watershed is an area where the natural environment and human activities interact strongly [1,2], and is also the main area of human life and reproduction. Statistically, the population of the world's major basins is as high as 2.24 billion, accounting for about one-third of the world's population. Human activities have resulted in dramatic land use transitions (LUTs), which have seriously affected the ecosystem services value (ESV). Then it would bring severe ecological problems that threaten sustainable human development [3–5]. It could be seen that the balance of the watershed ecosystem was easily affected by LUT, which posed a serious threat to the regional ecological environment. Therefore, it is necessary to model and analyze the impact of watershed ecosystem services. As the second-longest river in China and the fifth-longest river in the world [6,7], the Yellow River occupies important strategic positions both in social-economic development and as an ecological barrier in China [8]. However, the ecological environment of the Yellow River Basin (YRB, for short) is extremely fragile [9,10], which has been the focus and difficulty of watershed management in China since ancient times [6,11,12]. Notably, China's Central Government has incorporated ecological protection and high-quality development of the YRB into the major national strategy [8]. Thus far, the YRB has not been well protected, mainly since the land use was not reasonable and the ESV was also ignored.

Land use/cover change (LUCC) is a vital part of global environmental change and sustainable development, which has long been a question of great interest in a wide range of fields [13,14]. As one of the manifestations of LUCC, LUT has been introduced as a new way to research land use change in different stages of socio-economic development, which attracted extensive attention of the academic community [15–18]. Recently, an increasing number of empirical studies have explored LUT, and the research scales are mostly concentrated in the whole country [19,20], urban agglomeration [5,7], city [21,22], and county [23]. The research content included theories and hypotheses, the transition of overall land use pattern and structure, driving mechanisms, the environmental effects, and the relationship between LUT and related socio-economic activities [15–17,24–26]. Therefore, future research on LUT would focus more on the impact on social economy and ecological environment, to compensate for the existing research on the ecological function of land use. Additionally, LUT analyses in previous studies have usually been based on the transfer matrix to yield the quantitative changes of land use and obtained a spatial distribution through overlay analysis of the land use data in different periods [27]. The geo-informatic Tupu method could record composite spatio-temporal information of land use change using Tupu units. Moreover, the spatial pattern and time sequence characteristics could also be quantitatively expressed under the multiple spatio-temporal conditions [4,5], which has gradually become an effective method to LUT.

Ecosystem services refer to the benefits that human beings directly or indirectly obtain from the ecosystem, including supplying services, regulating services, cultural services, and support services [28–30]. The ecosystem services are related to human well-being, and are the basis for human survival and social-economic development [28]. LUT was considered to be one of the main driving forces to change ecosystem services at regional and global levels [31,32], which had a significant effect on the regional natural environment and ecosystem [33,34]. Statistically, the loss of ESV was estimated between 4.3 to 20.2 trillion USD/year from 1997 to 2011 at the global level due to LUT [29]. In China as well, ESVs have been decreased due to high resource consumption and the city's rapid expansion under economic development [22]. For example, ESV decreased by 0.45% (1988–2000) and 0.10% (2000–2008) [35]. Ecosystem services have been widely evaluated in the world [30,36]. Both the ESV and its comprehensive framework and principle for integrated assessment and evaluation have been studied in recent decades [13,28,37,38]. At present, three main approaches have been widely applied to assess ecosystem services, including equivalent factors, productivity, and biomass [28,39]. Among them, the equivalent factor method is more intuitive and easier to use, with fewer data requirements, comprehensive evaluation, and high comparability, and especially suitable for regional and global scale assessment [39].

Watershed, a natural catchment area, is a relatively independent and complete ecological unit with water as the core element [1,8]. It has become a basic consensus of the international community to study the ecological environment change from the overall level of the basin [40,41]. To explore the characteristics of LUT and its eco-environmental effects can provide scientific reference for ecological protection [42]. In the YRB, land use changes are a pervasive and common phenomenon [43], especially the continuous expansion of urban and rural construction land, and the space of agricultural and ecological land is squeezed [44], making the internal relationship of the natural ecosystem uncoordinated [45,46]. From the overall scale of the YRB, studying the response of ESV to LUT can provide an entry point for land management opportunities in the future. Furthermore, studies conducted on LUT in the YRB focus on the dynamics of cover changes and their causes [6,11,43] with little attention to address the impacts of such changes on the ecosystem services aspect.

The objective of this paper is threefold. The first is to assess the dynamic patterns of LUT in the YRB using the geo-information Tupu method. We use the change ratio and space separating degree to determine the spatio-temporal characteristics of Tupu change. The second is to evaluate the changes in ESV caused by LUT using the equivalent factor method. Thus, the ESV Tupu method was applied to further examine the response of ESV to LUT. The third is to discuss how to incorporate ecosystem services into land use policy implications. This paper contributes to the literature in several ways: (1) As for the research scale, this paper selects the special type of region as the research area. It could provide scientific guidance for the cross-regional collaborative governance by analyzing the ecological environment of the YRB from a systematic and holistic perspective. (2) As for the geographical space, this paper records the spatio-temporal composite information of land use change applying the geo-information Tupu method, and then systematizes and dynamizes the process of LUT and reveals its internal laws.

## 2. Materials and Methods

### 2.1. Study Area

The study site was the Yellow River Basin (YRB), located on China's north at 95°53′ E–119°05′ E and 32°10′ N–41°50′ N, as shown in Figure 1. It is the second largest river in China, with a total length of 5464 km and a total area of $7.95 \times 10^5$ km$^2$, accounting for 8.28% of the total area of China [47]. From west to east, the YRB flows through nine provinces and regions, namely Qinghai, Sichuan, Gansu, Ningxia, Inner Mongolia, Shaanxi, Shanxi, Henan, and Shandong. The topography in the YRB is dominated by hills, mountains, and plains, and the terrain is high in the west and low in the east. The YRB is dominated by a temperate continental monsoon climate with four distinct seasons and rich natural resources. In 2018, the total population of the YRB (nine provinces and autonomous regions) was $4.2 \times 10^8$, accounting for 30.3% of China, and the GDP was $2.39 \times 10^{13}$ yuan, accounting for 26.5% of the total national GDP. Referring to previous research [11], it was divided into upstream (Qinghai, Inner Mongolia, Ningxia, Gansu, Sichuan), midstream (Shaanxi and Shanxi), and downstream (Shandong and Henan).

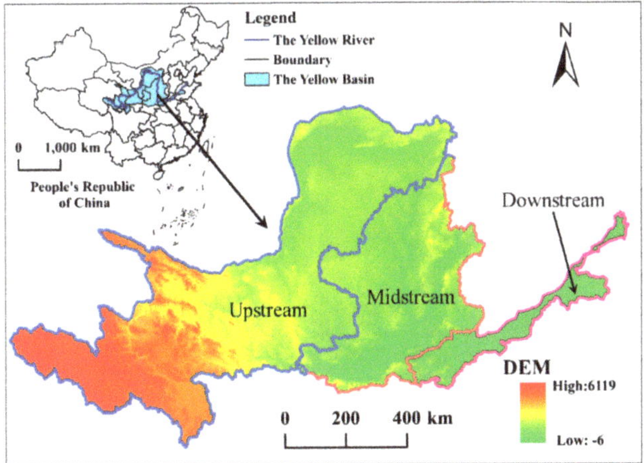


**Figure 1.** Location of the Yellow River Basin.

*2.2. Data Sources and Processing*

Taking the administrative division in 2019 as the standard, the data of each year were unified to the administrative unit. There are there main sources of data: (1) The boundaries of the YRB, and provincial and municipal administrative units were based on the 1:250,000 basic geographic data provided by the Data Center for Resources and Environmental Sciences, Chinese Academy of Sciences (RESDC) (http://www.resdc.cn), including 9 provincial units and 76 municipal administrative units. (2) The land use data of the YRB in 1990, 2000, 2010, and 2018 were collected from the Data Center for Resources and Environmental Sciences, Chinese Academy of Sciences (RESDC) (http://www.resdc.cn) [19,48,49]. The data processing method was as follows. Firstly, the production of the data set was mainly based on Landsat Multi-Spectral Scanner (MSS), Thematic Mapper (TM)/ Enhance Thematic Mapper (ETM), and Landsat 8 remote sensing images as the data sources. With the help of ArcGIS, the geographical elements were mathematically and physically interpreted by manual visual interpretation, and the land use data with a spatial resolution of 30 × 30 m were generated. The original data included 6 major categories and 25 secondary subcategories [19]. After strict quality control, the overall identified accuracies of the primary type of land use reached 94.3%, and the classification accuracies of the secondary type reached more than 91.2% [19,49], which met the requirement of user mapping accuracy on the 1: 100,000 scale. Secondly, the vector data of the boundary of the YRB were used to mask the 30 m raster land use data of the whole country, and the spatial database of land use in the study area was constructed. The reclassify tool was used to reclassify the land use types, and the value of the extracted land type was Set 1, and that of other land types were Set No-data. (3) The grain yield data were derived from the statistical yearbooks of each province in the YRB from 1991 to 2019, and the grain price data were derived from the "China Agricultural Produce Survey Yearbook" in 2019.

Referring to the national standard of land use classification (GB/T21010-2007) in China, the land use types were classified into six types: cultivated land, forestland, grassland, water area, construction land, and unused land. Then, the land use data were recoded in ArcGIS 10.6. The cultivated land, forestland, grassland, water area, construction land, and unused land were set as 1, 2, 3, 4, 5, and 6, respectively, with a unified classification standard, as shown in Figure 2.

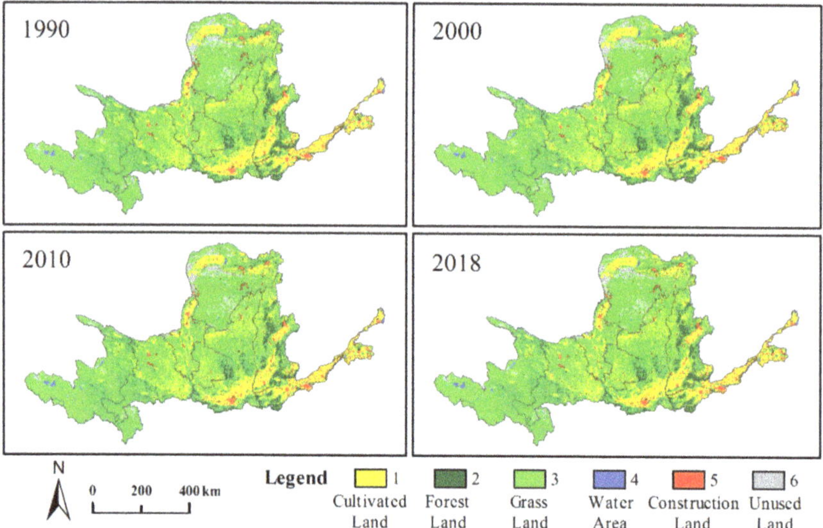


**Figure 2.** Land use distribution of the Yellow River Basin (YRB) from 1990 to 2018.

### 2.3. Geo-Information Tupu Methods

Tupu is a means to express content or transmit information by analyzing comprehensive maps, images, or tables [50]. It could effectively reflect the characteristics of spatial structure and spatio-temporal change. Geo-information Tupu is a geographic space–time analysis methodology [51], which could express various geographical elements through abstract mathematical forms [52]. Moreover, it could combine "TU expressing the characteristics of spatial unit" and "PU presenting the starting point and process of events", which makes up for the deficiency of traditional data mining methods in thinking and space [53]. According to the theory and method of geo-information Tupu, based on land use data of four periods in the YRB, the paper employs a "Raster Calculator" to conduct map algebraic operations and builds series Tupu models for LUT.

#### 2.3.1. Build the Process Tupu of LUT and ESV

The map algebraic superposition for LUTs' Tupu unit was performed in ArcGIS 10.6 to integrate the spatial information of Tupu code [54]. Specifically, the codes of adjacent two-phase grid elements were selected for an algebraic operation to obtain the LUT Tupu value, seen in Formula (1):

$$C = 10 \times A + B \tag{1}$$

where $C$ represents the Tupu code of the LUT during the research stage, $A$ represents the land use unit code value in the previous stage, and $B$ represents the land use unit code value in the later stage. Thus, the LUT Tupu of the YRB in 1990–2000, 2000–2010, and 2010–2018 can be obtained. Referring to existing studies [4,55], the Tupu of changes in ESV was obtained by multiplying the Tupu of LUT from 1990 to 2018 by the ESV per unit area of spatial grid. According to the change of ESV, the region was defined as an ecological preservation zone, ecological appreciation zone, and ecological impairment zone.

#### 2.3.2. Statistics of the Tupu Characteristic

Characteristics of the "TU" are mainly the spatial representation of the sequential changing process of the land use Tupu unit, and the quantitative features of the Tupu unit type during the two sampling periods are the characteristics of the "PU" of the LUT [4,5]. In this study, the characteristics of "TU" in LUT are quantitatively presented by the degree of spatial separation, and the visual observation of the Tupu process in different stages. The characteristics of "PU" are represented by the sorted Tupu unit table. The specific formula of change ratio and spatial separation index [23,54,56] are as follows:

$$A_{ij} = N_{ij} \times 100\% / \sum_{i=1}^{n} \sum_{j=1}^{n} N_{ij} (i \neq j) \tag{2}$$

$$S_{ij} = \frac{1}{2} \times \frac{\sqrt{F_{ij} / \sum_{i=1}^{n} \sum_{j=1}^{n} N_{ij}}}{N_{ij} / \sum_{i=1}^{n} \sum_{j=1}^{n} N_{ij}} \tag{3}$$

where $A_{ij}$ represents change ratio. It presents the ratio of the transformed land use Tupu unit area to the total area of all transformed land use Tupu units. $S_{ij}$ represents the spatial separating degree, and reflects the degree of dispersion of the Tupu unit. $F_{ij}$, $N_{ij}$ refers to the number of Tupu units and area of land use types converted from the land use type $i$ at the initial stage to the land use type $j$ at the last stage, and $n$ represents the number of land use types.

### 2.4. Calculation of ESV

#### 2.4.1. Revision of Value Coefficient

Since Costanza et al. [28] scientifically expounded the principles and methods of estimating ESV published in *Nature*, relevant studies had gradually become a hot topic in academia and widely used around the world [40,57]. However, there are some defects in the direct application in China, such as the estimation of the farmland is too low but the wetland is too high [35,58]. Therefore, from combined expert knowledge of more than 700 ecologists, Xie et al. [38] revised the ecosystem services classification and ecosystem equivalent table in China. The 17 ecosystem services proposed by Costanza et al. [28] were classified into four categories and nine sub-categories [38]. The equivalent value (EV) per unit area of food production of cultivated land was set to 1, and the EV of other ecosystem services can be quantified by comparing with the standard value of 1. Although some problems were still unsolved, it did not influence this method from being widely used by many scholars [35,58]. Therefore, this study adopted this method to further modify the ecological service coefficient according to the actual situation of the YRB.

As the land use types cannot correspond to ecosystem types one by one, this study chooses the closest land use type for equivalent valuation. The water area in this study included a water body and wetland, so this paper used the average equivalent ESV coefficients of water bodies and wetlands [38] to calculate the ESV. In this study, the ESV of construction land was assigned as zero [4]. The net profit of land use type production is regarded as the production value that the land use type can provide, and the net profit of food production per unit area of cultivated land is regarded as the ESV of one standard equivalent factor [39]. Generally, the economic value provided by natural ecosystems without human input is about 1/7 of the economic value of food provided by existing farmland per unit area [38]. According to the statistical yearbook of the provinces, the average grain yield of the YRB from 1990 to 2018 was 3987.98 kg/ (ha·a), and the grain purchase price was 0.47 USD/kg in 2018. Thus, the economic value of farmland grain yield provided by a standard equivalent factor in the YRB was calculated as about 267.76 (=3987.98 × 0.47 ÷ 7) USD/(ha·a). The value coefficients of ecological services provided by the modified ecosystems are shown in Table 1. The equations are as follows:

$$ESV = \sum_{i=1}^{n} ESV_i = \sum_{i=1}^{n} P_i \times A_i \tag{4}$$

where $ESV$ indicates the total ESV in the research area, $i$ is the number of land use types ($i = 1$–6). $ESV_i$ indicates the ESV of the $i$th land use type, $P_i$ and $A_i$ indicate the area and the ESV of the $i$th land use type, respectively.

**Table 1.** The ecosystem service value (ESV) per land use type after correction in the YRB.

| Primary-Types | Secondary-Types | Cultivated Land | Forestland | Grassland | Water Area | Unused Land |
| --- | --- | --- | --- | --- | --- | --- |
| SuyS | Food production | 267.76 | 88.36 | 115.14 | 119.15 | 5.36 |
| | Raw material | 104.43 | 797.92 | 96.39 | 78.99 | 10.71 |
| RegS | Gas regulation | 192.79 | 1156.72 | 401.64 | 390.93 | 16.07 |
| | Climate regulation | 259.73 | 1089.78 | 417.71 | 2089.87 | 34.81 |
| | Hydrological regulation | 206.18 | 1095.14 | 407.00 | 4312.27 | 18.74 |
| | Waste treatment | 372.19 | 460.55 | 353.44 | 3915.99 | 69.62 |
| SutS | Soil formation and retention | 393.61 | 1076.40 | 599.78 | 321.31 | 45.52 |
| | Biodiversity protection | 273.12 | 1207.60 | 500.71 | 953.23 | 107.10 |
| CulS | Recreation and culture | 45.52 | 556.94 | 232.95 | 1222.32 | 64.26 |
| Total | | 2115.30 | 7529.41 | 3124.76 | 13,404.07 | 372.19 |

#### 2.4.2. Spatial Analysis of ESV

To further analyze the spatial distribution of ESV in the YRB, the prefecture level cities were taken as the basic units to count the ESV of the YRB. Then combined with Formula (5), the average ESV (AESV) of each unit could be calculated. Using spatial analysis in ArcGIS 10.6, the AESV was divided into five grades from high to low, with the first level being the lowest and the fifth level being the highest. The spatial distribution of AESV was obtained. According to Formula (6), the change rate of average ESV in the study area was calculated and divided into significant increase, general increase, weak increase, stable, weak decrease, general decrease, and significant decrease, and the spatial distribution of the change rate of AESV was obtained.

$$AESV = ESV/A = (\sum_{i=1}^{n} P_i \times A_i / \sum_{i=1}^{n} P_i) \tag{5}$$

$$C = \frac{ASEV_{t_2} - ASEV_{t_1}}{ASEV_{t_1}} \times 100\% \tag{6}$$

where $AESV$ is the average ESV, $C$ indicates the change rate of AESV, $AESV_{t1}$ and $AESV_{t2}$ indicate the $AESV$ at $t_1$ and $t_2$. The other variables are the same as Formulas (3) and (4).

## 3. Results

### 3.1. Land Use Change in the YRB from 1990 to 2018

The primary land use type was grassland, as shown in Figure 2 and Table 2, accounting for 47.38%, 47.18%, 47.54%, and 47.51% of the total YRB area, respectively, which was mainly distributed in Qinghai, Inner Mongolia, Gansu, and Shaanxi. Cultivated land was the second, accounting for 26.83%, 27.06%, 26.27%, and 25.73%, which was mainly distributed in northern Henan and Shandong. Forestland area accounted for 13% of the total YRB area, and was mainly distributed in central and southern Shaanxi, northern Henan, and most areas of Shanxi. Construction land was mainly distributed along the lower reaches of the Yellow River and the estuary. Unused land was mainly distributed in Inner Mongolia, Qinghai, Sichuan, and northern Ningxia. The lowest proportion of the total area was the water area, which mainly distributed in the source of the YRB.

**Table 2.** Land use change in the YRB from 1990 to 2018.

| | | Cultivated Land | Forestland | Grassland | Water Area | Construction Land | Unused Land |
|---|---|---|---|---|---|---|---|
| | 1990 | 217,048 | 103,537 | 383,220 | 14,181 | 17,505 | 73,406 |
| Area (km²) | 2000 | 218,884 | 103,436 | 381,622 | 13,654 | 18,994 | 72,307 |
| | 2010 | 212,492 | 106,372 | 384,572 | 14,108 | 25,659 | 65,700 |
| | 2018 | 208,104 | 106,466 | 384,238 | 14,758 | 31,395 | 63,752 |
| | 1990 | 26.83 | 12.80 | 47.38 | 1.75 | 2.16 | 9.07 |
| Proportion (%) | 2000 | 27.06 | 12.79 | 47.18 | 1.69 | 2.33 | 8.94 |
| | 2010 | 26.27 | 13.15 | 47.54 | 1.74 | 3.17 | 8.12 |
| | 2018 | 25.73 | 13.16 | 47.51 | 1.82 | 3.88 | 7.88 |
| | 1990–2000 | 0.85 | −0.10 | −0.42 | −3.72 | 8.51 | −1.50 |
| Change percentage (%) | 2000–2010 | −2.92 | 2.84 | 0.77 | 3.33 | 35.09 | −9.14 |
| | 2010–2018 | −2.07 | 0.09 | −0.09 | 4.61 | 22.35 | −2.96 |
| | 1990–2018 | −4.12 | 2.83 | 0.27 | 4.07 | 79.35 | −13.15 |

As for the evolution of various land use types, the grassland area showed a trend of fluctuation, decreasing by −0.42% during 1990–2000, increasing by 0.77% during 2000–2010, and decreasing by −0.09% during 2010–2018. A faint decline (−0.10%) was found in the forestland during 1990–2000, while a continuous increase of 2.84% and 0.09% was observed in the following two stages. An apparent decrease in cultivated land was documented, while a remarkable increase in construction land was observed. Specifically, cultivated land decreased by −4.12%, and the construction land increased by 79.35% during 1990–2018. The water area decreased by −3.72% during 1990–2000, while a constant increase (3.33% during 2000–2010 and 4.61% during 2010–2018) was found during the last two periods. The area of unused land continued to decrease by 13.15%.

### 3.2. Tupu Analysis of LUTs in the YRB from 1990 to 2018

### 3.2.1. Spatial Distribution of Tupu Units from 1990 to 2018

(1)  *Spatial distribution of Tupu units from 1990 to 2000*

Figure 3A shows the most significant change of Tupu units is "grassland→cultivated land" (31), mainly distributed in the upstream of the YRB. The Tupu units of "unused land→grassland" (63) and "grassland→unused land" (36) were also remarkable, which were widely distributed in the upstream and midstream. The total number of Tupu units of "construction land→cultivated land" (51) was very small, far less than the Tupu units of "cultivated land→construction land" (15), and the spatial distribution of which was relatively scattered.

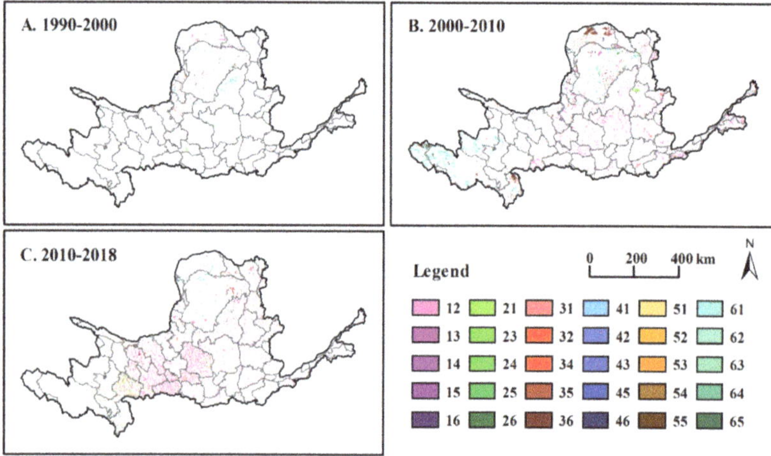

**Figure 3.** Process Tupu of LUTs during 1990–2018 in the YRB. (Note: **A**, **B** and **C** denotes the Tupu of LUTs during 1990–2000, 2000–2010, and 2010–2018, respectively. The No. 1–6 represent cultivated land, forestland, grassland, water area, construction land, and unused land, respectively. The Tupu units' codes could be seen in Table A1. For example, Code 12 represents cultivated land converted to forestland, and the other codes follow the same rule.)

The spatial separation degree was applied to represent the "Tu" features of the LUT in a quantitative manner. It could reflect the discrete degree of spatial distribution for LUT Tupu units. The higher the spatial separation degree, the greater dispersed the spatial distribution of Tupu units. Based on the spatial separation degree of the LUT Tupu units, we can obtain a visual display of the "Tu" of the LUT, as shown in Figure 4A. The spatial separation degree of Tupu units of Type 31, 63, 36, 15, 13, 23, 41, 32, and 61 were all very low, indicating that these Tupu units were more concentrated in space. Specifically, the spatial separation degree of Tupu units of "grassland→cultivated land" (31) and "unused land→grassland" (63) were 0.33 and 0.43, respectively. Moreover, the spatial distribution of Tupu units of Type 36 (0.56) and 13 (0.58) presented a spatial agglomeration phenomenon. However, the spatial separation degree of Tupu units of "construction land→unused land" was the largest (56.07), followed by the Tupu units of Type 52 and 54 (49.41 and 43.28), indicating that these Tupu units were very dispersed.

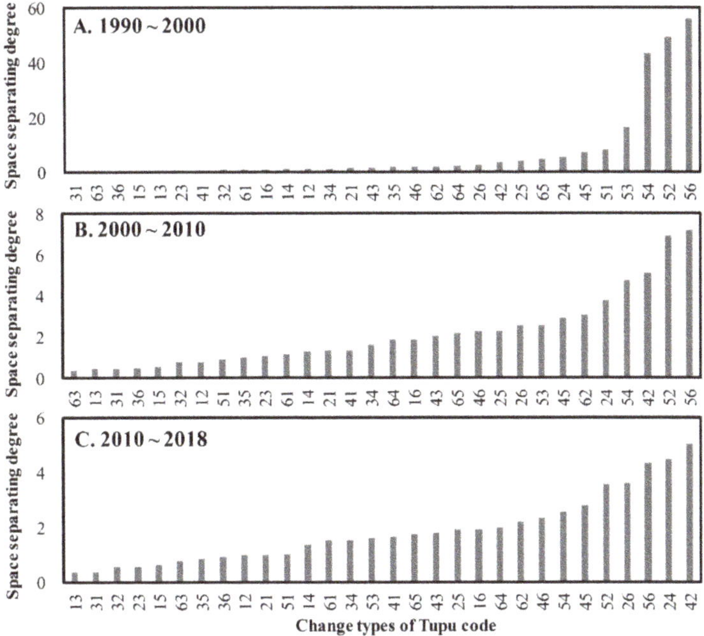

**Figure 4.** Spatial separation degree of Tupu units from 1990 to 2018 in the YRB. (Note: **A**, **B** and **C** denotes the spatial separation degree of Tupu units during 1990–2000, 2000–2010, and 2010–2018, respectively. No. 1–6 has the same meaning as Figure 3. The Tupu units' codes can be seen in Table A1).

(2)  *Spatial Distribution of Tupu Units from 2000 to 2010*

The spatial pattern distribution of Tupu units at this stage showed clustered distribution of some types, as shown in Figure 3B, which were more complicated than the previous stage. The mutual conversions between grassland and cultivated land, and grassland and unused land were still the most significant Tupu units at this stage, which showed a continuous expansion in the midstream and clustered distribution in the upstream. Meanwhile, the mutual transformation between cultivated land and construction land was relatively obvious, "cultivated land→construction land" (15) was mainly distributed in the downstream of the YRB, and also widely distributed in Shanxi, southern Shaanxi, and northwest Yinchuan of Ningxia. While the cultivated land occupation for construction land was relatively dispersed, "construction land→cultivated land" (51) was mainly distributed in the downstream, southern Shanxi, and Bayannaoer of Inner Mongolia. The process of occupying water area for construction land was intensified, and Tupu units of 45 were mainly distributed in Shanxi (Jinzhong, Luliang, and Changzhi), Henan (Luoyang and Zhengzhou), and coastal areas of Shandong.

As shown in Figure 4B, overall separation degree value of this stage was 64.88, which was much smaller than that of the previous stage (222.31), indicating that the spatial pattern of LUT in the YRB was less dispersed. The degrees of spatial separation in Tupu units (codes 63, 13, 31, 36, 15, 32, 12, and 51) were all less than 1, indicating that these Tupu units were intensively distributed in space. The degree of spatial separation of Tupu unit for Type 56 was the highest (7.22), followed by the Tupu units with codes of 52, 42, 54, 24, and 62, indicating that these Tupu units were relatively dispersed. We found that the conversions between construction land and other land use types were scattered throughout the YRB.

(3)   *Spatial distribution of Tupu units from 2010 to 2018*

The most important type of transition was the mutual transformation between cultivated land and grassland, which continued to expand in the midstream of the YRB and gradually developed the cluster distribution in the upstream, as shown in Figure 3C. The interconversion between grassland and forestland changed from scattered distribution to centralized distribution, but the distribution area changed little. The distribution of Type 15 was still relatively concentrated, except in Henan and Shandong, and the same situation also appeared in Gansu (northern Lanzhou). In addition, grassland erosion of forestland and unused land were concentrated, and "forestland→grassland" (23) and "unused land→grassland" (63) were concentrated in Inner Mongolia and Gansu.

The overall separation degree value of Tupu units of LUT was 55.97 during 2010–2018, which was further reduced compared with the previous two stages, as shown in Figure 4C. The spatial separation of Tupu units of Type 13, 31, 32, 23, 15, 63, 35, 36, 12, and 21 were all less than 1. Specifically, spatial separation degrees of Tupu units of Type 13, 31, 32, and 63 were 0.35, 0.37, 0.57, and 0.78, respectively, which were slightly lower than those in the previous stage. However, the spatial separation degrees of Tupu units of Type 15, 36, and 12 were higher than the previous stage. The spatial separation degree of Tupu units for Type 51 increased, indicating that although land reclamation activities are promoted continuously, it is still difficult to make up for the decrease in the total volume. Relatively speaking, Type 15 distribution has a higher spatial concentration, which further verifies that human activities occupy cultivated land violently. The spatial separation degree of Type 42 was the largest and slightly decreased compared with the previous stage. Similarly, the spatial distribution of Tupu units with codes of 56, 26, and 56 were discrete, with the spatial separation degree exceeding 3.

### 3.2.2. Quantity Change of Tupu Units from 1990 to 2018

The total area of land use change in 1990–2000, 2000–2010, and 2010–2018 was 14,872.17 km$^2$, 65,364.84 km$^2$, and 64,945.99 km$^2$, respectively, as shown in Table 3. During 1990–2000, the change rate of transformation between grassland and cultivated land, and unused land and grassland was 40.46%, which were the main transformation types. In the process of mutual transition between cultivated land and grassland, the area occupied by grassland (3757.86 km$^2$) was much larger than the area supplemented by cultivated land (1222.70 km$^2$). Overall, the accumulative change rate of grassland transfer-out in this stage was 41.19%, which was higher than 30.44% of grassland transfer-in, with a net decrease of grassland quantity. In addition, the scale occupied of cultivated land was also very large, but the area of other land use types transforming into the cultivated land (3452.68 km$^2$) was less than the area of cultivated land they occupy (5288.40 km$^2$). The conversion between construction land and water area, and forestland and unused land was less; the change rate was less than 0.01%.

During 2000–2010, the Tupu unit change rate of Type 63 was 19.85%, which was the most important LUT type and increased 4.66% compared with the previous stage, as shown in Figure 5. Tupu units of 13 rose to second place from fifth place of the previous stage and cultivated land became a grassland transfer-in type, ranking only second to unused land, which was mainly due to the implementation of the policy of returning farmland to grassland. The Tupu unit of 31 and 36 were other important transformation types, accounting for 12.59% and 11.75% of the converted land-use types, respectively. Among grassland transfer-in types, Type 23 (3143.63 km$^2$, 4.81%) and 43 (433.07 km$^2$, 0.66%) slightly decreased compared with the previous stage, and the number of Tupu units of 13 (9184.49 km$^2$, 14.05%), 53 (275.65 km$^2$, 0.42%), and 63 (12,976.90 km$^2$, 19.85%) all increased to a different extent. The conversion between cultivated land and construction land became increasingly prominent; a total of 6606.34 km$^2$ of cultivated land was converted into construction land, while only 2229.56 km$^2$ of construction land was converted into cultivated land. The accumulative change rate of cultivated land transfer-out was 30.99%, which was larger than cultivated land transfer-in (21.21%), with a net cultivated land decrease of 6392.91 km$^2$.

**Table 3.** Tupu unit sequence of land use transitions (LUTs) from 1990 to 2018 in the YRB.

| Sequence | 1990–2000 | | | 2000–2010 | | | 2010–2018 | | |
|---|---|---|---|---|---|---|---|---|---|
| | Type | Area (km²) | Change Ratio (%) | Type | Area (km²) | Change Ratio (%) | Type | Area (km²) | Change Ratio (%) |
| 1 | 31 | 3757.86 | 25.27 | 63 | 12,976.90 | 19.85 | 13 | 14,580.50 | 22.45 |
| 2 | 63 | 2258.58 | 15.19 | 13 | 9184.49 | 14.05 | 31 | 13,310.50 | 20.49 |
| 3 | 36 | 1333.30 | 8.97 | 31 | 8227.12 | 12.59 | 32 | 5538.27 | 8.53 |
| 4 | 15 | 1317.42 | 8.86 | 36 | 7680.71 | 11.75 | 23 | 5345.11 | 8.23 |
| 5 | 13 | 1222.70 | 8.22 | 15 | 6606.34 | 10.11 | 15 | 4733.88 | 7.29 |
| 6 | 23 | 885.09 | 5.95 | 32 | 3143.63 | 4.81 | 63 | 2998.93 | 4.62 |
| 7 | 41 | 861.78 | 5.79 | 12 | 2867.07 | 4.39 | 35 | 2622.12 | 4.04 |
| 8 | 32 | 642.81 | 4.32 | 51 | 2229.56 | 3.41 | 36 | 2207.89 | 3.40 |
| 9 | 61 | 490.99 | 3.30 | 35 | 1754.52 | 2.68 | 12 | 1886.48 | 2.90 |
| 10 | 16 | 365.24 | 2.46 | 23 | 1594.11 | 2.44 | 21 | 1844.11 | 2.84 |
| 11 | 14 | 274.22 | 1.84 | 61 | 1380.05 | 2.11 | 51 | 1714.97 | 2.64 |
| 12 | 12 | 273.10 | 1.84 | 14 | 1073.97 | 1.64 | 14 | 975.63 | 1.50 |
| 13 | 34 | 266.72 | 1.79 | 21 | 1021.60 | 1.56 | 61 | 761.38 | 1.17 |
| 14 | 21 | 171.74 | 1.15 | 41 | 1005.42 | 1.54 | 34 | 754.34 | 1.16 |
| 15 | 43 | 159.54 | 1.07 | 34 | 710.22 | 1.09 | 53 | 691.23 | 1.06 |
| 16 | 35 | 125.18 | 0.84 | 64 | 539.28 | 0.83 | 41 | 662.85 | 1.02 |
| 17 | 46 | 108.95 | 0.73 | 16 | 524.79 | 0.80 | 65 | 590.64 | 0.91 |
| 18 | 62 | 105.12 | 0.71 | 43 | 433.07 | 0.66 | 43 | 565.57 | 0.87 |
| 19 | 64 | 88.25 | 0.59 | 65 | 391.03 | 0.60 | 25 | 482.38 | 0.74 |
| 20 | 26 | 55.91 | 0.38 | 46 | 354.19 | 0.54 | 16 | 480.06 | 0.74 |
| 21 | 42 | 31.37 | 0.21 | 25 | 353.85 | 0.54 | 64 | 450.09 | 0.69 |
| 22 | 25 | 27.60 | 0.19 | 26 | 278.37 | 0.43 | 62 | 368.80 | 0.57 |
| 23 | 65 | 19.35 | 0.13 | 53 | 275.65 | 0.42 | 46 | 327.44 | 0.50 |
| 24 | 24 | 13.36 | 0.09 | 45 | 216.24 | 0.33 | 54 | 279.56 | 0.43 |
| 25 | 45 | 7.86 | 0.05 | 62 | 193.09 | 0.30 | 45 | 233.45 | 0.36 |
| 26 | 51 | 6.03 | 0.04 | 24 | 127.59 | 0.20 | 52 | 142.83 | 0.22 |
| 27 | 53 | 1.57 | 0.01 | 54 | 80.30 | 0.12 | 26 | 138.99 | 0.21 |
| 28 | 52 | 0.17 | 0.00 | 42 | 69.04 | 0.11 | 56 | 96.85 | 0.15 |
| 29 | 54 | 0.22 | 0.00 | 52 | 37.84 | 0.06 | 24 | 89.87 | 0.14 |
| 30 | 56 | 0.13 | 0.00 | 56 | 34.82 | 0.05 | 42 | 71.28 | 0.11 |
| Total | | 14,872.17 | 100.00 | | 65,364.84 | 100.00 | | 64,945.99 | 100.00 |

Note: No. 1–6 has the same meaning as Figure 3. The Tupu units' codes can be seen in Table A1.

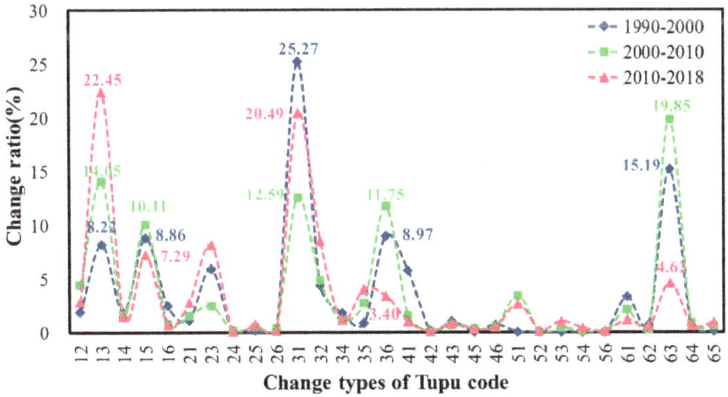


**Figure 5.** Change ratio of Tupu units from 1990 to 2018 in the YRB. (Note: No. 1–6 has the same meaning as Figure 3. The Tupu units' codes can be seen in Table A1.)

From 2010 to 2018, mutual transformation between cultivated land and grassland was the main LUT type, with an accumulative change rate of 22.45%. Among these types, the change rate of Type 13 was 22.45%, which increased significantly compared with the previous two stages; Type 31 decreased relative to the stage of 1990–2000, indicating that the quantity of supplementary grassland during 2010–2018 was large. However, the grassland transfer-in area was less than the transfer out area, and the accumulative change rate of grassland transfer-out (37.62%) was slightly higher than the accumulative change rate of grassland transfer-in (37.23%), with a net decrease of 251.78 km$^2$. The problem of cultivated land occupied by construction land was still outstanding, with an area of 4733.88 km$^2$. Compared with the previous two stages, the change rate of cultivated land transfer-out increased (34.89%), and transfer-in decreased (28.17%). It was mainly due to the fact that a large amount of cultivated land was occupied by grassland and construction land. Totally, the accumulative change rate of cultivated land transition (CLT) gradually increased, and reached 63.06% from 2010 to 2018, indicating that CLT has gradually become the main type of LUT in the YRB.

*3.3. The Impact of LUTs on ESV*

3.3.1. Changes in ESV from 1990 to 2018

The total ESV showed a fluctuating trend of decreasing first, increasing later, and then decreasing again—decreasing from 2653.56 × 10$^8$ USD in 1990 to 2644.23 × 10$^8$ USD in 2000, then increasing to 2665.65 × 10$^8$ USD in 2010, and decreasing to 2664.03 × 10$^8$ USD in 2018, which was higher than that in the year of 1990, as shown in Table 4. During the study period, grassland contributed to most of the total ESV (more than 45%), followed by forestland (30%), and unused land contributed the least to the total ESV. The ESV provided by forestland and water decreased during 1990–2000 and increased during the following two periods, while the ESV provided by cultivated land increased during 1990–2000 and decreased during the following two periods. The ESV provided by grassland decreased (−4.99 × 10$^8$ USD) during 1990–2000, increased (9.21 × 10$^8$ USD) during 2000–2010, and decreased (−1.04 × 10$^8$ USD) during 2010–2018. Notably, the ESV provided by unused land was in a decreasing state (−3.59 × 10$^8$ USD). Overall, the area of cultivated land decreased, resulting in the ESV's loss of 18.92 × 10$^8$ USD. However, the growth of forestland area in the same period brought an ESV increase of 22.06 × 10$^8$ USD, and eventually led to the overall rise of ESV, which verified the effectiveness of the implementation of the policy of returning farmland to forest and a series of protective forest protection plans in the YRB.

**Table 4.** Changes in total ESV from 1990 to 2018 (×10$^8$ USD).

| Region | Land Use Types | Codes | ESV in (10$^8$ USD) | | | | ESV Change | | | |
|---|---|---|---|---|---|---|---|---|---|---|
| | | | 1990 | 2000 | 2010 | 2018 | 1990–2000 | 2000–2010 | 2010–2018 | 1990–2018 |
| Yellow River Basin | Cultivated land | 1 | 459.12 | 463.01 | 449.49 | 440.2 | 3.89 | −13.52 | −9.29 | −18.92 |
| | Forestland | 2 | 779.57 | 778.81 | 800.92 | 801.63 | −0.76 | 22.11 | 0.71 | 22.06 |
| | Grassland | 3 | 1197.47 | 1192.48 | 1201.69 | 1200.65 | −4.99 | 9.21 | −1.04 | 3.18 |
| | Water area | 4 | 190.08 | 183.02 | 189.1 | 197.82 | −7.06 | 6.08 | 8.72 | 7.74 |
| | Unused land | 6 | 27.32 | 26.91 | 24.45 | 23.73 | −0.41 | −2.46 | −0.72 | −3.59 |
| | **Total** | | **2653.56** | **2644.23** | **2665.65** | **2664.03** | **−9.33** | **21.42** | **−1.62** | **10.47** |
| Upstream | Cultivated land | 1 | 186.65 | 190.13 | 189.7 | 184.94 | 3.48 | −0.43 | −4.76 | −1.71 |
| | Forestland | 2 | 325.3 | 323.86 | 335.91 | 338.26 | −1.44 | 12.05 | 2.35 | 12.96 |
| | Grassland | 3 | 950.38 | 943.66 | 952.39 | 953.1 | −6.72 | 8.73 | 0.71 | 2.72 |
| | Water area | 4 | 121.81 | 122.14 | 124.66 | 130.45 | 0.33 | 2.52 | 5.79 | 8.64 |
| | Unused land | 6 | 24.87 | 24.97 | 25.47 | 21.96 | 0.1 | 0.5 | −3.51 | −2.91 |
| | **Total** | | **1609.01** | **1604.76** | **1628.13** | **1628.71** | **−4.25** | **23.37** | **0.58** | **19.7** |
| Midstream | Cultivated land | 1 | 198.18 | 198.15 | 186.4 | 183.81 | −0.03 | −11.75 | −2.59 | −14.37 |
| | Forestland | 2 | 380.94 | 381.97 | 393.42 | 391.85 | 1.03 | 11.45 | −1.57 | 10.91 |
| | Grassland | 3 | 227.82 | 229.92 | 234.71 | 232.95 | 2.1 | 4.79 | −1.76 | 5.13 |
| | Water area | 4 | 33.19 | 32.1 | 29.76 | 30.12 | −1.09 | −2.34 | 0.36 | −3.07 |
| | Unused land | 6 | 2.26 | 1.78 | 1.79 | 1.68 | −0.48 | 0.01 | −0.11 | −0.58 |
| | **Total** | | **842.39** | **843.92** | **846.08** | **840.41** | **1.53** | **2.16** | **−5.67** | **−1.98** |
| Downstream | Cultivated land | 1 | 74.3 | 74.73 | 73.39 | 71.45 | 0.43 | −1.34 | −1.94 | −2.85 |
| | Forestland | 2 | 73.34 | 72.98 | 71.59 | 71.52 | −0.36 | −1.39 | −0.07 | −1.82 |
| | Grassland | 3 | 19.23 | 18.84 | 14.59 | 14.59 | −0.39 | −4.25 | 0 | −4.64 |
| | Water area | 4 | 34.94 | 28.64 | 34.15 | 36.77 | −6.3 | 5.51 | 2.62 | 1.83 |
| | Unused land | 6 | 0.18 | 0.15 | 0.05 | 0.09 | −0.03 | −0.1 | 0.04 | −0.09 |
| | **Total** | | **201.99** | **195.34** | **193.77** | **194.42** | **−6.65** | **−1.57** | **0.65** | **−7.57** |

The changes in ESV among the upstream, midstream, and downstream showed significant differences, as shown in Table 4. During the study period, the ESV provided by upstream increased ($19.7 \times 10^8$ USD), and the ESV provided by midstream and downstream decreased ($-1.98 \times 10^8$ USD, $-7.57 \times 10^8$ USD). In terms of stages, the ESV of upstream decreased during 1990–2000 ($-4.25 \times 10^8$ USD), increased greatly during 2000–2010 ($23.37 \times 10^8$ USD), and then increased slightly during 2010–2018 ($0.58 \times 10^8$ USD). The ESV of midstream increased during the first two periods and decreased during the following period, while the ESV of downstream showed an opposite tendency. In terms of the proportion of the ESV, the upstream, midstream, and downstream of the YRB were quite different. For upstream, the ESV provided by grassland, forestland, and cultivated land accounted for more than 90%, and the ESV of grassland accounted for 58.52%, which was in a dominant position. The difference between the midstream and the upstream was that the ESV of the forestland accounted for the largest proportion (46.05%); for downstream, cultivated land, forestland, and water area were the three land use types that provide more ESV, of which cultivated land accounted for the largest proportion (36.75%).

3.3.2. Spatial Distribution of ESV from 1990 to 2018

From 1990 to 2018, the distribution of average ESV (AESV) in the YRB had significant spatial heterogeneity, as shown in Figure 6. Overall, the high-value areas of AESV were mainly distributed in Qinghai, Gansu, central Shaanxi, northern Henan, and Shanxi. These areas were mountainous and hilly, not suitable for farming, forestland and grassland were widely distributed, and the ecological environment was relatively good. The low-value areas were mainly distributed in Ningxia, Inner Mongolia, and eastern Henan and Shandong, mainly due to the distribution of unused land and construction land in these areas, the ecological environment was relatively poor. Specifically, (1) Grade I was mainly distributed in the east of Henan Province (Xinxiang, Anyang, Puyang), and the west of Shandong (Liaocheng, Jining), and the western Inner Mongolia (Alxa League and Wuhai City). In addition, Zhengzhou of Henan and Bayannaoer of Inner Mongolia presented Grade I in 2018 and 2010, respectively. (2) Grade II was distributed in central and southern Ningxia, western Inner Mongolia, western Qinghai, northern Sichuan, and some cities in Gansu. In addition, northern Shaanxi (Yulin) and central Shandong (Jinan, Tai'an) also had distribution. (3) Grade III was mainly distributed in the upstream, including Lanzhou, Tianshui, Dingxi, Qingyang, Hohhot, Baotou, Ningxia, and some cities in Qinghai. (4) Grade IV was mainly distributed in the upstream (Qinghai, Gansu) and the midstream (southern Shaanxi and some cities in Shanxi). (5) Grade V was mainly distributed in the midstream and downstream—the midstream included Luliang, Taiyuan, Changzhi, Jincheng, Yan'an, and Tongchuan in Shaanxi, and the downstream included Luoyang, Sanmenxia, Jiyuan, etc. In addition, Wuwei of Gansu in the upstream was also Grade V. Notably, the AESV in Dongying of Shandong was upgraded from Grade II (1990) to Grade V in 2018.

From 1990 to 2018, the AESV in the YRB changed dramatically, as shown in Figure 7. Specifically, the change degree in 2000–2010 was the most severe, and the whole basin has changed in different degrees and directions. In 2010–2018, the changes were mainly concentrated in the north and southeast of the YRB, and the changes from 1990 to 2000 were relatively scattered. Among them, the AESV in Qinghai increased significantly (2000–2010), Yulin and Yan'an in Shaanxi also increased in 1990–2000 and 2000–2010, respectively. From 1990 to 2018, the AESV of Dongying showed a sustained and obvious upward trend. In addition, the AESV of Ningxia (Yinchuan, Shizuishan) decreased significantly, mainly in 2010–2018. The AESV of Shanxi (Luliang, Taiyang, Jinzhong, Jincheng, etc.) decreased significantly in 2000–2010, while Shaanxi (Xi'an, Weinan) decreased to varying degrees from 1990 to 2018, which led to a significant decrease in the whole study period. The AESV of Henan (Sanmenxia, Yuncheng, Xinxiang, Anyang, Puyang) and central Shandong (Jinan, Jining, Liaocheng, Tai'an) declined significantly. It is worth noting that Bayannaoer in Inner Mongolia decreased significantly from 2000 to 2010, but increased significantly from 2010 to 2018, which led to a general downward trend in the whole study period.

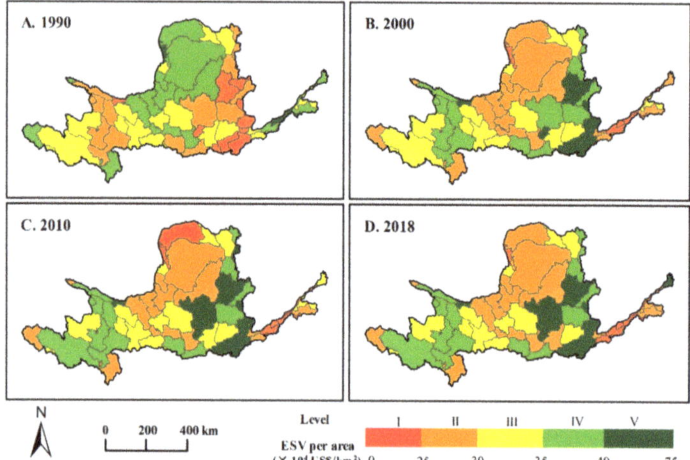



**Figure 6.** Spatial distribution of average ESV (AESV) in the YRB from 1990 to 2018.

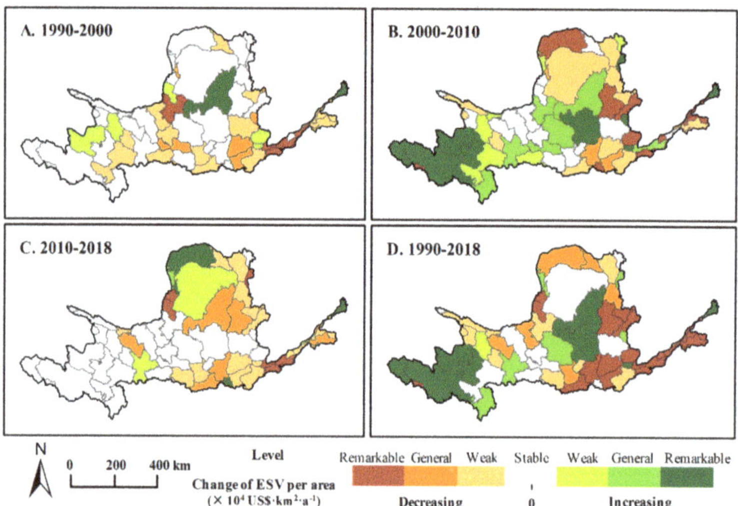



**Figure 7.** Change in spatial distribution of AESV in the YRB from 1990 to 2018.

### 3.3.3. Changes in ESV in Response to LUT

There are two effects of LUT on the ecological environment—improving (positive contribution rate) and reducing (negative contribution rate) ecosystem service functions [55], as shown in Table 5. From the perspective of each land use type transfer-out, the biggest ESV loss caused by forestland transfer-out was $48.047 \times 10^8$ USD, of which the loss from 2010 to 2018 was as high as $37.626 \times 10^8$ USD. Secondly, the ESV loss caused by the transformation of water area into other land use types was $39.639 \times 10^8$ USD, of which the losses caused by 2000–2010 and 2010–2018 were $23.722 \times 10^8$ USD and $21.112 \times 10^8$ USD, respectively. The unused land transfer-out brought the largest increase of ESV, which was $63.204 \times 10^8$ USD, followed by the cultivated land transfer-out, which was $35.351 \times 10^8$ USD. ESV increase resulting from construction land transfer-out was relatively small ($10.017 \times 10^8$ USD), which mainly occurred in 2000–2010.

**Table 5.** Changes in the total ESV in response to LUT in the YRB from 1990 to 2018.

| Type | 1990–2000 | | 2000–2010 | | 2010–2018 | | 1990–2018 | |
|---|---|---|---|---|---|---|---|---|
| | Changes of ESV ($\times 10^8$ USD) | Contribution Rate (%) | Changes of ESV ($\times 10^8$ USD) | Contribution Rate (%) | Changes of ESV ($\times 10^8$ USD) | Contribution Rate (%) | Changes of ESV ($\times 10^8$ USD) | Contribution Rate (%) |
| 12 | 1.479 | 2.943 | 15.523 | 7.639 | 10.213 | 5.227 | 22.947 | 6.374 |
| 13 | 1.234 | 2.455 | 9.271 | 4.563 | 14.718 | 7.532 | 20.550 | 5.708 |
| 14 | 3.096 | 6.160 | 12.124 | 5.967 | 11.014 | 5.637 | 17.042 | 4.734 |
| 15 | −2.787 | −5.546 | −13.974 | −6.877 | −10.014 | −5.125 | −23.900 | −6.639 |
| 16 | −0.637 | −1.267 | −0.915 | −0.450 | −0.837 | −0.428 | −1.287 | −0.358 |
| 21 | −0.930 | −1.851 | −5.531 | −2.722 | −9.984 | −5.110 | −12.370 | −3.436 |
| 23 | −3.899 | −7.758 | −7.021 | −3.455 | −23.543 | −12.049 | −29.065 | −8.074 |
| 24 | 0.078 | 0.156 | 0.750 | 0.369 | 0.528 | 0.270 | 1.356 | 0.377 |
| 25 | −0.208 | −0.414 | −2.664 | −1.311 | −3.632 | −1.859 | −5.336 | −1.482 |
| 26 | −0.400 | −0.796 | −1.992 | −0.980 | −0.995 | −0.509 | −2.632 | −0.731 |
| 31 | −3.793 | −7.547 | −8.305 | −4.087 | −13.436 | −6.876 | −20.564 | −5.712 |
| 32 | 2.831 | 5.633 | 13.847 | 6.815 | 24.394 | 12.484 | 35.481 | 9.856 |
| 34 | 2.742 | 5.456 | 7.301 | 3.593 | 7.754 | 3.968 | 14.866 | 4.130 |
| 35 | −0.391 | −0.778 | −5.482 | −2.698 | −8.193 | −4.193 | −12.519 | −3.478 |
| 36 | −3.670 | −7.303 | −21.142 | −10.405 | −6.077 | −3.110 | −26.487 | −7.358 |
| 41 | −9.728 | −19.357 | −11.350 | −5.586 | −7.483 | −3.830 | −19.829 | −5.508 |
| 42 | −0.184 | −0.366 | −0.406 | −0.200 | −0.419 | −0.214 | −0.661 | −0.184 |
| 43 | −1.640 | −3.263 | −4.452 | −2.191 | −5.814 | −2.975 | −8.331 | −2.314 |
| 45 | −0.105 | −0.209 | −2.898 | −1.426 | −3.129 | −1.601 | −4.994 | −1.387 |
| 46 | −1.420 | −2.826 | −4.616 | −2.272 | −4.267 | −2.184 | −5.824 | −1.618 |
| 51 | 0.013 | 0.026 | 4.716 | 2.321 | 3.628 | 1.857 | 5.670 | 1.575 |
| 52 | 0.001 | 0.002 | 0.285 | 0.140 | 1.075 | 0.550 | 0.585 | 0.163 |
| 53 | 0.005 | 0.010 | 0.861 | 0.424 | 2.160 | 1.105 | 1.603 | 0.445 |
| 54 | 0.003 | 0.006 | 1.076 | 0.530 | 3.747 | 1.918 | 2.141 | 0.595 |
| 56 | 0.000 | 0.000 | 0.013 | 0.006 | 0.036 | 0.018 | 0.018 | 0.005 |
| 61 | 0.856 | 1.703 | 2.406 | 1.184 | 1.327 | 0.679 | 3.713 | 1.031 |
| 62 | 0.752 | 1.496 | 1.382 | 0.680 | 2.640 | 1.351 | 4.286 | 1.191 |
| 63 | 6.217 | 12.371 | 35.720 | 17.579 | 8.255 | 4.225 | 44.918 | 12.477 |
| 64 | 1.150 | 2.288 | 7.028 | 3.459 | 5.866 | 3.002 | 10.653 | 2.959 |
| 65 | −0.007 | −0.014 | −0.146 | −0.072 | −0.220 | −0.113 | −0.366 | −0.102 |

Note: No. 1–6 has the same meaning as Figure 3. The Tupu units' codes can be seen in Table A1.

The transition of Type 12, 13, 14, 32, 34, 51, 52, 53, 54, 56, 61, 62, 63, and 64 contributed to the increase in ESV. In terms of improving ecological environment, the transition of Type 63 brought the ESV for the largest increase (44.918 × 10$^8$ USD), with the highest contribution rate of 12.477%. This was mainly due to the contribution rate of 12.371% and 17.579%, respectively, from 1990 to 2010. The contribution rate of grassland to forestland was the second (9.856%), of which the contribution rate from 2010 to 2018 was 12.484%. This reflected the transformation and utilization of barren grassland, which made the grassland ecosystem in some areas evolve to a higher level of forestland ecosystem. The third was the conversion of Type 12 and 13, with contribution rates of 6.374% and 5.708%, respectively, all of which occurred from 2000 to 2018. In addition, the transition of Type 14 also had a positive effect on the ecological environment, with a contribution rate of 4.734%, and the contribution rate of the three stages decreased slightly, which was 6.160%, 5.967%, and 5.637%, respectively.

The transition of Type 15, 16, 21, 23, 24, 25, 26, 31, 35, 36, 41, 42, 43, 45, 46, and 65 caused the ESV decline. In terms of ESV loss, the transition of Type 23 caused the largest decline of ESV (29.065 × 10$^8$ USD), with a contribution rate of −8.074%, of which the contribution rate from 2010 to 2018 was the highest (−12.049%). The conversion of Type 36 took second place, with a contribution rate of −7.358%, of which the contribution rate from 2000 to 2010 was the highest (−10.405%). The third was the conversion of Type 15, with a contribution rate of −6.639%, and the contribution rates of the three stages were similar, which were −5.546%, −6.877%, −5.125%, respectively. Due to the small transition area, the ESV decline caused by Type 65 was only 0.366 × 10$^8$ USD, but the contribution rate of the three stages showed an increasing trend.

Figure 8 shows the spatial distribution of ESV changes in the YRB has significant characteristics during 1990–2018. Ecological appreciation zones were widespread in the YRB, particularly in Qinghai, eastern Gansu, Bayannaoer of Inner Mongolia, and central and northern Shaanxi. Ecological impairment zones were mainly concentrated in Sichuan, southwest Gansu, northern Ningxia, Inner Mongolia, southern Shaanxi, and lower reaches of the YRB. In addition, ecological impairment zones were also scattered in other areas.

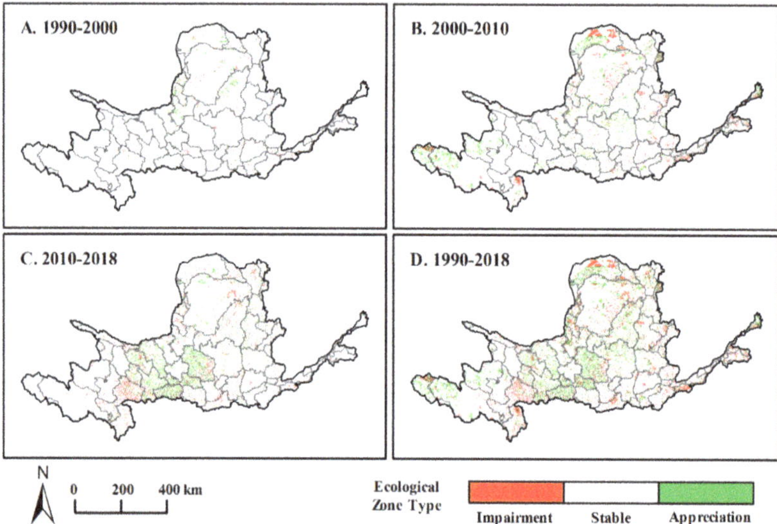

**Figure 8.** Tupu of changes in ESV from 1990 to 2018 in the YRB.

The range of ecological appreciation was little in 1990–2000, mainly concentrated in Ningxia, northern Inner Mongolia, and northern Shaanxi. During this period, the scope of ecological impairment was relatively large, mainly distributed in Gansu, Inner Mongolia, central and southern Shaanxi, central Shanxi, Henan, and Shandong. Furthermore, the ESV change in 1990–2000 was more severe than that in 2000–2010. The range of ecological appreciation in Shaanxi was expanded, but the range of ecological impairment was reduced. The range of ecological impairment in Inner Mongolia, Shanxi, Henan, and Shandong Province increased, and the range of ecological impairment in central Shanxi and coastal areas of Shandong tended to expand. At this stage, ecological appreciation zones were also distributed in Qinghai, Gansu, and Sichuan. From 2010 to 2018, the range of ecological appreciation zones decreased, and the range of ecological impairment zones increased. Compared with the first two stages, the ecological appreciation zones were widely distributed in Gansu, and the ecological impairment zones were transferred from east to west of the YRB.

## 4. Discussion

### 4.1. Interpretation of LUTs

As an important ecological functional area in China, the land use has changed dramatically in the YRB over the past 30 years [6]. Thereinto, the grassland and cultivated land were found to be the dominant land use types, and the conversions between these two types were also the most frequent. Notably, the grassland area changed by +1598 km$^2$ (1990–2000), +2950 km$^2$ (2000–2010), and −334 km$^2$ (2010–2018), respectively. Grassland increased significantly in 2000–2010, mainly due to the implementation of the Grain-for-Green policy since 2003 [59]. This policy aimed to withdraw the cultivated land that was not suitable for farming and to turn it into grassland. Therefore, the implementation of this policy has greatly increased the area of grassland. It shows that the protection of the ecologically fragile watershed by the project is worth learning from other countries and regions. Grassland area decreased from 2010 to 2018, interestingly, and different scholars have different judgments on the leading factors of such change in this period. Some studies believed that climate change was the primary factor [12], while others think that human activities were the leading factor [60]. It is worth noting that China's economy developed at a high speed since 2008, with the average annual GDP growth rate at around 10%. There is no doubt that a large number of high-intensity and unreasonable human activities have intensified the LUT, such as extensive urbanization construction, coal mining, and industrial processing [61], which have also caused direct losses of ESV. China's Central Government was also aware of this problem [8]. On the one hand, China's economy has begun to shift from rapid development to high-quality development. On the other hand, China has been vigorously promoting the construction of ecological civilization, and has implemented projects such as the protection and restoration of degraded grassland since 2012 [62]. Additionally, the transitions between grassland and unused land, and forestland were also relatively frequent, mainly due to the land exploitation, the Natural Forest Conservation Program, and deforestation [5]. During the study periods, the cultivated land area decreased, while the area of construction land continued to increase. This increase in construction land mainly came from the supplement of cultivated land, while the rural residential land was required to be reclaimed as new cultivated land. This was mainly due to the urban construction land increase vs. rural construction land decrease policy and the cultivated land balance policy [63]. This is the regulation and control policy of cultivated land and construction land with Chinese characteristics, which aims to adjust and utilize the unreasonable, inadequate, and abandoned rural construction land. In this way, it could effectively tap the full potential of existing resources of construction land in urban and rural areas and solve the shortage of construction land in cities and industrial parks. This policy is of great significance to the YRB, which could realize the coordination between economic development and ecological environment protection in the YRB. Additionally, the total area of cultivated land in China is large, but there is little cultivated land per capita and a serious shortage of reserve resources, especially in the YRB. These policies' implementation could promote new construction land to occupy as less cultivated land as possible, so as to ensure food security of the YRB or even China.

*4.2. Changes in ESV of Response to LUTs*

The ESV increased by $10.47 \times 10^8$ USD with a growth rate of 0.39% during 1990–2018, mainly due to the increase of water area and forestland area. According to the proportion of each ESV, the value of cultivated land, forestland, and grassland accounted for more than 90% of the total value, indicating that these land use types were of great significance to the ecological security of the YRB. Overall, the decline of ESV in the YRB was mainly due to the reduction of cultivated land and unused land area, resulting in $22.51 \times 10^8$ USD of ecological value loss in the study area, which should be taken as the object of key protection and restoration. However, the ESV of forestland showed different growth, which reflected the effect of ecological restoration policies, such as returning farmland to forest, natural forest protection plans, and three North Shelterbelt programs [64].

The ESV variation in the YRB showed different features in the upper, middle, and lower reaches. Specifically, cultivated land area had a decreasing trend in the upstream, midstream, and downstream, which had a bad impact on the ecological environment in the YRB. Additionally, it could cause a great loss of the ecological service value in the basin. Meanwhile, the water area in the midstream decreased slightly, which made the loss of ESV in the midstream. The area of water in downstream had increased at a slow speed, and the increase of ESV can hardly offset the loss of ESV caused by other land types. The water area in upstream increased rapidly, especially in 2010–2018. Then, the increase of ESV brought by the increase of water area offset the loss of ESV, which was caused by the decrease of cultivated land and unused land area. Therefore, it could make the ESV in the upstream increase generally and keep the stability of the upstream ecosystem.

In some sense, the impact of LUTs on ESV could be understood from the perspective of land use quantity transfer. However, the change of ESV affected by LUTs also showed obvious spatial differences due to the spatial difference of land use change [5]. Thereinto, the high-value areas of AESV mainly concentrated in the upstream and midstream of the YRB, while the low-value areas mainly concentrated in the downstream areas, which were anti-geographical gradient characteristics with an economic development level. In the upstream of the YRB, ESV in Qinghai, Lanzhou, and Sichuan had high value, but the economic development in these areas was backward and human activities were less.

In addition, these areas were mostly located at high altitudes, where the perennial melting water of ice and snow nourished the local grassland [54,55]. On the other hand, global warming also intensified the decline of the snow line [56]. The ESV of Shanxi in the midstream was relatively high. Previously, Shanxi Province was an important energy center in China. However, after 2013, Shanxi began to attach importance to transformation development and ecological environment protection and abandoned the extensive and high-intensity coal industry, thus achieving a high ESV effect at present. This also means that energy development and ecological and environmental protection in the YRB can be paid equal attention to, but the key premise lay in their coordination. From further analysis of the change Tupu of ESV in the YRB from 1990 to 2018, we found that the impairment areas of ecosystem services were mainly concentrated in the middle and lower reaches of the YRB, that is, the ecosystem protection pressure of the midstream and downstream was relatively greater than that of the upstream provinces [65].

Interestingly, the change of ESV in the YRB was also closely related to the positioning of the ecosystem service function in the upper and lower reaches of the YRB. For example, the main reason for the decline of ESV in Henan, Shandong, and other provinces in the lower reaches of the Yellow River was that these provinces have flat terrain, convenient irrigation conditions, and are suitable for farming, and they were mostly located as provinces with high yield, so many grasslands and water areas converted into cultivated land. However, Ningxia, Qinghai, Gansu, and other provinces in the upper reaches of the Yellow River have high terrain and steep slope, and they are mainly responsible for water and soil conservation and other ecological services [66]. In the past 30 years, the areas of grassland and forestland have increased significantly, and their service values have been significantly improved. Therefore, the ecological and functional differences between the upper, middle, and lower reaches of the YRB should be fully considered. To sum up, the ecosystem condition in most areas of the YRB has improved in the past 30 years, and the improvement extent is greater than the deterioration extent, which reflects the effect of the Yellow River Regulation to a certain extent. However, the pressure of ecosystem protection and restoration still exists in the YRB, especially in the areas where ESV declines.

### 4.3. Policy Implications

The LUT and the change of ESV urgently need more appropriate policy guidance, to better realize the protection and development of the YRB. Combined with our research, the following policy recommendations are suggested.

(1) Sustainable land management (SLM) claims to minimize the negative impacts of land degradation [67], which otherwise results in the deprivation of human welfare [68]. However, our research found that the unsustainable use of natural resources was still widespread in the YRB. Owing to the YRB existing across several of China's administrative provinces, it is urgent to break down the administrative districts and to establish strategic and participatory land use planning, including environmental and social impact assessments. In fact, such schemes of SLM are very tough to implement because of the multiple institutional interests from different sectors at different scales. Therefore, it is necessary for China's Central Government to establish a unified SLM organization for the YRB for the whole basin. One of the main functions of this organization is to perfect land use planning on the scale of the YRB, so as to make the LUT more scientific and reasonable. Moreover, the medium- and long-term governance blueprint of the YRB can be planned with reference to the advanced experience of the Rhine River or other watershed areas [67], which could ensure the sustainability, integrity, and clarity of the governance path.

(2) In the process of the land use transformation and its management in the YRB, there was an obvious absence in the power of enterprise organizations, social institutions, and the public. On the one hand, these non-governmental organizations have not been well developed, and their strength was still very weak; on the other hand, these social forces lack effective channels to participate. Therefore, the social cooperative governance mechanism is urgent to speed up the establishment and improvement, and let social forces fully participate in the management of the

YRB. Additionally, it is necessary to form a situation of social cooperation and co-governance by: (a) clarifying the boundary of responsibility among various social subjects, (b) building an efficient coordination and cooperation mechanism, and (c) establishing a multi-subject governance pattern in the YRB.

(3)  The annual per capita ESV of the YRB is only 628 USD and per capita GDP and ESV in 2018 is 12:1, reflecting that the YRB provided very low ESV per capita. Therefore, it is suggested to introduce measurement evaluation of ESV, and integrate the ecosystem services into the decision-making of land use and ecological protection. As we all know, land use for economic growth is unsustainable, so we must make the environmental value decision of land use. However, China's current land use planning and land use policies do not fully reflect the concept of sustainable land use. A large number of studies have focused on ecosystem services [3,34,55,69–71]; how to integrate ecosystem services into land use and ecological protection decisions has always been the focus of discussion [42,72]. Therefore, taking ESV as a quantitative indicator to measure the ecological effect of land use-related policies is of great significance to promote land use decision-making, urban management, and ecosystem protection.

(4)  The upper, middle, and lower reaches are the ecological center, energy center, and economic center in China, respectively. Therefore, it is necessary to fully consider the differences of the eco-environment in the upper, middle, and lower reaches, and classify the watershed according to the different protection priorities of the region. The main contradiction of its governance lies in how to balance the relationship between development and protection [66]. Thus, we suggest: (a) exploring the ecological compensation mechanism for carrying out land utilization in the YRB; (b) balancing the economic benefit of different areas and the principal part of land utilization in the upstream, midstream, and downstream of the YRB; and (c) coordinating the interesting relationship between economic construction and ecological protection. Internationally, Payment for Watershed Ecosystem Services (PWES) replaces the concept of Watershed Ecosystem Services [36]. Thus, to eliminate the negative impact of land use on the environment, some suggestions was purposed as follows: (a) taking ESV as the foundation for determining the ecological compensation standard; (b) exploring the establishment of an ecological compensation mechanism for different regions and different principal parts of land utilization in the upstream, midstream, and downstream; and (c) weighing the benefit difference brought by different land use types.

(5)  Due to the influence of ecosystem services preference in different land use management types, one or several types of specific ecosystem services were pursued, which could intentionally or unintentionally affect the provision of other ecosystem services [5]. This pattern has led to trade-offs and synergies in ecosystem services [61,73]. Therefore, carrying out in-depth research for the influence that the LUT exerts on ESV can provide decision references for further optimizing land use policies. In order to ensure the coordinated development of ecological, economic, and social benefits in the process of rapid urbanization, measures such as delineating the "three zones and three lines" (three zones—ecological zone, agricultural zone, and urban zone; three lines—permanent basic farmland red line, urban development boundary, and ecological red line) should be promoted [74]. At the same time, it is necessary to strengthen ecological protection and restoration, do more to repair ecological damage, and strive to achieve a good balance between the natural ecosystem and human activities.

*4.4. Uncertainties and Challenges for Future Research*

This research provided a novel approach, which integrated the geo-information Tupu method with spatial analysis to unravel the effect of LUT on ESV. To some extent, it could make up for the lack of comprehensive quantitative assessment of the ecosystem services of the entire YRB and other deficiencies in the existing research. The results could offer a new perspective into the relationship between land use transition, ecosystem services, and land use management widely recognized for understanding the complex human–ecosystem interactions in ecological topics. Moreover, we improved the previous research by revealing the internal process and geographical pattern of LUTs at a more refined raster data level (30 × 30 m). It can capture more information about LUT than that of 1000 × 1000 m resolution land use raster data [4].

Some deficiencies, however, hinder the study and more efforts can be made to improve. Firstly, this paper used the first-class classification standard to classify the land use in the YRB, and the estimated results are slightly rough. Future work could refine the land use types and improve the accuracy of the estimation results. Secondly, this paper has not studied the factors influencing the change of ESV in the YRB and has ignored the ecological value provided by construction land, which is also the content of research in the future.

## 5. Conclusions

This work revealed the spatio-temporal pattern of LUT by using the geo-information Tupu method and assessed the change in ESV caused by different Tupu units based on the equivalent factor method. The main conclusions were as follows. (1) The mutual conversions between grassland and unused land, and cultivated land and forestland were the primary Tupu units, which were distributed in the upstream and midstream. The conversion of cultivated land to construction land was also remarkable, being mainly distributed in the lower reaches of the basin. The overall separation degree value became increasingly smaller in these three periods, indicating that these Tupu units were more concentrated in space. The spatial separation degree in the Tupu units between construction land and other land use types was obviously higher than that of other Tupu units, indicating the spatial distribution of construction land transition was relatively scattered. Cultivated land transition has gradually become the main type of LUT in the YRB, with a proportion of 63.06% from 2010 to 2018. (2) The total ESV presented a fluctuated upward tendency ($+10.47 \times 10^8$ USD), but it was periodic, with grassland (45%) and forestland (30%) contributing the most to the total ESV. ESV presented the watershed characteristics of the upstream region > the midstream region > the downstream region of the economic development level. The ESV of upstream increased, while ESV of midstream and downstream decreased. From the perspective of spatial distribution, the high-value AESV were mainly distributed in Qinghai, Gansu, central Shaanxi, northern Henan, and most areas of Shanxi, while the low-value AESV were mainly distributed in Ningxia, Inner Mongolia, eastern Henan, and Shandong. Among them, the changes in AESV were the most drastic from 2000 to 2010. (3) LUT had a certain impact on quantity variation of ESV, and the impact of land type transfer-in and transfer-out on ESV change was quite different. Tupu units 63 and 32 contributed to most of the increase in the ESV, while Tupu units 23 and 36 contributed to most of the decrease in the ESV during 1990–2018. Additionally, the range of ecological appreciation zones were widely distributed and dispersed, and the range of ecological impairment zones were gradually expanded.

**Author Contributions:** Conceptualization, X.L. and D.Y.; methodology, D.Y.; software, D.Y.; validation, H.Y., G.L. and J.Z.; formal analysis, D.Y. and H.Y.; investigation, D.Y.; resources, X.L. and H.Y.; data curation, D.Y.; writing—original draft preparation, D.Y.; writing—review and editing, X.L., G.L., J.Z., and H.Y.; visualization, D.Y. and H.Y.; supervision, X.L. and H.Y.; project administration, X.L.; funding acquisition, X.L. All authors have read and agreed to the published version of the manuscript.

**Funding:** This study was supported by the National Natural Science Foundation of China (Grant No.71704177 and No.71874192), Jiangsu Social Science Youqing and Key Fund Project (19GLA006), Special Fund Project for Basic Scientific Research Business Expenses of Central University, and Open Foundation of Key Laboratory of the Coastal Zone Exploitation and Protection, Ministry of Land and Resource (2019CZEPK10).

**Acknowledgments:** The authors gratefully acknowledge the editors and anonymous reviewers for their constructive comments on our manuscript. We also thank the data support from the "Data Center for Resources and Environmental Sciences, Chinese Academy of Sciences (RESDC) (http://www.resdc.cn)".

**Conflicts of Interest:** The authors declare no conflict of interest.

## Appendix A

**Table A1.** The Tupu units' codes of land use transitions (LUTs) in the YRB.

| | Transition | In | | | | | |
|---|---|---|---|---|---|---|---|
| | | Cultivated Land | Forestland | Grassland | Water Area | Construction Land | Unused Land |
| | Cultivated land | / | 12 | 13 | 14 | 15 | 16 |
| | Forestland | 21 | / | 23 | 24 | 25 | 26 |
| Out | Grassland | 31 | 32 | / | 34 | 35 | 36 |
| | Water area | 41 | 42 | 43 | / | 45 | 46 |
| | Construction land | 51 | 52 | 53 | 54 | / | 56 |
| | Unused land | 61 | 62 | 63 | 64 | 65 | / |

## References


1. Schmidt, P.; Morrison, T.H. Watershed management in an urban setting: Process, scale and administration. *Land Use Policy* **2012**, *29*, 45–52. [CrossRef]
2. Liu, Y.; Bi, J.; Lv, J. Classification of ecosystem services and a reclassification framework of watershed ecosystem services. *Resour. Sci.* **2019**, *41*, 1189–1200. [CrossRef]
3. Long, H.; Liu, Y.; Hou, X.; Li, T.; Li, Y. Effects of land use transitions due to rapid urbanization on ecosystem services: Implications for urban planning in the new developing area of China. *Habitat Int.* **2014**, *44*, 536–544. [CrossRef]
4. Lu, X.; Shi, Y.; Chen, C.; Yu, M. Monitoring cropland transition and its impact on ecosystem services value in developed regions of China: A case study of Jiangsu Province. *Land Use Policy* **2017**, *69*, 25–40. [CrossRef]
5. Chen, W.; Zhao, H.; Li, J.; Zhu, L.; Wang, Z.; Zeng, J. Land use transitions and the associated impacts on ecosystem services in the Middle Reaches of the Yangtze River Economic Belt in China based on the geo-informatic Tupu method. *Sci. Total Environ.* **2020**, *701*, 134690. [CrossRef] [PubMed]
6. Wang, S.; Liu, J.; Ma, T. Dynamics and changes in spatial patterns of land use in Yellow River Basin, China. *Land Use Policy* **2010**, *27*, 313–323. [CrossRef]
7. Fang, C. Spatial Organization Pattern and High-Quality Development of Urban Agglomeration in the Yellow River Basin. *Econ. Geogr.* **2020**, *40*, 1–8. [CrossRef]
8. Lu, D.; Sun, D. Development and management tasks of the Yellow River Basin: A preliminary understanding and suggestion. *Acta Geogr. Sin.* **2019**, *74*, 2431–2436. [CrossRef]
9. Zhang, W.; Wang, L.; Xiang, F.; Qin, W.; Jiang, W. Vegetation dynamics and the relations with climate change at multiple time scales in the Yangtze River and Yellow River Basin, China. *Ecol. Indic.* **2020**, *10*, 105892. [CrossRef]
10. Wohlfart, C.; Mack, B.; Liu, G.; Kuenzer, C. Multi-faceted land cover and land use change analyses in the Yellow River Basin based on dense Landsat time series: Exemplary analysis in mining, agriculture, forest, and urban areas. *Appl. Geogr.* **2017**, *85*, 73–88. [CrossRef]
11. Zhang, B.; Miao, C. Spatiotemporal changes and driving forces of land use in the Yellow River Basin. *Resour. Sci.* **2020**, *42*, 460–473. [CrossRef]
12. Xu, S.; Yu, Z.; Yang, C.; Ji, X.; Zhang, K. Trends in evapotranspiration and their responses to climate change and vegetation greening over the upper reaches of the Yellow River Basin. *Agric. For. Meteorol.* **2018**, *263*, 118–129. [CrossRef]
13. Daily, G.C.; Soderqvist, T.; Aniyar, S.; Arrow, K.; Dasgupta, P.; Paul, R. The Value of Nature and the Nature of Value. *Science* **2000**. [CrossRef] [PubMed]
14. Perring, M.P.; De Frenne, P.; Baeten, L.; Maes, S.L.; Depauw, L.; Blondeel, H.; Caron, M.M.; Verheyen, K. Global environmental change effects on ecosystems: The importance of land-use legacies. *Glob. Chang. Biol.* **2016**, *22*, 1361–1371. [CrossRef]
15. Lambin, E.F.; Meyfroidt, P. Land use transitions: Socio-ecological feedback versus socio-economic change. *Land Use Policy* **2010**, *27*, 108–118. [CrossRef]
16. Long, H.; Li, T. The coupling characteristics and mechanism of farmland and rural housing land transition in China. *J. Geogr. Sci.* **2012**, *22*, 548–562. [CrossRef]

17. Tsai, Y.; Zia, A.; Koliba, C.; Bucini, G.; Guilbert, J.; Beckage, B. An interactive land use transition agent-based model (ILUTABM): Endogenizing human-environment interactions in the Western Missisquoi Watershed. *Land Use Policy* **2015**, *49*, 161–176. [CrossRef]

18. Long, H.; Qu, Y. Land use transitions and land management: A mutual feedback perspective. *Land Use Policy* **2018**, *74*, 111–120. [CrossRef]

19. Liu, J.; Kuang, W.; Zhang, Z.; Xu, X.; Qin, Y.; Ning, J.; Zhou, W.; Zhang, S.; Li, R.; Yan, C.; et al. Spatiotemporal characteristics, patterns, and causes of land-use changes in China since the late 1980s. *J. Geogr. Sci.* **2014**, *24*, 195–210. [CrossRef]

20. Ning, J.; Liu, J.; Kuang, W.; Xu, X.; Zhang, S.; Yan, C.; Li, R.; Wu, S.; Hu, Y.; Du, G.; et al. Spatio-temporal patterns and characteristics of land-use change in China during 2010–2015. *Acta Geogr. Sin.* **2018**, *28*, 547–561. [CrossRef]

21. Sun, P.; Xu, Y.; Wang, S. Terrain gradient effect analysis of land use change in poverty area around Beijing and Tianjin. *Trans. Chin. Soc. Agric. Eng.* **2014**, *30*, 277–288. [CrossRef]

22. Wang, C.; Wang, Y.; Wang, R.; Zheng, P. Modeling and evaluating land-use/land-cover change for urban planning and sustainability: A case study of Dongying city, China. *J. Clean Prod.* **2018**, *172*, 1529–1534. [CrossRef]

23. Liu, J.; Wang, P.; Li, J.; Xu, J.; Liu, X. An Algorithm for Land-Use Pattern Index and Its Application. *Geogr. Geo-Inf. Sci.* **2009**, *25*, 107–109.

24. Long, H.; Li, X. Analysis on regional land use transition: A case study in Transect of the Yangtze River. *J. Nat. Resour.* **2002**, *17*, 144–149.

25. Song, X. Discussion on land use transition research framework. *Acta Geogr. Sin.* **2017**, *72*, 471–487.

26. Nizalov, D.; Thornsbury, S.; Loveridge, S.; Woods, M.; Zadorozhna, O. Security of property rights and transition in land use. *J. Comp. Econ.* **2016**, *44*, 76–91. [CrossRef]

27. Liu, Y.; Long, H.; Li, T.; Tu, S. Land use transitions and their effects on water environment in Huang-Huai-Hai Plain, China. *Land Use Policy* **2015**, *47*, 293–301. [CrossRef]

28. Costanza, R.; Arge, R.; Groot, R.D.; Farberk, S.; Belt, M.V.D. The value of the world's ecosystem services and natural capital. *Nature* **1997**, *387*, 253–260. [CrossRef]

29. Costanza, R.; de Groot, R.; Sutton, P.; van der Ploeg, S.; Anderson, S.J.; Kubiszewski, I.; Farber, S.; Turner, R.K. Changes in the global value of ecosystem services. *Glob. Environ. Chang.* **2014**, *26*, 152–158. [CrossRef]

30. Redford, K.H.; Adams, W.M. Payment for Ecosystem Services and the Challenge of Saving Nature. *Conserv. Biol.* **2009**, *23*, 785–787. [CrossRef]

31. Gascoigne, W.R.; Hoag, D.; Koontz, L.; Tangen, B.A.; Shaffer, T.L.; Gleason, R.A. Valuing ecosystem and economic services across land-use scenarios in the Prairie Pothole Region of the Dakotas, USA. *Ecol. Econ.* **2011**, *70*, 1715–1725. [CrossRef]

32. Rounsevell, M.D.A.; Pedroli, B.; Erb, K.; Gramberger, M.; Busck, A.G.; Haberl, H.; Kristensen, S.; Kuemmerle, T.; Lavorel, S.; Lindner, M.; et al. Challenges for land system science. *Land Use Policy* **2012**, *29*, 899–910. [CrossRef]

33. Foley, J.A. Global Consequences of Land Use. *Science* **2005**, *309*, 570–574. [CrossRef] [PubMed]

34. Tolessa, T.; Senbeta, F.; Kidane, M. The impact of land use/land cover change on ecosystem services in the central highlands of Ethiopia. *Ecosyst. Serv.* **2017**, *23*, 47–54. [CrossRef]

35. Song, W.; Deng, X. Land-use/land-cover change and ecosystem service provision in China. *Sci. Total Environ.* **2017**, *576*, 705–719. [CrossRef]

36. Hauck, J.; Görg, C.; Varjopuro, R.; Ratamäki, O.; Jax, K. Benefits and limitations of the ecosystem services concept in environmental policy and decision making: Some stakeholder perspectives. *Environ. Sci. Policy* **2013**, *25*, 13–21. [CrossRef]

37. De Groot, R.S.; Alkemade, R.; Braat, L.; Hein, L.; Willemen, L. Challenges in integrating the concept of ecosystem services and values in landscape planning, management and decision making. *Ecol. Complex* **2010**, *7*, 260–272. [CrossRef]

38. Xie, G.; Zhen, L.; Lu, C.; Xiao, Y.; Chen, C. Expert Knowledge Based Valuation Method of Ecosystem Services in China. *J. Nat. Resour.* **2008**, *23*, 911–919.

39. Xie, G.; Zhang, C.; Zhang, C.; Xiao, Y.; Lu, C. The value of ecosystem services in China. *Resour. Sci.* **2015**, *37*, 1740–1746.

40. Chaikaew, P.; Hodges, A.W.; Grunwald, S. Estimating the value of ecosystem services in a mixed-use watershed: A choice experiment approach. *Ecosyst. Serv.* **2017**, *23*, 228–237. [CrossRef]

41. De Oliveira, V.A.; de Mello, C.R.; Beskow, S.; Viola, M.R.; Srinivasan, R. Modeling the effects of climate change on hydrology and sediment load in a headwater basin in the Brazilian Cerrado biome. *Ecol. Eng.* **2019**, *133*, 20–31. [CrossRef]

42. Liu, J.; Jin, X.; Xu, W.; Fan, Y.; Ren, J.; Zhang, X.; Zhou, Y. Spatial coupling differentiation and development zoning trade-off of land space utilization efficiency in eastern China. *Land Use Policy* **2019**, *85*, 310–327. [CrossRef]

43. Wang, G.; Wang, S.; Chen, Z. Land-use /land-cover changes in the Yellow River basin. *J. Tsinghua Univ. Sci. Technol.* **2004**, *44*, 1218–1222.

44. Jiang, W.; Yuan, L.; Wang, W.; Cao, R.; Zhang, Y.; Shen, W. Spatio-temporal analysis of vegetation variation in the Yellow River Basin. *Ecol. Indic.* **2015**, *51*, 117–126. [CrossRef]

45. He, Z.; He, J. Remote Sensing on Spatio-temporal Evolution of Vegetation Cover in the Yellow River Basin during 1982—2013. *Trans. Chin. Soc. Agric. Mach.* **2017**, *48*, 179–185. [CrossRef]

46. Omer, A.; Elagib, N.A.; Zhuguo, M.; Saleem, F.; Mohammed, A. Water scarcity in the Yellow River Basin under future climate change and human activities. *Sci. Total Environ.* **2020**, *749*, 141446. [CrossRef]

47. Tang, Y.; Tang, Q.; Tian, F.; Zhang, Z.; Liu, G. Responses of natural runoff to recent climatic variations in the Yellow River basin, China. *Hydrol. Earth Syst. Sci.* **2013**, *17*, 4471–4480. [CrossRef]

48. Liu, J.; Liu, M.; Zhuang, D.; Zhang, Z.; Deng, X. Study on spatial pattern of land-use change in China during 1995—2000. *Sci. China Ser. D Earth Sci.* **2003**, *46*, 373–384.

49. Liu, J.; Zhang, Z.; Xu, X.; Kuang, W.; Zhou, W.; Zhang, S.; Li, R.; Yan, C.; Yu, D.; Wu, S.; et al. Spatial Patterns and Driving Forces of Land Use Change in China in the Early 21st Century. *Acta Geogr. Sin.* **2009**, *64*, 1411–1420. [CrossRef]

50. Liao, K. The Discussion and Prospect for Geo-Informatic Tupu. *Geo-Inf. Sci.* **2002**, *3*, 14–20.

51. Wan, J.; Zhu, G.; Peng, Q. Land Use Information Tupu and Its Application. *Geomat. Inf. Sci. Wuhan Univ.* **2005**, *30*, 355–358. [CrossRef]

52. Chen, S.; Yue, T.; Li, H. Studies on Geo-Informatic Tupu and its application. *Geogr. Res. Aust.* **2000**, *19*, 337–343.

53. Ye, Q.; Liu, G.; Lu, Z.; Gong, Z.; Marco, M. Research of TUPU on Land Use/ Land Cover Change Based on GIS. *Prog. Geogr.* **2002**, *21*, 349–357. [CrossRef]

54. Wang, J.; Shao, J.; Li, Y. Geo-Spectrum Based Analysis of Crop and Forest Land Use Change in the Recent 20 Years in the Three Gorges Reservoir Area. *J. Nat. Resour.* **2015**, *30*, 235–247. [CrossRef]

55. Gong, W.; Yuan, L. Analysis of Land Use and Its Ecosystem Service Value Change by Graphic Information: A Case Study of Harbin Section of Songhuajiang Watershed. *Environ. Technol.* **2010**, *33*, 200–205. [CrossRef]

56. Yang, A.; Zhu, L.; Chen, S.; Jin, H.; Xia, X. Geo-informatic spectrum analysis of land use change in the Manas River Basin, China during 1975–2015. *Chin. J. Appl. Ecol.* **2019**, *30*, 3863–3874.

57. Nahuelhual, L.; Vergara, X.; Kusch, A.; Campos, G.; Droguett, D. Mapping ecosystem services for marine spatial planning: Recreation opportunities in Sub-Antarctic Chile. *Mar. Policy* **2017**, *81*, 211–218. [CrossRef]

58. Wang, Y.; Dai, E.; Yin, L.; Ma, L. Land use/land cover change and the effects on ecosystem services in the Hengduan Mountain region, China. *Ecosyst. Serv.* **2018**, *34*, 55–67. [CrossRef]

59. Wang, B.; Yang, Q.; Liu, Z. Effect of conversion of farmland to forest or grassland on soil erosion intensity changes in Yanhe River Basin, Loess Plateau of China. *Front. For. China* **2009**, *4*, 68–74. [CrossRef]

60. Kong, D.; Miao, C.; Wu, J.; Duan, Q. Impact assessment of climate change and human activities on net runoff in the Yellow River Basin from 1951 to 2012. *Ecol. Eng.* **2016**, *91*, 566–573. [CrossRef]

61. Haase, D.; Schwarz, N.; Strohbach, M.; Kroll, F.; Seppelt, R. Synergies, Trade-offs, and Losses of Ecosystem Services in Urban Regions: An Integrated Multiscale Framework Applied to the Leipzig-Halle Region, Germany. *Ecol. Soc.* **2012**, *17*. [CrossRef]

62. Yu, H.; Bian, Z.; Chen, F. Mechanism of Land Ecological Restoration in Mining Area: An Analy-sis based on Land Degradation Neutrality Framework. *China Land Sci.* **2020**, *34*, 86–95. [CrossRef]

63. Shen, X.; Wang, L.; Wu, C.; Lv, T.; Lu, Z.; Luo, W.; Li, G. Local interests or centralized targets? How China's local government implements the farmland policy of Requisition–Compensation Balance. *Land Use Policy* **2017**, *67*, 716–724. [CrossRef]

64. Yang, J.; Xie, B.; Zhang, D. Spatio-temporal variation of water yield and its response to precipitation and land use change in the Yellow River Basin based on InVEST model. *Chin. J. Appl. Ecol.* **2020**, *31*, 2731–2739. [CrossRef]

65. Liu, G.; Yang, Q.; Huang, J. The Change Characteristics and Influence Factors of Ecosystem Services Valuation of the Yellow River Basin from 2000 to 2015. *Environ. Conform. Assess.* **2020**, *12*, 90–97. [CrossRef]

66. Yu, H.; Bian, Z.; Mu, S.; Yuan, J.; Chen, F. Effects of Climate Change on Land Cover Change and Vegetation Dynamics in Xinjiang, China. *Int. J. Environ. Res. Public Health* **2020**, *17*, 4865. [CrossRef] [PubMed]

67. Vogl, A.L.; Bryant, B.P.; Hunink, J.E.; Wolny, S.; Apse, C.; Droogers, P. Valuing investments in sustainable land management in the Upper Tana River basin, Kenya. *J. Environ. Manag.* **2017**, *195*, 78–91. [CrossRef]

68. Siegmund-Schultze, M.; Ppel, J.K.; Sobral, M.D.C. Unraveling the water and land nexus through inter- and transdisciplinary research: Sustainable land management in a semi-arid watershed in Brazil's Northeast. *Reg. Environ. Chang.* **2018**, *15*, 2005–2017. [CrossRef]

69. Song, W.; Deng, X.; Liu, B.; Li, Z.; Jin, G. Impacts of Grain-for-Green and Grain-for-Blue Policies on Valued Ecosystem Services in Shandong Province, China. *Adv. Meteorol.* **2015**, *2015*, 1–10. [CrossRef]

70. Schirpke, U.; Kohler, M.; Leitinger, G.; Fontana, V.; Tasser, E.; Tappeiner, U. Future impacts of changing land-use and climate on ecosystem services of mountain grassland and their resilience. *Ecosyst. Serv.* **2017**, *26*, 79–94. [CrossRef]

71. Cao, Y.; Li, G.; Tian, Y.; Fang, X.; Li, Y.; Tan, Y. Linking ecosystem services trade-offs, bundles and hotspot identification with cropland management in the coastal Hangzhou Bay area of China. *Land Use Policy.* **2020**, *97*, 104689. [CrossRef]

72. Liu, W.; Zhan, J.; Zhao, F.; Yan, H.; Zhang, F.; Wei, X. Impacts of urbanization-induced land-use changes on ecosystem services: A case study of the Pearl River Delta Metropolitan Region, China. *Ecol. Indic.* **2019**, *98*, 228–238. [CrossRef]

73. Bennett, E.M.; Peterson, G.D.; Gordon, L.J. Understanding relationships among multiple ecosystem services. *Ecol. Lett.* **2009**, *12*, 1394–1404. [CrossRef] [PubMed]

74. Lian, X. Review on Advanced Practice of Provincial Spatial Planning: Case of a Western, Less Developed Province. *Int. Rev. Spat. Plan. Sustain. Dev.* **2018**, *6*, 185–202. [CrossRef]

**Publisher's Note:** MDPI stays neutral with regard to jurisdictional claims in published maps and institutional affiliations.

*Article*

# Habitat Quality Effect and Driving Mechanism of Land Use Transitions: A Case Study of Henan Water Source Area of the Middle Route of the South-to-North Water Transfer Project

Meijing Chen [1,2], Zhongke Bai [1,3,4], Qingri Wang [2,*] and Zeyu Shi [1]

[1] School of Land Science and Technology, China University of Geosciences (Beijing), Beijing 100083, China; 3012200009@cugb.edu.cn (M.C.); baizk@cugb.edu.cn (Z.B.); 3012200007@cugb.edu.cn (Z.S.)
[2] China Land Surveying and Planning Institute, Beijing 100035, China
[3] Key Laboratory of Land Consolidation and Rehabilitation, Ministry of Natural Resources, Beijing 100035, China
[4] Technology Innovation Center of Ecological Restoration Engineering in Mining Area, Ministry of Natural Resources, Beijing 100083, China
* Correspondence: wangqingri@mail.clspi.org.cn

**Citation:** Chen, M.; Bai, Z.; Wang, Q.; Shi, Z. Habitat Quality Effect and Driving Mechanism of Land Use Transitions: A Case Study of Henan Water Source Area of the Middle Route of the South-to-North Water Transfer Project. *Land* **2021**, *10*, 796. https://doi.org/10.3390/land10080796

Academic Editors: Hualou Long, Xiangbin Kong, Shougeng Hu and Yurui Li

Received: 29 June 2021
Accepted: 27 July 2021
Published: 29 July 2021

**Publisher's Note:** MDPI stays neutral with regard to jurisdictional claims in published maps and institutional affiliations.

**Abstract:** Accelerating urbanization and industrialization have had substantial impacts on economic and social activities, changed the surface environment of the earth, and affected global climate change and biodiversity. If reasonable and effective management measures are not implemented in time, unchecked urbanization and industrialization will damage the structure and function of the ecosystem, endanger human and biological habitats, and ultimately lead to difficulties in achieving sustainable development. This study investigates the habitat quality effect of land use transition and analyzes the cause and mechanism of such changes from an economic–social–ecological complex system perspective in the Henan Water Source (HWS) area of the Middle Route of the South-to-North Water Transfer Project (MRP). The study comprehensively examines the characteristics of land use transition from 2000 to 2020. The results indicate that the habitat quality of the HWS area of the MRP decreased slowly over the past 20 years, with a more obvious decrease in the past 10 years. Specifically, the proportion of high quality habitat areas is relatively large and stable, and the medium and low quality habitat areas increase significantly. Analyzing the change degree of the proportion of different levels of habitat quality area in each county, revealed that Dengzhou City had the most dramatic change, followed by the Xichuan and Neixiang counties; other counties did not undergo obvious change. The results of habitat quality factor detection by GeoDetector showed that land use transition plays a decisive role in the change of habitat quality. The types of land use with high habitat suitability compared to those with low habitat suitability will inevitably lead to a decrease in habitat quality. Additionally, elevation, slope, landform type, and annual precipitation are important factors affecting the habitat quality in the HWS area of the MRP, indicating that ecological factors determine the background conditions of habitat quality. The gross domestic product (GDP) per capita, the proportion of agricultural output value, grain yield per unit area in economic factors, population density, and urbanization rate in social factors affect the spatial differentiation of habitat quality to a certain extent. Soil type, annual mean temperature, vegetation type, and NDVI index have weak effects on habitat quality, while road network density and slope aspect have no significant effect on habitat quality. The results of this study provide a basis for the improvement of habitat quality, ecosystem protection and restoration, land resource management, and related policies in the HWS area of the MRP. They also provide references for the research and practice of the habitat quality effects of land use transition in other regions.

**Keywords:** land use transition; habitat quality effect; driving mechanism; the Middle Route of the South-to-North Water Transfer Project (MRP); Henan Water Source (HWS) area

439

## 1. Introduction

The accelerating process of urbanization and industrialization has led to the rapid change of land use patterns in China, and the conflict between the social economic system and natural ecosystem is thus increasing. Since the Reform and Opening Up in 1978, urbanization and industrialization have been promoted extensively in China [1], and great achievements have been made in economic and social development. Moreover, the land use pattern has changed dramatically. Construction utilized substantial amounts of cultivated land from 1997 to 2009, which led to the loss of approximately 8.2 million hm$^2$ of arable land in China [2], and threatened national food security. Moreover, to ensure food production, the Chinese government has formulated a strict system of farmland occupation and compensation balance, which results in a large number of ecological spaces, such as forestland, grassland, and unused land, being reclaimed for crop production. This practice has made the ecosystem function declining and climate change [3–5]. It is estimated that since 1850, land use/cover change has caused a 145 PgC loss in the global terrestrial ecosystem [6]. Since the beginning of the 21st century, China's ecological and environmental problems have become increasingly prominent. The government has gradually realized the importance of ecosystem protection for maintaining ecosystem functions, protecting biodiversity, and realizing regional sustainable development. The government, therefore, has taken the ecological civilization as an important development strategy to actively promote the improvement and protection of habitat quality and other ecosystem protection and restoration efforts.

Investigating the effect of land use transition on habitat quality has important scientific and practical significance in the current period of China's ecological civilization construction. As a new method in Land Use/Cover Change (LUCC) research, land use transition has been widely investigated by scholars in recent years [7–13]. Land use transition refers to the change in land use patterns over time corresponding to the transition of the economic and social development stage. With the improvement in economic and social development, regional land use pattern conflicts gradually weaken [9], which can effectively reflect the change of the natural environment, and social and economic development process [14]. Since Grainger [15,16] published research on land use transition in forest countries [9], scholars have extensively discussed the concept, connotation, mechanism [17], and methods of land use transition [18] and analyzed the land use transition characteristics of typical regions [19–23] or typical land types (e.g., rural homestead [9,24,25], industrial land [26,27], cultivated land [28–32]). Some scholars have discussed the environmental effect caused by regional land use transition [14,33–36], which has enhanced the theory, methodology, and empirical research of land us e transition. Regarding the environmental effect of land use transition, the change of ecosystem service value or environmental quality index has obtained an increasing amount of attention from scholars. However, the change in habitat quality has received less attention. Habitat quality refers to the ability of an ecosystem to provide suitable survival and development conditions for individuals and populations, which is an important basic condition to determine biodiversity [37]. It can effectively reflect the health degree of ecosystems [38–40], and is an important embodiment of ecosystem function [41,42]. Investigating the effect of land use transition on habitat quality and analyzing the changing trend, can understand the conflict between urban construction and environment, especially the conflict between human development and biodiversity protection. Investigating the driving mechanism of the habitat quality effect, can have a deeper understanding of the causes and mechanism of this conflict, which can provide a decision-making basis for sustainable management of the environment, the maintenance of biodiversity, and the realization of harmonious coexistence between humans and nature in the future.

## 2. Materials and Methods

### 2.1. Study Area

The Middle Route of the South-to-North Water Transfer Project (MRP) is an important inter basin water transfer project in China that began on 30 December 2003, and opened on 12 December 2014. It provides water for production, daily life, industry, and agriculture to more than 20 large and medium-sized cities in Henan, Hebei, Beijing, and Tianjin. By December 2020, the MRP had delivered 34.8 billion $m^3$ of water, and more than 69 million people had directly benefited from the project, which is of great strategic significance to optimizing China's water resources allocation pattern and promoting regional coordinated development. The water source area of the MRP is an important ecological function protection area for water conservation in China. It covers seven prefecture-level cities in Henan, Hubei, and Shaanxi provinces. Xichuan County, Xixia County, Neixiang County, Luanchuan County, Lushi County, and Dengzhou City are in Henan Province. The total administrative area of the six counties (cities) is 17,312 $km^2$, and the overall geographical location of the six counties is 32°22″ N–34°23′ N, 110°34′ E–112°20″ E (Figure 1). This area is the transition zone from the second to third steps of China's terrain. It has various geomorphic types, including mountains, hills, and plains, and is mainly composed of medium and large undulating mountains with medium altitude. There are many rivers in the Henan Water Source (HWS) area of the MRP, and the primary tributaries include Danjiang River, Guanhe River, and Xihe River. Xichuan County is the main distribution area of the Danjiang Reservoir area, which is the water source of the MRP. The research area has important biological habitats, such as Funiu Mountain National Nature Reserve, Baotianman National Nature Reserve, Dinosaur Egg Fossil Group National Nature Reserve, Danjiang Wetland National Nature Reserve, and Xixia Giant Salamander Provincial Nature Reserve, as well as several rare animal and plant resources, such as the endangered species of Chinese merganser, peach blossom jellyfish, etc.

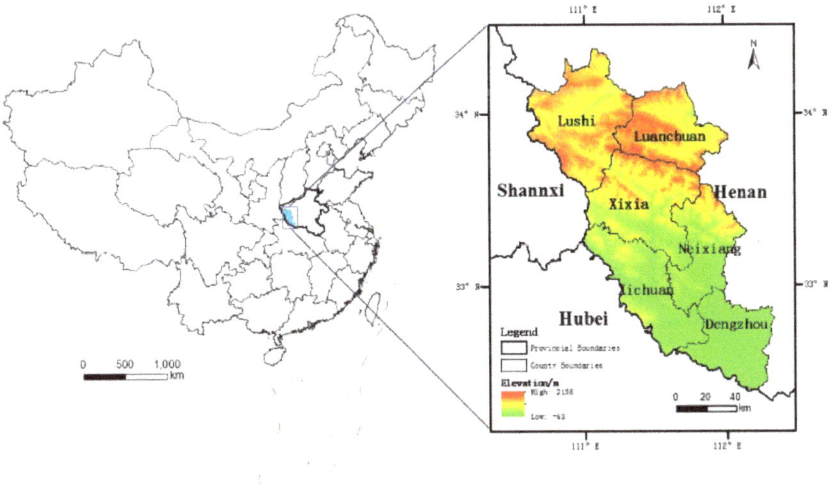


**Figure 1.** Location of Henan Water Source area of the Middle Route of the South-to-North Water Transfer Project.

### 2.2. Data Sources

The analysis of this study is mainly based on the data of land use, Digital Elevation Model (DEM), administrative boundaries, environmental properties, and economy and society. The sources of the involved data are specified as follows. (1) Land use data derived from the global 30 m land cover data (http://www.globallandcover.com/, 14

January 2021) of National Geomatics Center of China was obtained from 2000, 2010, and 2020 to represent before, during, and after the implementation of the first phase of the MRP. The land use types were divided into seven categories: Cultivated land, forest land, grassland, construction land, wetland, water area, and other land. (2) DEM data were derived from Geospatial Data Cloud platform with a spatial resolution of 30 m. (3) The vector data of administrative boundaries, highways, and railways were derived from the National Catalogue Service for Geographic Information (https://www.webmap.cn/, 26 January 2021). (4) Landform, vegetation, soil type, temperature, precipitation, and NDVI data were derived from the Resource and Environment Science and Data Center of the Chinese Academy of Science (https://www.resdc.cn/, 23 February 2021). (5) Economic and social data, namely, per capita GDP, population density, urbanization rate, grain output per unit area, etc., were derived from the *Luoyang Statistical Yearbook, Nanyang Statistical Yearbook, Sanmenxia Statistical Yearbook*, statistical bulletin of national economic and social development, and government work reports of each county. Some missing data were calculated and obtained using the moving average method. All the above data were transformed to the same coordinate system by projection (WGS_1984_UTM_ Zone_49n) and cut according to the administrative boundary of the study area.

*2.3. Methods*

In this study, the land use transfer matrix was used to study the characteristics of land use transition in the HWS area of the MRP. Then, the InVEST model was used to analyze the changes in habitat quality and degradation degree of the study area. Finally, the GeoDetector model was used to analyze the driving factors and mechanisms of regional habitat quality (Figure 2).

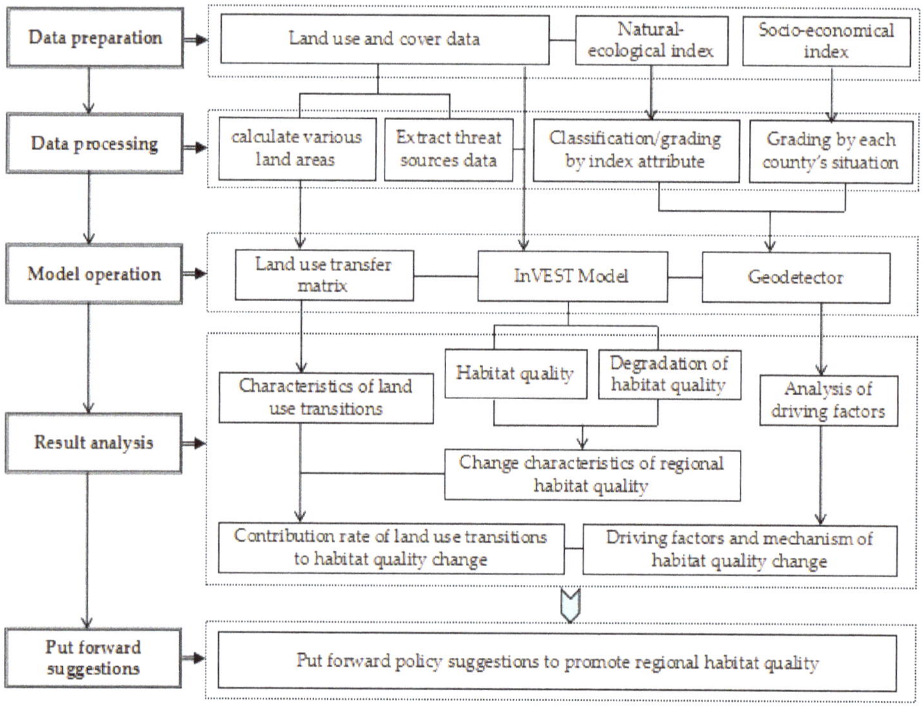


**Figure 2.** Research logic and process of this study.

### 2.3.1. Land Use Transfer Matrix

The land use transfer matrix reflects the conversion between different land use types in different periods of a certain region. This matrix can quantitively characterize the change in regional land use [14]. The land use transfer matrix is shown in Equation (1).

$$A_{ij} = \begin{bmatrix} A_{11} & A_{12} & \cdots & A_{1n} \\ A_{21} & A_{22} & \cdots & A_{2n} \\ \cdots & \cdots & \cdots & \cdots \\ A_{n1} & A_{n2} & \cdots & A_{nn} \end{bmatrix} \tag{1}$$

where $A$ is the area of land type, $i$ and $j$ are the land types before and after the transfer, respectively, $n$ is the number of land types, and $A_{ij}$ is the transfer area from land type $i$ to land type $j$. Each row represents the flow direction information from land type $i$ to other land types, and each column represents the source information from other land types to land type $j$.

### 2.3.2. Habitat Quality Module of the InVEST Model

The Habitat Quality module of the InVEST model can calculate the regional habitat quality index, and its spatial distribution by analyzing the land use cover map and the threat factors. Habitat quality index is a comprehensive index to evaluate the habitat suitability and degradation degree of land use type [43], and is calculated using Equation (2).

$$Q_{xj} = H_j \left[ 1 - \left( \frac{D_{xj}^z}{D_{xj}^z + K^z} \right) \right] \tag{2}$$

where $Q_{xj}$ is the habitat quality index of grid unit $x$ of land use type $j$; $H_j$ is the habitat suitability score of land use type $j$, with a range of 0–1; $z$ is the scale constant, generally 2.5; $K$ is the semisaturation constant, which was 0.5 in this study; and $D_{xj}$ is the habitat degradation index, which indicates the degradation degree of habitat under stress.

$$D_{xj} = \sum_{r=1}^{R} \sum_{y=1}^{Y_r} \left( \frac{\omega_r}{\sum_{r=1}^{R} \omega_r} \right) r_y i_{rxy} \beta_x S_{jr} \tag{3}$$

where $R$ is the number of stress factors, $Y_r$ is the total number of grid cells of stress factors, $\omega_r$ is the weight, $r_y$ is the number of stress factors on the grid cell, $\beta_x$ is the accessibility level of grid $x$, $S_{jr}$ is the sensitivity of land use type $j$ to stress factors, and the value range is 0–1, and $i_{rxy}$ is the influence distance of stress factors, which can be divided into linear and exponential decline.

$$i_{rxy} = 1 - \left( \frac{d_{xy}}{d_{r\,max}} \right), \text{ if linear} \tag{4}$$

$$i_{rxy} = \exp\left( -\left( \frac{2.99}{d_{r\,max}} \right) d_{xy} \right), \text{ if exponential} \tag{5}$$

Based on previous studies [44–48], the InVEST model user's guide [43], and the reality of HWS area, this study constructed an evaluation table of habitat threat factors and threat degree (as shown in Table 1) and the sensitivity of land use types to the threat factors (Table 2).

### 2.3.3. GeoDetector

GeoDetector is a statistical method used to detect spatial differentiation and reveal the driving factors behind it [49,50]. GeoDetector has been applied in many fields of natural and social sciences. The factor detection tool can detect the extent to which the independent variable $x$ explains the spatial differentiation of dependent variable $y$. The calculation formula of q value is as follows (Equation (6)):

$$q = 1 - \frac{\sum_{h=1}^{L} N_h \sigma_h^2}{N \sigma^2} = 1 - \frac{SSW}{SST}$$

$$SSW = \sum_{h=1}^{L} N_h \sigma_h^2, \ SST = N \sigma^2$$

(6)

where $h$ ($h = 1, 2, \ldots, l$) is the stratification of dependent variable $y$ or independent variable $x$, i.e., classification or partition; $N_h$ and $N$ are the unit numbers of layer $h$ and the whole region, respectively; $\sigma_h^2$ and $\sigma^2$ are the variances of layer $h$ and region $Y$, respectively; $SSW$ and $SST$ are the sum of variances within the layer and the total variances of the whole region, respectively; $q \in [0, 1]$, where the larger the value, the stronger the explanatory power of independent variable $x$ to dependent variable $y$, and vice versa.

**Table 1.** Habitat threat factors and threat degree.

| Threat Factors | Farthest Threat Distance (km) | Threat Degree | Declining Type |
|---|---|---|---|
| Cultivated land | 4 | 0.5 | linear |
| Construction land | 8 | 1.0 | exponential |
| Main traffic arteries | 6 | 0.9 | linear |
| Bare land | 5 | 0.8 | linear |

**Table 2.** Sensitivity of land use types to the threat factors.

| Land Use Types | Habitat Suitability | Cultivated Land | Construction Land | Main Traffic Arteries | Bare Land |
|---|---|---|---|---|---|
| Cultivated land | 0.4 | 0 | 0.8 | 0.6 | 0.4 |
| Forest land | 1.0 | 0.6 | 0.7 | 0.8 | 0.3 |
| Grassland | 0.8 | 0.8 | 0.6 | 0.5 | 0.4 |
| Wetland | 0.7 | 0.7 | 0.6 | 0.6 | 0.2 |
| Water area | 0.6 | 0.5 | 0.4 | 0.4 | 0.2 |
| Construction land | 0 | 0 | 0 | 0 | 0 |
| Other land | 0.2 | 0.5 | 0.7 | 0.2 | 0 |

## 3. Results

### 3.1. Land Use Transitions

#### 3.1.1. Changes in Land Types and Degree of Change

From 2000 to 2020, a total land area of 1522.66 km$^2$ changed, accounting for 8.9% of the research area. The change in the first 10 years was more intense with 921.08 km$^2$ of land changing, and slowed down in the past 10 years, during which the altered area was 814.86 km$^2$. All types of land changed by varying degrees (Figure 3).

Overall, the land types with greater changes were cultivated land, construction land, and water area. A new land type of "other land," mainly bare land, appeared in the study area from 2010 to 2020. During the study period, cultivated land, grassland, and forest land were the most reduced land types, while construction land, cultivated land, and forest land were the most increased land types (Table 3).

#### 3.1.2. Drastic Changes in Construction Land and Cultivated Land

During the study period, the land types with the largest change area were cultivated land and construction land (Table 4), with a net decrease of 437.74 km$^2$ in cultivated land and a net increase of 301.88 km$^2$ in construction land. (1) A large area of cultivated land was converted into construction land and water areas. From 2000 to 2010, 159.09 km$^2$ of cultivated land was converted into construction land, accounting for 17.27% of the total changed area, and 107.52 km$^2$ of cultivated land was converted into water areas, accounting for 11.67% of the total changed area. From 2010 to 2020, the area of cultivated land occupied by construction reached 263.60 km², accounting for 32.35% of cultivated land reduction. During this period, the area of cultivated land converted into water areas decreased slightly, reaching 103.42 km² and accounting for 12.70% of cultivated land reduction. (2) The cultivated land occupied by construction was obvious in the southeast plain. Owing to the flat terrain and rapid economic development, the cultivated land occupied by construction was primarily distributed in the southeast plain area (Figure 4), including most areas of Dengzhou City, the south of Neixiang County, and the southeast of Xichuan County. This particular land type was mainly scattered in the surrounding areas of the original urban and rural construction land with low altitudes and was relatively concentrated around the county town. (3) The area of forest land converted to cultivated land was larger than the total area cultivated land returned to forest land. From 2000 to 2010, 150.87 km² of forest land was converted into cultivated land, while 128.38 km² of cultivated land was converted into forest land in the same period. From 2010 to 2020, the conversion area of forest land to cultivated land was 5.07 km² larger than the conversion area of cultivated land to forest land.

### 3.1.3. Water Area

The water area of the study area increased by 274.38 km², with a growth rate of 155.12%, of which 121.18 km² increased in the first 10 years, and 153.20 km² increased in the second 10 years. This reflects the long-term efforts of the water source area to ensure water quality and quantity. To increase the sustainable water supply capacity of the MRP, Danjiangkou Reservoir implemented the dam heightening project in 2002 and passed the acceptance in 2013. The dam height increased from 162 m to 176.6 m, the normal water level increased from 157 m to 170 m, and the reservoir capacity increased from 17.45 billion m³ to 29.05 billion m³. A large number of farmland and villages around the reservoir area had been inundated. Among them, Xichuan County has an inundated area of 137 km² with about 150,000 people resettled. From 2000 to 2010, 79.96% of the total amount of water area transferred from cultivated land, followed by wetland (11.56%), grassland (5.58%), and forest land (2.60%), and the proportion of construction land which transitioned to water was relatively small. From 2010 to 2020, the area of cultivated land converted to water areas decreased slightly, accounting for 64.56% of the total converted water area. The proportion of forest land and grassland converted to water area increased significantly to 17.87% and 12.86%, respectively, and the proportion of wetland converted to water area decreased to 3.16%.

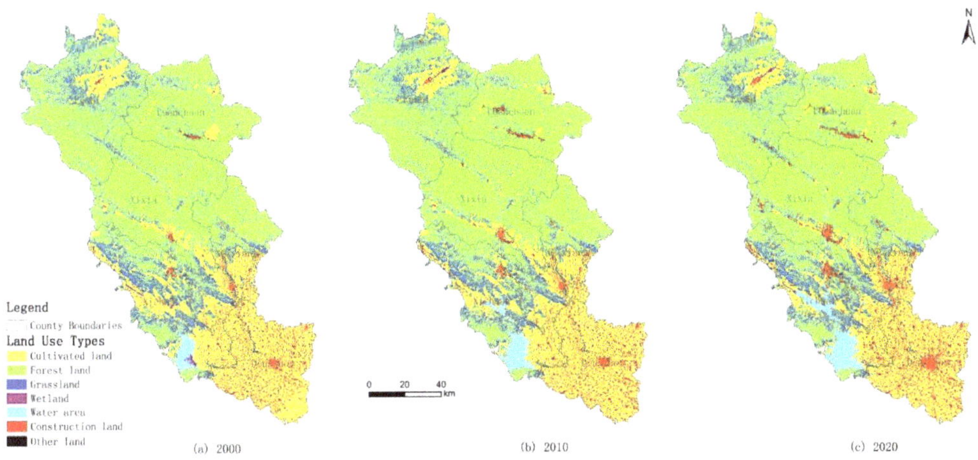


**Figure 3.** Land use maps in 2000, 2010, and 2020.

**Table 3.** Land use transfer matrix (unit: km$^2$).

| 2000 | 2020 | | | | | | | Reduction |
|---|---|---|---|---|---|---|---|---|
| | Grassland | Cultivated Land | Construction Land | Forest Land | Wetland | Water Area | Other Land | |
| Grassland | 1461.63 | 64.77 | 20.28 | 173.44 | 0.80 | 37.74 | 2.74 | 299.76 |
| Cultivated land | 47.48 | 4286.88 | 376.26 | 140.25 | 12.56 | 213.35 | 1.70 | 791.60 |
| Construction land | 2.58 | 111.71 | 481.20 | 0.66 | 0.05 | 1.92 | 0.59 | 117.51 |
| Forest land | 72.50 | 171.71 | 21.75 | 9146.06 | 0.19 | 15.24 | 1.71 | 283.09 |
| Wetland | 0.54 | 1.11 | 0.01 | 0.10 | 1.68 | 17.53 | 0.00 | 19.29 |
| Water area | 1.61 | 4.56 | 1.09 | 2.79 | 1.36 | 165.47 | 0.00 | 11.41 |
| Increase | 124.72 | 353.86 | 419.39 | 317.23 | 14.95 | 285.79 | 6.73 | |
| Change | −175.05 | −437.74 | 301.88 | 34.13 | −4.34 | 274.38 | 6.73 | |

**Table 4.** Changes of land use structure.

| Land Use Type | 2000 | | 2010 | | 2020 | |
|---|---|---|---|---|---|---|
| | Area/km$^2$ | % | Area/km$^2$ | % | Area/km$^2$ | % |
| Grassland | 1761.39 | 10.32 | 1614.21 | 9.46 | 1586.35 | 9.30 |
| Cultivated land | 5078.48 | 29.76 | 4901.65 | 28.72 | 4640.74 | 27.19 |
| Construction land | 598.71 | 3.51 | 715.48 | 4.19 | 900.59 | 5.28 |
| Forestland | 9429.16 | 55.25 | 9517.76 | 55.77 | 9463.29 | 55.45 |
| Wetland | 20.96 | 0.12 | 18.41 | 0.11 | 16.62 | 0.10 |
| Water area | 176.88 | 1.04 | 298.06 | 1.75 | 451.26 | 2.64 |
| Other land | 0.00 | 0.00 | 0.00 | 0.00 | 6.73 | 0.04 |

### 3.1.4. Ecological Land

The area of forest land, grassland, wetland, and water area with strong ecological functions in the study area was relatively large. In 2000, the area of the four land types was 11,388.39 km$^2$, accounting for 66.73% of the total research area, increasing by 60.06 km$^2$ in 2010, accounting for 67.09%. In 2020, the ecological land area continued to increase to 11,517.51 km$^2$, accounting for 69.07%. Among them, forest land initially increased rapidly then decreased slowly, grassland and wetland decreased continuously, and water area increased continuously and rapidly. This phenomenon fully reflects that many engineering measures, and ecological protection and restoration strategies were implemented to ensure the regional water supply capacity.

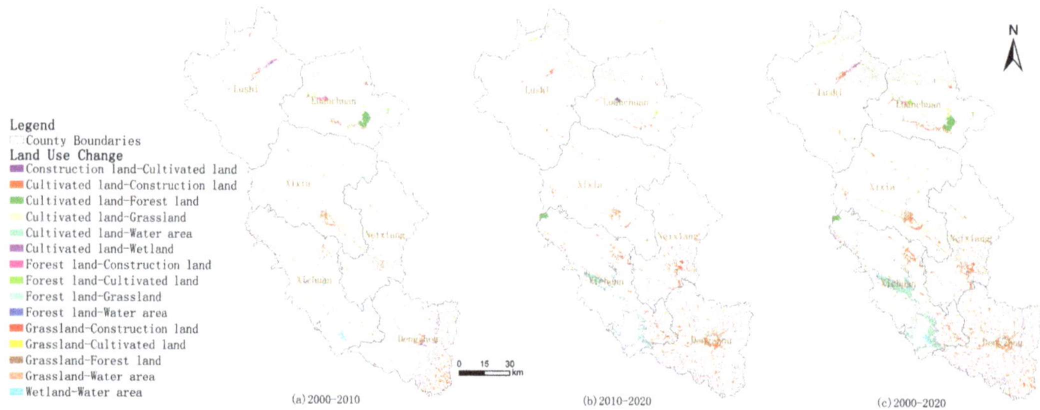


**Figure 4.** Spatial distribution of land use transitions.

### 3.1.5. Land Use Changes in Different Counties

By analyzing the change of land use types in each county (Table 5), it can be concluded that: The land use change in Xichuan County was the most severe, followed by Dengzhou City. The most obvious land types in Xichuan County were cultivated land and water area, of which cultivated land reduced by 244.66 km$^2$ over the past 20 years, with the proportion reducing by 8.70%, and the decrease between 2020 and 2010 was 30.16 km$^2$ more than that between 2010 and 2000. The water area increased from 5.53% of the total land in 2000 to 14.03% in 2020. The increase in water area between 2020 and 2010 was 12.54 km$^2$ more than that between 2010 and 2000. The most substantially changed land types in Dengzhou City were cultivated land and construction land, of which cultivated land decreased by 100.43 km$^2$ over 20 years and the construction land area increased by 98.07 km$^2$. The change range of land use in other counties was not significant, and the change proportions of different land types were all within 2.1%.

**Table 5.** Land use changes of each county (unit: km$^2$).

| County | Year | Grassland | Cultivated Land | Construction Land | Forestland | Wetland | Water Area | Other Land |
|---|---|---|---|---|---|---|---|---|
| Lushi | 2000 | 545.22 | 499.01 | 8.66 | 2598.88 | 0.00 | 5.03 | 0.00 |
| | 2010 | 495.89 | 489.43 | 22.84 | 2636.99 | 7.30 | 4.35 | 0.00 |
| | 2020 | 493.79 | 486.61 | 35.73 | 2627.94 | 7.99 | 4.73 | 0.00 |
| Luanchuan | 2000 | 137.39 | 185.87 | 15.94 | 2130.46 | 0.00 | 1.35 | 0.00 |
| | 2010 | 113.71 | 167.78 | 34.66 | 2152.64 | 0.76 | 1.48 | 0.00 |
| | 2020 | 113.50 | 168.13 | 49.09 | 2134.21 | 0.74 | 3.24 | 2.11 |
| Xixia | 2000 | 218.62 | 452.49 | 28.32 | 2743.72 | 0.00 | 4.56 | 0.00 |
| | 2010 | 201.70 | 458.53 | 48.54 | 2730.33 | 0.00 | 8.63 | 0.00 |
| | 2020 | 204.64 | 426.18 | 79.10 | 2725.24 | 0.00 | 12.55 | 0.00 |
| Xichuan | 2000 | 548.44 | 1197.63 | 81.73 | 810.69 | 19.85 | 155.58 | 0.00 |
| | 2010 | 506.39 | 1090.38 | 92.50 | 845.45 | 10.34 | 268.87 | 0.00 |
| | 2020 | 492.14 | 952.97 | 126.13 | 837.91 | 7.88 | 394.69 | 2.21 |
| Neixiang | 2000 | 285.57 | 781.75 | 105.81 | 1127.89 | 0.56 | 3.53 | 0.00 |
| | 2010 | 273.17 | 776.07 | 116.48 | 1132.96 | 0.00 | 6.44 | 0.00 |
| | 2020 | 262.92 | 745.57 | 154.22 | 1122.57 | 0.00 | 14.42 | 2.41 |
| Dengzhou | 2000 | 25.77 | 1960.55 | 358.17 | 15.70 | 0.55 | 6.80 | 0.00 |
| | 2010 | 23.01 | 1918.30 | 400.39 | 17.55 | 0.00 | 8.27 | 0.00 |
| | 2020 | 19.03 | 1860.12 | 456.24 | 13.56 | 0.00 | 18.58 | 0.00 |

### 3.2. Changes of Habitat Quality

The grid data of cultivated land, construction land, and bare land in 2000, 2010, and 2020 were extracted by ArcGIS 10.6, and the buffer area was set for the vector data of main traffic arteries in each period of the study area. After overlaying the land use maps, the corresponding grid data were extracted; then, the land use maps, various threat source data, habitat threat factors and threat degree, and the sensitivity evaluation table were input into the Habitat Quality module of InVEST 3.9.0. Subsequently, habitat quality distribution maps of the study area in 2000, 2010, and 2020 were obtained and divided into high, medium-high, medium, medium-low, and low grades using the equidistant method (Figure 5).

### 3.2.1. Overall Habitat Quality

From 2000 to 2020, the mean habitat quality value of the whole region decreased from 0.756 in 2000 to 0.755 in 2010, and then decreased to 0.750 in 2020, with a total converted area of 1520.81 km$^2$. The area of habitat quality improved was 709.83 km$^2$, and the area of habitat quality degraded was 810.98 km$^2$ (Figure 6). Overall, low and medium grade increasing and medium-low and medium-high grade decreasing trends were observed.

Among them, the medium-low habitat quality area reduction was the largest converted area, with a decrease of 473.74 km$^2$, and the low habitat quality area increased the most, by 308.61 km$^2$ (Table 6).

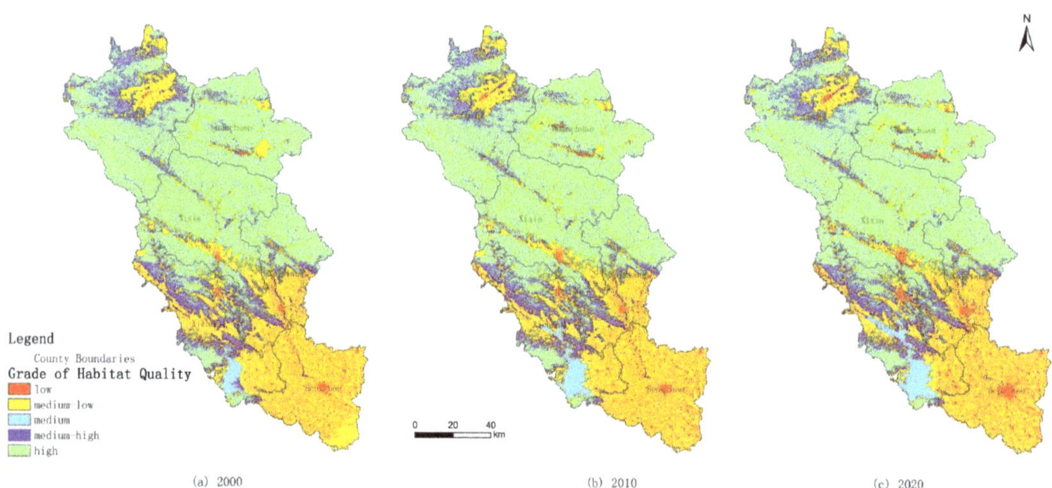


**Figure 5.** Spatio-temporal patterns of habitat quality.

### 3.2.2. High Quality Habitat Areas

The proportion of high quality habitat area continued to be about 55% during the study period. High quality habitat areas were mainly distributed in the northern and central regions, including most areas of Xixia County, Luanchuan County, Lushi County, and some areas of Xichuan County and Neixiang County. In 2000, the high quality habitat area was 9429.27 km$^2$, which increased to 9517.76 km$^2$ in 2010, and then decreased to 9463.29 km$^2$ in 2020. Overall, the areas of high quality habitat only changed less than 1% in the past 20 years.

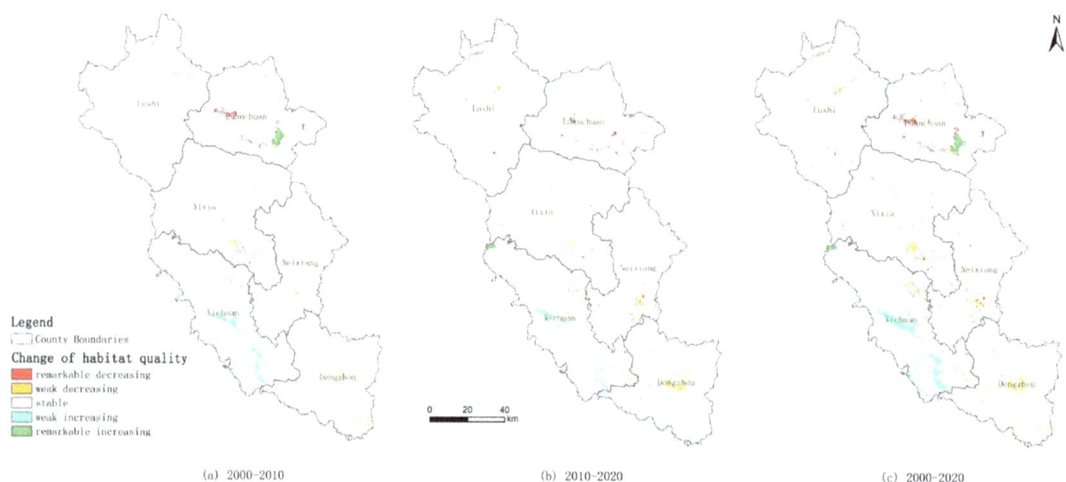


**Figure 6.** Spatial distribution of the changes of habitat quality.

**Table 6.** Spatial transfer matrix of habitat quality in the study area (unit: km$^2$).

| 2000 | 2020 | | | | | Reduction |
|---|---|---|---|---|---|---|
| | Low | Medium-Low | Medium | Medium-High | High | |
| Low | 481.79 | 111.71 | 1.92 | 2.63 | 0.66 | 116.92 |
| Medium-low | 377.97 | 4286.88 | 213.35 | 60.04 | 140.25 | 791.60 |
| Medium | 1.09 | 4.56 | 165.47 | 2.97 | 2.79 | 11.41 |
| Medium-high | 23.02 | 65.88 | 55.28 | 1464.55 | 173.52 | 317.69 |
| High | 23.46 | 171.71 | 15.24 | 72.79 | 9146.08 | 283.19 |
| Increase | 425.53 | 353.86 | 285.79 | 138.42 | 317.21 | |
| Change | 308.61 | −437.74 | 274.38 | −179.27 | 34.02 | |

### 3.2.3. Medium and Low Quality Habitat Areas

In 2000, the medium quality habitat area was 176.88 km$^2$, accounting for 1.04%. In 2010, the ratio increased to 1.75%, an increase of 121.18 km$^2$, with a growth rate of 68.51%. In 2020, the proportion was 2.64%, and the growth rate was 51.40%, with a total increase of 274.38 km$^2$ in 20 years. The low quality habitat area increased by 51.55% in the past 20 years, and the proportion increased from 3.51% in 2000 to 5.32% in 2020, mainly distributed in Neixiang County and Dengzhou City.

### 3.2.4. Changes of Habitat Quality in Different Counties

The change of habitat quality in Xichuan County was the most severe, followed by Dengzhou City and Neixiang County. Of all the counties in the study area, the mean habitat quality value only increased in Xichuan County (by 0.009 in 20 years) and decreased in other counties. Among them, the mean habitat quality value in Dengzhou City decreased the most, from 0.345 in 2000 to 0.328 in 2020, a decrease of 0.017 in 20 years. Neixiang County and Xixia County had the second largest decreases in mean habitat quality of 0.011 and 0.010, respectively; Luanchuan County and Lushi County exhibited decreases of 0.008 and 0.003, respectively.

The change of habitat quality in Xichuan County showed that the area of medium-low grade decreased significantly, which decreased by 244.665 km$^2$ in 20 years, while the area of medium grade increased rapidly, which increased by 239.11 km$^2$. In Dengzhou City and Neixiang County, the area of low quality habitat area both increased, while the proportion of medium-low grade area both decreased. The change range of Dengzhou City was larger than Neixiang County. The variation in the changes of habitat quality in other counties were not very significant and were all less than 1.2% (Table 7).

**Table 7.** Changes of habitat quality area in different grades of counties (unit: km$^2$).

| County | Year | low | Medium-Low | Medium | Medium-High | High |
|---|---|---|---|---|---|---|
| Lushi | 2000–2010 | 14.19 | −9.58 | −0.68 | −41.92 | 38.00 |
| | 2010–2020 | 12.89 | −2.82 | 0.38 | −1.40 | −9.05 |
| Luanchuan | 2000–2010 | 18.71 | −18.10 | 0.13 | −22.92 | 22.18 |
| | 2010–2020 | 16.54 | 0.36 | 1.76 | −0.23 | −18.43 |
| Xixia | 2000–2010 | 20.21 | 6.04 | 4.07 | −7.93 | −13.39 |
| | 2010–2020 | 30.57 | −32.35 | 3.93 | −6.06 | −5.08 |
| Xichuan | 2000–2010 | 10.77 | −107.25 | 113.28 | −51.55 | 34.75 |
| | 2010–2020 | 35.84 | −137.41 | 125.82 | −16.72 | −7.54 |
| Neixiang | 2000–2010 | 10.66 | −5.68 | 2.90 | −12.96 | 5.07 |
| | 2010–2020 | 40.17 | −30.35 | 11.01 | −10.31 | −10.57 |
| Dengzhou | 2000–2010 | 42.21 | −42.24 | 1.48 | −3.31 | 1.86 |
| | 2010–2020 | 55.85 | −58.19 | 10.31 | −3.99 | −3.99 |

### 3.2.5. Degree of Habitat Degradation

The obtained grid map of habitat quality degradation degree was reclassified using the equidistant method and divided into five categories: Weak, medium-weak, medium, medium-strong, and strong (Figure 7). The habitat quality degradation grades in the study area were mainly medium-weak, weak, and medium degradation. Overall, the degree of habitat degradation in the study area was reduced. The highest habitat degradation value was 0.200 in 2000, which increased slowly to 0.201 in 2010, and then decreased to 0.155 in 2020. In terms of spatial distribution, the areas with strong habitat quality degradation were mainly concentrated in the central and eastern low altitude regions, and the areas with medium-strong degradation were mainly concentrated in Dengzhou City and Xichuan County, for which, a significant increasing trend over the past 10 years was observed. Meanwhile, the areas with weak degradation were mainly concentrated in the south of Lushi County and Luanchuan County and the east of Xixia County. The medium-strong and medium-weak degradation areas showed an increasing trend, and the weak degradation area showed a decreasing trend, especially in the past 10 years.

### 3.3. Habitat Quality Effect of Land Use Transitions

### 3.3.1. Driving Factors of Habitat Quality Change

The spatial differentiation of regional habitat quality is restricted by different factors, such as ecological factors that affect the natural background conditions of the biological habitat environment; economic factors that determine the strength of regional economic development, reflecting the manner and degree of human interference with the biological habitat environment; and social factors that reflect human concern and awareness of habitat quality, as well as the protection and management ability. Based on the natural endowment characteristics of the study area, such as mountainous, undulating, a subtropical to the temperate transition zone, and a monsoon continental humid and semihumid climate, as well as the social and economic development characteristics, such as mountainous counties, relatively regressive economic development, and agricultural production dominance, the driving mechanism of the spatial pattern of habitat quality in 2020, was studied using 16 indicators (Table 8) including elevation, slope, geomorphology type, annual precipitation, vegetation type, land use type, per capita GDP, the proportion of agricultural output value, and urbanization rate.

In ArcGIS 10.6, the elevation, slope, annual precipitation, annual average temperature, and NDVI index of the study area were divided into 9 grades by using the natural breakpoint method, and the aspect, geomorphology type, soil type, vegetation type, and land use type were divided into 9, 10, 9, 8, and 7 categories, respectively, according to their classification standards and combined with the actual situation of the study area. The fishing net creating tool of ArcGIS 10.6 was used to generate $1 \times 1$ km grid data (18,297 evaluation units in total) of the study area. The road network density was calculated according to the ratio of the road length in the grid to the grid area and was divided into eight categories using the natural breakpoint method. Economic and social data were identified according to the spatial grid and divided into six categories according to the index values of each county. Based on the grid data of habitat quality and driving factors in the study area, the center point of $1 \times 1$ km grid was used as the sampling point (17,647 sampling points in total), the corresponding $X$ and $Y$ attribute values were extracted, and the generated data table was input into the GeoDetector for operation.

The results of factor detection (Figure 8) indicated that land use type was the most influential factor on habitat quality in the study area, with a q value as high as 0.99, followed by elevation, slope, geomorphology type, and annual precipitation, with a q value between 0.4 and 0.6. The q values of per capita GDP, the proportion of agricultural output value, grain yield per unit area, population density, and urbanization rate were all ~0.39, while the q values of soil type, annual mean temperature, vegetation type, and NDVI index were between 0.18 and 0.23, and the q values of road network density and aspect were the lowest. The results of risk detection reveal the suitable range or types of influencing factors of regional high quality habitat and provide a decision-making basis for the protection and restoration of the ecological system. According to the detection results (Table 8), the areas with high quality habitat were mostly distributed in forest land, with altitude >1503 m and slope of 33.48–40.06; additionally, the slope aspect was in the north, and the geomorphology type was dominated by medium altitude and large undulating mountains. The annual precipitation and annual average temperature were 554–591 mm and 12.30–13.40 °C, respectively; the main vegetation types were swamp and grass, the NDVI index was 0.08–0.29, and the road network density was <0.54.

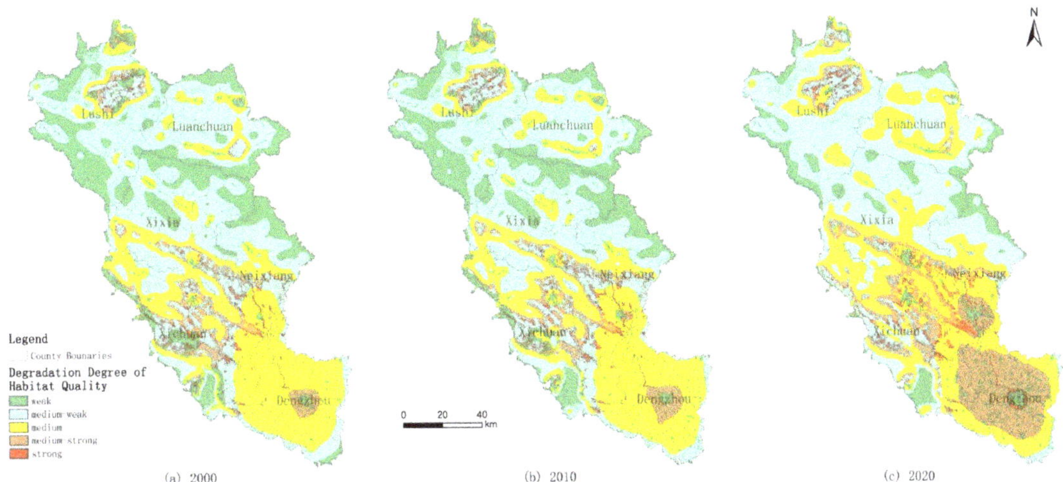


**Figure 7.** Spatio-temporal distribution of the degradation degree of habitat quality.

**Table 8.** Driving factors of habitat quality change and dominant range/type.

| | **Driving Factors** | | **Unit** | **Range/Type** |
|---|---|---|---|---|
| **Ecological factors** | Topography | elevation $x1$ | m | >1503 |
| | | slope $x2$ | ° | 33.48–40.06 |
| | | slope aspect $x3$ | — | North |
| | Geomorphology | geomorphology type $x4$ | — | Middle elevation relief mountains |
| | Soil | soil type $x5$ | — | Calcareous soil |
| | Climate | annual precipitation $x6$ | mm | 554–591 |
| | | annual average temperature $x7$ | °C | 12.30–13.40 |
| | Vegetation | vegetation type $x8$ | — | Swamp, grass |
| | | NDVI index $x9$ | — | 0.08–0.29 |
| | LUCC | land use type $x10$ | — | Forestland |
| **Economic factors** | GDP | per capita GDP $x11$ | Yuan/person | 55,716–57,676 |
| | Industry | proportion of agricultural output value $x12$ | % | 13.49–20.50 |
| | | grain yield per unit area $x13$ | kg/hm$^2$ | 4271–4532 |
| **Social factors** | Carrying capacity | population density $x14$ | Person/km$^2$ | 137.78–146.72 |
| | Development degree | urbanization rate $x15$ | % | 50.00–50.14 |
| | | road network density $x16$ | km/km$^2$ | <0.54 |

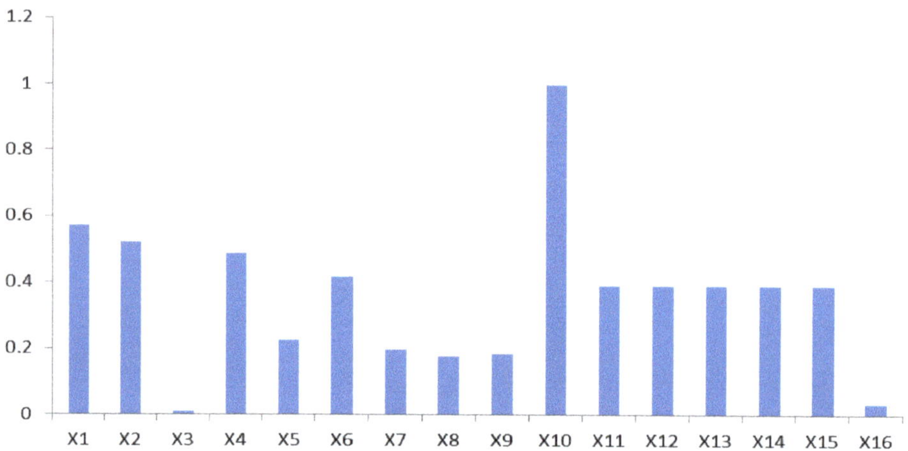

**Figure 8.** $Q$ value of each factor's influence on habitat quality.

It can be concluded that: (1) Land use type determines the regional habitat quality, and its analysis can be used to identify the core driving effect of land use transition on changing habitat quality; (2) ecological factors, such as elevation, slope, geomorphology type, and precipitation constitute the background conditions of biological habitat, which have an important impact on the quality of the habitat; (3) GDP per capita, the proportion of agricultural output value, grain yield per unit area, and population density, and urbanization rate are economic and social factors, respectively, which affect the spatial differentiation of habitat quality to a certain extent; (4) moreover, among the ecological factors, soil type, annual mean temperature, vegetation, type, and NDVI index (Normalized Difference Vegetation Index) have a weak effect on habitat quality, whereas road network density and slope aspect have no significant effect on the habitat quality.

### 3.3.2. Contribution of Land Use Transition to Habitat Quality Effect

As mentioned in Section 3.3.1, land use type is the core factor that determines the quality of regional habitat. In the process of regional economic and social development, human economic production activities and social management behavior jointly determine the direction and characteristics of land use transition. The economic and social activities in the study area, including agricultural planting, industrial development, and human living, contributed to the expansion of construction land with low habitat suitability in the southern plains and the surrounding areas of low altitude cities and towns, and constantly occupied cultivated land, woodland, and grassland with high habitat suitability. From 2000 to 2020, the total area of land use conversion with decreased habitat suitability (811.69 km$^2$) was larger than the area of land use conversion with increased habitat suitability (710.99 km$^2$) (Table 9), which led to a continuous decline of habitat quality in the study area; however, a series of measures have been implemented to curb the continuous degradation of habitat quality. In the past 20 years, to ensure the water supply capacity of the MRP to cities in northern China, the government increased the water area of the study area by increasing dams, and the increased water area mainly came from cultivated land. In the design of this study, the habitat suitability of water area (0.6) was greater than that of cultivated land (0.4); therefore, the habitat quality of Xichuan County, as the core distribution area of the reservoir, showed a gradually increasing trend, which differed from that of other counties.

**Table 9.** Changes in habitat suitability of different land types in the study area from 2000 to 2020.

| Conversion of Land Types with Declining Habitat Suitability | Area/km$^2$ | % | Conversion of Land Types with Improving Habitat Suitability | Area/km$^2$ | % |
|---|---|---|---|---|---|
| Cultivated land—Construction land | 376.26 | 46.36 | Cultivated land—Water area | 213.35 | 30.01 |
| Forest land—Cultivated land | 171.71 | 21.15 | Grassland—Forest land | 173.44 | 24.39 |
| Forest land—Grassland | 72.50 | 8.93 | Cultivated land—Forest land | 140.25 | 19.73 |
| Grassland—Cultivated land | 64.77 | 7.98 | Construction land—Cultivated land | 111.71 | 15.71 |
| Grassland—Water area | 37.74 | 4.65 | Cultivated land—Grassland | 47.48 | 6.68 |
| Forest land—Construction land | 21.75 | 2.68 | Cultivated land—Wetland | 12.56 | 1.77 |
| Grassland—Construction land | 20.28 | 2.50 | Water area—Forest land | 2.79 | 0.39 |
| Wetland—Water area | 17.53 | 2.16 | Construction land—Grassland | 2.58 | 0.36 |
| Forest land—Water area | 15.24 | 1.88 | Construction land—Water area | 1.92 | 0.27 |
| Water area—Cultivated land | 4.56 | 0.56 | Water area—Grassland | 1.61 | 0.23 |
| Grassland—Other land | 2.74 | 0.34 | Water area—Wetland | 1.36 | 0.19 |
| Forest land—Other land | 1.71 | 0.21 | Construction land—Forest land | 0.66 | 0.09 |
| Cultivated land—Other land | 1.70 | 0.21 | Construction land—Other land | 0.59 | 0.08 |
| Wetland—Cultivated land | 1.11 | 0.14 | Wetland—Grassland | 0.54 | 0.08 |
| Water area—Construction land | 1.09 | 0.13 | Wetland—Forest land | 0.10 | 0.01 |
| Grassland—Wetland | 0.80 | 0.10 | Construction land—Wetland | 0.05 | 0.01 |
| Forest land—Wetland | 0.19 | 0.02 | | | |
| Total | 811.68 | 100 | Total | 710.99 | 100 |

## 4. Discussion

### 4.1. Mechanism of Land Use Transitions Affecting Habitat Quality Change

The essence of land use transition is changing land use form in the process of economic and social development. The fundamental reason for land use transition is because of the change of land use type caused by human economic and social activities on the natural ecosystem. Therefore, it is necessary to analyze the mechanism of land use transition affecting habitat quality change from an economic–social–ecological complex system perspective (Figure 9).

In terms of the natural ecosystem, ecological background factors largely determine the quality of living conditions. For example, lush forests can provide animals with good hiding conditions, and abundant precipitation and suitable temperature can provide them with sufficient food. High altitude and steep mountains are difficult for human activities to reach; therefore, they are less disturbed and suitable for plant and animal habitats and reproduction areas. It can be inferred from the above mentioned detection results that each natural ecological element does not have the same effect on the habitat quality of the study area. First, land use type is the core determinant of habitat quality, indicating the research from land use transition to habitat quality change. Second, the influence of elevation and slope is strong, reflecting the important influence of the degree of human interference on the quality of habitat. Third, the geomorphology type and annual precipitation have an important impact on the spatial differentiation of habitat quality, indicating that these factors largely affect the quality conditions of biological habitats. Fourth, soil type, annual average temperature, vegetation type, and NDVI index have a certain effect on the habitat quality, indicating that they are also important influences on the habitat conditions of organisms. Fifth, the effect of the slope aspect is weak, indicating that the spatial distribution and changes of habitat quality are almost not affected by aspect conditions.

In terms of economic systems, humans engage in production and business activities, and not only do they obtain many resources needed for survival from the natural ecosystem and damage the stability of the original ecosystem, but they also change the types of land cover, which have important impacts on the natural environment. The impact of economic activities on the natural ecosystem is spatially manifested as changes in land use patterns, including spatial changes in the structure and distribution and temporal changes in the orientation and degree. In the process of land use transition, due to different habitat suitability, a change of land use type directly leads to a change of habitat suitability—which finally leads to a change of regional habitat quality. Land use types with lower habitat suitability not only affect their own habitat quality, but also negatively affect the habitat quality of the surrounding land use types. For example, the habitat quality of forest land adjacent to construction land is different from that adjacent to grassland, due to different potential threats, although they have the same habitat suitability under the two conditions.

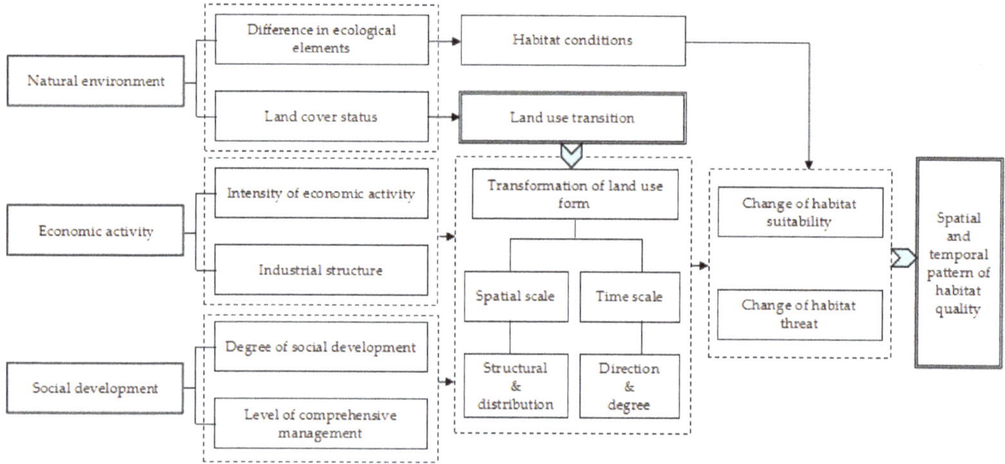


**Figure 9.** Mechanism of land use transitions affecting habitat quality change.

Regarding the social system, on the one hand, the higher the degree of social development, the more construction land and cultivated land with lower habitat suitability are needed to meet people's demands. Population growth and demand for agricultural products are important factors driving land system changes [51]. On the other hand, the higher the level of social development, the stronger the binding force of people on their own activities and behaviors. As human cognition improves, the concept of sustainable development of the harmonious coexistence between humans and nature will dominate social development, continuously using scientific management methods to minimize the impact of human activities on the natural environment. The degree of disturbance and destruction of habitat quality follows the Environmental Kuznets Curve. When the level of economic and social development is low, the habitat quality shows a trend of deepening with economic development. Then, with the improvement of people's cognitive ability and management level, the degree of disturbance and destruction will gradually decrease, and the habitat quality will gradually improve.

Therefore, to effectively improve habitat quality, we should study the endogenous factors and mechanisms of habitat quality change from an economic–social–ecological complex system perspective. From the above mentioned analysis, the core concept of controlling habitat quality change is to control the change of land cover type, which requires studying the dynamic mechanism of promoting land use change. According to the theory of human–earth system science, the interaction between humans activities and the earth's environment is the main driving force of the evolution of modern earth's surface system. In the coupling process of the human–earth system, the social and economic systems are the main bodies of human activities and the main causes of driving environmental changes [52]. Therefore, adjusting and optimizing the allocation of economic and social system elements and adopting reasonable control and management measures will help establish a coordinated and sustainable relationship between humans and land and realize the coordinated development of natural ecosystems and economic and social systems.

*4.2. Suggestions on Improving Regional Habitat Quality*

We recommend that the following points should be taken into consideration by policy-makers:

(1) The territorial and spatial planning must be strengthened, and water source areas must be regulated. On the basis of reasonable delineation of the "three areas and three lines," the local government should strictly regulate territorial and spatial use to prevent the extensive use and disorderly expansion of construction land caused by urban expansion; additionally, the government should continue to promote the return of farmland to forest and grassland and reasonably increase the quantity and quality of ecological space, improving the functions of water source area ecosystems, such as carbon sequestration, water conservation, and biodiversity conservation.

(2) Research, monitoring, and evaluation on environmental quality should be continued. An all-round, full-time, and long-period comprehensive monitoring system for the environment of the water source area should be established; research, monitoring, and evaluation should be conducted on water quality, water quantity, climate, vegetation, biodiversity, and other factors; changes of adverse factors affecting habitat quality should be reduced; and positive countermeasures for sudden ecological security incidents should be implemented, including the timely elimination or reduction of the impact.

(3) Environmental protection and restoration should be actively promoted. According to the theory of "landscape, forest, field, lake, and grass" community life, combined with the ecological space planning and control policy of water source areas, the core ecological protection area should be designated in the middle and north areas with high habitat quality, and the occupation and interference of human activities on the ecological space should be reduced in the southeast plain area with significantly declined habitat quality and strong habitat degradation. Based on the degree of ecosystem damage, different artificial support methods, such as conservation, natural restoration, assisted regeneration, and ecological reconstruction, should be adopted to conduct ecological restoration activities [53] in water source areas.

(4) A scientific and effective compensation mechanism for ecological protection should be established. The industrial development of water source areas is limited by the objective of environment protection, which leads to serious losses in local finance and people's income. The principles of clear authority and responsibility, overall coordination, and overall planning should be followed based on scientific research on quantitative accounting of ecological compensation for the MRP, and the authority and responsibility of government departments at all levels of the water source and receiving area should be clarified. The relevant industrial policies and laws, and regulations should be improved to form a long-term ecological compensation operation mechanism.

(5) Feasible paths to achieve green and sustainable development should be explored. Using the theory of "green water and mountains are also golden and silver mountains" as a guide, the government should explore the ecological resource asset accounting of water source areas and realize the ecological product value; actively cultivate and develop ecotourism, green agriculture, a special agricultural products processing industry, and other green industrial systems which rely on the local rich mountain landscape and biological resources; and form an endogenous mechanism for achieving high quality development of the ecology, economy, and society in the water source area.

## 5. Conclusions

Research on the habitat quality effect of land use transition can effectively reveal changes in the ecosystem under the influence of human activities, facilitate identification of the change characteristics and change trend, and control the change direction. Exploring the driving mechanism of habitat quality change can provide reasonable decision-making and action basis for the effective protection of biological habitats and construction of an ecological security pattern of harmonious coexistence between humans and nature. Using the HWS area of the MRP as an example, this study investigated the habitat quality effect and driving mechanisms of land use transition. Our work can serve as a guide for local governments aiming to effectively control regional land use transition, improve the environment, enhance water conservation capacity, and other ecosystem functions. The research on the driving mechanism of land use transition to habitat quality change from an ecological–economic–social complex system perspective proposed in this study can also further enrich the theory of land use transition and human–land system science and provide a reference for research on human–land system coupling and sustainable development. In future studies, the construction of an ecological security pattern of water source areas under the background of land use transition should be focused on, including identifying important ecological protection sources, strengthening green infrastructure construction, and improving the quality of ecological space, to provide a decision-making reference for the ecological protection and restoration of water source areas. A long-term mechanism to realize the value of ecological products in water source areas should be established under the guidance of the theory of human–earth system science to produce a low carbon, green, and sustainable development model in line with the actual situation of the region.

**Author Contributions:** Conceptualization, M.C. and Z.B.; methodology, Q.W.; software, M.C.; validation, Q.W. and Z.S.; formal analysis, M.C.; data curation, Z.B. and Z.S.; writing—original draft preparation, M.C.; writing—review and editing, Q.W.; funding acquisition, Z.B. All authors have read and agreed to the published version of the manuscript.

**Funding:** This study was funded by the National Key Research and Development Program of China (2018YFB2100703).

**Institutional Review Board Statement:** Not applicable.

**Informed Consent Statement:** Not applicable.

**Data Availability Statement:** The data presented in this study are available on request from the corresponding author.

**Conflicts of Interest:** The authors declare no conflict of interest.

## References


1. Lu, X.; Shi, Y.; Chen, C.; Yu, M. Monitoring cropland transition and its impact on ecosystem services value in developed regions of China: A case study of Jiangsu Province. *Land Use Policy* **2017**, *69*, 25–40. [CrossRef]
2. Xu, Y.; Mcnamara, P.; Wu, Y.; Dong, Y. An econometric analysis of changes in arable land utilization using multinomial logit model in Pinggu district, Beijing, China. *J. Environ. Manag.* **2013**, *128*, 324–334. [CrossRef] [PubMed]
3. Qu, Y.; Long, H. The economic and environmental effects of land use transitions under rapid urbanization and the implications for land use management. *Habitat Int.* **2018**, *82*, 113–121. [CrossRef]
4. Bongaarts, J. IPBES, 2019. Summary for policymakers of the global assessment report on biodiversity and ecosystem services of the Intergovernmental Science-Policy Platform on Biodiversity and Ecosystem Services. *Popul. Dev. Rev.* **2019**, *45*, 680–681. [CrossRef]
5. Oliver, T.H.; Heard, M.S.; Isaac, N.; Roy, D.B.; Procter, D.; Eigenbrod, F.; Freckleton, R.; Hector, A.; Orme, C.; Petchey, O.L. Biodiversity and resilience of ecosystem functions. *Trends Ecol. Evol.* **2015**, *30*, 673–684. [CrossRef]
6. Houghton, R.A.; Nassikas, A.A. Global and regional fluxes of carbon from land use and land cover change 1850–2015. *Glob. Biogeochem. Cycles* **2017**, *31*, 456–472. [CrossRef]
7. Perring, M.P.; De Frenne, P.; Baeten, L.; Maes, S.L.; Depauw, L.; Blondeel, H.; Caron, M.M.; Verheyen, K. Global environmental change effects on ecosystems: The importance of land-use legacies. *Glob. Chang. Biol.* **2016**, *22*, 1361–1371. [CrossRef] [PubMed]
8. Lambin, E.F.; Meyfroidt, P. Land use transitions: Socio-ecological feedback versus socio-economic change. *Land Use Policy* **2010**, *27*, 108–118. [CrossRef]
9. Long, H.; Li, T. The coupling characteristics and mechanism of farmland and rural housing land transition in China. *J. Geogr. Sci.* **2012**, *22*, 548–562. [CrossRef]
10. Tsai, Y.; Zia, A.; Koliba, C.; Bucini, G.; Guilbert, J.; Beckage, B. An interactive land use transition agent-based model (ILUTABM): Endogenizing human-environment interactions in the Western Missisquoi Watershed. *Land Use Policy* **2015**, *49*, 161–176. [CrossRef]

11. Long, H.; Qu, Y. Land use transitions and land management: A mutual feedback perspective. *Land Use Policy* **2018**, *74*, 111–120. [CrossRef]

12. Long, H.; Qu, Y.; Tu, S.; Zhang, Y.; Jiang, Y. Development of land use transitions research in China. *J. Geogr. Sci.* **2020**, *30*, 1195–1214. [CrossRef]

13. Long, H. *Land Use Transitions and Rural Restructuring in China*; Springer: Singapore, 2020.

14. Yang, Y.; Bao, W.; Li, Y.; Wang, Y.; Chen, Z. Land use transition and its eco-environmental effects in the Beijing–Tianjin–Hebei urban agglomeration: A production–living–ecological perspective. *Land* **2020**, *9*, 285. [CrossRef]

15. Grainger, A. *The Future Role of the Tropical Rain Forests in the World Forest Economy*; University of Oxford: Oxford, UK, 1986.

16. Grainger, A. The forest transition: An alternative approach. *Area* **1995**, *27*, 242–251.

17. Liu, Y.; Long, H. Land use transitions and their dynamic mechanism: The case of the Huang-Huai-Hai Plain. *J. Geogr. Sci.* **2016**, *26*, 515–530. [CrossRef]

18. Chen, R.; Ye, C.; Cai, Y.; Xing, X.; Chen, Q. The impact of rural out-migration on land use transition in China: Past, present andtrend. *Land Use Policy* **2014**, *40*, 101–110. [CrossRef]

19. Xu, J.; Sharma, R.; Fang, J.; Xu, Y. Critical linkages between land-use transition and human health in the Himalayan region. *Environ. Int.* **2008**, *34*, 239–247. [CrossRef] [PubMed]

20. Bae, J.; Joo, R.; Kim, Y. Forest transition in South Korea: Reality, path and drivers. *Land Use Policy* **2012**, *29*, 198–207. [CrossRef]

21. Yeo, I.; Huang, C. Revisiting the forest transition theory with historical records and geospatial data: A case study from Mississippi(USA). *Land Use Policy* **2013**, *32*, 1–13. [CrossRef]

22. Qu, Y.; Jiang, G.; Li, Z.; Tian, Y.; Wei, S. Understanding rural land use transition and regional consolidation implications in China. *Land Use Policy* **2019**, *82*, 742–753. [CrossRef]

23. Huang, H.; Zhou, Y.; Qian, M.; Zeng, Z. Land Use Transition and Driving Forces in Chinese Loess Plateau: A Case Study from Pu County, Shanxi Province. *Land* **2021**, *10*, 67. [CrossRef]

24. Wen, Y.; Zhang, Z.; Liang, D.; Xu, Z. Rural residential land transition in the Beijing-Tianjin-Hebei region: Spatial-temporal patterns and policy implications. *Land Use Policy* **2020**, *96*, 104700. [CrossRef]

25. Long, H.; Heilig, G.K.; Li, X.; Zhang, M. Socio-economic development and land-use change: Analysis of rural housing land transition in the Transect of the Yangtse River, China. *Land Use Policy* **2007**, *24*, 141–153. [CrossRef]

26. Tong, W.; Kazak, J.; Qi, H.; Vries, B.D. A framework for path-dependent industrial land transition analysis using vector data. *Eur. Plan. Stud.* **2019**, *27*, 1391–1412.

27. Zhu, F.; Zhang, F.; Ke, X. Rural industrial restructuring in China's metropolitan suburbs: Evidence from the land use transition of rural enterprises in suburban Beijing. *Land Use Policy* **2018**, *74*, 121–129. [CrossRef]

28. Lyu, L.; Gao, Z.; Long, H.; Wang, X.; Fan, Y. Farmland use transition in a typical farming area: The case of Sihong County in the Huang-Huai-Hai Plain of China. *Land* **2021**, *10*, 347. [CrossRef]

29. Lou, Y.; Yin, G.; Xin, Y.; Xie, S.; Li, G.; Liu, S.; Wang, X. Recessive transition mechanism of arable land use based on the perspective of coupling coordination of input–output: A case study of 31 provinces in China. *Land* **2021**, *10*, 41. [CrossRef]

30. Zhang, Y.; Long, H.; Ma, L.; Ge, D.; Tu, S.; Qu, Y. Farmland function evolution in the Huang-Huai-Hai Plain: Processes, patterns and mechanisms. *J. Geogr. Sci.* **2018**, *28*, 759–777. [CrossRef]

31. Ge, D.; Long, H.; Zhang, Y.; Ma, L.; Li, T. Farmland transition and its influences on grain production in China. *Land Use Policy* **2018**, *70*, 94–105. [CrossRef]

32. Zhang, L.; Lu, D.; Li, Q.; Lu, S. Impacts of socioeconomic factors on cropland transition and its adaptation in Beijing, China. *Environ. Earth Sci.* **2018**, *77*, 575. [CrossRef]

33. Yin, D.; Li, X.; Li, G.; Zhang, J.; Yu, H. Spatio-temporal evolution of land use transition and its eco-environmental effects: A case study of the Yellow River Basin, China. *Land* **2020**, *9*, 514. [CrossRef]

34. Qiu, L.; Pan, Y.; Zhu, J.; Amable, G.S.; Xu, B. Integrated analysis of urbanization-triggered land use change trajectory and implications for ecological land management: A case study in Fuyang, China. *Sci. Total Environ.* **2018**, *660*, 209–217. [CrossRef]

35. Long, H.; Liu, Y.; Hou, X.; Li, T.; Li, Y. Effects of land use transitions due to rapid urbanization on ecosystem services: Implications for urban planning in the new developing area of China. *Habitat Int.* **2014**, *44*, 536–544. [CrossRef]

36. Liu, Y.; Long, H.; Li, T.; Tu, S. Land use transitions and their effects on water environment in Huang-Huai-Hai Plain, China. *Land Use Policy* **2015**, *47*, 293–301. [CrossRef]

37. Newbold, T.; Hudson, L.N.; Hill, S.; Contu, S.; Lysenko, I.; Senior, R.A.; Börger, L.; Bennett, D.J.; Choimes, A.; Collen, B.; et al. Global effects of land use on local terrestrial biodiversity. *Nature* **2015**, *520*, 45–50. [CrossRef] [PubMed]

38. Fellman, J.B.; Eran, H.; William, D.; Sanjay, P.; Meador, J.P. Stream physical characteristics impact habitat quality for pacific salmon in two temperate coastal watersheds. *PLoS ONE* **2015**, *10*, e0132652. [CrossRef]

39. Hillard, E.M.; Nielsen, C.K.; Groninger, J.W. Swamp rabbits as indicators of wildlife habitat quality in bottomland hardwood forest ecosystems. *Ecol. Indic.* **2017**, *79*, 47–53. [CrossRef]

40. Riedler, B.; Lang, S. A spatially explicit patch model of habitat quality, integrating spatio-structural indicators. *Ecol. Indic.* **2017**, *94*, 128–141. [CrossRef]

41. Zhang, J.; Hull, V.; Huang, J.; Yang, W.; Zhou, S.; Xu, W.; Huang, Y.; Ouyang, Z.; Zhang, H.; Liu, J. Natural recovery and restoration in giant panda habitat after the Wenchuan earthquake. *For. Ecol. Manag.* **2014**, *319*, 1–9. [CrossRef]

42. Lin, Y.; Lin, W.; Wang, Y.; Lien, W.; Huang, T.; Hsu, C.; Schmeller, D.S.; Crossman, N.D. Systematically designating conservation areas for protecting habitat quality and multiple ecosystem services. *Environ. Model. Softw.* **2017**, *90*, 126–146. [CrossRef]
43. Sharp, R.; Douglass, J.; Wolny, S.; Arkema, K.; Bernhardt, J.; Bierbower, W.; Chaumont, N.; Denu, D.; Fisher, D.; Glowinski, K.; et al. *InVEST 3.8.7. User's Guide*; Collaborative Publication by the Natural Capital Project, Stanford University, University of Minnesota, the Nature Conservancy, World Wildlife Fund; Stanford University: Stanford, CA, USA, 2020.
44. Zhang, X.; Zhou, J.; Li, M. Analysis on spatial and temporal changes of regional habitat quality based on the spatial pattern reconstruction of land use. *Acta Geogr. Sin.* **2020**, *75*, 160–178.
45. Sallustio, L.; Toni, A.D.; Strollo, A.; Febbraro, M.D.; Gissi, E.; Casella, L.; Geneletti, D.; Munafò, M.; Vizzarri, M.; Marchetti, M. Assessing habitat quality in relation to the spatial distribution of protected areas in Italy. *J. Environ. Manag.* **2017**, *201*, 129–137. [CrossRef] [PubMed]
46. Sun, X.; Jiang, Z.; Liu, F.; Zhang, D. Monitoring spatio-temporal dynamics of habitat quality in Nansihu Lake basin, eastern China, from 1980 to 2015. *Ecol. Indic.* **2019**, *102*, 716–723. [CrossRef]
47. Hack, J.; Molewijk, D.; Beiler, M.R. A conceptual approach to modeling the geospatial impact of typical urban threats on the habitat quality of river corridors. *Remote Sens.* **2020**, *12*, 1315. [CrossRef]
48. Wu, L.; Sun, C.; Fan, F. Estimating the characteristic spatiotemporal variation in habitat quality using the InVEST model—A Case Study from Guangdong–Hong Kong–Macao Greater Bay Area. *Remote Sens.* **2021**, *13*, 1008. [CrossRef]
49. Wang, J.; Li, X.; Christakos, G.; Liao, Y.; Zhang, T.; Gu, X.; Zheng, X. Geographical detectors-based health risk assessment and its application in the neural tube defects study of the Heshun region, China. *Int. J. Geogr. Inf. Sci.* **2010**, *24*, 107–127. [CrossRef]
50. Wang, J.; XU, C. Geodetector: Principle and prospective. *Acta Geogr. Sin.* **2017**, *72*, 116–134.
51. Edmonds, J.A.; Link, R.; Waldhoff, S.T.; Cui, R. A global food demand model for the assessment of complex human-earth systems. *Clim. Chang. Econ.* **2017**, *8*, 1750012. [CrossRef]
52. Liu, Y. Modern human-earth relationship and human-earth system science. *Sci. Geogr. Sin.* **2020**, *40*, 1221–1234.
53. Bai, Z.; Shi, X.; Zhou, W.; Wang, J.; Zhao, Z.; Cao, Y. How does artificiality support and guide the natural restoration of ecosystems. *China Land Sci.* **2020**, *34*, 1–9.

 *land*

*Article*

# Land Use Transitions under Rapid Urbanization in Chengdu-Chongqing Region: A Perspective of Coupling Water and Land Resources

Yuchen Pan [1], Li Ma [1,*], Hong Tang [2], Yiwen Wu [1] and Zhongjian Yang [2]

[1] School of Public Affairs, Chongqing University, Chongqing 400044, China; 20183258@cqu.edu.cn (Y.P.); 20182887@cqu.edu.cn (Y.W.)

[2] Sichuan Center for Rural Development Research, College of Management, Sichuan Agricultural University, Chengdu 611130, China; tanghong@sicau.edu.cn (H.T.); yangzj.19@scu.edu.cn (Z.Y.)

* Correspondence: mal.1991@cqu.edu.cn

**Abstract:** Land resources and water resources are the important material basis of economic and social development, and their pattern determines the pattern of development. Based on the panel data of the Chengdu–Chongqing Economic Circle from 2011 to 2018, this paper evaluates the matching degree of water and land resources, and their respective matching degrees with the economic development in the Chengdu–Chongqing Economic Circle with the Gini coefficient method. Based on the two-way fixed effect model, an extended Cobb–Douglas production function model is established to analyze the sensitivity of economic growth to land and water factors. In addition, the restriction degree of water and land resources to the economic development of the Chengdu–Chongqing Economic Circle is measured quantitatively by using the growth damping coefficient. The results show that the water and land resources and economic development of the Chengdu–Chongqing Economic Circle have a high matching degree, but the inner cities have a great difference. The contribution of water resources to economic growth is greater than that of land resources. Both of them have a little growth drag, which shows that industrial development has disposed of the dependence of water and land resources. The development of the Chengdu–Chongqing Economic Circle needs to play the role of technological progress in promoting economic growth, and at the same time optimize the use of water and land resources to reduce its constraints on the economic growth. Finally, the policy suggestions of matching water and land resources and economic growth in different regions are put forward.

**Keywords:** water and land resources management; sustainable development; economic impacts; land use transitions; Chengdu–Chongqing economic circle



**Citation:** Pan, Y.; Ma, L.; Tang, H.; Wu, Y.; Yang, Z. Land Use Transitions under Rapid Urbanization in Chengdu-Chongqing Region: A Perspective of Coupling Water and Land Resources. *Land* **2021**, *10*, 812. https://doi.org/10.3390/land10080812

Academic Editors: Zahra Kalantari and Dietrich Schmidt-Vogt

Received: 30 June 2021
Accepted: 27 July 2021
Published: 2 August 2021



## 1. Introduction

Natural resources are a key factor in the economic development of all countries, of which water and land resources are the most basic natural resources. Daly points out that land is a key factor in all aspects of production [1]. The area, quality and development degree of land resources determine the production efficiency, which is the basis of human survival and economic activities. The endowment of water resources and the rationality of their development and utilization influence the utilization of other resources to a great extent [2,3]. Therefore, the abundance of water and land resources not only determines the regional ecological environment quality and population carrying capacity, but also affects the speed of the regional economic growth. The Heckscher–Ohlin theory and the bulk product theory also believe that the use of resource advantages can effectively promote regional economic development. However, with the continuous growth of the population, the shortage of water resources has become an important factor restricting the sustainable development of the environment and society [4]. The conflict between limited water resources and the rising water demand highlights the need for a rational and effective water allocation [5]. Large amounts of water and land have long been exploited for fuel, rapid economic growth and an increasing demand has intensified water and soil loss, posing huge risks to the economy [6–8].

China's land resources are seriously wasted. According to the Ministry of Land and Resources of the People's Republic of China, about 32 million acres of arable land are used for new construction each year, but much of that is wasted due to a lack of proper planning and use [9]. At the same time, China also faces the contradiction between the serious shortage of water resources and the rapid economic development. China is short of 50 billion cubic meters of water each year, ranking as one of the 13 countries with the worst water shortage by the United Nations [10]. On the other hand, the exploitation of water resources also poses a severe challenge to the sustainable development of the economy and society [11]. If the necessary industrial structure adjustment and effective engineering and non-engineering measures are not taken, the potential of water resources development and utilization will become smaller and smaller, and may even result in negative growth [12]. To make matters worse, the spatial and temporal distribution of water resources in China is highly uneven [4]. For example, the Yangtze River basin and its southern region hold only 37% of the land but 81% of China's water resources [13]. Urbanization and industrialization have led to a reduction in arable land and an increase in the imbalance between water supply and demand, with an even greater reduction in arable land and labor [14]. The contradiction between water and land resources and economic development intensifies, which restricts economic development. It is urgent to evaluate and grasp the matching condition of water and land resources and optimize the allocation and utilization of limited water and land resources. However, how can this constraint be measured quantitatively? How can the matching degree between water and land resources and economy be revealed, and how can the interactive mechanism between water and land resources and economic development be explored? Therefore, it is of great strategic significance to formulate macro-level optimal control policies and measures of water and land resources for ensuring a sustainable utilization of resources and promoting regional coordinated development.

As an important growth pole of the western development, the Chengdu–Chongqing region has an important strategic position in the national and regional development. The natural resources in this area are well endowed, but the level of urban economic development is not balanced, and many cities are relatively backward. Owing to the characteristics of the Chengdu–Chongqing area, it is typical and representative to explore the matching degree between water and land resources and economic development. Based on the above understanding, the objectives of this paper are to:

(1) Reveal the matching degree of water and land resources and economic development in the Chengdu–Chongqing Economic Circle.

(2) Probe into the reason of regional difference of different economic growth in the Chengdu–Chongqing Economic Circle.

(3) Put forward a policy suggestion of optimizing the allocation of water and land resources and promoting the development of the regional economy.

The study not only enriches the theoretical research on the relationship between water and land resources and economic development, but also provides scientific decision support for the coordinated development of the regional economy and the utilization of water and land resources.

## 2. Literature Review

Throughout the relevant research, we can see that the role of water and land resources in economic development has long attracted scholars' attention. Current research on the relationship between water resources and regional economic development mainly focuses on water resources allocation and economic output [15–18], using water resources utilization indexes such as degree and efficiency to describe the relationship between the level of water resources utilization and the level of economic development. Davies and Simonovic proposed the integration of global water resources and social–economic–environmental systems [19]. Qi et al. proposed a comprehensive indicator of the social and economic consumption level of water resources, and determined the red line of regional water resources utilization through the description of the pressure of water resources utilization [12]. Generally speaking, if there is no external influence, the total amount of water resources in an area is basically stable, and the total water consumption in the area will increase with economic development [20]. Therefore, it is urgent to promote water resource utilization by coordinating relevant factors and optimizing the allocation of limited agricultural water resources [5].

The land is a key factor in production in all aspects, and is the basis of human survival and economic activities [1]. Among all types of land, arable land plays a vital role. Ji et al. adopted the Nested IOA method to deal with the allocation of direct and embodied arable land of an urban economy under the background of economic globalization [21]. In addition, construction land has also made an important contribution to promoting social and economic growth [22–24]. Many existing theoretical and practical studies have shown that there is a significant interaction between the economy and changes in construction land [25–27]. Some scholars have conducted a detailed analysis of the dynamic relationship between economy and construction land. For example, a study on the relationship between urban construction land expansion and economic growth in the Yangtze River Economic Belt shows that urban construction land expansion in the Yangtze River Economic Belt has a significant impact on economic growth [28]. However, the inefficient use of construction land is also common [29].

The matching of water and land resources with socio-economic factors is the goal of the rational allocation of water and land resources. Many scholars have conducted a series of studies on the matching degree of water and land resources with socio-economic factors. Saway et al. used remote sensing and other methods to analyze the potential of local land and water resources [30]. Based on the coupling degree model and the water and land resource matching model, some scholars have analyzed the coupling effect between economic development and water and land resources development and the characteristics of temporal and spatial differentiation [31–33]. With the development of the economy and the scarcity of natural resources, scholars have begun to pay attention to the constraints of natural resources on the economy and the extent to which the development of land and space is restricted by water and land resources. Nordhaus et al. incorporated natural resources, including land, into the Solow model, and established two neoclassical economic growth models with and without resource constraints [34]. Based on the difference in the growth rate of output per capita between the two under steady-state conditions, Nordhaus innovatively proposed the concept of "growth drag" and used this model to examine the impact of resources and land on the US economy. Bruvoll et al. pointed out that the constraints of the natural environment will lead to the social cost of environmental governance, and used the dynamic resource environment application model to calculate and predict the degree of welfare loss caused by the environmental tail effect in Norway [35]. Romer defined the difference between the economic growth rate without resource constraints and the constrained growth rate as "growth drag ", and proposed a specific method to measure "growth drag" using the Cobb–Douglas production function [36].

With the rapid development of China's urbanization process and the increasing scarcity of water and land resources, many Chinese scholars have gradually established a framework for analyzing China's problems based on Romer's model. Some Chinese scholars have calculated China's growth drag: the growth drag of land resources from 1978 to 2002 was approximately 1.75% per year [37], while the growth drag of water and land resources in China was 0.1397% and 1.3201%, respectively, from 1981 to 2001 [38]. The growth drag based on China's provincial and municipal data proves that China's economic development is affected by water and land resources [39,40] and there are large regional differences. Different regions face different challenges due to different resource endowments and geographical characteristics [41]. Some scholars build other models. Song et al. established an urban spatial econometric model from the perspective of land resources on economic growth, not only obtaining the contribution of land resources to economic growth, but also obtaining the impact on economic fluctuations [42]. Wang and Li explained the relationship between urban industrial water use and economic growth by constructing a new decoupling model, and found that the effect of economic scale drives the use of total industrial water use and economic growth to weaken decoupling [43].

The limitations and scarcity of water and land resources have caused resistance to economic development, which has reached a consensus in the academic community. Scholars have determined more fruitful research results in the theories, methods, and countermeasures terms. However, there are still some deficiencies. In previous studies, land resources, water resources and economic development are often split into two relatively independent relationships, and the three are not analyzed and integrated into a unified system. In addition, previous studies have mostly explored from a holistic and macro perspective, and paid little attention to the local differentiation characteristics of different types of regions, and lacked representativeness and typicality. Owing to this, this study uses the Gini coefficient to evaluate the matching degree of water resources–land resources–GDP in Chengdu–Chongqing region, and calculates the growth drag of the development of the Chengdu–Chongqing Economic Circle on the basis of Romer's improved classic Cobb–Douglas production function; that is, it quantitatively measures the constraint of water and land resources on economic growth and the level of the regional matching degree and reveals the restraint mechanism of water and land resources on economic development. The research results have a strong scientific support for the theory and practice of enriching the matching degree of water and land resources and economic development.

### 3. A Theoretical Analysis Framework of Water, Land and Economic Development

New growth theory proposes that in any country and region, the economic development process will inevitably consume resources. According to Cobb–Douglas, the factors of production are labor force and capital, so the influence of labor force and capital on economic development should be included in the theory when discussing the relationship between water and land resources and economic development (Figure 1). Romer extended the Cobb–Douglas production function to land and water resources, forming a neoclassical growth model in which the factors of production are capital inputs, labor, and water and land resources, and their elasticity can be derived separately. Some Chinese scholars have pointed out that because of the limited resources, the consumption of resources in the previous stage must lead to the investment of economic growth in the next stage. This phenomenon is known as the "Growth tail effect" [38]. Social and economic development cannot be separated from water and land resources. If the growth rate of water and land resources is lower than the growth rate of the labor force, then the per capita ownership of water and land resources will decrease, thus, reducing the growth rate of per capita output, that is to say, creating growth drag and restricting economic development [44–46].

The population is one of the key factors that determine economic development. With the development of society and the increase in population, the demand for resources increases, and resources become more and more scarce. When some scholars study the dynamic relationship between economy and construction land, they often consider the change of population. Population change and economic growth are the main drivers of construction land expansion [47–49]. The contribution of the population to economic development is mainly reflected in the labor force. With the acceleration of urbanization, the opportunity cost of farming for farmers has gradually increased, which has accelerated the transfer of agricultural labor to non-agricultural industries, and further triggered the transformation of farmers' livelihoods and land-use patterns [50–52]. On the one hand, the transfer of agricultural labor to cities is conducive to promoting the transformation of land-use patterns, realizing the large-scale and intensive use of land in advantageous areas, and improving the economic benefits of agricultural production [53,54]. On the other hand, because farmers' production activities are profit-oriented, high-intensity land use may have a negative impact on the ecological environment, leading to a series of problems such as a reduction in biodiversity, soil pollution, and the deterioration of water quality [55–57]. Therefore, the impact of demographic changes on the economic development has two sides and is affected by multiple complex factors.

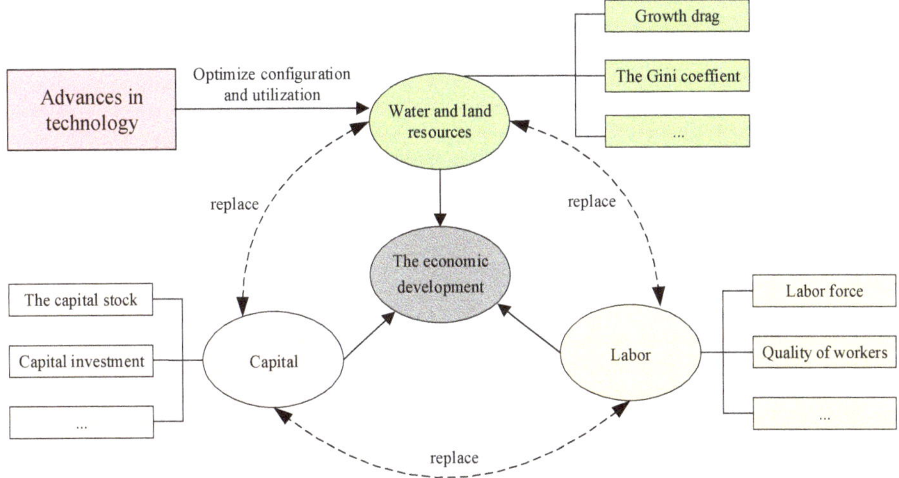


**Figure 1.** Analysis framework of economic development driving factors.

Red lines, such as land and environment, as well as the constraints of water and land resources on the economic development, have promoted technological innovation. Technological progress can optimize the use of existing water and land resources, and can develop more undeveloped water and land resources, such as the rational allocation of water resources, land development, rehabilitation, and intensive use of land [27,58–60]. Capital, labor force and water and land resources have substitutability in the production function. Through technological progress, effective labor and capital elements can replace the role of water and land resources to reduce absolute usage [61,62]. The government can change the extensive economic growth model, adjust the industrial structure, develop capital or technology-intensive industries in a targeted manner, and explore the potential for structural water-saving effects to relatively transfer and reduce the use of land and water resources [63–65]. An effective land resource protection policy, combined with the substitution effect of technological progress and other factors on land resources, is the key to maintaining a steady economic growth [66,67].

## 4. Study Site and Methodology

### 4.1. Study Site and Data Sources

The Chengdu–Chongqing Economic Circle is located in the hinterland of southwest China, including Chengdu, Zigong, Luzhou, Deyang, Mianyang (except Beichuan County and Pingwu County), Suining, Neijiang, Leshan, Nanchong, Meishan, Yibin, Guang'an, and Dazhou (except Wanyuan City), Ya'an (except Tianquan County and Baoxing County), Ziyang and another 15 cities, as well as Chongqing City (Figure 2). The topography is relatively complex, with four types of landforms, plateaus, mountains, hills and plains. It has prominent location advantages and an important strategic position, which is located at the junction of the "Belt and Road" and the Yangtze River Economic Belt. Aiming at the Cobb–Douglas production function, this study takes labor, capital, water resources, and land resources as basic input indicators, and economic output value as output indicators. Economic growth uses gross domestic product (GDP) to measure; labor (L) uses social employment data; capital (K) uses the perpetual inventory method to estimate. In order to eliminate the impact of inflation and maintain data consistency, the GDP and capital data are based on the national deflator, and comparable price adjustments were determined.

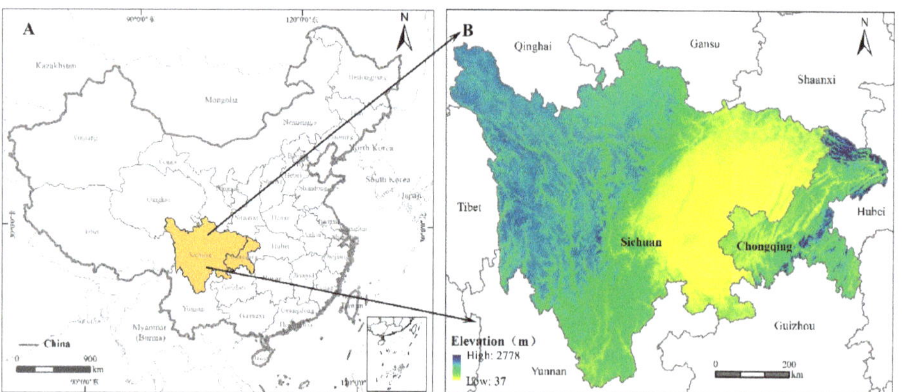


**Figure 2.** The location of study area. (**A**) depicts the location of Sichuan and Chongqing Provinces, and (**B**) presents the elevation information, which indicates that nearly half of the Sichuan Province is occupied by plain areas and the Chongqing is a typical hilly and mountainous region).

Based on the principles of science, reliability, and data availability, this study selects the sum of industrial water, agricultural water, and domestic water to measure the input of water resources. Water resources data (2011–2018) were obtained the Chongqing and Sichuan Water Resources Bulletins. Scholars differ greatly in the choice of land resource indicators. Some scholars use the sum of the three types of land areas, namely, the area of arable land, the area of forestry land, and the area of usable grassland as the land resource input. Considering the contribution of land resources to the output growth of the industrial sector, this study combined the characteristics of the land use structure in the Chengdu–Chongqing region and expressed the total amount of land sources by the sum of the construction land and the arable land area. The land data are from the survey data of land use changes in the relevant cities and prefectures over the years. Economic and other data (2011–2018) were obtained from China City Statistical Yearbook, Sichuan Provincial Statistical Yearbook, Chongqing Municipal Statistical Yearbook and other data.

### 4.2. Methods

#### 4.2.1. Estimation of Capital Stock

This study used the perpetual inventory method to estimate the fixed capital stock $K$ [68]. The basic formula is:

$$K_t = K_{t-1}(1-\delta) + I_t \tag{1}$$

Among them, $K_t$ is the stock of fixed capital in period $t$; $K_{t-1}$ is the stock of fixed capital in period $t-1$; $\delta$ is the discount rate; $I_t$ is the new fixed asset investment in period $t$.

Since no data on newly added fixed assets have been published in each city, Wang and Fan multiply the fixed asset investment in the whole society by the fixed asset investment delivery rate to calculate the newly added fixed assets before 1980 [69]. However, the length of the construction period of fixed asset investment projects is different, which makes the delivery and utilization rate of fixed asset investment vary. Therefore, when estimating new fixed assets, it was necessary to determine the period of urban investment and construction in China. Based on the actual situation in China, Ke and Xiang calculated that the weighted average construction period of fixed asset investment in the whole society is 3 years [70]. Additionally, then obtained:

$$K_t = K_{t-1}(1 - \delta) + \frac{I_t + I_{t-1} + I_{t-2}}{3} \tag{2}$$

$I_{t-1}$ is the newly added fixed capital investment during $t - 1$; $I_{t-2}$ is the newly added fixed capital investment during $t - 2$. Assuming the fixed capital depreciation rate is 0.05, the initial capital stock is estimated [71]:

$$K_0 = I_0(\frac{1 + g}{g + \delta}) \tag{3}$$

Among them, $I_0$ is the initial annual fixed capital investment; $g$ is the average growth rate of constant-price investment $I_0$.

### 4.2.2. Calculation of Matching Coefficient of Water and Land Resources

The water–land matching coefficient refers to the average amount of water resources per hectare of arable land in the region, which reflects the combination of water resources and arable land resources and the degree of water satisfaction to arable land. The greater the water–land matching coefficient, the richer the agricultural water resources that can be used in the region, the higher the satisfaction degree of the arable land in the region, and the more favorable the grain production of the arable land; on the contrary, the smaller the coefficient, the less water is available for agriculture. As the basic supporting condition of agricultural production, water resources are less matched in the region, which often becomes a restrictive condition for the healthy development of agriculture.

The calculation formula of water–land matching coefficient is:

$$R_i = \frac{W_i}{L_i}(i = 1, 2, \cdots n) \tag{4}$$

In the formula: $R_i$ is the water–land matching coefficient of the $i$-th prefecture and city; $W_i$ is the agricultural water use of the $i$-th prefecture and city, $L_i$ is the arable land area of the $i$-th prefecture and city; $n$ is the number of prefectures and cities.

The calculation formula of the regional scale water–land matching coefficient is:

$$R = \frac{\sum\limits_{n=1}^{n} W_i}{\sum\limits_{n=1}^{n} L_i} \tag{5}$$

In the formula: $R$ is the water–land matching coefficient of the Chengdu–Chongqing Economic Circle.

### 4.2.3. Calculation of Gini Coefficient of Water and Land Resources

In economics, the Gini coefficient can reflect the overall unfairness, and can decompose the total difference into the difference of different factors, so as to analyze the influence of different factors on the total difference. To this end, this study attempted to introduce the Gini coefficient as an indicator to assess the matching status between water and land resources and economic development, and to characterize the degree of inequality between the economy and water and land resources. Through the Gini coefficient, it was possible to explore whether the distribution of water and land resources in the region was compatible with the local economic growth, and whether the distribution was uneven in time and space.

The water–land Gini coefficient takes land resources as the basic matching object and water resources as the matching grading object. By calculating the water–land matching coefficient, the regions are sorted from low to high. In addition, the cumulative percentage of water and land resources in each region is calculated. The horizontal axis is the cumulative proportion of water resources, the vertical axis is the cumulative proportion of land resources, and the Lorentz curve is fitted. Water and land resources-economic Gini coefficient selects the total amount of water and land resources as the basic matching object, and uses GDP as the matching grading object to draw the Lorentz curve. Then, use definite integral to obtain the area A of the Figure sandwiched by the 0–1 Gini curve and the 45° line, and the area B of the Figure sandwiched by the 0–1 x-axis and the 45° line, the Gini coefficient can be obtained:

$$G = \frac{A}{A+B} \tag{6}$$

In the formula: $G$ is the Gini coefficient, $A$ is the area of the graph between the 0–1 Gini curve and the 45° line, and $B$ is the area of the graph between the x-axis and the 45° line.

Concerning related research results, this paper divided the matching degree of water and land resources in the Chengdu–Chongqing Economic Circle into 5 levels [72], as shown in Table 1.

**Table 1.** Classification of the matching degree between land and water resources and GDP.

| Gini Coefficient Interval | (0, 0.2) | (0.2, 0.3) | (0.3, 0.4) | (0.4, 0.5) | (0.5, 1) |
|---|---|---|---|---|---|
| The matching degree | Highly matching | Relative matching | General matching | Less matching | Extremely mismatching |

### 4.2.4. Growth Drag Model of Water and Land Resources

The "growth drag" model reflects the degree of restriction on economic development when a certain element is restricted. Romer incorporated natural resources based on the Solow model, and established a neoclassical growth model with and without asset constraints. The difference between the steady-state per capita output growth rate obtained by the two models was defined as the growth damping of natural resources, its expression is as follows:

$$Y(t) = K(t)^{\alpha} W(t)^{\beta} S(t)^{\gamma} [A(t)L(t)]^{(1-\alpha-\beta-\gamma)} \tag{7}$$

Among them, $Y(t)$ represents the value of economic output; $K(t)$ represents the capital stock; $W(t)$ represents the amount of water resources; $S(t)$ represents the amount of land resources; $A(t)$ represents the rate of technological progress; $L(t)$ represents the number of labors. $t$ represents time; $\alpha$, $\beta$, and $\gamma$ represent the elasticity of capital production, water resource production, and land resource production, respectively.

Based on the Solow model:

$$\begin{aligned} \overline{K}(t) &= sY(t) - \delta K(t) \\ \overline{L}(t) &= nL(t) \\ \overline{A}(t) &= gA(t) \end{aligned} \tag{8}$$

Among them, $s$ represents the savings rate, $\delta$ represents the capital depreciation rate, $n$ represents the labor force growth rate, and $g$ represents the technological progress rate.

Take the logarithm of (7) to obtain:

$$\ln Y(t) = \alpha \ln K(t) + \beta \ln W(t) + \gamma \ln S(t) + (1-\alpha-\beta-\gamma)[\ln A(t) + \ln L(t)] \tag{9}$$

Take the derivative of time on both sides of Formula (9). Since the derivative of the logarithm of a variable with respect to time is its growth rate, the growth rate function can be obtained, expressed by $g_Y(t)$, $g_K(t)$, $g_W(t)$, $g_S(t)$, $g_A(t)$ and $g_L(t)$, respectively, and the economic growth rate can be obtained:

$$g_Y(t) = \alpha g_K(t) + \beta g_W(t) + \gamma g_S(t) + (1-\alpha-\beta-\gamma)[g_A(t) + g_L(t)] \tag{10}$$

When the economy is on the path of balanced growth, to ensure that the growth rate remains unchanged, $\Delta K_t/k_{t-1} = Y_{t-1}/K_{t-1}$ and the growth rates of $Y(t)$ and $K(t)$ should be consistent. Simplify (10) to:

$$g_Y(t) = \frac{\beta g_W(t) + \gamma g_S(t) + (1 - \alpha - \beta - \gamma)(g + n)}{1 - \alpha} \tag{11}$$

Assuming that there are no natural resource constraints, the growth rates of water resources and land resources are both $n$. Therefore, the economic growth rate is:

$$g_Y(t) = \frac{(\beta + \gamma)n + (1 - \alpha - \beta - \gamma)(g + n)}{1 - \alpha} \tag{12}$$

In the presence of natural resource constraints, assuming that the growth rate of water and land resources is 0, then:

$$g_Y(t) = \frac{(1 - \alpha - \beta - \gamma)(g + n)}{1 - \alpha} \tag{13}$$

The growth drag of water and land resources growth can be obtained by subtracting the two formulas:

$$drag_W = \frac{\beta n}{1 - \alpha} \tag{14}$$

$$drag_S = \frac{\gamma n}{1 - \alpha} \tag{15}$$

For the calculation of labor force growth rate, this study uses the comprehensive method to calculate [73], namely:

$$rate = \sqrt[1+2+\ldots+n]{\frac{x_1 \cdot x_2 \ldots x_n}{x_0^n}} - 1 \tag{16}$$

In the formula: *rate* is the labor force growth rate, $x_0$ is the base period, and $x_n$ is the reporting period.

4.2.5. Panel Model Regression Test

This study selected the balanced short panel data of the Chengdu–Chongqing region from 2011 to 2018, and used the expanded Cobb–Douglas production function to establish a model and perform regression. First, the individual effect needed to be tested. Since there are generally autocorrelation disturbance items between the data of the same city (district) in different years, and the ordinary standard error is about half of the robust standard error of the cluster, the "least squares dummy variable model (LSDV)" was selected for the identification. At the same time, the mixed regression was tested, and the result was that the model had individual effects, and mixed regression should not be used. Furthermore, the joint significance test of individual effects and random effects was carried out. Since there was no strong theoretical reason to support which type of model was more suitable for estimation, this study selected the fixed effects model based on the results of Hausman's test. In consideration of the uncertainty brought by time to variables, a two-way fixed effect model was adopted for both fixed time and individuals.

## 5. Results

*5.1. Analysis of Matching Degree of Water and Land Resources*

From 2011 to 2018, the average water–land matching coefficient of the cities in the Chengdu–Chongqing Economic Circle was 2256.78 m$^3$/hm$^2$. From Table 2, the water–land matching coefficients of the six cities in Chengdu, Deyang, Mianyang, Leshan, Ya'an, and Meishan were all greater than the average of the Chengdu–Chongqing Economic Circle, the water–land matching coefficients of the remaining 10 cities were less than the average value. Compared with the amount of agricultural water resources, the spatial distribution of the water–land matching coefficients in various cities had obvious spatial differences (Figure 3). Chongqing, Dazhou, Yibin, Zigong, and Luzhou had the worst matching degree of water resources with arable land resources. This is due to the relative shortage of water resources and the relative abundance of arable land resources. From 2011 to 2018, Chengdu had the largest multi-year average water–land matching coefficient in the Chengdu–Chongqing Economic Circle (7625.93 m$^3$/hm$^2$), and Chongqing had the smallest (1053.35 m$^3$/hm$^2$), which showed a spatial difference compared with the provincial scale water and land matching coefficient of 2256.78 m$^3$/hm$^2$.

**Table 2.** The matching coefficient of water and land resources in Chengdu–Chongqing region.

| City | 2011 | 2015 | 2018 | Average |
|---|---|---|---|---|
| Chengdu | 7526.19 | 7712.14 | 5982.96 | 7625.93 |
| Zigong | 1407.41 | 1731.83 | 2043.39 | 1830.86 |
| Luzhou | 1516.74 | 1500.37 | 1589.72 | 1712.06 |
| Deyang | 8039.77 | 5143.17 | 5098.10 | 6136.44 |
| Mianyang | 3591.47 | 2841.30 | 2870.97 | 3457.47 |
| Suining | 1772.84 | 1934.89 | 2298.45 | 2168.10 |
| Neijiang | 1727.28 | 1700.93 | 1789.72 | 2191.84 |
| Leshan | 6111.82 | 2638.41 | 2463.10 | 4164.31 |
| Yibin | 1451.84 | 1560.41 | 1203.83 | 1769.47 |
| Nanchong | 2081.74 | 1579.75 | 1460.26 | 1960.55 |
| Dazhou | 1228.23 | 892.63 | 1364.74 | 1272.82 |
| Ya'an | 7470.44 | 4303.06 | 3239.12 | 6020.50 |
| Guang'an | 1969.50 | 2018.15 | 1176.55 | 2238.07 |
| Meishan | 6525.05 | 1334.70 | 3439.19 | 3143.44 |
| Ziyang | 2010.42 | 1707.24 | 1711.77 | 1906.99 |
| Chongqing | 1138.37 | 1064.41 | 1086.16 | 1053.35 |

Since the decline in 2014, the average water–land matching coefficient of the Chengdu–Chongqing Economic Circle has been maintained at 2000–2100, indicating that the matching degree of water and land resources has declined in recent years and the distribution of water and land is uneven (Figure 4).

*5.2. Analysis of Water and Land Resources—Economic Matching Degree*

This study sorts out and processes the data of water usage, land resource usage, GDP and other data in each region of the Chengdu–Chongqing Economic Circle from 2011 to 2018, and calculates the GDP corresponding to unit water resources and land resources in each region. The Lorentz curve is drawn according to the construction method of the Gini coefficient, and the current situation of matching between water and land resources and GDP was obtained.

From the Lorentz curve in Figure 5, it can be calculated that the average Gini coefficient of water and land resources in the Chengdu–Chongqing Economic Circle from 2011 to 2018 was 0.32, which is within a reasonable range, indicating that the overall spatial distribution of water and land resources is relatively balanced, but there is a big gap from the degree of high matching.

In Figure 6, the area formed by the water resources–GDP Lorentz curve and the 45° line in 2018 was 0.0944. According to the meaning of the Gini coefficient, the regional Gini coefficient of water resources and GDP in the Chengdu–Chongqing Economic Circle in 2018 was 0.1888. This value indicates that the utilization of water resources in the national economy of the region was highly matched. Similarly, the regional Gini coefficient of land resources and GDP was 0.3231, which indicates that the land resource utilization in the national economy of the region was relatively matched and even. In the two Lorentz curves, the land resources–GDP Lorentz curve is farther away from the absolute average line, which means that the balance of water resources was better than that of land resources, and the degree of matching of land resources to economic growth was lower.

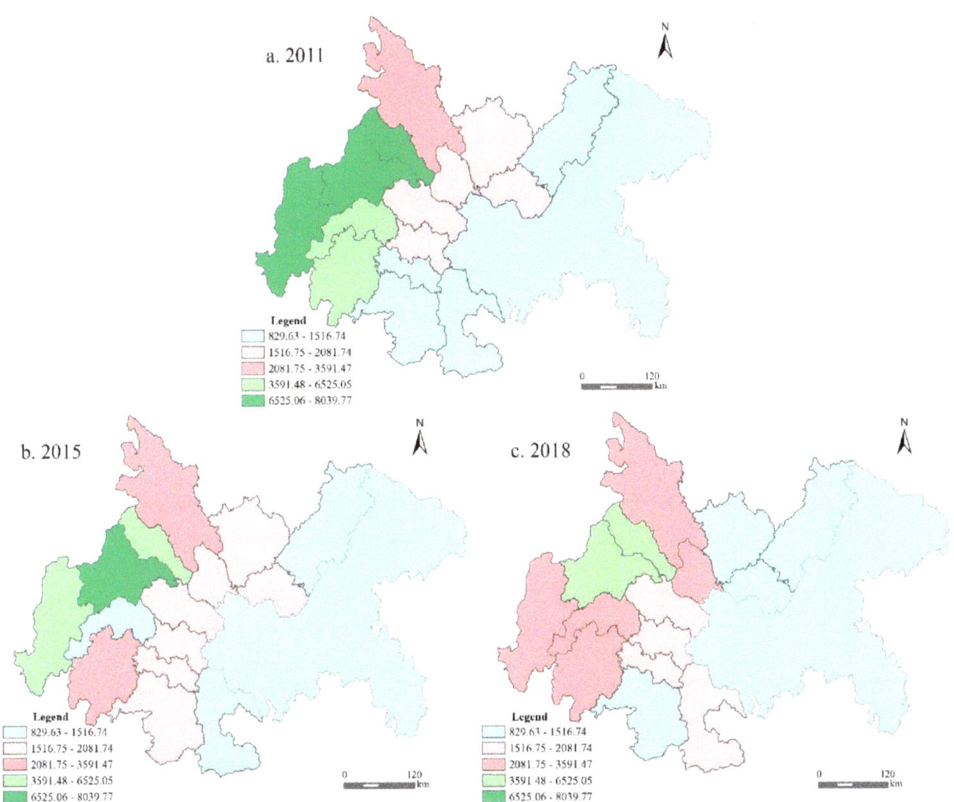



**Figure 3.** The spatio-temporal distribution characteristics of the matching coefficient of water and land resources in the Chengdu–Chongqing region from 2011 to 2018.

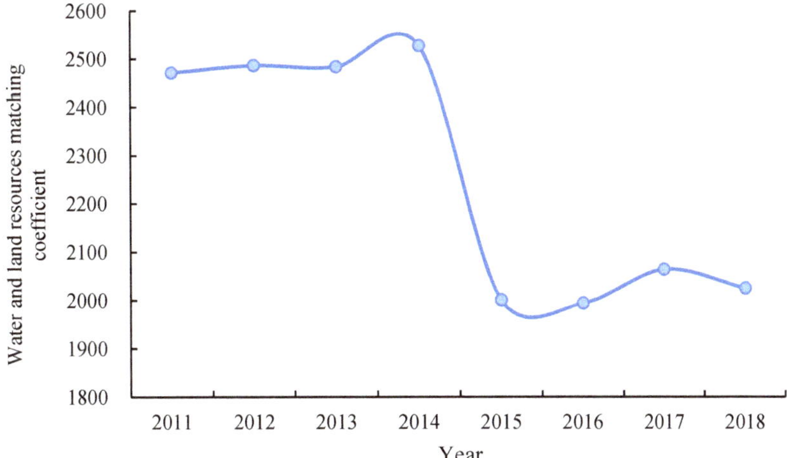



**Figure 4.** Change trend of water and land resources matching coefficient from 2011 to 2018.

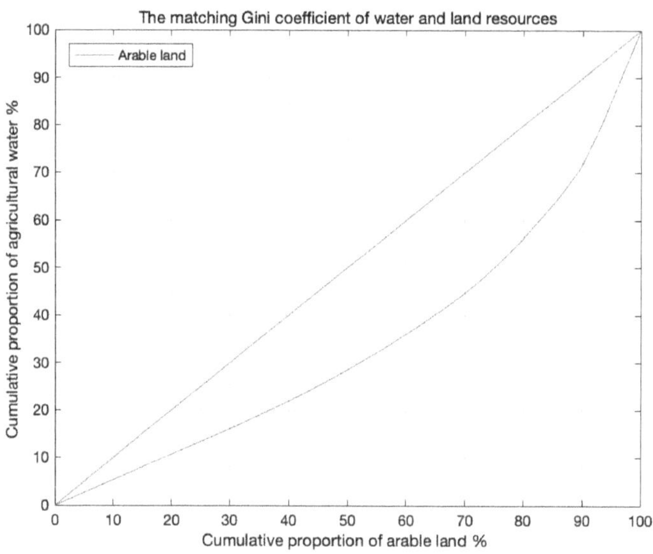



**Figure 5.** Lorentz Curve of Water and Land Resources Matching in Chengdu–Chongqing Economic Circle in 2018.

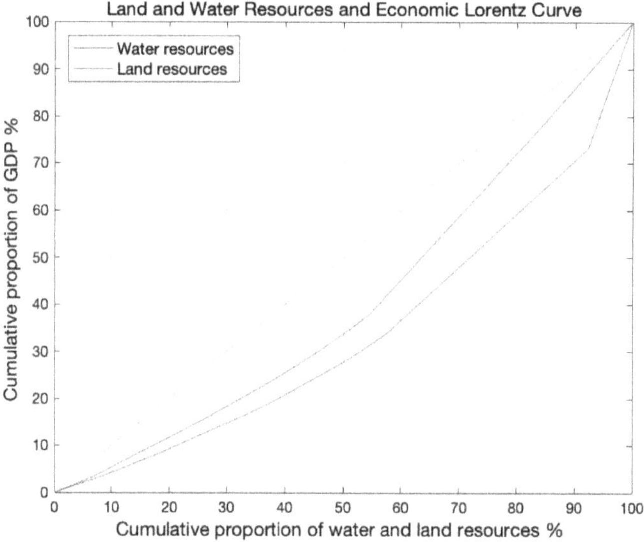



**Figure 6.** Water resources, land resources and GDP Lorenz curve in the Chengdu–Chongqing Economic Circle in 2018.

Further analysis of the Gini coefficient of water and land resources and the economic development from 2011 to 2018 (Figure 7) found that the Gini coefficient of water resources was relatively stable, rising slightly in 8 years and fluctuating between 0.1491 and 0.1888. It can be seen that water resources were highly matched with GDP in recent years, well adapted to the needs of the industrial development, and the distribution was relatively reasonable and even. The Gini coefficient of land resources was at a relatively high level and fluctuated greatly. In 2011–2014, land resources and GDP were in a relative matching degree, and the matching degree showed a downward trend after 2014. Its matching degree was lower than that of water resources, which may be because the complex and diverse terrain environment of the Chengdu–Chongqing Economic Circle had a restrictive effect on the degree of regional land resource utilization.

**Figure 7.** Gini Index of water resources, land resources and GDP in Chengdu–Chongqing Economic Circle from 2011 to 2018.

*5.3. Analysis of the Growth Drag of Water and Land Resources*

5.3.1. Panel Model Regression Results

According to Equations (7)–(9): according to the expanded Cobb–Douglas production function, the logarithm of GDP, labor force, capital, water resources and land resources of each city in Chengdu–Chongqing Economic Circle from 2011 to 2018 were taken to carry out a panel regression analysis.

The function fitting effect was better, labor and capital were significant at 1%, and water and land resources were significant at 5% (Table 3). It can be seen from Table 2 that the function fitting effect was better, labor and capital were significant at 1%, and water and land resources were significant at 5%. The results show that the four variables were closely related to output, and water and land resources can explain changes in GDP to a certain extent. Among them, the coefficients of labor and capital were relatively large, 0.891 and 0.307, respectively. It can be seen that the GDP was more sensitive in terms of capital stock and labor, which means that the development of the region still relies more on capital and labor input. In particular, the increase in labor was of great importance to economic development, which reflects that its economic development stage was still relatively low.

The biggest reason for labor flexibility was that most of the Chengdu–Chongqing Economic Circle was still dominated by labor-intensive industries, and there was a large demand for labor. Second, due to the continuous expansion of the industrial scale in the Chengdu–Chongqing region, the demand for labor continues to increase. Due to its high demand for labor, there has been an obvious phenomenon of population return in recent years. From 2012 to 2018, the permanent population of Sichuan Province increased by 2.99 million, returning to the level of 2000, and the population showed a trend of first decreasing and then increasing.

The elastic coefficients of water resources and land resources are relatively small, with 0.136 and 0.0894, respectively. It can be seen that the development and utilization of water resources contribute more to the economic development of the Chengdu–Chongqing Economic Circle. The Sichuan–Chongqing region relies on the Yangtze and Jialing Rivers, and water resources play an important role in the development of their industries. Areas with a relatively advanced level of economic development, such as the plains of Western

Sichuan, the hills of Central Sichuan, and Southern Sichuan, are relatively scarce of water resources and still have great potential. However, as far as land resources are concerned, due to the natural geographic characteristics of plain areas, developable land resources have been converted into current construction land earlier, and the land potential is not much reserved. Therefore, the further development and utilization of water resources has a more significant effect on the economic development, while the impact of land resources is relatively weak.

**Table 3.** Results of the extended Cobb–Douglas production function model.

|  | *fe* | *re* |
|---|---|---|
| ln$w$ | 0.136 * <br> (0.0577) | 0.125 * <br> (0.0566) |
| ln$s$ | 0.0894 * <br> (0.0382) | 0.115 ** <br> (0.0364) |
| ln$l$ | 0.891 *** <br> (0.157) | 0.730 *** <br> (0.0782) |
| ln$k$ | 0.307 *** <br> (0.0217) | 0.286 *** <br> (0.0210) |
| *cons* | −3.565 *** <br> (0.958) | −2.670 *** <br> (0.509) |
| $N$ | 128 | 128 |

Standard errors in parentheses. * $p < 0.05$, ** $p < 0.01$, *** $p < 0.001$.

### 5.3.2. Water and Land Resource Damping Coefficient

The rate of change of water and land resources was calculated by the comprehensive method, and combined with the panel regression results, the growth drag of water and land resources was measured, and the two drags were added to obtain the total damping of the area. The result is shown in Figure 8.

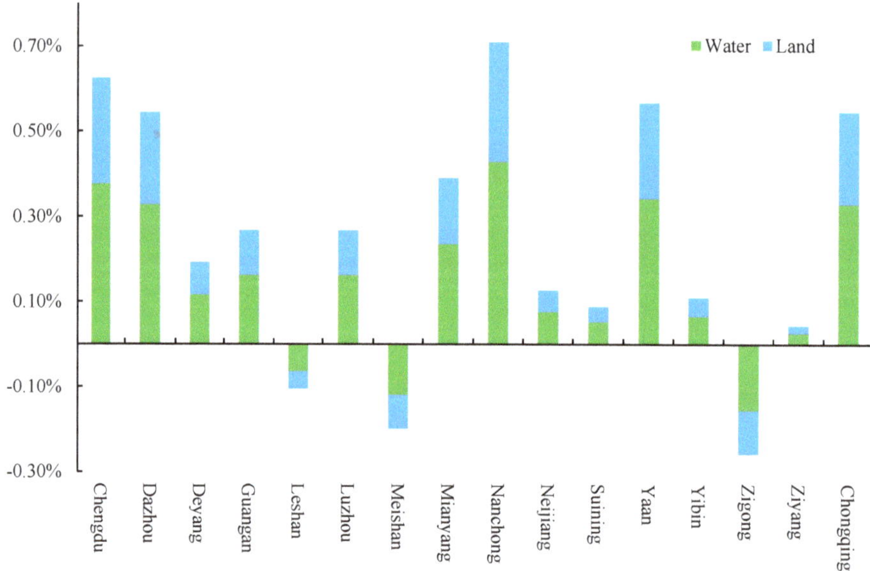


**Figure 8.** Water and land resources growth drag of cities in Chengdu-Chongqing Economic Circle.

In general, the damping coefficients of water and land resources in all regions of the Chengdu–Chongqing Economic Circle were relatively low, all below 0.7%. The first reason is that the elasticity of the capital output was relatively small. At present, the Chengdu–Chongqing Economic Circle has fewer capital-intensive industries, with labor-intensive industries dominating. Secondly, due to the generally low growth rate of labor, the Economic Circle is still in the early stage of development, and the industrial development is still immature. Most areas are mountainous and hilly, so the ability to attract labor is limited. In addition, the low elasticity of water and land resources is also the reason for low damping. At present, the development of various regions is relatively weakly dependent on natural resources, and the impact of land and water resources on the economy is relatively small. In contrast, the damping of water resources is greater than that of land resources. It can be seen that water resources are still an important factor restricting economic development.

There are large regional differences in various cities. From Figure 8, the damping of water and land resources in Leshan, Meishan and Zigong areas were all negative values. This means that the incremental supply of urban land was not only unable to promote urban economic growth, but was also being suppressed. The reason was that the labor force growth rate in these regions was negative, that is, the net population outflow. Therefore, even if the growth rate of natural resources was zero, it would not hinder economic growth. The main reasons for the outflow of the population in these areas are their small initial scale of economy and population, a low administrative level and urbanization level, and poor development foundation. The pains of industrial transformation have led to a weakening of the ability to absorb the local labor force, shrinking the original large-scale labor-intensive low-end industry jobs, especially in the third-tier construction of industrial cities represented by Leshan City, which will inevitably lead to population loss. The mountainous terrain conditions of Leshan lead to the frequent occurrence of geological disasters, and the shortage of water and land resources in Zigong has also led to the outflow of labor.

The labor transfer and relatively slow capital inflow have caused the marginal return of urban land development and utilization to be lower than its marginal cost, making the utilization and development of urban land have an "abnormal" negative effect on urban economic growth. The results show that the resources in these areas are in a surplus stage, and the outflow of population leads to a waste of resources.

Except for the above three regions, the damping coefficients of water and land resources in the other regions were all positive, indicating that natural resources restrict economic development to a certain extent. Among them, Nanchong, Chengdu and Ya'an had the largest total damping capacity. The reason is that the labor force growth rate in these areas is relatively high. Nanchong and Chengdu are rich in water and land resources, which have better development conditions and a greater ability to attract a population. Moreover, Nanchong and Chengdu have a relatively mature industrial development, a high degree of urbanization, and more labor inflows. The main reasons for the rapid growth of the labor force in Ya'an are: On the one hand, its population base and labor force are small. In the early stage of economic development, its output mainly relied on agriculture, and most of it was mountainous and hilly, leading to economic backwardness. Based on a low level, a smaller population increase can bring about a larger growth rate. On the other hand, in recent years, the government has actively promoted poverty alleviation policies and industrial reforms in Ya'an, so the population has grown to a certain extent.

## 6. Discussion and Conclusions

### 6.1. Discussion and Limitation

What factors can reduce the constraints of land and water resources on economic growth, and what can we do to reduce growth drag?

Growth drag is directly proportional to the elastic coefficients of water and land resources, that is, reducing the elastic coefficient can reduce the "tail effect" of economic growth. Its economic significance is to reduce the role of land in the economy. This also provides another explanation for economic growth that does not rely excessively on resources, but technological progress. A similar explanation can be made for the elasticity coefficient of the capital. As the elasticity coefficient of the capital decreases, so does the growth drag. Therefore, economic growth cannot rely on the increase in capital stock, and technological progress is the key. It is necessary to transfer economic growth to technological progress. Damping is directly proportional to the labor force growth rate. The smaller the labor growth rate, the slower the rate of reduction in per capita natural resources, the smaller the growth rate of the per capita average output on the path of balanced growth, and the smaller the resistance of natural resources to economic growth. The population continues to rise steadily. On the one hand, it provides a wealth of labor for economic growth; on the other hand, economic growth will absorb labor, and the labor force growth rate will increase accordingly. If you want to reduce the increase in the damping effect caused by the increase in the number of employees, you must maintain a moderate population size, while improving the quality of the population and increasing effective labor [74].

There is no doubt that we should pay attention to the restrictions and constraints of water and land resources on economic development, but we should not exaggerate the growth drag of water and land resources. It can be seen that the development of the Chengdu–Chongqing Economic Circle is not greatly constrained by water and land resources. The overall water and land resources are relatively matched with the economy, and there are large internal differences. Romer, when analyzing the complexity of resources and economic growth, believed that the share of land is the product of the real rental price of land and the land-output ratio [36]. Although the real rental price rarely decreases, the land–GDP ratio has been steadily declining; therefore, the share of land has also fallen, and the actual situation in China is also the same. Similarly, the share of water resources is also declining, and the decline in the share of land and water means a decline in "growth drag". The Chinese scholar Lu also determined a similar judgment [75]. He pointed out that the role of natural resources in China's economic growth is declining. The fact that the share of water and land resources has declined also means that the substitution elasticity between water and land resource inputs and other inputs is greater than 1. Therefore, the economy has shifted to those production methods that relatively seldom use water and land resources, so as to deal with the increasing scarcity of water and land resources.

From a short-term perspective, technological progress can give full play to the role of factor substitution and alleviate resource constraints; however, in the long run, resource constraints can be finally solved only when factor substitution and industrial structure adjustment work together to promote the development of non-resource-based industries [76]. Although it was concluded from the above analysis that factor substitution can alleviate resource constraints in the short term, land and water resources, as a basic input factor, are difficult to be effectively replaced by factors such as capital and labor when the development level of China's primary and secondary industries and the level of urbanization are relatively low. At the same time, due to the strict assumptions on the elasticity of factor substitution, it is difficult to meet the actual situation, so the substitution relationship between factors needs to be further explored and studied.

In addition, there are still some shortcomings in the study: (1) The regulation mechanism of water and land resources needs to be further deepened. The water consumption and land resource requirements of different industries vary greatly, and the industrial structure is more affected by the local resource endowments, so it is not possible to improve the utilization efficiency of water and land resources by completely changing the industrial structure. (2) Due to the availability of data and the choice of indicators, the analysis in this study failed to fully consider resource constraints, and only the area of arable land resources and construction land was considered in the land resources.

*6.2. Conclusions and Implications*

6.2.1. Conclusions

Based on the theory of economic growth, this study combined the panel regression test method with the Gini coefficient method, and used the panel fixed effect model to build the "growth drag" model of water and land resources in the Chengdu–Chongqing Economic Circle. The damping effect of water and land resources on the economic growth and the matching degree of water and land resources and the economy in the Chengdu–Chongqing Economic Circle were studied in depth. The following conclusions were drawn:

Water and land resources have a certain restraining effect on economic development, and their matching degree with economic development is relatively low. The restraining effect is often realized through the high contribution of capital in the industry, the high growth of the population and the dependence of the economy on water and land resources. The greater the dependence on capital stock, the faster the growth rate of the labor force, the more the industry depends on water and land resources, and the greater the growth damping of water and land resources. The growth damping of the Chengdu–Chongqing Economic Circle is relatively small. The damping of water resources is greater than the damping of land resources, and the matching degree between water resources and land resources, water and land resources and economic development is relatively high.

There are large differences within the region. The damping of water and land resources in the southwestern part of the Chengdu–Chongqing Economic Circle is negative in Leshan, Meishan, and Zigong, while the damping in Chengdu and Nanchong in the west is relatively large. The water and land resources in Chengdu–Chongqing Economic Circle are relatively rich, but for resource-scarce regions, the mechanism and approach of how to use a market mechanism to utilize internal and external resources, especially foreign resources, to alleviate resource constraint still need to be studied.

6.2.2. Policy Implications

The Chengdu–Chongqing Economic Circle is a central city cluster in Southwest China. It plays an important role in the country's strategy of improving regional cooperation mechanisms and creating coordinated development. The matching and balance of water resources and land resources is one of the key factors for achieving sustainable economic growth. Based on the above research, this study proposes the following policy measures to promote the optimal allocation of land and water resources, economic growth, and the sustainable use of resources.

From the perspective of water and land resources, expanding the total amount of water and land resources and increasing supply should be taken to solve the problem of the damping effect of water and land resources from the source, to realize sustainable economic development. Governments should increase investment in the construction of water conservancy facilities in the precipitation-rich areas of the Chengdu–Chongqing Economic Circle to store more water resources. For land resources, governments should scientifically carry out land development and consolidation, optimize the structure of land use, and take the road of intensification. Strengthening land consolidation and reclamation, prioritizing plans and arrangements for land that can be used after transformation, and effectively using plain land resources should also be taken. The proportion of unused land in Sichuan Province is relatively large. As the pressure on land resources becomes greater, corresponding measures must be taken for unused land or land that does not produce economic benefits to make it effective.

Change the way of resource utilization and give full play to the role of technology in economic growth. Due to the limitation of natural supply resources, as the population continues to grow and the scale of cities continues to expand, resources will still be an important factor restricting economic growth. At the same time, the dependence on economic growth must be transferred to technological progress to give full play to the mitigation effect of technological progress on resource constraints and to reduce the economy's dependence on resources. The government should increase investment in technology development, such as researching and developing water-saving varieties and adjusting the industrial structure, to play the role of factor substitution and improve the efficiency of factor allocation. The government can also set up incentive funds to encourage water conservation and the protection of water resources.

In terms of labor-related policies, it is necessary to rely on the resource advantages of universities to accelerate the optimization of the employment structure, and reduce the damping effect by promoting the rational and orderly transfer of the employed population from the primary industry to the secondary and tertiary industries. For areas with different damping, different policies can be adopted. The negative damping region: The damping values of water and land resources in Leshan, Meishan and Zigong are all negative. This shows that the resources in these areas are surplus, and the outflow of the population has led to the waste of resources. Therefore, for these regions, the introduction of talents and labor is the key to promoting economic development. Natural resources should be used to develop advantageous industries, and with the development of industries, talents will be attracted to further promote economic development and form a virtuous circle. The positive damping region: It reflects that natural resources restrict economic development to a certain extent. For areas such as Nanchong and Chengdu with higher damping and higher levels of economic development, an appropriate scale of labor should be maintained, while the quality of the population and effective labor should be improved. In addition, the distribution of labor should be consistent with the distribution of industries to further promote economic growth.

**Author Contributions:** Conceptualization, Y.P. and L.M.; methodology, Y.P.; software, H.T.; validation, Y.P., L.M. and H.T.; formal analysis, Y.W.; investigation, Z.Y.; resources, Y.P.; data curation, H.T.; writing—original draft preparation, L.M.; writing—review and editing, Y.P.; visualization, L.M.; supervision, Y.W.; project administration, H.T.; funding acquisition, L.M. and H.T. All authors have read and agreed to the published version of the manuscript.

**Funding:** Founded by China Postdoctoral Science Foundation (No. 2021M693709); Social Science Planning Research Doctoral Program of Chongqing Social Science Planning Office (Grant No. 2020BS43); Humanity and Social Science Youth Foundation of Ministry of Education Project (Grant No. 17YJC630136); The Fundamental Research Funds for the Central Universities (Grant No. 2021CDJSKJC02).

**Institutional Review Board Statement:** Not applicable.

**Informed Consent Statement:** Not applicable.

**Data Availability Statement:** Not applicable.

**Conflicts of Interest:** The authors declare no conflict of interest.

## References

1. Daly, H.E. When smart people make dumb mistakes. *Ecol. Econ.* **2000**, *1*, 1–2. [CrossRef]
2. Peng, L.; Deng, W.; Tan, J.; Lin, L. Matching relationship between land and water resources utilization and economic growth in Hengduan Mountain. *Acta Geogr. Sin.* **2020**, *75*, 1996–2008.
3. Wei, X.; Wang, J.; Wu, S.; Xin, X.; Wang, Z.; Lin, W. Comprehensive evaluation model for water environment carrying capacity based on VPOSRM framework: A case study in Wuhan, China. *Sustain. Cities Soc.* **2019**, *50*, 101640. [CrossRef]
4. Liu, G.; Hu, F.; Wang, Y.; Wang, H. Assessment of Lexicographic Minimax Allocations of Blue and Green Water Footprints in the Yangtze River Economic Belt Based on Land, Population, and Economy. *Int. J. Environ. Res. Public Health* **2019**, *16*, 643. [CrossRef] [PubMed]
5. Zhao, J.; Li, M.; Guo, P.; Zhang, C.; Tan, Q. Agricultural water productivity oriented water resources allocation based on the coordination of multiple factors. *Water* **2017**, *9*, 490. [CrossRef]
6. Liu, Y.; Wang, S.; Chen, B. Water–Land Scarcity Risk Nexus in the National Trade System Based on Multiregional Input–Output and Ecological Network Analyses. DEStech Transactions on Environment, Energy and Earth Sciences 2018. Available online: https://www.dpi-proceedings.com/index.php/dteees/article/view/27893 (accessed on 10 December 2020).
7. Bacca, E.J.M.; Knight, A.; Trifkovic, M. Optimal land use and distributed generation technology selection via geographic-based multicriteria decision analysis and mixed-integer programming. *Sustain. Cities Soc.* **2020**, *55*, 102055. [CrossRef]

8.    Moein, M.; Asgarian, A.; Sakieh, Y.; Soffianian, A. Scenario-based analysis of land-use competition in central Iran: Finding the trade-off between urban growth patterns and agricultural productivity. *Sustain. Cities Soc.* **2018**, *39*, 557–567. [CrossRef]

9.    *2005 China Land and Resources Bulletin*; National Land & Resources Information, 2006; pp. 26–30. Available online: http://www.gov.cn/jrzg/2006-04/28/content_268770.htm (accessed on 10 December 2020).

10.   UNESCO. *The United Nations World Water Development Report 2017*; UNESCO: Paris, France, 2017.

11.   Jia, Z.; Cai, Y.; Chen, Y.; Zeng, W. Regionalization of water environmental carrying capacity for supporting the sustainable water resources management and development in China. *Resour. Conserv. Recycl.* **2018**, *134*, 282–293. [CrossRef]

12.   Qi, Z.; Xiao, C.; Zhang, B.; Liang, X. Generalized index of regional socio-economic consumption level of water resource. *J. Water Supply Res. Technol.* **2020**, *69*, 113–121. [CrossRef]

13.   Song, X.; Shi, P.; Jin, R. Analysis on the contradiction between supply and demand of water resources in China owing to uneven regional distribution. *Arid Zone Res.* **2005**, *22*, 162–166.

14.   Zhao, M.; Chen, Z.; Zhang, H.; Xue, J. Impact Assessment of Growth Drag and Its Contribution Factors: Evidence from China's Agricultural Economy. *Sustainability* **2018**, *10*, 3262. [CrossRef]

15.   Cai, X. Implementation of holistic water resources-economic optimization models for river basin management–reflective experiences. *Environ. Model. Softw.* **2008**, *23*, 2–18. [CrossRef]

16.   Birol, E.; Koundouri, P.; Kountouris, Y. Assessing the economic viability of alternative water resources in water-scarce regions: Combining economic valuation, cost-benefit analysis and discounting. *Ecol. Econ.* **2010**, *69*, 839–847. [CrossRef]

17.   Xuan, W.; Quan, C.; Shuyi, L. An optimal water allocation model based on water resources security assessment and its application in Zhangjiakou Region, northern China. *Resour. Conserv. Recycl.* **2012**, *69*, 57–65. [CrossRef]

18.   Kragt, M.E. Hydro-economic modelling in an uncertain world: Integrating costs and benefits of water quality management. *Water Resour. Econ.* **2013**, *4*, 1–21. [CrossRef]

19.   Davies EG, R.; Simonovic, S.P. Global water resources modeling with an integrated model of the social–economic–environmental system. *Adv. Water Resour.* **2011**, *34*, 684–700. [CrossRef]

20.   Wu, F.; Zhan, J.; Zhang, Q.; Sun, Z.; Wang, Z. Evaluating impacts of industrial transformation on water consumption in the Heihe River basin of northwest China. *Sustainability* **2014**, *6*, 8283–8296. [CrossRef]

21.   Ji, X.; Han, M.; Ulgiati, S. Optimal allocation of direct and embodied arable land associated to urban economy: Understanding the options deriving from economic globalization. *Land Use Policy* **2020**, *91*, 104392. [CrossRef]

22.   Tan, S.K.; Rao, Y.X.; Zhu, X.B. Study on the influence of land investment on regional economic growth. *China Popul. Resour. Environ.* **2012**, *22*, 61–67.

23.   Long, H. *Land Use Transitions and Rural Restructuring in China*; Springer: Berlin/Heidelberg, Germany, 2020.

24.   Liang, X.; Jin, X.; Ren, J.; Gu, Z.; Zhou, Y. A research framework of land use transition in Suzhou City coupled with land use structure and landscape multifunctionality. *Sci. Total Environ.* **2020**, *737*, 139932. [CrossRef]

25.   Liu, Y.; Li, J.; Yang, Y. Strategic adjustment of land use policy under the economic transformation. *Land Use Policy* **2018**, *74*, 5–14. [CrossRef]

26.   Li, T.; Liao, H.; Yang, W.; Zhang, W.; Shi, J. Urbanization quality over time and space as well as coupling coordination of land, population and industrialization in Chongqing. *Econ. Geogr.* **2015**, *35*, 65–71.

27.   Ge, D.; Zhou, G.; Qiao, W.; Yang, M. Land use transition and rural spatial governance: Mechanism, framework and perspectives. *J. Geogr. Sci.* **2020**, *30*, 1325–1340. [CrossRef]

28.   Xie, H.; Zhu, Z.; Wang, B.; Liu, G.; Zhai, Q. Does the expansion of urban construction land promote regional economic growth in China? Evidence from 108 cities in the Yangtze River Economic Belt. *Sustainability* **2018**, *10*, 4073. [CrossRef]

29.   Du, G.; Cai, Y. Technical Efficiency of Built-up Land in China's Economic Growth during 1997–2007. *Progr. Geogr.* **2010**, *6*, 12.

30.   Sawaya, K.E.; Olmanson, L.G.; Heinert, N.J.; Brezonik, P.L.; Bauer, M.E. Extending satellite remote sensing to local scales: Land and water resource monitoring using high-resolution imagery. *Remote Sens. Environ.* **2003**, *88*, 144–156. [CrossRef]

31.   Dong, W.; Yang, Y.; Zhang, Y.F. Coupling effect and spatiotemporal differentiation between oasis city development and water-land resources. *Resour. Sci.* **2013**, *35*, 1355–1362.

32.   Wang, W.; Lu, N.; Wang, X. Spatial Matching Pattern of Water and Soil Resources in the Yellow River Delta. *J. Water Resour. Water Eng.* **2014**, *25*, 66–70.

33.   Zhang, J.H.; Li, J.; Tang, Y. Analysis of the spatio-temporal matching of water resource and economic development factors in China. *Resour. Sci.* **2012**, *34*, 1546–1555.

34.   Nordhaus, W.D.; Stavins, R.N.; Weitzman, M.L. Lethal model 2, The limits to growth revisited. *Brook. Pap. Econ. Act.* **1992**, *1992*, 1–59. [CrossRef]

35.   Bruvoll, A.; Glomsrød, S.; Vennemo, H. Environmental drag: Evidence from Norway. *Ecol. Econ.* **1999**, *30*, 235–249. [CrossRef]

36.   Romer, D. *Advanced Macroeconomics*, 2nd ed.; Shanghai University of Finance &Economics Press: Shanghai, China, 2001; pp. 30–38.

37.   Xue, J.; Wang, Z.; Zhu, J.; Wu, B. Analysis of the "tail effect" of China's economic growth. *Financ. Res.* **2004**, *30*, 5–14.

38.   Xie, S.; Wang, Z.; Xue, J. Analysis of the "Tail-effect of Growth" of Soil and Water Resources in China's Economic Development. *Manag. World* **2005**, *7*, 22–25.

39.   Cao, X.; Jin, X.; Zhou, Y. The "growth drag" of resources and environment on Nanjing's economic development. *China Popul. Resour. Environ.* **2011**, *21*, 558–561.

40.   Liu, Y. Economic growth drag in the Central China: Evidence from a panel analysis. *Appl. Econ.* **2013**, *45*, 2163–2174. [CrossRef]

41. Tian, X.; Geng, Y.; Viglia, S.; Bleischwitz, R.; Buonocore, E.; Ulgiati, S. Regional disparities in the Chinese economy. An emergy evaluation of provincial international trade. *Resour. Conserv. Recycl.* **2017**, *126*, 1–11. [CrossRef]

42. Song, M.; Ma, X.; Shang, Y.; Zhao, X. Influences of land resource assets on economic growth and fluctuation in China. *Resour. Policy* **2020**, *68*, 101779. [CrossRef]

43. Wang, S.; Li, R. Toward the coordinated sustainable development of urban water resource use and economic growth: An empirical analysis of Tianjin City, China. *Sustainability* **2018**, *10*, 1323. [CrossRef]

44. Ma, L.; Long, H.; Chen, K.; Tu, S.; LZhang, Y.; Liao, L. Green growth efficiency of Chinese cities and its spatio-temporal pattern. *Resour. Conserv. Recycl.* **2019**, *146*, 441–451. [CrossRef]

45. Zhang, Y.; Long, H.; Ma, L.; Tu, S.; Liao, L.; Chen, K.; Xu, Z. How does the community resilience of urban village response to the government-led redevelopment? A case study of Tangjialing village in Beijing. *Cities* **2019**, *95*, 102396. [CrossRef]

46. Long, H.; Qu, Y.; Tu, S.; Zhang, Y.; Jiang, Y. Development of land use transitions research in China. *J. Geogr. Sci.* **2020**, *30*, 1195–1214. [CrossRef]

47. Gao, X.; Su, L.; Liu, Y. Optimal Allocation of Construction Land Based on GIS. In Proceedings of the 2009 First International Workshop on Database Technology and Applications, Wuhan, China, 25–26 April 2009.

48. Liu, Y.; Luo, T.; Liu, Z.; Kong, X.; Li, J.; Tan, R. A comparative analysis of urban and rural construction land use change and driving forces: Implications for urban–rural coordination development in Wuhan, Central China. *Habitat Int.* **2015**, *47*, 113–125. [CrossRef]

49. Zhang, Y.; Long, H.; Tu, S.; Ge, D.; Ma, L.; Wang, L. Spatial identification of land use functions and their tradeoffs/synergies in China: Implications for sustainable land management. *Ecol. Indic.* **2019**, *107*, 105550. [CrossRef]

50. Song, X.; Huang, Y.; Wu, Z.; Ouyang, Z. Does cultivated land function transition occur in China? *J. Geogr. Sci.* **2015**, *25*, 817–835. [CrossRef]

51. Long, H.; Ge, D.; Zhang, Y.; Tu, S.; Qu, Y.; Ma, L. Changing man-land interrelations in China's farming area under urbanization and its implications for food security. *J. Environ. Manag.* **2018**, *209*, 440–451. [CrossRef]

52. Qu, Y.; Jiang, G.; Li, Z.; Tian, Y.; Wei, S. Understanding rural land use transition and regional consolidation implications in China. *Land Use Policy* **2019**, *82*, 742–753. [CrossRef]

53. Long, H.; Zhang, Y.; Tu, S. Rural vitalization in China: A perspective of land consolidation. *J. Geogr. Sci.* **2019**, *29*, 517–530. [CrossRef]

54. Lambin, E.F.; Meyfroidt, P.; Rueda, X.; Blackman, A.; Börner, J.; Cerutti, P.O.; Dietsch, T.; Jungmann, L.; Lamarque, P.; Lister, J.; et al. Effectiveness and synergies of policy instruments for land use governance in tropical regions. *Glob. Environ. Chang.* **2014**, *28*, 129–140. [CrossRef]

55. Lu, X.; Shi, Y.; Chen, C.; Yu, M. Monitoring cropland transition and its impact on ecosystem services value in developed regions of China: A case study of Jiangsu Province. *Land Use Policy* **2017**, *69*, 25–40. [CrossRef]

56. Lambin, E.F.; Meyfroidt, P. Land use transitions: Socio-ecological feedback versus socio-economic change. *Land Use Policy* **2010**, *27*, 108–118. [CrossRef]

57. Meyfroidt, P. Trade-offs between environment and livelihoods: Bridging the global land use and food security discussions. *Glob. Food Secur.* **2018**, *16*, 9–16. [CrossRef]

58. Li, Y.; Fan, P.; Liu, Y. What makes better village development in traditional agricultural areas of China? Evidence from long-term observation of typical villages. *Habitat Int.* **2019**, *83*, 111–124. [CrossRef]

59. Liu, Y.; Zhou, G.; Liu, D.; Yu, H.; Zhu, L.; Zhang, J. The interaction of population, industry and land in process of urbanization in China: A case study in Jilin Province. *Chin. Geogr. Sci.* **2018**, *28*, 529–542. [CrossRef]

60. Ma, L.; Long, H.; Tu, S.; Zhang, Y.; Zheng, Y. Farmland transition in China and its policy implications. *Land Use Policy* **2020**, *92*, 104470. [CrossRef]

61. Bruggeman, D.; Meyfroidt, P.; Lambin, E.F. Forest cover changes in Bhutan: Revisiting the forest transition. *Appl. Geogr.* **2016**, *67*, 49–66. [CrossRef]

62. Ge, D.; Long, H.; Zhang, Y.; Ma, L.; Li, T. Farmland transition and its influences on grain production in China. *Land Use Policy* **2018**, *70*, 94–105. [CrossRef]

63. Li, Y.; Westlund, H.; Liu, Y. Why some rural areas decline while some others not: An overview of rural evolution in the world. *J. Rural Stud.* **2019**, *68*, 135–143. [CrossRef]

64. Ma, L.; Long, H.; Tang, L.; Tu, S.; Zhang, Y.; Qu, Y. Analysis of the spatial variations of determinants of agricultural production efficiency in China. *Comput. Electron. Agric.* **2021**, *180*, 105890. [CrossRef]

65. Zhang, Y.; Long, H.; Li, Y.; Ge, D.; Tu, S. How does off-farm work affect chemical fertilizer application? Evidence from China's mountainous and plain areas. *Land Use Policy* **2020**, *99*, 104848. [CrossRef]

66. Yang, Y.; Wu, C.; Luo, G.; Wei, S. Effects of soil and water resources on economic growth damping in China. *Econ. Geogr.* **2007**, *27*, 529–532.

67. Long, H.; Qu, Y. Land use transitions and land management: A mutual feedback perspective. *Land Use Policy* **2018**, *74*, 111–120. [CrossRef]

68. Ye, Y. Calculation and analysis of total factor productivity in China and provinces. *Economists* **2002**, *3*, 115–121.

69. Wang, X.; Fan, G. *The Sustainability of China's Economic Growth—A Cross-century Review and Prospect*; Economic Science Press: Beijing, China, 2000.

70. Ke, S.; Xiang, J. Estimation of the fixed capital stocks in Chinese cities for 1996–2009. *Stat. Res.* **2012**, *29*, 19–24.

71. Marshall, R.; Mariam, C. *Measurement of Capital Stocks, Consumption of Fixed Capital, and Capital Services*; Central American Ad Hoc Group on National Accounts: Santo Domingo, Dominican Republic, 2005.

72. Jantzen, R.T.; Volpert, K. On the mathematics of income inequality: Splitting the gini index in two. *Am. Math. Mon.* **2012**, *119*, 824–837. [CrossRef]

73. Lin, F. *Economic Statistical Analysis Methods*; Social Sciences Academic Press: Beijing, China, 2003; pp. 57–69.

74. Filippini, R.; Mazzocchi, C.; Corsi, S. The contribution of Urban Food Policies toward food security in developing and developed countries: A network analysis approach. *Sustain. Cities Soc.* **2019**, *47*, 101506. [CrossRef]

75. Lu, D.D. New factors and new patterns of regional development in China. *Geogr. Res.* **2003**, *22*, 261–271.

76. Li, H.; Xiong, Z.; Xie, Y. Resource tax reform and economic structure transition of resource-based economies. *Resour. Conserv. Recycl.* **2018**, *136*, 389–398. [CrossRef]

*land*

MDPI

*Article*

# Spatially Explicit Evaluation and Driving Factor Identification of Land Use Conflict in Yangtze River Economic Belt

Jiaxing Cui [1], Xuesong Kong [2], Jing Chen [1], Jianwei Sun [3] and Yuanyuan Zhu [1,*]

[1] Hubei Provincial Key Laboratory for Geographical Process Analysis and Simulation, Academy of Wuhan Metropolitan Area, College of Urban and Environmental Sciences, Central China Normal University, Wuhan 430079, China; cuijiaxing@ccnu.edu.cn (J.C.); jing.chen@ccnu.edu.cn (J.C.)
[2] School of Resource and Environmental Sciences, Wuhan University, Wuhan 430079, China; xuesongk@whu.edu.cn
[3] School of Geographical and Environmental Sciences, Guizhou Normal University, Guiyang 550025, China; sunjianwei@whu.edu.cn
* Correspondence: zhuyy990@mail.ccnu.edu.cn

**Abstract:** Regional land use transitions driven by the adaptive reconciliation of existing land use conflict with socioeconomic development can lead to positive economic effects as well as new land use conflict. Although research on land use transition has progressed considerably, limited studies have explored the spatiotemporal dynamic pattern of land use conflict during the land use transition period. Previous evaluation approaches on land use conflict that mainly focus on status or potential conflict lack conflict intensity evaluation during the land use transition process. A new spatially explicit evaluation framework of land use conflict that directly examines three aspects of conflict, namely, ecological and agricultural (EAC), agricultural and construction (ACC), and ecological and construction (ECC) land conflicts based on ecological quality and agricultural suitability, is proposed in this study. The spatiotemporal dynamic pattern and driving factors of land use conflict in the Yangtze River Economic Belt of China in the period of 2000–2018 are evaluated. The results indicated that comprehensive land use conflict (CLUC) intensity slightly decreased by 9.91% and its barycenter showed a trend toward the west during 2000–2018. ACC is the most drastic conflict among the three aspects of conflict. The mean intensity of ACC reduced remarkably by 38.26%, while EAC increased by 33.15% and ECC increased by 28.28% during the research periods. The barycenter of EAC moved toward the east while the barycenter of ACC and ECC moved toward the west. The changes in the intensity and spreading pattern of land use conflict indices demonstrated the changes in the pattern of territorial space development. Total population, population density, per capita GDP, number of mobile phone users, and road density were strong drivers that influenced the land use conflict of territorial space. Multiple policy recommendations including improving territorial space planning and governance ability, and improving land use efficiency, were proposed to manage and resolve the land use conflict of territorial space. The results and conclusions of this study will help improve future regional land use policies and reduce land use conflict.

**Keywords:** land use conflict; land use transition; ecological value; agricultural land suitability; Yangtze River Economic Belt



**Citation:** Cui, J.; Kong, X.; Chen, J.; Sun, J.; Zhu, Y. Spatially Explicit Evaluation and Driving Factor Identification of Land Use Conflict in Yangtze River Economic Belt. *Land* **2021**, *10*, 43. https://doi.org/10.3390/land10010043

Received: 13 December 2020
Accepted: 1 January 2021
Published: 5 January 2021

**Publisher's Note:** MDPI stays neutral with regard to jurisdictional claims in published maps and institutional affiliations.



## 1. Introduction

Along with increasingly extensive and intensive human activities, the exceptional pace, magnitude, and spatial reach of human alterations of Earth's surface since the industrial revolution, or even earlier, has taken us to a new geological era called the Anthropocene [1]. Given that humans have increasingly become the predominant environmental force apart from geophysical sciences, the concept of the Anthropocene has been enthusiastically received [2,3]. The controversial idea of the "Age of Humans" has been extensively investigated by scholars from numerous other disciplines and public media and promoted

481

the consideration of the human–nature relation [4]. Land use and land cover change (LUCC) is an important research topic that tackles the changing human–nature relation and a core project of IGBP (International Geosphere Biosphere Programme)and crossover research of IHDP (International Human Dimensions Programme on Global Environmental Change) [4]. The competition among different land use types over limited land use resources with remarkable urbanization and economic development, such as expansion of mankind landscape at the expense of ecological space, has been a primary feature of LUCC [5,6]. LUCC and the trade-off between different land use types has formed serious land use conflicts represented by environmental damage, including habitat loss, biotic diversity decline, climate change, and soil degradation [4,7,8], as well as social problems, such as the vulnerability of places and people to climatic and economic or sociopolitical perturbations [9–12]. In addition, many land use transitions have taken place over wild areas both from the dominant morphology and recessive morphology with the quantity, spatial structure, and function change of land use and land cover [13–15]. On the one hand, land use transitions are largely driven by different land use types representing the benefits of different departments conflicting in space and will lead to a new balance of regional land use morphology patterns [16,17].On the other hand, land use transitions may trigger new land use conflicts during transformations between different land use types and allocation of land resources due to policy and institutional failures [14,18]. Thus, exploring the spatiotemporal characteristics of land use conflicts is an important research topic to enhance the understanding of the human–nature relation and provide critical insights into the dynamics of land use transitions.

Land use conflicts occur whenever land use stakeholders have incompatible interests about how land is used in local areas [19]. Land disputes are the manifestation of land use conflicts when incompatible interests related to certain land units occur between specific stakeholders, such as developers, farmers, and local residents. For example, the controversial land compensation through large land investments and land acquisitions that led to a large number of land disputes between residents and developers, can pose a risk to social stability [20]. Another manifestation of land use conflict is the presence of the environmental land use conflict between land users and the public interest in ecological protection that typically occurs during the competition and compromise between different objectives for all land uses within a limited space [19,21]. Human land use (agriculture, exploitation of mineral resources, industrial production, and living activities) commonly have negative ecological and environmental impacts, such as consumption of highly productive agricultural land, occupation of rural residential communities, and destruction of ecological spaces that lead to growing conflicts between economic gains, social objectives (food security), and environmental goals (biodiversity conservation, ecosystem services provision, habitat integrity, and biodiversity) [22]. In addition, the implementation of some environmental protection policies can conflict with native residents, such as pastoral evictions due to the implementation of green economy in Tanzania [23]. Rapid urbanization and economic development cause extensive environmental land use conflicts. This complicated problem has attracted considerable research attention. Some scholars have carried out in-depth investigations on the ecological environmental impacts of land use conflicts [24,25]. Other scholars have focused on the potential land use conflict over current and proposed uses [26,27]. Another issue that has drawn considerable research attention is conflict resolution, and varied approaches have been proposed to solve this complicated problem [28–30]. However, comprehensive investigations on the spatiotemporal characteristics of land use conflict from the perspective of ecology based on a spatially explicit evaluation approach around a large-scale range are lacking, and studies on the driving factors of environmental conflicts are inadequate.

Land use conflicts are complex processes and various aspects must be considered. Several valuable approaches on land use conflict evaluation from an ecological perspective have been developed. Some scholars evaluate land use conflict through the degree of deviation between the actual land use from the most adequate land use standing on a capability evaluation (called natural use), while the ruggedness number is widely used to evaluate land capability [24,25]. The evaluation approach, based on the ruggedness number, is suitable for identifying the land use conflict in rural areas filled with farmlands and forest spots but unsuitable for rapidly urbanized areas. Some scholars proposed a linear conceptual model considering the complexity, vulnerability, and stability of land use units, while some scholars proposed another linear conceptual model considering spatial type, spatial structure, and spatial process conflict indices, to calculate the land use conflict in urbanized areas [21,31]. In addition, the comprehensive assessment from three aspects, including conflicts over land use structure, land conversion, and landscape pattern, were proposed to calculate the land use conflict [19]. These two evaluation approaches can properly evaluate the status of land use conflict but fail to examine the land use conflict occurring among land use transitions within a period. Furthermore, multicriteria analysis and multisuitability evaluation of construction, agricultural, and ecological lands were used to calculate potential land use conflict [32,33]. However, to date, analyses that directly examine the actual occurrence of land use conflict during land use transition processes are limited. A spatially explicit land use conflict evaluation approach that simultaneously considers the transition among ecological, agricultural, and construction lands triggered by human activities in rapidly urbanized areas is proposed in this study.

China has experienced rapid urbanization and economic development in the last 40 years since its opening and reform [34]. Acceleration of industrialization and urbanization as well as population and industrial agglomeration has intensified land development and utilization intensity [35]. Large-scale land development and utilization has led to alterations of local ecological landscapes and increasing food demand of the growing population driven by agricultural land reclamation. Chaotic land use patterns due to planning control failures and inefficient land utilization cause land use conflict in many regions across China [36,37]. Territorial space suffers from wastage and unreasonable use of land resources as well as environmental and ecological degradation [38,39]. Optimization of territorial space development has recently attracted considerable research attention under ecological civilization construction. Coordination and optimization of territorial space, which are crucial in effective territorial space development, have become important research topics [40]. However, scientific understanding of the spatiotemporal dynamic pattern and driving factors of land use conflicts is still limited, especially on a large regional scale [41]. The Yangtze River Economic Belt (YREB), a rapid economic development region in China with accelerated industrialization and urbanization, causes serious urban expansion and land use change [42]. Economic development, economic activities, infrastructure, and residences have clearly extended the scope and depth of urban areas to their surrounding virgin territories [43,44]. Natural land covers have been increasingly converted into urban impervious surfaces, with urban expansion exhibiting very fragmented characteristics [45,46]. The complexity of this phenomenon and its effects influence many ecological and environmental impacts, such as increased soil erosion [47], poor water quality [48], and reduced aquatic biodiversity [49], from local to global scales. The scarcity of land resource has led to the accumulation of land use conflicts over time. The spatial conflict of land use in small regions has been explored in many studies while investigations on spatial conflict in large regions and its driving forces are relatively rare [50,51]. Research on spatiotemporal patterns and land use conflict of the territorial space in YREB can help identify policies and practices to mitigate conflicts. This work examines issues that planners, land use managers, and practitioners must address when dealing with the complexity of land use change in the area.

This study primarily aims to (1) explore the spatiotemporal characteristics of land use conflict in YREB between 2000 and 2018, (2) develop a step-wise multiple linear regression model for identifying socioeconomic driving forces of spatial conflicts, and (3) propose some policy implications of the spatial control of territorial space within the study area while estimating land use conflict. Our study will provide a spatially explicit evaluation framework and contribute to the understanding of underlying driving forces of land use conflict.

## 2. Research Area and Data Sources

### 2.1. Research Area

The coastal economic belt and YREB are the top two largest economic and territorial space developments in China. As shown in Figure 1, YREB is located in the land belt adjoining the Yangtze River that stretches across three major regions of China (i.e., eastern, central, and western China). YREB covers nine provinces (i.e., Jiangsu, Zhejiang, Anhui, Jiangxi, Hubei, Hunan, Sichuan, Yunnan, and Guizhou) and two municipalities (i.e., Shanghai and Chongqing) with a population of 599 million inhabitants, which is 42.9% of China's population, in its total area of approximately 2.05 million km$^2$, which accounts for 21.4% of the country's land mass. Benefiting from the golden waterway of the Yangtze, the high level of economic development in YREB contributed 46.2% of the nation's gross domestic product (GDP), which reached CNY 45.78 trillion, in 2019.

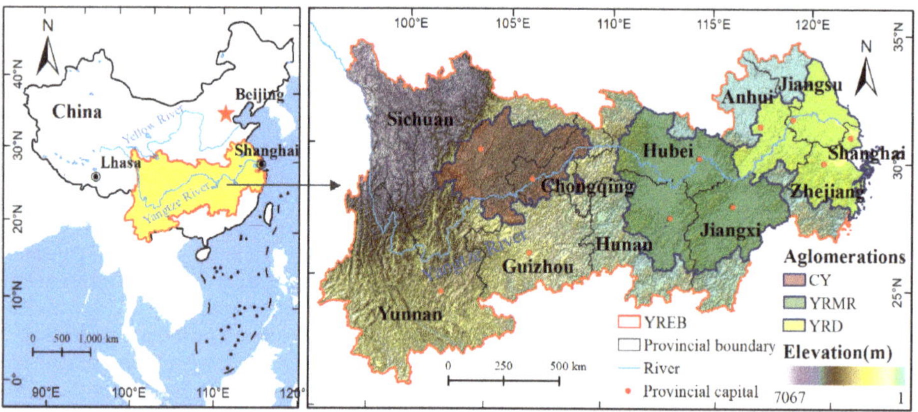


**Figure 1.** Geographical location of the study area.

Remarkable differences among regions in terms of natural conditions, regional transportation location, and socioeconomic development are due to the vast area and large span that crosses from east to west. Three urban agglomerations, namely, Cheng-Yu (CY), Yangtze River Middle Reaches (YRMR), and Yangtze River Delta (YRD), span the upper, middle, and lower reaches of YREB and represent the least developed western, less developed central, and most developed eastern regions in China, respectively. The development of YREB as a national strategy in 2014 has become one out of three regional development approaches in China that will change the country's national landscape both economically and environmentally. YREB has remarkable advantages due to its geographical location and is important in the coordinated development of the three plates of the eastern, middle, and western parts of China and the opening up of areas along the Yangtze River. Despite its high level of economic development, YREB has received considerable attention in ecological civilization construction. The State Council has called for efforts to facilitate the development of YREB by promoting appropriately coordinated environmental conservation and avoiding excessive development activities (*gong zhua da bao hu, bu gao da kai fa*) because its ecological protection is critical to the sustainable development of the Yangtze River Basin, which serves as an important ecological security barrier area. The optimization of land use and development pattern is crucial in the coordinated development of the ecosystem, economy, and society because of the serious problem of urban overexpansion and uncoordinated land development in YREB. Hence, exploring the spatiotemporal dynamic pattern and land use conflict of the territorial space in YREB is necessary.

### 2.2. Data Sources

Land use data in 2000, 2010, and 2018 (spatial resolution: 30 m) were provided by the Resource and Environment Science and Data Center of the Chinese Academy of Sciences (http://www.resdc.cn/) and then validated using nationwide field surveys [52–54]. Land use raster data demonstrated a resolution of 30 × 30 m. These data consisted of six classes (cropland, forest, grassland, waterbody, built-up land, and unused land) and 25 subclasses with a classification accuracy of 90% [55].

Defense Meteorological Satellite Program/Operational Linescan System (DMSP/OLS) nonradiative calibration nighttime stable lighting data of 2000, 2010, and 2018 were downloaded from National Earth System Science Data Center (http://www.geodata.cn/). The resolution of nonradio-calibrated DMSP/OLS night-stabilized lighting data is 30 arc seconds and approximately 1 km at the equator. Data were strictly processed to remove effects of fire, sunlight, moonlight, clouds, and aurora. Stable lighting data include lights from cities, towns, and other places with long-lasting light sources with the removal of background noise [56].

Moderate resolution imaging spectroradiometer (MODIS) normalized difference vegetation index (NDVI) data of 2000, 2010, and 2018 were obtained from Resource and Environment Science and Data Center. The NDVI measures changes in chlorophyll content and spongy mesophyll within the vegetation canopy and is widely used to represent the vigor and photosynthetic capacity of vegetation canopy. The value of NDVI lies between $-1$ and 1, with high values typically representing large vegetation cover and growth. The annual maximum NDVI value generated from 16-d composite MODIS NDVI product was used in this study.

Soil quality data, including physical soil fertility (represented by area weighted soil organic carbon), reference soil depth, and soil texture (calculated by the clay fraction, gravel content, sand fraction, and silt fraction), were accessed from the harmonized world soil database [57–59]. The harmonized world soil database is a 30 arc-second raster database with over 15,000 soil mapping units that combines existing regional and national updates of soil information worldwide (Soil and Terrain database programme, European Soil Database, soil map of China, and World Inventory of Soil property Estimates) with the information contained within the 1:5,000,000-scale FAO-UNESCO (Food and Agriculture Organization of the United Nations- United Nations Educational, Scientific, and Cultural Organization) soil map of the world (FAO, 1971–1981) [59]. Explicit temporal component is absent in the data set. Data for the year 2000 are nominal [58].

Meteorological data were processed using gauge-based analysis of daily precipitation product from China Meteorological Administration [60]. DEM (Digital elevation model) was processed using the SRTM (Shuttle Radar Topography Mission) product from United States Geological Survey. Administrative boundary and road maps were downloaded from the National Catalogue Service for Geographic Information (http://www.webmap.cn/main.do?method=index).

The basic geographic information base map including national boundary, sea land border, and major rivers was obtained from the Standard Map Service System maintained by the Map Technology Review Center, Ministry of Natural Resources (http://bzdt.ch.mnr.gov.cn/index.html). Provincial boundary, Prefecture-level city boundary, and road map were extracted from 1:250,000 national basic geographic database provided by the National Catalogue Service for Geographic Information (https://www.webmap.cn/commres.do?method=result25W).

Socioeconomic data used in this study were mainly derived from statistical records and government publications, including statistical yearbooks of Chinese cities and provinces, statistical bulletins of each prefectural city, and government work reports at various administrative levels.

A detailed description of the data collected is in Table 1. All geographic data were organized and reprojected to a uniform geographic projection using ArcGIS software. SPSS was applied to carry out regression analysis during the study period.

**Table 1.** Data sources and descriptions.

| Data Name | Data Type | Time Period | Resolution | Data Source |
| --- | --- | --- | --- | --- |
| Land use/land cover data | Grid | 2000, 2010, 2018 | 30 m × 30 m | Resource and Environment Science and Data Center |
| DMSP/OLS Night Light Data | Grid | 2000, 2010, 2018 | 30 arc-seconds | National Geophysical Data Center |
| NDVI | Grid | 2000, 2010, 2018 | 1000 m × 1000 m | Resource and Environment Science and Data Center |
| Meteorological data | Vector | 2000, 2010, 2018 | 0.25° × 0.25° | China Meteorological Administration |
| DEM | Grid | 2003 | 90 m × 90 m | United States Geological Survey (USGS) |
| Soil quality data | Grid | 2000 | 30 arc-seconds | Harmonized World Soil Database |
| National boundary, sea land border, and major rivers | Vector | 2019 | 1:60,000,000 | Standard Map Service System |
| Administrative boundary and road map | Vector | 2015 | 1:250,000 | National Catalogue Service for Geographic Information |
| Socioeconomic data | Txt | 2018 | County level | Statistical Yearbook of Chinese Cities |

**DMSP/OLS:** Defense Meteorological Satellite Program/Operational Linescan System; NDVI: normalized difference vegetation index; **DEM:** Digital Elevation Model.

## 3. Methodology

The research framework is presented in Figure 2. The land use classification system was established according to main functions of each piece of land. A land use conflict evaluation approach based on ecological quality and agricultural suitability was then constructed to estimate the land use conflict from the three aspects of ecological and agricultural (EAC), agricultural and construction (ACC), and ecological and construction (ECC) land conflicts. After evaluating the land use conflict of YREB according to the evaluation approach, temporal and spatial characteristics of land use conflict in the area as well as driving factors of land use conflict are explored. Finally, general policy suggestions for territorial space regulation are discussed in this study.

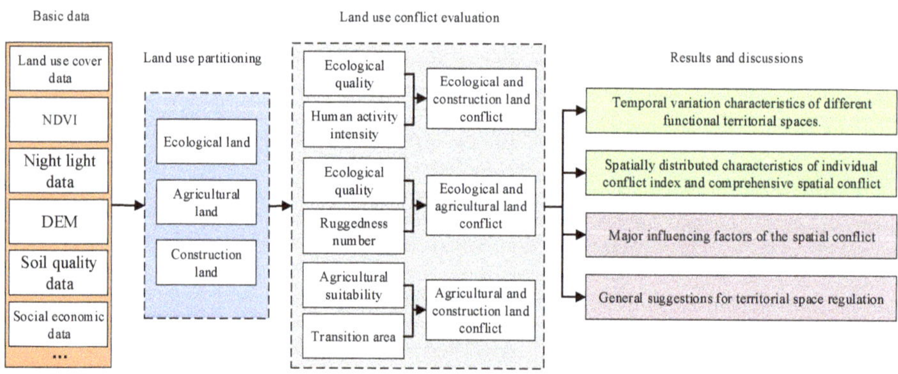


**Figure 2.** Research framework.

### 3.1. Land Use Classification System

Land use functions (LUFs) refer to the capacity of providing private and public products or services through diversified land use types and patterns [61]. Territorial space is a multifunctional comprehensive system and each piece of land provides one or combines several space functions in various forms [62]. According to the land use and functional statuses of various land use types in the study area, this article established a spatial classification system of construction, agricultural, and ecological areas [63]. Construction land mainly provides living, industrial production, and service industry functions. Agricultural land mainly provides agricultural production function. Ecological land mainly provides ecological functions such as climate regulation, gas regulation, water regulation, soil retention, and biodiversity. The development intensity of ecological, agricultural and construction lands and its disturbance to the natural ecosystems increase in turn. The classification system in this article can comprehensively reflect the change of land use functions and the transition of land use and development intensity. In addition, the land resources were classified as three land use types in the comprehensive zoning of land utilization [64]. Table 2 presents three classes of the territorial space.

**Table 2.** Land use classification system.

| Classes | Land Use Type | Description |
|---|---|---|
| Ecological land | Forest, grasslands, rivers, lakes, shoals, reservoirs, ponds, glaciers, and unutilized land | These land use types provide numerous ecological functions, such as climate regulation, gas regulation, water regulation, soil retention, and biodiversity. |
| Construction land | Urban and rural residential lands; construction land, including industrial, mining, and storage lands and roads. | These land use types mainly composed of impervious surface can provide industrial and mineral production as well as living functions, such as residence, shopping, education, and medical treatment. |
| Agricultural land | Paddy fields, irrigated land, and arid land. | The primary function of these land use types is agricultural production. |

### 3.2. Land Use Conflict

A new evaluation system for land use conflict based on real problems of land use practices and previous studies is proposed in this study (Figure 2). Our proposed index system includes three aspects, namely, ecological and construction (ECC), ecological and agricultural (EAC), and agricultural and construction (ACC) land conflicts. (1) ECC is defined by the deterioration of the ecosystem due to the intensification of human activities and construction land expansion during a certain period. Economic development and urbanization cause the construction land to expand with the occupation of large-scale ecological space. The shrinking ecological space through overexploitation generally indicates the deteriorating ecological environment and decreasing ecological service values [65,66]. Furthermore, increasing human activities may impose massive externalities to ecological land around the construction land [67]. Thus, both the increase and decrease in the total scale of ecological and construction land, as well as the negative externalities of construction land were considered in calculating ECC. (2) EAC is defined by the expanded agricultural land which unreasonably occupied a large area of ecological land during a certain period. The occupation of ecological land by agricultural land will reduce existing habitats, increase soil erosion, cause land degradation, and affect water quality, especially when the agricultural land is located in unsuitable areas [24,25,68]. (3) ACC is defined by the construction land expansion at the expense of agricultural land during a certain period. Decreasing agricultural land resources causes agricultural production risks and the threat of high food prices [69,70]. The increasing fertility of occupied agricultural land increases the conflict. The CLUC was calculated by considering the three kinds of land use conflict indicators.

ECC is calculated from the two aspects of ecological land loss index by construction expansion and threat indicator (TI) from human activities.

The ecological land loss index by construction expansion considers both the area and ecological quality of the ecological land loss as follows:

$$EL_{i,construction} = \frac{\sum_j^n A_{ij} \times EV_{ij}}{TA_i},\tag{1}$$

where $EL_{i,\,construction}$ is the ecological land loss index of region $i$ caused by construction expansion; $A_{ij}$ is the area of ecological land pixel $j$; $EV_{ij}$, the ecological value of ecological land pixel $j$ represented by its NDVI, is set to 1 when the land use class of pixel $j$ is a water body or shoal [67]; $n$ is the total number of ecological land pixels occupied by construction land in region $i$, and $TA_i$ is the total area of region $i$.

This article estimated the level of threat posed by human activities to ecological land using the following modified *TI*:

$$TI_i = \frac{\sum_j^m \sum_r^R A_{ij} \times EV_{ij} \times IC_r \times \left(1 - \frac{d_{jr}}{d_{jmax}}\right)}{TA_i}\tag{2}$$

where $TI_i$ is the threat level of human activities from construction land to ecological land in region $i$, R is the total number of construction land pixels within a certain distance to pixel $j$, $IC_r$ is the influence coefficient of the construction land pixel $r$ measured by the standardized value of night light brightness, $d_{jr}$ is the distance between ecological land pixel $j$ and construction land pixel $r$, and $d_{jmax}$ is the maximum effective distance of human threat reach across space, that was set to 5 km [71].

ECC is calculated by integrating ecological land loss index and *TI* of construction land. For the convenience of calculation, the average value of these two indexes is computed to express the following *ECC*:

$$ECC_i = \frac{EL_{i,construction} + TI_i}{2}\tag{3}$$

where $ECC_i$ is the ECC in region $i$.

EAC is calculated according to the ecological land loss in the course of a progressive invasion of ecological land by the agricultural land and the ecological risk caused by the agricultural land reclamation. The main consequence ecological risk and hazards of agricultural land reclamation include water erosion and soil structure decline. The ecological risk was evaluated via a combination of basin relief and drainage network density called the ruggedness number (RN) [24,25]:

$$RN = H_t \times D_d\tag{4}$$

where $H_t$ is the basin relief and defined as the elevation difference between lowest and highest points within the buffer area and $D_d$ is the drainage network density calculated by the kernel density of the drainage network.

EAC is calculated via the accumulated ecological risk by progressive invasion of ecological land by agriculture land as follows:

$$EAC_i = \frac{\sum_j^G A_{ij} \times RN_{ij}}{TA_i}\tag{5}$$

where $EAC_i$ is the $EAC$ of region $i$, G is the total number of ecological pixels occupied by agriculture land in region $i$, and $RN_{ij}$ is the RN of pixel $j$ in region $i$.

*ACC* is calculated via the agricultural land loss index by using a combination of the agricultural land loss area and the suitability of the lost agricultural land. The high suitability of the lost agricultural land increases ACC. The agricultural land suitability is evaluated through GIS-based multicriteria analysis [72] with selected criteria (Table 3).

**Table 3.** Agricultural land suitability evaluation criteria.

| Criteria | Aptitude, impact, and feasibility class [a] | | | |
|---|---|---|---|---|
| | $S_1$ | $S_2$ | $S_3$ | N |
| Length of dry season (months) | 0–2 | 2–3 | 3–4 | >4 |
| Mean annual temperature (°C) | >22 | 22–20 | 20–18 | <18 |
| Mean annual maximum temperature (°C) | >27 | 27–24 | 24–22 | <22 |
| Slope (%) | 0–8 | 8–16 | 16–30 | >30 |
| Drainage | Good | Moderate | Imperfect | Poor |
| Soil texture [b] | C, SC, SCL | SL | LS | S |
| Soil depth (cm) | >100 | 100–70 | 70–50 | <50 |
| Distance to the road (m) | <500 | 500–1000 | 1000–2000 | >2000 |

[a] Each factor was expressed using four suitability classes corresponding to a high (S1), moderate (S2), and marginal (S3) suitability, as well as unsuitable (N) conditions. [b] C: clay; SC: sandy clay; SCL: sandy clay loam; SL: sandy loam; LS loamy sand; S: sand.

ACC is calculated using the accumulated suitability index of the lost agricultural land as follows:

$$ACC_i = \frac{\sum_j^H A_{ij} \times AS_{ij}}{TA_i} \tag{6}$$

where $ACC_i$ is the ACC of region $i$, $H$ is the total number of agricultural pixels occupied by construction land in region $i$, and $AS_{ij}$ is the agricultural land suitability of pixel $j$ in region $i$.

CLUC is calculated using the combination of EAC, ACC, and ECC.

*3.3. Selecting Potentially Important Driving Factors*

Identifying major underlying factors of spatiotemporal patterns of the land use conflict is necessary to recognize evolution rules and the internal mechanism of the spatial conflict indices. Previous studies showed that demography, economy, and residents living in the area are important factors that drive structural changes and the pattern evolution of the territorial space [13]. We hypothesized in this study that three kinds of driving forces, namely demographic, economic, and life factors, would be crucial to the land use conflict of the territorial space. Specifically, this article selected eight socioeconomic variables covering three kinds of factors, namely, total population, population density, per capita GDP, investment in fixed assets, total retail sales of consumer goods, number of hospital beds, number of mobile phone users, and road density, to examine their relation to the spatial conflict of the territorial space (Table 4). However, historical data on YREB covering 130 cities (including 2 municipalities, 109 prefecture cities, 15 autonomous prefectures, 3 province-governing counties, and 1 forest district) in the study's time span from 2010 to 2018 are lacking. Therefore, given the regional characteristics of YREB and the difficulty in acquiring relevant data, this article gathered socioeconomic factors in 2018 to explore the driving forces of the land use conflict.

**Table 4.** Socioeconomic variables considered in this study.

| Variables | Abbreviation | Definitions |
| --- | --- | --- |
| Demographic factors | TPOP | Total population of prefectural city |
| | POPD | Population density of prefectural city |
| Economic factors | GDPPC | per capita GDP |
| | IFA | Investment in fixed assets |
| | TRSCG | Total retail sales of consumer goods |
| Life factors | HBN | Number of hospital beds |
| | MPUN | Number of mobile phone users |
| | RD | Road density |

### 3.4. Statistical Analysis

Both Pearson correlation and step-wise linear regression were used in this study. Eight socioeconomic variables were examined via Pearson correlation analysis to determine their individual correlation with conflict indices in 2018. This article also performed step-wise linear regressions in which EAC, ACC, ECC, and CLUC were taken as dependent variables while nine socioeconomic driving factors were used as independent variables to verify their significant correlation with conflict indices when other factors are held constant.

In addition, this article computed standardized regression coefficients to compare significant socioeconomic factors and determine which ones are statistically significant in the step-wise linear regression. Standardized regression coefficients represent the amount of change in the dependent variable in response to a change of one standard deviation in an independent variable. Thus, the large absolute value of the standardized regression coefficient indicates the importance of the independent variable. Statistical analyses were performed using SPSS for Windows (version 19.0).

## 4. Results

### 4.1. Variation Characteristics of Land Use in YREB

The land use area and its rates in YREB from 2000 to 2018 were obtained according to the land use classification system (Table 5). The ecological land is the largest region that accounts for more than 66% of the territorial area. This finding indicates the satisfactory environmental background of YREB that benefits from its suitable latitude and humid climate. The large area of ecological land provides a livable environment for human beings and also lays an acceptable foundation for an environmentally friendly economy. Agricultural land is the second most dominant class in YREB that accounts for more than 29% of the total area. YREB's long history of farming is because of its suitable climate, abundant labor force, and large plains. The high-level agricultural development leads to the large amount of agricultural population and high grain output. The Chinese government set up several agricultural production bases in YREB, including the area between Yangtze and Huai Rivers (*jianghuaidiqu*); Taihu, Poyang, and Dongting Lake Plains, and Jianghan and Chengdu Plains. Known as the "rice barn" of China, YREB plays an important role in China's food security. Strengthening the protection of agricultural land in YREB and controlling the land use conflict between agricultural land and other land use categories are important [73]. However, fine planting and excessive pursuit of yield demonstrated a negative effect on the surrounding ecology, including nonpoint source pollution from fertilizers and pesticides. The land use conflict between ecological and agricultural lands must be considered to realize satisfactory ecological services in the region. The share of construction land is relatively low, but these areas demonstrating the maximum human activity also receive considerable attention. YREB is a primary development belt of China that demonstrates high-intensity progress. Rapid urbanization and industrialization has caused high-intensity land use conflict in YREZ.

**Table 5.** Area of each land use class in The Yangtze River Economic Belt (YREB) from 2000 to 2018.

| Class | | 2000 | 2010 | 2018 |
|---|---|---|---|---|
| Ecological land | Area (million ha) | 136.15 | 136.19 | 135.84 |
| | Rate (%) | 66.41% | 66.43% | 66.26% |
| Agricultural land | Area (million ha) | 64.02 | 62.08 | 60.88 |
| | Rate (%) | 31.23% | 30.28% | 29.70% |
| Construction land | Area (million ha) | 4.84 | 6.73 | 8.28 |
| | Rate (%) | 2.35% | 3.28% | 4.04% |

The annual variation of each land use category is presented in Table 6. The three land classes changed in varying degrees. The construction land demonstrating the most remarkable changes increased rapidly by 3,448,260 ha during 2000–2018. Its ascending speed reached 18.51% per year during 2000–2010 and then decreased to 13.86% during 2010–2018. This finding evidently shows the land use front via urbanization and economic development, which accelerated at the expense of land use. By contrast, agricultural land reduced rapidly by 1,939,000 and 1,196,240 ha during 2000–2010 and 2010–2018, respectively, because the rapid urbanization and economic development used a considerable amount of arable land. Its decline speed first achieved −0.30% during 2000–2010 and then decreased to −0.24% during 2010–2018. The reduced consumption of agricultural land was due to the gradual tightening of arable land management policies concerning food security. The ecological land increased by 37,700 ha during 2000–2010 because of the Grain for Green policy and excavation of reservoirs and ponds for irrigation and fish culture by residents. However, ecological land reduced by 350,720 ha during 2010–2018 due to the strict farmland protection system. The agricultural land occupied by construction land is usually supplemented from the ecological land to accomplish the difficult task of minimizing farmland areas.

**Table 6.** Average annual change of land use area in YREB during the study periods.

| Class | | 2000–2010 | 2010–2018 | Sum | Average |
|---|---|---|---|---|---|
| ecological land | Annual variation (thousand ha) | 3.77 | −43.84 | −313.02 | −17.39 |
| | Rate(%) | 0.02% | 0.08% | 0.86% | 0.05% |
| agricultural land | Annual variation (thousand ha) | −193.9 | −149.53 | −3135.24 | −174.18 |
| | Rate(%) | −0.30% | −0.24% | −4.92% | −0.27% |
| construction land | Annual variation (thousand ha) | 190.13 | 193.37 | 3448.26 | 191.57 |
| | Rate(%) | 18.51% | 13.86% | 295.98% | 16.44% |

*4.2. Spatiotemporal Dynamic Analysis of Each Land Use Conflict Indicator*

4.2.1. Temporal Variability in Land Use Conflict during 2000–2010 and 2010–2018

The conflict indices of YREB during 2000–2010 and 2010–2018 are shown in Figure 3. The aggregative indicator of CLUC decreased slightly throughout the study period. The slight decrease in CLUC throughout the entire study period indicated that economic development and urbanization gradually reduce land type conversion and result in the reduction in the land function transformation. By comparison, ACC is the most serious conflict indicator among the three aspects of conflict indicators. ACC reduced considerably by 38.26% during the last decade due to strict arable land protection policies and the pressure of food security that effectively prevented land development from occupying arable land. EAC and ECC both demonstrated a trend of rising slowly due to the lack of spatial regulation and governance zoning of ecological land. The unlimited expansion of construction land occupying large ecological land and the arable land reclamation of ecological land that compensates for the lost arable land finally resulted in remarkable EAC and ECC. The intensifying EAC and ECC and still high ACC indicate the high demand of our spatial government ability.

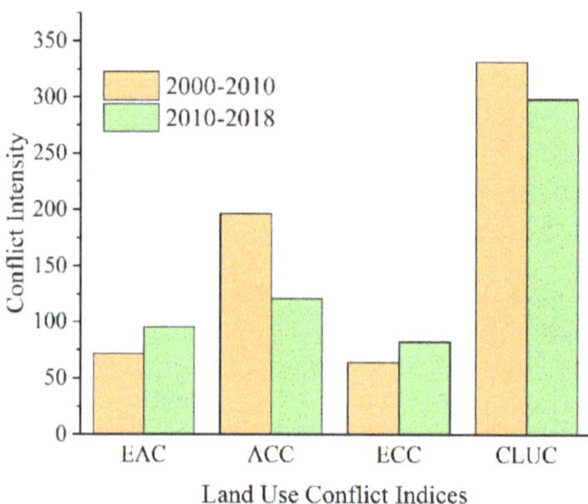


**Figure 3.** Variation of the average conflict indices of YREB.

### 4.2.2. Spatiotemporal Dynamic Pattern of Land Use Conflict Indicators

Spatial distribution patterns of land use conflict indicators during 2000–2010 and 2010–2018 are illustrated in Figure 4. This article classified the area into five types of regions, namely, low, semilow, medium, semihigh, and high value regions, according to natural breaking points. Different land use conflict indicators demonstrate unique distribution characteristics. High and semihigh values of the EAC are spread out over a large area. The midwest region exhibited serious conflict, especially in Yunnan, Hubei, Guizhou, and Eastern Sichuan. Eastern provinces, including Jiangsu, Zhejiang, Anhui, and Shanghai, demonstrated low-scale conflict values. The spatial distribution pattern of the EAC was controlled by multiple factors in which the economic development and topographic condition were the most evident. Relatively high values of the EAC were concentrated in interlaced zones of ecological and agricultural lands. Large populations strongly depend on farming for their living due to the area's poor economic development. Residents expand the agricultural land to increase production and ecological land was compressed and perturbed by agricultural land which led to serious EAC, while increasing the vulnerability of the entire territory space. The period of 2010–2018 generally demonstrated an intensified trend of EAC compared to the period of 2000–2010. The area exhibited high, semihigh, and medium values from 40,100 km$^2$ to 75,600 km$^2$ and its mean conflict intensity increased by 33.15%. Hence, the agricultural development considerably affected the ecosystem in YREB, especially in the last decade. Responding to problems that cause damage to the environment and restricting the agricultural sustainable development, reform, and promotion of agricultural development are necessary. The status of the distributing pattern still presented the dispersion type. The EAC in Yunnan is still serious and the EAC in Guizhou and Hunan increased considerably in the second period.

Both periods demonstrated similar spatial distribution patterns of the ACC index. Areas with high and semihigh values concentrated around the three urban agglomerations indicated that the urbanization and economic development of urban agglomerations cause considerable ACC. The intensity and scope of ACC of YRD are considerably larger than other urban agglomerations along the upstream and midstream of the Yangtze River mainly because of the higher intensity of urban expansion at the expense of agricultural land in YRD than the findings in the two other urban agglomerations. The period of 2010–2018 demonstrated a considerably reduced ACC than the period of 2000–2010. The area with relatively higher and high values reduced from 50,700 km$^2$ to 20,800 km$^2$ and its mean conflict intensity reduced by 38.26%, thereby indicating that the overall agricultural land loss due to human activities and overall deterioration considerably reduced. The agricultural land protection policy played a crucial role in controlling ACC.

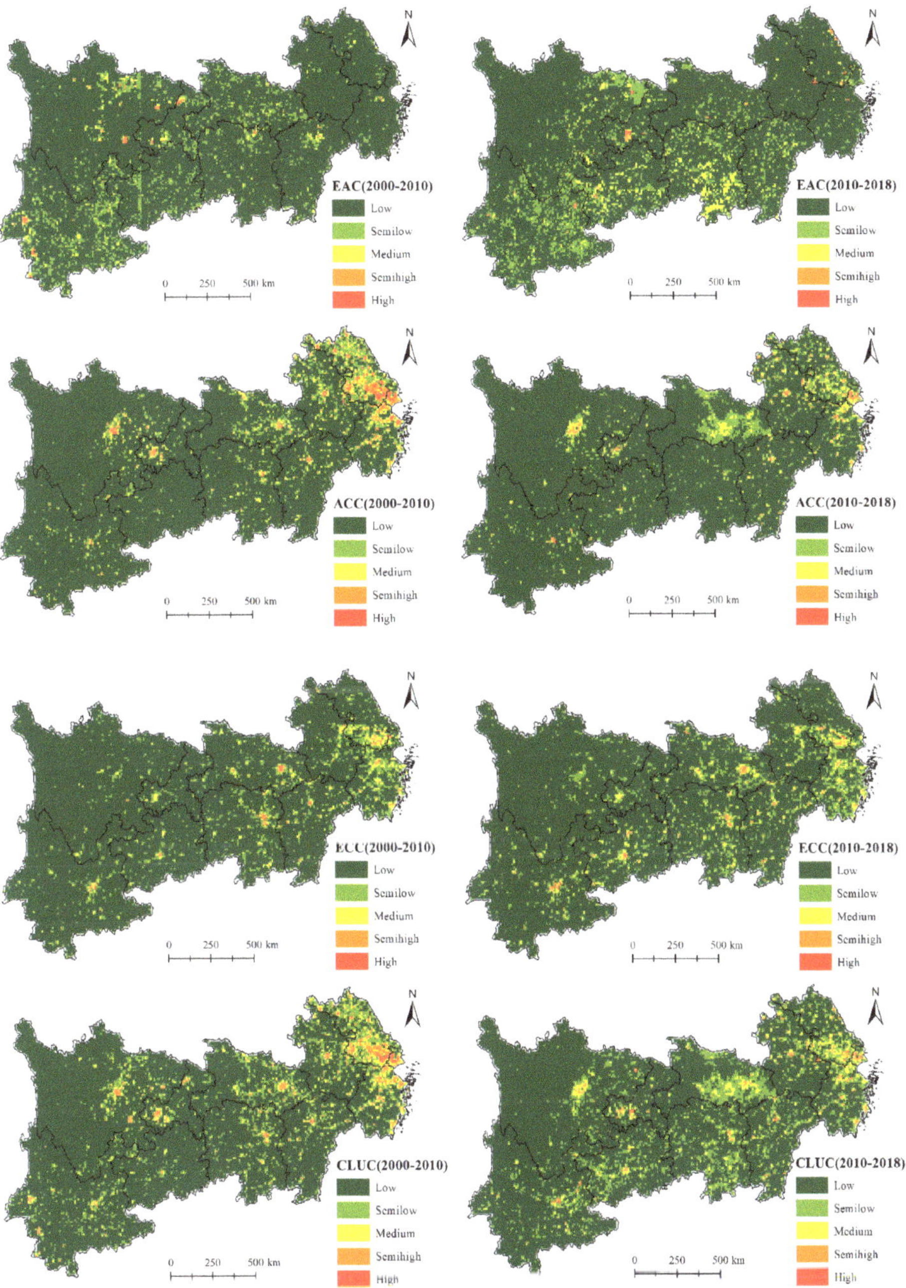


**Figure 4.** Spatial distribution pattern of land use conflict indicators during 2000–2010 and 2010–2018.

The ECC demonstrated evident east high and west low trend, as shown in Figure 4. High and semihigh value areas in YRD spread over a wide area. By comparison, high and semihigh value areas in the middle and western parts of the region concentrated in economically developed large cities. The high population density and rapid economic development in these regions required large amounts of construction land at the expense of ecological land. Meanwhile, the intensive production and living activities around these economic development zones interfere with the surrounding ecosystem and lead to severe land use conflict between humans and wild nature. The overall trend of the ECC increased both in size and distribution. The high, semihigh, and medium value areas increased from 61,500 km$^2$ to 79,800 km$^2$ and the average conflict intensity increased by 28.82%. Meanwhile, the distribution spread out from central cities to a wide area.

CLUC of YREB was calculated by integrating the three aspects of EAC, ACC, and ECC. As shown in Figure 4, CLUC in the eastern coastal region, especially YRD, is relatively high over a wide area. CLUC in middle and western inland regions is mainly concentrated in central cities. Wuhan, Chengdu, Chongqing, Changsha, and Nanchang are the five main conflict centers in the middle and western regions and are largely consistent with the range of the three national urban agglomerations. The rapid economic development and urbanization process in these areas stimulated the function transformation among ecological, agricultural, and construction lands and resulted in intense land use conflict. Areas with low values were located in western Hubei, western Sichuan, and Guizhou during 2000–2010. The overall trend of CLUC slightly declined by 9.91% during the research period.

### 4.2.3. Spatiotemporal Variation Pattern and Standard Deviation Ellipse Analysis of Land Use Indicators

The explicit variation and weighted standard deviation ellipse (SDE) of the conflict was calculated to further explore the variation of land use conflict indicators between the two periods (Figure 5).

Areas with increased EAC are mainly spread over middle and western regions. SDE locations of EAC spread along the western part of the study area confirmed the serious EAC in middle and western parts of YREB. Weighted mean centers revealed an evident shift in the direction of north by east due to the increase in EAC in the middle region and slight decrease in some areas in the western region of the study area. The increased degree of dispersion of EAC is depicted by the enlarged long and short axes of SDEs.

Areas with remarkably increasing ACC were mainly located in peripheral regions of Wuhan and Chengdu. These were star cities with rapid development in the past decade. At the same time, several cities, such as Kunming in Yunnan, Guiyang in Guizhou, Hefei in Anhui, and Nanchang in Jiangxi, also demonstrated considerably increasing ACC. By comparison, ACC evidently decreased in YRD and some regions in the central urban area of midwestern cities. This finding indicated that the urbanization process of YRD noticeably slowed down after reaching a high level. The weighted SDE of ACC was delineated to reveal its orientation and spatiotemporal development trends further. The weighted mean center of ACC clearly distributed along the east side indicated that ACC of the eastern part was considerably more serious than the western part. Furthermore, changes in the SDE location revealed a considerable shift in the direction of south by west. Hence, the land development intensity transferred from the eastern coastal area to the inland midwest because of the midwest regions accepting the industrial transfer from the eastern coastal area and the active economic development of the midwest region.

The comparison of distribution patterns of ECC in both periods showed that the intensity of ECC increases throughout the study area, except for a small region in Jiangsu and other big cities. The region development accelerated in the west, especially in Kunming and Guiyang, and caused a serious threat to the ecosystem. SDEs in both periods also demonstrated that ECC transferred to the west.

A large imbalance exists between western and oriental CLUC variance. CLUC decline regions are mainly located in the eastern coastal region and some urban centers of several big cities in middle and western inland regions, such as Wuhan, Chengdu, Changsha, and Nanchang. Regions with increasing CLUC were scattered around a wide area in the midwest. The level of dispersion of CLUC was enhanced considerably. SDEs of CLUC spread over northeast of the study area indicated the strong CLUC intensity in the east but weak CLUC intensity in the west. The overall CLUC transferred from east to west.

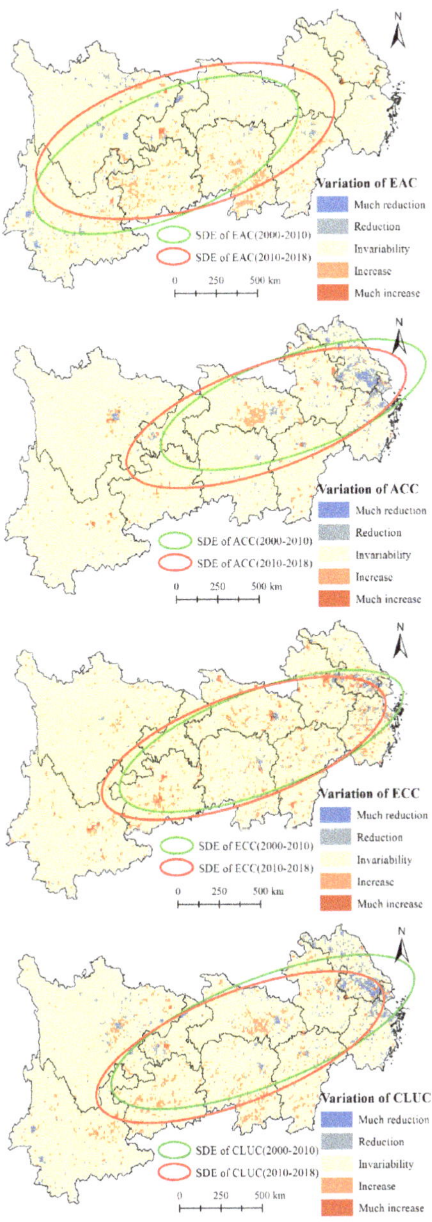


**Figure 5.** Variation and standard deviation ellipse (SDEs) during study periods of land use conflict indicators.

### 4.3. Driving Factors

Our correlation analysis showed that all selected socioeconomic variables were significantly correlated with land use conflict indices (with at least one indicator) although values of their Pearson correlation coefficients differed significantly (Table 7). Pearson correlation coefficients of total population of prefectural city (TPOP), number of hospital beds (HBN), and road density (RD) were larger than 0.60 and six of the variables were approximately 0.4 and less in the EAC conflict. High Pearson correlation coefficients of 0.807 between RD and EAC indicated that road extension leads to agricultural land reclamation through the destruction of ecological land. A Pearson correlation coefficient of $-0.189$ in POPD demonstrated that EAC occurs mainly in remote areas far from the city. Pearson correlation coefficients of TPOP, investment in fixed assets (IFA), total retail sales of consumer goods (TRSCG), HBN, and number of mobile phone users (MPUN) were larger than 0.7 with ACC. POPD, per capita GDP (GDPPC), and RD were approximately 0.3 and more with ACC. This finding indicated that population growth, economic development, and lifestyle improvement all generated ACC and exerted strong pressure on the protection of agricultural land. All driving factors demonstrated Pearson correlation coefficients of less than 0.6 with ECC. High Pearson correlation coefficients of more than 0.6 were observed in TPOP, IFA, TRSCG, HBN, MPUN, RD, and LUC. TPOP exhibited the highest Pearson correlation coefficient of 0.839 with CLUC, followed by HBN and MPUN with values of 0.824 and 0.790, respectively.

**Table 7.** Pearson correlation coefficients between the land use conflict indices and socioeconomic factors.

| Variables | EAC | ACC | ECC | CLUC |
|---|---|---|---|---|
| TPOP | 0.534 ** | 0.800 ** | 0.470 ** | 0.839 ** |
| POPD | $-0.189$ * | 0.573 ** | — | 0.217 * |
| GDPPC | — | 0.428 ** | 0.349 ** | 0.244 ** |
| IFA | 0.438 ** | 0.736 ** | 0.530 ** | 0.775 ** |
| TRSCG | 0.245 ** | 0.780 ** | 0.471 ** | 0.677 ** |
| HBN | 0.559 * | 0.729 ** | 0.505 ** | 0.824 ** |
| MPUN | 0.389 ** | 0.826 ** | 0.517 ** | 0.790 ** |
| RD | 0.807 ** | 0.376 ** | 0.419 ** | 0.743 ** |

Note: * Significant at the 0.05 level, ** Significant at the 0.01 level, and—Not significant.

The step-wise multiple linear regression results revealed that conflict indices can be appropriately explained by socioeconomic factors. Table 8 indicates that R-square values of Models 1, 2, 3, and 4 can reach 0.653, 0.734, 0.360, and 0.803, respectively. Unlike the results of the Pearson correlation analysis, our step-wise multiple linear regression showed that only some of the selected socioeconomic factors are statistically significant. TPOP represents demographic variables with a significantly positive relationship with EAC, ACC, and CLUC. By comparison, the negative coefficient of POPD, EAC, and CLUC indicated that land efficiency improvement effectively reduces land use conflict. Economic factors showed significantly positive coefficients with EAC, ACC, ECC, and CLUC in different aspects. GDPPC is the most important driving factor and positively correlated with EAC, ECC, and CLUC. IFA mainly drives the land exploitation and is thus positively correlated with ACC and ECC. TRSCG mainly drives the increase in ACC.

The lifestyle transition of residents also significantly affects the land use conflict. The significantly positive relationship of the coefficient of HBN and EAC indicated that the public service upgrading intensifies the land use conflict. The positive influence of MPUN on ACC indicated that lifestyle improvement affects the land use conflict. The positive correlation of RD with EAC and CLUC at 0.448 and 0.512, respectively, implied that the transportation development seriously disturbs the ecological system. However, RD showed a significant negative relationship with ECC because the growth in the living standard of residents improves the awareness of ecological protection and reduces the conflict.

**Table 8.** Standardized regression coefficients and coefficients of determination ($R^2$).

| Variables | EAC | ACC | ECC | CLUC |
|---|---|---|---|---|
| | Model 1 | Model 2 | Model 3 | Model 3 |
| TPOP | 0.002 * | 0.002 ** | — | 0.005 ** |
| POPD | −35.81 * | — | — | −42.61 ** |
| GDPPC | 2.214 ** | — | 0.068 ** | 1.812 ** |
| IFA | — | 0.99 * | 1.43 ** | — |
| TRSCG | — | 0.347 ** | — | — |
| HBN | 0.16 * | — | — | — |
| MPUN | — | 0.001 ** | — | — |
| RD | 0.448 ** | — | −0.315 ** | 0.512 * |
| R-squared | 0.653 | 0.734 | 0.360 | 0.803 |

Note: * Significant at the 0.05 level, ** Significant at the 0.01 level, and—Not significant. CLUC: comprehensive land use conflict; ecological and agricultural (EAC), agricultural and construction (ACC), and ecological and construction (ECC) land conflicts.

## 5. Discussion

### 5.1. Major Influencing Factors of Land Use Conflict in YREB

Our study demonstrated that the major influencing factors of land use conflict depend on the statistical methods used (i.e., Pearson correlation analysis vs. step-wise linear regression analysis in this case). The results of the Pearson correlation analysis indicated that all selected socioeconomic factors, except POPD, are significantly correlated with conflict indices individually (Table 6). This finding suggested that each factor can contribute to the land use conflict by enhancing human activities and intensifying land development. Particularly, the total population demonstrated the strongest positive correlation with all conflict indices. Population is a main driver of land development and urban expansion due to the requirement of sufficient production and living spaces [74,75]. People gathered together will generate various human activities and cause serious ecological space disturbance by encroaching on the natural habitat of animals and plants or polluting the water and soil [76,77]. For example, urbanization and population are important driving forces of the morphological change of lakes and the conflict between rapid urban growth and the maintenance of water landscape is increasingly intensified [78]. The increasing population causes a sharp rise in the land use conflict.

Road density is another important driver of land use conflict of the territorial space. On the one hand, roads split the natural environment and destroy the ecosystem integrity to some extent while exerting pressure on ecological and eco-production lands [79]. On the other hand, roads open the land for resource extraction and other human activities while increasing accessibility and mobility, thereby extending the level of human disturbance on many ecological processes [80,81]. Road networks affect the spatial structure of urban landscapes and exert additional widespread influences on the regional ecological environment with continuous expansion [82]. Hence, road density is an important driver of land use conflict, especially for EAC.

Other economic factors, such as per capita GDP, investment in fixed assets, and total retail sales of consumer goods all have significant relationships with conflict indices. The prevalence of the extensive use of land in economic development leads to the large consumption of land resources at the expense of ecological and eco-production land loss [83]. Overexploitation and utilization of the land and lack of land order administrative policies in economic development lead to remarkable changes in the territorial space. China is still in the period of rapid industrialization and urbanization, and massive land development will continue to intensify land use conflict and impose pressure on both ecological and food security. This trend is especially evident in many megacities, such as Beijing, and coastal regions of China, such as YRD and Pearl River Delta regions [51]. Accelerating transformations in economic development and land use modes is necessary in future developments.

Lifestyle improvements represented by the number of hospital beds and mobile phone users are also important driving factors of spatial conflicts. The rapid economic development that accompanied the industrialization and urbanization in China increased the demand of lifestyle experience. The industrialized lifestyle will consume additional energy and land and cause remarkable spatial conflict. Green and healthy lifestyles should be promoted seriously and become the habit and attitude of modern humans.

The step-wise linear regression analysis provided an efficient prediction model of spatial conflict using only socioeconomic variables with independent information. Hence, a small set of socioeconomic variables were statistically significant in this analysis due to the correlation between many variables. For example, the economic development represented by local financial revenue was strongly related to per capita GDP; hence, local financial revenue was excluded from the list of major influencing factors.

### 5.2. Policy Implications

Land use conflict is a serious problem during the land use transition process [14,84]. Promoting coordinated land use by exploring measures from the land use policy perspective of land use transitions is important. Land use policies play an important role in shaping the pattern of territorial space development and adjusting the land use conflict. China has introduced many land use policies that have serious effects on land use conflict since the 1980s (Table 9). For example, China's farmland protection system set the state policy for protecting farmlands with a series of measures and regulations, such as cultivated land reserve index of each administrative region assigned by the land administration department and tax levy on farmland occupation. Basic farmland protection and farmland dynamic balance systems have been established to protect farmlands from human construction occupation. These farmland protection policies play a crucial role in the adjustment of ACC.

Meanwhile, China introduced several ecological control policies, such as nature reserve regulation, the Grain for Green Project, ecological red line plans, and national parks. The first nature reserve was initiated in 1956 in Zhaoqing, Guangdong Province, and the nature reserve regulation was issued in 1994 to regulate the institution and administration of nature reserves. Since then, many nature reserve areas have played a considerable role in protecting biodiversity, preserving natural heritage, improving the quality of the ecological environment, and safeguarding national ecological security. At present, China has 2750 nature reserves that cover approximately 15% of the entire land area. However, serious problems, such as overlapping extent, multiple management, and obscure boundaries, were observed in the past institution and administration of nature reserves [85]. The state intends to establish a system of protected natural areas dominated by national parks to solve such problems through natural area integration, marginal adjustment, and clear functional position. The ecological red line policy is another control plan for protecting areas with special and important ecological functions. The ecological red line includes more territory space than nature reserves and covers more than a quarter of the land area of the entire country. The Grain for Green Project is a policy implemented for actively restoring farmland to forests and livestock pastures to natural grasslands. The Grain for Green policy has been carried out in more than 33 million ha to expand the ecological land effectively and consolidate ecological security in the last 20 years. These ecological control policies play a remarkable role in the adjustment of EAC and ECC.

Our findings have several implications for policy implementation and land use management in YREB. Land use conflict in YREB is still serious despite these implemented land use policies. The coordination of ecological, agricultural, and construction lands should be seriously considered in existing territorial space planning. For example, territorial space planning should solve problems of crossing and overlapping of basic farmlands and ecological red lines. Human activities and infrastructure projects should be strictly limited and gradually exist from the core region of nature reserves.

**Table 9.** Major governmental policies that influence the land use conflict of the territorial space in China from the 1980s to 2010s.

| Year | Policy/Decree | Main Contents |
| --- | --- | --- |
| 1986 | Farmland protection system | Cherish and rationally use every inch of land; effectively protecting farmlands is the basic state policy that our country must adhere to for a long time |
| 1989 | Basic farmland | The state established the basic farmland protection system and provided special protection to basic farmlands. |
| 1994 | Nature reserve regulation | The state formulated regulations to strengthen the construction and management of nature reserves and protect the natural environment and resources. |
| 1997 | Farmland dynamic balance system | The state established the farmland dynamic balance system in which farmlands are supplemented with no fewer than the land occupied by construction. |
| 1999 | Grain for Green | The state formulated policies to bring forward large-scale efforts to return farmlands to forests and restore livestock pastures to natural grasslands. |
| 2012 | Ecological red line | Areas with special and important ecological functions all over the country were strictly delineated into the ecological red line policy for special protection. |
| 2019 | National park | The state established a system of protected natural areas dominated by national parks. |

Development concept and governance ability must be improved to realize territorial spatial pattern optimization, accelerate ecological civilization, and promote high-quality development in the implementation of territorial space planning. Territorial space planning should be formulated with a comprehensive understanding of evolution context and regularity feature of the regional physical geographical environment and land use spatial patterns. Evaluating the resource environment carrying capacity, suitability of territorial space development, and disaster risk will build a solid foundation for territorial space planning. The harmonizing and processing rule of land use conflict should be investigated further on the basis of coordinated relations among urban development, population, land use, industry, and resource environment background to optimize the territorial spatial pattern. The economic development and population growth both demonstrate significant relationships with spatial conflicts. Hence, policies and regulations should be introduced to coordinate the economic development, population growth, and land development and utilization, especially in highly urbanized and flourishing economic areas. Road density is another important variable that can influence the land use conflict. Road construction must demonstrate reasonable layout planning and reduce the segmentation in the ecological land. The few remaining roadless and low-traffic areas in YREB should be considered and additional conservation measures must be included in laws and policies [81]. Systems of organizational scheme, technical standard, implementation supervision, as well as regulations and policies related to territorial space planning should be completed. Additional detailed and specific land control measures and regulations should be introduced to build coordinated territorial space patterns.

The land use efficiency and functional quality of construction land should be improved via intensive utilization and compact development. The scale of urban and rural construction land should be strictly controlled through the delineation of urban development boundary. We should overall allocate the industrial distribution and promote industrial clustering and centralization. The coupling relationship of production and living lands should be optimized according to local conditions by using the new development model of city and industry integration. Food security is a critical mission for agricultural land. Ensuring agricultural land scale and committing to the guarantee of food supply are realistic tasks that can be achieved by allocating the agricultural land on the basis of the "no reduction in quantity, no loss of mass, improvement in ecology, and optimization in layout" strategy. The high-quality agricultural land should be protected and measures must be taken to facilitate the large-scale industrialization and modernization development of agriculture. The concept of ecological priority must be established for ecological land. Ecological protection space should be delineated first during the formulation of territorial space planning. Ecological corridors and networks should be constructed to form a complete regional pattern for ecological security. Comprehensive land adjustment and ecological remediation must be promoted to improve land quality, optimize space function, and enhance ecological services.

## 6. Conclusions

Land use conflict is an important problem in land use transitions and territorial space development. This article proposed a new framework to evaluate the land use conflict spatially and explicitly from the three aspects of EAC, ACC, and ECC. This article then examined dynamic changes of land use conflict and performed Pearson correlation versus step-wise linear regression to determine factors that influence land use conflict during 2000–2018 in YREB, which is an important economic belt in China. The period of 2010–2018 demonstrated slightly lower compressive land use conflict than the period of 2000–2010. However, considerable differences exist in various conflict types. ACC reduced remarkably by 38.26%, while EAC and ECC showed a slowly increasing trend. The spatiotemporal dynamic pattern of each conflict indicator exhibits unique features. Higher and high values of the EAC were relatively decentralized and mainly spread out over middle and western inland regions. The period of 2010–2018 demonstrated higher intensity of EAC, with its center shifting to the east, compared with the period of 2000–2010. Areas with higher and high values of ACC are concentrated around three urban agglomerations, with YRD exhibiting the highest intensity. Compared with the period of 2000–2010, ACC evidently reduced and shifted to the west during 2010–2018. The ECC trend was evidently high in the east and low in the west. Compared with the period of 2000–2010, the overall ECC trend increased both in size and distribution and transferred to the west during 2010–2018. CLUC was mainly concentrated in central cities in middle and western inland regions and widespread in YRD. The overall CLUC transferred from east to west. Variables from three aspects of demographic, economic, and life factors significantly affected land use conflict. Policies, such as improvement of the territorial space planning system and promotion of comprehensive land adjustment, were put forward to adjust land use conflict.

This work has the following limitations. The land use conflict only considers transitions of land use categories but ignores the landscape change of the region. In addition, the driving factor analysis includes incompetence socioeconomic factors due to the data limitation. The exploration of the spatiotemporal dynamic pattern of land use conflict is a significant piece of work which can enhance the understanding of the dynamics of land use transitions and reveal the problems of agricultural land shrinking and environmental and ecological degradation during rapid urbanization and industrialization. Therefore, it is important to constantly monitor the land use conflict to identify the state of regional territorial space development and constantly adjust governance measures. The evaluation approach of land use conflict needs further improvement to provide deeper insight and more details of the analysis results. Future studies can focus on improving the analysis framework of land use conflict and further explore additional driving factors.

**Author Contributions:** Conceptualization, J.C. (Jiaxing Cui); methodology, Y.Z.; software, J.S.; validation, Y.Z.; formal analysis, J.C. (Jiaxing Cui); investigation, J.S.; resources, J.C. (Jing Chen); data curation, J.C. (Jing Chen.); writing—original draft preparation, J.C. (Jiaxing Cui); writing—review and editing, Y.Z.; visualization, X.K.; supervision, X.K.; project administration, J.C. (Jiaxing Cui); funding acquisition, J.C. (Jiaxing Cui). All authors have read and agreed to the published version of the manuscript.

**Funding:** This research was funded by the National Natural Science Foundation of China, grant number 41901201 and 41961031, and the Fundamental Research Funds for the Central Universities, grant number CCNU20QN034.

**Institutional Review Board Statement:** Not applicable.

**Informed Consent Statement:** Not applicable.

**Data Availability Statement:** Publicly available datasets were analyzed in this study. Land use data can be found here: [http://www.resdc.cn/]. Soil quality data can be found here: [http://westdc.westgis.ac.cn/zh-hans]. Nighttime lighting data can be found here: [http://www.geodata.cn/]. The basic geographic information base map can be found here: [http://bzdt.ch.mnr.gov.cn/index.html. Digital elevation model data can be found here: [http://www.webmap.cn/main.do?method=index].

**Acknowledgments:** We are grateful to Ruihao Li, Qiuxia Li, Shenyang Li, Zhe Li, Yating Ma, Ran Sun, Xinyi Dai, Zixuan Dai, and Huajie Luo for their help in part of the data processing.

**Conflicts of Interest:** The authors declare no conflict of interest.

## References


1. Crutzen, P.J. Geology of mankind. *Nature* **2002**, *415*, 23. [CrossRef]
2. Thomson, G.; Newman, P. Cities and the Anthropocene: Urban governance for the new era of regenerative cities. *Urban Stud.* **2020**, *57*, 1502–1519. [CrossRef]
3. Trischler, H. The Anthropocene: A Challenge for the History of Science, Technology, and the Environment. *NTM* **2016**, *24*, 309–335. [CrossRef] [PubMed]
4. Grimm, N.B.; Faeth, S.H.; Golubiewski, N.E.; Redman, C.L.; Wu, J.; Bai, X.; Briggs, J.M. Global change and the ecology of cities. *Science* **2008**, *319*, 756–760. [CrossRef]
5. Tu, Y.; Chen, B.; Yu, L.; Xin, Q.; Gong, P.; Xu, B. How does urban expansion interact with cropland loss? A comparison of 14 Chinese cities from 1980 to 2015. *Landsc. Ecol.* **2020**. [CrossRef]
6. Curtis, P.G.; Slay, C.M.; Harris, N.L.; Tyukavina, A.; Hansen, M.C. Classifying drivers of global forest loss. *Science* **2018**, *361*, 1108–1111. [CrossRef]
7. Zhang, W.; Villarini, G.; Vecchi, G.A.; Smith, J.A. Urbanization exacerbated the rainfall and flooding caused by hurricane Harvey in Houston. *Nature* **2018**, *563*, 384–388. [CrossRef]
8. Pielke, R.A. Land Use and Climate Change. *Science* **2005**, *310*, 1625. [CrossRef]
9. Jianchu, X.; Fox, J.; Vogler, J.B.; Yongshou, Z.P.; Lixin, Y.; Jie, Q.; Leisz, S. Land-use and land-cover change and farmer vulnerability in Xishuangbanna prefecture in southwestern China. *Environ. Manag.* **2005**, *36*, 404–413. [CrossRef]
10. McNeeley, S.M.; Even, T.L.; Gioia, J.B.M.; Knapp, C.N.; Beeton, T.A. Expanding vulnerability assessment for public lands: The social complement to ecological approaches. *Clim. Risk Manag.* **2017**, *16*, 106–119. [CrossRef]
11. Lioubimtseva, E. A multi-scale assessment of human vulnerability to climate change in the Aral Sea basin. *Environ. Earth Sci.* **2014**, *73*, 719–729. [CrossRef]
12. Huang, X.; Huang, X.; He, Y.; Yang, X. Assessment of livelihood vulnerability of land-lost farmers in urban fringes: A case study of Xi'an, China. *Habitat Int.* **2017**, *59*, 1–9. [CrossRef]
13. Liu, Y.; Fang, F.; Li, Y. Key issues of land use in China and implications for policy making. *Land Use Policy* **2014**, *40*, 6–12. [CrossRef]
14. Qu, Y.; Long, H. The economic and environmental effects of land use transitions under rapid urbanization and the implications for land use management. *Habitat Int.* **2018**, *82*, 113–121. [CrossRef]
15. Liang, X.; Jin, X.; Ren, J.; Gu, Z.; Zhou, Y. A research framework of land use transition in Suzhou City coupled with land use structure and landscape multifunctionality. *Sci. Total Environ.* **2020**, *737*, 139932. [CrossRef]
16. Long, H.; Qu, Y. Land use transitions and land management: A mutual feedback perspective. *Land Use Policy* **2018**, *74*, 111–120. [CrossRef]
17. Wang, J.; Li, Y.; Wang, Q.; Cheong, K.C. Urban–Rural Construction Land Replacement for More Sustainable Land Use and Regional Development in China: Policies and Practices. *Land* **2019**, *8*, 171. [CrossRef]
18. Huang, L.; Hu, S.; Li, S.; Fu, Z. Nonmarketization Bargaining and Actual Compensation Level for Land Requisition: A Qualitative Comparative Analysis of China's Land Requisition Conflict Events. *Sustainability* **2019**, *11*, 6084. [CrossRef]
19. Ma, W.; Jiang, G.; Chen, Y.; Qu, Y.; Zhou, T.; Li, W. How feasible is regional integration for reconciling land use conflicts across the urban–rural interface? Evidence from Beijing–Tianjin–Hebei metropolitan region in China. *Land Use Policy* **2020**, *92*, 104433. [CrossRef]
20. Wu, Y.; Heerink, N. Foreign direct investment, fiscal decentralization and land conflicts in China. *China Econ. Rev.* **2016**, *38*, 92–107. [CrossRef]
21. Zhou, D.; Lin, Z.; Lim, S.H. Spatial characteristics and risk factor identification for land use spatial conflicts in a rapid urbanization region in China. *Environ. Monit. Assess.* **2019**, *191*, 677. [CrossRef] [PubMed]
22. Liu, Z.; Liu, Y.; Baig, M.H.A. Biophysical effect of conversion from croplands to grasslands in water-limited temperate regions of China. *Sci. Total Environ.* **2019**, *648*, 315–324. [CrossRef] [PubMed]
23. Bergius, M.; Benjaminsen, T.A.; Maganga, F.; Buhaug, H. Green economy, degradation narratives, and land-use conflicts in Tanzania. *World Dev.* **2020**, *129*, 104850. [CrossRef]

24. Pacheco, F.A.L.; Varandas, S.G.P.; Sanches Fernandes, L.F.; Valle Junior, R.F. Soil losses in rural watersheds with environmental land use conflicts. *Sci. Total Environ.* **2014**, *485–486*, 110–120. [CrossRef]

25. Valle Junior, R.F.; Varandas, S.G.; Sanches Fernandes, L.F.; Pacheco, F.A. Groundwater quality in rural watersheds with environmental land use conflicts. *Sci. Total Environ.* **2014**, *493*, 812–827. [CrossRef]

26. Moore, S.A.; Brown, G.; Kobryn, H.; Strickland-Munro, J. Identifying conflict potential in a coastal and marine environment using participatory mapping. *J. Environ. Manag.* **2017**, *197*, 706–718. [CrossRef]

27. Kim, I.; Arnhold, S. Mapping environmental land use conflict potentials and ecosystem services in agricultural watersheds. *Sci. Total Environ.* **2018**, *630*, 827–838. [CrossRef]

28. Helwege, A. Challenges with resolving mining conflicts in Latin America. *Extr. Ind. Soc.* **2015**, *2*, 73–84. [CrossRef]

29. Helbron, H.; Schmidt, M.; Glasson, J.; Downes, N. Indicators for strategic environmental assessment in regional land use planning to assess conflicts with adaptation to global climate change. *Ecol. Indic.* **2011**, *11*, 90–95. [CrossRef]

30. Chen, S.; Yi, Z.-F.; Campos-Arceiz, A.; Chen, M.-Y.; Webb, E.L. Developing a spatially-explicit, sustainable and risk-based insurance scheme to mitigate human–wildlife conflict. *Biol. Conserv.* **2013**, *168*, 31–39. [CrossRef]

31. He, Y.; Tang, C.; Zhou, G.; He, S. The Analysis of Spatial Conflict Measurement in Fast Urbanization Region from the Perspective of Geography——A Case Study of Changsha-Zhuzhou-Xiangtan Urban Agglomeration. *J. Nat. Resour.* **2014**, *29*, 1660–1674.

32. Zou, L.; Liu, Y.; Wang, J.; Yang, Y.; Wang, Y. Land use conflict identification and sustainable development scenario simulation on China's southeast coast. *J. Clean. Prod.* **2019**, *238*, 117899. [CrossRef]

33. Iojă, C.I.; Niță, M.R.; Vânău, G.O.; Onose, D.A.; Gavrilidis, A.A. Using multi-criteria analysis for the identification of spatial land-use conflicts in the Bucharest Metropolitan Area. *Ecol. Indic.* **2014**, *42*, 112–121. [CrossRef]

34. You, H.; Yang, X. Urban expansion in 30 megacities of China: Categorizing the driving force profiles to inform the urbanization policy. *Land Use Policy* **2017**, *68*, 531–551. [CrossRef]

35. Long, H.; Qu, Y.; Tu, S.; Zhang, Y.; Jiang, Y. Development of land use transitions research in China. *J. Geogr. Sci.* **2020**, *30*, 1195–1214. [CrossRef]

36. Xu, Y.; McNamara, P.; Wu, Y.; Dong, Y. An econometric analysis of changes in arable land utilization using multinomial logit model in Pinggu district, Beijing, China. *J. Environ. Manag.* **2013**, *128*, 324–334. [CrossRef] [PubMed]

37. Feng, Y.; Liu, Y.; Tong, X. Spatiotemporal variation of landscape patterns and their spatial determinants in Shanghai, China. *Ecol. Indic.* **2018**, *87*, 22–32. [CrossRef]

38. Liu, Y.; Peng, J.; Zhang, T.; Zhao, M. Assessing landscape eco-risk associated with hilly construction land exploitation in the southwest of China: Trade-off and adaptation. *Ecol. Indic.* **2016**, *62*, 289–297. [CrossRef]

39. Li, Y.; Sun, X.; Zhu, X.; Cao, H. An early warning method of landscape ecological security in rapid urbanizing coastal areas and its application in Xiamen, China. *Ecol. Model.* **2010**, *221*, 2251–2260. [CrossRef]

40. Wang, D.; Jiang, D.; Fu, J.; Lin, G.; Zhang, J. Comprehensive Assessment of Production–Living–Ecological Space Based on the Coupling Coordination Degree Model. *Sustainability* **2020**, *12*, 2009. [CrossRef]

41. Lin, G.; Jiang, D.; Fu, J.; Cao, C.; Zhang, D. Spatial Conflict of Production–Living–Ecological Space and Sustainable-Development Scenario Simulation in Yangtze River Delta Agglomerations. *Sustainability* **2020**, *12*, 2175. [CrossRef]

42. Jin, G.; Deng, X.; Zhao, X.; Guo, B.; Yang, J. Spatiotemporal patterns in urbanization efficiency within the Yangtze River Economic Belt between 2005 and 2014. *J. Geogr. Sci.* **2018**, *28*, 1113–1126. [CrossRef]

43. Ma, Q.; He, C.; Wu, J. Behind the rapid expansion of urban impervious surfaces in China: Major influencing factors revealed by a hierarchical multiscale analysis. *Land Use Policy* **2016**, *59*, 434–445. [CrossRef]

44. Mantas, V.M.; Marques, J.C.; Pereira, A.J.S.C. A geospatial approach to monitoring impervious surfaces in watersheds using Landsat data (the Mondego Basin, Portugal as a case study). *Ecol. Indic.* **2016**, *71*, 449–466. [CrossRef]

45. Tong, L.; Hu, S.; Frazier, A.E. Hierarchically measuring urban expansion in fast urbanizing regions using multi-dimensional metrics: A case of Wuhan metropolis, China. *Habitat Int.* **2019**, *94*, 102070. [CrossRef]

46. Huang, X.; Xia, J.; Xiao, R.; He, T. Urban expansion patterns of 291 Chinese cities, 1990–2015. *Int. J. Digit. Earth* **2017**, *12*, 62–77. [CrossRef]

47. Mitsova, D. Coupling Land Use Change Modeling with Climate Projections to Estimate Seasonal Variability in Runoff from an Urbanizing Catchment Near Cincinnati, Ohio. *ISPRS Int. J. Geo-Inf.* **2014**, *3*, 1256–1277. [CrossRef]

48. Du, N.; Ottens, H.; Sliuzas, R. Spatial impact of urban expansion on surface water bodies—A case study of Wuhan, China. *Landsc. Urban Plan.* **2010**, *94*, 175–185. [CrossRef]

49. Yao, X.; Christensen, M.J.; Bao, G.; Zhang, C.; Li, X.; Li, C.; Nan, Z. A toxic endophyte-infected grass helps reverse degradation and loss of biodiversity of over-grazed grasslands in northwest China. *Sci. Rep.* **2015**, *5*, 18527. [CrossRef]

50. Montanari, A.; Londei, A.; Staniscia, B. Can we interpret the evolution of coastal land use conflicts? Using Artificial Neural Networks to model the effects of alternative development policies. *Ocean Coast. Manag.* **2014**, *101*, 114–122. [CrossRef]

51. Valle Junior, R.F.; Varandas, S.G.P.; Sanches Fernandes, L.F.; Pacheco, F.A.L. Environmental land use conflicts: A threat to soil conservation. *Land Use Policy* **2014**, *41*, 172–185. [CrossRef]

52. Liu, J.; Kuang, W.; Zhang, Z.; Xu, X.; Qin, Y.; Ning, J.; Zhou, W.; Zhang, S.; Li, R.; Yan, C.; et al. Spatiotemporal characteristics, patterns, and causes of land-use changes in China since the late 1980s. *J. Geogr. Sci.* **2014**, *24*, 195–210. [CrossRef]

53. Liu, J.; Liu, M.; Tian, H.; Zhuang, D.; Zhang, Z.; Zhang, W.; Tang, X.; Deng, X. Spatial and temporal patterns of China's cropland during 1990–2000: An analysis based on Landsat TM data. *Remote Sens. Environ.* **2005**, *98*, 442–456. [CrossRef]

54. Li, D.; Xu, E.; Zhang, H. Influence of ecological land change on wind erosion prevention service in arid area of northwest China from 1990 to 2015. *Ecol. Indic.* **2020**, *117*, 106686. [CrossRef]

55. Ning, J.; Liu, J.; Zhao, G. Spatio-temporal characteristics of disturbance of land use change on major ecosystem function zones in China. *Chin. Geogr. Sci.* **2015**, *25*, 523–536. [CrossRef]

56. Zhong, Y.; Lin, A.; He, L.; Zhou, Z.; Yuan, M. Spatiotemporal Dynamics and Driving Forces of Urban Land-Use Expansion: A Case Study of the Yangtze River Economic Belt, China. *Remote Sens.* **2020**, *12*, 287. [CrossRef]

57. Wang, H.; Meng, X. *Siol Map Based Harmonized World Soil Database (v1.2)*; National Tibetan Plateau Data Center, Institute of Tibetan Plateau Research: Beijing, China, 2018.

58. Wieder, W. *Regridded Harmonized World Soil Database v1.2*; ORNL Distributed Active Archive Center: Oak Ridge, TN, USA, 2014.

59. Fischer, G.; Nachtergaele, F.; Prieler, S.; van Velthuizen, H.T.; Verelst, L.; Wiberg, D. *Global Agro-Ecological Zones Assessment for Agriculture (GAEZ 2008)*; IIASA: Laxenburg, Austria; FAO: Rome, Italy, 2008.

60. Shen, Y.; Xiong, A. Validation and comparison of a new gauge-based precipitation analysis over mainland China. *Int. J. Climatol.* **2016**, *36*, 252–265. [CrossRef]

61. Zhang, Y.; Long, H.; Tu, S.; Ge, D.; Ma, L.; Wang, L. Spatial identification of land use functions and their tradeoffs/synergies in China: Implications for sustainable land management. *Ecol. Indic.* **2019**, *107*, 105550. [CrossRef]

62. Liao, G.; He, P.; Gao, X.; Deng, L.; Zhang, H.; Feng, N.; Zhou, W.; Deng, O. The Production–Living–Ecological Land Classification System and Its Characteristics in the Hilly Area of Sichuan Province, Southwest China Based on Identification of the Main Functions. *Sustainability* **2019**, *11*, 1600. [CrossRef]

63. Zhao, X.; Li, S.; Pu, J.; Miao, P.; Wang, Q.; Tan, K. Optimization of the National Land Space Based on the Coordination of Urban-Agricultural-Ecological Functions in the Karst Areas of Southwest China. *Sustainability* **2019**, *11*, 6752. [CrossRef]

64. Gao, L.; Ma, C.; Wang, Q.; Zhou, A. Sustainable use zoning of land resources considering ecological and geological problems in Pearl River Delta Economic Zone, China. *Sci. Rep.* **2019**, *9*, 16052. [CrossRef]

65. Yin, C.; Kong, X.; Liu, Y.; Wang, J.; Wang, Z. Spatiotemporal changes in ecologically functional land in China: A quantity-quality coupled perspective. *J. Clean. Prod.* **2019**, *238*, 117917. [CrossRef]

66. Long, H.; Liu, Y.; Hou, X.; Li, T.; Li, Y. Effects of land use transitions due to rapid urbanization on ecosystem services: Implications for urban planning in the new developing area of China. *Habitat Int.* **2014**, *44*, 536–544. [CrossRef]

67. Chi, Y.; Shi, H.; Zheng, W.; Sun, J.; Fu, Z. Spatiotemporal characteristics and ecological effects of the human interference index of the Yellow River Delta in the last 30 years. *Ecol. Indic.* **2018**, *89*, 880–892. [CrossRef]

68. Valera, C.A.; Valle Junior, R.F.; Varandas, S.G.P.; Sanches Fernandes, L.F.; Pacheco, F.A.L. The role of environmental land use conflicts in soil fertility: A study on the Uberaba River basin, Brazil. *Sci. Total Environ.* **2016**, *562*, 463–473. [CrossRef]

69. Li, T.; Long, H.; Zhang, Y.; Tu, S.; Ge, D.; Li, Y.; Hu, B. Analysis of the spatial mismatch of grain production and farmland resources in China based on the potential crop rotation system. *Land Use Policy* **2017**, *60*, 26–36. [CrossRef]

70. Ge, D.; Long, H.; Zhang, Y.; Ma, L.; Li, T. Farmland transition and its influences on grain production in China. *Land Use Policy* **2018**, *70*, 94–105. [CrossRef]

71. Yang, J.; Yang, J.; Luo, X.; Huang, C. Impacts by expansion of human settlements on nature reserves in China. *J. Environ. Manag.* **2019**, *248*, 109233. [CrossRef] [PubMed]

72. Nguyen, T.T.; Verdoodt, A.; Van, Y.T.; Delbecque, N.; Tran, T.C.; Van Ranst, E. Design of a GIS and multi-criteria based land evaluation procedure for sustainable land-use planning at the regional level. *Agric. Ecosyst. Environ.* **2015**, *200*, 1–11. [CrossRef]

73. Li, J.; Jiang, H.; Bai, Y.; Alatalo, J.M.; Li, X.; Jiang, H.; Liu, G.; Xu, J. Indicators for spatial–temporal comparisons of ecosystem service status between regions: A case study of the Taihu River Basin, China. *Ecol. Indic.* **2016**, *60*, 1008–1016. [CrossRef]

74. Deng, X.; Zhao, C.; Yan, H. Systematic Modeling of Impacts of Land Use and Land Cover Changes on Regional Climate: A Review. *Adv. Meteorol.* **2013**, *2013*, 1–11. [CrossRef]

75. Kuang, W.; Liu, J.; Dong, J.; Chi, W.; Zhang, C. The rapid and massive urban and industrial land expansions in China between 1990 and 2010: A CLUD-based analysis of their trajectories, patterns, and drivers. *Landsc. Urban Plan.* **2016**, *145*, 21–33. [CrossRef]

76. Viglia, S.; Civitillo, D.F.; Cacciapuoti, G.; Ulgiati, S. Indicators of environmental loading and sustainability of urban systems. An emergy-based environmental footprint. *Ecol. Indic.* **2018**, *94*, 82–99. [CrossRef]

77. Jackson, C.R.; Bahn, R.A.; Webster, J.R. Water Quality Signals from Rural Land Use and Exurbanization in a Mountain Landscape: What's Clear and What's Confounded? *JAWRA J. Am. Water Resour. Assoc.* **2017**, *53*, 1212–1228. [CrossRef]

78. Chen, K.; Wang, X.; Li, D.; Li, Z. Driving force of the morphological change of the urban lake ecosystem: A case study of Wuhan, 1990–2013. *Ecol. Model.* **2015**, *318*, 204–209. [CrossRef]

79. Feng, Y.; Yang, Q.; Tong, X.; Chen, L. Evaluating land ecological security and examining its relationships with driving factors using GIS and generalized additive model. *Sci. Total Environ.* **2018**, *633*, 1469–1479. [CrossRef]

80. Hu, X.; Wu, Z.; Wu, C.; Ye, L.; Lan, C.; Tang, K.; Xu, L.; Qiu, R. Effects of road network on diversiform forest cover changes in the highest coverage region in China: An analysis of sampling strategies. *Sci. Total Environ.* **2016**, *565*, 28–39. [CrossRef]

81. Selva, N.; Kreft, S.; Kati, V.; Schluck, M.; Jonsson, B.-G.; Mihok, B.; Okarma, H.; Ibisch, P.L. Roadless and Low-Traffic Areas as Conservation Targets in Europe. *Environ. Manag.* **2011**, *48*, 865–877. [CrossRef]

82. Mo, W.; Wang, Y.; Zhang, Y.; Zhuang, D. Impacts of road network expansion on landscape ecological risk in a megacity, China: A case study of Beijing. *Sci. Total Environ.* **2017**, *574*, 1000–1011. [CrossRef]

83. Tan, M.; Li, X.; Xie, H.; Lu, C. Urban land expansion and arable land loss in China—a case study of Beijing–Tianjin–Hebei region. *Land Use Policy* **2005**, *22*, 187–196. [CrossRef]
84. Liu, Y.; Long, H.; Li, T.; Tu, S. Land use transitions and their effects on water environment in Huang-Huai-Hai Plain, China. *Land Use Policy* **2015**, *47*, 293–301. [CrossRef]
85. Wang, L.; Pan, Y.; Cao, Y.; Li, B.; Wang, Q.; Wang, B.; Pang, W.; Zhang, J.; Zhu, Z.; Deng, G. Detecting early signs of environmental degradation in protected areas: An example of Jiuzhaigou Nature Reserve, China. *Ecol. Indic.* **2018**, *91*, 287–298. [CrossRef]

*Article*

# The Evolution of the Interactive Relationship between Urbanization and Land-Use Transition: A Case Study of the Yangtze River Delta

Bo Niu [1], Dazhuan Ge [1,*], Rui Yan [2], Yingyi Ma [3], Dongqi Sun [2], Mengqiu Lu [4] and Yuqi Lu [1]

[1] School of Geography, Nanjing Normal University, Nanjing 210023, China; 201302052@nnu.edu.cn (B.N.); luyuqi@njnu.edu.cn (Y.L.)
[2] Institute of Geographic Sciences and Natural Resources Research, Chinese Academy of Sciences, Beijing 100101, China; yanrui20@mails.ucas.ac.cn (R.Y.); sundq@igsnrr.ac.cn (D.S.)
[3] School of Architectural Engineering, Jinling Institute of Technology, Nanjing 211169, China; mayingyi@jit.edu.cn
[4] School of International Economics and Trade, Nanjing University of Finance and Economics, Nanjing 210023, China; mqlu3201@nufe.edu.cn
* Correspondence: gedz@njnu.edu.cn; Tel.: +86-25-8589-1347

**Abstract:** In recent years, the impact of land-use systems on global climate change has become increasingly significant, and land-use change has become a hot issue of concern to academics, both within China and abroad. Urbanization, as an important socioeconomic factor, plays a vital role in promoting land-use transition, which also shows a significant spatial dependence on urbanization. This paper constructs a theoretical framework for the interaction relationship between urbanization and land-use transition, taking the Yangtze River Delta as an example, and measures the level of urbanization from the perspective of population urbanization, economic urbanization and social urbanization, while also evaluating the level of land-use morphologies from the perspective of dominant and recessive morphologies of land-use. We construct a PVAR model and coupled coordination model based on the calculated indexes for empirical analysis. The results show that the relationship between urbanization and land-use transition is not a simple linear relationship, but tends to be complex with the process of urbanization, and reasonable urbanization and land-use morphologies will promote further benign coupling in the system. By analyzing the interaction relationship between urbanization and land-use transition, this study enriches the study of land-use change and provides new pathways for thinking about how to promote high-quality urbanization.

**Keywords:** urbanization; land-use transition; interactive relationship evolution; PVAR model; coupled coordination

**Citation:** Niu, B.; Ge, D.; Yan, R.; Ma, Y.; Sun, D.; Lu, M.; Lu, Y. The Evolution of the Interactive Relationship between Urbanization and Land-Use Transition: A Case Study of the Yangtze River Delta. *Land* **2021**, *10*, 804. https://doi.org/10.3390/land10080804

Academic Editor: Ileana Pătru-Stupariu

Received: 28 June 2021
Accepted: 28 July 2021
Published: 30 July 2021

**Publisher's Note:** MDPI stays neutral with regard to jurisdictional claims in published maps and institutional affiliations.

## 1. Introduction

Land-use change is a significant form of interaction and connection between human activities and the natural environment [1]. With the development of the economy, the impact of the land-use system on global climate change and global environmental change has become increasingly significant. In recent years, land-use change has become a hot issue [2–6], both in China and abroad. Grainger described land-use morphology as the general morphology of the actual land cover over a certain period of time, and proposed the concept of land-use transition (the changes in the land-use morphology over a certain period of time) [7], inspired by the "forest transition" hypothesis put forward by Mather [8]. At the beginning of the 21st century, the concept of land-use transition was introduced into China by Hualou Long, and it has been explored as a new means of land-use/cover change (LUCC) [9]. At the beginning of the introduction of land-use transition in China, it mainly referred to the time-series changes in land-use morphology affected by social and economic development [10]. With the deepening of the related research, the connotations of land-use morphology became enriched, and its meaning expanded from the structure of land-use types to encompass the dominant and recessive morphologies of land-use. The former refers to the structure of the composition of the main land-use types in a region during a specific period of time, and the latter refers to the land-use morphology, relying on land-use dominant morphology, such as properties, functions, input and output [11,12]. The connotations of land-use transition have also been further deepened. At present, under the theoretical framework of the dominant and recessive morphologies of land-use, studies on land-use dominant morphology transition mainly focus on the problems of the evolution of quantitative land-use structural change (changes in the proportion of different types of land-use) [13] and space–time morphological characteristics (characteristics of land-use morphologies in terms of time series and spatial differences) [14]. Studies of land-use recessive morphology transition mainly focus on transitions in land-use function (main uses of the land) [15–17], land-use efficiency (output efficiency per unit of the land) [18], and land-use intensity (input per unit of the land) [19].

Multiple studies have identified different natural environmental and socioeconomic factors that function as the driving forces of land-use transition, and have carried out analyses of the driving mechanisms [20–22]. Among these factors, urbanization, as an important force promoting social and economic development, has a profound impact on the change of land-use morphology. Additionally, land-use transition also shows a significant spatial dependence on urbanization [23]. China's rapid urbanization began in the 1990s [24]. Under the current trend of economic globalization, urbanization not only promotes the rapid development of the social economy, but also remodels the form and morphology of land-use. The change in urban land-use structure, and the orderliness and rationality of land-use, are closely related to the urbanization process [23]. A review of related research on urbanization and land-use transition in recent years shows that current studies mainly focus on the role of urbanization in certain dimensions of land-use, such as land-use structure [25,26], land-use efficiency [27] and land-use intensity [28,29]. The effects of land-use transition mainly include the socioeconomic effect [30–32] and the eco-environmental effect [33–35]. However, few studies start from a systems theory perspective to undertake a comprehensive analysis of the two-way interaction and the mechanism between the urbanization system and the land-use transition system. Most studies only focus on the one-way impact of urbanization on land-use transition [23], or the impact of land-use transition [30–35]. The results of the study of Hualou Long and Yi Qu show that there is mutual feedback between land-use transition and land management, but this study does not address the interrelationships between land-use transition and urbanization [36].

An in-depth assessment of urbanization will reveal that economic efficiency is no longer the only goal of urbanization; urbanization is beginning to shift to high-quality, sustainable development. In this context, the relationship between urbanization and land-use also changes. Analyzing the two-way role between urbanization and the land-use system from the perspective of systems theory is of great significance to further understand both urbanization and land-use transition. From the perspective of land-use morphology, this paper argues that there is a two-way interaction relationship between urbanization and land-use transition (urbanization shapes land-use morphologies, and land-use morphologies feed back into and influence the process of urbanization), and that the interaction relationship tends to become more complex as urbanization progresses. Based on this, we focus on the two-way role between the urbanization system and land-use transition. The difficulty lies in how to reasonably measure and simulate the land recessive morphology [37] and how to quantify the two-way interaction between the urbanization system and the dominant and recessive morphology land-use systems. In view of this, this study takes the Yangtze Triangle—which has the highest densities of economic activity, population, and cities in China [38]—as its research object, and constructs a comprehensive index system to quantitatively describe its urbanization and the dominant and recessive morphologies of its land-use. This study also establishes panel vector autoregression models (PVAR) and a coupled coordination model to describe the urbanization system and land-use morphology system from 2000 to 2020. In order to further enrich the theory of land-use morphology transition and clarify the relationship between urbanization and land-use transition, this paper analyzes the interactive relationships and mechanisms among the subsystems of land-use morphology.

## 2. Theoretical Framework

### 2.1. The Effect of Urbanization on Land-Use Transition

Land-use transition refers to a change in land-use morphology over a certain period of time [10], usually corresponding to a stage of transformation of social and economic development [11]. The driving forces of land-use transition can be divided into endogenous natural driving factors and exogenous economic driving factors according to their sources, wherein exogenous economic driving factors are generally considered the main factors driving changes in land-use morphology. Among the economic factors driving land-use transition, urbanization plays a significant role [39]. The main forms of urbanization include the spatial agglomeration of the population in cities (population urbanization), the development of social and economic industries to a higher stage (economic urbanization), and changes in the lifestyle and quality of life of urban residents (social urbanization). The above urbanization process has brought about corresponding changes in the areas of land-use structures, land-use forms, etc. (as well as urbanization). The expansion in terms of area of urbanized construction land causes the spatial morphology of land-use to change significantly [40]. In addition, changes in the recessive morphology of land-use, such as ownership, use, mode of operation, and input–output, are also reflected in agricultural land acquisition and conversion, and the transfer of rural collective construction land [41]. Therefore, urbanization causes land-use type change by promoting the expansion of urban space. The spatial morphology of land-use tends to be complicated, and the value attributes of land are also closely related to it. In addition, land-use morphology is also closely related to the stage of urbanization development, which makes the land-use morphology different in different regions.

### 2.2. The Effect of Land-Use Transition on Urbanization

The response of urbanization to land-use transition is related to the environmental effects of land-use transition, and the evolution of the spatial structure and the function of land-use. With the acceleration of urbanization, dramatic land-use transition in a region will inevitably lead to regional economic, social, and environmental changes [30–35,42]. Urbanization is an integrated process that encompasses higher stages of development in economic, ecological and social dimensions. Land-use changes in spatial structure, efficiency, and other dominant and recessive morphological aspects will directly affect the efficiency and quality of the development of society and the economy, which will also have a significant impact on the promotion of urbanization. Taking the problem of Urban Villages (A general term for the villages in the city) in China's urbanization as an example, to a certain extent, this is the result of unsuccessful land-use transition. This also reveals that inefficient urban and rural land-use and fragmental land spatial morphology will hinder the transformation of urbanization into the stage of high-quality development. The effects of land-use transition on resources and the environment are an important driving factor in promoting the transition to urbanization. An unreasonable land-use transition process (in the context of rapid urbanization, the rapid expansion of land for urban construction has put the regional ecology under duress) will worsen the resource and environmental problems, and lead to the aggravation of the ecological and environmental crisis [33–35], which will hinder the transition of the urban development strategy and the construction of resource/environment-friendly urbanization.

### 2.3. Interactive Relationship between Urbanization and Land-Use Transition

The relationship between urbanization and land-use transition is not a simple one-way relationship, but an interactive process. In terms of land-use dominant morphology, rapid urbanization inevitably leads to the expansion of urban space, which is mainly reflected in the rapid increase in the land area under urban construction, which aggravates the trend of land fragmentation in the absence of rational planning and control. The efficient utilization and sustainable transition of land-use are key to ensuring the orderly progress of urbanization. However, an unreasonable land-use structure and function system will become a significant obstacle inhibiting the development of urbanization. The urbanization process is closely related to the structural system and quantitative characteristics of land-use dominant morphology. The effectiveness of spatial expansion control in urbanization directly determines the direction and trend of construction land expansion, and significantly changes the quantitative relationship and structural characteristics of land-use dominant morphology. In the process of rapid urbanization, the structural transformation of construction land and agricultural land ensnares strong institutional and financial support for urbanization, which is an important guarantee to ensure the rapid advancement of urbanization. The above analysis shows that the dominant morphological change in land-use is an intuitive reflection of the spatial projection of the urbanization process, and it also affects the quality and process of urbanization development to a certain extent.

In terms of land-use recessive morphology, changes in land-use efficiency and intensity, and in the level of land-use function, effectively reflect the quality of urbanization. Relevant studies point out that the overall efficiency of land-use shows a downward trend against the background of rapid urbanization [43], reflecting that in the extensive urbanization mode, the value of land is not fully activated, and the problems of land recessive morphology transition and rapid urbanization become increasingly prominent. Similarly, disordered land-use patterns and management modes, inefficient land function development status, and extensive land inputs and outputs all become obstacles to further urbanization. In contrast, rational land-use structures and efficient land development patterns constitute an orderly land-use system and are conducive to the formation of a benign interactive relationship with the urbanization system, which promotes and coordinates the transformation between the two systems.

This study analyzes the interactive relationship between urbanization and land-use transition from the perspective of systems theory and constructs the theoretical framework in Figure 1, drawing on the concepts of the dominant and recessive morphologies of land-use to construct the interactive relationship between urbanization and land-use transition. Urbanization has reshaped land-use morphology and has an important impact on the dominant and recessive morphologies of land-use, which is manifested in changes in the spatial morphology, quantitative structure, efficiency, and function of land-use. Land-use morphology responds to urbanization and produces land-use transition, which, in turn, affects urbanization and its sustainable development. Based on this, the interaction mode between the urbanization system and the land-use system is discussed, and its interactive characteristics will become an important basis for promoting the development of urbanization quality. The coordinated coupling of land-use morphology and urbanization status is conducive to promoting the construction of high-quality urbanization. In contrast, a coupled antagonistic state of the two systems is a significant obstacle inhibiting the benign development of urbanization. Unreasonable urbanization will strengthen the conflict in the pattern of regional land-use morphology and become an important driving force promoting transition in the dominant and recessive morphologies of land-use.

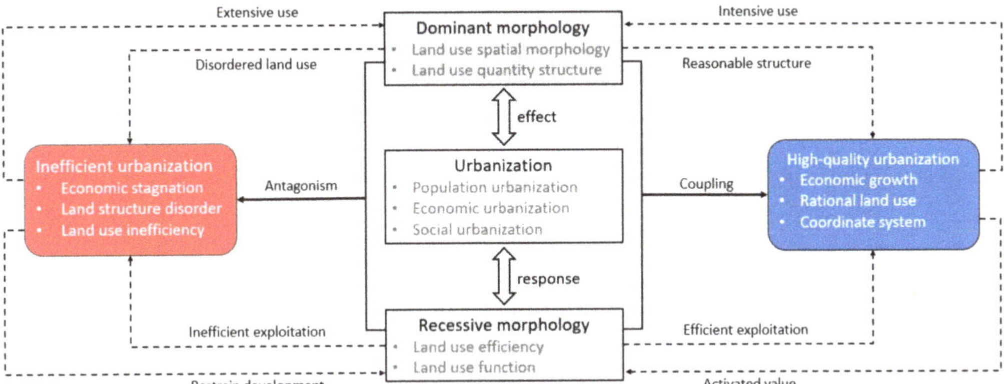



**Figure 1.** Analytical framework of interactive relationship between urbanization and land-use transition.

## 3. Data and Methods

### 3.1. Data Sources

To measure the level of urbanization, referring to the index system of related studies [44–46], this paper constructs a comprehensive evaluation system of urbanization and calculates the urbanization index (UI) of cities in the Yangtze River Delta region from the three dimensions of population urbanization (the spatial agglomeration level of the population in cities), economic urbanization (the development level of social and economic industries), and social urbanization (the quality of life of urban residents). The index data are from the China Urban Statistical Yearbook (2000–2020) and the local statistical yearbook (some of the indicators are not listed in the China Urban Statistical Yearbook, which are collated in the local statistical yearbooks of each prefecture-level city, such as the number of students in colleges and universities per 10,000 people, the number of beds in hospitals and health centers per 10,000 people, and the green coverage rate of built-up areas). It is worth noting that, as the relevant data from the 2020 Statistical Yearbook have not yet been released, the social and economic data from each local statistical yearbook in 2019 are applied to replace them. Due to the differences in the caliber of the statistical methods used to measure the urban population in some cities in earlier years, the urbanization rate of some cities increased nearly four times over five years. Therefore, this research applies the linear regression method to fit the urbanization rate of some cities in Anhui Province in 2000 to make the data stable.

As for the land-use data, in recent years, land-use remote sensing data have become an important support for LUCC-related studies [47]. This paper uses grid data for land-use (with a resolution of 1 km) in 2000, 2005, 2010, 2015, and 2020, which are available through the Resource and Environment Science and Data Center of the Chinese Academy of Sciences. On the basis of preprocessing the raster data, we calculate the relevant landscape pattern indexes using Fragstats to measure the land-use dominant morphology index (LUDMI). Combined with the raster data and socioeconomic data, we calculate the land-use recessive morphology index (LURMI).

### 3.2. Research Methods

Based on the main analysis methods and mathematical models of the article, we have drawn the analysis process into a flow plot (Figure 2).

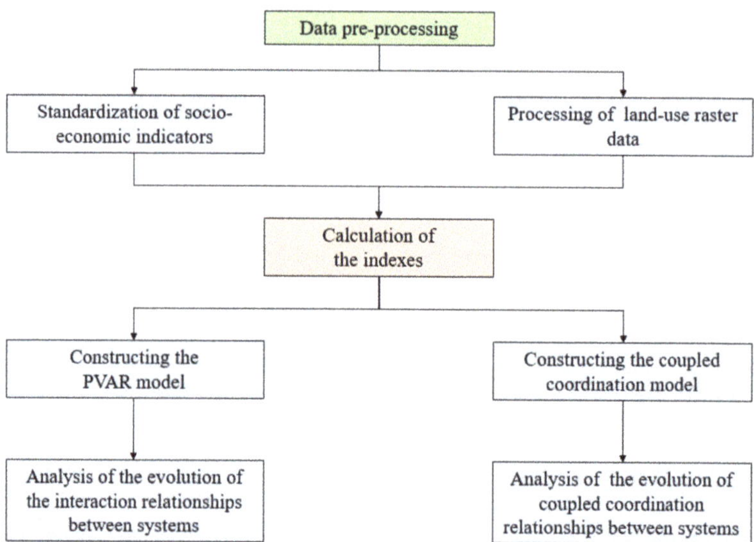

**Figure 2.** The process of data analysis.

### 3.2.1. Index System Construction and Weight Calculation

The purpose of this study is to explore the interactive relationship between urbanization and land-use transition, and the evolutionary mechanism that governs it. Based on the existing research [44], we select indicators from the three dimensions of population urbanization, economic urbanization and social urbanization to evaluate the comprehensive level of urbanization and then calculate the UI. From the perspective of land-use morphology, we study the dominant and recessive morphologies of land-use to describe the transformation of land-use quantitatively. Specifically, land-use dominant morphology (LUDM) is measured from the perspective of land spatial morphology (LSM) and land quantitative structure (LQS). By using the two aspects of land-use efficiency (LUE) and land-use function (LUF), this paper quantitatively simulates land-use recessive morphology (LURM).

Regarding the selection of the indicators, in the measurement of LUDM, we calculate the landscape fragmentation index using Fragstats to measure land-use spatial morphology from the perspective of landscape ecology. We also define the construction land structure index to evaluate the quantitative structure of urban land-use.

The calculation formula is $I_j = S_{con} / S_{non-con}$.

In the formula, $I_j$ is the construction land structure index of each city; $S_{con}$ is the area of urban construction land; $S_{non-con}$ is the area of urban non-construction land.

In terms of the measurement of LURM, we quantify LURM from dimensions of LUE and LUF. LUE is measured by calculating the proportion of the built-up area, investment in fixed assets per square kilometer of land, and the index of the comprehensive degree of land-use. Among these indexes, the comprehensive degree index of land-use characterizes the comprehensive degree of land development and utilization in the region by giving different types of land energy levels [48].

The calculation formula is $D_{ij} = 100 \times \sum_{r=1}^{n} A_{ij} \times C_r$.

In the formula, $D_{ij}$ is a comprehensive index of land-use degree in different years for each city; r means different land-use types; $A_{ij}$ is the proportion of different land-use types in the total land area; $C_r$ is the energy level of different land-use types. Furthermore, we quantify LUF using the aspects of the ecological, economic, and social functions of land-use. On the basis of the selection and calculation of the above indexes, the weight of each index is calculated using the entropy method, and then LUDMI and LURMI are calculated.

In terms of index assignment, the main methods can be divided into subjective and objective methods. To avoid the influence of subjective factors on the study, this paper applies the entropy method to determine the weight. The entropy value method refers to the definition of entropy in thermodynamics, which describes the degree of disorder of the system state with information entropy and calculates the index weight according to the value of information entropy. The main calculation steps are as follows [49]:

- Index standardization—Because the dimensions of different indicators are different, it is necessary to standardize the indicators. $x'_{\theta ij} = x_{\theta ij}/x_{max}$ is applied for the standardization of positive indicators, and $x'_{\theta ij} = x_{min}/x_{\theta ij}$ is used for the standardization of negative indicators. In the formula, $\theta$ represents the year, i represents the city, and j represents the indicator;
- Calculate the proportion of the index value—$Y_{\theta ij} = x'_{\theta ij} / \sum_{i=1}^{m} \sum_{\theta=1}^{n} x'_{\theta ij}$;
- Calculate the entropy of the index information $e_j = -k \sum_{i=1}^{m} \sum_{\theta=1}^{n} \left( Y_{\theta ij} \times \ln Y_{\theta ij} \right)$, in this formula, k is a constant term, and k =ln(mn);
- Calculate the redundancy of information entropy $d_j = 1 - e_j$;
- Calculate the weight of indicators—$a_j = d_j / \sum_{j=1}^{r} d_j$.

The weight of each measurement index of the urbanization system and land-use morphology is calculated by the above calculation formula, as shown in the Tables 1 and 2. After the index weight is obtained, the comprehensive measurement index of urbanization and land-use morphology is obtained by weighted summation.

**Table 1.** Comprehensive measurement index system for the level of urbanization.

| Dimension | Index | Index Weight |
|---|---|---|
| Population Urbanization | Urban population density | 0.093 |
| | Urbanization rate | 0.096 |
| | Per capita GDP | 0.090 |
| Economic Urbanization | Proportion of tertiary industry in GDP | 0.096 |
| | Total investment in fixed assets | 0.087 |
| | Regional passenger volume | 0.091 |
| | Total amount of social consumer goods per capita | 0.090 |
| Social Urbanization | Average wage of employees | 0.092 |
| | Education expenditure per capita | 0.082 |
| | Number of students in colleges and universities per 10,000 people | 0.088 |
| | Number of beds in hospitals and health centers per 10,000 people | 0.095 |

**Table 2.** Land-use measurement index system.

| LUMI | Dimension | Index | Index Weight |
|---|---|---|---|
| LUDMI | LSM | Landscape fragmentation index | 0.49 |
| | LQS | Construction land structure index | 0.51 |
| LURMI | LUE | Proportion of built-up area | 0.16 |
| | | Investment in fixed assets per square kilometer of land | 0.18 |
| | | Comprehensive land-use index | 0.16 |
| | LUF | Green coverage rate of built-up area | 0.18 |
| | | GDP per square kilometer of land | 0.15 |
| | | Population density | 0.17 |

### 3.2.2. PVAR Model

The traditional VAR model, proposed by Sims [50], is applied to predict and analyze the impact of random disturbances on variables in the time series, but the model does not consider the problems brought by panel data. To address this issue, Holtz-Eakin [51] proposed the panel vector autoregression (PVAR) model. In recent years, following work to develop it by Love [52] and Lian [53], the model has matured and has been widely applied. The expression of the PVAR model is as follows:

$$Y_{it} = \sum_{j=1}^{m} \beta_j Y_{it-j} + \varphi_i + \omega_t + \varepsilon_{it} \tag{1}$$

In the formula, $Y_{it}$ is the column vector of UI and the land-use explicit and implicit form index; i and t represent the city and time, respectively; m is the order of delay; $\beta_j$ is the coefficient matrix of each lag term, which represents the degree of interpretation of $Y_{it}$; $\varphi_i$ is introduced to indicate the individual fixed effect, and reflects the heterogeneity of the cities in the Yangtze River Delta; $\omega_t$ represents the specific impact effect of each period and is the time effect vector; $\varepsilon_{it}$ is a random perturbation term.

### 3.2.3. Coupling Coordination Model

Coupling is originally a physical concept, which describes the phenomenon whereby two (or more) systems influence each other through interaction. The degree of coupling describes the degree of interaction between systems or elements, whereas the degree of coordination of coupling describes the trend of the system from disorder to order. The model's expression is as follows:

$$C = \sqrt{\frac{2 \times f(x) \times f(y)}{[f(x) \times f(y)]^2}} \tag{2}$$

$$T = \alpha f(x) \times \beta f(y) \tag{3}$$

$$D = \sqrt{C \times T} \tag{4}$$

In the formula above, C is the value of the degree of coupling, and f(x) and f(y) are the comprehensive evaluation scores of the two systems; T is the comprehensive evaluation index of the development of the two systems, $\alpha$ and $\beta$ are the undetermined coefficients; D is the coupling coordination index. To calculate the degree of coupling coordination, this paper divides it into three categories and six subcategories for evaluation and the classification results are summarized in Table 3.

**Table 3.** Classification of degree of coupling coordination.

| Category | D | | Subcategory |
|---|---|---|---|
| Coordinated type | 0.80–1.0 | T1 | High coordination |
| | 0.60–0.80 | T2 | Suboptimal coordination |
| Transitional type | 0.50–0.59 | T3 | Barely coordinated |
| | 0.40–0.50 | T4 | On the verge of disorder |
| Disordered type | 0.15–0.40 | T5 | Mild disorder |
| | 0–0.15 | T6 | Serious disorder |

## 4. Empirical Analysis of the Interactive Evolution

### 4.1. Two-Way Interaction Analysis Based on the PVAR Model

#### 4.1.1. Index System Construction and Weight Calculation

In order to ensure the validity of the model results and avoid the pseudo-regression problem, and because the panel data applied in this paper only contain five periods, the HT test method, which is suitable for the unit root test of short panel data, is used to test the stationarity of the panel data.

The test results (Table 4) show that all variables except UI pass the HT test and that the data are stable. Based on the analytic ideas of this paper, two PVAR models need to be built to analyze the relationships between urbanization and land-use morphological systems (Model 1), and between urbanization systems and land-use morphological subsystems

(Model 2), respectively. Therefore, UI, LUDMI, and LURMI, as well as UI and the land-use morphological subsystem measures index, were tested covariantly using the Kao test. The test results (Table 5) show a long-term equilibrium relationship between UI and the remaining variables, i.e., the construction of the PVAR model can be conducted with the current data.

**Table 4.** HT unit root test.

| Variable | Stat. (Prob.), t |
|---|---|
| UI | 0.7642 (0.9997) |
| LUDMI | −0.4939 (0.0000) *** |
| LURMI | −0.6038 (0.0000) *** |
| LSMI | −0.8604 (0.0000) *** |
| LQSI | −0.7914 (0.0000) *** |
| LUEI | −0.6096 (0.0000) *** |
| LUFI | −0.3661 (0.0003) *** |

*** denotes the 1% statistical significance levels.

**Table 5.** Co-integration test based on the Kao test.

| | Model 1 Stat. (Prob.) | Model 2 Stat. (Prob.) |
|---|---|---|
| Modified Dickey–Fuller t | 4.8641 (0.0000) *** | 5.4363 (0.0000) *** |
| Dickey–Fuller t | 3.5435 (0.0002) *** | 5.1615 (0.0000) *** |
| Augmented Dickey–Fuller t | 1.6939 (0.0451) ** | 1.4211 (0.0776) * |
| Unadjusted modified Dickey–Fuller t | 1.3858 (0.0829) * | 1.3963 (0.0813) * |
| Unadjusted Dickey–Fuller t | −1.3953 (0.0815) * | −1.1374 (0.1277) |

*, **, and *** denote the 10%, 5%, and 1% statistical significance levels, respectively.

### 4.1.2. Model Order Determination and Granger Causality Test

The AIC test, BIC test, and HQIC test are also performed to determine the optimal lag order before establishing the PVAR model, and the test results are shown in Table 6. The results of the three tests show that the optimal lag order of Model 1 and Model 2 are all of order 1, so in this paper, order 1 is selected to construct the PVAR Model.

**Table 6.** Model lag order test.

| | AIC | BIC | HQIC |
|---|---|---|---|
| | −10.9054 * | −7.8874 * | −9.67947 * |
| Model 1 | −9.97569 | −5.8373 | −8.31419 |
| | −5.42103 | 0.848137 | −3.13814 |
| | −21.1123 * | −15.8537 * | −18.9762 * |
| Model 2 | −19.8522 | −12.3679 | −16.8474 |
| | 19.6186 | 31.3211 | 23.88 |

* denotes the optimal lag order calculated by each test method.

Granger causality tests are further performed for each variable in Model 1 versus Model 2 on the basis of determining model order (Table 7). The test results, shown below, show that only the Granger causality test of LSM for the remaining system variables does not pass the significance test, indicating that the fragmentation trend of land-use is not the Granger causality for the remaining variables, but that there is significant Granger causality between the remaining variables, which also further illustrates the rationality of the theoretical framework constructed in this paper.

**Table 7.** Granger causality test for variables.

| Equation | Excluded | Chi$^2$ | Df | Prob > chi$^2$ |
|---|---|---|---|---|
|  | LUDMI | 70.778 | 1 | 0.225 |
| UI | LURMI | 1.4742 | 1 | 0.000 *** |
|  | All | 71.282 | 2 | 0.000 *** |
|  | UI | 4.7969 | 1 | 0.029 ** |
| LURMI | LUDMI | 14.925 | 1 | 0.000 *** |
|  | All | 15.049 | 2 | 0.001 *** |
|  | UI | 0.56121 | 1 | 0.454 |
| LUDMI | LURMI | 0.89521 | 1 | 0.344 |
|  | All | 32.018 | 2 | 0.000 *** |
|  | LUEI | 6.7271 | 1 | 0.009 *** |
|  | LUFI | 4.0217 | 1 | 0.045 ** |
| UI | LSMI | 16.244 | 1 | 0.000 *** |
|  | LQSI | 21.881 | 1 | 0.000 *** |
|  | All | 29.718 | 4 | 0.000 *** |
|  | UI | 5.6744 | 1 | 0.017 ** |
|  | LUFI | 2.6833 | 1 | 0.101 |
| LUEI | LSMI | 8.7802 | 1 | 0.003 *** |
|  | LQSI | 13.131 | 1 | 0.000 *** |
|  | All | 17.048 | 4 | 0.002 *** |
|  | UI | 6.2896 | 1 | 0.012 ** |
|  | LUEI | 1.9892 | 1 | 0.158 |
| LUFI | LSMI | 6.196 | 1 | 0.013 ** |
|  | LQSI | 6.5609 | 1 | 0.010 ** |
|  | All | 8.7349 | 4 | 0.068 * |
|  | UI | 0.0082 | 1 | 0.928 |
|  | LUEI | 1.4912 | 1 | 0.222 |
| LSMI | LUFI | 1.2134 | 1 | 0.271 |
|  | LQSI | 1.7681 | 1 | 0.184 |
|  | All | 4.4417 | 4 | 0.350 |
|  | UI | 4.2785 | 1 | 0.039 ** |
|  | LUEI | 6.7495 | 1 | 0.009 *** |
| LQSI | LUFI | 4.2216 | 1 | 0.040 ** |
|  | LSMI | 9.9698 | 1 | 0.002 *** |
|  | All | 21.192 | 4 | 0.000 *** |

*, **, and *** denote the 10%, 5%, and 1% statistical significance levels, respectively.

### 4.1.3. Pulse Response Analysis

To further analyze the mechanism and path between the urbanization system and land-use transition, this paper conducted pulse response analysis using Stata 15.0 and took the trend of the pulse response function for further exploration.

Figure 3 presents the results of the pulse response analysis between the urbanization system and the land-use morphology system. Figure 3a,d shows that the elevation of LUDMI negatively inhibits the enhancement of the level of urbanization, that is, in both the LSM and land quantity structure, the enhancement of the land fragmentation level and increase in the structural ratio of construction land will hinder the enhancement of the comprehensive level of urbanization. The land-use pattern under rapid urbanization has resulted in an increase in the fragmentation of the landscape and the structural proportion of construction land, which will affect the urbanization system. According to the results (Table 8) of the cumulative pulse response, urbanization plays a positive role in LUDM, and the enhancement at the level of LUDM plays a negative role in urbanization.

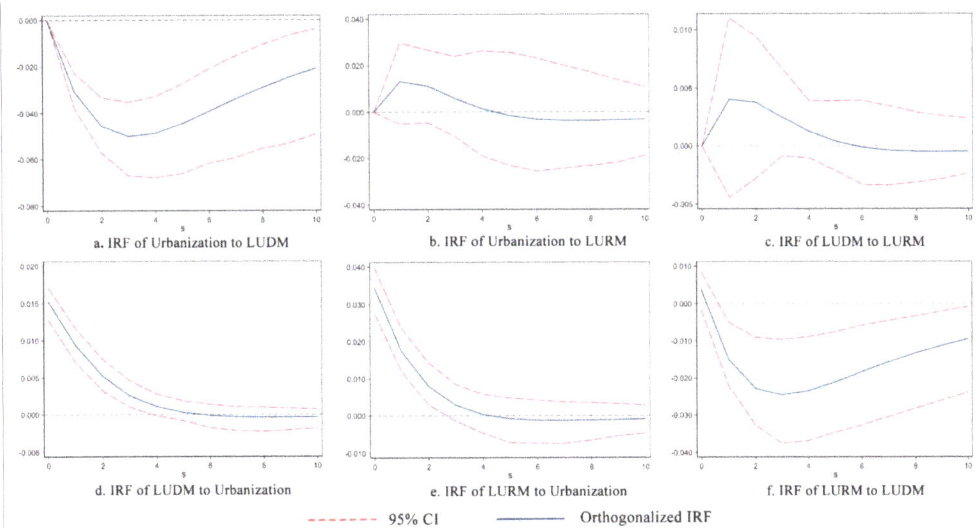

**Figure 3.** Pulse response results of Model 1.

**Table 8.** Cumulative impulse response values of Model 1.

| Response to | Urbanization | LUDM | LURM |
| --- | --- | --- | --- |
| Urbanization | 0.1079 | −0.3669 | 0.0133 |
| LUDM | 0.0325 | −0.0318 | 0.01 |
| LURM | 0.0569 | −0.1707 | 0.0613 |

Figure 3b,e show the relationship between urbanization and LURM. The results show that the relationship between urbanization and LURM manifests mutual promotion in the early stage, but that the positive promotion of one gradually weakens the other. However, after period 4, the interaction between the two turns into a weakly negative mutual inhibition relationship, and the impulse response of both becomes stable. Overall, the relationship between the urbanization system and the LURM system is positive and mutually beneficial.

According to the results of Figure 3c,f, the enhancement at the level of LURM, that is, the enhancement at the level of LUE and LUF, promotes landscape fragmentation and manifest a high structural ratio of construction land. However, after period 6, further increases at the level of LURM start to contribute to a trend toward the fragmentation of land and the continued growth of construction land, and LUDM shows a persistent negative effect on the enhancement of LURM. From the results of the cumulative pulse response, overall, the level of elevation of LURM promotes the level of elevation of LUDM, and LUDM transition, in turn, acts on LURM, inhibiting further improvements in it.

The above results show that the relationship between urbanization and land transition is not a simple linear relationship, but a complex non-linear relationship. When the system develops to a certain stage, the forces acting within the systems will change, e.g., the relationship between the urbanization system and LURM will change from one of mutual promotion to one of mutual inhibition.

Based on the results of Model 1, the relationships among urbanization, LUDM, and LURM are analyzed. In order to further discuss the specific mechanism of each system, Model 2 is established under the analysis framework of this paper.

According to the results of the impulse response analysis and the cumulative impulse response value of Model 2 (Figure 4 and Table 9), in general, in addition to the positive effect of LUE on the development of urbanization, the cumulative effects of the fragmentation trend of LSM, the rapid expansion of construction land, and the improvement of LUF on urbanization are negative. Specifically, the enhancement of landscape fragmentation and the rapid expansion of construction land exerts a sustained negative effect on the enhancement of urbanization levels. LUE promotes the increase in urbanization level in the early period, but the positive effect gradually weakens and turns to a negative effect after period 6. There are fluctuations in the role of LUF in the level of urbanization, which, although initially shown as having an inhibitory effect on urbanization, continues to have a weak and stable positive promotional effect beyond period 4. In addition, urbanization systems have similar characteristics regarding the forces of each land-use morphological subsystem. Urbanization consistently promotes the fragmentation of regional land-use spatial morphology, and the forces gradually weaken, decaying to 0 after period 5. Moreover, the effects of urbanization on LQS, LUE, and LUF are all characterized by an early-period promotion and a late-period depression. Specifically, urbanization promotes the rapid expansion of constructed land-uses and the enhancement of LUE at the beginning, but restrains the development of both subsystems by the beginning of period 4; urbanization significantly promotes LUF in periods 0 to 2, but rapidly changes to have a continuously negative effect in period 3, suppressing the rising level of LUF.

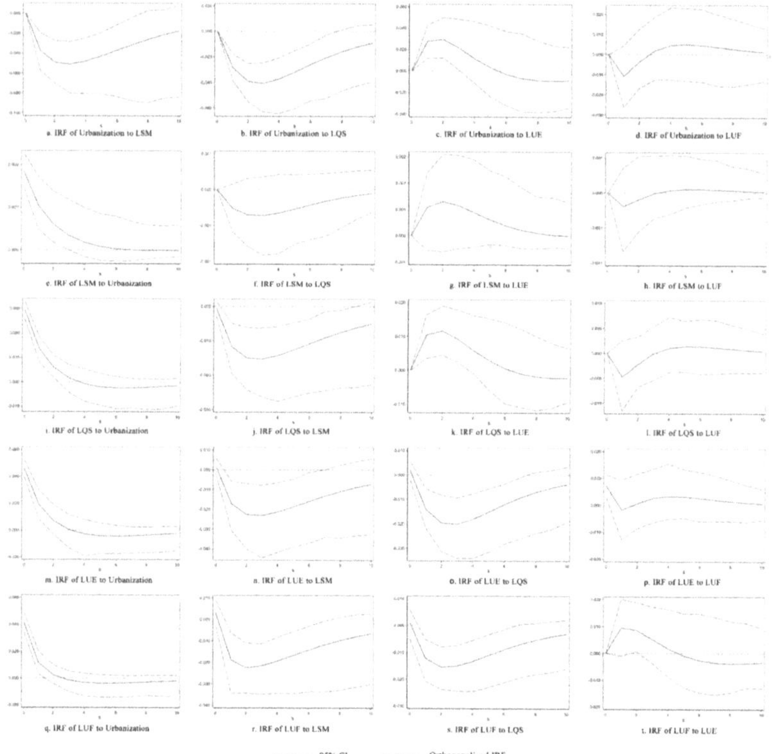

**Figure 4.** Pulse response results of Model 2.

Table 9. Cumulative impulse response values of Model 2.

| Response to | Urbanization | LSM | LQS | LUE | LUF |
|---|---|---|---|---|---|
| Urbanization | 0.0986 | −0.3602 | −0.2605 | 0.0577 | −0.0045 |
| LSM | 0.004 | 0.0054 | −0.0022 | 0.0028 | −0.0009 |
| LQS | 0.0376 | −0.1081 | −0.0413 | 0.0321 | −0.0089 |
| LUE | 0.0489 | −0.1574 | −0.1225 | 0.1018 | −0.0041 |
| LUF | 0.0337 | −0.1456 | −0.0945 | 0.0213 | 0.0537 |

Based on the exploration of the interactive relationship among urbanization and each land-use morphological subsystem described above, the relationship between the land-use morphological subsystems is analyzed by Model 2 in this paper (Figure 4). Inside the LUDM system, there appears to be an antagonism between the trend toward fragmentation of land and the expansion of construction land, suggesting that the expansion of construction land can, to some extent, slow the trend toward fragmentation in regional land in a landscape sense, and that the fragmentation of land is not conducive to the expansion of construction land. Inside the LURM system, where LUF is overall negative compared to LUE, the forces fluctuate and are not significant; LUE acts positively on LUF in the initial period, but changes to have a negative effect from period 5; this negative effect shows no sign of abating until stage 10. However, according to the results of the cumulative pulse response, the active force of LUE on LUF is positively promoted overall.

In terms of the effects of the LUDM subsystems on the LURM subsystems, the fragmentation of LSM and the expansion of construction land both negatively contribute to the enhancement of the levels of LUE and LUF; the effects peak in period 2 and gradually weaken, but persist thereafter. In terms of the effects of LURM on LUDM, increasing levels of LUF exert an overall weak inhibitory effect on land fragmentation versus the expansion of construction land, but there is some fluctuation in this effect; the enhancement of LUE makes the land morphology more fragmented, and the expansion of construction land is promoted in the early period and converted to a negative suppressing effect after period 6.

### 4.2. Analysis of the Coupling Coordination of Urbanization and Land-Use Transition

In the theoretical framework of this paper, we identify the direction, magnitude, and evolution of the relationships among urbanization, LUDM, and LURM by building PVAR models for pulse response analysis. In order to further analyze the evolution of the relationships among the three systems, this paper conducts a specific time and spatial analysis via a coupling coordination model.

#### 4.2.1. Time Series Analysis of Coupling Coordination

Based on the calculation of the degree of coupling coordination among urbanization, LUDM, and LURM for a more intuitive view of the area under study, we plot the degree of coupling coordination among the systems as a scatter plot (Figure 5), with the degree of coupling coordination between urbanization and LUDM and between urbanization and LURM as the X-axis and Y-axis, respectively, and those between LUDM and LURM as the Z-axis.

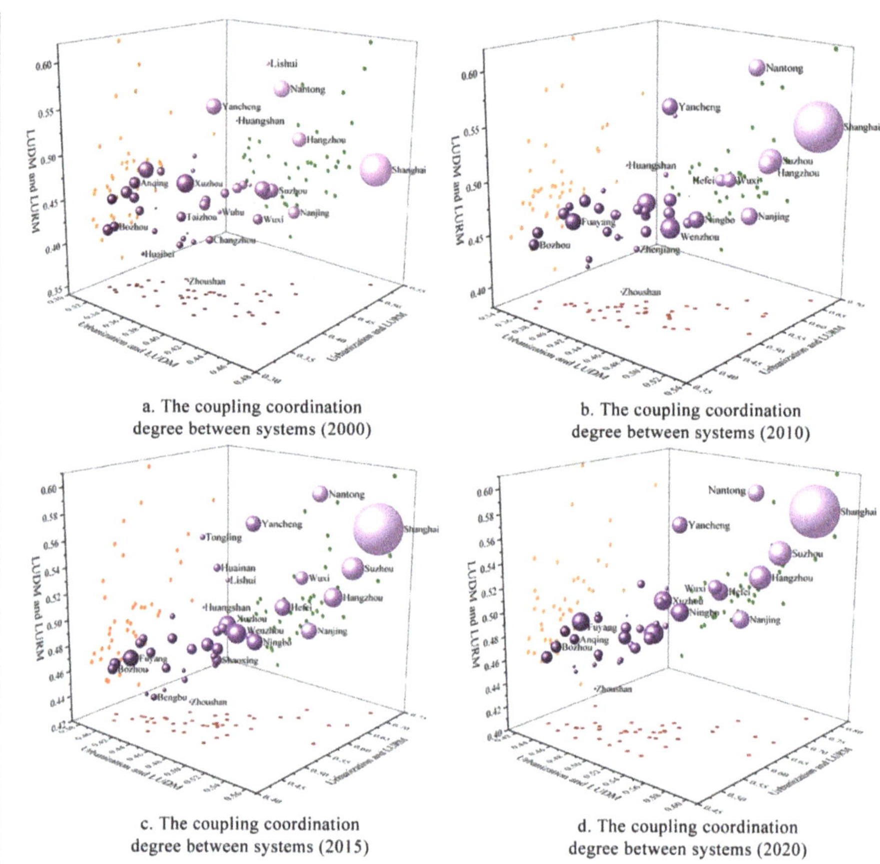

a. The coupling coordination
degree between systems (2000)

b. The coupling coordination
degree between systems (2010)

c. The coupling coordination
degree between systems (2015)

d. The coupling coordination
degree between systems (2020)

**Figure 5.** Time evolution of the coupling coordination relationship between urbanization and land-use morphology in the Yangtze River Delta.

In Figure 5, the size of the points represents the population size of each city, and the positioning of the point closer to the origin indicates a lower overall coupling coordination of the systems in that city, the relationship between systems is more antagonistic, and the development of system becomes moribund. The results show that the level of coordination between the urbanization system and the land-use morphology system varied among cities in the long triangle region in 2000, presenting discrete states in the plots. However, as time progressed, some regional hub cities showed large gains in both city size and system coordination, such as Shanghai, Suzhou, and Hangzhou. As can be seen from the projection of the scatter plots in the XY, XZ, and YZ planes, an overall increase in the level of coordination in the coupling between urbanization and land-use morphological systems in the long triangle occurred over the past 20 years, but the magnitude of the overall increase was not substantial, with only a few regional hub cities such as Shanghai and Suzhou having increased with respect to all aspects.

Furthermore, it is interesting to see from the graph that since 2000, there has been a convergence trend in the scatter plot, affected by the degree of coupling and coordination between the urbanization system and the land-use morphology system. This further shows that the urbanization mode and land-use mode play an important role in promoting and influencing the coordinated relationship between urbanization and land-use patterns, with results similar to the Matthew effect. A reasonable urbanization path selection and land-use mode can allow some cities to not only improve their level of urbanization, but also to expand their scale; the development of each system also tends to be coupled and coordinated, realizing a benign interaction between the systems.

### 4.2.2. Spatial Analysis of the Coupling Coordination

To further discuss the spatial pattern of the coupling and coordination relationship between urbanization and land-use morphology, the computational results are visualized in this paper using ArcGIS, and Figure 6 shows the spatial evolution of coupling coordination between the urbanization and land-use morphology systems in the Yangtze River Delta from 2000 to 2020.

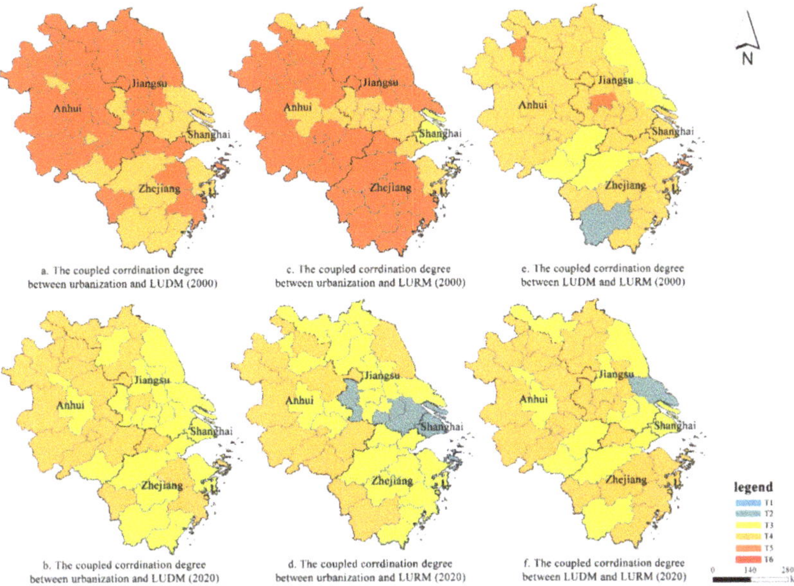


**Figure 6.** Spatial evolution of the coupling coordination relationship between urbanization and land-use morphology in the Yangtze River Delta.

In this paper, six types of degrees of coupling coordination are classified according to the calculation results. Figure 6a,b shows the spatial evolution pattern of the coupling coordination relationship between urbanization and LUDM in the Yangtze River Delta. In 2000, the coupling coordination relationships between urbanization and LUDM in over half of the cities in the Yangtze River Delta were mildly disordered (T5), while the rest were on the verge of disorder (T4). Specifically, cities belonging to T5 were mainly located in economically underdeveloped areas, such as Anhui Province and northern Jiangsu Province. Subsequently, the coordination relationship between the urbanization and LUDM of cities in the Yangtze River Delta began to moderate, and was advanced from coast to inland and from center to periphery.

Figure 6c,d shows the spatial evolution pattern of the coupling coordination relationship between urbanization and LURM. In 2000, almost all the cities' coordination relationships between urbanization and LURM in the Yangtze River Delta were in a state of antagonistic disorder, except Shanghai, which belonged to T3, and the cities in southern Jiangsu, eastern Anhui, and northeastern Zhejiang, which belonged to T4. In the subsequent development, the relationship between urbanization and LURM in the Yangtze River Delta area was gradually promoted from the east to the west. In the past 20 years, the level of coupling coordination between urbanization and LURM has been greatly improved in

Shanghai, southern Jiangsu, and eastern Zhejiang. In particular, the relationships of coupling coordination between urbanization and LURM in Shanghai, Nanjing, Suzhou, Wuxi, and Zhoushan were in a state of suboptimal coordination in 2020. The relationship between LUDM and LURM (Figure 6c,f) is not significantly related to the level of urbanization and economic development of cities. In the past 20 years, the coupling coordination relationship between LUDM and LURM in the Yangtze River Delta has not changed significantly. At present, the relationship between LUDM and LURM in more than half of the cities is on the verge of disorder.

## 5. Discussion

### 5.1. Interactive Evolution Analysis of Urbanization and Land-Use Transition

Based on the analysis of the relationship between urbanization and land-use transition in the Yangtze River Delta over the past 20 years by the PVAR model and coupling coordination model, we summarize the development process, as shown in Figure 7. This paper divides the evolution of the interactive relationship between urbanization and land-use transition into two main stages. S1 is the stage of early rapid urbanization, and S2 is the stage of accelerated rapid urbanization. On this basis, we put forward the stage of high-quality urbanization (S3), in which urbanization and land-use transition interact benignly.

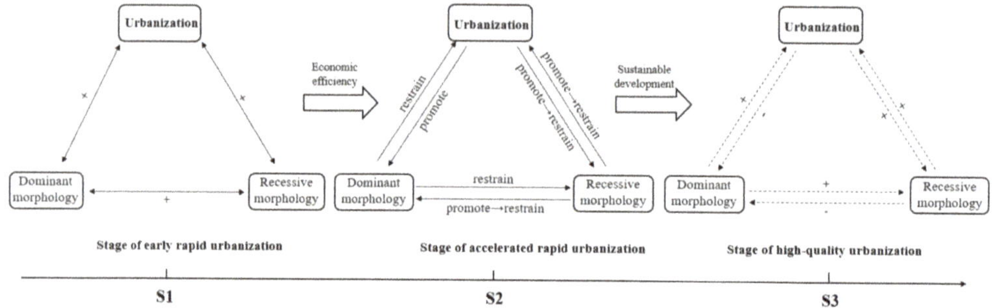


**Figure 7.** The bidirectional interactive response relationship between urbanization and land-use morphology.

In S1, the development level of each system is low, the relationship between the systems is simple, and the systems exhibit a synergistic promotional relationship with each other. Although urbanization promotes the rapid expansion of construction land and enhances the trend of land fragmentation, resulting in an increase in LUDMI, in the early stage, the extensive land-use model also brings some economic benefits, such that the level of urbanization increases. LUDM and LURM show a mutually promoting relationship. Meanwhile, a positive mutual promotion relationship is also observed between urbanization and LURM. However, it cannot be ignored that the coupling coordination relationship among the systems in this stage is mildly disordered or on the verge of disorder, which further illustrates that extensive urbanization brings about extensive land-use patterns that are unfavorable to the high-quality development of urbanization in the long run.

As rapid urbanization progresses, the relationship between urbanization and land-use transition moves into S2. In parallel with increasing urbanization, there is an urbanization orientation that solely targets short-term economic benefits, with the relationships among systems tending to be complex. For example, the trend toward land fragmentation caused by rapid urbanization is ultimately fed back to the urbanization system to inhibit further improvements in the urbanization level. The rapid expansion of construction land and the encroachment on other land do bring economic benefits and promote the improvement of the urbanization level in a certain period of time, but at a certain stage, this brings structural problems, which inhibit the development of urbanization in turn. In this stage, the interaction between urbanization and LURM has changed from positive to negative, which also indicates that, in the long run, rapid urbanization will further affect the improvement of LUE and LUF, thus hindering the progress of spatial urbanization. Similarly, land fragmentation and the rapid expansion of construction land have a negative inhibiting effect on LURM, indicating that the unreasonable land-use structure is not conducive to the activation of land value. The effect of LURM on LUDM changes from promotion to inhibition, which further illustrates that the inefficient and low-functional land-use mode will aggravate the trend of urban construction land expansion and land fragmentation. In S2, the coordination relationship between the urbanization system and the land-use system improves, but this mainly reflects the improvement of the coordination relationship between urbanization and LURM. The levels of coordination relationships between urbanization and LUDM, as well as those between LUDM and LURM, do not improve.

This paper holds that urbanization and land-use transition are adapted to each other, and that there is a long-term balanced relationship between the two systems. A reasonable urbanization path and land-use pattern will form a benign interactive relationship, which will further deepen urbanization. In the long run, the extensive development mode brought about by the pursuit of short-term economic benefits will cause the system to fall out of order, show antagonistic effects among the systems, and hinder the efficient development of the social economy. Based on this, we have constructed an ideal stage of high-quality urbanization with sustainable development as the orientation. In S3, the patterns of urbanization and land-use have changed, and the structure of land-use is orderly and reasonable. Due to the intensive and efficient land-use under the high-quality urbanization mode, land fragmentation and the rapid expansion of construction land are restrained; urbanization effectively promotes the improvement of the LUF level and land output efficiency, which will also further affect urbanization and continue to promote its level.

Through the analysis of the evolution of the interactive relationship between urbanization and land-use transition, it is not difficult to see that the relationship between urbanization and land-use transition is not a simple linear relationship; the two impact and affect each other, and the relationship tends to become more complex with the development of the social economy to a higher stage. In order to realize the coordination relationship of systems in S3, urbanization path selection and land-use mode decision-making are the key issues. A reasonable urbanization path and an efficient, intensive land-use mode not only promote the sustainable development of urbanization to a higher stage, but also promote benign coupling in the relationship between systems.

### 5.2. Interactive Feedback between Rapid Urbanization and Land-Use Transition

Against the background of rapid urbanization, the extensive economic development and land-use pattern in the pursuit of short-term benefits are not only the result of this stage, but also factors influencing urbanization itself. Most cities in the Yangtze River Delta still have the problem that their land-use pattern is not suitable for higher stages of urbanization, which is not only an urban development problem in the Yangtze River Delta area, but also a common problem caused by rapid urbanization. In recent years, some cities in China have started to shrink for various reasons [54]. However, urban planning in China has long been established at the top level of the growth doctrine [55], and urban space still tends to grow more disordered. In addition, inefficient land-use, unreasonable land structures, and the occupation of agricultural land by urban construction land have further widened the urban–rural gap. The value of urban and rural land has not been fully activated. Further exploration of efficient utilization patterns is needed to tap the potential of land for regional coordinated development. These problems of urbanization and land-use are the real problems brought about by the rapid urbanization process. As the analytical logic of this article shows, these problems are both the results and the factors affecting the next stage of development. Therefore, how to achieve the benign development of high-quality urbanization between urbanization and land-use transition is a problem worthy of further discussion. As mentioned above, the process of rapid urbanization in China started in the 1990s [24], and with the aim of increasing the speed of urban development, the rough development and land-use patterns have brought about urban problems such as Urban Villages and land-use fragmentation [56–59]. The results of this paper reveal the interaction between the urbanization system and the land-use system, and that land is not only a container and carrier of the city. From a systems theory perspective, the relationship between land and city is one of feedback and mutual influence. Taking a rational view of China's urbanization, as the understanding of urbanization deepens, the quality of urbanization becomes the first goal of development. How to achieve high-quality urbanization is the main issue. According to this study, we believe that land is not only a carrier of cities, but its participation in socio-economic development as a spatial element directly interacts with urbanization. Therefore, land-use transformation by means of land management and spatial remediation, and the formation of a positive interactive relationship with the urbanization system, are the keys to solving the problems brought about by China's rapid urbanization process, and achieving high-quality urbanization transformation.

In terms of the methodology, from a systems theory perspective, this paper takes the Yangtze River Delta as an example, creates two PVAR models and a coupled coordination model to analyze the evolution of the interactive relationship between urbanization and land-use transition, and holds that there is a long-term balanced relationship between the two whereby they influence and adapt to each other, which ultimately has a profound impact on the sustainable development of the social economy.

In terms of the results, this paper analyzes the evolutionary process of the interaction between land-use transformation and urbanization through the case of the Yangtze River Delta region, and analyzes the inter-system action mechanism. On the one hand, this study explores the interaction between urbanization and land-use transition from the perspective of the dominant and recessive morphologies of land-use, which enriches the connotation of land-use research. On the other hand, by dividing the interactive evolution stages of urbanization and land-use transition and analyzing the internal mechanism, we further clarify the importance of urbanization path selection and land-use mode decision-making, and provide a reference for urban development decision-making. In addition, the question still remains of how to achieve the coordinated transition of urbanization and land-use proposed in this paper, and ultimately achieve sustainable and high-quality urbanization—this issue has not been analyzed in detail in this paper, and is also worthy of further discussion.

## 6. Conclusions

This paper analyzes the interactive relationship between urbanization and land-use transition in the Yangtze River Delta from 2000 to 2020 from a systems theory perspective, and holds that there is a long-term equilibrium relationship between urbanization and land-use transition and that the two cause and affect each other, which ultimately has a profound impact on urbanization and the sustainable development of the social economy. The main conclusions of the empirical analysis of the Yangtze River Delta are as follows:

1. With the rapid development of urbanization in the Yangtze River Delta, the interaction between urbanization and land-use transition has changed from a simple positive interaction to negative inhibition between systems, and the interaction between systems has become more complex. Specifically, rapid urbanization intensifies the trend toward land fragmentation and promotes the rapid expansion of construction land, which hinders the further development of urbanization;

2. The structural problems brought about by rapid urbanization also make the interactive relationship between urbanization and LURM change. The relationship between them will inhibit both when it develops to a certain stage, which hinders the promotion of the overall level. This further reflects that the extensive development mode of rapid urbanization is not conducive to the improvement of land function level and LUE in the long run. Ultimately, urbanization itself will also be affected;

3. Although the degree of coupling coordination between the urbanization system and the land-use system in the Yangtze River Delta region increased from 2000 to 2020, the overall level of improvement was not significant, and the system relationship of most cities was still on the verge of disorder. This indicates that, in the long run, the land-use transition problems brought about by the rapid urbanization mode will hinder the benign development of the system relationship;

4. The coupling coordination relationship between urbanization and land-use transition in the Yangtze River Delta appears to be a convergence phenomenon, which also shows that a reasonable urbanization path and mode will promote benign coupling in the relationships between systems. This will ultimately make the city scale expand and the economy develop continuously; moreover, the systems will also achieve coordinated transition.

**Author Contributions:** Conceptualization, D.G.; methodology, B.N.; software, R.Y.; validation, B.N., Y.M. and M.L.; formal analysis, B.N. and D.G.; investigation, B.N. and R.Y.; resources, Y.L.; data curation, Y.M. and M.L.; writing—original draft preparation, B.N.; writing—review and editing, D.S.; visualization, B.N. and R.Y.; supervision, D.G. and D.S.; project administration, D.G. and M.L.; funding acquisition, D.G., Y.L. and M.L. All authors have read and agreed to the published version of the manuscript.

**Funding:** This research was funded by "the National Natural Science Foundation of China" (grant number 41901204, 42001125), "The Foundation of Humanity and Social Sciences of the Ministry of Education of China" (grant number 19YJCZH036, 20YJC790093), "China Postdoctoral Science Foundation" (grant number 2019M660109, 2021T140303), "Jiangsu Provincial Science Foundation" (grant number BK20190717), and "Jiangsu Provincial Social Science Foundation" (grant number 19GLC002).

**Institutional Review Board Statement:** Not applicable.

**Informed Consent Statement:** Not applicable.

**Data Availability Statement:** Not applicable.

**Conflicts of Interest:** The authors declare no conflict of interest.

# References


1. Mooney, H.A.; Duraiappah, A.; Larigauderie, A. Evolution of natural and social science interactions in global change research programs. *Proc. Natl. Acad. Sci. USA* **2013**, *110*, 3665–3672. [CrossRef]
2. Sterling, S.M.; Ducharne, A.; Polcher, J. The impact of global land-cover change on the terrestrial water cycle. *Nat. Clim. Chang.* **2012**, *3*, 385–390. [CrossRef]
3. Sun, P.; Zhou, L.; Ge, D.; Lu, X.; Sun, D.; Lu, M.; Qiao, W. How does spatial governance drive rural development China's farming areas? *Habitat Int.* **2021**, *109*, 102320. [CrossRef]
4. Ge, D.; Lu, Y. Rural spatial governance for territorial planning in China: Mechanisms and path. *Acta Geogr. Sin.* **2021**, *76*, 1422–1437.
5. Pei, T.; Sobolevsky, S.; Ratti, C.; Shaw, S.L.; Li, T.; Zhou, C. A new insight into land use classification based on aggregated mobile phone data. *Int. J. Geogr. Inf. Sci.* **2014**, *28*, 1998–2007. [CrossRef]
6. Cho, S.J.; Mccarl, B. Major united states land use as influenced by an altering climate: A spatial econometric approach. *Land* **2021**, *10*, 546. [CrossRef]
7. Grainger, A. National land use morphology: Patterns and possibilities. *Geography* **1995**, *80*, 235–245.
8. Mather, A.S. The forest transition. *Area* **1992**, *24*, 367–379.
9. Long, H. Land use transition: A new integrated approach of land use/cover change study. *Geogr. Geo-Inf. Sci.* **2003**, *19*, 87–90.
10. Long, H.; Li, X. Analysis on regional land use transition: A case study in Transect of the Yangtze River. *J. Nat. Resour.* **2002**, *17*, 144–149.
11. Long, H. Land use transition and land management. *Geogr. Res.* **2015**, *34*, 1607–1618.
12. Long, H.; Tu, S. Land use transition and rural vitalization. *China Land Sci.* **2018**, *32*, 1–6.
13. Guo, C.; Gao, J.; Fan, P.; Yao, F. Land use transition and hotpots detection in Yongcheng city based on the grid scale. *China Land Sci.* **2016**, *30*, 43–51.
14. Lambin, E.F.; Meyfroidt, P. Land use transitions: Socio-ecological feedback versus socio-economic change. *Land Use Policy* **2010**, *27*, 108–118. [CrossRef]
15. Song, X.; Wu, Z.; Ouyang, Z. Changes of cultivated land function in China since 1949. *Acta Geogr. Sin.* **2014**, *69*, 435–447.

16. Song, X.; Li, X. Theoretical explanation and case study of regional cultivated land use function transition. *Acta Geogr. Sin.* **2019**, *74*, 992–1010.

17. Thenail, C.; Baudry, J. Farm riparian land use and management: Driving factors and tensions between technical and ecological functions. *Environ. Manag.* **2005**, *36*, 640–653. [CrossRef]

18. Xie, H.; Chen, Q.; Lu, F.; Wu, Q.; Wang, W. Spatial-temporal disparities, saving potential and influential factors of industrial land use efficiency: A case study in urban agglomeration in the middle reaches of the Yangtze River. *Land Use Policy* **2018**, *75*, 518–529. [CrossRef]

19. Newbold, T.; Scharlemann, J.P.W.; Butchart, S.H.M.; Sekercioglu, C.H.; Alkemade, R.; Booth, H.; Purves, D.W. Ecological traits affect the response of tropical forest bird species to land-use intensity. *Proc. R. Soc. B Biol. Sci.* **2013**, *280*, 20122131. [CrossRef]

20. Liu, Y.; Long, H. Land use transitions and their dynamic mechanism: The case of the Huang-Huai-Hai Plain. *J. Geogr. Sci.* **2016**, *26*, 515–530. [CrossRef]

21. Chen, S.; Guo, B.; Zhang, R.; Zang, W.; Wei, C.; Wu, H.; Yang, X.; Zhen, X.; Li, X.; Zhang, D.; et al. Quantitatively determine the dominant driving factors of the spatial–temporal changes of vegetation NPP in the Hengduan Mountain area during 2000–2015. *J. Mt. Sci.* **2021**, *18*, 427–445. [CrossRef]

22. Peng, J.; Zhao, M.; Guo, X.; Pan, Y.; Liu, Y. Spatial-temporal dynamics and associated driving forces of urban ecological land: A case study in Shenzhen City, China. *Habitat Int.* **2017**, *60*, 81–90. [CrossRef]

23. Chen, W.; Li, J.; Ran, D. On the Spatial Relationship Between Land Use Transition and Urbanization in the Middle Yangtze River Urban Cluster. *Resour. Environ. Yangtze Basin* **2019**, *28*, 1036–1048.

24. Chen, B.; Hao, S.; Yang, X. The dynamic mechanism of the rapid development of urbanization in China. *Acta Geogr. Sin.* **2004**, *59*, 1068–1075.

25. Fernández-Nogueira, D.; Corbelle-Rico, E. Determinants of Land Use/Cover Change in the Iberian Peninsula (1990–2012) at Municipal Level. *Land* **2019**, *9*, 5. [CrossRef]

26. Liu, X.; Zhang, D.; Chen, B. Characteristics of China's Town-level Land Use in Rapid Urbanization Stage. *Acta Geogr. Sin.* **2008**, *63*, 301–310.

27. Siciliano, G. Urbanization strategies, rural development and land use changes in China: A multiple-level integrated assessment. *Land Use Policy* **2012**, *29*, 165–178. [CrossRef]

28. Yang, R.; Luo, X.; Xu, Q.; Zhang, X.; Wu, J. Measuring the impact of the multiple cropping index of cultivated land during continuous and rapid rise of urbanization in china: A study from 2000 to 2015. *Land* **2021**, *10*, 491. [CrossRef]

29. Kuang, B.; Lu, X.; Han, J.; Zuo, J. How urbanization influence urban land consumption intensity: Evidence from China. *Habitat Int.* **2020**, *100*, 102103. [CrossRef]

30. Ge, D.; Zhou, G.; Qiao, W.; Yang, M. Land use transition and rural spatial governance: Mechanism, framework and perspectives. *J. Geogr. Sci.* **2020**, *30*, 1325–1340. [CrossRef]

31. Ge, D.; Long, H.; Qiao, W.; Wang, Z.; Sun, D.; Yang, R. Effects of rural–urban migration on agricultural transformation: A case of Yucheng City, China. *J. Rural. Stud.* **2020**, *76*, 85–95. [CrossRef]

32. Qu, Y.; Zhang, F.; Guo, L.; Xu, Y. Estimation of farmland quality after rural residential land consolidation and its application. *Trans. Chin. Soc. Agric. Eng.* **2012**, *28*, 226–233.

33. Amin, A.; Fazal, S.; Mujtaba, A.; Singh, S.K. Effects of land transformation on water quality of Dal Lake, Srinagar, India. *J. Indian Soc. Remote. Sens.* **2014**, *42*, 119–128. [CrossRef]

34. Qu, Y.; Zhang, Y.; Zhu, W.; Wang, S. Pattern and mode of comprehensive land consolidation form the perspective of production-life-ecological function. *Mod. Urban Res.* **2021**, 33–39. [CrossRef]

35. Yang, Q.; Duan, X.; Wang, L.; Jin, Z. Land use transformation based on ecological-production-living spaces and associated eco-environment effects: A case study in the Yangtze Delta. *Sci. Geogr. Sin.* **2018**, *38*, 97–106.

36. Long, H.; Qu, Y. Land use transitions and land management: A mutual feedback perspective. *Land Use Policy* **2018**, *74*, 111–120. [CrossRef]

37. Long, H.; Qu, Y.; Tu, S.; Li, Y.; Ge, D.; Zhang, Y.; Ma, L.; Wang, W.; Wang, J. Land use transitions under urbanization and their environmental effects in the farming areas of China: Research progress and prospect. *Adv. Earth Sci.* **2018**, *33*, 455–463.

38. Cui, Y.; Liu, X.; Li, D.; Deng, Q.; Xu, J.; Shi, X.; Qin, Y. Urbanization spatial correlation characteristics and intrinsic mechanism in the Yangtze River Delta region. *Acta Geogr. Sin.* **2020**, *75*, 1301–1315.

39. Chen, Y.; Xie, B.; Li, X.; Deng, C.; Zhu, Y. The preliminary research on relationship between the change of land use and urbanization in Changsha from 2003 to 2013. *Econ. Geogr.* **2015**, *35*, 149–154.

40. Dadashpoor, H.; Azizi, P.; Moghadasi, M. Land use change, urbanization, and change in landscape pattern in a metropolitan area. *Sci. Total Environ.* **2019**, *655*, 707–719. [CrossRef] [PubMed]

41. Lv, X.; Huang, X.; Zhang, Q. A literature review on urban-rural construction land transition. *City Plan. Rev.* **2015**, *39*, 105–112.

42. Nuissl, H.; Haase, D.; Lanzendorf, M.; Wittmer, H. Environmental impact assessment of urban land use transitions—A context-sensitive approach. *Land Use Policy* **2009**, *26*, 414–424. [CrossRef]

43. Zhou, L.; Zhang, M.; Xu, J.; Sun, D. The measurement of efficiency and orderly of urban land use in Shangdong province in the rapid urbanization. *Areal Res. Dev.* **2014**, *33*, 135–140.

44. Chen, M.; Lu, D.; Zhang, H. Comprehensive evaluation and the driving factors of China's urbanization. *Acta Geogr. Sin.* **2009**, *64*, 387–398.

45. Wang, F.; Mao, A.; Li, H.; Jia, M. Quality measurement and regional difference of urbanization in Shandong Province based on the entropy method. *Sci. Geogr. Sin.* **2013**, *33*, 1323–1329.
46. Ou, X.; Zhen, F.; Qin, Y.; Zhu, L.; Wu, H. Study on compressive level and ideal impetus of regional urbanization: The case of Jiangsu Province. *Geogr. Res.* **2008**, *27*, 993–1002.
47. Chen, Y.; Yang, P. Recent progresses of international study on land use and land cover change (LUCC). *Econ. Geogr.* **2001**, *21*, 95–100.
48. Zhuang, D.; Liu, J. Study on the model of regional differentiation of land use degree in China. *J. Nat. Resour.* **1997**, *12*, 10–16.
49. Yang, L.; Sun, Z. The development of western new-type urbanization level evaluation based on entropy method. *Econ. Probl.* **2015**, 115–119. [CrossRef]
50. Sims, C.A. Comparison of interwar and postwar business cycles: Monetarism reconsidered. *Am. Econ. Rev.* **1980**, *70*, 250–257.
51. Holtz-Eakin, D.; Newey, W.; Rosen, H.S. Estimating vector autoregressions with panel data. *Econom. J. Econom. Soc.* **1988**, *56*, 1371–1395. [CrossRef]
52. Love, I.; Zicchino, L. Financial development and dynamic investment behavior: Evidence from panel VAR. *Q. Rev. Econ. Financ.* **2006**, *46*, 190–210. [CrossRef]
53. Lian, Y.; Cheng, J. Investment-cash flow sensitivity: Financial constraints or agency costs? *J. Financ. Econ.* **2007**, *33*, 37–46.
54. Du, Z.; Jin, L.; Ye, Y.; Zhang, H. Characteristics and influences of urban shrinkage in the exo-urbanization area of the Pearl River Delta, China. *Cities* **2020**, *103*, 102767. [CrossRef]
55. Li, X.; Wu, K.; Long, Y.; Li, Z.; Luo, X.; Zhang, X.; Wang, D.; Yang, D.; Kuai, Y.; Li, Y.; et al. Academic debates upon shrinking cities in China for sustainable development. *Geogr. Res.* **2017**, *36*, 1997–2016.
56. Yan, X.; Wei, L.; Zhou, R. Research on the coordination between urban and rural area in the rapid urbanization with the redevelopment of Guangzhou village-amid-the-city as a case. *City Plan. Rev.* **2004**, *28*, 30–38.
57. Long, H.; Qu, Y.; Tu, S.; Zhang, Y.; Jiang, Y. Development of land use transitions research in China. *J. Geogr. Sci.* **2020**, *30*, 1195–1214. [CrossRef]
58. Long, H.; Zhang, Y. Rural planning in China: Evolving theories, approaches, and trends. *Plan. Theory Pract.* **2020**, *21*, 782–786. [CrossRef]
59. Long, H.; Zhang, Y.; Tu, S. Rural vitalization in China: A perspective of land consolidation. *J. Geogr. Sci.* **2019**, *29*, 517–530. [CrossRef]

*Article*

# Urban Land Use Transitions and the Economic Spatial Spillovers of Central Cities in China's Urban Agglomerations

Hui Li [1], Kunqiu Chen [2,3,*], Lei Yan [4], Yulin Zhu [1], Liuwen Liao [5] and Yangle Chen [6]

[1]   School of Economics, Central South University of Forestry and Technology, Changsha 410004, China;
      t20172336@csuft.edu.cn (H.L.); t19970886@csuft.edu.cn (Y.Z.)
[2]   School of Public Administration, Hunan University, Changsha 410082, China
[3]   Institute of Geographic Sciences and Natural Resources Research, Chinese Academy of Sciences,
      Beijing 100101, China
[4]   School of Business, Central South University, Changsha 410083, China; yanlei36@csu.edu.cn
[5]   College of Economics and Management, Changsha University, Changsha 410022, China;
      liaoliuwen_12@163.com
[6]   College of Tourism, Hainan University, Haikou 570228, China; 991352@hainanu.edu.cn
*    Correspondence: chenkq.18b@igsnrr.ac.cn

**Abstract:** Urbanization and land use transformation are typical characteristics of China in recent decades. Studying the effects of urban land use transitions (ULUT) on the economic spatial spillovers of central cities (ESSCC) can provide a reference for China to optimize cities' land space layout and promote their coordinated development. Based on the direct and indirect effects of ULUT in central cities on the production factors and economic growth in other cities, this paper reveals the mechanisms underlying the influence of ULUT on ESSCC. Then, we usethe expanded geographical distance-weighted spatial Durbin model with the panel data of 152 Chinese urban agglomeration cities from 2003 to 2016 to empirically test it. The results show that, since 2003, the rate of urban land expansions, the level of urban land intensive use (ULIU), the degree of land marketization, and the urban land prices in China have increased substantially; and the proportionate supplies of industrial land, commercial land, and residential land have decreased. Moreover, ULUT between cities have significant spatial autocorrelations. The current ULUT have positive but small effects on ESSCC. Among them, ULIU has the greatest promotion effects on ESSCC. The impacts of ULUT on ESSCC vary greatly among urban agglomerations. The ULUT in central cities indirectly enhance the ESSCC, which mainly depend on the positive effects of ULUT on enterprise investment, infrastructure investment, labor and technological efficiency and the spatial spread effects of these production factors. This is the main intermediate mechanism by which the ULUT in central cities enhance the ESSCC. Continuing to strengthen ULIU, promote the improvement of land marketization, and establish and improve the coordination mechanism for the economic development of urban agglomerations will help to strengthen the ESSCC in urban agglomerations. The results provide evidence for how the Chinese government can enhance the ESSCC and promote the coordinated development of cities through ULUT under new urbanization.

**Keywords:** land use transitions; spatial effects; urban land expansions; land intensive use; land marketization; land structure; land prices; urbanization

 check for **updates**

**Citation:** Li, H.; Chen, K.; Yan, L.; Zhu, Y.; Liao, L.; Chen, Y. Urban Land Use Transitions and the Economic Spatial Spillovers of Central Cities in China's Urban Agglomerations. *Land* **2021**, *10*, 644. https://doi.org/10.3390/land10060644

Academic Editor: Adrianos Retalis

Received: 17 May 2021
Accepted: 14 June 2021
Published: 17 June 2021

**Publisher's Note:** MDPI stays neutral with regard to jurisdictional claims in published maps and institutional affiliations.



## 1. Introduction

Since China's reform and opening up, its economy and urban population have both been growing rapidly, leading to the rapid and large-scale expansion of urban land [1–3] and the continuous reduction inagricultural land [4–6]. From 2000 to 2005, the area of construction land in China increased by 1,705,300 hectares and cultivated land decreased by 686,500 hectares [7]; from 2010 to 2015, construction land increased by 2.46 million hectares, and cultivated land decreased by 490,000 hectares [8]. However, these extensive expansions of urban land have led to problems such as low land use efficiency, idle land, and waste [9]. As land resource constraints continue to tighten [10,11], urban land use transitions (ULUT) should shift away from the previous explicit expansion and move toward the implicit transition of intensive land use [12–14], land marketization [15], and optimization of the land supply structure to promote the transformation of economic development [16,17].

In order to accelerate the implicit transitions of urban land use, the Chinese government has issued several policies. The Ministry of Land and Resources promulgated the Regulations on the Transfer of State-owned Construction Land Use Rights by Bidding, Auction, and Listing in 2007. It stipulates that industrial, commercial, tourism, entertainment, residential, and other operating land, as well as the same parcel of land with more than two intentional users, shall be leased by bidding, auction, or listing [18].This policy has further promoted the marketization of urban land leasing in China. In 2008, the State Council issued the Outline of the National Land Use Master Plan, which proposed to economically and intensively use construction land, strictly control the newly added construction land, control the rapid expansions of urban industrial and mining land, and give priority to ensuring the construction land for foundation facilities, public service facilities, low-rent housing, affordable housing, and ordinary housing, to promote the intensive use of urban land, control the scale of newly added urban land, and optimize the supply structure of urban land [19]. In 2014, the Ministry of Land and Resources issued the Guiding Opinions of the Ministry of Land and Resources on Promoting Land Conservation and Intensive Utilization, which required a gradual reduction in the scale of newly added construction land, improvements in construction land use efficiency, adjustments in the rational proportion of construction land, and expansions of the paid use range of state-owned land [20]. At this point, the scale of newly added construction land in cities had gradually decreased; the proportion of industrial and mining land, commercial land, and residential land in urban land leasing has gradually declined [21]; the level of urban land intensive use (ULIU) has further improved [11,22]; and the proportion of leased land for bidding, auction, and listing has further increased.

With China's rapid urbanization over the past four decades, the scale of its cities' areas has continued to expand and the connections between cities have become increasingly close, leading to the formation of many urban agglomerations [23]. Urban agglomerations have become an important growth pole for the rapid development of the national economy and an important engine for the coordinated development of regional economies [24,25]. The National New Urbanization Plan aims to establish a coordination mechanism for the development of urban agglomerations, enhance the economic spatial spillovers of central cities (ESSCC), accelerate the development of small and medium-sized cities, prioritize the development of small towns, and promote the coordinated development of various cities [26]. The 2018 Central Economic Work Conference further stated that "it is necessary to strengthen the ESSCC to boost China's high-quality development".

ULUT interact with urban economic growth, population, and production factors [27,28]. On the one hand, the rapid growth of the urban economy and population has driven the continuous expansions of urban land and improved land use efficiency [29–32]. On the other hand, urban land expansions are also used as an important tool to obtain local fiscal revenue and regional economic growth [33,34].ULUT have an important influence on production factor changes and spatial flows [35], while production factors and economic growth have spatial effects [36]. Therefore, ULUT may also have impacts on the ESSCC in urban agglomerations. The development of urban agglomerations has become the key to driving regional economic development, and the ESSCC in urban agglomerations is particularly important. China is in the process of transforming its economic development by encouraging urbanization and promoting the high-quality, coordinated development of cities. In 2016, the State Council pointed out the need to improve and perfect the land use mechanism in Several Opinions of the State Council on Deepening the Construction of New Urbanization. In 2020, the Chinese government promulgated the Proposals of the Central Committee of the Communist Party of China on Formulating the Fourteenth

Five-Year Plan for National Economic and Social Development and Long-Term Goals for 2035, which further proposed "optimizing the layout of land and space and promoting coordinated regional development and new types of towns". Therefore, how to promote new urbanization and the coordinated development of cities through ULUT is particularly important [37].

Europe and the United States are dominated by private ownership of land. The previous related research included land as a production factor in the economic growth model and explored the impact of land on economic growth [38–41]. In recent years, the research on the relationship between land and economic growth under the land private ownership system has mainly discussed the impact of land use control and land landscape on housing prices [42–45], and the impact of urban land use on the agglomeration of factors [46,47]. However, China has implementedpublic ownership of land, and urban land is supplied by the government. Therefore, the transition of urban land quantity and utilization is often used as a tool to stimulate production factors and fiscal revenue, thereby stimulating regional economic growth. Research on the impact of ULUT on economic growth mainly focuses on explicit ULUT—that is, the impacts of urban land expansions on the urban economy and its production factors. Studies have shown that the expansions of urban land have significantly expanded local fiscal revenue and basic investment in China, and urban land commodification has become a main source of municipal finance and funding for urban maintenance and construction [48]. At the same time, urban land is used as a tool to attract investment and drive urban investment. Through the expansions of urban land to drive urban investment, investment-driven growth in China has been generated in the past decade [49].The expansion of industrial land directly stimulates economic growth, and urban land leasing also indirectly drives economic growth by attracting foreign direct investment and infrastructure investment [35].Increasing land leasing revenues will directly speed up economic growth [50]. However, in the process of attracting investment, local governments often adopt low prices to obtain enterprise investment, which leads to inefficient land use [49] and low input–output efficiency [48], and makes them sometimes unable to achieve the goal of stimulating economic growth. At the same time, the expansions of urban construction land to increase fiscal revenue have also led to an excess of residential land and an imbalance in the land supply structure in China. It is necessary to reverse the low-price urban land leasing policy to create a properly functioning land market [51].

Some scholars have also considered the spatial effects of urban land use on production factors and economic growth. He et al. pointed out that interregional competition would lead to the spatial dependence of land supply and land use changes in China [35,52]. The landscape between cities also has significant spatial correlation [53]. At the same time, urban land expansions may also affect the changes in production factors and economic growth in other cities. Wu et al. pointed out that in the process of attracting investment in China, competition drives down industrial land prices, which leads to the spatial correlation of urban land prices [54–56]. Wei et al. found that there were more urban land expansions opportunities in provincial-level central cities, which inhibited the economic growth of other cities and widened the development gap between cities in the province [57].

The current urban land use policies in China promote ULUT from explicit urban land expansion to the implicit transitions of land intensive use, land marketization, and urban land supply structure optimization. The implicit transitions of urban land use replace the explicit transition, and become the leading factors to promote the new urbanization. However, existing studies have neglected the impact of the implicit transitions of urban land use on production factors and economic growth. In addition, although existing studies have paid attention to the spatial correlation and spatial effects on production factors and economic growth of urban land expansion and urban land prices in China, there is a lack of research on the impact of ULUT on ESSCC. Then, what are the impacts of ULUT on ESSCC? How do these effects (if any) come into being? How canthe ULUT be optimized to enhance the ESSCC and promote the coordinated development of cities? Answering these questions can provide a basis for tightening constraints on land resources and China's ULUT under the background of new urbanization.

With urban land expansions representing the explicit ULUT, and ULIU, urban land supply structure (ULSS), urban land marketization and urban land prices representing the implicit ULUT, this paper first describes the characteristics of ULUT in China. Then, from the perspective of ULUT affecting the spatial flow and the spatial effect of production factors, we tease out the transmission mechanism of ULUT on the ESSCC in urban agglomerations. Finally, data on Chinese urban agglomerations from 2003 to 2016 were collected to estimate the influence of explicit and implicit ULUT on ESSCC in urban agglomerations and its intermediate mechanism.

## 2. Influence Mechanism of ULUT on ESSCC

ULUT include explicit transitions and implicit transitions. Explicit ULUT refer to land use structure in a certain region over a specific period, with features such as the quantity and spatial pattern of land use types. In contrast, implicit ULUT are more profound and refer to land use that depends on the explicit morphology and requires analysis, testing, detection, and investigation. Implicit transitions have multiple attributes such as quality, property rights, operating methods, utilization efficiency, and functional structure [58–63]. In most countries such as the United States and Europe, land is privately owned, and land is used more as a factor of production to affect economic growth. Changes in urban land use are often more the result of urban economic growth. In China, land belongs to public ownership, and urban land is used more as a tool for local governments to drive the growth of production factors and stimulate economic growth. The ULUT in China areoften the cause of urban economic growth, or mutual causality. Therefore, the influence mechanism of ULUT on ESSCC in this article is especially applicable to China. ULUT affect the ESSCC in urban agglomerations in two ways. On the one hand, ULUT may affect the spatial flow of production factors and have a direct spatial effect on the economic growth of other cities; on the other hand, after ULUT affect the scale of a city's production factors, those production factors will affect the economic growth of other cities through their own spatial effects. This means that ULUT will have indirect spatial effects on the economic growth of other cities due to the spatial effects of the production factors (see Figure 1).

The effect of ULUT on the ESSCC in urban agglomerations can also be considered from two perspectives. First, ULUT strengthen the diffusion effect of the central cities in urban agglomerations and drive the economic growth of other cities. ULUT of the central cities in urban agglomerations drive the growth of local production factors and economic growth and then promote the economic growth of other cities through the spatial spillover effects of production factors and economic growth [36].At the same time, ULUT of central cities in urban agglomerations may also directly drive the growth of production factors in surrounding cities. Second, ULUT intensify the agglomeration and back flow effects in central cities in urban agglomerations and inhibit the economic growth of other cities. ULUT in central cities in urban agglomerations may prompt other urban production factors to directly agglomerate and flow to central cities. It is also possible that ULUT indirectly drive the production factors of other cities to collect in central cities, because of the agglomeration effects of production factors and economic growth, thus hindering the economic growth of other cities [64].

First, urban land expansions will affect the ESSCC in urban agglomerations, which may cause a diffusion effect or a siphon effect. The supply of urban land is used as a tool to attract investment, absorb labor, expand local fiscal revenue, increase infrastructure investment, and stimulate regional economic growth [65]. Investment, transportation infrastructure, labor, and economic growth have spatial spillover effects that may drive economic growth in other cities [64]. In this way, the supply of urban land in the central cities of urban agglomerations may indirectly enhance ESSCC. At the same time, if acity adopts the industrial chain investment of urban agglomerations, the urban land expansion in one city will also increase the enterprise investment in other cities [66], and enhance the

ESSCC. Infrastructure investment often supports investment between cities [67], which means that the expansion of urban land in central cities in urban agglomerations may also directly drive the growth of corporate investment and infrastructure investment in other cities and, in turn, drive their economic growth. However, larger scale central cities with stronger agglomeration effects may also attract production factors from surrounding cities through their newly added urban land, thereby having a siphon effect on other cities' economic growth in the urban agglomeration. At the same time, local governments have bottom-line competition in the process of attracting capital and will compete for more land leasing indicators. As a result, central cities that are more capable of fighting for land leasing indicators [57,68] may also crowd out the economic growth of other cities in urban agglomerations.

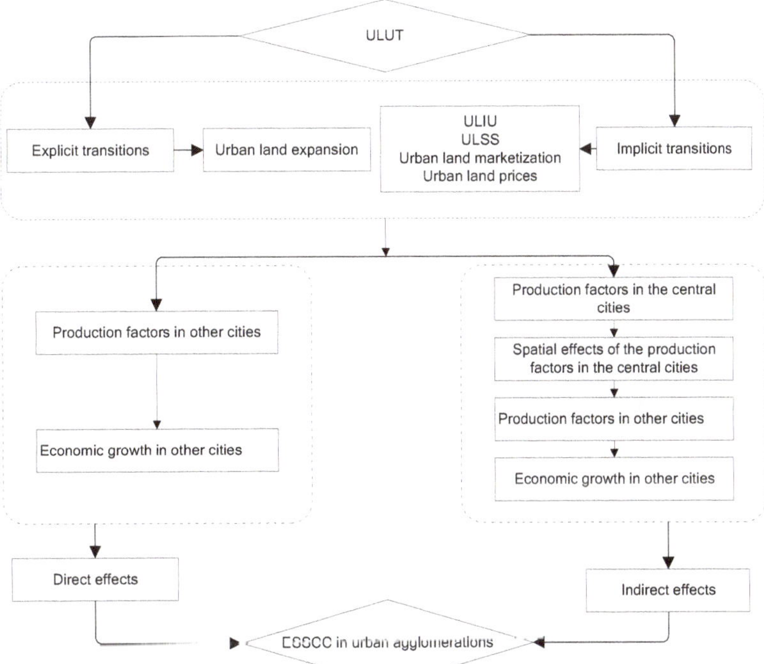


**Figure 1.** The influence mechanism of ULUT on the ESSCC in urban agglomerations.

Second, ULIU affects the ESSCC of urban agglomerations by influencing the industrial structure and its spatial emulation effect. Local governments continuously improve ULIU to ease the constraints of urban land resources on urban economic growth [69]. In the process of corporate investment, they gradually formulate assessments of land investment intensity and output intensity, to screen companies and industries. At the same time, they continue to revitalize the stock of construction land, eliminate enterprises and industries with low land use efficiency, and reintroduce enterprises with higher land use efficiency so as to improve land use efficiency and promote urban economic growth. The central government has increased ULIU, forcing local governments to continuously optimize the industrial structure to further stimulate urban economic growth. In the process of continuously improving the levels of ULIU, local governments also promote the flow of production factors, such as enterprise investment, labor force, and infrastructure investment, among cities, thus affecting the ESSCC of urban agglomerations. In addition, ULIU between cities has spatial learning and emulation effects that may also indirectly affect the ESSCC in urban agglomerations [70].

Third, the ULSS affects the urban investment structure and industrial structure, which in turn affect the ESSCC in urban agglomerations. The structure of urban land includes industrial and mining storage land, commercial land, residential land, and other lands. Industrial and mining storage land is used for industrial development, which mainly affects industrial investment; commercial land is used for commercial, financial, catering, hotels, and other operating service industries and their corresponding ancillary facilities, which mainly affects service industry investment; residential land mainly affects real estate investment; and other land includes public management, public services land, transportation land, water area, and water conservancy facilities land, which mainly affects infrastructure investment. With the continuous optimization of ULSS, urban investment structures and industrial structures have also changed [71]. Meanwhile, different industrial types also have different effects driving economic radiation. Therefore, ULSS will affect the ESSCC in urban agglomerations.

Fourth, urban land marketization affects economic growth through financing effects and resource allocation effects, which further affects the ESSCC in urban agglomerations. The ULUT away from planned allocation to market-oriented allocation are an important part of China's economic market reform. Urban land leasing has changed from planned allocation to agreement to the current transfer method based on bidding, auctioning, and listing. In order to attract investment, local governments often resort to depressing land prices, or even offering land free of charge to enterprises [51,55]. As the degree of land marketization continues to increase, cities with less economic competitiveness have no way to attract investment by lowering land prices. The growing marketization of urban land can not only increase the degree of land capitalization and promote the expansion of the production scale by increasing urban financing [72], but it can also improve the efficiency of resource allocation through more effective uses of land price signals and more effective combinations of production factors. Ultimately, the increase in the marketization of urban land is conducive to the promotion of urban economic growth, and it can further affect the ESSCC of urban agglomerations through the spatial effects of production factors and economic output.

Fifth, urban land prices affect the ESSCC in urban agglomerations through bottom-line competition and the enterprise screening effect [54]. Economic growth competition between regions will cause local governments to continuously lower their land leasing prices and the quality of corporate investment in order to obtain investment [51,55]. The competition will not necessarily drive regional economic growth and may reduce the ESSCC in urban agglomerations. However, increasing urban land prices helps to screen companies by favoring those enterprises with high technical efficiency and high profit margins and squeezing out those with low profit margins and low technical efficiency, thus helping to strengthen the ESSCC [73]. However, the rapid increase in urban land prices in urban agglomerations' central cities may also crowd out corporate investment, thus forcing some companies to migrate to surrounding cities. For example, due to the rapid increase in urban land prices and labor prices in China's coastal cities, a large number of manufacturing companies have moved to Southeast Asia and the mid-western parts of China. At the same time, rising urban areas have brought about a rapid increase in housing prices, which may also cause labor to flow from central cities to surrounding cities. In this way, an increase in the urban land prices of a central city may directly drive the economic growth of other cities [74]. However, land prices are often spatially correlated. An increase in land prices in central cities will lead to the land prices rising in the surrounding cities as well, which will directly crowd out investment and labor in other cities and inhibit their economic growth.

## 3. Methodology

### 3.1. Spatial Autocorrelation Analysis

ULUT may have spatial autocorrelation, due to land use competition, imitation and technology spillover effect of land use among cities. This paper uses Moran I to test this spatial autocorrelation, and the calculation formula is as follows:

$$I = \frac{\sum_{i=1}^{n} \sum_{j \neq i}^{n} W_{ij}(X_i - \overline{X})(X_j - \overline{X})}{S^2 \sum_{i=1}^{n} \sum_{j \neq i}^{n} W_{ij}} \tag{1}$$

where $X_i$ and $X_j$ refer to the variables of ULUT of the cities $i$ and $j$, respectively; $n$ is numbers of the samples. The sample of this paper is the panel data of 152 cities from 2003 to 2016, so $n$ equals 2128. $\overline{X}$ is the average value of each variable. $S^2$ is the variance of each variable.

$W_{ij}$ is the spatial distance weight matrix. We take spherical distance as the geographical distance of two cities. The spatial effect tends to diminish with distance [36,75], so we use the inverse of the geographical distance square as the element of the spatial weight matrix. The calculation is as follows:

$$w_{ij} = \begin{cases} = 0 & (i = j) \\ \frac{1}{d_{ij}^2} & (i \neq j) \end{cases} \tag{2}$$

where $w_{ij}$ is the element of the spatial weight matrix; $d_{ij}$ refers to spherical distance of two cities, $d_{ij} = R \times \arccos(\cos(\alpha_i - \alpha_j)\cos\beta_i\cos\beta_j + \sin\beta_i\sin\beta_j)$. In the formula for $d_{ij}$, $R$ is the equatorial radius of the earth, determined to be 6378 km; $\alpha$ and $\beta$ are longitude and latitude, respectively.

This paper uses the software stata.14 to estimate the spatial autocorrelation panel model. First, the spatial weight matrix is generated according to Equation (2). Then, based on this spatial weight matrix, we use the "spatgsa" command to estimate global spatial autocorrelation, and perform a 2-tailed test on the exponential significance.

### 3.2. The Econometric Model to Examine the Impacts of ULUT on ESSCC

Based on the traditional measurement model, considering the spatial differences and the spatial correlation between variables, a spatial measurement model has been developed [76]. Spatial measurement models help to test the spatial influence between variables. The spatial lag model can test the spatial influence of the dependent variable. The spatial Durbin model can simultaneously test the spatial effects of the key independent variables and dependent variables [77,78]. This article attempts to explore the direct and indirect spatial effects of the ULUT in central cities on the economic growth and production factors of other cities, and it is more suitable to adopt the spatial Durbin model. Therefore, this article draws on the form of the spatial Durbin model, incorporates the spatial lag variables that examine the independent variables and dependent variables in the model, and constructs a measurement model as shown in Equations (3) and (4).The paper constructs a basic production function equation in which the output is the regional gross product and the input factors are the capital stock and labor scale. To examine the impact of ULUT on the urban economics, variables reflecting ULUT are included in the model. To test the direct spatial effect of ULUT incentral cities on ESSCC, the spatial lag variable of ULUT in central cities was introduced as an explanatory variable. To examine the indirect effects of ULUT in the central cities on ESSCC, we further introduce the spatial lag variable of the regional gross product of central cities. Assuming that the production function is in the form of Cobb–Douglas, after taking the log of both sides, we obtain the measurement model of the impacts of ULUT on the ESSCC as follows:

$$\ln y_{it} = \beta_0 + \beta_1 \ln k_{it} + \beta_2 \ln n_{it} + \lambda_1 WC \ln ule_{it} + \lambda_2 WC \ln uliu_{it} + \lambda_3 WCpil_{it}$$
$$+ \lambda_4 WCpsl_{it} + \lambda_5 WCphl_{it} + \lambda_6 WCulm_{it} + \lambda_7 WC \ln ulp_{it}$$
$$+ \rho(WC \ln y)_{it} + \alpha_1 \ln ule_{it} + \alpha_2 \ln uliu_{it} + \alpha_3 pil_{it} + \alpha_4 psl_{it} \qquad (3)$$
$$+ \alpha_5 phl_{it} + \alpha_6 ulm_{it} + \alpha_7 \ln ulp_{it} + \varepsilon_{it}$$

where $\ln y$ represents the gross domestic product (GDP) in a city, and its log is used as the dependent variable. The control variables are $\ln k$ and $\ln n$, as the logs of capital stock and labor scale, respectively; $\ln ule$, $\ln uliu$, $ulm$ and $\ln ulp$ are variables, representing urban land expansions, ULIU, urban land marketization and urban land prices, respectively; *pil*, *psl*, and *phl* stand for the proportion of industrial and mining storage land, commercial service land, and residential land in the leased land, and together, they reflect the ULSS; $WC \ln y$, $WC \ln ule$, $WC \ln uliu$, $WCpil$, $WCpsl$, $WCphl$, $WCulm$, and $WC \ln ulp$ are the spatial lag variables of the corresponding variables of the central cities in the urban agglomerations; $i$ and $t$ represent the city and year, respectively; $\varepsilon$ is the residual error; $\beta$, $\lambda$, $\rho$ and $\alpha$ are the coefficients to be estimated. According to $\lambda$, the direct effects of ULUT on the ESSCC can be tested, which means that when a certain aspect of ULUT in a central city changes by 1%, the GDP of other cities will change by $\lambda$%; by combining $\rho$ and $\alpha$, we can determine the indirect effects of ULUT on the ESSCC, which means that every 1% change of ULUT in the central city will result in economic changes $\left(\rho \times \alpha \times \frac{1}{d^2}\right)$% in other cities $d$ kilometers away from the central city.

The spatial lag variables of the corresponding variables of the central cities in the urban agglomerations are constructed as follows: first, set the dummy variable of the central city. Central cities include the central cities in the urban agglomeration development plan approved by China's State Council. If a city is the central city in the urban agglomerations, it is given a value of 1, and 0 otherwise. Take the urban agglomeration in the Middle Reaches of the Yangtze River, for example. According to the urban agglomeration plan, the central cities are Wuhan, Changsha, and Nanchang. The urban agglomeration planning scope covers 28 cities above the prefecture level in the provinces of Hubei, Hunan, and Jiangxi. Additionally, the cities affected by ULUT of a central city are the other 27 cities in addition to the central city itself. Then, multiply each city's corresponding variables by the dummy variable of the central cities, and finally, multiply by the spatial distance weight matrix $W$, to obtain the spatially lagged variables of the central cities in the urban agglomerations.

### 3.3. The Econometric Model to Examine the Intermediate Mechanism

Theoretical analysis shows that ULUT directly or indirectly affect the ESSCC in urban agglomerations by affecting production factors and their spatial effects. This paper selects five main production factors—enterprise investment, fiscal expenditure, infrastructure investment, labor, and technical efficiency—as the explained variables to test the intermediate mechanism of urban agglomeration central cities' ULUT affecting the ESSCC. The econometric model is as follows:

$$f_{it} = a_0 + \sum_{n=1} a_n x_{n,it} + z_1 WC \ln ule_{it} + z_2 WC \ln uliu_{it} + z_3 WCpil_{it}$$
$$+ z_4 WCpsl_{it} + z_5 WCphl_{it} + z_6 WCulm_{it} + z_7 WC \ln ulp_{it}$$
$$+ \theta WCf_{it} + b_1 \ln ule_{it} + b_2 \ln uliu_{it} + b_3 pil_{it} + b_4 psl_{it} \qquad (4)$$
$$+ b_5 phl_{it} + b_6 ulm_{it} + b_7 \ln ulp_{it} + \sigma_{it}$$

where $f_{it}$ is the explained variable. When testing the enterprise investment mechanism, fiscal expenditure mechanism, infrastructure investment mechanism, labor mechanism, and technical efficiency mechanism, the log of corporate investment, the log of fiscal expenditure, the log of infrastructure investment, the log of labor, and technical efficiency are used as explained variables, respectively. The control variable is $x_n$. In the enterprise investment mechanism test, the control variables include the log of GDP lagged by one period, the log of capital stock lagged by one period, the log loan balance at the end of the year, and the log of the total population at the end of the year. In the fiscal expenditure mechanism test, the control variables include the log of GDP, the log of fiscal revenue, and the log of the total population at the end of the year. In the infrastructure investment mechanism test, the control variables include the log of fiscal expenditure and the log of the built-up area. In the labor mechanism test, the control variables include the log of the total population at the end of the year, the log of the number of hospitals per 10,000 people, the log of the number of pupils per 10,000 people, and the log of the built-up area. In the technical efficiency mechanism test, the control variables are the log of enterprise investment, the log of the number of college students per 10,000 people, and the log of foreign direct investment. $\sigma$ is the residual error term. $a$, $z$, $\theta$, and $b$ are the coefficients to be estimated. The direct and indirect effects of ULUT of central cities in urban agglomerations on the production factors of other cities can be measured by the coefficient $z$ and combining $\theta$ and $b$, respectively.

### 3.4. Solution of Endogeneity and Collinearity

The ULIU is obtained by dividing the city's GDP by the urban construction land area. As an explanatory variable, it has a correlation with the explained GDP and various production factors. This makes a potentially endogenous variable in the two econometric models. Therefore, this paper adopts the two-stage estimation method of instrumental variables of the fixed effects model. We select the log of the total land leasing area of other cities in the province or the log of the sum of the land leasing areas of other provinces and cities with the exception of this municipality, the log of the built-up area, and the proportion of the urban construction land in the built-up area, the area of the built-up area, the age of the party secretary, the number of years as the party secretary, and the dummy variable of whether the tenure of the party secretary is within its first two years, as instrumental variables. The reasons are as follows: First, the land leasing area is often strictly limited to a certain period of time [79]. Therefore, the sum of the land leasing scale in other cities will affect the land leasing scale of the city, thus affecting the ULIU of the city, but it does not directly affect the GDP and production factors of the city. Second, existing studies have shown that the younger the party secretary, the fewer the number of years serving or when in the first two years serving as a party secretary, the party secretary will more impulsively use land to stimulate economic growth [80]. Third, the built-up area is often determined by the city in its long-term spatial evolution and can be regarded as an exogenous variable. When the area of urban construction land occupies a large area of the built-up area, it will force the ULIU, but it will not directly affect GDP and production factors of the city.

The explanatory variables in each estimation model of our Equations (3) and (4) are first analyzed by a correlation coefficient. The correlation coefficient test results show that the correlation between the independent variables in the six regression equations is not very high, and the absolute value of the highest correlation coefficient does not exceed 0.5. Then, after the mixed regression least squares estimation of the above equation, the independent variables' variance inflation factors are analyzed. The results show that the variance inflation factors of most variables are below 10. Although the variance inflation factors of $L1. \ln k$, $\ln gy$, $\ln ge$, $\ln y$ and $l. \ln y$ exceed 10, these variables are only control variables, and their estimated coefficients are not the focus of this article. Although the variance inflation factors of key observation variables such as $wc \ln ulp$, $wc \ln uliu$, $wc \ln ule$, $wculm$, and $wc \ln n$ exceed 10, they are not large. Additionally, because this is just the variance inflation factors in the mixed regression model, after adopting the panel data model, it helps to reduce collinearity. Therefore, in the two-stage least squares estimation process of the panel data's instrumental variables used in this paper, the collinearity of these observed variables has little effect on the accuracy of its estimated coefficients. Based on the above analysis, this article believes that the collinearity problem in the estimation equation is acceptable.

### 3.5. Variables and Data

Urban land expansions, ULIU, and urban land prices are measured by the newly added construction land area in the state-owned construction land leased, the GDP of the unit construction land, and the land leasing price, respectively; urban land marketization is calculated by the proportion of land leased by bidding, auction, and listing in the leased land. The proportion of industrial and mining storage land, commercial service land, and residential land in the leased land, together, reflects the ULSS. The labor is calculated by adding the number of employees to the number of self-employed private workers. Enterprise investment is obtained by subtracting investment in fixed assets for urban construction from investment in fixed assets for society as a whole. Drawing on the methods of Ke [81], the capital stock in the initial year 2003 is estimated by the average investment growth rate, depreciation rate, and investment in the initial year 2003, and then, the perpetual inventory method is used to calculate the capital stock of cities above the prefecture level from 2004 to 2016, for which the depreciation rate is 9.6%. Urban infrastructure investment is represented by urban construction fixed asset investment. Technical efficiency is estimated by using the stochastic frontier production function model with the total social capital stock and labor as input elements and GDP as output [82,83]. The relevant data affected by prices are transformed as comparable prices.

In this paper, 152 cities above the prefectural level in urban agglomerations are selected as samples (see Figure 2). The urban agglomerations in the samples, which are approved by the State Council of China as national urban agglomerations, include the Beijing–Tianjin–Hebei (including 11 cities), Yangtze River Delta (including 27 cities), Pearl River Delta (including 16 cities), Middle Reaches of the Yangtze River (including 28 cities), Central Plains (including 26 cities), Guanzhong Plain (including 5 cities), Lanzhou–Xining(including 4 cities), Hohhot–Baotou–Ordos–Yulin (including 4 cities), Chengdu–Chongqing (including 16 cities), Beibu Gulf (including 6 cities), and Harbin–Changchun (including 9 cities) urban agglomerations. Then, based on China's regional differences, we select 8 urban agglomerations for typical analysis. Among them, we chose the Beijing–Tianjin–Hebei, Yangtze River Delta, and Pearl River Delta urban agglomerations in eastern China; the Middle Reaches of the Yangtze River and Central Plains urban agglomerations in central China; Guanzhong Plain urban agglomerations in northwestern China; the Chengdu–Chongqing urban agglomerations in southwestern China; the Harbin–Changchun urban agglomerations in northeastern China.

Data on newly added construction land leased; urban construction land area; land leasing area; bidding, auction, and listing leased area; land transaction price from 2003 to 2016; and the land leasing area distinguished by land type from 2009 to 2016 come from the China Land and Resources Statistical Yearbook. The China Land and Resources Statistical Yearbook did not record data on the land leasing area distinguished by land type from 2003 to 2008. Therefore, this paper matches the 398,706 land leases signed from 1 January 2003 to 31 December 2008 in the China Land Market Net to the prefecture level and sorts out and calculates the leased area of each city according to land type from 2003 to 2008. Data on urban built-up area and urban infrastructure investment are from the Statistical Yearbook of Urban Construction in China. The ages of the party secretaries of cities above the prefectural level and the number of years in office are manually collected from the Internet. The remaining data come from the China City Statistical Yearbook.

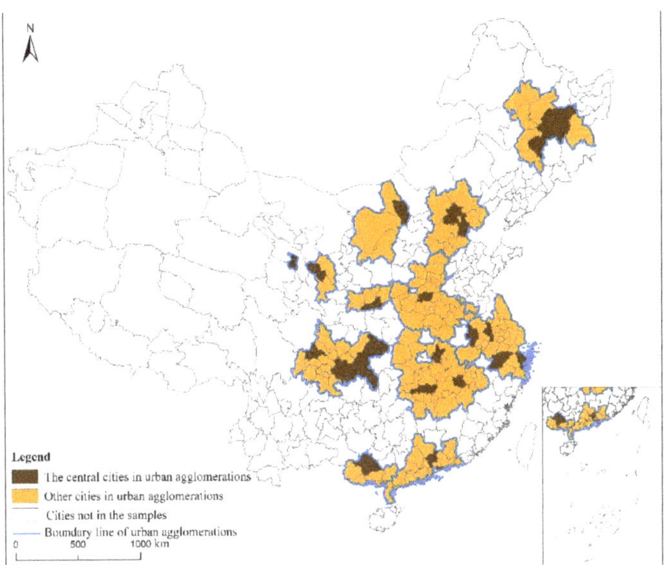


**Figure 2.** Samples.

## 4. Results

### 4.1. ULUT in China during 2003–2016

Based on the mean values of ULUT indicators in 152 cities above the prefecture level in China from 2003 to 2016 (see Table 1), this paper analyzes the temporal change characteristics of ULUT in China. The rate of urban land expansion has increased. In 2003, the average size of newly added construction land in the land leased was 324.22 hectares, and thishad expanded to 573.87 hectares by 2016. The ULIU has increased significantly. In 2003, the GDP per hectare of urban construction land was 31.683 billion yuan, which had increased by 121.31% to reach 70.118 billion. Moran's I is used to analyze the spatial autocorrelation of ULUT in Chinese urban agglomerations (see Table 2). The Moran's I of urban land expansions, the ULIU, the proportion of industrial land in leased land, urban land marketization, and the urban land prices are positive, and the corresponding z statistic is relatively large; Moran's I are significant at the 1% level. This indicates urban land expansions, the ULIU, the proportion of industrial land in leased land, urban land marketization, and the urban land prices have significantly positive spatial autocorrelations.

**Table 1.** Mean value of ULUT.

| Variables | Year | Total | Beijing–Tianjin–Hebei | Yangtze River Delta | Pearl River Delta | Middle Reaches of Yangtze River | Central Plain | Guanzhong Plain | Chengdu–Chongqing | Harbin–Changchun |
|---|---|---|---|---|---|---|---|---|---|---|
| ule (hm$^2$) | 2003 | 324.22 | 583.53 | 846.62 | 92.55 | 146.22 | 125.41 | 261.56 | 74.53 | 99.96 |
| | 2016 | 573.87 | 884.33 | 645.44 | 439.09 | 598.77 | 524.65 | 573.49 | 569.41 | 294.58 |
| uliu | 2003 | 316.83 | 284.37 | 344.14 | 644.72 | 245.05 | 209.08 | 247.85 | 264.79 | 314.08 |
| (10$^2$millionyuan/hm$^2$) | 2016 | 701.18 | 692.05 | 768.95 | 1462.23 | 653.03 | 421.83 | 557.46 | 526.34 | 481.01 |
| pil (%) | 2003 | 32.28 | 41.94 | 37.28 | 24.09 | 40.22 | 25.97 | 23.63 | 26.62 | 35.46 |
| | 2016 | 26.01 | 29.17 | 25.07 | 19.76 | 27.35 | 30.60 | 18.66 | 17.66 | 35.10 |
| psl (%) | 2003 | 16.31 | 20.79 | 16.69 | 21.88 | 11.70 | 14.27 | 26.60 | 19.08 | 14.36 |
| | 2016 | 7.52 | 8.80 | 7.65 | 4.81 | 8.10 | 8.29 | 7.51 | 7.11 | 8.83 |
| phl (%) | 2003 | 24.40 | 18.91 | 24.35 | 23.09 | 23.21 | 27.74 | 27.52 | 27.26 | 22.07 |
| | 2016 | 17.38 | 23.08 | 16.69 | 13.93 | 16.34 | 20.96 | 14.93 | 20.05 | 13.04 |
| ulm (%) | 2003 | 35.79 | 33.26 | 32.68 | 53.48 | 37.76 | 36.38 | 33.04 | 32.24 | 38.38 |
| | 2016 | 92.52 | 91.87 | 97.42 | 84.58 | 94.81 | 94.48 | 93.96 | 95.55 | 86.58 |
| ulp (10 thousand | 2003 | 256.12 | 297.66 | 300.85 | 233.58 | 243.72 | 228.91 | 178.39 | 277.22 | 188.86 |
| yuan/hm$^2$) | 2016 | 1386.86 | 1657.22 | 2299.41 | 2417.77 | 789.72 | 618.45 | 522.07 | 953.41 | 476.21 |

**Table 2.** Spatial autocorrelation of ULUT.

| Variables | Total | Beijing–Tianjin–Hebei | Yangtze River Delta | Pearl River Delta | Middle Reaches of Yangtze River | Central Plain | Guanzhong Plain | Chengdu–Chongqing | Harbin–Changchun |
|---|---|---|---|---|---|---|---|---|---|
| ule | 0.222 *** | 0.321 *** | 0.321 *** | 0.042 | 0.170 | 0.270 *** | 0.678 *** | 0.077 * | 0.12 * |
| | (3.54) | (5.55) | (9.43) | (0.74) | (1.00) | (8.84) | (4.25) | (1.94) | (1.80) |
| uliu | 0.389 *** | 0.418 *** | 0.388 *** | 0.010 | −0.327 | 0.249 *** | 0.400 *** | 0.329 *** | −0.023 |
| | (6.23) | (6.98) | (11.54) | (0.23) | (1.89) | (8.06) | (2.53) | (7.73) | (0.20) |
| pil | 0.323 *** | 0.064 | 0.104 *** | 0.094 | 0.631 *** | 0.080 *** | 0.022 | 0.204 *** | 0.067 |
| | (5.14) | (1.14) | (3.10) | (1.56) | (3.69) | (2.64) | (0.21) | (4.84) | (1.01) |
| psl | 0.002 | 0.065 | 0.089 *** | 0.022 | −0.337 * | 0.071 *** | 0.482 *** | 0.234 *** | −0.033 |
| | (0.04) | (1.22) | (2.72) | (0.44) | (1.97) | (2.43) | (3.12) | (5.79) | (0.35) |
| phl | 0.036 | 0.002 | 0.092 *** | −0.085 | −0.006 | 0.029 | 0.121 | 0.113 *** | 0.044 |
| | (0.58) | (0.13) | (2.77) | (1.29) | (0.02) | (1.02) | (0.83) | (2.75) | (0.72) |
| ulm | 0.757 *** | 0.839 *** | 0.842 *** | 0.517 *** | 0.880 *** | 0.751 *** | 0.822 *** | 0.815 *** | 0.736 *** |
| | (12.02) | (13.85) | (24.51) | (8.18) | (5.13) | (24.11) | (5.10) | (18.89) | (10.02) |
| ulp | 0.276 *** | 0.048 | 0.198 *** | 0.468 *** | 0.269 | 0.320 *** | −0.116 | 0.334 *** | 0.019 |
| | (4.47) | (1.07) | (6.08) | (7.77) | (1.60) | (10.62) | (0.65) | (7.85) | (0.49) |

Note: The values not in brackets areMoran's I, and the values in brackets are Moran's I's z statistics. * and *** respectively indicate that the estimated coefficients are significant at the level of 1% and 10%.

However, the differential analysis of the spatial correlations of ULUT of cities in typical urban agglomerations shows that there are great differences in the spatial autocorrelations of ULUT in urban agglomerations. The urban land expansions in the Beijing–Tianjin–Hebei, Yangtze River Delta, Central Plains, Guanzhong Plain, and Chengdu–Chongqing urban agglomerations have significant positive spatial autocorrelations, but are not significant in the Pearl River Delta and Middle Reaches of the Yangtze River Delta urban agglomerations. There are significant positive spatial autocorrelations in the ULIU in the Beijing–Tianjin–Hebei, Yangtze River Delta, Central Plains, Guanzhong Plain, and Chengdu–Chongqing urban agglomerations, but these are not significant in the Pearl River Delta, Middle Reaches of the Yangtze River, and Harbin–Changchun urban agglomerations; the spatial autocorrelation of the ULSS in urban agglomerations varies greatly. The proportion of industrial land in leased land, commercial land in leased land, and residential land in leased land has a significant positive spatial autocorrelation in the Yangtze River Delta and Chengdu–Chongqing urban agglomerations. However, the spatial autocorrelation of the above three types of urban land supply in the Beijing–Tianjin–Hebei, Pearl River Delta, and Harbin–Changchun urban agglomerations is not significant. The proportion of industrial land leased in the Middle Reaches of the Yangtze River urban agglomeration has a positive spatial autocorrelation, while the proportion of commercial land leased has a negative spatial autocorrelation. There is a significant positive spatial autocorrelation in the proportion of industrial land leased and commercial land leased in the Central Plains urban agglomeration. In the Guanzhong Plain urban agglomeration, the proportion of commercial land leased has a significant positive spatial autocorrelation, but the proportion of industrial land leased and residential land leased has no significant spatial autocorrelation.

### 4.2. The Impacts of ULUT on the ESSCC in Urban Agglomerations

Based on the estimated effects of ULUT on ESSCC (see Table 3), we conclude that: (1) Urban land expansion and ULIU directly and indirectly promote ESSCC in urban agglomerations; (2) ULSS, urban land marketization and urban land prices indirectly enhance ESSCC in urban agglomerations (see Figure 3). In the following figures, "→" refers to significantly positive effects, "→" refers to significantly negative effects, and ""refers to insignificant effects.

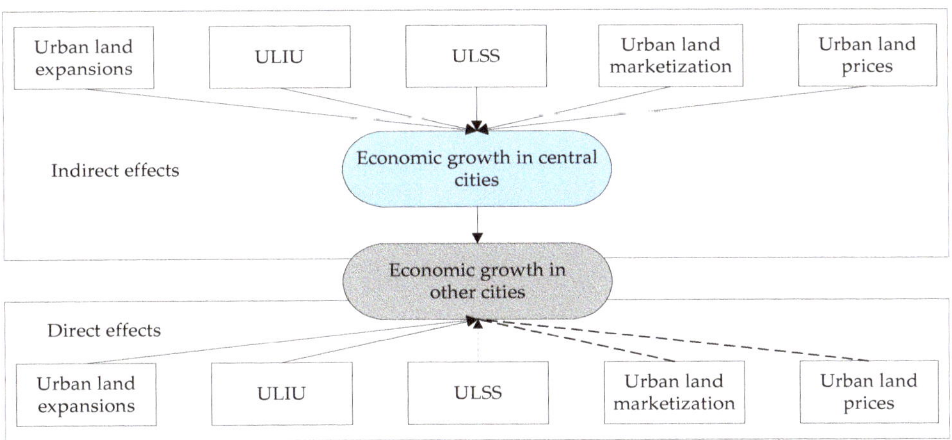


**Figure 3.** The impacts of ULUT on ESSCC in urban agglomerations.

**Table 3.** Regression estimate results for impacts of ULUT on ESSCC.

| Variables | Total | Beijing Tianjin Hebei | Yangtze River Delta | Pearl River Delta | Middle Reaches of the Yangtze River | Central Plains | Guanzhong Plain | Chengdu Chongqing | Harbin Changchun |
|---|---|---|---|---|---|---|---|---|---|
| wclnule | 3.785 *** | 42.327 | 5.632 | 5.308 ** | 0.482 | 13.483 ** | 5.721 | −7.551 * | 3.461 |
| | (3.81) | (1.10) | (1.56) | (2.09) | (0.22) | (2.28) | (0.51) | (1.93) | (0.06) |
| wclnuliu | 7.709 * | 469.958 *** | −21.237 | −22.800 | −0.505 | −166.681 ** | 55.958 | 48.209 | −1081.976 |
| | (1.71) | (3.32) | (0.56) | (1.00) | (0.14) | (2.26) | (1.32) | (0.37) | (1.52) |
| wcpil | −7.594 | 114.457 | 81.976 * | 70.497 | −9.224 | 67.758 | 16.321 | 87.596 | 614.922 |
| | (1.50) | (0.54) | (1.92) | (1.24) | (1.44) | (1.16) | (0.28) | (0.90) | (0.77) |
| wcpsl | −7.533 | 286.438 | −165.571 * | −165.238 | −16.225 | 5.1 | −3.311 | 420.252 ** | 119.619 |
| | (0.54) | (0.83) | (1.67) | (1.24) | (0.88) | (0.04) | (0.03) | (2.34) | (0.06) |
| wcphl | −9.386 | 508.635 * | 90.394* | 135.784 ** | −10.862 | −59.918 | −66.129 | 9.942 | −56.784 |
| | (1.13) | (1.71) | (1.69) | (2.16) | (1.06) | (0.64) | (0.73) | (0.11) | (0.07) |
| wculm | 0.671 | −354.712 ** | 137.364 *** | 6.782 | 9.909 | 150.606 ** | 68.478 | 194.865 | 2659.256 ** |
| | (0.13) | (2.46) | (3.58) | (0.15) | (1.33) | (2.56) | (0.97) | (1.61) | (2.59) |
| wclnulp | −0.470 | 79.048 | −29.839 | 22.916 | −6.690 | −30.321 | −26.007 | −62.948 | 543.726 |
| | (0.12) | (0.99) | (1.38) | (1.36) | (1.31) | (0.75) | (0.68) | (1.54) | (1.03) |
| wclny | 45.670 * | 185.569 | 50.292 | 52.938 | −59.757 | 760.988 *** | −81.858 * | −992.131 *** | 723.047 |
| | (1.89) | (1.06) | (0.51) | (0.94) | (0.51) | (4.15) | (1.81) | (−5.06) | (0.45) |
| lnule | 0.004 *** | 0.003 | 0.007 ** | 0.001 | 0.002 | 0.007* | 0.014 ** | −0.002 | 0.004 |
| | (3.72) | (0.19) | (2.52) | (0.41) | (0.76) | (1.92) | (2.37) | (1.07) | (0.66) |
| lnuliu | 1.028 *** | 0.230 * | −0.029 | −0.046 | −0.219 *** | 0.132 | −0.015 | 0.345 * | −0.032 |
| | (3.92) | (1.75) | (0.84) | (0.91) | (3.74) | (1.30) | (0.14) | (1.70) | (0.17) |
| pil | 0.001 *** | $9 \times 10^{-4}$ | $3 \times 10^{-4}$ | $-7 \times 10^{-4}$ | $-1 \times 10^{-4}$ | $5 \times 10^{-4}$ | $7 \times 10^{-4}$ | $-6 \times 10^{-4}$ | 0.003 ** |
| | (4.11) | (1.52) | (0.65) | (1.09) | (0.15) | (1.49) | (1.02) | (1.26) | (2.54) |
| psl | $2 \times 10^{-4}$ | $8 \times 10^{-4}$ | −0.002 *** | 0.003 ** | $2 \times 10^{-4}$ | $4 \times 10^{-4}$ | −0.002 | $-6 \times 10^{-4}$ | −0.002 |
| | (0.57) | (1.12) | (2.81) | (2.54) | (0.15) | (0.70) | (1.09) | (0.66) | (1.03) |
| phl | 0.000 | $-2 \times 10^{-4}$ | $-6 \times 10^{-4}$ | $8 \times 10^{-4}$ | $-9 \times 10^{-4}$ | $-4 \times 10^{-4}$ | 0.001 | −0.001 | 0.001 |
| | (0.12) | (0.25) | (1.29) | (1.08) | (1.41) | (0.82) | (1.01) | (1.46) | (0.66) |
| ulm | 0.001 *** | 0.001 ** | $-4 \times 10^{-4}$ | $-4 \times 10^{-4}$ | 0.002 *** | $4 \times 10^{-4}$ | $3 \times 10^{-4}$ | 0.003 *** | $-6 \times 10^{-4}$ |
| | (6.23) | (2.36) | (−0.91) | (0.73) | (2.69) | (0.78) | (0.40) | (3.93) | (0.47) |
| lnulp | 0.054 *** | 0.030 | 0.061 *** | −0.002 | 0.149 *** | 0.098 *** | 0.016 | −0.040 * | 0.069 |
| | (6.49) | (1.23) | (3.10) | (0.07) | (4.49) | (4.20) | (0.36) | (1.65) | (1.54) |
| id-effects | yes | yes | yes | yes | yes | yes | yes | yes | yes |
| contorls | yes | yes | yes | yes | yes | yes | yes | yes | yes |
| sargan-p | 0.11 | 0.11 | 0.20 | 0.13 | 0.13 | 0.29 | 0.13 | 0.24 | 0.25 |
| Obs | 2128 | 154 | 378 | 224 | 392 | 364 | 70 | 224 | 126 |
| R-squared | 0.92 | 0.95 | 0.95 | 0.92 | 0.91 | 0.95 | 0.98 | 0.97 | 0.87 |

Note: The values not in brackets are coefficients, and the values in brackets are corresponding *t*-statistics. *, ** and *** respectively indicate that the estimated coefficients are significant at the level of 1%, 5% and 10%.

### 4.2.1. The Impacts of Urban Land Expansions on the ESSCC in Urban Agglomerations

From the empirical results (see Table 3), we can conclude that urban land expansions in central cities are helpful to enhance the ESSCC. On the one hand, the urban land expansions in central cities can directly drive the economic growth of other cities. Each 1% increase in urban land area in central cities will drive the economic growth of other cities in a radius of 100 km from central cities by 0.0004%. The calculation process is that the estimated coefficient of $wc \ln ule$ is 3.785 multiplied by the reciprocal of the square of the distance 100, because this article chooses the reciprocal of the square of the distance as the element of the spatial weight matrix. The calculation process of direct influence effects below is similar to this. On the other hand, urban land expansions in central cities can indirectly drive the economic growth of other cities in the urban agglomerations, because of the significantly positive effects of urban expansions on urban economic growth and the significant spatial diffusion effect of central cities' economics. Each 1% increase in the urban land expansions in central cities can significantly drive economic output growth by 0.0039% and indirectly drive the economic growth of other cities within a radius of 100 km from central cities by 0.000018%. The calculation process is to multiply $\ln ule$'s estimated coefficient which is 0.004 by $wc \ln y$'s estimated coefficient which is 45.670, and then, multiply by the reciprocal of the square of 100, because the reciprocal of the square of the distance is an element of the spatial weight matrix in this paper. The calculation process of the indirect influence effect below is similar to this.

The empirical results of the urban land expansions affecting the ESSCC show that, although the urban land expansions in central cities can help to enhance the ESSCC, its direct and indirect effects are very small. Enhancing the ESSCC cannot rely on the expansions of urban land in central cities. At present, the Chinese central government has scientifically and rationally delineated urban boundaries through its territorial spatial planning and placed restrictions on the expansion of urban areas [84], which will have little negative impacts on the improvement of the ESSCC.

The estimation results of the typical agglomerations sample show that the impacts of urban land expansions on ESSCC vary greatly. The expansions of urban land in the Beijing–Tianjin–Hebei, Yangtze River Delta, Middle Reaches of the Yangtze River, Central City of Harbin–Changchun urban agglomerations did not have a significant impact on ESSCC. The urban land expansions of central cities in the Pearl River Delta can directly drive the economic growth of other cities in the urban agglomeration. The urban land expansions of central cities in the Central Plains urban agglomeration can directly and indirectly promote the economic growth of other cities in the urban agglomeration, but the urban land expansions of the central cities in the Chengdu–Chongqing urban agglomeration can directly inhibit the economic growth of other cities in the urban agglomerations.

### 4.2.2. The Influences of the ULIU on the ESSCC in Urban Agglomerations

ULIU in central cities can significantly enhance the ESSCC. Each 1% increase in ULIU in central cities can directly drive the economic growth of other cities in a radius of 100 km from central cities by 0.0007%. Additionally, each 1% increase in ULIU in central cities can drive the central cities' economic growth by 1.0283% (see Table 3), and then, the GDPs of other cities in a radius of 100 km from central cities in the urban agglomeration can indirectly grow by 0.0047% through the diffusion effect of the central cities' economic growth.

China's rapid urban expansions in recent decades have led to the conversion of a large amount of agricultural land into urban construction land. Land resource constraints are becoming increasingly tight, and ULIU has become inevitable [68–70]. In general, ULIU has significantly promoted ESSCC, which is much higher than the positive effect of urban land expansions. This shows that ULIU should be prioritized over urban land expansions to promote ESSCC. However, since the diffusion effects of the economic growth of central cities are small, the current intensive use of urban land in central cities only has a small positive effect on the economic growth of other cities. It is necessary to further enhance the diffusion effects of the economic growth of the central cities in urban agglomerations to increase the positive impacts of ULIU on the ESSCC.

The estimation results show that the Pearl River Delta, Yangtze River Delta, Middle Reaches of the Yangtze River, Guanzhong Plain, and Harbin–Changchun urban agglomerations' ULIU in central cities has no significant impact on the ESSCC. The ULIU in the central cities of the Beijing–Tianjin–Hebei urban agglomeration will directly drive the economic growth of other cities in the urban agglomeration. However, the ULIU in the central cities of the Central Plains urban agglomeration will directly inhibit the economic growth of other cities. Additionally, the intensive land use in the central cities of the Chengdu–Chongqing urban agglomeration will indirectly inhibit the economic growth of other cities in the urban agglomeration (see Table 3).

### 4.2.3. The Effects of the ULSS on the ESSCC in Urban Agglomerations

In the supply structure of urban land, only the proportion of industrial land in the leased land of central cities can indirectly significantly increase the GDP of other cities in urban agglomerations. However, the proportion of commercial land and the proportion of residential land in the leased land has no significant effect on ESSCC (see Table 3). This also confirms the local economic development model in which local governments boost the economy through industrial land leasing and increase fiscal revenue through commercial land leasing [85]. Industrial land leasing can drive economic growth through the introduction of industrial investment and then promote the economic growth of related industries in other cities in the urban agglomeration through diffusion effects such as industrial chains. However, it is relatively difficult to form a spatial correlation between commercial land and residential land in the cities and their spatial influence on economic growth.

The estimation results on typical urban agglomerations show that the increase in the proportion of residential land in the central cities of the Beijing–Tianjin–Hebei urban agglomeration can help to significantly drive the economic growth of other cities. The increase in the proportion of industrial land and the proportion of residential land in leased land can directly increase the ESSCC in the Yangtze River Delta, but an increase in the proportion of commercial land in leased land can inhibit the economic growth of other cities. The increase in the proportion of residential land in leased land of central cities in the Pearl River Delta can directly drive the economic growth of other cities in the urban agglomeration. The increase in the proportion of commercial land in leased land in the central cities of the Chengdu–Chongqing urban agglomeration can directly enhance its role in driving central cities' economic spatial spillovers. However, the ULSS in the Middle Reaches of the Yangtze River, the Central Plains, the Guanzhong Plain, and the Harbin–Changchun urban agglomerations did not have significant impacts on the ESSCC (see Table 3).

### 4.2.4. The Impacts of Urban Land Marketization on the ESSCC in Urban Agglomerations

The marketization of urban land indirectly increases ESSCC in urban agglomerations. From 2003 to 2016, the proportion of urban land leased by bidding, auction, and listing increased by 60% in China. Assuming that the proportion of land leased by bidding, auction, and listing in central cities also increased by 60%, this will drive the central cities' economic growth by 0.0660% and then indirectly drive the GDP of other cities in a radius of 100 km from central cities in urban agglomerations to increase by 0.0003% through the diffusion effect of the economic growth of the central cities (see Table 3). Overall, the degree of urban land marketization in China has greatly improved [86], but the marketization of urban land has less of a driving effect on economic growth. In addition, the diffusion effect of central cities is small, which leads to small positive effects of land marketization on the ESSCC.

The estimation results of typical urban agglomerations show that the marketization of urban land in the central cities of the Yangtze River Delta, Central Plains, and Harbin–Changchun urban agglomeration can help to enhance the ESSCC. However, the marketization of urban land in central cities in the Chengdu–Chongqing urban agglomeration indirectly enhances its siphon effect, leading to a decline in economic growth in other cities in the urban agglomeration. The urban land marketization of central cities in Beijing–Tianjin–Hebei can directly crowd out the economic growth of other cities. The urban land marketization of central cities in the Pearl River Delta, the Middle Reaches of the Yangtze River, and the Guanzhong Plain urban agglomerations have no significant impacts on ESSCC.

### 4.2.5. The Influences of the Urban Land Prices on the ESSCC in Urban Agglomerations

Rising urban prices indirectly strengthen the ESSCC. Each 1% increase in urban land prices drives the economic growth of central cities by 0.0537%, and then, indirectly promotes the economic growth of other cities in a radius of 100 km from central cities by 0.0002% through the diffusion effects of economic growth in the central cities. The direct effects of urban land prices on the ESSCC are non-significant (see Table 3).

The estimation results show that China's urban land prices rose rapidly, and effectively stimulated urban economic growth. On the one hand, the increase in urban land prices has compelled enterprises to increase their investment, thereby driving economic growth [73]. On the other hand, as urban prices rise, it may be conducive to the expansion of fiscal revenue to stimulate economic growth through fiscal expenditures and infrastructure investment [50]. Finally, through the spatial spillovers of the economic growth in central cities, urban land prices indirectly promote the economic growth of other cities in the urban agglomerations.

The estimated results of the urban agglomerations show that an increase in the urban land prices in the central cities of the Central Plains urban agglomeration can indirectly enhance ESSCC. However, urban land prices' increase in the central cities of the Chengdu–Chongqing urban agglomeration can weaken its siphon effect. The urban land prices in other representative urban agglomerations do not have significant impacts on the ESSCC.

### 4.3. The Intermediate Mechanism of the ULUT Affect the ESSCC in Urban Agglomerations

The estimation results show that there are significant spatial diffusion effects on enterprise investment, infrastructure investment, labor, and technical efficiency in the central cities in urban agglomerations. Only the spatial diffusion effects of fiscal expenditure are not significant (see Table 4). The spatial diffusion effects of the production factors in central cities also give economic growth a significant radiating driving effect.

### 4.3.1. The Influences of the Urban Land Expansions on the Production Factors and Their Spatial Effects in Urban Agglomerations

The urban land expansions in central cities can drive the growth of labor therein and then, indirectly drive the growth of labor in other cities through the spatial linkage effect of labor (see Table 4 and Figure 4). However, the urban land expansions in central cities will directly inhibit the growth of labor and technical efficiency in other cities in the urban agglomeration. Urban land expansions have no direct significant impact on corporate investment, fiscal expenditure, and infrastructure investment in other cities.

**Table 4.** Estimation results for the intermediate mechanism of the ULUT affect the ESSCC in urban agglomerations.

| Variables | Enterprise Investment | Fiscal Expenditure | Infrastructure Investment | Labor | Technical Efficiency |
|---|---|---|---|---|---|
| wclnule | 0.430 | −1.168 | 1.701 | −0.090 ** | −0.009 *** |
| | (0.21) | (0.98) | (0.34) | (2.54) | (2.83) |
| wclnuliu | 0.941 ** | −2.421 | 13.986 | −0.258 | $1 \times 10^{-4}$ |
| | (2.05) | (0.58) | (0.96) | (0.17) | (0.02) |
| wcpil | 6.250 | 11.637 * | −36.089 | −5.049 | −0.004 |
| | (0.55) | (1.92) | (1.40) | (1.11) | (0.26) |
| wcpsl | 37.611 | 5.831 | −18.192 | −6.335 | −0.053 |
| | (1.14) | (0.28) | (0.26) | (0.80) | (1.15) |
| wcphl | 11.470 | 15.214 | −1.742 | 30.476 | −0.052* |
| | (0.71) | (1.54) | (0.04) | (1.40) | (1.88) |
| wculm | −3.369 | 7.469 | 37.428 | −65.053 *** | 0.007 |
| | (0.29) | (1.05) | (1.42) | (5.03) | (0.43) |
| wclnulp | −8.398 | −9.814 ** | −49.653 *** | −9.959 | 0.019 |
| | (1.21) | (2.23) | (2.64) | (1.20) | (1.50) |
| wcf | 10.765 ** | 2.303 | 1.093 ** | 12.731 ** | 179.882 *** |
| | (2.29) | (0.68) | (2.29) | (2.03) | (4.28) |
| lnule | 0.003 | 0.004 *** | −0.106 | 39.797 *** | 0.000 |
| | (1.46) | (2.77) | (0.90) | (4.42) | (1.22) |
| lnuliu | 1.039 *** | 0.081 ** | 0.014 ** | $2 \times 10^{-4}$ | $1.65 \times 10^{-5}$ *** |
| | (2.82) | (2.45) | (2.54) | (0.11) | (3.14) |
| pil | $2 \times 10^{-4}$ | 0.000 | −0.001 | $-2 \times 10^{-4}$ | $-1.49 \times 10^{-6}$ ** |
| | (0.47) | (0.14) | (1.36) | (0.73) | (2.39) |
| psl | $-3 \times 10^{-4}$ | −0.001 *** | $13 \times 10^{-4}$ | −0.001 ** | $-3.60 \times 10^{-6}$ *** |
| | (0.44) | (2.78) | (0.82) | (−2.13) | (3.37) |
| phl | $7 \times 10^{-4}$ | $-3 \times 10^{-4}$ | $3 \times 10^{-4}$ | −0.001 *** | $-4.40 \times 10^{-6}$ *** |
| | (1.64) | (1.07) | (0.25) | (2.85) | (6.09) |
| ulm | 0.001 *** | $14 \times 10^{-4}$*** | $14 \times 10^{-4}$ | 0.002 *** | $7.36 \times 10^{-6}$ *** |
| | (3.01) | (6.06) | (1.48) | (6.75) | (12.21) |
| lnulp | 0.032 ** | −0.015 | 0.069 | 0.133 *** | $1 \times 10^{-4}$ *** |
| | (1.98) | (1.48) | (1.63) | (9.93) | (4.32) |
| id-effects | yes | yes | yes | yes | yes |
| controls | yes | yes | yes | yes | yes |
| sargan-p | 0.29 | 0.34 | 0.19 | 0.41 | 0.33 |
| Obs | 1976 | 2128 | 2128 | 2128 | 2128 |
| R-squared | 0.88 | 0.95 | 0.31 | 0.59 | 0.86 |

Note: The values not in brackets arecoefficients, and the values in brackets are corresponding *t*-statistics. *, ** and *** respectively indicate that the estimated coefficients are significant at the level of 1%, 5% and 10%.

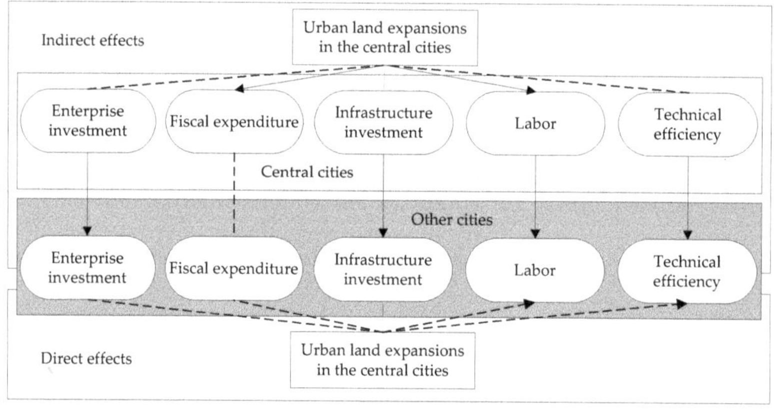


**Figure 4.** The impacts of urban land expansions on the production factors in urban agglomerations.

### 4.3.2. The Influences of the ULIU on the Production Factors and Their Spatial Effects in Urban Agglomerations

Intensive of urban land use in central cities has significantly driven the growth of production factors in other cities. The ULIU in central cities hassignificant positive effects on enterprise investment, infrastructure investment, and technical efficiency, and it can indirectly drive the growth of these production factors in other cities through their spatial diffusions (see Table 4 and Figure 5). Simultaneously, the ULIU in central cities can also directly drive enterprise investment in other cities. Thus, strengthening ULIU is currently an effective way to strengthen the ESSCC in China [84].

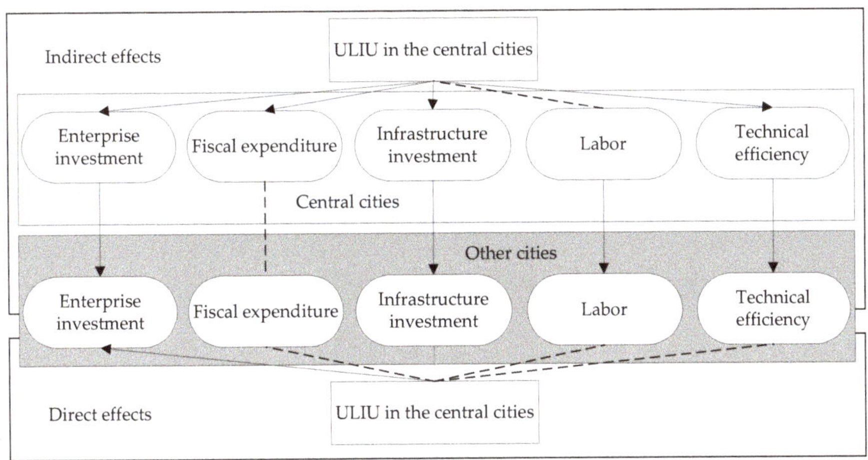


**Figure 5.** The impacts of ULIU on the production factors in urban agglomerations.

### 4.3.3. The Influences of the ULSS on the Production Factors and Their Spatial Effects in Urban Agglomerations

Increasing the proportion of industrial land, commercial land, and residential land in leased land in central cities can indirectly squeeze out labor in other cities and also inhibit the improvement of technical efficiency in other cities. An increase in the proportion of commercial land and the proportion of residential land in leased land in central cities can squeeze out labor and indirectly reduce the labor in other cities through the spatial linkage effect of labor. At the same time, an increase in the proportion of industrial land, commercial land, and residential land in leased land in central cities can inhibit the improvement of urban technical efficiency in other cities indirectly (see Table 4 and Figure 6). Therefore, when optimizing the urban spatial structure and reducing the proportion of industrial land in China, the input–output benefit of existing industrial land should be increased and technical efficiency should be improved to mitigate the damage of the reduction in the proportion of industrial land on the improvement of technical efficiency.

### 4.3.4. The Influences of the Urban Land Marketization on the Production Factors and Their Spatial Effects in Urban Agglomerations

Urban land marketization indirectly promotes the radiating and leading role of production factors in central cities (see Table 4 and Figure 7). Urban land marketization effectively prevents the introduction of low investment and low-efficiency enterprises, which may be caused by negotiated leasing and the "bottom line" competition among local governments. Additionally, it also effectively promotes the growth of corporate investment, labor, and technical efficiency [73,85], and then drives the increase in these production factors in other cities through their spatial diffusion effect.

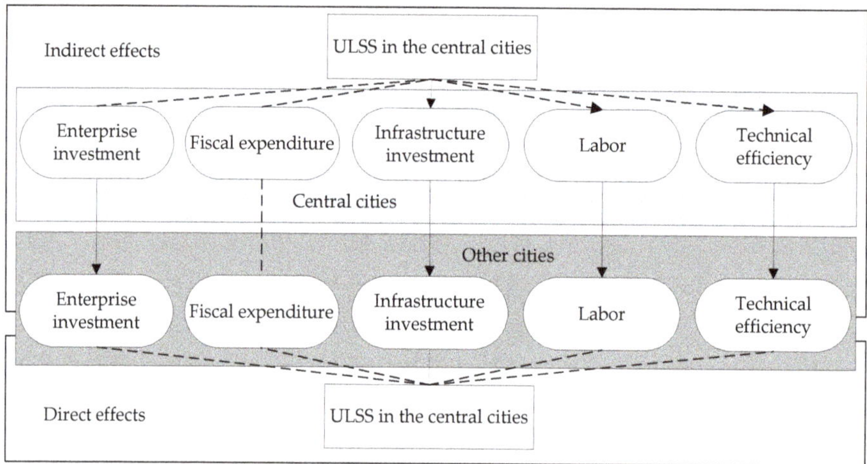

**Figure 6.** The impacts of ULSS on the production factors in urban agglomerations.

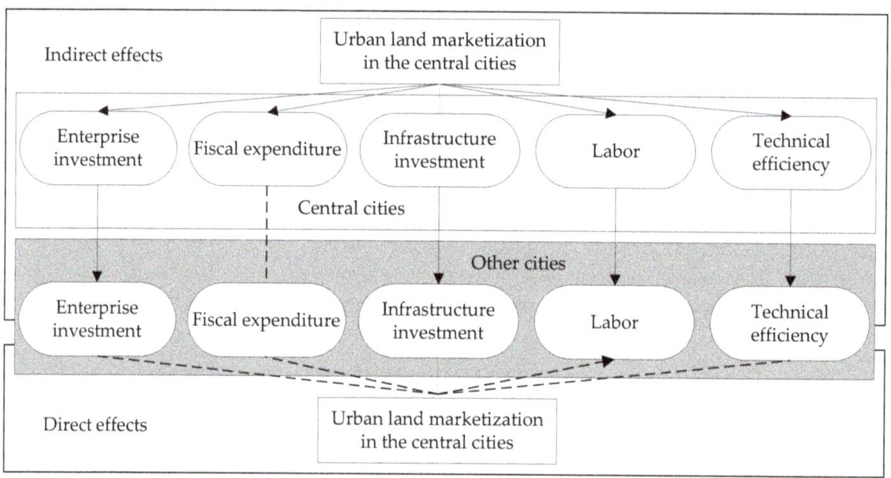

**Figure 7.** The impacts of urban land marketization on the production factors in urban agglomerations.

### 4.3.5. The Influences of the Urban Land Prices on the Production Factors and Their Spatial Effects in Urban Agglomerations

Rising urban land prices in central cities will directly crowd out the fiscal expenditures and infrastructure investment of other cities, but this will indirectly stimulate the growth of corporate investment, labor and technical efficiency in other cities as well (see Table 4 and Figure 8). The estimated results of the impacts of rising urban land prices on corporate investment show that rising urban land prices have not only failed to crowd out corporate investment, labor, and technical efficiency, but they have spurred them instead. The reason is that as the urban land prices continue to rise, companies must increase their investment intensity to obtain higher returns per unit of land [73]. At the same time, the rising land prices also have a screening effect on enterprises. Only companies with higher technical efficiency and better input–output efficiency can earn profits in that environment. Rising urban land prices have also forced technical efficiency improvements.

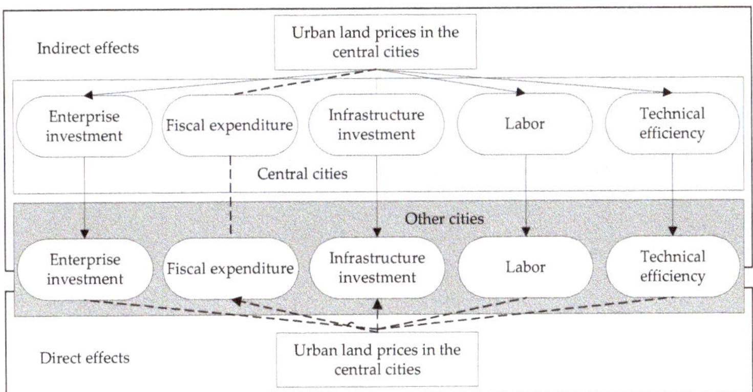



**Figure 8.** The impacts of urban land prices on the production factors in urban agglomerations.

## 5. Discussion

### 5.1. Why Are the Positive Effects of ULUT on the ESSCC Small?

The study finds that in general, China's ULUT has enhanced the ESSCC, but the positive effects are small. From the estimation results, on the one hand, urban land expansions, the adjustment of ULSS, urban land marketization, and urban land prices have positive but small effects on economic growth; on the other hand, the spatial spillovers of economic growth and production factors in central cities in China's urban agglomerations are very small.

Regional economic growth and urbanization are important assessment tools for the promotion of local officials in China [87]. Therefore, in China, urban land expansions are often used as a tool in the "local officials promotion tournament" [49]. However, there exists regional "bottom line competition" in the process of attracting investment through urban land expansion, which achieves the purpose of introducing enterprise investment projects by lowering the land leasing price and reducing the quality of enterprise investment [55,80]. After enterprises are introduced, there may be problems relating to land hoarding, low investment intensity, and low profitability that reduce the driving effects of urban land expansions on enterprise investment, fiscal revenue, and economic growth [51]. In view of this, the Chinese government has issued a series of policies to improve urban land use efficiency by assessing land investment intensity and output intensity through bidding, auction, and listing to lease land and optimize the supply structure of urban land [17,84]. This study found that strengthening the intensive use of urban land and promoting urban land marketization in China are indeed effective measures through which to enhance the ESSCC in urban agglomerations.

On the basis of urbanization, the Chinese government has approved many urban agglomerations with a view to enhancing the ESSCC and promoting the coordinated development of urban agglomerations through their development planning [66]. However, there are many problems with urban agglomerations, such as their high industrial homogeneity, weak industrial correlation, prevailing local protectionism, serious market segmentation, and unsmooth circulation of production factors. More resources and factors are directed to central cities, and urban agglomerations lack systematic economic development planning and a perfect regional cooperation mechanism [88]. These problems result in the impact of the ESSCC remaining small, and the ESSCC in some urban agglomerations is not significant or even negative.

### 5.2. Suggestions for ULUT to Promote the ESSCC

The results suggest that enhancing the ESSCC, strengthening the intensive use of land, and deepening the land marketization reform are important measures to promote the economic radiation of central cities by strengthening the ULUT. Therefore, eliminating local protectionism and barriers to the flow of factors between cities, enhancing collaboration between cities, further strengthening the assessment of land use efficiency, prohibiting extensive expansion of construction land, and local governments using more market-oriented methods to supply land will contribute to the positive effect of ULUT on ESSCC.

It is necessary to establish and improve the coordinated development mechanism among cities in urban agglomerations [66,88]. At present, the spatial spillovers of economic growth and production factors in the central cities in China's urban agglomerations are relatively small. Establishing and perfecting a coordinated development mechanism between cities in urban agglomerations can effectively enhance the ESSCC, e.g., by establishing relevant mechanisms for inter-citycommunication and coordination, planning docking, technical support, and industrial division and cooperation to promote practical cooperation between cities in infrastructure construction, public utilities development, and industrial division of labor and breaking industry monopolies and regional blockades to promote the free flow and orderly competition of commodities and various elements. For cross-city project construction, industry transfer, and investment activities, etc., joint construction, demutualization operation, and other ways and means should be adopted to share benefits. By standardizing the financial transfer payment system and establishing an inter-city compensation system and a special fund for urban coordinated development, the surrounding cities will be compensated for their loss of interests in the process of coordinated development. The formulation of laws and regulations that promote the coordinated development of cities should also be made quicker.

The intensive use of urban land should also continue to be strengthened, as the ULIU has directly and indirectly significantly enhanced the ESSCC. China needs to further reduce the scale of newly added urban land used for construction in cities, focus on revitalizing the existing land used for construction, strictly enforce the regulations on recovering idle land or collecting idle land fees in accordance with the law, and encourage the redevelopment of inefficient land in cities. Moreover, it should intensify the development of urban land, increase land investment, and overall control the per capita land use index so as to improve the input–output efficiency of urban land and optimize the supply structure of urban land. The coordination and connection between industrial development planning and land space planning should be strengthened to promote the migration of some industries in the central urban areas of megacities to satellite cities to further enhance the positive effect of ULIU on the ESSCC.

The urban land market mechanism needs to be further improved [12,89]. The increase in the marketization level of urban land has indirectly significantly enhanced the ESSCC. It is necessary to further improve the urban land market mechanism, deepen the reform of the system of the paid use of state-owned construction land, expand the scope of paid use of state-owned land, gradually implement the paid use of land for business infrastructure and social undertakings, reduce the scope of land allocation, and improve the secondary market for land lease, transfer, and mortgage to promote the coordinated development of cities in urban agglomerations. It isnecessary to speed up the formation of a sound and healthy price mechanism, improve the implementation policies related to the minimum price standard for industrial land leasing, establish an effective mechanism for adjusting the reasonable price ratio between industrial land and residential land, and constrain extensive land use by leveraging prices to encourage intensive use of urban land.

### 5.3. Limitations and Future Research

From the perspective of the spatial effects of ULUT, this paper combs the data to estimate their impacts on the ESSCC in urban agglomerations and uncover its intermediate mechanism, to meaningfully supplement the existing research on urban land use and economic growth in China. It also provides a valuable reference for urban land use and resource regulations, territorial space planning, and new urbanization.

However, this article also has limitations. First, due to data availability, this paper only contains a sample of cities above the prefecture level in the urban agglomerations and does not contain cities at the county level. Therefore, the impacts of the ULUT in central cities on the production factors and economic growth of county-level cities, and the impacts of ULUT in prefecture-level cities in urban agglomerations on the production factors and economic growth in county-level cities have not been identified. Second, this paper selects the urban land expansions to describe the explicit transition, and the ULIU, the ULSS, the urban land marketization, and urban land prices to describe the implicit transition. However, the content of urban land use transition is richer than this [58–63]. Third, this paper identifies the effect and intermediate mechanism of ULUT on the ESSCC. However, the factors affecting the effects of ULUT on the ESSCC are very complicated, and it is difficult to conduct systematic empirical tests on all of the possible factors in a single paper. In the future, we will focus our quantitative and empirical research on the factors of ULUT affecting corporate investment structure, input–output efficiency, and labor mobility.

## 6. Conclusions

There have been relevant studies in the context of private land ownership, mainly from the perspective of land as a limited production factor, discussing the impact of land on economic growth [38–41]; China's land system is publicly owned, and land is often used by local governments as a tool to achieve economic growth goals. Urban land expansion is used to drive investment and stimulate economic growth [33–35]; in the context of the public ownership of land in China, this article accepts land as a tool for economic growth of local governments. Based on the perspective of the flow of production factors, this paper studies the effect of ULUT on the ESSCC. Compared with the existing research results, it is found that in addition to urban land expansion, explicit ULUT will affect production factors and thus, affect economic growth [33,34], and implicit ULUT, such as ULIU, urban land supply structure, and urban land marketization, urban land prices, etc., will also affect production factors and thus, economic growth. Moreover, strengthening the implicit ULUT is more conducive to enhancing the ESSCC. Since 2003, China's urban land use has undergone rapid transitions that havemanifested in the rapid expansions of urban land, the increase in ULIU, the gradual decline in the proportion of urban industrial land, commercial land, and residential land in leased land, urban land marketization, the substantial increase in land prices, and other implicit transition features. Moreover, there are significantly positive spatial autocorrelations in ULUT.

The ULUT in China has enhanced the ESSCC in urban agglomerations. On the one hand, ULUT indirectly enhances ESSCC through their positive effects on economic growth and the spatial spillovers of economic growth; on the other hand, both the urban land expansions and ULIU in central cities can directly drive the economic growth in other cities in the agglomerations, thus enhancingthc ESSCC. The ULUT in central cities indirectly promotes the growth of enterprise investments, infrastructure investments, labor, and technical efficiencies in other cities in the agglomerations, through its positive effects on these production factors and their spatial diffusion effects. This is the main intermediate mechanism by which the ULUT in central cities enhances the ESSCC. However, the positive effect of ULUT on ESSCC in the agglomerations is small, as its promoting effects on economic growth and the spatial spillovers of economic growth in central cities are small. In addition, there are great differences in the impact and formation mechanisms of ULUT in the central cities of major urban agglomerations on the ESSCC.

China has experienced nearly 40 years of rapid economic growth and urbanization, which has brought about large-scale expansion of urban land. As land resources continue to tighten, China is undergoing ULUT, from the original extensive urban land expansion to implicit ULUT, such as optimizing urban land supply structure, promoting urban land marketization, and strengthening ULIU, etc. A new urbanization plan has been issued to enhance the ESSCC and promote coordinated development between cities. This paper combines ULUT with ESSCC, and the results obtained can provide valuable references for enhancing the ESSCC by optimizing ULUT. Based on the results, continuing to strengthen ULIU and improve the urban land market are conducive to enhancing the ESSCC in urban agglomerations. However, only by establishing the coordination mechanism of the economic development of urban agglomerations can it be more helpful to exert the positive impact of ULUT on the ESSCC.

**Author Contributions:** Conceptualization, H.L., K.C., Y.C. and Y.Z.; methodology, H.L., K.C. and L.L.; validation, H.L., K.C. and L.Y.; formal analysis, H.L. and L.Y.; data curation, H.L. and L.L.; writing—original draft preparation, H.L., K.C. and L.Y.; writing—review and editing, H.L., K.C., L.Y., Y.Z., L.L. and Y.C.; visualization, K.C., Y.Z. and Y.C.; supervision, K.C.; funding acquisition, H.L. All authors have read and agreed to the published version of the manuscript.

**Funding:** This study is supported by the National Social Science Foundation of China (No. 18CJY033), National Natural Science Foundation of China (No. 41971216) and the Provincial Social Science Foundation of Hunan (No.18YBQ132).

**Institutional Review Board Statement:** Not applicable.

**Informed Consent Statement:** Not applicable.

**Data Availability Statement:** The data came from the China City Statistical Yearbook, the China Land and Resources Statistical Yearbook, the Statistical Yearbook of Urban Construction in China, and China Land Market Net.

**Conflicts of Interest:** The authors declare no conflict of interest.

## References


1. Deng, X.; Huang, J.; Rozelle, S.; Uchida, E. Economic Growth and the Expansion of Urban Land in China. *Urban Stud.* **2009**, *47*, 813–843. [CrossRef]
2. Ji, Y.; Guo, X.; Zhong, S. Land Financialization, Uncoordinated Development of Population Urbanization and Land Urbanization, and Economic Growth: Evidence from China. *Land* **2020**, *9*, 481. [CrossRef]
3. Xiao, J.; Shen, Y.; Ge, J.; Tateishi, R.; Tang, C.; Liang, Y.; Huang, Z. Evaluating urban expansion and land use change in Shijiazhuang, China, by using GIS and remote sensing. *Landsc. Urban Plan.* **2006**, *75*, 69–80. [CrossRef]
4. Yu, Y.; Xu, T.; Wang, T. Outmigration Drives Cropland Decline and Woodland Increase in Rural Regions of Southwest China. *Land* **2020**, *9*, 443. [CrossRef]
5. Yan, L.; Hong, K.; Chen, K.; Li, H.; Liao, L. Benefit distribution of collectively-owned operating construction land entering the market in rural China: A multiple principal–agent theory-based analysis. *Habitat Int.* **2021**, *109*, 102328. [CrossRef]
6. Yan, L.; Hong, K.; Li, H. Transfer of Land Use Rights in Rural China and Farmers' Utility: How to Select an Optimal Payment Mode of Land Increment Income. *Land* **2021**, *10*, 450. [CrossRef]
7. Ning, J.; Liu, J.; Kuang, W.; Xu, X.; Zhang, S.; Yan, C.; Li, R.; Wu, S.; Hu, Y.; Du, G.; et al. Spatiotemporal patterns and characteristics of land-use change in China during 2010–2015. *J. Geogr. Sci.* **2018**, *28*, 547–562. [CrossRef]
8. Zhang, X.; Wu, Y.; Shen, L. An evaluation framework for the sustainability of urban land use: A study of capital cities and municipalities in China. *Habitat Int.* **2011**, *35*, 141–149. [CrossRef]
9. Long, H.; Li, Y.; Liu, Y.; Woods, M.; Zou, J. Accelerated restructuring in rural China fueled by 'increasing vs. decreasing balance' land-use policy for dealing with hollowed villages. *Land Use Policy* **2012**, *29*, 11–22. [CrossRef]
10. Yu, J.; Zhou, K.; Yang, S. Land use efficiency and influencing factors of urban agglomerations in China. *Land Use Policy* **2019**, *88*. [CrossRef]
11. Lu, X.; Ke, S. Evaluating the effectiveness of sustainable urban land use in China from the perspective of sustainable urbanization. *Habitat Int.* **2018**, *77*, 90–98. [CrossRef]
12. Wang, J.; Chen, Y.; Shao, X.; Zhang, Y.; Cao, Y. Land-use changes and policy dimension driving forces in China: Present, trend and future. *Land Use Policy* **2012**, *29*, 737–749. [CrossRef]
13. Gong, J.; Chen, W.; Liu, Y.; Wang, J. The intensity change of urban development land: Implications for the city master plan of Guangzhou, China. *Land Use Policy* **2014**, *40*, 91–100. [CrossRef]
14. Wei, Y.H.D. Restructuring for growth in urban China: Transitional institutions, urban development, and spatial transformation. *Habitat Int.* **2012**, *36*, 396–405. [CrossRef]
15. Liu, T.; Cao, G.; Yan, Y.; Wang, R.Y. Urban land marketization in China: Central policy, local initiative, and market mechanism. *Land Use Policy* **2016**, *57*, 265–276. [CrossRef]
16. Gao, J.; Wei, Y.D.; Chen, W.; Chen, J. Economic transition and urban land expansion in Provincial China. *Habitat Int.* **2014**, *44*, 461–473. [CrossRef]
17. Long, H. Land use policy in China: Introduction. *Land Use Policy* **2014**, *40*, 1–5. [CrossRef]
18. The Ministry of Land and Resources of China. Regulations on the Transfer of State-Owned Construction Land Use Rights by Bidding, Auction and Listing. Available online: http://www.gov.cn/gongbao/content/2008/content_970300.htm (accessed on 28 September 2007).

19. State Council of China. Outline of National Overall Land Use Planning in China (2006–2020). Available online: http://www.gov.cn/zxft/ft149/content_1144625.htm (accessed on 24 October 2008).

20. The Ministry of Land and Resources of China. Guiding Opinions of the Ministry of Land and Resources on Promoting Land Conservation and Intensive Utilization. Available online: http://www.gov.cn/xinwen/2014-09/26/content_2756852.htm (accessed on 26 September 2014).

21. Zhou, Y.; Huang, X.; Chen, Y.; Zhong, T.; Xu, G.; He, J.; Xu, Y.; Meng, H. The effect of land use planning (2006–2020) on construction land growth in China. *Cities* **2017**, *68*, 37–47. [CrossRef]

22. Chen, Y.; Chen, Z.; Xu, G.; Tian, Z. Built-up land efficiency in urban China: Insights from the General Land Use Plan (2006–2020). *Habitat Int.* **2016**, *51*, 31–38. [CrossRef]

23. Yang, Y.; Bao, W.; Li, Y.; Wang, Y.; Chen, Z. Land Use Transition and Its Eco-Environmental Effects in the Beijing–Tianjin–Hebei Urban Agglomeration: A Production–Living–Ecological Perspective. *Land* **2020**, *9*, 285. [CrossRef]

24. Allen, J. *Global City-Regions: Trends, Theory, Policy*; Oxford University Press: Oxford, UK, 2001.

25. Zhen, F.; Cao, Y.; Qin, X.; Wang, B. Delineation of an urban agglomeration boundary based on SinaWeibo microblog 'check-in' data: A case study of the Yangtze River Delta. *Cities* **2017**, *60*, 180–191. [CrossRef]

26. The Central Committee of the Communist Party of China, State Council of China. The National New Urbanization Plan (2014–2020). Available online: http://www.gov.cn/zhengce/2014-03/16/content_2640075.htm (accessed on 16 March 2014).

27. Long, H.; Tang, G.; Li, X.; Heilig, G.K. Socio-economic driving forces of land-use change in Kunshan, the Yangtze River Delta economic area of China. *J. Environ. Manag.* **2007**, *83*, 351–364. [CrossRef]

28. Ding, C.; Lichtenberg, E. Land and Urban Economic Growth in China. *J. Reg. Sci.* **2010**, *51*, 299–317. [CrossRef]

29. Seto, K.C.; Kaufmann, R.K. Modeling the Drivers of Urban Land Use Change in the Pearl River Delta, China: Integrating Remote Sensing with Socioeconomic Data. *Land Econ.* **2003**, *79*, 106–121. [CrossRef]

30. Long, H.; Heilig, G.; Li, X.; Zhang, M. Socio-economic development and land-use change: Analysis of rural housing land transition in the Transect of the Yangtze River, China. *Land Use Policy* **2007**, *24*, 141–153. [CrossRef]

31. Wang, J.; He, T.; Lin, Y. Changes in ecological, agricultural, and urban land space in 1984–2012 in China: Land policies and regional social-economical drivers. *Habitat Int.* **2018**, *71*, 1–13. [CrossRef]

32. Kuang, W.; Liu, J.; Dong, J.; Chi, W.; Zhang, C. The rapid and massive urban and industrial land expansions in China between 1990 and 2010: A CLUD-based analysis of their trajectories, patterns, and drivers. *Landsc. Urban Plan.* **2016**, *145*, 21–33. [CrossRef]

33. Lin, G.; Zhang, A. Emerging spaces of neoliberal urbanism in China: Land commodification, municipal finance and local economic growth in prefecture-level cities. *Urban Stud.* **2015**, *52*, 2774–2798. [CrossRef]

34. Lichtenberg, E.; Ding, C. Local officials as land developers: Urban spatial expansion in China. *J. Urban Econ.* **2009**, *66*, 57–64. [CrossRef]

35. He, C.; Huang, Z.; Wang, R. Land use change and economic growth in urban China: A structural equation analysis. *Urban Stud.* **2014**, *51*, 2880–2898. [CrossRef]

36. Ke, S. Agglomeration, productivity, and spatial spillovers across Chinese cities. *Ann. Reg. Sci.* **2009**, *45*, 157–179. [CrossRef]

37. Han, W.; Zhang, X.; Zheng, X. Land use regulation and urban land value: Evidence from China. *Land Use Policy* **2020**, *92*, 104432. [CrossRef]

38. Nichols, A. Land and economic growth. *Am. Econ. Rev.* **1970**, *60*, 332–340.

39. McCain, A. Land in Fellner's model of economic growth: Comment. *Am. Econ. Rev.* **1970**, *60*, 495–499.

40. Drazen, A.; Eckstein, Z. On the organization of rural markets and the process of economic development. *Am. Econ. Rev.* **1988**, *78*, 431 443.

41. Rhee, C. Dynamic Inefficiency in an Economy with Land. *Rev. Econ. Stud.* **1991**, *58*, 791–797. [CrossRef]

42. Ihlanfeldt, R. The effect of land use regulation on housing and land prices. *J. Urban Econ.* **2007**, *61*, 420–435. [CrossRef]

43. Kok, N.; Monkkonen, P.; Quigley, J.M. Land use regulations and the value of land and housing: An intra-metropolitan analysis. *J. Urban Econ.* **2014**, *81*, 136–148. [CrossRef]

44. Irwin, G.; Bockstael, N. The problem of identifying land use spillovers: Measuring the effects of open space on residential property values. *Am. J. Agric. Econ.* **2001**, *83*, 698–704. [CrossRef]

45. Hui, E.; Liang, C. Spatial spillover effect of urban landscape views on property price. *Appl. Geogr.* **2016**, *72*, 26–35. [CrossRef]

46. Glaeser, E.; Henderson, V. Urban economics for the developing world: An introduction. *J. Urban Econ.* **2017**, *98*, 1–5. [CrossRef]

47. Anas, A.; Kim, I. General equilibrium models of polycentric urban land use with endogenous congestion and job agglomeration. *J. Urban Econ.* **1996**, *40*, 232–256. [CrossRef]

48. Tian, L.; Ma, W. Government intervention in city development of China: A tool of land supply. *Land Use Policy* **2009**, *26*, 599–609. [CrossRef]

49. Liu, M.; Tao, R.; Yuan, F.; Cao, G. Instrumental land use investment-driven growth in China. *J. Asia Pac. Econ.* **2008**, *13*, 313–331. [CrossRef]

50. Gao, H. Public land leasing, public productive spending and economic growth in Chinese cities. *Land Use Policy* **2019**, *88*, 104076. [CrossRef]

51. Zhu, J. Changing Land Policy and its Impact on Local Growth: The Experience of the Shenzhen Special Economic Zone, China, in the 1980s. *Urban Stud.* **1994**, *31*, 1611–1623. [CrossRef]

52. Liu, J.; Guo, Q. A spatial panel statistical analysis on cultivated land conversion and Chinese economic growth: Integrating, monitoring, assessment and management. *Ecol. Indic.* **2015**, *51*, 20–24. [CrossRef]

53. Overmars, K.; Koning, G.; Veldkamp, A. Spatial autocorrelation in multi-scale land use models. *Ecol. Model.* **2003**, *164*, 257–270. [CrossRef]

54. Wu, Y.; Zhang, X.; Skitmore, M.; Song, Y.; Hui, E.C. Industrial land price and its impact on urban growth: A Chinese case study. *Land Use Policy* **2014**, *36*, 199–209. [CrossRef]

55. Huang, Z.; Du, X. Strategic interaction in local governments' industrial land supply: Evidence from China. *Urban Stud.* **2017**, *54*, 1328–1346. [CrossRef]

56. Li, C.; He, J.; Duan, X. Modeling the Collaborative Evolution of Urban Land Considering Urban Interactions under Intermediate Intervention, in the Urban Agglomeration in the Middle Reaches of the Yangtze River in China. *Land* **2020**, *9*, 184. [CrossRef]

57. Wei, Y.D.; Li, H.; Yue, W. Urban land expansion and regional inequality in transitional China. *Landsc. Urban Plan.* **2017**, *163*, 17–31. [CrossRef]

58. Long, H.; Li, T. The coupling characteristics and mechanism of farmland and rural housing land transition in China. *J. Geogr. Sci.* **2012**, *22*, 548–562. [CrossRef]

59. Long, H.; Liu, Y.; Hou, X.; Li, T.; Li, Y. Effects of land use transitions due to rapid urbanization on ecosystem services: Implications for urban planning in the new developing area of China. *Habitat Int.* **2014**, *44*, 536–544. [CrossRef]

60. Long, H.; Qu, Y. Land use transitions and land management: A mutual feedback perspective. *Land Use Policy* **2018**, *74*, 111–120. [CrossRef]

61. Chen, K.; Long, H.; Liao, L.; Tu, S.; Li, T. Land use transitions and urban-rural integrated development: Theoretical framework and China's evidence. *Land Use Policy* **2020**, *92*, 104465. [CrossRef]

62. Long, H.; Qu, Y.; Tu, S.; Zhang, Y.; Jiang, Y. Development of land use transitions research in China. *J. Geogr. Sci.* **2020**, *30*, 1195–1214. [CrossRef]

63. Long, H. *Land Use Transitions and Rural Restructuring in China*; Springer: Singapore, 2020; pp. 3–29.

64. Ke, S.; Feser, E. Count on the Growth Pole Strategy for Regional Economic Growth? Spread–Backwash Effects in Greater Central China. *Reg. Stud.* **2009**, *44*, 1131–1147. [CrossRef]

65. Tao, R.; Su, F.; Liu, M.; Cao, G. Land Leasing and Local Public Finance in China's Regional Development: Evidence from Prefecture-level Cities. *Urban Stud.* **2010**, *47*, 2217–2236.

66. Ye, C.; Zhu, J.; Li, S.; Yang, S.; Chen, M. Assessment and analysis of regional economic collaborative development within an urban agglomeration: Yangtze River Deltaas a case study. *Habitat Int.* **2019**, *83*, 20–29. [CrossRef]

67. Yu, N.; Jong, M.; Storm, S.; Mi, J. Spatial spillover effects of transport infrastructure: Evidence from Chinese regions. *J. Transp. Geogr.* **2013**, *28*, 56–66. [CrossRef]

68. Hui, E.C.; Wu, Y.; Deng, L.; Zheng, B. Analysis on coupling relationship of urban scale and intensive use of land in China. *Cities* **2015**, *42*, 63–69. [CrossRef]

69. Zhang, P.; Yang, D.; Qin, M.; Jing, W. Spatial heterogeneity analysis and driving forces exploring of built-up land development intensity in Chinese prefecture-level cities and implications for future Urban Land intensive use. *Land Use Policy* **2020**, *99*, 104958. [CrossRef]

70. Cheng, X.; Shao, H.; Li, Y.; Shen, C.; Liang, P. Urban Land Intensive Use Evaluation Study Based on Nighttime Light—A Case Study of the Yangtze River Economic Belt. *Sustainability* **2019**, *11*, 675. [CrossRef]

71. Lai, Y.; Chen, K.; Zhang, J.; Liu, F. Transformation of Industrial Land in Urban Renewal in Shenzhen, China. *Land* **2020**, *9*, 371. [CrossRef]

72. Lin, G.C.S.; Yi, F. Urbanization of Capital or Capitalization on Urban Land? Land Development and Local Public Finance in Urbanizing China. *Urban Geogr.* **2011**, *32*, 50–79. [CrossRef]

73. Du, J.; Thill, J.-C.; Peiser, R.B. Land pricing and its impact on land use efficiency in post-land-reform China: A case study of Beijing. *Cities* **2016**, *50*, 68–74. [CrossRef]

74. Rong, Z.; Wang, W.; Gong, Q. Housing price appreciation, investment opportunity, and firm innovation: Evidence from China. *J. Hous. Econ.* **2016**, *33*, 34–58. [CrossRef]

75. Ansenlin, L. Spatial externalities, spatial multipliers, and spatial econometrics. *Int. Reg. Sci. Rev.* **2003**, *26*, 153–166. [CrossRef]

76. Anselin, L. Thirty years of spatial econometrics. *Pap. Reg. Sci.* **2010**, *89*, 3–25. [CrossRef]

77. Corrado, L.; Fingleton, B. Where is the economics in spatial economics? *J. Reg. Sci.* **2011**, *52*, 210–239. [CrossRef]

78. Pinkse, J.; Slade, M.E. The Future of Spatial Econometrics. *J. Reg. Sci.* **2010**, *50*, 103–117. [CrossRef]

79. Feng, J.; Lichtenberg, E.; Ding, C. Balancing act: Economic incentives, administrative restrictions, and urban land expansion in China. *China Econ. Rev.* **2015**, *36*, 184–197. [CrossRef]

80. Yang, Q.; Zhuo, P.; Yang, J. Industrial land transfer and competition quality bottom line competition: An empirical study based on panel data of prefecture-level cities in China from 2007 to 2011. *Manag. World* **2014**, *11*, 24–34. (In Chinese)

81. Ke, S.; Xiang, J. Estimation of the fixed capital stocks in Chinese cities for 1996–2009. *Stat. Res.* **2012**, *29*, 19–24. (In Chinese)

82. Battese, G.E.; Coelli, T.J. Frontier production functions, technical efficiency and panel data: With application to paddy farmers in India. *J. Prod. Anal.* **1992**, *3*, 153–169. [CrossRef]

83. Battese, G.E.; Coelli, T.J. A model for technical inefficiency effects in a stochastic frontier production function for panel data. *Empir. Econ.* **1995**, *20*, 325–332. [CrossRef]

84. Liu, Y.; Li, J.T.; Yang, Y. Strategic adjustment of land use policy under the economic transformation. *Land Use Policy* **2018**, *74*, 5–14. [CrossRef]
85. Yuan, F.; Wei, Y.; Xiao, W. Land marketization, fiscal decentralization, and the dynamics of urban land prices in transitional China. *Land Use Policy* **2019**, *89*, 104208. [CrossRef]
86. Fan, X.; Qiu, S.; Sun, Y. Land finance dependence and urban land marketization in China: The perspective of strategic choice of local governments on land transfer. *Land Use Policy* **2020**, *99*, 105023. [CrossRef]
87. Shu, C.; Xie, H.; Jiang, J.; Chen, Q. Is Urban Land Development Driven by Economic Development or Fiscal Revenue Stimuli in China? *Land Use Policy* **2018**, *77*, 107–115. [CrossRef]
88. Li, H.; Liu, Y.; He, Q.; Peng, X.; Yin, C. Simulating Urban Cooperative Expansion in a Single-Core Metropolitan Region Based on Improved CA Model Integrated Information Flow: Case Study of Wuhan Urban Agglomeration in China. *J. Urban Plan. Dev.* **2018**, *144*, 05018002. [CrossRef]
89. Li, L.; Bao, H.X.; Robinson, G.M. The return of state control and its impact on land market efficiency in urban China. *Land Use Policy* **2020**, *99*, 104878. [CrossRef]

MDPI

St. Alban-Anlage 66

4052 Basel

Switzerland

Tel. +41 61 683 77 34

Fax +41 61 302 89 18

www.mdpi.com

*Land* Editorial Office

E-mail: land@mdpi.com

www.mdpi.com/journal/land